Handbook of
Experimental Pharmacology

Volume 143

Editorial Board

G.V.R. Born, London
P. Cuatrecasas, Ann Arbor, MI
D. Ganten, Berlin
H. Herken, Berlin
K. Starke, Freiburg i. Br.
P. Taylor, La Jolla, CA

Springer

*Berlin
Heidelberg
New York
Barcelona
Hong Kong
London
Milan
Paris
Singapore
Tokyo*

Nitric Oxide

Contributors

J.-L. Balligand, T.R. Billiar, C. Bogdan, B. Brüne, H. Bult,
V. Burkart, R. Busse, A. Costa, T.M. Dawson, V.L. Dawson,
V.J. Dzau, M.G. Espey, K. Falke, M. Feelisch, I. Fleming,
U. Förstermann, A. Friebe, J. Fukuto, J. Garthwaite, H. Gerlach,
S. Ghosh, M.B. Grisham, A.B. Grossman, E. Hackenthal,
D. Keh, M.M. Kockx, D. Koesling, G. Kojda, H. Kolb,
P.A. MacCarthy, W. Martin, K.E. Matthys, L. McNaughton,
K.M. Miranda, J.B. Mitchell, S. Moncada, P. Navarra,
J. Parkinson, A. Radomski, M.W. Radomski, D. Rees,
K. Sandau, M. Sasaki, G. Sawicki, H.H.H.W. Schmidt,
A.M. Shah, D.J. Stuehr, P. Vallance, H.E. von der Leyen,
A. von Knethen, E.R. Werner, D.A. Wink, R. Zamora

Editor
B. Mayer

Springer

Universitäts-Professor Dr. BERND MAYER
Department of Pharmacology and Toxicology
Karl-Franzens-University of Graz
Universitätsplatz 2
8010 Graz
AUSTRIA
e-mail: mayer@kfunigraz.ac.at

With 73 Figures and 34 Tables

ISBN 3-540-66122-0 Springer-Verlag Berlin Heidelberg New York

Library of Congress Cataloging-in-Publication Data
Nitric oxide / Contributors, J.-L. Balligand . . . [et al.]; editor, B. Mayer.
  p. cm. — (Handbook of experimental pharmacology; v. 143)
  Includes bibliographical references and index.
  ISBN 3540661220 (hbk.: alk. paper)
  1. Nitric oxide—Physiological effect. 2. Nitric oxide—Pathophysiology. I. Mayer, B.
(Bernd), 1959– II. Balligand, J.-L. (Jean-Luc) III. Series.
QP535.N1 N5417 2000   99-087335
572'.54 – dc21

This work is subject to copyright. All rights are reserved, whether the whole or part of the material is concerned, specifically the rights of translation, reprinting re-use of illustrations, recitation, broadcasting, reproduction on microfilms or in any other way, and storage in data banks. Duplication of this publication or parts thereof is permitted only under the provisions of the German Copyright Law of September 9, 1965, in its current version, and permission for use must always be obtained from Springer-Verlag. Violations are liable for Prosecution under the German Copyright Law.

Springer-Verlag is a company in the BertelsmannSpringer publishing group
© Springer-Verlag Berlin Heidelberg 2000
Printed in Germany

The use of general descriptive names, registered names, etc. in this publication does not imply, even in the absence of a specific statement, that such names are exempt from the relevant protective laws and regulations and free for general use.

Product liability: The publishers cannot guarantee the accuracy of any information about dosage and application contained in this book. In every individual case the user must check such information by consulting the relevant literature.

Coverdesign: design & production GmbH, Heidelberg

Typesetting: Best-set Typesetter Ltd., Hong Kong

SPIN: 10673334   27/3020 – 5 4 3 2 1 0 – printed on acid-free paper

Preface

It must be hard, nowadays, for students in biomedical disciplines to escape knowing something about nitric oxide (NO). Even among the general public, it must be becoming one of the better-known molecules. It is hard to imagine that little more than a decade ago, NO was the preserve of a small handful of scientists. In the late 1980s, probably about half a dozen laboratories contributed essentially to the discovery that NO was a naturally occurring signal molecule in blood vessels. Three of the leaders of these groups, Robert Furchgott, Louis Ignarro, and Ferid Murad, were awarded the 1998 Nobel Prize for Medicine. Salvador Moncada, another of these pioneers, has kindly contributed the Introduction to this volume.

It is a strong testimony to those discoveries that a book like this can only spare a few opening pages for them. The cascade of research that they set off does not leave much time for dwelling on history. The known roles of NO soon expanded to include its production by neurons and activated macrophages. A wide spectrum of researchers became motivated to investigate what NO might be doing in their systems. The catalogue of effects of NO grew rapidly, and with it, the wish to manipulate NO-related processes for therapeutic reasons. It emerged early on that NO has harmful as well as beneficial effects. Thus, strategies to boost inadequate NO signalling, on the one hand, and to inhibit excessive NO production, on the other, are each indicated by a variety of disease states.

Maintaining an overview of NO research has become very difficult for a single individual. Thus it has become more important than ever to collect the perspectives of multiple experts. This volume is one such collection. The first part of the book deals with the basic chemistry and enzymology of NO, thus laying a molecular basis for what follows. The middle section surveys the physiological roles of NO under normal conditions. The concluding section explores the relevance of NO to disease, both as a pathogenic factor and a therapeutic target.

I would like to thank the authors for the time and effort involved in preparing their contributions. Although claims of completeness would be out of place, I am confident that the result is a useful cross-section of the field. Last but not least, I would also like to thank Ms. Doris Walker and the staff of Springer-Verlag for their help and support.

January 2000   B. Mayer

List of Contributors

BALLIGAND, J.-L., Department of Medicine, Unit of Pharmacology and Therapeutics, University of Louvain Medical School, FATH 5349, Tour Pasteur +2, 53 avenue E. Mounier, B-1200 Brussels, Belgium
e-mail: balligand@mint.ucl.ac.be

BILLIAR, T.R., Department of Surgery, University of Pittsburgh, 676 Scaife Hall, 3550 Terrace Street, Pittsburgh, PA 15261, USA
e-mail: billiartr@msx.upmc.edu

BOGDAN, C., Institut für Klinische Mikrobiologie, Immunologie und Hygiene, Universität Erlangen, Wasserturmstr. 3-5, D-91054 Erlangen, Germany
e-mail: christian.bogdan@mikrobio.med.uni-erlangen.de

BRÜNE, B., University of Erlangen-Nürnberg, Faculty of Medicine, Department of Medicine IV-Experimental Division, Loschgestr. 8, D-91054 Erlangen, Germany
e-mail: mfm423@rzmail.uni-erlangen.de

BULT, H., Division of Pharmacology, Department of Medicine, University of Antwerp (UIA), B-2610 Wilrijk (Antwerp), Belgium
e-mail: bult@uia.ua.ac.be

BURKART, V., Diabetes Research Institute at the Heinrich Heine University of Düsseldorf, Auf'm Hennekamp 65, D-40225 Düsseldorf, Germany
e-mail: burkart@dfi.uni-duesseldorf.de

BUSSE, R., Institut für Kardiovaskuläre Physiologie, Klinikum der Johann Wolfgang Goethe-Universität, Theodor-Stern-Kai 7, D-60590 Frankfurt am Main, Germany
e-mail: r.busse@em.uni-frankfurt.de

COSTA, A., Institute of Neurology "C.Mondio", University of Pavia, Pavia, Italy

DAWSON, T.M., Departments of Neurology and Neuroscience,
  Johns Hopkins University School of Medicine, 600 N. Wolfe Str.,
  Carnegie 214, Baltimore, MD 21287, USA

DAWSON, V.L., Departments of Neurology, Neuroscience and Physiology,
  Johns Hopkins University School of Medicine, 600 N. Wolfe Str.,
  Carnegie 214, Baltimore, MD 21287, USA
  e-mail: vdawson@jhmi.edu

DZAU, V.J., Cardiovascular Research, Department of Medicine,
  Harvard Medical School and Brigham and Women's Hospital, Boston,
  MA 02115, USA

ESPEY, M.G., Tumor Biology Section, Radiation Biology Branch,
  National Cancer Institute, Bldg. 10, Rm. B3-B69, Bethesda,
  MD 20892, USA

FALKE, K., Charité, Campus Virchow-Klinikum,
  Klinik für Anaesthesiologie und operative Intensivmedizin,
  Augustenburger Platz 1, D-13353 Berlin, Germany

FEELISCH, M., The Wolfson Institute for Biomedical Research,
  University College London, St. Martins House,
  140 Tottenham Court Rd., London, W1P 9LN, United Kingdom

FLEMING, I., Institut für Kardiovaskuläre Physiologie, Klinikum der
  Johann Wolfgang Goethe-Universität, Theodor-Stern-Kai 7,
  D-60590 Frankfurt am Main, Germany

FÖRSTERMANN, U., Department of Pharmacology,
  Johannes Gutenberg University, Obere Zahlbacher Str. 67,
  D-55101 Mainz, Germany
  e-mail: ulrich.forstermann@uni-mainz.de

FRIEBE, A., Abt. für Pharmakologie und Toxikologie,
  Medizinische Fakultät MA N1/39, Ruhr-Universität Bochum,
  D-44780 Bochum, Germany
  e-mail: andreas.friebe@ruhr-uni-bochum.de

FUKUTO, J., Department of Molecular Pharmacology,
  University of California, Los Angeles, CA 90269, USA

GARTHWAITE, J., The Wolfson Institute for Biomdical Research,
  The Cruciform Building, University College London, Gower Street,
  London WC1E 6BT, United Kingdom
  e-mail: john.garthwaite@ucl.ac.uk

List of Contributors

GERLACH, H., Charité, Campus Virchow-Klinikum, Klinik für
  Anaesthesiologie und operative Intensivmedizin,
  Augustenburger Platz 1,
  D-13353 Berlin, Germany

GHOSH, S., Department of Immunology, Lerner Research Institute,
  Cleveland Clinic Foundation, 9500 Euclid Avenue, Cleveland,
  OH 44195, USA

GRISHAM, M.B., Department of Molecular and Cellular Physiology,
  Louisiana State University Medical Center, Shreveport, LA 71130, USA

GROSSMAN, A.B., Department of Endocrinology, St. Bartholomew's Hospital,
  London EC1A 7BE, United Kingdom
  e-mail: A.B. grossman@mds.qmw.ac.uk

HACKENTHAL, E., Pharmakologisches Institut, Universität Heidelberg,
  Im Neuenheimer Feld 366, D-69120 Heidelberg, Germany
  e-mail: eberhard.hackenthal@urz.uni-heidelberg.de

KEH, D., Charité, Campus Virchow-Klinikum, Klinik für Anaesthesiologie
  und operative Intensivmedizin Augustenburger Platz 1,
  D-13353 Berlin, Germany
  e-mail: dkeh@CHARITE.DE

KOCKX, M.M., Division of Pharmacology, Department of Medicine,
  University of Antwerp (UIA), B-2610 Wilrijk (Antwerp), Belgium

KOESLING, D., Abt. für Pharmakologie und Toxikologie,
  Medizinische Fakultät MA N1/39, Ruhr-Universität Bochum,
  D-44780 Bochum, Germany
  e-mail: koesling@iname.com

KOJDA, G., Institut für Pharmakologie und Klinische Pharmakologie,
  Heinrich Heine-Universität
  Düsseldorf, Moorenstr. 5, D-40225 Düsseldorf, Germany
  e-mail: kojda@uni-duesseldorf.de

KOLB, H., German Diabetes Research Institute at the Heinrich Heine
  University of Düsseldorf, Auf'm Hennekamp 65, D-40225 Düsseldorf,
  Germany
  e-mail: burkart@dfi.uni-duesseldorf.de

MACCARTHY, P.A., Department of Cardiology,
  University of Wales College of Medicine, Heath Park,
  Cardiff CF4 4XN, United Kingdom

MARTIN, W., Division of Neuroscience and Biomedical Systems,
   Institute of Biomedical & Life Sciences, University of Glasgow,
   Glasgow G12 8QQ, Scotland, United Kingdom
   e-mail: W.Martin@bio.gla.ac.uk

MATTHYS, K.E., Division of Pharmacology, Department of Medicine,
   University of Antwerp (UIA), B-2610 Wilrijk (Antwerp),
   Belgium

MCNAUGHTON, L., Department of Pharmacology,
   9-50 Medical Sciences Building,University of Alberta,
   Edmonton, AB, Canada T6G 2H7

MIRANDA, K.M., Tumor Biology Section, Radiation Biology Branch,
   National Cancer Institute, Bldg. 10, Rm. B3-B69, Bethesda, MD 20892,
   USA

MITCHELL, J.B., Tumor Biology Section, Radiation Biology Branch,
   National Cancer Institute, Bldg. 10, Rm. B3-B69, Bethesda, MD 20892,
   USA

MONCADA, S., The Wolfson Institute for Biomedical Research,
   University College London, 140 Tottenham Court Rd.,
   London W1P 9LN, United Kingdom
   e-mail: s.moncada@ucl.ac.uk

NAVARRA, P., Institute of Pharmacology, Catholic University of Rome,
   Rom, Italy

PARKINSON, J., Immunology Department, Berlex Biosciences Inc.,
   Richmond, CA 94804, USA
   e-mail: john_parkinson@berlex.com

RADOMSKI, A., Department of Pharmacology,
   9-50 Medical Sciences Building,University of Alberta,
   Edmonton, AB, Canada T6G 2H7

RADOMSKI, M.W., Department of Pharmacology,
   9-50 Medical Sciences Building,University of Alberta,
   Edmonton, AB, Canada T6G 2H7
   e-mail: marek.radomski@ualberta.ca

REES, D., Centre for Clinical Pharmacology,
   The Wolfson Institute for Biomedical Research, 140 Tottenham
   Court Rd., London W1P 9LN, United Kingdom

List of Contributors

SANDAU, K., University of Erlangen-Nürnberg, Faculty of Medicine,
  Department of Medicine IV-Experimental Division, Loschgestr. 8,
  D-91054 Erlangen, Germany

SASAKI, M., Departments of Neurology, Neuroscience and Physiology,
  Johns Hopkins University School of Medicine,
  600 N. Wolfe Str., Carnegie 214, Baltimore, MD 21287, USA

SAWICKI, G., Department of Pharmacology,
  9-50 Medical Sciences Building,University of Alberta,
  Edmonton, AB, Canada T6G 2H7

SCHMIDT, H.H.H.W., Rudolf-Buchheim-Institute for Pharmacology,
  Justus-Liebig-University, Frankfurter Str. 107, D-35392 Gießen,
  Germany
  e-mail: harald.schmidt©pharma.med.uni-giessen de

SHAH, A.M., Department of Cardiology, GKT School of Medicine,
  King's College London, Denmark Hill Campus, Bessemer Road,
  London SE5 9PJ, United Kingdom
  e-mail: ajay.shah@kcl.ac.uk

STUEHR, D.J., Department of Immunology, Lerner Research Institute,
  Cleveland Clinic Foundation, 9500 Euclid Avenue, Cleveland,
  OH 44195, USA
  e-mail: stuehrd@ccf.org

VALLANCE, P., Centre for Clinical Pharmacology, University College London,
  Rayne Institute, 5 University St,
  London W1C 655, United Kingdom
  e-mail: patrick.vallance@ucl.ac.uk

VON DER LEYEN, H.E., Cardiogene AG, Max-Planck-Str. 15A,
  D-40699 Erkrath, Germany
  e-mail: vdleyen@cardiogene.de

VON KNETHEN, A., University of Erlangen-Nürnberg, Faculty of Medicine,
  Department of Medicine IV-Experimental Division, Loschgestr. 8,
  D-91054 Erlangen, Germany

WERNER, E.R., Institut für Medizinische Chemie und Biochemie der
  Leopold-Franzens Universität Innsbruck, Fritz-Pregl-Str. 3,
  A-6020 Innsbruck, Austria
  e-mail: ernst.r.werner@uibk.ac.at

WINK, D.A., Tumor Biology Section, Radiation Biology Branch,
National Cancer Institute, Bldg. 10, Rm. B3-B69, Bethesda,
MD 20892
e-mail: wink@box-w.nih.gov

ZAMORA, R., Department of Surgery, University of Pittsburgh,
440 Scaife Hall, Pittsburgh, PA 15261, USA

Contents

Introduction
S. MONCADA 1

Section I: Chemistry

CHAPTER 1

The Chemical Biology of Nitric Oxide. Balancing Nitric Oxide with Oxidative and Nitrosative Stress
D.A. WINK, K.M. MIRANDA, M.G. ESPEY, J.B. MITCHELL,
M.B. GRISHAM, J. FUKUTO, and M. FEELISCH. With 8 Figures 7

A. Introduction 7
B. Direct Effects 9
   I. Reactions Between NO and Metal Complexes 9
   II. Interaction of NO with Metal–Oxygen and Metal–Oxo Complexes 11
   III. The Reaction of NO with Radical Species 13
C. Indirect Effects 14
D. Nitrosative Stress 15
E. Oxidative Stress 18
F. NO/O_2^- Chemistry 21
G. Conclusion 23
References 24

Section II: Biochemistry and Pharmacology of NO Synthesis and Action

CHAPTER 2

Enzymology of Nitric Oxide Synthases
D.J. STUEHR and S. GHOSH. With 14 Figures 33

A. Introduction 33
B. NOS Structure–Function 33
   I. Domain Organization 33

II. NOS Oxygenase Domains and Mutagenesis	35
1. Arg-Binding Site	36
2. H$_4$biopterin-Binding Site	38
3. N-Terminal Hairpin Loop	41
4. NOS Cysteines and Metal Binding	42
III. NOS Reductase Domains	43
1. General Features	43
2. Catalytic Properties and Response to CaM	44
3. Mutagenesis	45
IV. CaM Activation of NOS	47
1. Mechanism of Action	47
2. Structural Determinants of CaM Binding	47
V. NOS Domain Interactions	48
C. Catalysis of NO Synthesis from L-Arg	49
I. Heme-NO Complex Formation	49
1. NOS Partitioning into an NO-Bound Form During Catalysis	49
2. Impact of NO Complex Formation on NOS Catalysis	49
3. The NO Complex and NOS O$_2$ Response	50
II. The Active Catalytic Cycle	51
1. Steps involved, O$_2$, Binding and Activation	51
2. NOS Heme Iron Reduction	52
3. Control of Heme Reduction by H$_4$B and Arg	53
III. Enzyme Structural Features that may Impact on NO Synthesis	54
IV. Roles for Heme and H$_4$B	56
D. Control Mechanisms and Targeting	58
I. NOS Dimerization	58
1. Stepwise Assembly Mechanism	58
2. Positive and Negative Regulation	59
II. Is NOS Oxygenase Domain Structure Modified by Dimerization, H$_4$B Binding, or Both?	59
III. Another Type of NO Inhibition	60
IV. Interactions Between NOS and Other Proteins	60
1. PDZ, PIN	60
2. Caveolins	61
3. Heat Shock Proteins	62
4. Kalirin	62
References	62

CHAPTER 3

Regulation of Nitric Oxide Synthase Expression and Activity
U. FÖRSTERMANN 71

A. Introduction	71
B. Nitric Oxide Synthase I	73
I. Cellular Expression of NOS-I	73
II. Regulation of NOS-I Expression	73
III. Regulation of NOS-I Activity	74
C. Nitric Oxide Synthase II	75
I. Cellular Expression of NOS-II	75
II. Regulation of NOS-II Expression	75
III. Regulation of NOS-II Activity	78
D. Nitric Oxide Synthase III	79
I. Cellular Expression of NOS-III	79
II. Regulation of NOS-III Expression	79
III. Regulation of NOS-III Activity	81
E. Summary and Conclusions	82
References	83

CHAPTER 4

Enzymology of Soluble Guanylyl Cyclase
D. KOESLING and A. FRIEBE. With 4 Figures 93

A. Introduction	93
B. Regulation of sGC	94
I. NO, the Physiological Activator of sGC	94
II. Mechanism of Activation of sGC by NO	96
III. Termination of the NO-Induced Activation	96
IV. CO: a Physiological Activator of sGC?	98
V. Redox Regulation of sGC?	98
VI. Modulators of sGC	99
1. ODQ: An Inhibitor of the Stimulated Activity of sGC	99
2. YC-1: A Novel Activator of sGC	99
C. Structure of sGC	101
I. Isoforms and Tissue Distribution	101
II. Primary Structure and Homology among the Subunits of sGC	102
III. The Regulatory Heme-Binding Domain	102
IV. Catalytic Domain	103
D. Conclusions	105
References	105

CHAPTER 5

Nitric Oxide Synthase Inhibitors I:
Substrate Analogs and Heme Ligands
J.F. PARKINSON. With 4 Figures 111

A. Introduction	111
I. Therapeutic Concepts for NOS Inhibitors	111
II. NOS-Knockout Mice	112
B. Mechanism-Based NOS Inhibitors	114
I. Substrate-Based NOS Inhibitors	114
1. Arginine Analogs	114
2. Amidine-Containing Inhibitors	119
3. Summary for Substrate Analogs	124
II. Heme Ligands	124
1. Summary for Heme Ligands	127
III. Towards Rational Design of NOS Inhibitors	127
References	129

CHAPTER 6

Nitric-Oxide-Synthase Inhibitors II – Pterin Antagonists/ Anti-Pterins
E.R. WERNER and H.H.H.W. SCHMIDT. With 6 Figures 137

A. Introduction	137
B. H_4B Dependence of the NOS Reaction	137
I. NOS-Associated H_4B	138
II. Allosteric and Stabilising Effects	138
III. Possible Electron-Transfer Role	138
IV. The Pterin-Binding Site	140
C. Pterin-Based Inhibition OF NOS	140
I. Manipulating Intracellular H_4B Levels	140
II. Approaches to Pterin Antagonists	141
III. 4-Amino-H_4B	142
1. Effects of 4-Amino-H_4B on Purified Enzymes	142
2. Effects of 4-Amino-H_4B on Cultured Cells	144
3. Effects of 4-Amino-H_4B in Animals	144
IV. Further 4-Aminopteridines	145
1. The 4-Amino Function	145
2. The 2, 5 and 7 Positions	145
3. The C6 Side Chain and Pterin Exosite	147
4. Conclusion	149
V. 4-Oxopteridines as Inhibitors of NOS	149
1. Specificity and the Anti-Pterin-Binding Domain	151
2. Type-I and -II Anti-Pterins	151
3. 4-Oxo Anti-Pterins in Intact Cells	153
4. Conclusions	153
D. Outlook	154
References	155

Contents

CHAPTER 7

Mechanisms of Cellular Resistance Against Nitric Oxide
B. BRÜNE, ANDREAS VON KNETHEN, and K. SANDAU.
With 3 Figures 159

A. Introduction 159
    I. Cell Death: Apoptosis Versus Necrosis 159
    II. NO: Formation and Signaling 161
B. Cytotoxicity of Nitric Oxide 161
    I. NO-Mediated Cytotoxicity/Apoptosis 161
    II. Apoptotic-Signal Transduction: p53 Accumulation and
        Caspase Activation 162
C. Resistance Against NO$^{\bullet}$-Mediated Toxicity 164
    I. Antagonism by Bcl-2-Family Members 164
    II. Protection by NO$^{\bullet}$ and O_2^- Co-Generation 164
    III. Protective Protein Expression 166
    IV. cGMP Formation and Protein Thiol Modification 167
D. Conclusions 169
References 171

Section III: Physiological Functions of NO

CHAPTER 8

Nitric Oxide and Regulation of Vascular Tone
R. BUSSE and I. FLEMING. With 3 Figures 179

A. Regulation of Vascular Tone 179
B. Endothelial Nitric Oxide Synthase 180
    I. Ca^{2+}-Dependent eNOS Activation 181
        1. The Interaction of eNOS with CaM 181
        2. The Interaction of eNOS with Caveolin-1 182
        3. Other Modulators of eNOS Activity 183
            a) Endothelial NOS-Associated Protein-1 183
            b) Hsp90 183
            c) Phosphorylation 184
    II. Ca^{2+}-Independent eNOS Activation 185
    III. The Link Between Fluid Shear Stress and
        NO Production 186
C. Mechanisms of Action of NO on Vascular Smooth Muscle 187
    I. Effects of NO on $[Ca^{2+}]_i$ 187
    II. Effects of NO on Cyclic Nucleotide Phosphodiesterase
        III 189

    III. Effects of NO on Other Systems Involved in the Control
         of Vascular Tone 190
         1. Endothelin-1 190
         2. Noradrenaline 190
         3. NO and Iron-Containing Proteins 191
         4. NO and Mitochondrial Respiration 191
    IV. Dinitrosyl Iron Complexes, Nitrosothiol-Containing Proteins
         and Vascular Tone 192
D. NO and the Control of Blood Flow 193
    I. Interaction between NO and O_2^- 195
    II. NO and 20-HETE 197
References 198

CHAPTER 9

Regulation of Cardiac Function by Nitric Oxide
J.-L. BALLIGAND. With 1 Figure 207

A. Introduction 207
B. Specifics on Cardiac NOS Biology 207
    I. Which Isoform(s)? 207
    II. How are they Regulated? 209
         1. Endothelial Nitric Oxide Synthase 209
             a) Expressional Control 209
             b) Acute Regulation of Activity 210
                 α. Mechanical Forces 210
                 β. Beating Rate 210
                 γ. β-Adrenergic Agonists 211
                 δ. Muscarinic Cholinergic Agonists 211
                 ε. Acute Effect of Cytokines 212
         2. Inducible Nitric Oxide Synthase 213
             a) Expressional Control 213
             b) Acute Regulation of Activity 213
C. Intracellular Mechanisms of Action of NO in
   Cardiac Muscle Cells 214
    I. Cyclic GMP-Dependent Mechanisms 214
         1. Contraction-Enhancing Mechanisms 214
         2. Contraction-Decreasing Mechanisms 217
    II. Cyclic GMP-Independent Mechanisms 218
         1. Contraction-Enhancing Mechanisms 218
         2. Contraction-Decreasing Mechanisms 218
D. Regulation of Cardiac Function by eNOS 219
    I. Basal Systolic and Diastolic Function 219
    II. Regulation of β-Adrenergic Response 220
    III. Regulation of Muscarinic Cholinergic Response 221

E. Regulation of Cardiac Function by iNOS	222
I. Basal Contractile Function	223
II. Regulation of β-Adrenergic Response	224
III. iNOS and Cardiomyocyte Biology	224
F. Conclusion and Perspectives	225
References	226

CHAPTER 10

Regulation of Platelet Function
L. McNaughton, A. Radomski, G. Sawicki,
and M.W. Radomski 235

A. Introduction	235
I. Platelet Rheology	235
II. Platelet Control	235
B. Nitric Oxide	236
I. NO in Platelets: the Quest	236
II. Molecular Biology of Platelet NOS	236
III. Regulation of NO Generation in Platelet Microenvironment	237
1. Cell Activation	237
2. Role of Substrate	238
3. Role of Co-Factors	238
4. Rheology	238
IV. Physiological Effects of NO on Platelets	238
1. Effects of NO on Platelet Function In Vitro	238
2. Effects of NO on Platelet Function In Vivo	239
3. NO in Synergistic Regulation of Platelet Function	239
V. The Mechanisms of NO Action on Platelets	239
C. The Role of NO in the Pathogenesis of Vascular Disorders Associated with Platelet Activation	241
I. Pathomechanism	241
II. Atherosclerosis, Thrombosis and Hypertension	242
III. Diabetes Mellitus and Stress	242
IV. Pre-Eclampsia	243
V. Septicaemia	243
VI. Uraemia	244
VII. Cancer	244
D. Pharmacological Modulation of Formation and Action of NO on Platelets	244
I. L-Arginine	244
II. Stimulators of NOS	245
III. Inhibitors of NOS and NO Scavengers	245
IV. NO Gas	246
V. NO Donors	246

VI. Novel NO Donors 248
VII. NO-Independent Activators of GC-S 249
E. Conclusions 249
References 249

CHAPTER 11

The Physiological Roles of Nitric Oxide in the Central Nervous System
J. GARTHWAITE 259

A. Introduction 259
B. Acute Actions of NO 261
   I. Synaptic Transmission 261
   II. Gap Junctions 263
   III. Local Cerebral Blood Flow 264
   IV. Glial Cells 265
C. NO and Synaptic Plasticity 266
   I. Short-Term Plasticity 266
   II. Long-Term Potentiation 267
   III. Long-Term Depression 268
D. NO and Developmental Plasticity 268
E. Concluding Remarks 270
References 270

CHAPTER 12

The Role of Nitric Oxide in the Peripheral Nervous System
W. MARTIN 277

A. Introduction 277
   I. Nomenclature 277
   II. Historical Perspective 277
   III. The Concept of Non-Adrenergic, Non-Cholinergic Neurotransmission 278
   IV. The Concept of Nitrergic Nerves 279
B. Properties of Nitrergic Nerves 281
   I. Properties of nNOS 281
   II. Localisation of nNOS in Nitrergic Nerves 281
   III. Anatomical Distribution and Physiological Functions of Nitrergic Nerves 282
   IV. Unitary Transmission, Dual Transmission and Co-Transmission 283
C. Nature of the Nitrergic Neurotransmitter 284
   I. Predicted Differences in the Effects of Drugs on Nerve-Derived and Bath-Applied NO 285

	II. Evidence that the Nitrergic Neurotransmitter is a NO-Like or NO-Releasing Molecule	285
	III. Evidence that NO is the Nitrergic Neurotransmitter and is Protected from Inactivation	287
D.	Pre-Junctional Mechanisms	289
	I. Activation of Nitrergic Nerves	289
	II. Role of Ca^{2+} in Activation of Nitrergic Nerves	289
	III. Pre-Junctional Augmentation of Nitrergic Transmission	290
	IV. Blockade of Nitrergic Transmission by Inhibition of NOS	291
E.	Nerve–Nerve Interactions	293
	I. Nitrergic–Adrenergic Interactions	293
	II. Nitrergic–Cholinergic Interactions	293
	III. Nitrergic–NANC Interactions	294
F.	Junctional and Post-Junctional Mechanisms	295
	I. Scavengers of NO	295
	II. Blockade of Soluble Guanylate Cyclase	295
	III. Post-Junctional Potentiation of Nitrergic Transmission	296
G.	Post-Junctional Transduction Pathway	297
	I. Role of Cyclic GMP	297
	II. Inhibition of Calcium Mobilisation	297
	III. Role of Membrane Hyperpolarisation	298
H.	Concluding Remarks	299
References		299

CHAPTER 13

Nitric Oxide and Neuroendocrine Function
P. NAVARRA, A. COSTA, and A. GROSSMAN. With 4 Figures 315

A.	Introduction	315
B.	NO Biosynthesis in the Hypothalamus: Relationship Between Localization and Function	315
C.	Physiology of Hypothalamic NO	317
	I. Vasopressin and Oxytocin	317
	II. Corticotrophin-Releasing Hormone and the Hypothalamo–Pituitary–Adrenal Axis	318
	III. Hypothalamo–Pituitary–Gonadal Axis	320
	IV. Other Hormonal Systems	322
References		323

CHAPTER 14

The Role of Nitric Oxide in Kidney Function
E. HACKENTHAL. With 6 Figures 329

A. Introduction 329
B. Nitric Oxide Synthase Isoforms in the Kidney 329
C. Distribution of NOS in the Kidney 330
    I. Distribution of NOS in the Renal Vasculature 330
    II. Distribution of NOS in Renal Tubules 331
    III. Distribution of NOS in Renal Nerves 333
D. Physiological Roles of NO 333
    I. Role of NO in the Regulation of Renal Blood Flow 333
        1. Endogenous Mediators of NO Release 333
        2. Inhibitors of NOS 334
    II. Role of NO in Glomerular Circulation 335
    III. Role of NO in Renal Autoregulation 336
        1. The Myogenic Response and NO 338
        2. NO and Tubuloglomerular Feedback 338
    IV. Role of NO in the Control of Medullary Blood Flow
        and Pressure Natriuresis 341
E. Tubular Functions of NO 342
F. NO, Renin Secretion and Renin Synthesis 343
    I. NO as a Stimulator of Renin Secretion 344
    II. NO and Pressure Control of Renin Release 344
    III. NO, Renal Nerves and Renin Release 347
    IV. NO and Macula-Densa-Mediated Renin Secretion 348
    V. NO, Prostaglandins and Renin Synthesis 350
G. Concluding Remarks 352
References 353

Section IV: The Role of Pharmacological Action of NO in Human Disease

CHAPTER 15

Therapeutic Importance of Nitrovasodilators
G. KOJDA. With 3 Figures 365

A. Introduction 365
B. Mechanisms of Action 367
C. Hemodynamic Actions 368
    I. Preferential Venodilation 368
    II. Vessel-Size-Selective Coronary Vasodilation 369
    III. Effects on Blood Pressure 370
    IV. Other Effects on Hemodynamics 371

V. Effects on Platelets	371
D. Pharmacokinetics	372
E. Clinical Use	373
I. Effects in Stable Angina	373
1. Treatment and Short-Term Prevention of Anginal Attacks	374
2. Long-Term Management of Chronic Stable Angina	374
II. Effects in Unstable Angina	374
III. Effects in Acute Myocardial Infarction	375
IV. Effects in Heart Failure	375
V. Effects in Gastrointestinal Disorders	376
VI. Effects on the Uterus	376
F. Nitrate Tolerance	377
G. Side Effects and Contraindications	378
References	378

CHAPTER 16

Therapeutic Potential of NOS Inhibitors in Septic Shock
P. VALLANCE, D. REES, and S. MONCADA. With 5 Figures 385

A. Introduction	385
B. Clinical Features of Sepsis	385
I. Cardiovascular Changes	386
II. Tissue Oxygenation	386
III. Tissue and Organ Damage	386
C. NO in Experimental Models of Shock	387
I. Cardiovascular Changes	388
II. Tissue Oxygenation	388
III. Tissue and Organ Damage	389
D. NO in Clinical Sepsis	390
I. iNOS Induction in Humans	391
II. Cardiovascular Changes	391
III. Tissue Oxygenation	393
IV. Tissue and Organ Damage	393
E. Outcome Studies	394
F. Conclusions	394
References	395

CHAPTER 17

Inhalation Therapy with Nitric Oxide Gas
D. KEH, H. GERLACH, and K. FALKE. With 7 Figures 399

A. Introduction	399
B. Therapy with NO Gas	400

	I. NO Inhalation in ARDS Patients	400
	1. Introduction	400
	2. Acute Effects of NO Inhalation in Patients with ARDS	401
	3. NO Inhalation and Non-Cardiogenic Pulmonary Edema	404
	4. Dose–Response Relationship of NO Inhalation	404
	5. Effects of NO Inhalation on Right Heart Function	408
	6. NO Non-Responders	409
	7. NO Dependency	409
	8. Recent Studies of NO Inhalation in ARDS	410
	II. NO Inhalation in PPHN	412
	III. NO Inhalation in Other Diseases	414
	IV. NO Autoinhalation	415
C.	NO Metabolism, Toxicology, and Adverse Effects	417
	I. NO Uptake and Clearance	417
	II. NO and Nitrogen Dioxide	417
	III. NO, Superoxide, and Peroxynitrite	419
	IV. NO and S-Nitrosothiols	422
	1. Methemoglobin	422
D.	NO Administration	424
	I. The NO/Nitrogen Gas Mixture	424
	II. Delivery of NO	425
	III. Monitoring of NO Inhalation	429
	1. Chemiluminescence	429
	2. Electrochemical Analyzers	430
References		432

CHAPTER 18

The Function of Nitric Oxide in the Immune System
C. BOGDAN ... 443

A. Introduction	443
B. Type-2 NOS (NOS-II, iNOS) and the Immune System	444
I. Cell Types	444
II. Induction and Regulation	444
1. Overview	444
2. Transcriptional Regulation	447
3. Positive and Negative Regulation of NOS-II by Cytokines, Ligand–Receptor Interactions, and Microbial Products	449
a) Cytokines	449
b) Cross-Linking of Cell-Surface Receptors	451
c) Microbial Products	451

III. Functions	453
1. Overview	453
2. Antimicrobial Functions	453
a) Results from Host-Cell-Free Experiments and Studies in Rodents	453
b) NO as an Antimicrobial Molecule in Humans	456
c) Interaction Between NO and Other Antimicrobial Effector Pathways	457
3. Anti-Tumor Function	459
4. Autotoxic Functions	460
5. Regulatory Functions	461
a) Regulation of Proliferation, Apoptosis and Survival, and Cytotoxic Activity of Lymphocytes	462
b) Modulation of Cytokine Responses	463
α. NO and IL-12	465
c) Leukocyte Chemotaxis and Adhesion	466
d) Immune (T-Helper Cell) Deviation	466
C. Other NOS Isoforms and Perspective	467
References	468

CHAPTER 19

Nitric Oxide: A True Inflammatory Mediator
R. ZAMORA and T.R. BILLIAR 493

A. Introduction	493
I. Biosynthesis of NO	493
B. NO and Inflammation	494
I. The Chemical Mediators of the Vascular Response	495
II. NO and the Vascular Response to Injury	496
III. NO in Acute Inflammatory Responses	498
IV. NO and Inflammatory Cytokines	499
V. NO and Arachidonic Acid Metabolites	501
C. NO in Immunity and Chronically Inflammatory Diseases	502
I. NO and the Immune Response	502
II. NO and Chronic Inflammatory Processes	506
III. Induced NO in Antimicrobial Defense Mechanisms	508
D. Conclusions	510
References	511

CHAPTER 20

Nitric Oxide in the Immunopathogenesis of Type 1 Diabetes
V. BURKART and H. KOLB 525

A. Introduction 525

B. Type 1 Diabetes 525
  I. Clinical Characteristics 525
  II. Studies on the Immunopathogenesis of Type 1
      Diabetes 526
  III. Cellular Immune Reactions Against Pancreatic Islet
       Cells 527
C. NO as a Major Pathogenetic Factor in Immune-Mediated
   Diabetes 528
   I. Cellular Sources of β-Cell-Damaging NO 528
      1. Macrophages 528
      2. Endothelial Cells 529
      3. β Cells 529
   II. Primary Target Structures of NO in the β Cell 530
       1. Mitochondria 530
       2. Nuclear DNA 530
   III. Pathways of NO-Induced β-Cell Death 531
        1. Mitochondrial Damage 531
        2. Apoptotic Pathway 531
        3. Poly(Adenosine Diphosphate–Ribose)Polymerase-
           Dependent Pathway 532
D. Open Issues 533
E. Strategies to Protect Islet Cells from NO-Induced Damage 534
   I. Suppression of NO Formation 534
   II. Improvement of β-Cell Defense Mechanisms 535
   III. Inhibition of the PARP-Dependent Pathway 536
   IV. Regulation of Th1/Th2 Balance in Islet Inflammation 536
F. Concluding Remarks 537
References 538

CHAPTER 21

The Role of Nitric Oxide in Cardiac Ischaemia–Reperfusion
P.A. MacCarthy and A.M. Shah. With 2 Figures 545

A. Introduction 545
B. Consequences of Myocardial Ischaemia–Reperfusion 546
C. Interaction Between NO and ROS 547
D. Potential Ways in Which NO and ONOO⁻ May Influence
   Myocardial Ischaemia–Reperfusion 547
   I. Changes in Coronary Blood Flow and Vessel–Blood Cell
      Interactions 550
   II. Direct Effects of NO and ONOO⁻ on Myocardium 550
E. Experimental Studies 551
   I. Post-Ischaemic Endothelial Dysfunction 551
   II. Myocardial Function 552

	1. NO as a Beneficial Agent	552
	a) Post-Ischaemic Contractile Function	552
	α. Buffer-Perfused Preparations	556
	β. Blood/Neutrophil-Perfused Preparations	557
	b) Myocardial Infarction	557
	c) Reperfusion-Induced Arrhythmia	558
	2. NO as a Deleterious Agent	559
F.	Reasons for Conflicting Experimental Results	561
G.	NO and Ischaemic Preconditioning	562
H.	Summary and Conclusions	563
References		564

CHAPTER 22

Nitric Oxide and Atherosclerosis
H. Bult, K. E. Matthys, and M.M. Kockx 571

A.	Introduction	571
B.	Stages of Intimal Thickening and Atherosclerosis	571
	I. The Physiological Intima: the Soil for Atherosclerosis	571
	II. Successive Stages of Atherosclerosis	572
	III. Accelerated Atherosclerosis	572
C.	Pathogenic Mechanisms	573
	I. The Initiation of Atherosclerosis	573
	II. Remodeling of the Artery	574
	III. Plaque Stability	575
D.	Dysfunction of eNOS Signaling in Atherosclerosis	575
	I. Impaired Relaxation in Isolated Arteries	575
	II. In vivo Studies of the eNOS Defect in Atherosclerotic Arteries	576
	III. The Systemic Nature of the Defective eNOS Signaling	576
E.	Explanations for the Defective eNOS-Signaling Pathway	577
	I. Endothelial Receptor Dysfunction	577
	II. Expression of eNOS mRNA and Protein	578
	III. THB Deficiency	579
	IV. Arginine Availability	579
	1. Conduit Arteries with Atherosclerosis	579
	2. Conduit Arteries Without Overt Atherosclerosis	580
	3. Arterioles Without Overt Atherosclerosis	580
	4. Possible Explanations for the Arginine Paradox	580
	V. Endogenous NOS Antagonists	581
	VI. Negative Feedback by NO Derived from iNOS	582
	VII. Superoxide Anion Inactivates NO	582
F.	Expression of iNOS	584
	I. iNOS Expression in Atherosclerosis	584

II. Mechanical Injury and iNOS Expression	585
G. NO: a Radical with Anti-Atherogenic Properties	585
I. In Vitro Studies	585
1. Interference with Oxidative Processes	585
2. Maintenance of Endothelial Barrier Function	587
3. Interference with Leukocyte Recruitment	587
4. Antiproliferative Action of NO	587
5. Antiplatelet Effects of NO	588
II. In Vivo Studies	588
1. Inhibition of Experimental Atherosclerosis	588
2. Inhibition of Intimal Thickening by NO	590
a) Neointima Formation after Balloon Denudation ...	590
b) Intimal Hyperplasia Due to Perivascular Manipulation	591
3. Inhibition of Intimal Hyperplasia in Vein Grafts ..	592
4. Inhibition of Intimal Hyperplasia Induced by Balloon Angioplasty	592
5. Stimulation of Compensatory Remodeling	592
H. NO: a Radical Promoter of Atherosclerosis	593
I. Peroxynitrite Formation	593
II. LDL Oxidation	594
III. Oxidative Cell Injury	594
IV. NO and Apoptosis	595
1. NO as an Inhibitor of Apoptosis in the Normal Arterial Wall	595
2. NO as an Inducer of Apoptosis	596
a) PARP- and NO-Induced DNA Repair and Apoptosis	596
b) p53/p21 and NO-Induced DNA Repair and Apoptosis	596
3. NO, Apoptosis and Plaque Stability	597
V. Matrix Breakdown	598
I. Summary	598
References	599

CHAPTER 23

Nitric Oxide in Brain Ischemia/Reperfusion Injury
M. SASAKI, T.M. DAWSON, and V.L. DAWSON. With 2 Figures 619

A. Introduction	619
B. Neuronal NOS	619
C. Endothelial NOS	623
D. Immunologic NOS	626
E. The Role of NO in Focal Ischemic Brain Damage	627

F.	Targets of NO	629
G.	Summary	631
	References	631

CHAPTER 24

Therapeutic Potential of Nitric Oxide Synthase Gene Manipulation
H.E. VON DER LEYEN and V.J. DZAU. With 1 Figure 639

A.	General Principles of Gene Therapy	639
B.	Gain of Function	640
	I. Overexpression of the NOS Gene	640
	1. Overexpression of Endothelial Constitutive NOS	641
	2. Overexpression of Inducible NOS	644
C.	Loss of Function	646
	I. Inhibition of NOS by Antisense Technology	646
D.	Transgenic Animals with Disrupted NOS Gene	647
E.	Potential Therapeutic Applications of NOS Gene Transfer	649
	References	649
	Subject Index	655

Introduction

S. MONCADA

It gives me great pleasure to write the Introduction to this book on nitric oxide (NO). Many thanks are due to Bernd Mayer, who approached me over a year ago with a request to give an overview of the discovery of NO for a book to be sponsored by the Springer-Verlag publishing company. Much has already been said about the discovery of NO, and we have written about it extensively, most recently for the National Heart, Lung, and Blood Institute (MONCADA and HIGGS 1999). However, I would like this to be a vehicle for some more general considerations.

The discovery that NO plays a fundamental role in biological phenomena has developed to the stage where, at present, there is hardly any area of research that does not relate in some way to this molecule. Furthermore, its large variety of actions in physiology and pathophysiology and the obvious consequences that follow the pharmacological or genetic manipulation of its generation indicate that the knowledge acquired will be used to invent and develop new approaches to the prevention and treatment of diseases. It is remarkable that this has happened in little more than a decade, and the disbelief of the scientific community when the initial evidence was presented has now been transformed into general understanding and into frequent proposals of daring hypotheses that are now acceptable within the framework of NO biology.

Progress in science proceeds in an uneven fashion. Long periods of slow accumulation of data, often unrelated, are followed by a rapid qualitative leap in understanding. This changes the nature of the research and creates a novel field, leading to the development of a paradigm within which new and old facts are then incorporated. This breakthrough is usually recognized as discovery. The very nature of this process is such that the slow phase of quantitative growth can only be visualized once the discovery has been made. The story of NO is no exception and, with hindsight, different lines can be drawn from the breakthrough point to almost any piece of prior research at any time in the past.

Some 10 years ago, while writing the introduction to the book for the proceedings of the first international meeting on NO, I made an attempt to acknowledge some of those observations that were significant and that had "unknowingly paved the way to the discovery that NO, synthesized from L-arginine, plays a role as a regulator of cell function and communication"

(MONCADA 1990). That meeting was structured to include most of the major contributors to the NO story and, in addition, researchers from allied fields that the discovery had highlighted and brought together, such as work on the soluble guanylate cyclase and/or work on L-arginine.

The main steps in the discovery of NO as a biological mediator were: the seminal observation of endothelium-dependent relaxation and endothelium-derived relaxing factor (EDRF); the appearance (among others) of the hypothesis that EDRF might be NO or a related substance; the identification of NO and EDRF as the same substance; the demonstration of the release of NO by vascular endothelial cells, and the finding that L-arginine was the precursor for the synthesis of NO. This last observation represented a major part of the discovery, because it revealed the existence of a biochemical pathway leading to the generation of NO. N^G-monomethyl-L-arginine provided an invaluable biochemical and pharmacological tool for studies of the role of NO in biology.

The explosion of research in the 10 years that followed these findings has not exhausted the possibilities of new and exciting observations, and the present book is witness to this. A number of investigators with new approaches, assisted by the tools of modern biology, are now contributing to the further expansion of the field. Thus, the book contains chapters on topics including the enzymology of the soluble guanylate cyclase, the development of novel NO-synthase inhibitors, the use of NO-synthase gene therapy and the role of NO in cardiac function, inflammatory disease and brain ischemia/reperfusion injury. These chapters are among many that give an overview of the current research on NO.

A highly significant part of research on the biology of NO in recent years has been the increasing recognition of two of the main aspects of its nature as a biological mediator: its function as a physiological regulator and the part it plays in pathophysiological processes. In the latter respect, attention has focused not only on the actions of NO itself but also on the consequences of its interactions with a host of other molecules. Chief among these is superoxide which, in 1986, was discovered to interact with EDRF, an observation that was instrumental in the identification of EDRF as NO. The interaction of superoxide with NO, which leads to the formation of the powerful oxidant peroxynitrite, is now known to be a key pathophysiological step, the consequences of which are not completely understood.

I have been lucky to be part of this endeavor from its inception and look forward to being able to follow it for some years to come. The field is very robust. At this stage, we cannot see any major weaknesses in the paradigm of NO biology and, with little exception, every hypothesis can be fully or partially accommodated. We have witnessed very significant additional contributions in the last few years, and no doubt there are more to come. However, as in all fields of biological research, minor inconsistencies will doubtless appear in the future, leading to significant new observations that will be impossible to explain with NO alone. Other breakthroughs will then herald the development of another new area of research that will put the work on NO, its discoveries and contributions, into their proper historical perspective.

References

Moncada S (1990) Introduction. In: Moncada S, Higgs EA (eds) Nitric oxide from L-arginine: a bioregulatory system. Elsevier Science B.V., Amsterdam, pp 1–4

Moncada S, Higgs EA (1999) Discovery of biological roles of nitric oxide in the cardiovascular system. In: Panza, JA, Cannon RO III (eds) Endothelium, nitric oxide and atherosclerosis. Futura, New York, pp 13–27

Section I
Chemistry

CHAPTER 1
The Chemical Biology of Nitric Oxide. Balancing Nitric Oxide with Oxidative and Nitrosative Stress

D.A. WINK, K.M. MIRANDA, M.G. ESPEY, J.B. MITCHELL, M.B. GRISHAM, J. FUKUTO, and M. FEELISCH

A. Introduction

Since the discovery of nitric oxide (NO) as an endogenous mediator of numerous physiological processes ranging from regulation of cardiovascular function to memory formation, there has been some question as to the effect its chemistry has on biology (IGNARRO 1989; MONCADA et al. 1991; DAWSON et al. 1992; FELDMAN et al. 1992). In the immune system, this diatomic radical is involved in numerous anti-pathogenic and tumoricidal processes (HIBBS 1991; MACMICKING et al. 1997). Yet, despite these properties critical to maintaining homeostasis, NO has been implicated as a participant in different pathophysiological conditions (GROSS and WOLIN 1995; WINK et al. 1998c). To further complicate definition of the exact roles of NO in vivo, both protective and deleterious effects have been attributed to NO, even in the same biological event. Mechanistic explanations to account for these differences are still being pursued.

Unlike other biological mediators, the chemistry of NO determines its biological properties. NO can undergo numerous reactions, including many that result in the formation of biologically reactive nitrogen oxide species (RNOS). The large variety of potential chemical reactions related to NO makes it difficult to determine which are pertinent in vivo. In an attempt to categorize these diverse reactions in terms of their biological significance, we have developed the concept of the "chemical biology of NO" (WINK et al. 1996a, 1996c, 1997; WINK and MITCHELL 1998). This concept separates the chemical reactions into the two basic categories of direct and indirect effects (Fig. 1).

Direct chemical reactions are defined as those in which NO interacts with biological targets. The most common reaction is between NO and heme-containing proteins. Such reactions are generally rapid, requiring low concentrations of NO and most likely account for the majority of the physiological effects of NO. Conversely, indirect effects involve RNOS derived from the reaction between NO and either O_2 or O_2^-, rather than NO itself. These RNOS can react with and modulate critical intracellular molecules. It appears that NO produced at low concentrations for short periods of time primarily mediates direct effects, while higher, sustained NO concentrations induce indirect effects.

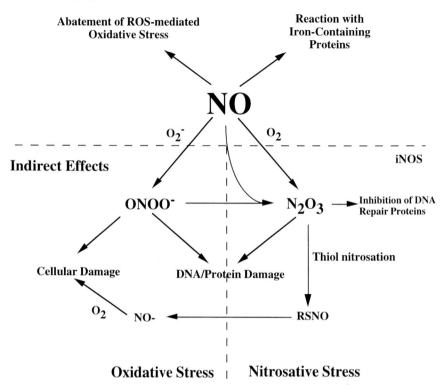

Fig. 1. Chemical reactivity in the chemical biology of nitric oxide

The reactive intermediates that are responsible for the indirect effects of NO can undergo various chemical reactions that result in either nitrosative or oxidative stress (WINK and MITCHELL 1998), depending on the species involved. Nitrosation in vivo appears to be effected primarily by N_2O_3, while oxidation is mediated by peroxynitrite ($ONOO^-$) as well as the nitroxyl ion, NO^- (WINK et al. 1997). Several studies suggest that nitrosative stress is orthogonal to oxidative stress with respect to RNOS chemistry (WINK et al. 1997; WINK and MITCHELL 1998). This implies that nitrosative and oxidative stress are produced by different chemical mechanisms. NO can also interact with reactive oxygen species (ROS) that mediate oxidative stress. Biological effects, such as cytotoxicity, occur by different pathways under nitrosative and oxidative stress (WINK and MITCHELL 1998). This suggests that a balance exists between these two types of stresses and that a shift in this balance may determine the functional outcome in responses ranging from signal transduction to cell death.

Separation of chemical reactions into direct and indirect effects is analogous to the concept of constitutive and inducible isozymes of nitric oxide synthase (NOS). This enzyme, which catalyzes the oxidation of arginine to citrulline and NO, can be found constitutively in a variety of cells, such as endothelial cells (eNOS) and neurons (nNOS) (GRIFFITH and STUEHR 1995; NATHAN and XIE 1994). Constitutively expressed NOS is thought to generate low amounts of NO at concentrations only as high as submicromolar for a few minutes at the cell level. However, inducible NOS (iNOS) generates NO for prolonged periods of time to produce concentrations ranging from 1–10 μM in the cellular environment. Therefore, in general, the isozyme type will dictate the chemistry and the ultimate biological outcome of NO production. An additional consideration is the proximity of the biological target to the NO source. Both direct and indirect reactions occur close to cells, such as macrophages, that produce high levels of NO, while cells farther away primarily experience direct effects. In the following sections, we explore the different chemical reactions relevant to the chemical biology of NO in the context of oxidative and nitrosative stress.

B. Direct Effects

There are three major types of reactions between NO and biological metals: the direct reaction of NO with the metal center, redox reactions with dioxygen complexes and redox reactions with high-valence oxo complexes (Fig. 2). These reactions, whose rates, in some cases, are near the diffusion-controlled limit, are important in numerous physiological and pathophysiological conditions.

I. Reactions Between NO and Metal Complexes

Although NO can react with a variety of metals to form metal nitrosyl complexes, the biological importance of each reaction must be considered. For instance, copper and cobalt will react with NO to form nitrosyls, but these reactions are too slow to be of major significance in vivo (COTTON and WILKINSON 1988). The most facile (and thus most physiologically relevant) reactions of this type are those between NO and heme proteins. Other biologically important transition metals, such as zinc, do not react with NO in vivo (COTTON and WILKINSON 1988).

The most notable protein that forms a metal nitrosyl complex in vivo is soluble guanylyl cyclase. Formation of the heme nitrosyl adduct results in decoupling of the distal histidine to produce a five-coordinate complex (STONE and MARLETTA 1994; Yu et al. 1994). This alteration in protein configuration activates the enzyme and leads to conversion of guanosine triphosphate to cyclic guanosine monophosphate (cGMP) in a different domain of the enzyme. The concentration of NO required to activate guanylyl cyclase is low

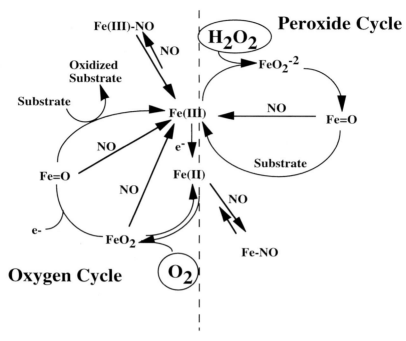

Fig. 2. Reactivity of nitric oxide with heme proteins in oxygen or peroxide reaction cycles

(EC_{50}~100 nM; FORSTERMANN and ISHII 1996). Synthesis of cGMP has many ramifications in a number of tissues, including vascular smooth muscle (MURAD 1994). The influence of NO on soluble guanylyl cyclase has profound effects on vascular tone, platelet function, neurotransmission and a variety of other cellular interactions.

In contrast to the activation of guanylyl cyclase, NO inhibits cytochrome $P450$ in addition to other heme monooxygenases (Fig. 2) (KHATSENKO et al. 1993; WINK et al. 1993c; STADLER et al. 1994). Endogenously generated NO has been postulated to play a role in regulation of hormone metabolism by controlling testosterone synthesis (ADAMS et al. 1992). In addition, inhibition of cytochrome $P450$ has some important pathophysiological sequelae. During chronic infection or septic shock, NO can be produced in copious amounts. Inhibition of liver cytochrome $P450$s (WINK et al. 1993c; STADLER et al. 1994) inhibits drug metabolism (KHATSENKO et al. 1993). However, the binding of NO to the heme domain of cytochrome $P450$ can result in release of free heme, which in turn activates heme oxygenase in hepatocytes (KIM et al. 1995). The activation of heme oxygenase may serve as a protective mechanism against a variety of pathophysiological conditions (STOCKER 1990; CHOI and ALAM 1996). Thus, the interaction of NO with $P450$ may involve regulatory physiological functions and produce positive or negative pathophysiological effects.

Another important role of heme nitrosyl formation is regulation of NOS activity. Catalysis of arginine oxidation to citrulline and NO occurs in the heme domain of NOS by a mechanism similar to substrate oxidation by cytochrome $P450$ (GRIFFITH and STUEHR 1995). However, it has been shown that NO will inhibit the oxidation of arginine, which suggests that NO levels produced by the enzyme are controlled by a negative-feedback mechanism analogous to that operating at cytochrome $P450$ (Fig. 2) (GRISCAVAGE et al. 1994; ABU-SOUD et al. 1995; GRISCAVAGE et al. 1995; HURSHMAN and MARLETTA 1995). Comparison of the different isozymes of NOS shows that eNOS and nNOS are more susceptible than iNOS to inhibition by NO (GRISCAVAGE et al. 1995). This indicates that significantly higher NO fluxes can be achieved with iNOS than with either eNOS or nNOS. The difference in NO-mediated inhibition of NOS activity is apparently due to the relative reactivity of NO and to the stability of the resultant Fe–NO complex within NOS. The stable Fe–NO complex restricts the potential concentration of NO that can be produced by nNOS and likely by eNOS. Even under conditions of elevated intracellular calcium concentrations, this feedback mechanism will prevent the production of significant amounts of RNOS from nNOS and eNOS. Therefore, the predominant source of RNOS (indirect effects) in vivo may be from iNOS.

The competitive inhibition of NOS activity through a heme nitrosyl complex can serve to regulate tissue blood flow depending on oxygen tension. It has been shown that oxidation of arginine under catalytic conditions results in the formation of an Fe-NO complex in nNOS which blocks the active site to O_2 binding and inhibits arginine oxidation (ABU-SOUD et al. 1995; HURSHMAN and MARLETTA 1995). Since NO binds to nNOS with a low K_m, the O_2 dependency is linear in the physiological range. This suggests that NOS may serve as an O_2 sensor and may attenuate the O_2 supply to tissue (ABU-SOUD et al. 1996). NOS catalyzes NO production in the lung alveoli in an O_2-dependent manner (DWEIK et al. 1998), which suggests that blood flow in the lung is regulated by O_2 tension. Since NO and O_2 compete for the heme binding site, the relative stability of the nitrosyl versus the dioxygen adduct determines the level of NO produced. It is thought that this mechanism may play a crucial role in regulating blood flow in different tissues.

II. Interaction of NO with Metal–Oxygen and Metal–Oxo Complexes

The reactivity of NO with metals is not limited simply to covalent interactions with metal ions. Various metal–oxygen complexes and metallo–oxo complexes rapidly react with NO (Fig. 2). The reaction between NO and oxyhemoglobin is a determinant of NO behavior in vivo as important as activation of guanylyl cyclase. Complexation between NO and oxyhemoglobin results in formation of methemoglobin and nitrate (DOYLE and HOEKSTRA 1981; FEELISCH 1991).

$$\text{oxyHb[Fe(II)-O2]} + \text{NO} \rightarrow \text{metHb[Fe(III)]} + \text{NO}_3^- \quad (1)$$

This reaction likely provides the primary endogenous mechanism to both eliminate NO and control the movement and concentration of NO in vivo (LANCASTER 1994). In addition, this reaction exerts an important control over the progression of indirect effects.

NO also reacts rapidly with metal–oxo and metal–peroxo adducts (WINK et al. 1997; WINK and MITCHELL 1998). Highly reactive metal complexes of this type are formed from oxidation by agents like hydrogen peroxide (H_2O_2) and can lead to cellular damage, such as lipid peroxidation (PUPPO and HALLIWELL 1988).

$$\text{Fe(II)} + H_2O_2 \rightarrow \text{Fe(IV)} = O + H_2O \quad (2)$$

The oxidative chemistry of these high-valence metal complexes can be abated by reaction with NO to reduce the hypervalent heme complex to a normal valent state (KANNER et al. 1991; GORBUNOV et al. 1995; WINK et al. 1994b).

$$\text{Fe(IV)} = O + \text{NO} \rightarrow \text{Fe(II)} + NO_2^- \quad (3)$$

The antioxidant properties of NO may be a primary mechanism by which NO protects tissue from peroxide-mediated damage (Wink et al. 1994b). Another important reaction is that between NO and catalase. Cytokine-stimulated hepatocytes have been demonstrated to reduce catalase activity due to the production of NO (Kim et al. 1995), and similar inhibition of H_2O_2 consumption was observed using NO donors (Wink et al. 1996b). Inhibition of catalase by NO could play a role in the tumoricidal activity of macrophages (Farias-Eisner et al. 1996). Catalase can be inhibited by NO through either metal–nitrosyl formation or interaction with metal–oxo species. NO binds to the heme moiety to form a nitrosyl complex with a rate constant of $3 \times 10^7 \text{ M}^{-1}\text{s}^{-1}$ and a kdis of $1 \times 10^5 \text{ M}^{-1}$ (Hoshino et al. 1993). An Fe–NO adduct would prevent the binding of H_2O_2 to the metal ion by occupying the coordination site in analogy to the mechanism for the inhibition of P450 and NOS. It is estimated that a concentration of 10–15 μM NO will inhibit H_2O_2 consumption by 80% by this mechanism (Farias-Eisner et al. 1996). Since cells that express iNOS have reduced catalase activity, this suggests that local NO concentrations near these cells may be as high as 10 μM for prolonged periods of time.

H_2O_2 may also attenuate NO levels through the enzymatic action of catalase. Initially, H_2O_2 and catalase react to form complex I and H_2O (Fig. 3). Further reaction of this complex with H_2O_2 produces O_2. However, NO can also rapidly react with complex I to form complex II, which then rapidly reacts with an additional NO. This results in consumption of NO and retardation of H_2O_2 depletion through conversion of 2 moles of NO and 1 mole of H_2O_2 to 2 moles of HNO_2 (BROWN 1995). The K_i for NO in this reaction is 0.18 μM. Thus, at submicromolar levels of NO, such as those produced by constitutive NOS,

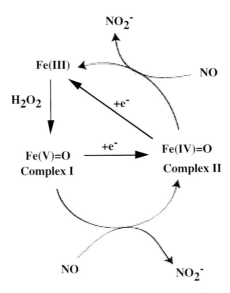

Fig. 3. The chemistry of nitric oxide and high-valence metal complexes

the steady-state concentration of H_2O_2 may be controlled, in part, by this mechanism.

The reaction of NO with high-valence heme derived from peroxide may attenuate NO levels in vivo. Reports have shown that elevation of the activity of glutathione peroxidase, which does not react with NO, increases the bioavailability of NO from eNOS (Li et al. 1992). This implies that H_2O_2, via a similar mechanism to that shown in Fig. 2, may play a crucial role in regulation of the direct effects of NO in vivo. However, the mechanism of Fe–NO formation would be important under conditions where NO concentrations are higher than those of H_2O_2. These two mechanisms may be important in sequences of NO and peroxide bursts under certain physiological and pathophysiological conditions.

III. The Reaction of NO with Radical Species

Another direct effect of NO is its reaction with other radical species. For instance, the tyrosyl radical formed in the catalytic turnover of ribonucleotide reductase reacts with NO, which results in enzyme inhibition (LEPOIVRE et al. 1990, 1991, 1992; KWON et al. 1991). Due to the suppression of DNA synthesis, inhibition of ribonucleotide reductase has been proposed to be a factor in the cytostatic properties of NO.

Lipid peroxidation is an important reaction in cell death and in different stages of the inflammatory process (HALLIWELL 1991). During lipid peroxidation, a variety of lipid–oxy and lipid–peroxy adducts are formed, which in turn

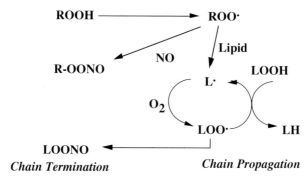

Fig. 4. Abatement of chain propagation in lipid peroxidation

perpetuate lipid oxidation, resulting in cell-membrane compromise (Fig. 4). NO reacts with oxidants formed from exposure to copper, xanthine oxidase or azo-bis-amidinopropane at near diffusion-controlled rates (Hogg et al. 1995; Rubbo et al. 1995).

$$LOO + NO \rightarrow LOONO \qquad (4)$$

This reaction protects cells by removing ROS and thus terminating lipid peroxidation (Padmaja and Huie 1993; Wink et al. 1994b, 1995; Gupta et al. 1997). Chain termination also prevents the oxidation of low-density lipoproteins in both endothelial (Struck et al. 1995) and macrophage cells (Hogg et al. 1995). The level of oxidized cholesterol may slow down the onset of atheroscleroses mediated by activated macrophages. Furthermore, other processes in inflammation, such as the production of leukotrienes, are affected by NO. Lipoxygenase, which mediates a variety of lipid oxidation steps, is also inhibited by NO.

C. Indirect Effects

The indirect effects of NO are often associated with pathophysiological conditions, and higher nitrogen oxides are thought to be the chemical species responsible for the etiology of numerous diseases related to NO. Although this may be true in some cases, direct-effect participation of NO cannot be ignored. For example, the interaction of NO and O_2^- can result in the formation of $ONOO^-$, which is a powerful oxidizer of a variety of macromolecules. Despite the potential for oxidative chemistry, the NO/O_2^- reaction in vivo is not likely to mediate oxidation. Instead, this reaction will primarily scavenge NO and, thus, inhibit direct effects. This is best illustrated by the role the NO/O_2^- reaction plays in regulation of blood pressure. If O_2^- is present, vasodilation is impeded not by reaction of $ONOO^-$ with a biological target but simply because NO has been consumed (Furchgott and Vanhoute 1989). The

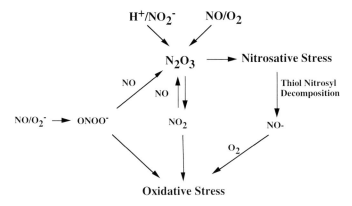

Fig. 5. The indirect effects of nitric oxide

indirect effects and the exact role of each of the intermediates must be cautiously evaluated.

The chemical relationship between the principle RNOS species formed under biological systems (N_2O_3, $ONOO^-$, NO^- and NO_2) is dependent upon the type of chemistry associated with the intermediates (Fig. 5). For instance, N_2O_3 is not a powerful oxidant, yet it readily nitrosates nucleophiles (WINK et al. 1996d). Conversely, $ONOO^-$ and NO^- do not nitrosate substrates directly but readily mediate oxidative chemistry (MILES et al. 1996; WINK et al. 1997; WONG et al. 1998). A comparison of the resultant thiol products in the presence of RNOS can illustrate this point. Aerobic NO solutions produce N_2O_3 and, when exposed to solutions of glutathione (GSH), form nearly 100% of the nitrosative product, S-nitrosoglutathione (WINK et al. 1994a). However, if GSH is exposed to $ONOO^-$, NO_2 or NO^-, the major products are oxidized – not nitrosated – thiols (PRYOR et al. 1982b; DOYLE et al. 1988; RADI et al. 1991; WONG et al. 1998). However, the reaction of NO with $ONOO^-$, NO_2 or NO^- (Fig. 5) inhibits thiol oxidation (PRYOR et al. 1982b; WINK et al. 1997) and, except in the case of NO^-, facilitates thiol nitrosation via N_2O_3 formation. Since thiols are the primary reactive sites for the indirect effects of NO within the cell, one can see a balance emerging between oxidative and nitrosative stress. With this in mind, in the following sections we discuss the sources and conditions that might lead to nitrosative or oxidative stress, the chemical reactions that are responsible for both types of stress and the likely biological targets.

D. Nitrosative Stress

Study of the chemistry of nitrosation dates back to the turn of the century (WILLIAMS 1988). Interest in the biological implications of nitrosation was increased in the mid 1970s, when concern arose that nitrosamines formed from

NO_3^--induced nitrosation of amines in the gastrointestinal tract were a potential source of carcinogens (BARTSCH et al. 1990). A decade later, the observation of NO_3^- and nitrosamine production under conditions of infection or by activated leukocytes was a critical link in elucidating the endogenous formation of NO in vivo (GREEN et al. 1981; STUEHR and MARLETTA 1985; MARLETTA 1988; HIBBS 1991). Further study showed that nitrosamines are produced during certain types of chronic infections, which implies that nitrosative stress does occur in vivo (LIU et al. 1991). Nitrosation of thiols and S-nitrosothiols themselves have been demonstrated to be physiologically relevant in processes ranging from cardiovascular function to cancer (STAMLER 1994; WINK et al. 1998d). The formation of these biologically active nitrosative products both in vitro and in vivo suggests that nitrosative stress is an important mediator of the biological effects of NO (Fig. 6), and this further complicates any mechanistic interpretation.

Nitrosation is defined as donation of a nitrosonium ion (NO^+) to a nucleophile by RNOS, while nitrosylation is formation of a nitrosyl adduct, such as those formed between the reaction of NO and metals. There are three potential sources of RNOS that mediate nitrosation: the autoxidation of NO, acidic NO_2^- and the NO/O_2^- reactions under fixed excess fluxes of NO (discussed below). Except in the gastric regions (pH < 1.5), the primary nitrosating intermediates are isomers of N_2O_3. The route of formation of this intermediate is important and determines where and when nitrosation occurs.

The chemical reaction most noted for the formation of N_2O_3 is the autoxidation of NO with O_2. This reaction has been studied for decades due to its

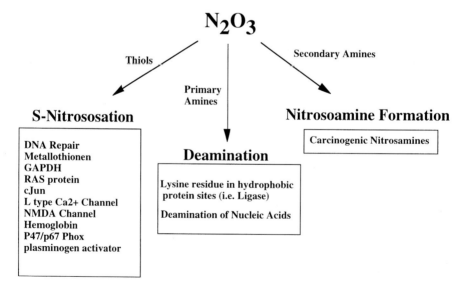

Fig. 6. The effects of nitrosative stress

importance in nitrogen oxide chemistry in the atmosphere (SCHWARTZ and WHITE 1983). The autoxidation of NO in aqueous, hydrophobic or gas phase has a third-order rate equation with second-order dependency on NO (SCHWARTZ and WHITE 1983)

$$2NO + O_2 \rightarrow 2NO_2 \quad -d[NO]/dt = k_{NO}[NO]^2[O_2] \tag{5}$$

Knowledge of this reaction resulted in initial confusion and disbelief that this potentially toxic species could participate in biological systems. Due to the instability of NO in the presence of O_2 and the formation of toxic chemical species such as N_2O_3 and NO_2, it was hard to envision why nature would choose NO as a physiological mediator. However, the lifetime of NO, which is inversely proportional to its concentration, is dictated by the second-order dependence of the autoxidation reaction (Eq. 5) (FORD et al. 1993; WINK et al. 1993a). Therefore, as NO diffuses away from the cellular source and is thereby diluted, its lifetime increases sufficiently to allow it to react with other biological targets, such as guanylyl cyclase, without interference from the autoxidation reaction and production of related RNOS. As NO levels increase, the formation of RNOS increases dramatically. Thus, it is likely that, in local regions of high NO output, intermediates associated with the autoxidation also occur in vivo.

This supposition leads to the question of where in the cell autoxidation will likely occur. The rate constants for the autoxidation of NO in hydrophobic and aqueous solutions are similar. This suggests that the rate of autoxidation is dependent not on the medium in which it occurs but on the solubility of both NO and O_2. The concentrations of both gases have been found to be 10–50 times higher in lipid membranes than in aqueous solution (DENICOLA et al. 1996). Thus, if a cell is exposed to a NO flux from either a chemical donor or another cell, NO would partition such that the NO levels would be at least 10 times greater in the membrane than in the surrounding aqueous solution. The autoxidation reaction should, therefore, occur much faster in membranes than in aqueous solution. A recent study has shown that, in the presence of micelles, the rate of autoxidation is increased 1500 times more than the rate in aqueous solution (LIU et al. 1998). This work suggests that nitrosation reactions mediated via the autoxidation reaction are likely to predominate in the membrane. Hence, membrane-bound proteins that are functionally and structurally dependent on certain thiols or amines would be most affected by nitrosative stress, and thiol nitrosation can be expected to play an important role in membrane-associated protein chemistry.

The next factor to consider in the chemistry of nitrosative stress is the mechanism of formation of N_2O_3. In the gas phase or in hydrophobic solvents, the autoxidation of NO occurs through the formation of NO_2 (Eq. 5) (SCHWARTZ and WHITE 1983). Further reaction between NO and NO_2 results in an equilibrium with N_2O_3.

$$NO_2 + NO \leftrightarrow N_2O_3 \tag{6}$$

In aqueous solution, an additional reaction results in the rapid conversion of N_2O_3 to nitrous acid.

$$N_2O_3 + H_2O \rightarrow 2HNO_2 \quad (7)$$

Thus, autoxidation in aqueous solutions, as in the gas phase, results in N_2O_3; however, there is no formation of "free" NO_2. Under acidic conditions, H_2ONO^+ is formed; this ion can further nitrosate an additional HNO_2 to form N_2O_3 (WILLIAMS 1988).

$$H_2ONO^+ + HNO_2 \rightarrow N_2O_3 + H_3O^+ \quad (8)$$

Disproportionation of N_2O_3 then produces NO_2 according to Eq. 6 to give a similar product mixture as in the gas- or lipid-phase reactions. Studies have suggested that different isomers of N_2O_3 are formed in aqueous solution and hydrophobic media (WINK et al. 1993a; WINK and FORD 1995). Competition reactions demonstrated that, in aqueous solution, NO_2 cannot be trapped. This suggests that NO_2 arising from equilibrium with N_2O_4 cannot escape the solvent cage before N_2O_4 reacts with two additional molecules of NO to produce N_2O_3. To best illustrate the difference, the nitrosation versus nitration of phenol and tyrosine was examined (PIRES et al. 1994; WINK et al. 1994a). The exposure of tyrosine to gas-phase synthesized RNOS or to acidified nitrite solutions gave nitrotyrosine exclusively through what is thought to be a NO_2-mediated reaction (WINK et al. 1994a). However, when the RNOS where formed from the autoxidation of NO in H_2O, nitrotyrosine was not produced. This suggests that the autoxidation of NO in cytosol will not produce nitrotyrosine whereas, in hydrophobic environments, such as in cell membranes, this may well occur.

The selectivity of the intermediates formed in the autoxidation reaction has been determined. Since N_2O_3 is hydrolyzed to NO_2^- with a half-life of 1 ms, only substrates that are present in high enough concentration and have sufficient affinity will react (WINK et al. 1996d). At neutral pH, thiol-containing peptides have an affinity 1000 times greater than any other amino acid (WINK et al. 1994a, 1996d). In addition, buffers, such as carbonate and phosphate, have affinities less than 400 times those of thiol-containing peptides (WINK et al. 1996d). These results suggest that the primary products of the NO/O_2 reaction in aqueous solution or biological media will be S-nitrosothiols, which are compounds that have been shown to have a variety of effects on biological functions (STAMLER 1994; FEELISCH and STAMLER 1996). Their detection in different biological fluids raises the possibility that this reaction does occur as a result of NOS-generated NO.

E. Oxidative Stress

Oxidation, or the removal of electrons from a substrate, does occur under normal physiological conditions. However, there is a significant difference between normal cellular redox chemistry and that associated with oxidative stress. Under conditions of oxidative stress, powerful oxidizing agents yield oxidized products not found under normal physiological conditions. For example, oxidation of nucleic acids results in DNA-strand breaks (HALLIWELL and GUTTERIDGE 1984). Similarly, oxidation of lipids leads to lipid peroxidation, while oxidation of proteins modifies structure and impedes processes such as enzymatic activity. These reactions have been associated with the onset of different pathophysiological conditions, which suggests that chronic oxidative stress is involved in the etiology of many disease states (HALLIWELL and GUTTERIDGE 1984).

The chemistry of NO can result in the formation of different RNOS that are capable of producing conditions of oxidative stress. The three major RNOS that mediate oxidative stress are NO_2, NO^- and $ONOO^-$. The primary sources of NO_2 are the same as those involved in nitrosative stress: autoxidation of NO, the NO/O_2^- reaction (as discussed below) and the H^+/NO_2^- reaction. Nitrogen dioxide can also nitrate substances such as tyrosine (WINK et al. 1994a) and so might serve as a source of nitrotyrosine observed in vivo. Although DNA does not appear to be altered by NO_2 through processes such as strand breaks (ROUTLEDGE et al. 1994), NO_2 can induce lipid peroxidation (PRYOR 1982b). Under conditions of excess NO, particularly in membranes, nitrosative stress can be induced by formation of N_2O_3 from the rapid reaction of NO with the autoxidation product NO_2 (Eq. 6) or with HOONO (Eqs. 9, 6) (WINK et al. 1997). Therefore, NO_2-mediated oxidation would most likely occur from the acidification of nitrite (Eq. 8) and so is probably limited in vivo.

The nitrogen oxide species NO^- is a chemical intermediate whose role is emerging in the biology of NO. It has been shown that formation of NO^- can result from different processes under a variety of biological situations. One primary source of NO^- is the decomposition of *S*-nitrosothiols (WINK and FEELISCH 1996). The nucleophilic attack of thiols to form *S*-nitrosothiols can result in NO^- and disulfide. Decomposition of dithiothreitol (DTT)-*S*-nitrosothiol results in the formation of NO^- and oxidized DTT (ARNELLE and STAMLER 1995). Other reports suggest that NO^- can be formed from the decomposition of iron dinitrosyls similar to those observed in tumor cells exposed to activated macrophage (BONNER and PEARSALL 1982). One intriguing possibility is that NO^- may be directly derived from NOS or from one of the resulting products (HOBBS et al. 1994; SCHMIDT et al. 1996). Several reports have suggested that NO^- is one of the initial products formed from the conversion of arginine to citrulline. Other reports proposed that oxidation of the catalytic intermediate in NOS activity, hydroxyarginine, may result in NO^- (PUFAHL et al. 1995). Taken together, these processes open up the possibility that NO^- may play a role in the biology of NO.

Several substances that release NO^- have been synthesized, and these have provided a method to study the effects of NO^- in biology (FEELISCH and STAMLER 1996). One of these, $Na_2N_2O_3$ (sodium trioxodinitrate) or Angeli's salt (AS), releases NO^- and NO_3^- at neutral pH. Nitroxyl, which is isoelectric to O_2, can exist in either the triplet or singlet state. It is believed that AS releases NO^- in the singlet state (BONNER and STEDMAN 1996). Intersystem crossing to the triplet state appears to be slower than the singlet-state chemical reactivity. Therefore, the singlet state is of more chemical importance than the ground state. This is probably true for most reactions of NO^-. Recent studies using AS have shown that NO^- is a potent toxic entity (WINK et al. 1998b). The cytotoxicity of AS was two orders of magnitude greater than that of NO-releasing compounds such as SIN-1 and the nitric oxide–nucleophile adduct 1,1-diethyl-2-hydroxy-2-nitroso-hydrazine and was comparable to that of H_2O_2 and alkylhydroperoxide. Hypoxia abates AS cytotoxicity, which suggests that the RNOS chemistry responsible for cell death requires a reaction between NO^- and O_2. Experiments using metal chelators indicate that ROS produced via Fenton-type reactions are not involved. It appears that addition of AS results in both a dramatic loss of GSH and in DNA double-strand breaks. The latter result suggests that the chemical intermediate is not H_2O_2 or $ONOO^-$, as neither of these compounds mediates double-strand breaks.

We have utilized several methods to examine the oxidative, nitrosative and nitrative properties of AS under aerobic and anaerobic conditions (WINK et al. 1998b). Two of the techniques involve the oxidation of nonfluorescent compounds to fluorescent products. Oxidation of dihydrorhodamine (DHR) to rhodamine is a two-electron process. Angeli's salt had a selectivity similar to that of $ONOO^-$ for DHR and was not quenched by azide, which is a N_2O_3 scavenger. Despite these similarities, the oxidative yield from AS was twice that of $ONOO^-$. In contrast to DHR, hydroxyphenylacetic acid (HPA) requires oxidation by only one electron to form a fluorescent dimer. AS produced little product, while $ONOO^-$ proved to be a very effective one-electron oxidant. Thirdly, hydroxylation of benzoic acid was more efficient with AS than with $ONOO^-$. Although nitration of HPA was not detected with AS, $ONOO^-$ readily nitrated HPA. Therefore, it appears that NO^-, in the presence of O_2, produces an intermediate which is distinct from $ONOO^-$.

Examination of O_2 consumption during the decomposition of AS shows that, for every mole of NO^- formed, one mole of O_2 is consumed (WINK et al. 1998a). The NO_2^-/NO_3^- ratio was 0.4. At low concentrations of AS (<100 μM), H_2O_2 was detected. This suggests that nitroxyl, in the presence of O_2, forms H_2O_2. We therefore propose that NO^- is in equilibrium with a hydrated form (H_2ONO^-) in aqueous solution. Singlet NO^- has been shown to react with thiols and amines in a similar manner. This hydrated NO^- adduct can further react with O_2 to form O_2NOO^{3-} (Fig. 6). This species, or a protonated form, can either decompose to form H_2O and NO_3^- or H_2O_2 and NO_2^-. It is O_2NOO^{3-} which is probably responsible for the oxidative chemistry and the cytotoxicity under aerobic conditions in cells.

There appear to be two types of reactivity of NO^- in biological systems: an oxygen-independent and an oxygen-dependent pathway (Fig. 7). Using DHR oxidation (which requires O_2) as a probe, it appears that amines, such

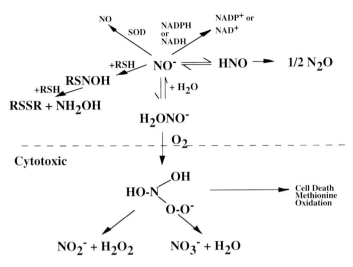

Fig. 7. The oxidative chemistry of niroxyl

as hydroxylamine and thiols, rapidly react with NO⁻ directly. In addition, NO, reduced nicotinamide adenine dinucleotide phosphate and superoxide dismutase (SOD) all react with NO⁻ (MURPHY and SIES 1991; WINK et al. 1998b). Nitroxyl also has been shown to effectively react with metalloproteins, such as myoglobin and catalase. However, oxidation reactions (such as those with methionine) and hydroxylation reactions require O_2. Therefore, there are two tiers of chemical reactivity. It does appear that, in a biological environment, the reaction of NO⁻ with O_2 leads to DNA damage and cell death.

F. NO/O_2^- Chemistry

The reaction between O_2^- and NO has been shown to be very important in the fundamental understanding of NO behavior in biology (BECKMAN et al. 1990; PRYOR and SQUADRITO 1996). Huie and Padmaja showed that NO and O_2^- react at nearly diffusion-controlled rates to form ONOO⁻ (HUIE and PADMAJA 1993). One of the first observations that suggested O_2^- might, in part, control NO concentrations was that SOD enhanced the effect of endothelium-derived hyperpolarizing factor (FURCHGOTT and ZAWADZKI 1980). Due to the fast rate constant for formation of the powerful oxidant ONOO⁻, this reaction could potentially play an important role in the effectiveness of NO on various pathophysiological conditions (BECKMAN et al. 1990). However, the relative pseudo-first-order rate constant is of equal importance in determining whether a reaction occurs in vivo. In other words, the impact a particular reaction will have in vivo is governed by the concentration of the reactants as much as by the rate constant.

The estimated cellular concentrations of O_2^- and NO, under normal conditions, are 1 pM and 0.1–1 μM, respectively. Thus, the location and amount of ONOO$^-$ formation from reactions between these two radicals is likely to be controlled by O_2^- production. Since O_2^- reacts with SOD and NO with similar rate constants, ONOO$^-$ formation will be dependent upon the competing reaction of O_2^- with SOD. The intracellular concentration of SOD is thought to be between 4 μM and 10 μM, while MnSOD is probably as high as 50 μM in the mitochondria, where most of the cellular O_2^- is produced. For 10% of the O_2^- formed to be converted to ONOO$^-$, the NO concentration would have to be 0.4–5 μM. In addition, other reaction partners of O_2^-, such as aconitase ($3 \times 10^7 M^{-1}s^{-1}$) and ferricytochrome c ($5 \times 10^6 M^{-1}s^{-1}$), could also play roles in the abatement of ONOO$^-$ production. Despite the high rate constant, the reaction of O_2^- with NO may be restrained to specific sites.

In neutral solution, ONOO$^-$ has been shown to be a powerful oxidant, in that it can oxidize thiols, initiate lipid peroxidation, nitrate tyrosine, cleave DNA, nitrate and oxidize guanosine and oxidize methionine. The oxidizing species is an excited state of peroxynitrous acid (HOONO*), which is in equilibrium with ONOO$^-$ (KOPPENOL et al. 1992). In the absence of adequate substrate, HOONO* simply rearranges to form nitrate. This isomerization can be considered as a detoxification pathway. However, in high enough concentrations, substrates, such as tyrosine (1 mM for a 50% yield), can react with HOONO* to give nitrotyrosine. It is thought that most of the nitration and oxidative chemistry proceeds through the HOONO* species (KOPPENOL et al. 1992). There are some species that can directly react with the anion. For example, metals can react with ONOO$^-$ at $6 \times 10^3 M^{-1}s^{-1}$ (BECKMAN et al. 1992), and the metals in Cu-SOD and Fe-ethylene diamine tetraacetic acid enhance nitration reactions. Heme-containing enzymes, such as myeloperoxidase ($6 \times 10^6 M^{-1}s^{-1}$), lactoperoxidase ($3.3 \times 10^5 M^{-1}s^{-1}$) and horseradish peroxidase ($3.2 \times 10^6 M^{-1}s^{-1}$), also react rapidly with OONO$^-$ (FLORIS et al. 1993). Reaction of OONO$^-$ with myeloperoxidase and lactoperoxidase resulted in formation of compound II, while compound I was formed from horseradish peroxidase. In fact, this report proposes an interesting mechanism by which compound I is initially formed and then rapidly oxidizes NO_2^- to NO_2. We postulate that NO_2 formation in the presence of NO leads to N_2O_3, which then facilities a number of nitrosation reactions.

Another point to consider is that NO or O_2^- can react with ONOO$^-$ to produce NO_2 (KOPPENOL et al. 1992; BECKMAN et al. 1994; MILES et al. 1996).

$$OONO^- + NO \xrightarrow{H+} NO_2 + NO_2^- \tag{9}$$
$$OONO^- + O_2^- \xrightarrow{H+} NO_2 + H_2O + O_2 \tag{10}$$

This places further restrictions on the chemistry mediated directly by ONOO$^-$. Several studies with NO donors and xanthine oxidase (XO) have examined the fluxes of NO and O_2^- in the context of nitrosative and oxidative chemistry (MILES et al. 1996). XO is considered to be a model for oxidative

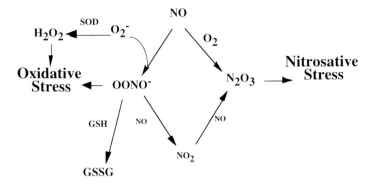

Fig. 8. Balance of oxidative and nitrosative stress in the NO/O_2^- reaction

stress and generates the ROSs O_2^- and H_2O_2 (Fig. 5). In the presence of a NO releasing agent, the amount of H_2O_2 produced by XO is unaffected (WINK et al. 1996b). However, the amount of O_2^- formed was dramatically reduced. It has been proposed that NO reacts with O_2^- formed from XO to form $ONOO^-$, which then isomerizes to form nitrate (CLANCY et al. 1992; WINK et al. 1993b; RUBBO et al. 1994; MILES et al. 1995).

$$NO + O_2^- \rightarrow OONO^- \tag{11}$$

$$OONO^- + H^+ \rightarrow NO_3^- \tag{12}$$

$$OONO^- + DHR \rightarrow oxidation \tag{13}$$

If the reaction is performed in the presence of DHR (MILES et al. 1996), an increase in oxidation is observed. Since $OONO^-$ oxidizes DHR (KOOY et al. 1994), the hypothesis that $ONOO^-$ is a RNOS generated from NO/XO is further supported (Eq. 13).

An intriguing aspect of this study is that the fluxes of NO and O_2^- varied relative to each other. Maximal oxidation through $ONOO^-$ was only achieved at a ratio of 1:1 while, under an excess of either radical, the chemistry of $ONOO^-$ was quenched. In the presence of excess NO or O_2^-, $ONOO^-$ is converted to NO_2 (Eqs. 9, 10) (KOPPENOL et al. 1992; BECKMAN et al. 1994; MILES et al. 1996). NO_2 can rapidly react with NO to form the nitrosating species, N_2O_3 (Eq. 6). Although nitrosation does not occur directly through $ONOO^-$ or $HOONO^*$, there are several mechanisms by which it can arise through the NO/O_2^- reaction. These mechanisms may be important in the biology of NO (Fig. 8).

G. Conclusion

The chemical biology of NO outlined in this chapter is meant to provide a road map for researchers who are investigating the cellular and molecular aspects of NO. The importance of concentration and timing with other ROS cannot be overstated. The direct effects predominate; however, it is the indirect effects that may govern the pathophysiological characteristics of NO. With temporal, stoichiometric, concentration and spatial considerations, the potential reactions of NO can be placed in the context of biological systems.

References

Abu-Soud HM, Wang J, Rousseau DL, Fukuto JM, Ignarro LJ, Stuehr DJ (1995) Neuronal nitric oxide synthase self-inactivates by forming a ferrous-nitrosyl complex during aerobic catalysis. J Biol Chem 270:22997–23006

Abu-Soud HM, Rousseau DL, Stuehr DJ (1996) Nitric oxide binding to the heme of neuronal nitric-oxide synthase links its activity to changes in oxygen tension. J Biol Chem 271:32515–32518

Adams ML, Nock B, Truong R, Cicero TJ (1992) Nitric oxide control of steroidogenesis: endocrine effects of N^G-nitro-L-arginine and comparisons to alcohol. Life Sci 50:PL35–40

Arnelle DR, Stamler JS (1995) NO+, NO, and NO- donation by S-nitrosothiols: implications for regulation of physiological functions by S-nitrosylation and acceleration of disulfide formation. Arch Biochem Biophys 318:279–285

Bartsch H, Ohshima H, Shuker DE, Pignatelli B, Calmel SS (1990) Exposure of humans to endogenous N-nitroso compounds: implications in cancer etiology. Mutat Res 238:255–267

Beckman JS, Beckman TW, Chen J, Marshall PH, Freeman BA (1990) Apparent hydroxyl radical production by peroxylnitrites: implications for endothelial injury from nitric oxide and superoxide. Proc Natl Acad Sci USA 87:1620–1624

Beckman JS, Ischiropoulos H, Zhu L, van der Woerd M, Smith C, Chen J, Harrison J, Martin JC, Tsai M (1992) Kinetics of superoxide dismutase- and iron-catalyzed nitration of phenolics by peroxynitrite. Arch Biochem Biophys 298:438–445

Beckman JS, Chen J, Ischiropoulos H, Crow JP (1994) Oxidative chemistry of peroxynitrite. Methods in Enz 233:229–240

Bonner FT, Pearsall KA (1982) Aqueous nitrosyliron(II) chemistry. 1. Reduction of nitrite and nitric oxide by iron(II) and (trioxodinitrato)iron(II) in acetate buffer. Intermediacy of nitrosyl hydride. Inorg Chem 21:1973–1978

Bonner FT, Stedman G (1996) The chemistry of nitric oxide and redox-related species (eds. Feelisch M, Stamler J), Methods in nitric oxide research. John Wiley and Sons, NY, pp 13–18

Brown GC (1995) Reversible binding and inhibition of catalase by nitric oxide. Eur J Biochem 232:188–191

Choi AM, Alam J (1996) Heme oxygenase-1: function, regulation, and implication of a novel stress-inducible protein in oxidant-induced lung injury. Am J Respir Cell Mol Biol 15:9–19

Clancy RM, Leszczynska-Piziak J, Abramson SB (1992) Nitric oxide, an endothelial cell relaxation factor, inhibits neutrophil superoxide anion production via a direct action on the NADPH oxidase. J Clin Invest 90:1116–1121

Cotton FA, Wilkinson G (1988) Advanced inorganic chemistry (eds Cotton FA, Wilkinson G) Wiley and Sons

Dawson TM, Dawson VL, Synder SH (1992) A novel neuronal messenger molecule in brain: the free radical, nitric oxide. Ann Neurol 32:297–311

Denicola A, Souza JM, Radi R, Lissi E (1996) Nitric oxide diffusion in membranes determined by fluorescence quenching. Arch Biochem Biophys 328:208–212

Doyle MP, Hoekstra JW (1981) Oxidation of nitrogen oxides by bound dioxygen in hemoproteins. J Inorg Biochem 14:351–356

Doyle MP, Mahapatro SN, Broene RD, Guy JK (1988) Oxidation and reduction of hemoproteins by trioxodinitrate(II). The role of nitrosyl hydride and nitrite. J Am Chem Soc 110:593–599

Dweik RA, Laskowski D, Abu-Soud HM, Kaneko F, Hutte R, Stuehr DJ, Erzurum SC (1998) Nitric oxide synthesis in the lung. Regulation by oxygen through a kinetic mechanism. J Clin Invest 101:660–666

Farias-Eisner R, Chaudhuri G, Aeberhard E, Fukuto JM (1996) The chemistry and tumoricidal activity of nitric-oxide hydrogen-peroxide and the implications to cell resistance susceptibility. J Biol Chem 271:6144–6151

Feelisch M (1991) The biochemical pathways of nitric oxide formation from nitrovasodilators: appropriate choice of exogenous NO donors and aspects of preparation and handling of aqueous NO solutions. J Cardiovasc Pharmacol 17:S25–S33

Feelisch M, Stamler JS (1996) Donors of nitrogen (eds. Feelisch M, Stamler J) Methods in nitric oxide research. John Wiley and Sons, NY, pp 71–118

Feldman PL, Griffith OW, Stuehr DJ (1992) The surprising life of nitric oxide. Chem Eng. News December 20:26–38

Floris R, Piersma SR, Yang G, Jones P, Wever R (1993) Interaction of myeloperoxidase with peroxynitrite. A comparison with lactoperoxidase, horse radish peroxidase and catalase. Eur J Biochem 215:767–775

Ford PC, Wink DA, Stanbury DM (1993) Autoxidation kinetics of aqueous nitric oxide. FEBS Lett 326:1–3

Furchgott RF, Vanhoute PM (1989) Endolthelium-derived relaxing and contracting factors. FASEB J 3:2007–2018

Furchgott RF, Zawadzki JV (1980) The obligatory role of endothelial cells in the relaxation of arterial smooth muscle by acetylcholine. Nature 288:373–376

Gorbunov NV, Osipov AN, Day BW, Zayas-Rivera B, Kagan VE, Elsayed NM (1995) Reduction of ferrylmyoglobin and ferrylhemoglobin by nitric oxide: a protective mechanism against ferryl hemoprotein-induced oxidations. Biochemistry 34:6689–6699

Green LC, Tannenbaum SR, Goldman P (1981) Nitrate synthesis in the germfree and conventional rat. Science 212:56–58

Griffith OW, Stuehr DJ (1995) Nitric oxide synthases: properties and catalytic mechanism. Annu Rev Physiol 57:707–736

Griscavage JM, Fukuto JM, Komori Y, Ignarro LJ (1994) Nitric oxide inhibits neuronal nitric oxide synthase by interacting with the heme prosthetic group. Role of tetrahydrobiopterin in modulating the inhibitory action of nitric oxide. J Biol Chem 269:21644–21649

Griscavage JM, Hobbs AJ, Ignarro LJ (1995) Negative modulation of nitric oxide synthase by nitric oxide and nitroso compounds. Adv Pharmacol 34:215–234

Gross SS, Wolin MS (1995) Nitric oxide: pathophysiological mechanisms. Annu Rev Physiol 57:737–769

Gupta MP, Evanoff V, Hart CM (1997) Nitric oxide attenuates hydrogen peroxide-mediated injury to porcine pulmonary artery endothelial cells. Am J Physiol 272:L1133–1141

Halliwell B (1991) Reactive oxygen species in living systems: source, biochemistry, and role in human disease. Am J Med 91:14S-22S

Halliwell B, Gutteridge JMC (1984) Oxygen toxicity, oxygen radicals, transition metals, and disease. Biochem J 219:1–14

Hibbs JB Jr (1991) Synthesis of nitric oxide from L-arginine: a recently discovered pathway induced by cytokines with antitumour and antimicrobial activity. Res Immunol 142:565–569

Hobbs AJ, Fukuto JM, Ignarro LJ (1994) Formation of free nitric oxide from L-arginine by nitric oxide synthase: direct enhancement of generation by superoxide dismutase. Proc Natl Acad Sci 91:10992–10996

Hogg N, Kalyanaraman B, Joseph J, Struck A, Parthasarathy S (1993) Inhibition of low-density lipoprotein oxidation by nitric oxide. Potential role in atherogenesis. FEBS Lett 334:170–174

Hogg N, Struck A, Goss SP, Santanam N, Joseph J, Parthasarathy S, Kalyanaraman B (1995) Inhibition of macrophage-dependent low density lipoprotein oxidation by nitric-oxide donors. J Lipid Res 36:1756–1762

Hoshino M, Ozawa K, Seki H, Ford PC (1993) Photochemistry of nitric oxide adducts of water-soluble iron(III) porphyrin and ferrihemoproteins studied by nanosecond laser photolysis. J Am Chem Soc 115:9568–9575

Huie RE, Padmaja S (1993) The reaction of NO with superoxide. Free Radic Res Commun 18:195–199

Hurshman AR, Marletta MA (1995) Nitric oxide complexes of inducible nitric oxide synthase: spectral characterization and effect on catalytic activity. Biochemistry 34:5627–5634

Ignarro LJ (1989) Endothelium-derived nitric oxide: pharmacology and relationship to the actions of organic esters. Pharm Res 6:651–659

Kanner J, Harel S, Granit R (1991) Nitric oxide as an antioxidant. Arch Biochem Biophys 289:130–136

Khatsenko OG, Gross SS, Rifkind AB, Vane JR (1993) Nitric oxide is a mediator of the decrease in cytochrome P450-dependent metabolism caused by immunostimulants. Proc Natl Acad Sci USA 90:11147–11151

Kim Y-M, Bergonia HA, Muller C, Pitt BR, Watkins WD, Lancaster JR (1995) Nitric oxide and intracellular heme. Adv Pharmacol 34:277–291

Kooy NW, Royall JA, Ischiropoulos H, Beckman JS (1994) Peroxynitrite-mediated oxidation of dihydrorhodamine 123. Free Radic Biol Med 16:149–156

Koppenol WH, Moreno JJ, Pryor WA, Ischiropoulus H, Beckman JS (1992) Peroxynitrite, a cloaked oxidant formed by nitric oxide and superoxide. Chem Res Toxicol 5:834–842

Kwon NS, Stuehr DJ, Nathan CF (1991) Inhibition of tumor cell ribonucleotide reductase by macrophage-derived nitric oxide. J Exp Med 174:761–767

Lancaster J (1994) Simulation of the diffusion and reaction of endogenously produced nitric oxide. Proc Natl Acad Sci USA 91:8137–8141

Lepoivre M, Chenais B, Yapo A, Lemaire G, Thelander L, Tenu JP (1990) Alterations of ribonucleotide reductase activity following induction of the nitrite-generating pathway in adenocarcinoma cells. J Biol Chem 265:14143–14149

Lepoivre M, Fieschi F, Coves J, Thelander L, Fontecave M (1991) Inactivation of ribonucleotide reductase by nitric oxide. Biochem Biophys Res Commun 179:442–448

Lepoivre M, Flaman JM, Henry Y (1992) Early loss of the tyrosyl radical in ribonucleotide reductase of adenocarcinoma cells producing nitric oxide. J Biol Chem 267:22994–23000

Li Y, Severn A, Rogers MV, Palmer RM, Moncada S, Liew EY (1992) Catalase inhibits nitric oxide synthesis and the killing of intracellular *Leishmania major* in murine macrophages. Eur J Immunol 22:441–446

Liu RH, Baldwin B, Tennant BC, Hotchkiss JH (1991) Elevated formation of nitrate and N-nitrosodimethylamine in woodchucks (*Marmota monax*) associated with chronic woodchuck hepatitis virus infection. Cancer Res 51:3925–3929

Liu X, Miller MJS, Joshi MS, Thomas DD, Lancaster JRJ (1998) Accelerated reaction of nitric oxide with O_2 within the hydrophobic interior of biological membranes. Proc Natl Acad Sci USA 95:2175–2179

Marletta MA (1988) Mammalian synthesis of nitrite, nitrate, nitric oxide and N-nitrosating agents. Chem Res Toxicol 1:249–257

MacMicking J, Xie QW, Nathan C (1997) Nitric oxide and macrophage function. Annu Rev Immunol 15:323–350

Miles AM, Gibson M, Krishna M, Cook JC, Pacelli R, Wink DA, Grisham MB (1995) Effects of superoxide on nitric oxide-dependent N-nitrosation reactions. Free Radical Res 233:379–390

Miles AM, Bohle DS, Glassbrenner PA, Hansert B, Wink DA, Grisham MB (1996) Modulation of superoxide-dependent oxidation and hydroxylation reactions by nitric oxide. J Biol Chem 271:40–47

Moncada S, Palmer RMJ, Higgs EA (1991) Nitric oxide: physiology, pathophysiology, and pharmacology. Pharmacol Rev 43:109–142

Murad F (1994) The nitric oxide-cyclic GMP signal transduction system for intracellular and intercellular communication. Recent Progress in Hormone Res 49:239–248

Murphy ME, Sies H (1991) Reversible conversion of nitroxyl anion to nitric oxide by superoxide dismutase. Proc Natl Acad Sci USA 88:10860–10864

Nathan C, Xie Q (1994) Regulation of biosynthesis of nitric oxide. J Biol Chem 269:13725–13728

Padmaja S, Huie RE (1993) The reaction of nitric oxide with organic peroxyl radicals. Biochem. Biophys. Res Commun 195:539–544

Pires M, Ross DS, Rossi MJ (1994) Kinetic and Mechanistic aspects of the NO oxidation by O_2 in Aqueous Phase. Int J Chem Kinet 26:1207–1227

Pryor WA (1982a) Lipid peroxides in biology and medicine (eds Yagi K) Academic, New York, pp 1–22

Pryor WA, Squadrito GL (1996) The chemistry of peroxynitrite and peroxynitrous acid: products from the reaction of nitric oxide with superoxide. Am J Phys 268:L699–L721

Pryor WA, Church DF, Govindan CK, Crank G (1982b) Oxidation of thiols by nitric oxide and nitrogen dioxide: synthetic utility and toxicological implications. J Org Chem 47:156–159

Pufahl RA, Wishnok JS, Marletta MA (1995) Hydrogen peroxide-supported oxidation of N^G-hydroxy-L-arginine by nitric oxide synthase. Biochemistry 34:1930–1941

Puppo A, Halliwell B (1988) Formation of hydroxyl radicals from hydrogen peroxide in the presence of iron: is haemoglobin a biological Fenton reagent? Biochem J 249:185–190

Radi R, Beckman JS, Bush KM, Freeman BA (1991) Peroxynitrite oxidation of sulfhydryls: the cytotoxic potential of superoxide and nitric oxide. J Biol Chem 266:4244–4250

Routledge MN, Mirsky FJ, Wink DA, Keefer LK, Dipple A (1994a) Nitrite-induced mutations in a forward mutation assay: influence of nitrite concentration and pH. Mutat Res 322:341–346

Routledge MN, Wink DA, Keefer LK, Dipple A (1994b) DNA sequence changes induced by two nitric oxide donor drugs in the supF assay. Chem Res Toxicol 7:628–632

Rubbo H, Radi R, Trujillo M, Telleri R, Kalyanaraman B, Barnes S, Kirk M, Freeman BA (1994) Nitric oxide regulation of superoxide and peroxynitrite dependent lipid peroxidation: formation of novel nitrogen containing oxidized lipid derivatives. J Biol Chem 269:26066–26075

Rubbo H, Parthasarathy S, Barnes S, Kirk M, Kalyanaraman B, Freeman BA (1995) Nitric oxide inhibition of lipoxygenase-dependent liposome and low-density lipoprotein oxidation: termination of radical chain propagation reactions and formation of nitrogen-containing oxidized lipid derivatives. Arch Biochem Biophys 324:15–25

Schmidt HH, Hofmann H, Schindler U, Shutenko ZS, Cunningham DD, Feelisch M (1996) No, NO from NO synthase. Proc Natl Acad Sci USA 93:14492–14497

Schwartz SE, White WH (1983) In: Trace atmospheric constituents. Properties, transformation and fates. John Wiley and Sons, pp 1–117

Stadler J, Trockfeld J, Shmalix WA, Brill T, Siewert JR, Greim H, Doehmer J (1994) Inhibition of cytochrome P450 1A by nitric oxide. Proc Natl Acad Sci USA 91:3559–3563

Stamler JS (1994) Redox signaling: nitrosylation and related target interactions of nitric oxide. Cell 78:931–936

Stocker R (1990) Induction of haem oxygenase as a defense against oxidative stress. Free Radic Res Commun 9:101–112

Stone JR, Marletta MA (1994) Soluble gunaylate cyclase from bovine lung: activation with nitric oxide and carbon monoxide and spectral characterization of the ferrous and ferric state. Biochemistry 33:5636–5640

Struck AT, Hogg N, Thomas JP, Kalyanaraman B (1995) Nitric oxide donor compounds inhibit the toxicity of oxidized low-density lipoprotein to endothelial cells. FEBS Lett 361:291–294

Stuehr DJ, Marletta MA (1985) Mammalian nitrate biosynthesis: mouse macrophages produce nitrite and nitrate in response to *Escherichia coli* lipopolysaccharide. Proc Natl Acad Sci USA 82:7738–7742

Williams DLH (1988) In: Nitrosation. Cambridge. Oxford

Wink DA, Ford PC (1995) nitric oxide reactions important to biological systems: a survey of some kinetics investigations. Methods: a companion to methods in enzymology 7:14–20

Wink DA, Feelisch M, (1996) Formation and detection of nitroxyl and nitrous oxide, (eds Feelisch M, Stamler J) Methods in nitric oxide research. John Wiley and Sons. NY, pp 403–412

Wink DA, Mitchell JB (1998) The chemical biology of nitric oxide: insights into regulatory, cytotoxic and cytoprotective mechanisms of nitric oxide. Free Rad Biol Med 25:434–456

Wink DA, Darbyshire JF, Nims RE, Saveedra JE, Ford PC (1993a) Reactions of the bioregulatory agent nitric oxide in oxygenated aqueous media: determination of the kinetics for oxidation and nitrosation by intermediates generated in the NO/O_2 reaction. Chem Res Toxicol 6:23–27

Wink DA, Hanbauer I, Krishna MC, DeGraff W, Gamson J, Mitchell JB (1993b) Nitric oxide protects against cellular damage and cytotoxicity from reactive oxygen species. Proc Natl Acad Sci USA 90:9813–9817

Wink DA, Osawa Y, Darbyshire JF, Jones CR, Eshenaur SC, Nims RW (1993c) Inhibition of cytochromes P450 by nitric oxide and a nitric oxide-releasing agent. Arch Biochem Biophys 300:115–123

Wink DA, Nims RW, Darbyshire JF, Christodoulou D, Hanbauer I, Cox GW, Laval F, Laval J, Cook JA, Krishna MC, DeGraff W, Mitchell JB (1994a) Reaction kinetics for nitrosation of cysteine and glutathione in aerobic nitric oxide solutions at neutral pH. Insights into the fate and physiological effects of intermediates generated in the NO/O_2 reaction. Chem Res Toxicol 7:519–525

Wink DA, Hanbauer I, Laval F, Cook JA, Krishna MC, Mitchell JB (1994b) Nitric oxide protects against the cytotoxic effects of reactive oxygen species. Ann NY Acad Sci 738:265–278

Wink DA, Cook JA, Krishna MC, Hanbauer I, DeGraff W, Gamson J, Mitchell JB (1995) Nitric oxide protects against alkyl peroxide-mediated cytotoxicty: further insights into the role nitric oxide plays in oxidative stress. Arch Biochem Biophys 319:402–407

Wink DA, Hanbauer I, Grisham MB, Laval F, Nims RW, Laval J, Cook JC, Pacelli R, Liebmann J, Krishna MC, Ford PC, Mitchell JB (1996a) The chemical biology of NO. Insights into regulation, protective and toxic mechanisms of nitric oxide. Curr Topics Cellular Regulation 34:159–187

Wink DA, Cook J, Pacelli R, DeGraff W, Gamson J, Liebmann J, Krishna M, Mitchell JB (1996b) Effect of various nitric oxide-donor agents on peroxide mediated toxicity. A direct correlation between nitric oxide formation and protection. Arch Biochem. Biophys 331:241–248

Wink DA, Grisham M, Mitchell JB, Ford PC (1996c) Direct and indirect effects of nitric oxide. Biologically relevant chemical reactions in biology of NO. Methods in Enz 268:12–31

Wink DA, Grisham MB, Miles AM, Nims RW, Krishna MC, Pacelli R, Teague D, Poore CMB, Cook JC (1996d) Methods for the determination of selectivity of the reactive nitrogen oxide species for various substrates. Methods in Enz 268:120–130

Wink DA, Cook JA, Kim S, Vodovotz Y, Pacelli R, Kirshna MC, Russo A, Mitchell JB, Jourd'heuil D, Miles AM, Grisham MB (1997) Superoxide modulates the oxidation and nitrosation of thiols by nitric oxide derived reactive intermediates. J Biol Chem 272:11147–11151

Wink DA, Feelisch M, Fukuto J, Chistodoulou D, Jourd'heuil D, Grisham MB, Vodovotz Y, Cook JA, Krishna M, DeGraff W, Kim S, Gamson J, Mitchell JB (1998a) The cytotoxic mechanism of nitroxyl: possible implications for the pathophysiological role of NO. Nitric Oxide 2:114

Wink DA, Feelisch M, Fukuto J, Chistodoulou D, Jourd'heuil D, Grisham MB, Vodovotz YB, Cook JA, Krishna M, DeGraff W, Kim S, Gamson J, Mitchell JB (1998b) The cytotoxic mechanism of nitroxyl: possible implications for the pathophysiological role of NO. Arch Biochem Biophys in press

Wink DA, Vodovotz Y, Laval J, Laval F, Dewhirst MW, Mitchell JB (1998c) The multifaceted roles of nitric oxide in cancer. Carcinogenesis 19:711–721

Wink DA, Feelisch M, Vodovotz Y, Fukuto J, Grisham MB (1999) The chemical biology of NO. An update, Reactive oxygen species in biological systems, in press

Wong PS, Hyun J, Fukuto JM, Shirota FM, DeMaster EG, Shoeman DW, Nagasawa HT (1998) Reaction between S-nitrosothiols and thiols: generation of nitroxyl (HNO) and subsequent chemistry. Biochemistry 37:5362–5371

Yu AE, Hu S, Spiro TG, Burstyn JN (1994) Resonance raman spectroscopy of soluble guanylyl cyclase reveals displacement of distal and proximal heme ligand by NO. J Am Chem Soc 116:4117–4118

Section II
Biochemistry and Pharmacology of NO Synthesis and Action

CHAPTER 2
Enzymology of Nitric Oxide Synthases

D.J. STUEHR and S. GHOSH

A. Introduction

Nitric oxide (NO) is synthesized in animals and simpler life forms by sequential oxidation of a terminal guanidino nitrogen of L-arginine (L-Arg) (KERWIN et al. 1995; GRIFFITH and STUEHR 1995) (Fig. 1). This pathway was first described in mouse macrophages and bovine endothelium in 1987, although enzymatic hydroxylation at other atoms in L-Arg had been previously demonstrated. The reaction is catalyzed by the NO synthases (NOSs), which all utilize reduced nicotinamide adenine dinucleotide phosphate (NADPH) and O_2 as cosubstrates. Three NOS isoforms have evolved to function in animals, and each gene is located on a different chromosome (NATHAN and XIE 1994). Although alternatively spliced NOSs are sometimes expressed, they are often missing structural elements critical for full function (BRENMAN et al. 1996; EISSA et al. 1998). Two of three NOSs are constitutively expressed in cells, and they synthesize NO in response to increased Ca^{2+} or, in some cases, in response to Ca^{2+}-independent stimuli such as shear stress (FLEMING et al. 1998). These NOSs function in signal transduction cascades by linking temporal changes in Ca^{2+} level to NO production, which serves as an activator of soluble guanylate cyclase (IGNARRO and MURAD 1995). The constitutive enzymes are designated nNOS and eNOS (or NOS I and III, respectively), after the cell types in which they were originally discovered (rat neurons and bovine endothelial cells). An inducible NOS (iNOS or NOS II) is constitutively expressed only in select tissues, such as lung epithelium (DWEIK et al. 1998), and is more typically synthesized in response to inflammatory or proinflammatory mediators (HIERHOLZER et al. 1998; KRONCKE et al. 1995). Expression of iNOS may be beneficial in host defense or in modulating the immune response, but its expression is also linked to a number of inflammatory diseases (reviewed in MACMICKING et al. 1997; KRONCKE et al. 1995; NATHAN 1997).

B. NOS Structure–Function

I. Domain Organization

Each NOS polypeptide is comprised of an N-terminal oxygenase domain and a C-terminal reductase domain, with an approximate 30-amino acid recogni-

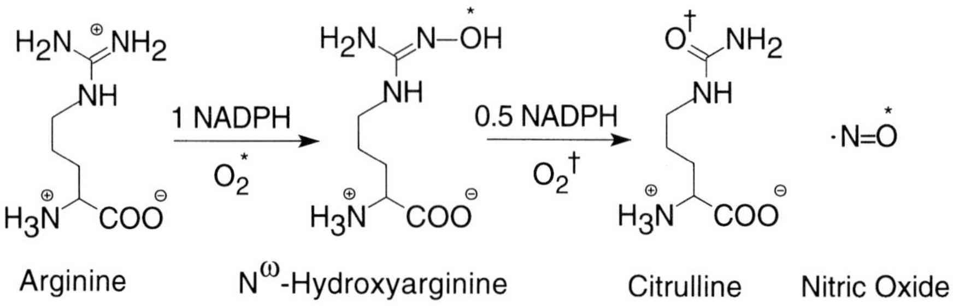

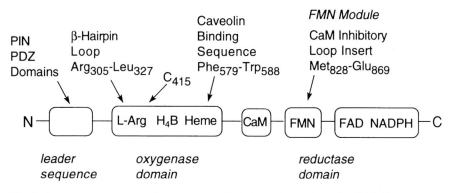

Fig. 1. Enzyme-catalyzed nitric oxide (NO) synthesis from L-Arg. Hydroxylation of L-Arg generates N^G-hydroxy-L-Arg (N^G-OH-L-arg) as an intermediate. The second step converts N^G-OH-L-arg to products NO and citrulline. Oxygen incorporation into products is noted by * or †

Fig. 2. Domain arrangement in rat neuronal nitric oxide synthase (nNOS). The enzyme consists of an approximate 220-amino acid N-terminal leader sequence that targets the enzyme in cells, a core oxygenase domain of approximately 500 amino acids, which forms the active catalytic site, an approximate 30-amino acid CaM-binding domain, and an approximate 700-amino acid reductase domain that is comprised of separate flavin mononucleotide (FMN) and reduced nicotinamide adenine dinucleotide phosphate (NADPH)–flavin adenine dinucleotide (FAD) modules and is responsible for electron import and transfer to the oxygenase domain. Sequence locations of the N-terminal hairpin loop, heme-binding cysteine (C_{415}), a caveolin binding consensus sequence, and putative CaM autoinhibitory loop are noted. The structural organization of iNOS and eNOS are similar to nNOS except that they do not contain an extensive leader sequence and the iNOS reductase domain is missing the CaM autoinhibitory loop. For additional description see HEMMENS and MAYER (1997)

tion sequence for the Ca^{2+}-binding protein, calmodulin (CaM), located between the two domains (reviewed in MASTERS et al. 1996; HEMMENS and MAYER 1997; STUEHR 1997) (Fig. 2). Sequence comparisons show that NOS oxygenase domains share a core region that starts at about proline 76 in mouse iNOS and extends approximately 500 residues to the CaM-binding sequence.

The core region binds heme, tetrahydrobiopterin (H_4B), L-Arg, and forms the active site where NO synthesis takes place. The N-terminal region that is located upstream from the oxygenase core varies in length among the NOSs, participates in cellular targeting, and in some cases may affect NOS structure or catalysis. The C-terminal reductase domains start at the end of the CaM-binding sequences and extend 570–625 residues. The reductase domains bind flavin mononucleotide (FMN), flavin adenine dinucleotide (FAD), and NADPH. During NO synthesis, the reductase flavins aquire electrons from NADPH and transfer them to the heme iron, which permits it to bind and activate O_2 and catalyze NO synthesis (MASTERS et al. 1996; HEMMENS and MAYER 1997; STUEHR 1997). In eNOS and nNOS, the flavin-to-heme electron transfer is triggered by CaM binding, which explains how Ca^{2+} and CaM can regulate NO synthesis by these constitutive NOSs. In mouse iNOS, CaM binding is essentially irreversible (CHO et al. 1992), consistent with iNOS being continuously active once assembled. When expressed in bacteria, eNOS and nNOS fold properly without CaM (RODRIGUEZ-CRESPO et al. 1996; ROMAN et al. 1995; RODRIGUEZ-CRESPO and ORTIZ DE MONTELLANO 1996).

All three NOSs are homodimers in their active forms (MASTERS et al. 1996; HEMMENS and MAYER 1997; STUEHR 1997). Only the oxygenase domains of each NOS appear essential to form the dimer (CHEN et al. 1996; MCMILLAN and MASTERS 1995; GHOSH et al. 1997), although some reports suggest that the reductase domains or CaM-binding site may help regulate the interaction in eNOS (LEE et al. 1995; HELLERMANN and SOLOMONSON 1997; VENEMA et al. 1997). The oxygenase and reductase domains of each NOS have been expressed separately in a number of different cell systems and, in general, retain their individual composition, structure, and catalytic properties (CHEN et al. 1996; MCMILLAN and MASTERS 1995; GHOSH et al. 1997; GACHHUI et al. 1996; BOYHAN et al. 1997; CUBBERLEY et al. 1997). Structural, biochemical, spectroscopic, mutagenic, and kinetic information obtained using the isolated domains or full-length enzymes is reviewed in the following sections.

II. NOS Oxygenase Domains and Mutagenesis

Crystal structures are available for monomeric (CRANE et al. 1997) and dimeric mouse iNOS oxygenase domains (CRANE et al. 1998) and for dimeric oxygenase domains of human iNOS (FISCHMANN et al. 1999), human eNOS (FISCHMANN et al. 1999), and bovine eNOS (RAMAN et al. 1998). Several point and deletion mutagenesis studies, mostly carried out prior to solution of the crystal structures, have also been reported. A review of certain pertinent information follows, but interested readers should consult the original articles for details.

To achieve correct crystallization, a portion of each oxygenase domain protein's N-terminus was deleted in all cases. The downstream CaM-binding site was either not present in the constructs or, if present, was not observable in the protein crystal structure (CRANE et al. 1998; RAMAN et al. 1998;

FISCHMANN et al. 1999). Together, the data show there is striking but not complete conservation of tertiary structure between all of the dimeric oxygenase proteins mentioned above (CRANE et al. 1998; RAMAN et al. 1998; FISCHMANN et al. 1999). The core structure of each oxygenase is made up of several overlapping or winged beta sheets. This distinguishes NOS from the cytochrome P_{450} superfamily, whose folds are mainly comprised of helical elements (POULOS and RAAG 1992; RAVICHANDRAN et al. 1993). The structural elements responsible for binding L-Arg, H_4B, and heme are located throughout the oxygenase domain, rather than being arranged in a clear linear series of subdomains along the polypeptide sequence. However, some subdomain structure within the oxygenase can still be inferred (FISCHMANN et al. 1999).

1. L-Arg-Binding Site

Each dimeric structure reveals how Arg and analogs bind within the active site. There are two identical substrate-binding channels that lead to either heme pocket. L-Arg binds in an elongated conformation, with only its guanidino nitrogens held close to the heme iron (Fig. 3), consistent with a role for heme in activating O_2 for substrate oxidation. One terminal and one bridging guanidino nitrogen of L-Arg form hydrogen bonds with the carboxylate oxygens of a conserved glutamate. Mutagenesis of this glutamate to Ala caused an absolute and specific loss of L-Arg binding in both bovine eNOS and mouse iNOS (GACHHUI et al. 1997; CHEN et al. 1997) (Table 1), indicating that it is essential for L-Arg binding. Other residues that help bind the guanidino group of L-Arg and its α-amino and carboxylate moieties are depicted in Fig. 3. Mutagenesis at some of these sites has been reported.

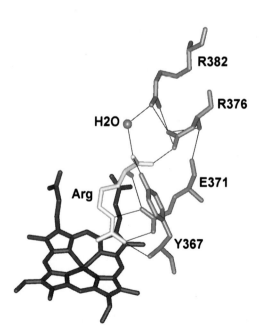

Fig. 3. L-Arg binding in mouse iNOS. An Arg molecule (*white*) binds in the active site with its guanidino group pointing down toward the heme (*black*). The L-Arg molecule forms hydrogen bonds with several protein residues (*gray*) that are located on the substrate binding helix. Hydrogen bonding that occurs between protein residues and L-Arg guanidino nitrogens or its α-amino and carboxyl groups are shown as *thin lines*. Adapted from CRANE et al. (1998)

Table 1. Properties of point mutants in nitric oxide synthase (NOS) oxygenase domains

NOS residue[a]	Phenotype	Reference
R80Aox	No Apparent effect	Ghosh et al. 1999; Crane et al. 1999
K82Aox	Defect in dimerization	Ghosh et al. 1999; Crane et al. 1999
N83Aox	Defect in dimerization	Ghosh et al. 1999; Crane et al. 1999
W84Aox	No apparent effect	Ghosh et al. 1999; Crane et al. 1999
D92ox, rnD314A	Defect in dimerization	Sagami and Shimizu 1998; Iwasaki et al. 1999; Ghosh et al. 1999; Crane et al. 1999
T93ox, rnT315A	Defect in dimerization	Sagami and Shimizu 1998; Iwassaki et al. 1999; Ghosh et al. 1999; Crane et al. 1999
H95Aox	Defect in dimerization	Ghosh et al. 1999; Crane et al. 1999
K97Aox	No apparent effect	Ghosh et al. 1999
C109Aox, heC99A, rnC331A	Decreased H_4B binding affinty, normal activity. Arginine binding is affected in case of bNOS	Ghosh et al. 1997; Chen et al. 1994; Chen et al. 1995; Martasek et al. 1998
G111Aox	No apparent effect	Ghosh et al. 1999
S112Aox	No apparent effect	Ghosh et al. 1999
M114Aox	No apparent effect	Ghosh et al. 1999
N115Aox	No apparent effect	Ghosh et al. 1999
P116Aox	No apparent effect	Ghosh et al. 1999
K117Aox	No apparent effect	Ghosh et al. 1999
C194A, heC184A, rnC415A, hiC200A, hiC200H	Defect in heme binding	Sari et al. 1996; Chen et al. 1994; Richards et al. 1996; Cubberley et al. 1997; Perry and Marletta 1998
[C205], hiC211A	No apparent effect	Cubberley et al. 1997
[C216], hiC222A	No apparent effect	Cubberley et al. 1997
[S245], he235A	No apparent effect	Cubberley et al. 1997
[C278], hiC284A	No apparent effect	Cubberley et al. 1997
[E352], heE342I	Normal	Chen et al. 1997
E357Aox, heE347L	Normal Arg and/or H_4B binding, lower activity in heNOS, oxygenase domain is normal in miNOS	Gachhui et al. 1997; Chen et al. 1997
[T370], heT360A	Normal	Chen et al. 1998
E371Aox, heE361Q	Absolute defect in Arg binding, inactive	Gachhui et al. 1997; Chen et al. 1997
[C372], hiC378A	No apparent effect	Cubberley et al. 1997
R375Aox, heR365L	Monomer in case of miNOS, active in heNOS	Ghosh et al. 1999; Chen et al. 1998
D376Aox	Defect in Arg binding and dimerization	Gachhui et al. 1997
[C378], heC368A	Normal	Chen et al. 1998
D379Aox, heD369I	Defect in heme, Arg and H_4B binding, inactive	Gachhui et al. 1997; Chen et al. 1998

Table 1. *Continued*

NOS residue[a]	Phenotype	Reference
[R382], heR372L	Same as above	CHEN et al. 1998
[Y383], heY373F	Normal	CHEN et al. 1998
E387Aox, heE377I	Normal	GACHHUI et al. 1997; CHEN et al. 1997
E388Aox, heD378E	Normal	GACHHUI et al. 1997; CHEN et al 1998
E396Aox	Normal	GACHHUI et al. 1997
E411Aox	Defect in dimerization, low activity	GACHHUI et al. 1997
D429Aox	Defect in heme, H_4B and/or Arg binding, low activity	GACHHUI et al. 1997
E435Aox	Normal	GACHHUI et al. 1997
E444Aox	Defect in heme, H_4B and/or Arg binding	GACHHUI et al. 1997
A447I	Normal	CHO et al. 1995
G450A	Absolute defect in dimerization	CHO et al. 1995
C451A, hiC457, heC441A	Normal	CHO et al. 1995; CUBBERLEY et al. 1997; CHEN et al. 1994
P452A	Normal	CHO et al. 1995
A453I	Absolute defect in dimerization	CHO et al. 1995
W455Aox	Defect in dimerization, inactive	GHOSH et al. 1999
W455Yox	Oxygenase domain normal, active in heterodimer	GHOSH et al. 1999
W455Fox	Same as above	GHOSH et al. 1999
W457Aox	Oxygenase domain inactive, partially active in heterodimer	GHOSH et al. 1999
W457Fox	Oxygenase domain normal, active in heterodimer	GHOSH et al. 1999
P461A	Defect in dimerization and H_4B binding	CHO et al. 1995
F470Aox	Oxygenase domain inactive, partially active in heterodimer	GHOSH et al. 1999
F470Yox	Oxygenase domain normal, active in heterodimer	GHOSH et al. 1999
F470Wox	Same as above	GHOSH et al. 1999

Species other than mouse iNOS are indicated by: *h* human; *r* rat; *i* inducible, *e* endothelial, *n* neuronal.
[a] The leftmost residue represents numbering for mouse iNOS residues. Brackets mean that the mutation has not been reported in mouse iNOS. The suffix "ox" means the mutation was studied using the NOS oxygenase domain, otherwise mutants were studied as full-length NOS.

2. H_4B-Binding Site

There are two identical H_4B-binding sites in all NOS oxygenase dimers, consistent with biochemical studies showing maximal activity requires that each subunit contain a H_4B (HEVEL and MARLETTA 1992; GORREN et al. 1996). The H_4B-binding site is near the dimer interface and utilizes protein structural

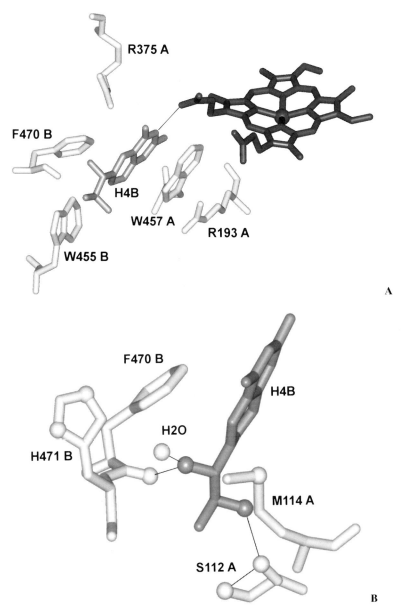

Fig. 4A,B. H$_4$B binding in mouse iNOS. **A** The view looks down on the distal side of the heme (*black*). H$_4$B (*gray*) is surrounded by the heme and five conserved protein residues (*white*), two of which are provided by the partner subunit of the dimer (designated with a *B*). A hydrogen bond (*thin line*) between one heme propionate oxygen and the N5 of H$_4$B is also shown. **B** Detailed view of the interactions between H$_4$B's dihydroxypropyl side chain and mouse iNOS residues contributed from the same subunit (*A*) or partner subunit (*B*). Adapted from CRANE et al. (1998)

Table 2. Properties of nitric oxide synthase (NOS) N-terminal deletion mutants

Mouse iNOS	Bovine eNOS	Rat bNOS	Phenotype
GHOSH et al. 1997	RODRIGUEZ-CRESPO et al. 1997	BOYHAN et al. 1997	
Δ 65 Oxy	Δ 52 FL	Δ 220 Oxy	Normal
Δ 114 Oxy	Δ 91 FL	Δ 349 Oxy	Defect in Arg and/or H_4B binding, mixture monomer-dimer
Δ 117 Oxy	Δ 105 FL		Defect in dimerization and Arg, H_4B binding

elements that are also involved in the dimeric interaction (CRANE et al. 1998; RAMAN et al. 1998; FISCHMANN et al. 1999). Some specific aspects of the H_4B-binding site in mouse iNOS are depicted in Fig. 4. The H_4B ring is positioned to the side of the heme, almost perpendicular to and somewhat below the heme plane. The N5 of H_4B hydrogen bonds with a carboxylate of a heme propionate side chain. In addition, there are five highly conserved residues that either hydrogen bond with H_4B or engage in pi cloud stacking with its pterin ring (CRANE et al. 1998). Surprisingly, the F470 and W455 residues derive from the partner polypeptide in the dimer, meaning residues are shared between subunits to make each H_4B site. This is consistent with the extensive structural integration in the dimer interface, and may help explain why H_4B can promote NOS subunit dimerization or stabilize an existing dimeric structure (STUEHR 1997; KLATT et al. 1995). The residues that participate in H_4B binding are also linked to the L-Arg binding site through an integrated network of hydrogen bonding, which may explain why these molecules bind in a cooperative manner to the NOS (LIST et al. 1997; KLATT et al. 1994; MAYER et al. 1997a).

The 6-dihydroxypropyl side chain of H_4B also interacts with several residues (Fig. 4). Its two hydroxyl groups hydrogen bond with the backbone carbonyl of F470, a hisitdine imidazole nitrogen, and a serine hydroxyl. The serine is conserved among the NOS and is part of an N-terminal structure whose presence is essenial for dimerization and high affinity H_4B binding (Table 2). Work with H_4B analogs that differ only in their side chain structure suggests that side-chain interactions with NOS are a major determinant of binding affinity (PRESTA et al. 1998a).

Roles of four conserved iNOS residues (R375, W455, W457, and F470) that interact directly with the H_4B ring (Fig. 4) were tested recently by means of site-directed mutagenesis (GHOSH et al. 1999; CHEN et al. 1998) (Table 1). In iNOS individual conversion to Ala generated heme-containing monomers at each position except for W457A. The monomers displayed an absolute defect in forming a homodimer but did form heterodimers that were either fully or partially active. Thus, R375, W455, F470, but not W457, are important for stabilizing the iNOS dimer, but in a dimeric setting their individual identities or aromatic character need not be maintained for H_4B binding or catalysis of NO synthesis. Interestingly, the R365A eNOS mutant (the same as

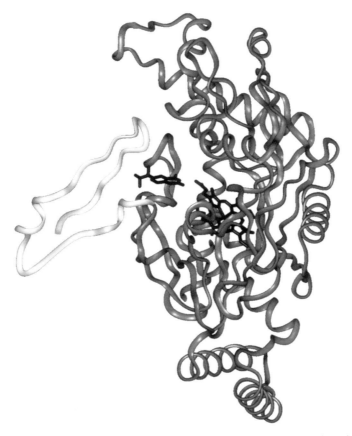

Fig. 5. The N-terminal hairpin loop. Ribbon diagram of one oxygenase domain subunit of a mouse iNOS dimer highlighting its N-terminal loop (residues Q77 to S100, *white*), bound H_4B and heme (*black*). Adapted from CRANE et al. (1998)

iNOS R375A) formed a homodimer that had reduced NO synthesis activity, suggesting that a difference exists between the two NOSs regarding this residue's importance for dimerization. Aromatic substitution at W455, W457, or F470 in iNOS generated monomers that could dimerize with H_4B and Arg. Their affinities toward Arg and H_4B were near normal, but they appeared to have some changes in the heme pocket and NO synthesis activities (Table 1). Further characterization should help us understand how these residues modulate the multiple effects of H_4B on NOS structure and catalysis.

3. N-Terminal Hairpin Loop

In all NOS oxygenase dimers, the N-termini of the core region (segment Q77-S105 in mouse iNOS) form a hairpin loop structure (Fig. 5). The loop structures are potentially significant for three reasons: (i) their primary sequences show considerable variation among the NOSs, suggesting they may impart characteristics unique to each isoform; (ii) they contain residues near or involved in H_4B binding; and (iii) they may reach out from the parent subunit to interact with the partner subunit of the dimer, a potentially dynamic means of dimer cross-talk (CRANE et al. 1998). Unfortunately, structural disintegrity in the crystal structures near the putative subunit overlap region prevented a definitive assignment regarding crossover among the N-terminal hooks until recently (CRANE et al. 1999). Progressive deletion analysis in iNOS, eNOS, and nNOS had already demonstrated the importance of the N-terminal loop structure in maintaining dimer stability or for binding Arg and H_4B (Table 2). Deletions preceding the loop structure had little or no effect on structure and catalysis, while deletions up to 40 residues into the core region progressively prevented dimerization and binding or response to L-Arg and H_4B, without affecting heme content or properties. For nNOS, it is important to note that this isoform contains a unique ~220-residue leader sequence upstream from the hairpin loop structure (Fig. 2). In one study, deletion of this N-terminal leader sequence appeared to lower activity by two-thirds by an unknown mechanism (BRENMAN et al. 1996), although, in a more recent study, a similar deletion did not affect nNOS heme incorporation or affinity toward H_4B and L-Arg (BOYHAN et al. 1997) (Table 2). An N-terminally deleted NOS expressed in testis also showed partial loss of activity (WANG et al. 1997). Point mutagenesis of residues in the hairpin loop region of iNOS and nNOS inhibits dimerization (IWASAKI et al. 1999; GHOSH et al. 1999) and affects NADPH oxidation and NO synthesis (SAGAMI and SHIMIZU 1998) (Table 1), and provides biochemical evidence for N-terminal hook swapping (GHOSH et al. 1999).

4. NOS Cysteines and Metal Binding

Eight Cys residues are conserved within the core oxygenase domains of the NOS. Individual mutagenesis of five conserved Cys to Ala had little or no effect on NOS structure and catalysis (CUBBERLEY et al. 1997; CHO et al. 1995; CHEN et al. 1994; RICHARDS and MARLETTA 1994) (Table 1). Mutagenesis and sequence analysis correctly predicted the heme-binding Cys in each NOS (MCMILLAN and MASTERS 1995; RENAUD et al. 1993; XIE et al. 1996; CHEN et al. 1994; RICHARDS and MARLETTA 1994). This Cys resides in a motif that is generally conserved among thiol-ligated hemeproteins, but there are some interesting differences in its orientation between NOSs and cytochrome $P450$s (CRANE et al. 1998).

The N-termini of all NOSs also contain a highly conserved dual Cys motif (C104XXXXC109 in mouse iNOS) that apparently serves to bind metals such as Zn^{2+} (RAMAN et al. 1998; FISCHMANN et al. 1999). Crystal structures show that each subunit of a dimer can contribute its dual Cys motif to generate a tetrahedral coordination of four Cys to one Zn^{2+} atom per dimer. In the absence of bound Zn^{2+}, the two Cys109 residues of mouse iNOS can form a disulfide bridge across the dimer interface (CRANE et al. 1998). Disulfide bond formation could represent a reversible modification in response to oxidant stress, and may promote N-terminal hook swapping between NOS subunits (CRANE et al. 1999).

Mutation of Cys109 to Ala in mouse iNOS and the analogous Cys residues in eNOS and nNOS have been reported (Table 1). Because this mutation removes two of the four essential metal ligands and is likely to prevent metal binding at the site, the mutant phenotypes can shed light on metal function. In general, the Ala mutants of eNOS and iNOS display lower H_4B affinity but otherwise have normal structure, L-Arg affinity, and catalytic activity (GHOSH et al. 1997; CHEN et al. 1994; CHEN et al. 1995). The nNOS Ala mutant displays a defect in Arg binding in the absence of H_4B, but is otherwise normal (MARTASEK et al. 1998). These results suggest that metal binding at this site likely stabilizes the local structure which, in turn, helps bind the H_4B dihydroxypropyl side chain and possibly stabilize the dimer, but otherwise does not impact on aspects of catalysis once H_4B is bound. Because Zn^{2+} has only been observed in recombinant NOS expressed in bacteria, it is presently unclear whether Zn^{2+} or other metals bind to NOS in intact animal cells.

Several other oxygenase point mutations affected both L-Arg and H_4B binding and, in some cases, also inhibited dimerization (Table 1). Such mixed phenotypes likely reflect the integration between binding sites for these molecules and the subunit interface (CRANE et al. 1998). In two cases, relatively conservative point mutations (G450A, A453I) completely prevented iNOS subunit dimerization and consequent L-Arg or H_4B binding (CHO et al. 1995). The crystal structures suggest that this phenotype could result from G450 and A453 residues providing backbone hydrogen bonding interactions that insure proper alignment between the C- and N-termini of adjacent oxygenase subunits in the iNOS dimer. Thus, defective dimerization in the G450A and A453I mutants, as well as in the N-terminal hairpin loop mutants noted above, all support a model where exchange occurs between subunit N-terminal hairpin loops in a NOS dimer (CRANE et al. 1998).

III. NOS Reductase Domains

1. General Features

The NOS reductase domain amino acid sequences are at least 50% homologous with other FMN- and FAD- containing reductases, such as cytochrome P_{450} reductase (PORTER 1991; WANG et al. 1997b), and data obtained thus far suggest the NOS reductase domains conserve the key features of this class of enzymes. As illustrated in the crystal structure of the related enzyme cytochrome P_{450} reductase (WANG et al. 1997b), the NOS reductase domain is likely comprised of two modules, one that binds NADPH and FAD, and the other that binds FMN (WANG et al. 1997b) (Fig. 2). Electron transfer typically procedes from the NADPH–FAD module to the FMN module, and from the FMN module to the heme acceptor which, in the case of NOS, is its oxygenase domain.

The NOS reductase domains can stabilize their one-electron reduced flavin semiquinone forms to varying degrees. However, the semiquinone electron is transferred slowly to the heme iron upon CaM binding (MCMILLAN and MASTERS 1995; GACHHUI et al. 1996; PRESTA et al. 1997; STUEHR and IKEDA-SAITO 1992). The reductase semiquinone can accept at least two additional electrons from NADPH to be shared between the FAD and FMN groups (MCMILLAN and MASTERS 1995; GACHHUI et al. 1996). Thus, it is likely that the three- and two-electron reduced forms of the reductase transfer electrons to the heme during NO synthesis. Indeed, stopped-flow analysis indicates that NOS flavin reduction by NADPH is faster than electron transfer to the heme (PRESTA et al. 1998a; GACHHUI et al. 1998), suggesting that flavins remain sufficiently reduced during NO synthesis and their reduction is not rate limiting.

Although the physical details of the electron transfer events are important, they have yet to be worked out for NOS or for any related reductase-hemeprotein pair. Recent crystallography work with a related enzyme cytochrome P_{450}BM3 reveals how its FMN module can contact its attached heme domain, providing a possible glimpse at what an electron transfer complex may look like (SEVRIOUKOVA et al. 1999). Interestingly, a portion of the FMN module that contacts the heme domain in the P_{450}BM3 structure also contacts the NADPH–FAD module in the cytochrome P_{450} reductase structure (WANG et al. 1997b), supporting an earlier proposal that the FMN module may rapidly shuttle between its adjacent upstream and downstream partners to perform electron transfer (SEVRIOUKOVA et al. 1999). Whether FMN module shuttling is a general feature shared among related flavoproteins is still unknown. In NOS, the fact that CaM binding activates electron transfer between the FMN and heme adds another level of complexity.

2. Catalytic Properties and Response to CaM

CaM plays a critical role in activating NOS because its binding triggers electron transfer from the reductase domain to the heme (ABU-SOUD et al. 1994). CaM binding alters the conformation of the reductase domain (GACHHUI et al. 1996; NARAYANASAMI et al. 1997; GALLI et al. 1996), increases the rate of electron transfer from NADPH into the flavins (GACHHUI et al. 1996; GACHHUI et al. 1998), and increases the rate at which the reductase transfers electrons to acceptors such as cytochrome c or ferricyanide (HEMMENS and MAYER 1997; RODRIGUEZ-CRESPO et al. 1996; ROMAN et al. 1995; CHEN et al. 1996; LIST et al. 1997; GACHHUI et al. 1996; MARTASEK et al. 1996; GACHHUI et al. 1998). Oxygen, however, is reduced relatively slowly by the NOS reductases (LIST et al. 1997; GACHHUI et al. 1996; ABU-SOUD et al. 1994; MILLER et al. 1997; XIA and ZWEIER 1997; VASQUEZ-VIVAR et al. 1998). Most of these changes also occur when CaM binds to eNOS, but to a smaller extent (CHEN et al. 1996; LIST et al. 1997; MARTASEK et al. 1996; PRESTA et al. 1997; VASQUEZ-VIVAR et al. 1998).

Work with a recombinant nNOS reductase domain that contains a functional N-terminal CaM-binding site showed that CaM activates electron transfer to external acceptors independent of the oxygenase domain (MCMILLAN

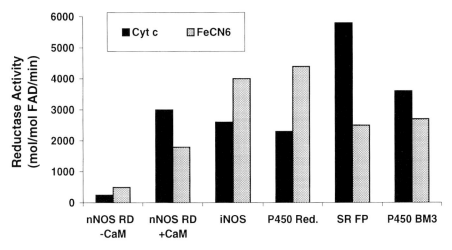

Fig. 6. Calmodulin (*CaM*) relieves repression of nNOS reductase domain catalysis. The cytochrome c and ferricyanide (FeCN6) reductase activities (expressed as turnover number per minute) of the rat nNOS reductase domain (*RD*) in the presence or absence of bound CaM are compared with activity values obtained from the literature for the related flavoproteins mouse iNOS, rat cytochrome P_{450} reductase (P_{450} *Red*), bacterial sulfite reductase flavoprotein (*SR FP*), and bacterial cytochrome P_{450}BM3

and MASTERS 1995; GACHHUI et al. 1996; GACHHUI et al. 1998; MILLER et al. 1997). In contrast, oxygenase domain properties and reactivity are either unchanged or affected to a much lesser extent upon CaM binding to the full-length NOSs (MATSUOKA et al. 1994; GERBER et al. 1997). Although this means that CaM must activate NO synthesis by influencing the reductase domain, which changes enable heme reduction during NO synthesis remain to be defined.

Comparing the nNOS reductase domain to related flavoproteins regarding their rates of cytochrome c or ferricyanide reduction shows that in the CaM-free state nNOS is much slower than all other flavoproteins of its class (Fig. 6). However, CaM binding increases the nNOS rates to comparable levels. This suggests that the nNOS flavoprotein evolved to supress electron flux in the basal state, and the supression is relieved by CaM. Because NADPH reduction of the FAD and FMN centers is also quickened when CaM binds to nNOS, this provides a mechanism where CaM relieves supression of electron transfer both into and out of the reductase domain (GACCHUI et al. 1998).

3. Mutagenesis

Mutational analysis of NOS reductase domains has not been as extensive as for NOS oxygenase domains. Conserved residues involved in NADPH binding have been identified at the extreme C-terminal end of the mouse iNOS reductase (XIE et al. 1994). More recently, two conserved acidic residues in the rat

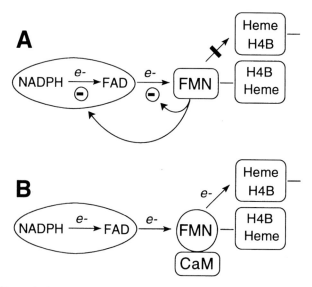

Fig. 7A,B. Key role for the flavin mononucleotide (FMN) module in controlling electron transfer in nNOS. **A** In the absence of calmodulin (CaM), the FMN module represses electron transfer from reduced nicotinamide adenine dinucleotide phosphate (NADPH) into flavin adenine dinucleotide (FAD), accepts electrons slowly from FAD, and is unable to transfer electrons to the heme. **B** CaM binding alters the structure of the FMN module and relieves repression of both upstream and downstream electron transfer events. Adapted from ADAK et al. (1999)

nNOS FMN module (D918, E919) and a neighboring conserved Phe (F892) were mutated to assess their role in reductase function (ADAK et al. 1999). This study was based on work in related flavoproteins that suggest a role for the analogous conserved acidic residues in flavoprotein–hemeprotein interaction and electron transfer (SHEN et al. 1999; JENKINS et al. 1997). Individual or dual conversion of D918 and D919 residues to Ala destabilized FMN binding and, thereby, inhibited heme reduction and NO synthesis, but otherwise did not affect other aspects of the nNOS. The effects were largely reversed by adding FMN. This indicates that the primary effect of mutation was on FMN binding affinity, implying that these residues function differently in nNOS than other flavoproteins.

Further study with the FMN-free nNOS mutants revealed that the effects of CaM on the reductase are probably restricted to and mediated by the FMN module. Surprisingly, the FMN module repressed upstream electron transfer from NADPH into the NADPH–FAD module, and the repression was relieved either by removing FMN or by CaM binding. Repression of upstream electron transfer is a new function for an FMN module in any flavoprotein. The effect appears to be unique to the NOS, and likely explains how reductase functions are supressed in the absence of bound CaM (Fig. 7).

IV. CaM Activation of NOS

1. Mechanism of Action

What structural elements of CaM are important for binding to nNOS and "derepressing" the various electron-transfer reactions? This has been investigated using CaM mutants whose Ca^{2+}-binding site in each of CaM's four domains was individually abolished (STEVENS-TRUSS et al. 1997), using various chimeras constructed from CaM and the homologous Ca^{2+}-binding protein cardiac troponin c (GACHHUI et al. 1998; SU et al. 1995; GEORGE et al. 1996) and using plant CaMs that bind to nNOS but either do or do not activate full NO synthesis (CHO et al. 1998). These studies reveal: (1) structural integrity of CaM domains 1, 3, and 4 are important to maintain full activation of NO synthesis; and (2) electron transfer to the heme may involve the latch region of CaM (formed by domains 1 and 3), and specific residues in the fourth domain of CaM. Importantly, several of the CaM proteins mentioned above increased rates of electron transfer into and out of the nNOS reductase domain, while triggering little or no heme reduction and NO synthesis. This indicates that the two processes are separable and implies that CaM must provide additional elements besides those that affect reductase domain electron transfer in order to support heme reduction and NO synthesis.

2. Structural Determinants of CaM Binding

Full-length NOSs or synthetic peptides that represent their canonical CaM-binding sequences have been used to determine CaM binding and dissociation kinetics and derive a structural model (ANAGLI et al. 1995; PERSECHINI et al. 1996; VORHERR et al. 1993; ZOCHE et al. 1996; STEVENS TRUSS and MARLETTA 1995; MATSUBARA et al. 1997). Selective deletion of the CaM-binding sequence has also been investigated in eNOS (VENEMA et al. 1996b). Whether CaM binds to the iNOS CaM-binding sequence in its Ca^{2+}-free form is controversial (STEVENS-TRUSS and MARLETTA 1995; MATSUBARA et al. 1997). In general, results using CaM-binding sequence peptides suggest that CaM affinity towards each NOS is largely determined by the canonical sequence. However, experiments in which the canonical binding segments were swapped among the NOSs indicate that at least two other regions in NOS help modulate CaM-binding affinity. In iNOS, a region essential for high-affinity CaM binding is located in the reductase domain between the CaM-binding sequence and the FMN subdomain (RUAN et al. 1996; VENEMA et al. 1996a). A distinct region in the oxygenase domain also appears to increase affinity (LEE and STULL 1998). However, CAM binding by nNOS and eNOS appears to be inhibited by a distinct structural element that is present as an insertion in the reductase domains of eNOS and nNOS but absent in iNOS (SALERNO et al. 1997). This apprixi-mate 40-amino acid insert is located within the FMN module (Fig. 2) and is thought to be an autoinhibitory loop based on function of similar inserts present in other CaM-responsive enzymes. Studies that further characterize these positive and negative control elements are awaited with interest.

V. NOS Domain Interactions

Oxygenase–oxygenase and oxygenase–reductase domain interactions have been examined using yeast two-hybrid analysis (VENEMA et al. 1997), nNOS–eNOS–iNOS chimeras in which reductase domains have been swapped (NISHIDA and ORTIZ de MONTELLANO 1998), and iNOS heterodimers formed by co-expressing dissimilar NOS polypeptides or NOS chimeras in cells (LEE et al. 1995; XIE et al. 1996; RUAN et al. 1996) or by combining dissimilar purified iNOS polypeptides in vitro (SIDDHANTA et al. 1996; SIDDHANTA et al. 1998).

The chimera work showed that each reductase can transfer electrons to support NO synthesis when linked to oxygenase domains from other NOS isoforms. This suggests the oxygenase–reductase surface interaction sites are conserved among the NOS. Some highlights of the yeast two-hybrid and heterodimer studies were:

1. The reductase domain of eNOS, but not of iNOS or nNOS, may play a role in dimer formation.
2. An iNOS mutant that could not form a homodimer with itself (for example, G450A) could still dimerize with wild-type subunits or with a variety of mutants that contained a dissimilar mutation, often to generate an active heterodimer.
3. A single iNOS reductase domain transfers electrons to only one of the two hemes in a dimer, and this is sufficient to support a normal rate of catalysis by that heme.
4. Electron transfer occurs between iNOS reductase and oxygenase domains that are located on adjacent subunits of the heterodimer.

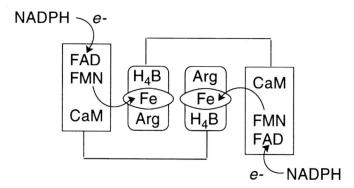

Fig. 8. Role for dimerization in completing electron transfer. In the iNOS dimer, each reductase domain appears to transfer electrons from flavin mononucleotide (FMN) to the heme located in the partner subunit. Adapted from SIDDHANTA et al. (1998)

Together, these results imply that a domain swap takes place in an iNOS dimer to complete electron transfer to the heme (Fig. 8). This model helps explain why NOS dimerization is critical for NO synthesis, and suggests co-evolution of dimeric structure and domain function. Whether eNOS and nNOS dimers also engage in cross-subunit electron transfer during NO synthesis, remains to be addressed. Further testing of this model is awaited with interest.

Isolated eNOS and iNOS reductase domains can support varying degrees of NO synthesis when mixed with their respective oxygenase domains (CHEN et al. 1996; GHOSH et al. 1995). In both cases, the rates are less than those observed for intact full-length enzymes. Poor catalysis in these systems is consistent with the fact that NOS domains are covalently linked and are also brought together by dimerization. These facets remove selective pressure to maintain working interactions between separated domains in solution.

C. Catalysis of NO Synthesis from L-Arg

I. Heme-NO Complex Formation

1. NOS Partitioning into an NO-Bound Form During Catalysis

The proposed mechanism of NO synthesis draws parallels from oxygenation reactions catalyzed by the cytochromes P_{450}: stepwise activation of O_2 at a heme center, leading to mixed function oxygenation of the substrate. However, NOS and cytochrome P_{450} catalysis differ in an important way: only in the NOS does product NO also bind to the heme during steady-state catalysis. This partitions the enzyme molecules between a NO-free, active form and a NO-bound, inactive form. Although the NO complexes are inherently inactive (they cannot bind O_2), they eventually decay to active ferric enzyme. However, for iNOS and nNOS, NO-complex decay is sufficiently slow such that about 70–90% of these enzymes exist as their NO complexes during the steady state (HURSHMAN and MARLETTA 1995; ABU-SOUD et al. 1995). The kinetics of NO binding to iNOS, eNOS, and nNOS hemes have been recently studied (ABU-SOUD et al. 1998; HUANG et al. 1999; NEGRERIE et al. 1999; SCHEELE et al. 1997, 1999).

2. Impact of NO Complex Formation on NOS Catalysis

NO complex formation is significant for at least two reasons:

1. It makes the NOS exhibit burst kinetics (Fig. 9, Table 3). When the reaction is initiated, all enzyme molecules are active but eventually an equilibrium is reached between free and NO-bound forms of the enzyme (i.e., the steady state). This takes about 2s for nNOS and less than 1 min for iNOS. Only the fraction of NOS molecules that are NO-free are active. Thus, NO synthesis activities for nNOS (and for iNOS in cases where NO is not scavenged by oxyhemoglobin) in the literature represent only a fraction

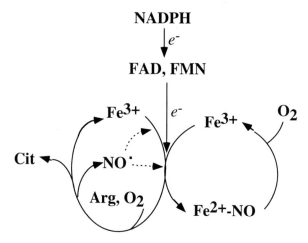

Fig. 9. Partitioning of nitric oxide synthase (NOS) into active and inactive cycles during nitric oxide (NO) synthesis. nNOS and iNOS cycle in an active, NO-generating cycle (*left*) and an inactive cycle (*right*) that involves formation and decay of a heme-NO complex. When product NO binds to ferric or ferrous NOS (*dotted arrows*), it shifts the NOS molecule to the inactive cycle. Decay of the NO complex enables the NOS molecule to join the active cycle. Adapted from ADAK et al. (1999)

Table 3. Iron–nitric oxide (Fe–NO) complex formation and nitric oxide synthase (NOS) catalysis

Reaction period	Burst phase	Steady state
Definition	NO synthesis initiated, Fe–NO complex forms	Equilibrium reached between NO complex formation and decay
Time frame	First 2–60s	After 2–60s
Major rate-limiting step	Heme reduction? Product release?	Decay of the Fe–NO complex

(approximately 10–30%) of the enzyme's true activity, which is only seen in the burst phase when all molecules are active.

2. It makes NOS circulate in two cycles during the steady state: an active cycle that generates NO and an inactive cycle that involves formation and decay of the NO complex (Fig. 9). All enzyme molecules start in the active cycle before NO complex formation occurs (the burst phase, Table 3). But once NO complex formation reaches equilibrium (the steady state), a given enzyme molecule can circulate in and between the active and inactive cycles many times, although enzyme cycling in the inactive cycle can take longer than in the active cycle.

3. NO Complex and the NOS O_2 Response

Partitioning NOS into two cycles fundamentally affects how activity is regulated. For example, each cycle has its own rate-limiting step (Table 4) which together determine the rate of NO synthesis that is observed in the steady state. Importantly, both cycles are O_2-dependent (Fig. 9), but have different concentration–response profiles for O_2 (ABU-SOUD et al. 1996). The active cycle's O_2 response is determined by the affinity of the ferrous NOS heme toward O_2, which for nNOS has an apparent K_mO_2 of about $30\mu M$. In contrast, the inactive cycle's O_2 response is determined by the oxygen sensitivity of the Fe–NO complex, whose rate of decay increases in a near linear relationship to O_2 concentration. Thus, O_2 response parameters from both cycles determine the observed activity versus O_2 concentration response for a NOS. However, in situations where more than 50% of the enzyme is present as a NO complex during the steady state (this is typical for nNOS and iNOS), or when O_2 is present at concentrations that nearly saturate the heme, it is the O_2 dependence of NO complex decay that essentially determines the observed activity versus O_2 concentration curve for the enzyme. In practice, this enables nNOS and iNOS to generate NO at a rate that is almost linearly related to the O_2 concentration across the physiologic O_2 concentration range and beyond (ABU-SOUD et al. 1996). Evidence that iNOS and nNOS in lung and carotid body tissues alter their O_2 response according to this mechanism is available (DWEIK et al. 1998; PROBHAKAR et al. 1993).

II. The Active Catalytic Cycle

1. Steps involved, O_2, Binding and Activation

The active cycle contains all the steps that lead to formation of NO. These include substrate binding, heme iron reduction by the flavin domain, O_2 binding, bond making, bond breaking, proton transfer, and electron transfer steps involved in oxygen activation or product formation, release of products, and any protein conformational changes that are associated with these steps. Although most of the details are still unclear, it is generally accepted that heme reduction leads to O_2 binding and a stepwise activation of heme-bound O_2 to generate the oxidant species that participate in NO synthesis (Fig. 10). Evidence suggests that distinct iron–oxygen species react in each step of NO synthesis (ABU-SOUD et al. 1997a, 1997b; BEC et al. 1998; SATO et al, 1998). As shown in Fig. 10, an electrophilic iron–oxo species (III), which is typically invoked in cytochrome P_{450} hydroxylation, is thought to hydroxylate L-Arg and form enzyme-bound N^G-OH-l-arg in the first step of NO synthesis. In the second step, the iron–dioxy species (I) formed by O_2 binding to ferrous heme is thought to act as an oxidant and abstract an H atom or electron from N^G-OH-L-arg. This generates a nucleophilic iron–peroxo species (II), which typically forms as an intermediate during cytochrome P_{450} oxygen activation. However, instead of the O–O bond of (II) cleaving to form water and species (III), in the second step the iron–peroxo species (II) is predicted to react directly with the N^G-OH-L-arg radical, generating a tetrahedral intermediate that rearranges to NO and citrulline (Fig. 10). These issues are discussed in detail in other papers (KERWIN et al. 1995; GRIFFITH and STUEHR 1995; BEC et al. 1998; MARLETTA et al. 1998).

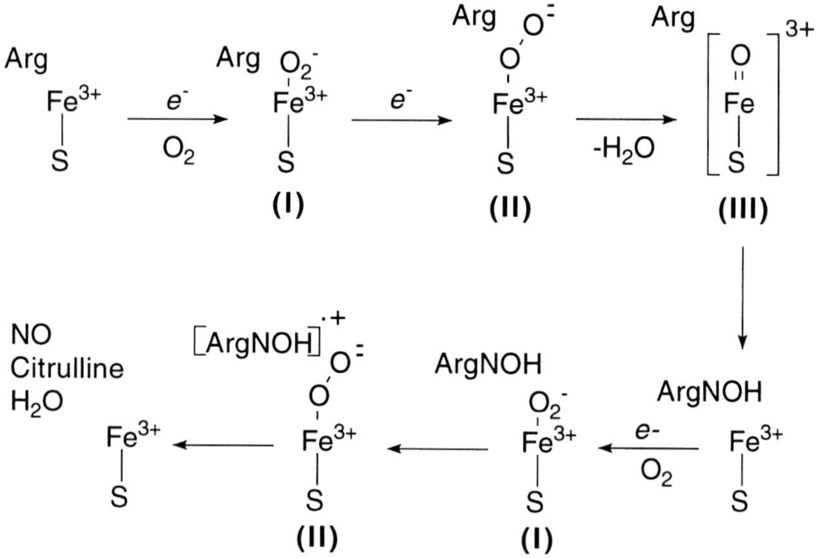

Fig. 10. Proposed mechanism of heme-based O_2 activation and reactivity of the heme-oxy complexes during NO synthesis by NOS. The iron oxo species (III) is envisioned to react with Arg, while the ferrous-O_2 species (I) reacts with the intermediate ArgNOH

2. NOS Heme Iron Reduction

Heme reduction is essential for oxygen activation during NO synthesis and is a potential rate-limiting step in the active cycle. Evidence suggests that heme reduction may occur at different rates in the eNOS, nNOS, and iNOS forms. For example, NOS chimeras displayed a rate of NO synthesis characteristic of the reductase they contained (NISHIDA and ORTIZ DE MONTELLANO 1998). This was particularly striking for the eNOS reductase–nNOS oxygenase chimera, whose rate of NO synthesis was slowed considerably compared with native nNOS. The authors proposed that electron transfer from the eNOS reductase may be slow enough to become rate limiting for NO synthesis.

Another study used CaM–troponin c chimeras to activate nNOS and directly compare rates of heme iron reduction and NO synthesis (GACHHUI et al. 1998). Chimeras that supported slower rates of heme iron reduction also had proportionally slower rates of NO synthesis. Similar results were obtained using CaM mutants that are defective in Ca^{2+} binding (STEVENS-TRUSS et al. 1997). This established that heme reduction does indeed become rate limiting in nNOS when it is slowed compared with wild type. However, it is currently unknown whether increasing the rate of heme reduction beyond wild-type would also increase the rate of NO synthesis.

3. Control of Heme Reduction by H_4B and Arg

Controlling heme reduction is important from the standpoint of reactive oxygen toxicity, because in the absence of substrate or H_4B, heme reduction in NOS leads to uncoupled oxidation of NADPH and production of superoxide and H_2O_2. Interestingly, rates of uncoupled NADPH oxidation under the various conditions differ among the NOS isoforms, suggesting their heme reduction is regulated differently by substrate and H_4B. Reduction potential measurements for the iNOS and nNOS hemes taken in the presence or absence of substrates and H_4B have been recently reported, and provide a thermodynamic basis for the different uncoupled NADPH oxidation rates (PRESTA et al. 1998b). As shown in Fig. 11 and Table 4, the heme in iNOS is held at a much lower potential than in nNOS in the absence of Arg and H_4B. This causes a thermodynamic block for heme iron reduction in iNOS, whereas, in nNOS, heme reduction can occur and leads to significant uncoupled NADPH oxidation. Individually, H_4B or Arg have either positive or negative effects on heme reduction potential in iNOS and nNOS, respectively, which are usually (but not always) mirrored by changes in NADPH oxidation under each circumstance (Table 4). Arg analogs also positively or negatively regulate heme reduction in iNOS and nNOS, and actually block heme reduction in some cases (S-ethylisothiourea, N-nitroArg) (PRESTA et al. 1998b).

It is remarkable that iNOS evolved for H_4B and Arg to thermodynamically regulate its heme reduction, whereas these molecules do not appear to be important for regulating heme iron reduction in nNOS. Perhaps this relates to different availabilities of H_4B and particularly Arg in cells or tissues that express iNOS versus nNOS (discussed in GRIFFITH and STUEHR 1995).

Table 4. Relationship between reduced nicotinamide adenine dinucleotide phosphate (NADPH) oxidation rate, heme iron reduction, and heme redox potential in iNOS and nNOS

Condition	NADPH oxidase[a] (nmol/min/mg enzyme)	Heme reduction[b] by NADPH	E_m (mV)[c]
iNOS			
$-H_4B, -Arg$	61 ± 5	No	-347
$-H_4B, +Arg$	165 ± 1	Yes	-235
$+H_4B, -Arg$	245 ± 1	Yes	-295
$+H_4B, +Arg$	1270 ± 140	Yes	-263
nNOS			
$-H_4B, -Arg$	540 ± 90	Yes	-239
$+H_4B, -Arg$	1450 ± 30	Yes	-257
$-H_4B, +Arg$	360 ± 10	Yes	-220
$+H_4B, +Arg$	880 ± 10	Yes	-248

Data from Presta et al. 1998.
[a] Measured with full-length NOS.
[b] Measured under anaerobic conditions with full-length NOS.
[c] Measured with the NOS oxygenase domains.

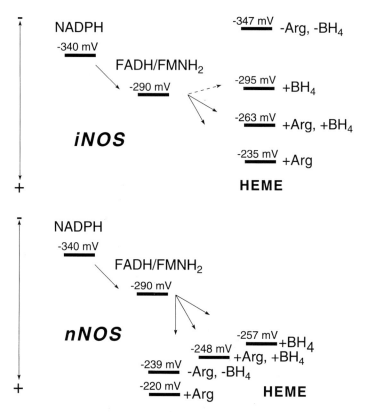

Fig. 11. Effect of L-Arg and H_4B on the thermodynamics of heme iron reduction in iNOS and nNOS. In the absence of H_4B and L-Arg, the iNOS heme is held at too low a potential to accept electrons from the flavins. Binding L-Arg or H_4B increases the iNOS heme potential into a range where electron transfer is thermodynamically possible. In contrast, the nNOS heme is held at a more positive potential whether L-Arg and H_4B are bound or not, and heme reduction is possible in all cases. Adapted from PRESTA et al. (1998)

III. Enzyme Structural Features that May Impact on NO Synthesis

Because NOS and cytochrome P_{450}s catalyze similar chemistry, the crystal structures provide an opportunity to identify important similarities and differences in their respective heme environments. Some issues, including their similar hydrophobic distal pockets, the lack of an obvious hydrogen bond donor in NOS for O_2 cleavage during NO synthesis, and implications for catalysis, have been discussed (CRANE et al. 1997, 1998).

Figure 12 shows two different views of the heme environment in the mouse iNOS dimer, highlighting three conserved aromatic residues that surround the NOS heme. F363 and especially W188 engage in aromatic stacking with the porphyrin ring in a manner that is seen in some peroxidases but is

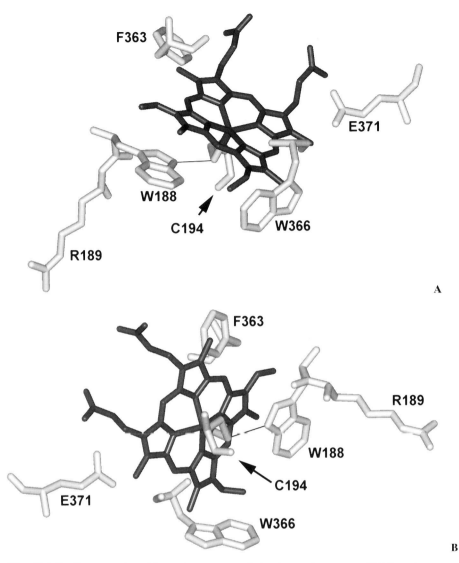

Fig. 12A,B. Conserved residues that surround the heme in mouse iNOS and may impact on its function. **A** Looking down from above the distal face of the heme. **B** Looking up from below the proximal face of the heme. The heme-binding cysteine (C194) and the residue responsible for holding Arg above the heme (E371) are included for orientation. The hydrogen bond formed between an indole nitrogen of W188 and the C194 sulfur is drawn as a *thin line*. Adapted from CRANE et al. (1997, 1998). See text for additional discussion

not typical for cytochrome P_{450}s (CRANE et al. 1998). The indole nitrogen of W188 also hydrogen bonds with the cysteine thiolate that serves as the fifth heme ligand in NOS. Strong hydrogen bonding to the heme thiolate is not generally seen in P_{450}s, but is present in chloroperoxidase, and in NOS is speculated to reduce the electronegativity of the heme iron through an inductive effect (CRANE et al. 1998). W366 is positioned near a portion of the heme that is exposed to solvent in the dimer, which has been proposed to be a potential site for reductase domain docking and electron transfer (CRANE et al. 1998). R189 also connects the protein surface to the proximal side of the NOS heme via W188. Similarly positioned positive residues are implicated in helping reductase proteins dock with cytochrome P_{450} for electron transfer to the heme (SEVRIOUKOVA et al. 1999; WANG et al. 1997). How these residues effect NO synthesis is a fascinating question and is under investigation.

IV. Roles for Heme and H_4B

The NOS dimer crystal structures all show that the hemes are positioned at a considerable distance from one another (Fig. 13). This is consistent with each heme functioning independently regarding its accepting electrons from the reductase and catalyzing NO synthesis (GORREN et al. 1996; GORREN et al. 1998; SIDDHANTA et al. 1996; 1998). The relative position of bound substrate to H_4B versus the heme iron clearly supports a heme-based chemistry for oxygen activation and NO synthesis, and rules out a direct role for H_4B in the reaction

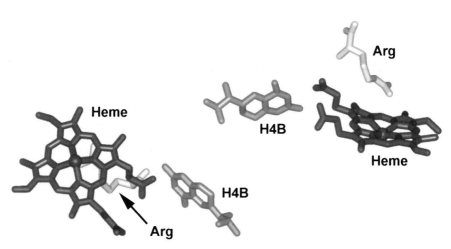

Fig. 13. Relative orientations of heme, L-Arg, and H_4B molecules in each active site of the iNOS dimer. Approximate distances in angstroms (Å) are: within the same subunit the heme iron to the guanidino nitrogen (4 Å), the guanidino nitrogen to the 4a carbon of the H_4B ring (12 Å); and between subunits are: heme edge to the heme edge in the partner subunit (24 Å), and heme iron to heme iron (34 Å). Adapted from CRANE et al. (1998)

Table 5. Functions of H_4B in nitric oxide synthase (NOS)

Function	Reference
Stabilizes dimer	Baek et al. 1993; McMillan and Masters 1995; Tzeng et al. 1995; Abu-Soud et al. 1995; Ghosh et al. 1996; Mayer et al. 1997
Protects against proteolysis	Ghosh et al. 1997
Prevents binding of large heme ligands	Ghosh et al. 1997; Renodon et al. 1998; Sennequier et al. 1999
Slows CO binding to the heme 100 fold (but not NO or O_2 binding)	Abu-Soud et al. 1997a, 1998; Scheele et al. 1997; 1999; Huang et al. 1999; Sato et al. 1998
Stabilizes Fe-S bond	Wang et al. 1995; Abu-Soud et al. 1998; Huang et al. 1999
Heme iron spin state change	McMillan and Masters 1993; Ghosh et al. 1997; Mayer et al. 1997; Gerber et al. 1997; Rusche et al. 1998
Alter heme reduction potential	Presta et al. 1998a, 1998b
Destabilize heme-O_2 complex	Abu-Soud et al. 1997

chemistry. In fact, one of the current challenges is to explain how H_4B can be essential for NO synthesis given its postion in the active site.

H_4B impacts several different aspects of NOS structure and function (Table 5). Most but not all of its effects can be mimicked by dihydro biopterin (Presta et al. 1998a), which does not support NO synthesis. The one effect that seems specific for H_4B is its destabilizing the ferrous–O_2 complex (Abu-Soud et al. 1997). H_4B could modulate reactivity of the ferrous–O_2 complex (species I in Fig. 10) through its hydrogen-bonding interactions with a heme propionate (Crane et al. 1998) or through less obvious electronic effects on the heme. This impacts directly on the second step of NO synthesis, because the ferrous–O_2 complex appears to react directly with bound N^G-OH-L-arg (Abu-Soud et al. 1997). In fact, the ability of H_4B to modulate heme ferrous–O_2 reactivity potentially explains the recent observation that H_4B-free iNOS can generate citrulline and nitroxyl from N^G-OH-L-arg rather than citrulline and NO (Rusche et al. 1998).

Work that examined an electron donating function of H_4B was done with nNOS in the absence of NADPH (Campos et al. 1995). Some conversion of L-Arg to N^G-OH-L-arg was observed and potentially attributed to H_4B-derived reducing equivalents, but the electron involved was eventually found to derive from the flavin semiquinone present in the nNOS reductase domain (Witteveen et al. 1998). The nNOS also exhibited a poor ability to reduce dihydro biopterin to H_4B (Witteveen et al. 1996). In contrast, a single turnover study of NO synthesis by nNOS suggested H_4B may provide an electron to the heme for conversion of L-Arg to N^G-OH-L-arg (Bec et al. 1997). However, this was not indicated in an earlier single turnover study with nNOS (Abu-Soud et al. 1997b). In any case, electron donation from H_4B remains an important possibility that needs to be fully examined before we can understand the role of H_4B in NO synthesis.

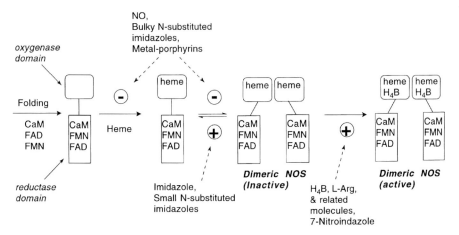

Fig. 14. Steps involved in assembly of an active NOS dimer. A functional reductase domain forms first, followed by heme incorporation, dimerization, and H₄B binding in the oxygenase domains. Several molecules can either promote or inhibit dimer assembly, at the points indicated. See text for details

D. Control Mechanisms and Targeting

I. NOS Dimerization

1. Stepwise Assembly Mechanism

The NOS polypeptide must incorporate FAD, FMN, heme, H$_4$B, and form a homodimer to become active. For iNOS, this is accomplished in stages (Fig. 14) (STUEHR 1997). As shown in Fig. 14, CaM binding and FAD and FMN incorporation occur first to form a functional reductase domain, followed by heme incorporation into the oxygenase domain, and dimerization of heme-containing monomers with stable incorporation of H$_4$B. In eNOS and nNOS, heme incorporation is also essential for dimer assembly, but H$_4$B may not be needed (RODRIGUEZ-CRESPO et al. 1996; RODRIGUEZ-CRESPO et al. 1997; HEMMENS et al. 1998; LIST et al. 1997; VENEMA et al. 1997). However, H$_4$B can stabilize the nNOS dimer once formed (KLATT et al. 1995). In animal cells, heme incorporation appears to be rate limiting for iNOS dimer assembly, because heme-free iNOS monomers accumulate in cells, whereas heme-containing monomers do not (ALBAKRI and STUEHR 1996). It is unclear whether this is also the case for nNOS and eNOS. However, when any NOS is expressed in bacteria, heme-containing monomers and dimers are typically present upon lysis. A study that investigated the abilities of various metal protoporphyrins to incorporate into heme-free nNOS monomers suggests that formation of a metal–thiolate bond is a critical step that alters nNOS conformation and enables the monomers to dimerize (HEMMENS et al. 1998).

2. Positive and Negative Regulation

As shown in Fig. 14, a number of molecules can promote or inhibit formation of the NOS dimer. H_4B and L-Arg, analogs of H_4B and Arg, and 7-nitroindazole typically promote formation of NOS dimers by binding to and stabilizing the heme-containing dimer once it has formed (SENNEQUIER and STUEHR 1996; PRESTA et al. 1998a; KLATT et al. 1995). In intact mammalian cells, H_4B was required for significant dimerization of human iNOS (TZENG et al. 1995). Small imidazoles also promote NOS dimerization, but act by a distinct mechanism that initially involves their binding to the heme-containing iNOS monomer (SENNEQUIER et al. 1999).

NO blocks cellular assembly of dimeric iNOS by inhibiting heme insertion into the monomer (ALBAKRI and STUEHR 1996; STUEHR 1997). This normally occurs in cells that are induced to express iNOS, results in accumulation of heme-free monomers, and can be largely prevented by adding a NOS inhibitor. Bulky N-substituted imidazoles, such as clotrimazole, also inhibit cellular assembly of the iNOS dimer through an effect on heme (SENNEQUIER et al. 1999). Certain proteins that interact with the N-terminus of iNOS also appear to inhibit dimer assembly (RATOVITSKI et al. 1999).

II. Is NOS Oxygenase Domain Structure Modified by Dimerization, H_4B Binding, or Both?

This question is important but still controversial. For iNOS, it has been addressed by aligning the hemes in the H_4B-free iNOS monomer and the L-Arg- and H_4B-replete iNOS dimer structures (CRANE et al. 1998). The analysis revealed structural differences in only certain protein elements (CRANE et al. 1998). Dimerization and H_4B binding appeared to recruit peripheral loop and helical elements from both the N- and C-terminal ends of the core oxygenase domain into a central region to participate in forming the dimer interface, H_4B binding site, and substrate channel. In particular, a disordered loop structure in the monomer was converted to a helical lariat in the dimer that provides most of the conserved residues that interact with H_4B. There also was a change in position of one helix, which exposed a heme edge on the solvent-accessible side of the dimer relative to the monomer. This analysis clearly shows that structural changes occur between heme-containing monomer and H_4B-saturated dimer, but cannot distinguish whether dimerization or H_4B incorporation is the cause. However, studies with the heme-containing, H_4B-free iNOS dimer isolated from bacteria suggest its heme environment is equivalent to that of the heme-containing monomer (ABU-SOUD et al. 1998). This, along with the many changes observed upon binding H_4B to the NOS dimer (Table 5), argue that H_4B binding may cause a greater degree of conformational change than subunit dimerization alone in the absence of H_4B.

Recent crystal structures of the eNOS oxygenase show very little structural difference between the H_4B-bound dimer and the H_4B-free, substrate-bound dimer (RAMAN et al. 1998). All of the structural changes noted for the iNOS dimer above were present in both forms of eNOS. This would indicate that dimerization rather than H_4B binding caused the structural changes in this case. However, a structure for the heme-containing eNOS monomer is not yet available. Moreover, in the H_4B-free eNOS dimer L-Arg or glycerol molecules were present in place of H_4B at its binding site, and it is possible that they could mimic H_4B in causing structural changes. Alternatively, eNOS and iNOS may have fundamentally difererent structural responses to dimerization and H_4B binding. Further insights are awaited with interest.

III. Another Type of NO Inhibition

NO inactivates NOS when its synthesis occurs in the presence of subsaturating H_4B (GRISCAVAGE et al. 1995). Under this condition a gradual inactivation takes place over several minutes, unlike the instant inhibition attributed to heme-NO complex formation as discussed above. In at least one case, the inhibited NOS was reactivated by adding excess H_4B (GRISCAVAGE et al. 1994). Although the mechanism of inactivation has not been directly investigated, it could involve uncoupled O_2 reduction that occurs in the absence of sufficient H_4B (KLATT et al. 1994; MAYER and WERNER 1995), dimer dissociation into monomers (STUEHR 1997), or NO binding to the heme and subsequent breakage of the cysteine thiolate–heme iron bond, as occurs for iNOS and nNOS treated with reagent NO in the absence of H_4B and L-Arg (ABU-SOUD et al. 1998; MIGITA et al. 1997).

The mechanism by which NO inhibits H_4B-free nNOS was recently investigated (HUANG et al. 1998). NO bound to the heme of H_4B-free nNOS and converted the enzyme to an inactive P_{420} species, which no longer maintained its cysteine thiol axial heme ligand. The protein reactivated over a 1-h time period when incubated with H_4B alone or plus L-Arg, and enzyme reactivation was associated with re-ligation of heme iron to the cysteine thiolate. This suggests H_4B and L-Arg perform an important role in preventing NO inactivation of NOS through the P_{420} conversion mechanism.

IV. Interactions Between NOS and Other Proteins

1. PDZ, PIN

The N-terminal approximate 220-amino acid leader sequence unique to nNOS contains PDZ-binding motifs that interact with several proteins, including syntrophin, post-synaptic density protein 93 (PSD 93), PSD 95 (FANNING and ANDERSON 1999), and phosphofructokinase M (FIRESTEIN and BREDT 1999). The general function of PDZ domains may be to localize their ligands to the appropriate plasma membrane domain or metabolic complex. Syntrophins are membrane-associated components of the dystrophin-associated protein complex in muscle cells that bind to nNOS via interaction between the PDZ domains in these proteins. Significantly, nNOS isoforms lacking a PDZ domain fail to localize to the plasma membrane (BRENMAN et al. 1996) and nNOS is diffusely distributed throughout the cytosol in mice lacking the syntrophin gene (KAMEYA et al. 1999). Thus, PDZ domains, like those found in syntrophin and nNOS, have the capacity to serve as adapters or linkers between transmembrane and cytosolic proteins, recruiting cytosolic proteins to specific subcellular domains.

The mammalian homologue of dynein light chain, called protein inhibitor of NO synthase (PIN), also interacts with the leader sequence of nNOS at a distinct site and was reported to destabilize its dimeric structure, thereby inhibiting the enzymatic activity (JAFFREY and SNYDER 1996; TOCHIO et al. 1998). The ability of PIN to inhibit various NOS in a purified system has also been investigated (HEMMENS et al. 1998).

2. Caveolins

The oxygenase domains of nNOS, eNOS and iNOS all contain consensus binding motifs for caveolins, which are a family of structural scaffolding proteins present in membrane caveolae (GARCIA-CARDENA et al. 1996; MICHEL 1997). In many tissues, caveolae may serve as sites for sequestration of signaling molecules such as receptors, G proteins and protein kinases, as well as eNOS (MICHEL 1997). Caveolae have not yet been identified in neuronal cells, although it has been speculated that there exists a neuron-specific caveolin isoform caveolin-3. Caveolae are certainly present in skeletal muscle, and it appears possible, if not likely, that nNOS expressed in skeletal muscle may interact with muscle-specific caveolin-3 isoform. Caveolin-3 binds to nNOS and inhibits its NO synthesis, the inhibition is reversed by CaM, and this process is thought to locate nNOS near other signal transduction proteins (VENEMA et al. 1997). Because the binding of caveolin and CaM may be mutually exclusive, it seems unlikely that iNOS, which binds CaM tightly, is regulated by interactions with caveolin or its targetting to caveolae. Futher work on the issues is awaited with interest.

Recent work on eNOS and caveolin-1 interactions suggests its physiological relevance as an inhibitor of NOS (FERON et al. 1999). Some work has been done to understand the mechanism of inhibition. Although caveolin-1 does not interact directly with CaM or with the CaM-binding motif on eNOS, adding CaM can reverse the inhibition by caveolin-1 in a competitive manner (MICHEL et al. 1997a; JU et al. 1997; GARCIA-CARDENA et al. 1997; MICHEL and FERON 1997; MICHEL et al. 1997b). Mutagenesis of the caveolin-1 binding motif prevented inhibition of eNOS by caveolin-1, suggesting it does indeed interact with a portion of the oxygenase domain (GARCIA-CARDENA et al. 1997). At this point, it is unknown how caveolin-1 inhibits and how CaM can reverse the inhibition. A recent report shows that caveolin-1 interacts directly with the eNOS reductase domain to inhibit its electron transfer function (GHOSH et al. 1998). This suggests that caveolins can bind to multiple sites in the NOSs and may prevent or slow electron transfer between the reductase and oxygenase domains.

3. Heat Shock Proteins

Recently, it has been shown that eNOS is activated by Hsp90 (GARCIA-CARDENA et al. 1998). This study demonstrated a direct activation of purified eNOS catalytic activity by purified Hsp90 in the absence of added heme or other co-factors, and in the absence of co-factors for the Hsp90-based chaperone machinery. On the basis of these results, they suggested that Hsp90 may act as an allosteric modulator of eNOS. This is an important finding because the specific activity of eNOS preparations have always been about three- to fivefold lower than comparable iNOS or nNOS preparations. Recently, a portion of nNOS was shown to exist as a molybdate-stabilized nNOS-Hsp90 heterocomplex in the cytosolic fraction of human embryonic kidney 293 cells stably transfected with rat nNOS (BENDER et al. 1999). However, heterocomplex assembly with Hsp90 is not required for increased heme binding and nNOS activation in this cell-free system. It is still unclear how Hsp90 might increase eNOS activity and/or regulate nNOS in vivo.

4. Kalirin

Screening a hipppocampal cDNA library revealed, using a yeast two-hybrid approach, that kalirin associates with iNOS in vitro and in vivo and may inhibit iNOS activity by destabilizing the dimer (RATOVITSKI et al. 1999). Continued work along these lines is likely to identify many other cellular proteins that interact with different NOS isoforms to locate them in cells and positively or negatively regulate their activity.

Acknowledgments. The authors thank Dennis Wolan for helping prepare the manuscript, Drs. Tainer, Getzoff, Crane, and Arvai from Scripps Research Institute for many insightful discussions, and those who provided manuscripts prior to publication. Supported in part by NIH grants CA53914, GM51491, HL58883, and a grant from Berlex Biosciences. DJS is an Established Investigator of the American Heart Association.

References

Abu-Soud HM, Feldman PL, Clark P, Stuehr DJ (1994a) Electron transfer in the nitric-oxide synthases. Characterization of L-arginine analogs that block heme iron reduction. J Biol Chem 269:32318–32326

Abu-Soud HM, Gachhui R, Raushel FM, Stuehr DJ (1997) The ferrous-dioxy complex of neuronal nitric oxide synthase. Divergent effects of L-arginine and tetrahydrobiopterin on its stability. J Biol Chem 272:17349–17353

Abu-Soud HM, Presta A, Mayer B, Stuehr DJ (1997) Analysis of neuronal NO synthase under single-turnover conditions: conversion of Nomega-hydroxyarginine to nitric oxide and citrulline. Biochemistry 36:10811–10816

Abu-Soud HM, Rousseau DL, Stuehr DJ (1996) Nitric oxide binding to the heme of neuronal nitric-oxide synthase links its activity to changes in oxygen tension. J Biol Chem 271:32515–32518

Abu-Soud HM, Wang J, Rousseau DL, Fukuto JM, Ignarro LJ, Stuehr DJ (1995) Neuronal nitric oxide synthase self-inactivates by forming a ferrous-nitrosyl complex during aerobic catalysis. J Biol Chem 270:22997–23006

Abu-Soud HM, Wu C, Ghosh DK, Stuehr DJ (1998) Stopped-flow analysis of CO and NO binding to inducible nitric oxide synthase. Biochemistry 37:3777–3786

Abu-Soud HM, Yoho LL, Stuehr DJ (1994b) Calmodulin controls neuronal nitric-oxide synthase by a dual mechanism. Activation of intra- and interdomain electron transfer. J Biol Chem 269:32047–32050

Adak S, Ghosh S, Abu-Soud HM, Stuehr DJ (1999) Role of Reductase Domain Cluster 1 Acidic Residues in Neuronal Nitric Oxide Synthase. Characterization of the FMN-free enzyme. J Biol Chem 274:(in press)

Albakri QA, Stuehr DJ (1996) Intracellular assembly of inducible NO synthase is limited by nitric oxide-mediated changes in heme insertion and availability. J Biol Chem 271:5414–5421

Anagli J, Hofmann F, Quadroni M, Vorherr T, Carafoli E (1995) The calmodulin-binding domain of the inducible (macrophage) nitric oxide synthase. Eur J Biochem 233:701–708

Baek KJ, Thiel BA, Lucas S, Stuehr DJ (1993) Macrophage nitric oxide synthase subunits. Purification, characterization, and role of prosthetic groups and substrate in regulating their association into a dimeric enzyme. J Biol Chem 268:21120–21129

Bec N, Gorren AC, Voelker C, Mayer B, Lange R (1998) Reaction of neuronal nitric-oxide synthase with oxygen at low temperature. Evidence for reductive activation of the oxy-ferrous complex by tetrahydrobiopterin. J Biol Chem 273:13502–13508

Bender AT, Silverstein AM, Demady DR, Kanelakis KC, Noguchi S, Pratt WB, Osawa Y (1999) Neuronal nitric-oxide synthase is regulated by the Hsp90–based chaperone system in vivo. J Biol Chem 274:1472–1478

Boyhan A, Smith D, Charles IG, Saqi M, Lowe PN (1997) Delineation of the arginine- and tetrahydrobiopterin-binding sites of neuronal nitric oxide synthase. Biochem J 323:131–139

Brenman JE, Bredt DS (1996) Nitric oxide signaling in the nervous system. Methods Enzymol 269:119–129

Campos KL, Giovanelli J, Kaufman S (1995) Characteristics of the nitric oxide synthase-catalyzed conversion of arginine to N-hydroxyarginine, the first oxygenation step in the enzymic synthesis of nitric oxide. J Biol Chem 270:1721–1728

Chang WJ, Iannaccone ST, Lau KS, Masters BS, McCabe TJ, McMillan K, Padre RC, Spencer MJ, Tidball JG, Stull JT (1996) Neuronal nitric oxide synthase and dystrophin-deficient muscular dystrophy. Proc Natl Acad Sci USA 93:9142–9147

Chen PF, Berka V, Tsai AL, Wu KK (1998) Effects of Asp-369 and Arg-372 mutations on heme environment and function in human endothelial nitric-oxide synthase. J Biol Chem 273:34164–34170

Chen PF, Tsai AL, Berka V, Wu KK (1996) Endothelial nitric-oxide synthase. Evidence for bidomain structure and successful reconstitution of catalytic activity from two separate domains generated by a baculovirus expression system. J Biol Chem 271:14631–14635

Chen PF, Tsai AL, Berka V, Wu KK (1997) Mutation of Glu-361 in human endothelial nitric-oxide synthase selectively abolishes L-arginine binding without perturbing the behavior of heme and other redox centers. J Biol Chem 272:6114–6118

Chen PF, Tsai AL, Wu KK (1994) Cysteine 184 of endothelial nitric oxide synthase is involved in heme coordination and catalytic activity. J Biol Chem 269:25062–25066

Chen PF, Tsai AL, Wu KK (1995) Cysteine 99 of endothelial nitric oxide synthase (NOS-III) is critical for tetrahydrobiopterin-dependent NOS-III stability and activity. Biochem Biophys Res Commun 215:1119–1129

Cho HJ, Martin E, Xie QW, Sassa S, Nathan C (1995) Inducible nitric oxide synthase: identification of amino acid residues essential for dimerization and binding of tetrahydrobiopterin. Proc Natl Acad Sci USA 92:11514–11518

Cho HJ, Xie QW, Calaycay J, Mumford RA, Swiderek KM, Lee TD, Nathan C (1992) Calmodulin is a subunit of nitric oxide synthase from macrophages. J Exp Med 176:599–604

Cho MJ, Vaghy PL, Kondo R, Lee SH, Davis JP, Rehl R, Heo WD, Johnson JD (1998) Reciprocal regulation of mammalian nitric oxide synthase and calcineurin by plant calmodulin isoforms. Biochemistry 37:15593–15597

Crane BR, Arvai AS, Gachhui R, Wu C, Ghosh DK, Getzoff ED, Stuehr DJ, Tainer JA (1997) The structure of nitric oxide synthase oxygenase domain and inhibitor complexes [see comments]. Science 278:425–431

Crane BR, Arvai AS, Ghosh DK, Wu C, Getzoff ED, Stuehr DJ, Tainer JA (1998) Structure of nitric oxide synthase oxygenase dimer with pterin and substrate. Science 279:2121–2126

Crane BR, Rosenfeld RJ, Arvai AS, Ghosh DK, Ghosh S, Tainer JA, Stuehr DJ, Getzoff ED (1999) N-terminal domain swapping and metal ion binding in nitric oxide synthase dimerization. EMBO J 18:6271–6281

Cubberley RR, Alderton WK, Boyhan A, Charles IG, Lowe PN, Old RW (1997) Cysteine-200 of human inducible nitric oxide synthase is essential for dimerization of haem domains and for binding of haem, nitroarginine and tetrahydrobiopterin. Biochem J 323:141–146

Dweik RA, Laskowski D, Abu-Soud HM, Kaneko F, Hutte R, Stuehr DJ, Erzurum SC (1998) Nitric oxide synthesis in the lung. Regulation by oxygen through a kinetic mechanism. J Clin Invest 101:660–666

Eissa NT, Yuan JW, Haggerty CM, Choo EK, Palmer CD, Moss J (1998) Cloning and characterization of human inducible nitric oxide synthase splice variants: a domain, encoded by exons 8 and 9, is critical for dimerization. Proc Natl Acad Sci USA 95:7625–7630

Feron O, Dessy C, Moniotte S, Desager JP, Balligand JL (1999) Hypercholesterolemia decreases nitric oxide production by promoting the interaction of caveolin and endothelial nitric oxide synthase. J Clin Invest 103:897–905

Firestein BL, Bredt DS (1999) Interaction of neuronal nitric-oxide synthase and phosphofructokinase-M. J Biol Chem 274:10545–10550

Fischmann TO, Hruza A, Niu XD, Fossetta JD, Lunn CA, Dolphin E, Prongay AJ, Reichert P, Lundell DJ, Narula SK, Weber PC (1999) Structural characterization of nitric oxide synthase isoforms reveals striking active-site conservation. Nat Struct Biol 6:233–242

Fleming I, Bauersachs J, Fisslthaler B, Busse R (1998) Ca2+-independent activation of the endothelial nitric oxide synthase in response to tyrosine phosphatase inhibitors and fluid shear stress. Circ Res 82:686–695

Gachhui R, Abu-Soud HM, Ghosha DK, Presta A, Blazing MA, Mayer B, George SE, Stuehr DJ (1998) Neuronal nitric-oxide synthase interaction with calmodulin-troponin C chimeras. J Biol Chem 273:5451–5454

Gachhui R, Ghosh DK, Wu C, Parkinson J, Crane BR, Stuehr DJ (1997) Mutagenesis of acidic residues in the oxygenase domain of inducible nitric-oxide synthase identifies a glutamate involved in arginine binding .Biochemistry 36:5097–5103

Gachhui R, Presta A, Bentley DF, Abu-Soud HM, McArthur R, Brudvig G, Ghosh DK, Stuehr DJ (1996) Characterization of the reductase domain of rat neuronal nitric oxide synthase generated in the methylotrophic yeast Pichia pastoris. Calmodulin response is complete within the reductase domain itself. J Biol Chem 271: 20594–20602

Galli C, MacArthur R, Abu-Soud HM, Clark P, Stuehr DJ, Brudvig GW (1996) EPR spectroscopic characterization of neuronal NO synthase [published erratum appears in Biochemistry 1996 Jun 4;35(22):7298]. Biochemistry 35:2804–2810

Garcia-Cardena G, Fan R, Shah V, Sorrentino R, Cirino G, Papapetropoulos A, Sessa WC (1998) Dynamic activation of endothelial nitric oxide synthase by Hsp90. Nature 392:821–824

Garcia-Cardena G, Martasek P, Masters BS, Skidd PM, Couet J, Li S, Lisanti MP, Sessa WC (1997) Dissecting the interaction between nitric oxide synthase (NOS) and caveolin. Functional significance of the nos caveolin binding domain in vivo. J Biol Chem 272:25437–25440

Garcia-Cardena G, Oh P, Liu J, Schnitzer JE, Sessa WC (1996) Targeting of nitric oxide synthase to endothelial cell caveolae via palmitoylation: implications for nitric oxide signaling. Proc Natl Acad Sci USA 93:6448–6453

George SE, Su Z, Fan D, Wang S, Johnson JD (1996) The fourth EF-hand of calmodulin and its helix-loop-helix components: impact on calcium binding and enzyme activation. Biochemistry 35:8307–8313

Gerber NC, Rodriguez-Crespo I, Nishida CR, Ortiz de Montellano PR (1997) Active site topologies and cofactor-mediated conformational changes of nitric-oxide synthases. J Biol Chem 272:6285–6290

Ghosh DK, Abu-Soud HM, Stuehr DJ (1995) Reconstitution of the second step in NO synthesis using the isolated oxygenase and reductase domains of macrophage NO synthase. Biochemistry 34:11316–11320

Ghosh DK, Abu-Soud HM, Stuehr DJ (1996) Domains of macrophage N(O) synthase have divergent roles in forming and stabilizing the active dimeric enzyme. Biochemistry 35:1444–1449

Ghosh DK, Crane BR, Ghosh S, Wolan D, Gachhui R, Crooks C, Presta A, Tainer JA, Getzoff ED, Stuehr DJ (1999) Inducible nitric oxide synthase: role of N-terminal β-hairpin hook and pterin binding segment in dimerization and tetrahydrobiopterin interaction. EMBO J 18:6260–6270

Ghosh DK, Stuehr DJ (1995) Macrophage NO synthase: characterization of isolated oxygenase and reductase domains reveals a head-to-head subunit interaction. Biochemistry 34:801–807

Ghosh DK, Wu C, Pitters E, Moloney M, Werner ER, Mayer B, Stuehr DJ (1997) Characterization of the inducible nitric oxide synthase oxygenase domain identifies a 49 amino acid segment required for subunit dimerization and tetrahydrobiopterin interaction. Biochemistry 36:10609–10619

Ghosh S, Gachhui R, Crooks C, Wu C, Lisanti MP, Stuehr DJ (1998) Interaction between caveolin-1 and the reductase domain of endothelial nitric-oxide synthase. Consequences for catalysis. J Biol Chem 273:22267–22271

Ghosh S, Wolan D, Adak S, Crane BR, Kwon NS, Tainer J, Getzoff E, Stuehr DJ (1999) Mutational Analysis of the Tetrahydrobiopterin Binding Site in Inducible Nitric Oxide Synthase. Tetrahydrobiopterin function in NO synthase. J Biol Chem (in press)

Gorren AC, List BM, Schrammel A, Pitters E, Hemmens B, Werner ER, Schmidt K, Mayer B (1996) Tetrahydrobiopterin-free neuronal nitric oxide synthase: evidence for two identical highly anticooperative pteridine binding sites. Biochemistry 35:16735–16745

Gorren ACF, Schrammel A, Schmidt K, Mayer B (1998) Effects of pH on the structure and function of neuronal nitric oxide synthase Biochem J 331:801–807

Griffith OW, Stuehr DJ (1995) Nitric oxide synthases: properties and catalytic mechanism. Annu Rev Physiol 57:707–736

Griscavage JM, Fukuto JM, Komori Y, Ignarro LJ (1994) Nitric oxide inhibits neuronal nitric oxide synthase by interacting with the heme prosthetic group. Role of tetrahydrobiopterin in modulating the inhibitory action of nitric oxide. J Biol Chem 269:21644–21649

Griscavage JM, Hobbs AJ, Ignarro LJ (1995) Negative modulation of nitric oxide synthase by nitric oxide and nitroso compounds. Adv Pharmacol 34:215–234

Hellermann GR, Solomonson LP (1997) Calmodulin promotes dimerization of the oxygenase domain of human endothelial nitric-oxide synthase. J Biol Chem 272:12030–12034

Hemmens B, Mayer B (1997) Methods in Molec Biol 100 (Titheradge, M A ed): 1–32, Humana Press, Totowa, NJ, USA

Hemmens B, Gorren AC, Schmidt K, Werner ER, Mayer B (1998a) Haem insertion, dimerization and reactivation of haem-free rat neuronal nitric oxide synthase. Biochem J 332:337–342

Hemmens B, Woschitz S, Pitters E, Klosch B, Volker C, Schmidt K, Mayer B (1998b) The protein inhibitor of neuronal nitric oxide synthase (PIN): characterization of its action on pure nitric oxide synthases. FEBS Lett 430:397–400

Hevel JM, Marletta MA (1992) Macrophage nitric oxide synthase: relationship between enzyme-bound tetrahydrobiopterin and synthase activity. Biochemistry 31:7160–7165

Hierholzer C, Harbrecht B, Menezes JM, Kane J, MacMicking J, Nathan CF, Peitzman AB, Billiar TR, Tweardy DJ (1998) Essential role of induced nitric oxide in the initiation of the inflammatory response after hemorrhagic shock. J Exp Med 187:917–928

Huang L, Abu-Soud HM, Hille R, Stuehr DJ (1999) Nitric oxide-generated P420 nitric oxide synthase: characterization and roles for tetrahydrobiopterin and substrate in protecting against or reversing the P420 conversion. Biochemistry 38:1912–1920

Hurshman AR, Marletta MA (1995) Nitric oxide complexes of inducible nitric oxide synthase: spectral characterization and effect on catalytic activity. Biochemistry 34:5627–5634

Ignarro L, Murad F ed (1995) Nitric Oxide: Biochemistry, Molecular Biology and therapeutic implications. Advances in Pharmacology 34: Academic Press, San Diego, CA

Iwasaki T, Hori H, Hayashi Y, Nishino T (1999) Modulation of the remote heme site geometry of recombinant mouse neuronal nitric-oxide synthase by the N-terminal hook region. J Biol Chem 274:7705–7713

Jaffrey SR, Snyder SH (1996) PIN: an associated protein inhibitor of neuronal nitric oxide synthase. Science 274:774–777

Jenkins CM, Genzor CG, Fillat MF, Waterman MR, Gomez-Moreno C (1997) Negatively charged anabaena flavodoxin residues (Asp144 and Glu145) are important for reconstitution of cytochrome P450 17alpha-hydroxylase activity. J Biol Chem 272:22509–22513

Ju H, Zou R, Venema VJ, Venema RC (1997) Direct interaction of endothelial nitric oxide synthase and caveolin-1 inhibits synthase activity. J Biol Chem 272:18522–18525

Kameya S, Miyagoe Y, Nonaka I, Ikemoto T, Endo M, Hanaoka K, Nabeshima Y, Takeda S (1999) alpha1-syntrophin gene disruption results in the absence of neuronal-type nitric-oxide synthase at the sarcolemma but does not induce muscle degeneration. J Biol Chem 274:2193–2200

Kerwin JF, Jr, Lancaster JR, Jr., Feldman PL (1995) Nitric oxide: a new paradigm for second messengers. J Med Chem 38:4343–4362

Klatt P, Schmid M, Leopold E, Schmidt K, Werner ER, Mayer B (1994) The pteridine binding site of brain nitric oxide synthase. Tetrahydrobiopterin binding kinetics, specificity, and allosteric interaction with the substrate domain. J Biol Chem 269:13861–13866

Klatt P, Schmidt K, Lehner D, Glatter O, Bachinger HP, Mayer B (1995) Structural analysis of porcine brain nitric oxide synthase reveals a role for tetrahydrobiopterin and L-arginine in the formation of an SDS-resistant dimer. EMBO J 14:3687–3695

Kroncke KD, Fehsel K, Kolb-Bachofen V (1995) Inducible nitric oxide synthase and its product nitric oxide, a small molecule with complex biological activities. Biol Chem Hoppe Seyler 376:327–343

Kroncke KD, Fehsel K, Sommer A, Rodriguez ML, Kolb-Bachofen V (1995) Nitric oxide generation during cellular metabolization of the diabetogenic N-methyl-N-nitroso-urea streptozotozin contributes to islet cell DNA damage. Biol Chem Hoppe Seyler 376:179–185

Lee CM, Robinson LJ, Michel T (1995) Oligomerization of endothelial nitric oxide synthase. Evidence for a dominant negative effect of truncation mutants. J Biol Chem 270:27403–27406

Lee SJ, Stull JT (1998) Calmodulin-dependent regulation of inducible and neuronal nitric-oxide synthase J Biol Chem 273:27430–27437

List BM, Klosch B, Volker C, Gorren AC, Sessa WC, Werner ER, Kukovetz WR, Schmidt K, Mayer B (1997) Characterization of bovine endothelial nitric oxide synthase as a homodimer with down-regulated uncoupled NADPH oxidase activity: tetrahydrobiopterin binding kinetics and role of haem in dimerization. Biochem J 323:159–165

MacMicking J, Xie QW, Nathan C (1997) Nitric oxide and macrophage function. Annu Rev Immunol 15:323–350

Marletta MA, Hurshman AR, Rusche KM (1998) Catalysis by nitric oxide synthase. Curr Opin Chem Biol 2:656–663

Martasek P, Liu Q, Liu J, Roman LJ, Gross SS, Sessa WC, Masters BS (1996) Characterization of bovine endothelial nitric oxide synthase expressed in E. coli. Biochem Biophys Res Commun 219:359–365

Martasek P, Miller RT, Liu Q, Roman LJ, Salerno JC, Migita CT, Raman CS, Gross SS, Ikeda-Saito M, Masters BS (1998) The C331A mutant of neuronal nitric-oxide synthase is defective in arginine binding. J Biol Chem 273:34799–34805

Masters BS, McMillan K, Sheta EA, Nishimura JS, Roman LJ, Martasek P (1996) Neuronal nitric oxide synthase, a modular enzyme formed by convergent evolution: structure studies of a cysteine thiolate-liganded heme protein that hydroxylates L-arginine to produce NO. as a cellular signal [published erratum appears in FASEB J 1996 Jul;10(9):1107]. FASEB J 10:552–558

Matsubara M, Hayashi N, Titani K, Taniguchi H (1997) Circular dichroism and 1H NMR studies on the structures of peptides derived from the calmodulin-binding domains of inducible and endothelial nitric-oxide synthase in solution and in complex with calmodulin. Nascent alpha-helical structures are stabilized by calmodulin both in the presence and absence of Ca2+. J Biol Chem 272:23050–23056

Matsuoka A, Stuehr DJ, Olson JS, Clark P, Ikeda-Saito M (1994) L-arginine and calmodulin regulation of the heme iron reactivity in neuronal nitric oxide synthase. J Biol Chem 269:20335–20339

Mayer B, Werner ER (1995) Why tetrahydrobiopterin? Adv Pharmacol 34:251–261

Mayer B, Wu C, Gorren AC, Pfeiffer S, Schmidt K, Clark P, Stuehr DJ, Werner ER (1997) Tetrahydrobiopterin binding to macrophage inducible nitric oxide synthase: heme spin shift and dimer stabilization by the potent pterin antagonist 4-amino-tetrahydrobiopterin. Biochemistry 36:8422–8427

McMillan K, Masters BS (1995) Prokaryotic expression of the heme- and flavin-binding domains of rat neuronal nitric oxide synthase as distinct polypeptides: identification of the heme-binding proximal thiolate ligand as cysteine-415. Biochemistry 34:3686–3693

Michel JB, Feron O, Sacks D, Michel T (1997) Reciprocal regulation of endothelial nitric-oxide synthase by Ca2+-calmodulin and caveolin. J Biol Chem 272:15583–15586

Michel JB, Feron O, Sase K, Prabhakar P, Michel T (1997) Caveolin versus calmodulin. Counterbalancing allosteric modulators of endothelial nitric oxide synthase. J Biol Chem 272:25907–25912

Michel T, Feron O (1997) Nitric oxide synthases: which, where, how, and why? J Clin Invest 100:2146–2152

Migita CT, Salerno JC, Masters BS, Martasek P, McMillan K, Ikeda-Saito M (1997) Substrate binding-induced changes in the EPR spectra of the ferrous nitric oxide complexes of neuronal nitric oxide synthase. Biochemistry 36:10987–10992

Miller RT, Martasek P, Roman LJ, Nishimura JS, Masters BS (1997) Involvement of the reductase domain of neuronal nitric oxide synthase in superoxide anion production. Biochemistry 36:15277–15284

Narayanasami R, Nishimura JS, McMillan K, Roman LJ, Shea TM, Robida AM, Horowitz PM, Masters BS (1997) The influence of chaotropic reagents on neuronal nitric oxide synthase and its flavoprotein module. Urea and guanidine hydrochloride stimulate NADPH-cytochrome c reductase activity of both proteins. Nitric Oxide 1:39–49

Nathan C (1997) Inducible nitric oxide synthase: what difference does it make? J Clin Invest 100:2417–2423

Nathan C, Xie QW (1994) Regulation of biosynthesis of nitric oxide. J Biol Chem 269:13725–13728

Negrerie M, Berka V, Vos MH, Liebl U, Lambry JC, Tsai AL, Martin JL (1999) Geminate recombination of nitric oxide to endothelial nitric oxide synthase and mechanistic implications, J Biol Chem 274:24694–24702

Nishida CR, Ortiz de Montellano PR (1998) Electron transfer and catalytic activity of nitric oxide synthases. Chimeric constructs of the neuronal, inducible, and endothelial isoforms. J Biol Chem 273:5566–5571

Persechini A, White HD, Gansz KJ (1996) Different mechanisms for Ca2+ dissociation from complexes of calmodulin with nitric oxide synthase or myosin light chain kinase. J Biol Chem 271:62–67

Porter TD (1991) An unusual yet strongly conserved flavoprotein reductase in bacteria and mammals. Trends Biochem Sci 16:154–158

Poulos TL, Raag R (1992) Cytochrome P450cam: crystallography, oxygen activation, and electron transfer. FASEB J 6:674–679

Prabhakar NR, Kumar GK, Chang CH, Agani FH, Haxhiu MA (1993) Nitric oxide in the sensory function of the carotid body. Brain Res 625:16–22

Presta A, Weber-Main AM, Stankovich MT, Stuehr DJ (1998) Comparative Effects of Substrates and Pterin Cofactor on the Heme Midpoint Potential in Inducible and Neuronal Nitric Oxide Synthases. J Am Chem Soc 120:9460–9465

Presta A, Liu J, Sessa WC, Stuehr DJ (1997) Substrate binding and calmodulin binding to endothelial nitric oxide synthase coregulate its enzymatic activity. Nitric Oxide 1:74–87

Presta A, Siddhanta U, Wu C, Sennequier N, Huang L, Abu-Soud HM, Erzurum S, Stuehr DJ (1998) Comparative functioning of dihydro- and tetrahydropterins in supporting electron transfer, catalysis, and subunit dimerization in inducible nitric oxide synthase. Biochemistry 37:298–310

Raman CS, Li H, Martasek P, Kral V, Masters BS, Poulos TL (1998) Crystal structure of constitutive endothelial nitric oxide synthase: a paradigm for pterin function involving a novel metal center. Cell 95:939–950

Ratovitski EA, Alam MR, Quick RA, McMillan A, Bao C, Kozlovsky C, Hand TA, Johnson RC, Mains RE, Eipper BA, Lowenstein CJ (1999) Kalirin inhibition of inducible nitric-oxide synthase. J Biol Chem 274:993–999

Ravichandran KG, Boddupalli SS, Hasermann CA, Peterson JA, Deisenhofer J (1993) Crystal structure of hemoprotein domain of P450BM-3, a prototype for microsomal P450's. Science 261:731–736

Renaud JP, Boucher JL, Vadon S, Delaforge M, Mansuy D (1993) Particular ability of liver P450s3A to catalyze the oxidation of N omega-hydroxyarginine to citrulline and nitrogen oxides and occurrence in no synthases of a sequence very similar to the heme-binding sequence in P450s. Biochem Biophys Res Commun 192:53–60

Renodon A, Boucher JL, Wu C, Gachhui R, Sari MA, Mansuy D, Stuehr DJ (1998) Formation of nitric oxide synthase-iron(II) nitrosoalkane complexes: severe restriction of access to the iron(II) site in the presence of tetrahydrobiopterin. Biochemistry 37:6367–6374

Richards MK, Marletta MA (1994) Characterization of neuronal nitric oxide synthase and a C415H mutant, purified from a baculovirus overexpression system. Biochemistry 33:14723–14732

Rodriguez-Crespo I, Gerber NC, Ortiz de Montellano PR (1996) Endothelial nitric-oxide synthase. Expression in Escherichia coli, spectroscopic characterization, and role of tetrahydrobiopterin in dimer formation. J Biol Chem 271:11462–11467

Rodriguez-Crespo I, Moenne-Loccoz P, Loehr TM, Ortiz de Montellano PR (1997) Endothelial nitric oxide synthase: modulations of the distal heme site produced by progressive N-terminal deletions. Biochemistry 36:8530–8538

Rodriguez-Crespo I, Ortiz de Montellano PR (1996) Human endothelial nitric oxide synthase: expression in Escherichia coli, coexpression with calmodulin, and characterization Arch Biochem. Biophys 336:151–156

Roman LJ, Sheta EA, Martasek P, Gross SS, Liu Q, Masters BS (1995) High-level expression of functional rat neuronal nitric oxide synthase in Escherichia coli. Proc Natl Acad Sci USA 92:8428–8432

Ruan J, Xie Q, Hutchinson N, Cho H, Wolfe GC, Nathan C (1996) Inducible nitric oxide synthase requires both the canonical calmodulin-binding domain and additional sequences in order to bind calmodulin and produce nitric oxide in the absence of free Ca2+. J Biol Chem 271:22679–22686

Rusche KM, Spiering MM, Marletta MA (1998) Reactions catalyzed by tetrahydrobiopterin-free nitric oxide synthase. Biochemistry 37:15503–15512

Sagami I, Shimizu T (1998) The crucial roles of Asp-314 and Thr-315 in the catalytic activation of molecular oxygen by neuronal nitric-oxide synthase. A site-directed mutagenesis study. J Biol Chem 273:2105–2108

Salerno JC, Harris DE, Irizarry K, Patel B, Morales AJ, Smith SM, Martasek P, Roman LJ, Masters BS, Jones CL, Weissman BA, Lane P, Liu Q, Gross SS (1997) An autoinhibitory control element defines calcium-regulated isoforms of nitric oxide synthase. J Biol Chem 272:29769–29777

Sari MA, Booker S, Jaouen M, Vadon S, Boucher JI, Pompon D, Mansuy D (1996) Expression in yeast and purification of functional macrophage nitric oxide synthase. Evidence for cysteine-194 as iron proximal ligand. Biochemistry 35:7204–7213

Sato H, Sagami I, Daff S, Shimizu T (1998) Autoxidation rates of neuronal nitric oxide synthase: effects of the substrates, inhibitors, and modulators. Biochem Biophys Res Commun 253:845–849

Scheele JS, Bruner E, Kharitonov VG, Martasek P, Roman LJ, Masters BS, Sharma VS, Magde D (1999) Kinetics of NO ligation with nitric-oxide synthase by flash photolysis and stopped-flow spectrophotometry. J Biol Chem 274:13105–13110

Scheele JS, Kharitonov VG, Martasek P, Roman LJ, Sharma VS, Masters BS, Magde D (1997) Kinetics of CO ligation with nitric-oxide synthase by flash photolysis and stopped-flow spectrophotometry. J Biol Chem 272:12523–12528

Sennequier N, Stuehr DJ (1996) Analysis of substrate-induced electronic, catalytic, and structural changes in inducible NO synthase. Biochemistry 35:5883–5892

Sennequier N, Wolan D, Stuehr DJ (1999) Antifungal imidazoles block assembly of inducible NO synthase into an active dimer. J Biol Chem 274:930–938

Sevrioukova IF, Li H, Zhang H, Peterson JA, Poulos TL (1999) Structure of a cytochrome P450–redox partner electron-transfer complex. Proc Natl Acad Sci USA 96:1863–1868

Shen AL, Sem DS, Kasper CB (1999) Mechanistic studies on the reductive half-reaction of NADPH-cytochrome P450 oxidoreductase. J Biol Chem 274:5391–5398

Siddhanta U, Presta A, Fan B, Wolan D, Rousseau DL, Stuehr DJ (1998) Domain swapping in inducible nitric-oxide synthase. Electron transfer occurs between flavin and heme groups located on adjacent subunits in the dimer. J Biol Chem 273:18950–18958

Siddhanta U, Wu C, Abu-Soud HM, Zhang J, Ghosh DK, Stuehr DJ (1996) Heme iron reduction and catalysis by a nitric oxide synthase heterodimer containing one reductase and two oxygenase domains. J Biol Chem 271:7309–7312

Stevens-Truss R, Beckingham K, Marletta MA (1997) Calcium binding sites of calmodulin and electron transfer by neuronal nitric oxide synthase. Biochemistry 36: 12337–12345

Stevens-Truss R, Marletta MA (1995) Interaction of calmodulin with the inducible murine macrophage nitric oxide synthase. Biochemistry 34:15638–15645

Stevens-Truss R, Beckingham K, Marletta MA (1997) Calcium binding sites of calmodulin and electron transfer by neuronal nitric oxide synthase. Biochemistry 36:12337–12345

Stricker NL, Christopherson KS, Yi BA, Schatz PJ, Raab RW, Dawes G, Bassett DE, Jr., Bredt DS, Li M (1997) PDZ domain of neuronal nitric oxide synthase recognizes novel C-terminal peptide sequences. Nat Biotechnol 15:336–342

Stuehr DJ (1999) Mammalian Nitric Oxide Synthases BBA Bioenergetics 1411:217–230

Stuehr DJ (1997) Structure-function aspects in the nitric oxide synthases. Annu Rev Pharmacol Toxicol 37:339–359

Stuehr DJ, Ikeda-Saito M (1992) Spectral characterization of brain and macrophage nitric oxide synthases. Cytochrome P-450-like hemeproteins that contain a flavin semiquinone radical. J Biol Chem 267:20547–20550

Su Z, Blazing MA, Fan D, George SE (1995) The calmodulin-nitric oxide synthase interaction. Critical role of the calmodulin latch domain in enzyme activation. J Biol Chem 270:29117–29122

Tochio H, Ohki S, Zhang Q, Li M, Zhang M (1998) Solution structure of a protein inhibitor of neuronal nitric oxide synthase. Nat Struct Biol 5:965–969

Tzeng E, Billiar TR, Robbins PD, Loftus M, Stuehr DJ (1995) Expression of human inducible nitric oxide synthase in a tetrahydrobiopterin (H4B)-deficient cell line: H4B promotes assembly of enzyme subunits into an active dimer. Proc Natl Acad Sci USA 92:11771–11775

Vasquez-Vivar J, Kalyanaraman B, Martasek P, Hogg N, Masters BS, Karoui H, Tordo P, Pritchard KA, Jr. (1998) Superoxide generation by endothelial nitric oxide synthase: the influence of cofactors. Proc Natl Acad Sci USA 95:9220–9225

Venema RC, Ju H, Zou R, Ryan JW, Venema VJ (1997a) Subunit interactions of endothelial nitric-oxide synthase. Comparisons to the neuronal and inducible nitric-oxide synthase isoforms. J Biol Chem 272:1276–1282

Venema RC, Sayegh HS, Kent JD, Harrison DG (1996) Identification, characterization, and comparison of the calmodulin-binding domains of the endothelial and inducible nitric oxide synthases. J Biol Chem 271:6435–6440

Venema VJ, Ju H, Zou R, Venema RC (1997b) Interaction of neuronal nitric-oxide synthase with caveolin-3 in skeletal muscle. Identification of a novel caveolin scaffolding/inhibitory domain. J Biol Chem 272:28187–28190

Vorherr T, Knopfel L, Hofmann F, Mollner S, Pfeuffer T, Carafoli E (1993) The calmodulin binding domain of nitric oxide synthase and adenylyl cyclase. Biochemistry 32:6081–6088

Wang J, Rousseau DL, Abu-Soud HM, Stuehr DJ (1994) Heme coordination of NO in NO synthase. Proc Natl Acad Sci USA 91:10512–10516

Wang J, Stuehr DJ, Rousseau DL (1997a) Interactions between substrate analogues and heme ligands in nitric oxide synthase. Biochemistry 36:4595–4606

Wang M, Roberts DL, Paschke R, Shea TM, Masters BS, Kim JJ (1997b) Three-dimensional structure of NADPH-cytochrome P450 reductase: prototype for FMN- and FAD-containing enzymes. Proc Natl Acad Sci USA 94:8411–8416

Wang Y, Goligorsky MS, Lin M, Wilcox JN, Marsden PA (1997c) A novel, testis-specific mRNA transcript encoding an NH2-terminal truncated nitric-oxide synthase. J Biol Chem 272:11392–11401

Witteveen CF, Giovanelli J, Yim MB, Gachhui R, Stuehr DJ, Kaufman S (1998) Reactivity of the flavin semiquinone of nitric oxide synthase in the oxygenation of arginine to NG-hydroxyarginine, the first step of nitric oxide synthesis. Biochem Biophys Res Commun 250:36–42

Wu C, Zhang J, Abu-Soud H, Ghosh DK, Stuehr DJ (1996) High-level expression of mouse inducible nitric oxide synthase in Escherichia coli requires coexpression with calmodulin. Biochem Biophys Res Commun 222:439–444

Xia Y, Zweier JL (1997) Direct measurement of nitric oxide generation from nitric oxide synthase. Proc Natl Acad Sci USA 94:12705–12710

Xie QW, Cho H, Kashiwabara Y, Baum M, Weidner JR, Elliston K, Mumford R, Nathan C (1994) Carboxyl terminus of inducible nitric oxide synthase. Contribution to NADPH binding and enzymatic activity. J Biol Chem 269:28500–28505

Xie QW, Leung M, Fuortes M, Sassa S, Nathan C (1996) Complementation analysis of mutants of nitric oxide synthase reveals that the active site requires two hemes. Proc Natl Acad Sci USA 93:4891–4896

Zoche M, Bienert M, Beyermann M, Koch KW (1996) Distinct molecular recognition of calmodulin-binding sites in the neuronal and macrophage nitric oxide synthases: a surface plasmon resonance study. Biochemistry 35:8742–8747

CHAPTER 3
Regulation of Nitric Oxide Synthase Expression and Activity

U. FÖRSTERMANN

A. Introduction

Nitric oxide (NO) can control vital functions, such as neurotransmission or vascular tone (via activation of soluble guanylyl cyclase), gene transcription and mRNA translation (via iron-responsive elements), and post-translational modifications of proteins (via adenosine diphosphate ribosylation) (FÖRSTERMANN et al. 1994). In higher concentrations, NO is capable of destroying parasites and tumor cells by inhibiting iron-containing enzymes or directly interacting with the DNA of these cells (NATHAN and HIBBS 1991). In view of this multitude of functions of NO, it is important to understand the mechanisms by which cells accomplish and regulate the production of this molecule.

In mammals, three isozymes of NO synthase [NOS: L-arginine, NADPH: oxygen oxidoreductase (nitric-oxide-forming); EC 1.14.13.39], have been identified, and the complementary DNAs (cDNAs) for these enzymes have been isolated. NOS-I [NOS1, neuronal NOS (nNOS)] is a low output enzyme that is constitutively expressed. The prototypical enzyme is present in neurons. NOS-II [NOS2, inducible NOS (iNOS)] is a high-output NOS whose expression is induced by cytokines and other agents. The prototypical enzyme is expressed by activated murine macrophages. NOS-III [NOS3, endothelial NOS (eNOS)] is also a low-output enzyme that is constitutively expressed. The prototypical enzyme is found in endothelial cells. The nature of NOS-I and NOS-III as low-output enzymes and NOS-II as a high-output enzyme depends not so much on the conversion rate of the different isozymes (Table 1) but rather reflects the short-lasting, pulsatile, Ca^{2+}-activated NO production of NOS-I and NOS-III versus the continuous, Ca^{2+}-independent NO production of NOS-II. All three isoforms of NOS utilize L-arginine as their substrate. Reported K_m values for L-arginine and V_{max} values for NO- or L-citrulline formation are shown in Table 1. Half-saturating L-arginine concentrations seem to be similar for the two constitutive forms of NOS, NOS-I and -III, and are somewhat higher for the inducible NOS-II (Table 1).

The amount of NO produced by a NOS depends both on the level of expression of the enzyme in a specific cell and on its enzymatic activity. Both parameters are regulated in a cell- and isozyme-specific fashion.

Table 1. Kinetic properties of the three isoforms of nitric oxide synthase (NOS)

Isoform	Protein source	Calcium dependence of activity (EC_{50}) (μM)	Calmodulin dependence of activity (EC_{50}) (nM)	K_m of L-arginine (μM)	V_{max} of citrulline or NO (μmol/mg protein/min)	References
NOS-I (nNOS)	Rat cerebellum	0.2	10	1.5	0.96	Bredt and Snyder 1990
	Rat cerebellum	0.35	3.5			Schmidt et al. 1991
	Porcine cerebellum	0.4	35–70	2.2	0.107	Mayer et al. 1990
	Rat NOS-I expressed in pCIS kidney cells	>80% Decrease in the absence of Ca^{2+}	>80% Decrease by trifluoperazine (100 μM)		0.73	Bredt et al. 1991
	Human NOS-I expressed in COS-1 cells	Activity abolished by EGTA	Activity abolished by trifluoperazine (100 μM)			Nakane et al. 1993
NOS-II (iNOS)	Mouse RAW 264.7 macrophages	<20% Increase at 2 mM Ca^{2+} vs no Ca^{2+}	None	2.8	1.3	Stuehr et al. 1991
	Mouse RAW 264.7 macrophages			16		Hevel et al. 1991
	Rat peritoneal macrophages	None	None	32.3	1.052	Yui et al. 1991
	Human DLD-1 adenocarcinoma cells	60% Decrease by EGTA	No effect of trifluoperazine (50 μM)			Sherman et al. 1993b
	Human NOS-II expressed in 293 embryonic kidney cells	50–70% Decrease by EGTA or EDTA	50% Decrease by trifluoperazine (50 μM)			Geller et al. 1993
NOS-III (eNOS)	Bovine aortic endothelial cells	0.3	3.5	2.9	0.015	Pollock et al. 1991
	Bovine NOS-III expressed in COS-1 cells	Total dependence	Total dependence	2.8		Sessa et al. 1992, 1993

EDTA, ethylene diamine tetraacetic acid; *EGTA*, ethylene glycol-bis(β-aminoethyl ether)-N,N,N',N'-tetraacetic acid; *eNOS*, endothelial NOS; *iNOS*, inducible NOS; *nNOS*, neuronal NOS; *NO*, nitric oxide.

B. Nitric Oxide Synthase I
I. Cellular Expression of NOS-I

NOS-I was first identified in the brain of various species. In brain, NOS-I is mainly a soluble enzyme which migrates with a molecular mass of 160 kDa on sodium dodecyl sulfate polyacrylamide gel electrophoresis (SDS-PAGE) (BREDT and SNYDER 1990; MAYER et al. 1990; SCHMIDT et al. 1991). The cDNA-deduced amino acid sequences predict proteins of 160 kDa and 161 kDa for the rat and human enzyme, respectively (BREDT et al. 1991; NAKANE et al. 1993).

NOS-I expression is not confined to neuronal cells. It has also been identified in certain areas of the spinal cord (DUN et al. 1992), in sympathetic ganglia and adrenal glands (DUN et al. 1993; SHENG et al. 1993), in peripheral nitrergic nerves (HASSALL et al. 1992; SAFFREY et al. 1992; SHENG et al. 1992), in epithelial cells of lung, uterus, and stomach (SCHMIDT et al. 1992a), including human lung epithelial cells (ASANO et al. 1994), in kidney macula densa cells (SCHMIDT et al. 1992a), and in pancreatic islet cells (SCHMIDT et al. 1992b). NOS-I is also present in skeletal muscle of humans, guinea pigs, and rats and in human rhabdomyosarcoma cell lines (NAKANE et al. 1993; KOBZIK et al. 1994; GATH et al. 1996, 1997).

II. Regulation of NOS-I Expression

Although constitutively expressed in several cell types, the expression level of NOS-I can be regulated dynamically. NOS-I mRNA upregulation seems to represent a response of neuronal cells to stress or injury induced by physical, chemical and biological agents. Examples include pain induced by formalin (LAM et al. 1996), brain injury caused by heat stress, axonal transection, colchicine treatment, or experimental allergic encephalomyelitis (CALZA et al. 1997; LUMME et al. 1997; SHARMA et al. 1997), and chronic electrical stimulation of skeletal muscle (REISER et al. 1997). A similar response is observed in the rat paraventricular nucleus and adrenal cortex during immobilization stress (CALZA et al. 1993; TSUCHIYA et al. 1996) and after mechanical or pathological lesions including spinal cord, axonal, or nerve injuries (HERDEGEN et al. 1993; LIN et al. 1997; VIZZARD 1997), hypophysectomy (VILLAR et al. 1994), or middle cerebral artery occlusion leading to focal ischemia (ZHANG et al. 1994). NOS-I protein expression increased in olfactory bulb neurons during infections with vesicular stomatitis virus (KOMATSU and REISS 1997). NOS-I expression appears to be regulated also by changes in neuronal activity; its expression was higher when rat cerebellar granule cells were kept in the presence of 10 mM K^+ compared with 25 mM K^+ (TASCEDDA et al. 1996). In the same cell type, the inhibition of the glutamatergic transmission drastically increased NOS-I expression (BAADER and SCHILLING 1996). NOS-I expression can also be triggered by steroid hormones. It has been demonstrated that estradiol and pregnancy could induce NOS-I expression in several tissues of the rat (WEINER et al. 1994a;

CECCATELLI et al. 1996; XU et al. 1996). In male rats, testosterone treatment has been described to stimulate the expression of the neuronal isoform in the penis (REILLY et al. 1997). Corticosterone treatment resulted in an upregulation of heme oxygenase-2 and a concomitant decrease of NOS-I transcription in rat brain (WEBER et al. 1994). Lithium and tacrine, a cholinesterase inhibitor currently used in the treatment of the symptoms of Alzheimer's disease, increased the expression of NOS-I synergistically in the hippocampus of the rat. This effect could be inhibited by corticosterone (BAGETTA et al. 1993). Recent evidence obtained in our laboratory with murine N1E-115 neuroblastoma cells indicated that glucocorticoids inhibit NOS-I expression by reducing the transcription of the gene (SCHWARZ et al. 1998).

A downregulation of NOS-I expression has been documented in guinea-pig skeletal muscle and rat brain after in vivo treatment with bacterial lipopolysaccharide (LPS) (LIU et al. 1996; GATH et al. 1997). Treatment of rats with LPS or interferon-γ (IFN-γ) also decreased the expression of NOS-I in brain, stomach, rectum, and spleen (BANDYOPADHYAY et al. 1997).

III. Regulation of NOS-I Activity

Neuron-derived NO can be either beneficial or detrimental, depending on the cellular context. Neuronal NOS-I must, therefore, be tightly regulated. NOS-I is a Ca^{2+}- and calmodulin-dependent enzyme (FÖRSTERMANN et al. 1990; SCHMIDT et al. 1991). Its activity is regulated by physiological changes in the intracellular Ca^{2+} concentration. Thus, hormones and neurotransmitters leading to increases in intracellular Ca^{2+} will activate the enzyme in neurons and probably other cells. As Ca^{2+} concentrations change rapidly in neuronal cells, the physiological NO production of these cells likely consists of short-lasting puffs of NO rather than a continuous NO formation.

In addition to this acute mechanism of regulation, other parameters are likely to be important for NO production by NOS-I. An important one may be the subcellular localization of NOS-I protein. In neurons, both soluble and particulate protein is found. Depending on the individual study, the particulate enzyme represents between 30 and 60% of the total neuronal NOS-I protein (HECKER et al. 1994; ARBONES et al. 1996; RODRIGO et al. 1997). In electron microscopy studies of kidney macula densa cells, the neuronal isoform has been seen associated mainly with small vesicles (TOJO et al. 1994). In skeletal muscle, NOS-I protein is mostly particulate (CHANG et al. 1996; GATH et al. 1996). The particulate localization of part of the NOS-I protein is most probably due to the PDZ/GLGF motif found in the N-terminal sequence of the NOS-I protein. This motif participates in protein–protein interactions with several other membrane-associated proteins (PONTING et al. 1997; SCHEPENS et al. 1997). In brain, synaptic association of NOS-I is mediated by the binding of the PDZ/GLGF motif to the postsynaptic density protein PSD-95 and/or to the related PSD-93 protein (BRENMAN et al. 1996). N-methyl-D-aspartate (NMDA) receptors are also known to be associated with PSD-95 (KORNAU et al. 1995). Thus, the complexed NOS-I may be the enzyme primarily activated during NMDA receptor-mediated Ca^{2+} influx into neuronal cells.

In skeletal muscle, the muscle-specific isoform (μNOS-I) is attached to the sarcolemma–dystrophin complex via the PDZ/GLGF motif and mainly interacts with α1-synthrophin (BRENMAN et al. 1995). High concentrations of this enzyme are found near motor endplates (GATH et al. 1996). NOS-I splice variants lacking a PDZ/GLGF domain, identified in NOS-I δ/δ mutant mice, do not associate with PSD-95 in brain or with skeletal muscle sarcolemma (BRENMAN et al. 1996). Also, this membrane association of NOS-I may facilitate activation of the enzyme during sarcolemmal depolarization.

Finally, NOS-I can be phosphorylated at serine and threonine residues by Ca^{2+}/calmodulin-dependent protein kinase II and protein kinases A, C, and G (NAKANE et al. 1991; DINERMAN et al. 1994b). In vitro, phosphorylation by Ca^{2+}/calmodulin kinase II and protein kinases A and G reduces the catalytic activity of the enzyme (NAKANE et al. 1991; DINERMAN et al. 1994b).

C. Nitric Oxide Synthase II
I. Cellular Expression of NOS-II

NOS-II can be induced in many cell types by cytokines, LPS, and a variety of other agents. The enzyme was first isolated from murine macrophages (HEVEL et al. 1991; STUEHR et al. 1991). On SDS-PAGE, NOS-II from macrophages has a molecular mass of 125–135 kDa and is a predominantly soluble enzyme (HEVEL et al. 1991; STUEHR et al. 1991). The cDNAs encoding NOS-II have been cloned by several groups from mouse, rat, and human cells (FÖRSTERMANN and KLEINERT 1995). The deduced amino acid sequences predict proteins of 130–131 kDa molecular mass.

Immunohistochemical localization of NOS-II in rats treated with *Propionibacterium acnes* and LPS demonstrated the enzyme in many cell types in addition to macrophages (BANDALETOVA et al. 1993). NOS-II immunoreactivity has also been reported in pancreatic islets of diabetic BB rats (KLEEMANN et al. 1993). KOBZIK et al. (1993) found strong NOS-II labeling of macrophages from LPS-treated rats. Also, human alveolar macrophages in areas of inflammation are NOS-II-immunoreactive (KOBZIK et al. 1993; TRACEY et al. 1994b). NOS-II immunoreactivity was also present in murine and human lung epithelial cells after cytokine stimulation (ASANO et al. 1994; ROBBINS et al. 1994).

II. Regulation of NOS-II Expression

Unlike NOS-I and NOS-III, expressional regulation of NOS-II represents the main mechanism of activation of NOS-II. In uninduced cells, expression of NOS-II is usually very low or undetectable. The first agents that were found to induce expression of this enzyme in macrophages and other cells were LPS and cytokines, such as interleukin-1 (IL-1), IFN-γ, and tumor necrosis factor-α (TNF-α). The type of cytokine (or the cytokine combination) that produces good NOS-II expression varies among species and among cell types within the same species. In addition, agents other than cytokines are efficacious inducers of NOS-II in some cell types. For example, in murine 3T3 fibroblasts and vascular smooth-muscle cells, NOS-II is expressed in response to cyclic adenosine monophosphate (cAMP)-elevating agents, such as forskolin or dibutyryl cAMP (GILBERT and HERSCHMAN 1993a; KOIDE et al. 1993); protein kinase C (PKC)-stimulating agents, such as tetradecanoyl phorbol-13-acetate (GILBERT and HERSCHMAN 1993a; HORTELANO et al. 1993); and growth factors, such as platelet-derived growth factor (PDGF) and fibroblast growth factor (FGF) (GILBERT and HERSCHMAN 1993a).

However, there is a large number of compounds known to prevent the induction of NOS-II by cytokines. First, there is the group of inhibitory cytokines and growth factors. IL-4 (BOGDAN et al. 1994), IL-8 (MCCALL et al. 1992), IL-10 (CUNHA et al. 1992), monocyte chemotactic protein-1 (ROJAS et al. 1993), and macrophage-deactivating factor (DING et al. 1990) are inhibitors of NOS-II induction in macrophages and neutrophils. NOS-II induction is also prevented by three isoforms of transforming growth factor-β (TGF-β1, 2, and 3) in macrophages and smooth-muscle cells (DING et al. 1990; FÖRSTERMANN et al. 1992; SCHINI et al. 1992; VODOVOTZ et al. 1993); PDGF-AB and -BB and insulin-like growth factor I in rat vascular smooth-muscle cells (SCHINI et al. 1992, 1994); and basic and acidic FGF in bovine retinal pigmented epithelial cells (GOUREAU et al. 1993). Compounds such as the tyrosine kinase inhibitors genistein, herbimycin A, and tyrphostin (DONG et al. 1993) and the inhibitors of nuclear factor-κB (NF-κB), pyrrolidine dithiocarbamate (PDTC) and diethyl-dithiocarbamate (MÜLSCH et al. 1993; SHERMAN et al. 1993a; XIE et al. 1994), prevent NOS-II induction in macrophages, indicating that tyrosine kinases and NF-κB are involved in induction. Glucocorticoids are effective inhibitors of NOS-II induction in endothelial cells, macrophages, fibroblasts, and smooth-muscle cells (DI ROSA et al. 1990; RADOMSKI et al. 1990; SCHINI et al. 1992; GILBERT and HERSCHMAN 1993a; KANNO et al. 1993).

We are only beginning to understand the molecular actions of compounds that stimulate or inhibit induction. In murine macrophages, LPS, IFN-γ, and other agents increase transcription of the NOS-II gene. Significant fragments of the 5'-flanking region of the murine NOS-II gene have been cloned (LOWENSTEIN et al. 1993; XIE et al. 1993). The promoter of the gene contains a "TATA box" and numerous consensus sequences for the binding of transcription factors (some in multiple copies), such as IFN-γ response element (IFN-γ-RE), NF-κB-binding motifs, nuclear factor IL-6 (NF-IL6)-binding sites, IFN-α-stimulated response element (RE), activating protein 1 (AP-1) site, and TNF-RE. Many of these sequences are associated with stimuli that induce NOS-II expression (XIE et al. 1993). To localize functionally important sequences of the regulatory region, mutants of the NOS-II 5'-flanking region were constructed and placed upstream of a reporter gene. The degree of expression of the reporter gene was dependent on two regulatory regions upstream of the putative "TATA box". The first region (positions −48 to −209) contains LPS-related response elements, such as the putative binding sites for NF-IL6 and NF-κB. This region was responsive to LPS, suggesting that it regulates LPS-induced expression of the gene (LOWENSTEIN et al. 1993). The NF-κB-binding site on the promoter sequence begins 55 bp upstream of the "TATA box" (XIE et al. 1994). Oligonucleotide probes containing the NF-κB site plus the 9 or 47 nucleotides downstream bound proteins that rapidly appeared in the nuclei of LPS-treated macrophages. The NF-κB inhibitor PDTC blocked both the activation of proteins binding to the NF-κB-binding site and the production of NO in LPS-treated macrophages, indicating that NF-κB activation is essential for transcription of the NOS-II gene in murine macrophages (XIE et al. 1994). The second region (positions −913 to −1029) mediated the potentiation of LPS induction by IFN-γ and, thus, probably is responsible for IFN-γ-mediated regulation of NOS-II induction (LOWENSTEIN et al. 1993). In positions −951 to −911, MARTIN et al. (1994) identified a cluster of four enhancer elements known to bind IFN-γ-responsive transcription factors, including an IFN regulatory factor binding site (IRF-E) at nucleotides −913 to −923. Site-specific mutagenesis of two nucleotides within the IRF-E abolished the enhancement of transcription by IFN-γ.

The human NOS-II gene is overexpressed in a number of human inflammatory diseases. In vitro, human cells generally require a cytokine mixture typically consisting of INF-γ, IL-1β, and TNF-α for NOS-II induction (SHERMAN et al. 1993b; KLEINERT et al. 1996). However, human chondrocytes (CHARLES et al. 1993) and human primary hepatocytes (GELLER et al. 1995) can be stimulated with IL-1β alone to express significant levels of NOS-II and, in human DLD-1 epithelial cells, INF-γ alone can produce about 30% of the maximal NOS-II induction (KLEINERT et al. 1998a, 1998c). Therefore, transcription factors induced by INF-γ alone are responsible for at least 30% of the transcription of the human NOS-II gene in DLD-1 cells. For maximal transcription, additional transcription factors regulated by TNF-α and IL-1β seem to be required. The human NOS-II promoter contains consensus sequences for the binding of transcription factors, including IFN-γ regulatory factor-1, signal transducer and activator of transcription (STAT) binding to γ-activation site, AP-1, NF-κB, and others (SPITSIN et al. 1996; LINN et al. 1997). Recent evidence suggests that the INF-γ–janus kinase-2–STAT1α pathway is an important pathway for NOS-II induction in human DLD-1 and A549/8 cells (KLEINERT et al. 1998c). Also NF-κB has been shown to be functionally important for the activity of the NOS-II promoter in human cells (KLEINERT et al. 1996; NUNOKAWA et al. 1996). The addition of NF-κB inhibitors significantly suppressed cytokine-stimulated NOS-II mRNA expression and NO synthesis in A549 epithelial cells and AKN-1 hepatocytes (KLEINERT et al. 1996; TAYLOR et al. 1998). The promoter activity of a 7-kb human NOS-II promoter fragment transfected into these cells revealed an about fivefold induction by cytokines (TAYLOR et al. 1998). Several putative NF-κB cis-regulatory transcription-factor-binding sites upstream of –5 kb seem to be required for cytokine-induced promoter activity in these cells (TAYLOR et al. 1998). Interestingly, the same 7-kb human NOS-II-promoter fragment was not significantly stimulated by cytokine induction when transfected into DLD-1 cells, and different inhibitors of NF-κB activation (PDTC, 3,4-dichloroisocoumarin, dexamethasone and panepoxydone) did not block cytokine-induced NOS-II mRNA expression in these cells (LINN et al. 1997; KLEINERT et al. 1998c). Therefore, in DLD-1 cells, the role of NF-κB may be that of a "basal" transcription factor of the human NOS-II promoter (similar to the "TATA box").

Paradoxically, some agents may stimulate NOS-II induction in one cell type and inhibit it in another. TGF-β and PDGF-AB and -BB, for example, are inhibitors of NOS-II induction in mouse macrophages and rat vascular smooth-muscle cells (DING et al. 1990; FÖRSTERMANN et al. 1992; SCHINI et al. 1992; GILBERT and HERSCHMAN 1993b) but stimulate induction in 3T3 fibroblasts (GILBERT and HERSCHMAN 1993a, 1993b). TGF-β also stimulates induction in bovine retinal pigmented epithelial cells (GOUREAU et al. 1993). cAMP-elevating agents prevent induction in primary rat astrocytes (FEINSTEIN et al. 1993) but produce induction in 3T3 cells and vascular smooth-muscle cells (GILBERT and HERSCHMAN 1993a; KOIDE et al. 1993). Thus, signal transduction pathways leading to NOS-II induction seem to differ markedly between cells.

In addition to transcriptional events, post-transcriptional phenomena can regulate the expression of NOS-II. WEISZ et al. (1994) reported that LPS, in addition to promoting NOS-II transcription, prolonged the half-life of NOS-II mRNA about fourfold, thereby contributing to enhanced NOS-II protein formation. In mouse peritoneal macrophages, three mechanisms have been described for the inhibition of NOS-II induction by TGF-β. The growth factor reduced NOS-II mRNA by decreasing its stability without affecting transcription. It also reduced NOS-II translation efficiency and accelerated the degradation of NOS-II protein (VODOVOTZ et al. 1993). Other inhibitory mechanisms have been reported for IL-4. This cytokine did not affect NOS-II mRNA stability but strongly reduced NOS-II mRNA and protein at later times (24–72h) of IFN-γ induction (BOGDAN et al. 1994).

III. Regulation of NOS-II Activity

Once expressed, no regulatory mechanisms are known for the activity of NOS-II. Interestingly, the amino acid sequence of the murine enzyme demonstrates a binding site for calmodulin (LYONS et al. 1992; XIE et al. 1992) despite the Ca^{2+} independence of its activity. The murine enzyme seems to tightly bind calmodulin even in the absence of Ca^{2+} (CHO et al. 1992). The induced NOS from human hepatocytes and DLD-1 adenocarcinoma cells lost some activity following Ca^{2+} chelation and/or exposure to a calmodulin antagonist (GELLER et al. 1993; SHERMAN et al. 1993b), suggesting that the human NOS-II sequence does not convey the same tight binding of calmodulin as the mouse enzyme and, therefore, the human NOS-II activity shows some limited degree of Ca^{2+} dependence.

D. Nitric Oxide Synthase III

I. Cellular Expression of NOS-III

NOS-III was first identified in endothelial cells (FÖRSTERMANN et al. 1991; POLLOCK et al. 1991). The endothelial NOS-III is more than 90% particulate. The solubilized enzyme shows a denatured molecular mass of 135 kDa (FÖRSTERMANN et al. 1991; POLLOCK et al. 1991). cDNAs encoding NOS-III have been isolated from bovine and human endothelial cells. The deduced amino acid sequences predict proteins of 133 kDa molecular mass for both species.

Immunohistochemical studies using a specific antibody to NOS-III located the enzyme in various types of arterial and venous endothelial cells (POLLOCK et al. 1993) and syncytiotrophoblasts of human placenta (MYATT et al. 1993), LLC-PK$_1$ kidney tubular epithelial cells (TRACEY et al. 1994a), interstitial cells of the canine colon (XUE et al. 1994), and neurons of the rat hippocampus and other brain regions (DINERMAN et al. 1994a).

II. Regulation of NOS-III Expression

NOS-III is found to be constitutively expressed in the cells mentioned above. However, several mechanisms modulate the expression of the NOS-II gene. Exercise training and shear stress produced by flowing blood upregulate NOS-III expression (NISHIDA et al. 1992; SESSA et al. 1994). In human primary pulmonary artery endothelial cells (ZIESCHE et al. 1996), cultured porcine pulmonary artery endothelial cells (DAI et al. 1995), and bovine pulmonary artery endothelial cells (LIAO et al. 1995), hypoxia reduced NOS-III mRNA and/or protein. At least in the bovine species, this was attributed to both a decreased rate of transcription and a destabilization of the NOS-III mRNA (LIAO et al. 1995). In non-pulmonary endothelial cells, the findings are more controversial. In bovine aortic endothelial cells (BAEC), ARNET et al. (1996) saw an upregulation of NOS-III mRNA and protein expression in cells incubated at low oxygen tension (1%). In this study, hypoxia did not change the stability of NOS-III mRNA. Nevertheless, the promoter activity of a 1.6-kb DNA fragment of the 5'-flanking sequence of the human NOS-III gene was enhanced by hypoxia (ARNET et al. 1996). However, the human NOS-III promoter contains no homology to the published binding sequence of the hypoxia-induced transcription factor HIF. Also, ZHANG et al. (1993) reported upregulation of endothelial NOS-III immunoreactivity in cerebral blood vessels during cerebral ischemia. Other reports, however, demonstrated reductions of NOS-III expression in human umbilical vein endothelial cells (HUVEC) and BAEC exposed to low oxygen tension (McQuillan et al. 1994; Phelan and Faller 1996).

It has been shown that NOS-III mRNA and protein are increased in growing versus resting BAEC (Arnal et al. 1994). This enhanced NOS-III expression was found to be the result of a greater stability of the NOS-III mRNA in proliferating compared with confluent BAEC (Arnal et al. 1994). Growth-arrested BAEC showed enhanced expression of a protein that interacts with the 3' untranslated region (3'-UTR) of the bovine NOS-III mRNA and destabilized the mRNA (Harrison 1997).

In BAEC and HUVEC, TNF-α downregulates NOS mRNA, protein, and activity (Lamas et al. 1992; Nishida et al. 1992). The downregulation of NOS-III mRNA expression by TNF-α in HUVEC has been ascribed to a destabilization of NOS-III mRNA with no effect on transcription (Yoshizumi et al. 1993). In bovine endothelial cells, this destabilization seems to result from a specific interaction of a TNF-α-induced protein with the 3'-UTR sequence of the NOS-III mRNA (Alonso et al. 1997).

There are several reports that estrogens can upregulate the expression of NOS-III mRNA and protein. In guinea pigs, near-term pregnancy and treatment with estradiol but not progesterone increased calcium-dependent NOS activity in various tissues (Weiner et al. 1994b). Both pregnancy and estradiol also enhanced NOS-III mRNA (along with NOS-I mRNA) in guinea-pig skeletal muscle (Weiner et al. 1994b). An increase in NOS-III mRNA has also been seen in the aortas of pregnant or estrogen-treated rats, but not progesterone- or testosterone-treated rats (Goetz et al. 1994). Also, the kidneys from female rats contain more NOS-III protein than those of male or oophorectomized female rats (Neugarten et al. 1997). However, in in vitro studies with ovine fetal pulmonary artery endothelial cells, estrogens enhance NOS-III activity, but no change in NOS-III mRNA expression was detected (Lantin-Hermoso et al. 1997). A study on bovine endothelial cells claimed that 17α-ethinyl estradiol did not enhance the expression of NOS-III but increased the release of bioactive NO by inhibiting superoxide anion production (Arnal et al. 1996). Studies performed in our own laboratory (Kleinert et al. 1998b) demonstrated that estrogens did enhance NOS-III mRNA and protein expression in permanent human endothelial EA.hy 926 cells. The increased NOS-III expression resulted from an increased NOS-III promoter activity with unchanged mRNA stability. In the absence of a bona fide estrogen-responsive element in the human NOS-III promoter (see below), the increased NOS-III-promoter activity may result from an enhanced binding activity of transcription factor Sp1 (which is essential for the human NOS-III promoter; see below) (Kleinert et al. 1998b).

Incubation of BAEC (Ohara et al. 1995) or human EA.hy 926 endothelial cells (Li et al. 1998) with phorbol esters enhanced NOS-III expression. Ohara et al. (1995) concluded from their results that downregulation of PKC by long-term incubation with phorbol esters (or PKC inhibition with staurosporine) enhances NOS-III expression. In contrast to these results, in our own experiments with human EA.hy 926 endothelial cells, the time course of phorbol-ester-induced enhancement of NOS-III expression paralleled PKC activation (LI et al. 1998). Also, specific PKC inhibitors, such as bisindolylmaleimide I, Gö-6976, Ro-31-8220, and chelerythrine, prevented the phorbol-ester-induced enhancement of NOS-III expression. Based on transfection experiments with a 3.5-kb human NOS-III-promoter fragment, the phorbol-ester-stimulated enhancement of NOS-III expression seems to be a transcriptional event (LI et al. 1998).

III. Regulation of NOS-III Activity

As a Ca^{2+}- and calmodulin-dependent enzyme (FÖRSTERMANN et al. 1991; POLLOCK et al. 1991), endothelial NOS-III can be activated by agonists as a consequence of an increase in the intracellular concentration of free Ca^{2+}. This activation is likely to represent the most important mechanism for acute changes in enzyme activity. However, recent evidence suggests that the enzyme can also be activated in a Ca^{2+}-independent fashion and that the subcellular targeting of the enzyme may be important for its activity.

In response to fluid shear stress, the continuous production of NO by endothelial cells is associated with only a transient and minimal increase in intracellular Ca^{2+}. In the absence of extracellular Ca^{2+} and in the presence of the calmodulin antagonist calmidazolium, shear stress still stimulated a continuous production of NO that was sensitive to the nonspecific kinase inhibitors staurosporine and calphostin C and the tyrosine kinase inhibitor erbstatin A. A similar activation of NOS-III could be induced by protein phosphatase inhibitors (such as phenylarsine oxide), suggesting that tyrosine phosphorylation of NOS III or an associated regulatory protein is crucial for its activation at basal levels of Ca^{2+}. Incubation of endothelial cells with phenylarsine oxide or exposure to fluid shear stress resulted in a time-dependent tyrosine phosphorylation of mainly Triton X-100-insoluble (cytoskeletal) proteins along with a parallel change in the detergent solubility of NOS-III such that the enzyme was recovered in the cytoskeletal fraction. Phenylarsine oxide also induced an immediate and maintained NO-mediated relaxation of isolated rabbit carotid arteries. This relaxation was insensitive to the removal of extracellular Ca^{2+} and the calmodulin antagonist calmidazolium but was abrogated by the tyrosine-kinase inhibitor erbstatin A. Thus, endothelial cells seem to be able to respond to mechanical and humoral stimuli, activating NOS-III by Ca^{2+}-dependent and -independent pathways (FLEMING et al. 1997, 1998).

NOS-III has been shown to be targeted to Golgi membranes and plasmalemmal caveolae by a complex process probably dependent on myristoylation, palmitoylation, and serine and tyrosine phosporylation of the enzyme and protein–protein interactions with caveolins (FERON et al. 1996; GARCIA-CARDENA et al. 1996a, 1996b; SHAUL et al. 1996). Myristoylation at the N-terminal glycine (glycine-2 before removal of the initiation methionine by a specific aminopeptidase) is probably important for the membrane association of the enzyme (POLLOCK et al. 1992). Mutation of the N-terminal myristoylation site (Gly 2) converted NOS-III into a 92% cytosolic enzyme (SESSA et al. 1993). Kinetic analysis of the wild-type and mutated enzymes in vitro revealed similar K_m values for L-arginine (2–4 μM), suggesting that the mutation did not alter the function of the NOS-III enzyme itself (SESSA et al. 1993). However, mislocalization of NOS-III caused by mutation of the N-terminal myristoylation or cysteine palmitoylation sites (Cys 15 and 26) reduced *cellular* NO production, suggesting that intracellular targeting of the NOS-III is critical for endothelial NO production (SESSA et al. 1995). Experiments with NOS-III green fluorescent protein (GFP) chimeras revealed that the first 35 amino acids of NOS-III are sufficient to target GFP to the Golgi region of NIH 3T3 cells (LIU et al. 1997).

NOS-III has also been shown to interact with caveolins, a family of transmembrane proteins that form a key structural component of the caveolae. In endothelial cells, the interaction between NOS-III and caveolin-1 was found to be regulated by Ca^{2+}/calmodulin. Addition of calmodulin disrupted the heteromeric complex formed between NOS-III and caveolin-1 in a Ca^{2+}-dependent fashion. In addition, overexpression of caveolin-1 markedly attenuated NOS-III enzyme activity, but this inhibition is reversed by purified calmodulin. Thus, the inhibitory NOS-III–caveolin-1 complex is disrupted by binding of Ca^{2+}/calmodulin to NOS-III, leading to enzyme re-activation (FERON et al. 1996; MICHEL et al. 1997).

NOS-III can undergo serine phosphorylation in response to bradykinin and a translocation from the particulate to the cytosolic fraction (MICHEL et al. 1993). Also, tyrosine phosphorylation has been demonstrated for NOS-III (Garcia-CARDENA et al. 1996a). Phophorylation of eNOS has also been described in response to shear stress (CORSON et al. 1993). Phosphorylation may influence NOS-III activity, subcellular trafficking and interaction with other proteins in caveolae. However, the exact physiological relevance of NOS-III phosphorylation remains to be determined.

E. Summary and Conclusions

Of the three established isozymes of NOS, NOS-I (nNOS) and NOS-III (eNOS) are constitutive, low-output, Ca^{2+}-activated enzymes whose physiological function is signal transduction. In addition to the intracellular Ca^{2+} level, post-translational modifications and the subcellular targeting of the enzymes determine their activity. Also, their expression levels can be modified in a cell-specific fashion. NOS-II (iNOS) is mainly regulated at the level of transcription, but post-transcriptional and -translational mechanisms that regulate protein expression of this isoform have also been described. The stimuli and conditions that determine NOS-II expression are cell- and species-specific. Once expressed, NOS-II does not seem to be subject to any significant regulation of its enzymatic activity. Thus, Mother Nature has invented a large array of regulatory mechanisms controlling the production of the pluripotent molecule NO.

References

Alonso J, Sanchez de Miguel L, Monton M, Casado S, Lopez-Farre A (1997) Endothelial cytosolic proteins bind to the 3' untranslated region of endothelial nitric oxide synthase mRNA: regulation by tumor necrosis factor alpha. Mol Cell Biol 17: 5719–5726

Arbones ML, Ribera J, Agullo L, Baltrons MA, Casanovas A, Riveros Moreno V, Garcia A (1996) Characteristics of nitric oxide synthase type I of rat cerebellar astrocytes. Glia 18:224–232

Arnal JF, Yamin J, Dockery S, Harrison DG (1994) Regulation of endothelial nitric oxide synthase mRNA, protein, and activity during cell growth. Am J Physiol 267:C1381–C1388

Arnal JF, Clamens S, Pechet C, Negre-Salvayre A, Allera C, Girolami JP, Salvayre R, Bayard F (1996) Ethinylestradiol does not enhance the expression of nitric oxide synthase in bovine endothelial cells but increases the release of bioactive nitric oxide by inhibiting superoxide anion production. Proc Natl Acad Sci USA 93: 4108–4113

Arnet UA, McMillan A, Dinerman JL, Ballermann B, Lowenstein CJ (1996) Regulation of endothelial nitric-oxide synthase during hypoxia. J Biol Chem 271: 15069–15073

Asano K, Chee C, Gaston B, Lilly CM, Gerard C, Drazen JM, Stamler JS (1994) Constitutive and inducible nitric oxide synthase gene expression, regulation, and activity in human lung epithelial cells. Proc Natl Acad Sci USA 91:10089–10093

Baader SL, Schilling K (1996) Glutamate receptors mediate dynamic regulation of nitric oxide synthase expression in cerebellar granule cells. J Neurosci 16: 1440–1449

Bagetta G, Corasaniti MT, Melino G, Paoletti AM, Finazzi Agro A, Nistico G (1993) Lithium and tacrine increase the expression of nitric oxide synthase mRNA in the hippocampus of rat. Biochem Biophys Res Commun 197:1132–1139

Bandaletova T, Brouet I, Bartsch H, Sugimura T, Esumi H, Ohshima H (1993) Immunohistochemical localization of an inducible form of nitric oxide synthase in various organs of rats treated with *Propionibacterium acnes* and lipopolysaccharide. Apmis 101:330–336

Bandyopadhyay A, Chakder S, Rattan S (1997) Regulation of inducible and neuronal nitric oxide synthase gene expression by interferon-γ and VIP. Am J Physiol 272: C1790–C1797

Bogdan C, Vodovotz Y, Paik J, Xie QW, Nathan C (1994) Mechanism of suppression of nitric oxide synthase expression by interleukin-4 in primary mouse macrophages. J Leukocyte Biol 55:227–233

Bredt DS, Snyder SH (1990) Isolation of nitric oxide synthetase, a calmodulin-requiring enzyme. Proc Natl Acad Sci USA 87:682–685

Bredt DS, Hwang PM, Glatt CE, Lowenstein C, Reed RR, Snyder SH (1991) Cloned and expressed nitric oxide synthase structurally resembles cytochrome P-450 reductase. Nature 351:714–718

Brenman JE, Chao DS, Xia H, Aldape K, Bredt DS (1995) Nitric oxide synthase complexed with dystrophin and absent from skeletal muscle sarcolemma in Duchenne muscular dystrophy. Cell 82:743–752

Brenman JE, Chao DS, Gee SH, McGee AW, Craven SE, Santillano DR, Wu Z, Huang F, Xia H, Peters MF, Froehner SC, Bredt DS (1996) Interaction of nitric oxide synthase with the postsynaptic density protein PSD-95 and α1-syntrophin mediated by PDZ domains. Cell 84:757–767

Calza L, Giardino L, Ceccatelli S (1993) NOS mRNA in the paraventricular nucleus of young and old rats after immobilization stress. Neuroreport 4:627–630

Calza L, Giardino L, Pozza M, Micera A, Aloe L (1997) Time-course changes of nerve growth factor, corticotropin-releasing hormone, and nitric oxide synthase isoforms and their possible role in the development of inflammatory response in experimental allergic encephalomyelitis. Proc Natl Acad Sci U S A 94:3368–3373

Ceccatelli S, Grandison L, Scott RE, Pfaff DW, Kow LM (1996) Estradiol regulation of nitric oxide synthase mRNAs in rat hypothalamus. Neuroendocrinology 64: 357–363

Chang WJ, Iannaccone ST, Lau KS, Masters BS, McCabe TJ, McMillan K, Padre RC, Spencer MJ, Tidball JG, Stull JT (1996) Neuronal nitric oxide synthase and dystrophin-deficient muscular dystrophy. Proc Natl Acad Sci U S A 93:9142–9147

Charles IG, Palmer RMJ, Hickery MS, Bayliss MT, Chubb AP, Hall VS, Moss DW, Moncada S (1993) Cloning, characterization, and expression of a cDNA encoding an inducible nitric oxide synthase from the human chondrocyte. Proc Natl Acad Sci U S A 90:11419–11423

Cho HJ, Xie QW, Calaycay J, Mumford RA, Swiderek KM, Lee TD, Nathan C (1992) Calmodulin is a subunit of nitric oxide synthase from macrophages. J Exp Med 176:599–604

Corson MA, Berk BC, Navas JP, Harrison DG (1993) Phosphorylation of endothelial nitric oxide synthase in response to shear stress. Circulation (Suppl) 88:I–183

Cunha FQ, Moncada S, Liew FY (1992) Interleukin-10 (IL-10) inhibits the induction of nitric oxide synthase by interferon-γ in murine macrophages. Biochem Biophys Res Commun 182:1155–1159

Dai A, Zhang Z, Niu R (1995) The effects of hypoxia on nitric oxide synthase activity and mRNA expression of pulmonary artery endothelial cells in pigs. Chung Hua Chieh Ho Ho Hu Hsi Tsa Chih 18:164–166

Di Rosa M, Radomski M, Carnuccio R, Moncada S (1990) Glucocorticoids inhibit the induction of nitric oxide synthase in macrophages. Biochem Biophys Res Commun 172:1246–1252

Dinerman JL, Dawson TM, Schell MJ, Snowman A, Snyder SH (1994a) Endothelial nitric oxide synthase localized to hippocampal pyramidal cells: implications for synaptic plasticity. Proc Natl Acad Sci USA 91:4214–4218

Dinerman JL, Steiner JP, Dawson TM, Dawson V, Snyder SH (1994b) Cyclic nucleotide dependent phosphorylation of neuronal nitric oxide synthase inhibits catalytic activity. Neuropharmacology 33:1245–1251

Ding A, Nathan CF, Graycar J, Derynck R, Stuehr DJ, Srimal S (1990) Macrophage deactivating factor and transforming growth factors $\beta 1$ $\beta 2$ and $\beta 3$ inhibit induction of macrophage nitrogen oxide synthesis by IFN-γ. J Immunol 145:940–944

Dong ZY, Qi XO, Xie KP, Fidler IJ (1993) Protein tyrosine kinase inhibitors decrease induction of nitric oxide synthase activity in lipopolysaccharide-responsive and lipopolysaccharide-nonresponsive murine macrophages. J Immunol 151:2717–2724

Dun NJ, Dun SL, Förstermann U, Tseng LF (1992) Nitric oxide synthase immunoreactivity in rat spinal cord. Neurosci Lett 147:217–220

Dun NJ, Dun SL, Wu SY, Förstermann U (1993) Nitric oxide synthase immunoreactivity in rat superior cervical ganglia and adrenal glands. Neurosci Lett 158:51–54

Feinstein DL, Galea E, Reis DJ (1993) Norepinephrine suppresses inducible nitric oxide synthase activity in rat astroglial cultures. J Neurochem 60:1945–1948

Feron O, Belhassen L, Kobzik L, Smith TW, Kelly RA, Michel T (1996) Endothelial nitric oxide synthase targeting to caveolae. Specific interactions with caveolin isoforms in cardiac myocytes and endothelial cells. J Biol Chem 271:22810–22814

Fleming I, Bauersachs J, Busse R (1997) Calcium-dependent and calcium-independent activation of the endothelial NO synthase. J Vasc Res 34:165–174

Fleming I, Bauersachs J, Fisslthaler B, Busse R (1998) Ca2+-independent activation of the endothelial nitric oxide synthase in response to tyrosine phosphatase inhibitors and fluid shear stress. Circ Res 82:686–695

Förstermann U, Kleinert H (1995) Nitric oxide synthase: expression and expressional control of the three isoforms. Naunyn Schmiedebergs Arch Pharmacol 352:351–364

Förstermann U, Gorsky LD, Pollock JS, Ishii K, Schmidt HH, Heller M, Murad F (1990) Hormone-induced biosynthesis of endothelium-derived relaxing factor/nitric oxide-like material in N1E-115 neuroblastoma cells requires calcium and calmodulin. Mol Pharmacol 38:7–13

Förstermann U, Pollock JS, Schmidt HHHW, Heller M, Murad F (1991) Calmodulin-dependent endothelium-derived relaxing factor/nitric oxide synthase activity is present in the particulate and cytosolic fractions of bovine aortic endothelial cells. Proc Natl Acad Sci USA 88:1788–1792

Förstermann U, Schmidt HHHW, Kohlhaas KL, Murad F (1992) Induced RAW 264.7 macrophages express soluble and particulate nitric oxide synthase: inhibition by transforming growth factor-beta. Eur J Pharmacol Mol Pharmacol 225:161–165

Förstermann U, Closs EI, Pollock JS, Nakane M, Schwarz P, Gath I, Kleinert H (1994) Nitric oxide synthase isozymes: characterization, molecular cloning and functions. Hypertension 23:1121–1131

Garcia-Cardena G, Fan R, Stern DF, Liu J, Sessa WC (1996a) Endothelial nitric oxide synthase is regulated by tyrosine phosphorylation and interacts with caveolin-1. J Biol Chem 271:27237–27340

Garcia-Cardena G, Oh P, Liu J, Schnitzer JE, Sessa WC (1996b) Targeting of nitric oxide synthase to endothelial cell caveolae via palmitoylation: implications for nitric oxide signaling. Proc Natl Acad Sci USA 93:6448–6453

Gath I, Closs EI, Gödtel-Armbrust U, Schmitt S, Nakane M, Wessler I, Förstermann U (1996) Inducible NO synthase II and neuronal NO synthase I are constitutively expressed in different structures of guinea pig skeletal muscle: Implications for contractile function. FASEB J 10:1614–1620

Gath I, Gödtel-Armbrust U, Förstermann U (1997) Expressional downregulation of neuronal-type NO synthase I in guinea pig skeletal muscle in response to bacterial lipopolysaccharide. FEBS Lett 410:319–323

Geller DA, Lowenstein CJ, Shapiro RA, Nussler AK, Di SM, Wang SC, Nakayama DK, Simmons RL, Snyder SH, Billiar TR (1993) Molecular cloning and expression of inducible nitric oxide synthase from human hepatocytes. Proc Natl Acad Sci U S A 90:3491–3495

Geller DA, Devera ME, Russell DA, Shapiro RA, Nüssler AK, Simmons RL, Billiar TR (1995) A central role for IL-1β in the in vitro and in vivo regulation of hepatic inducible nitric oxide synthase IL-1β induces hepatic nitric oxide synthesis. J Immunol 155:4890–4898

Gilbert RS, Herschman HR (1993a) Macrophage nitric oxide synthase is a glucocorticoid-inhibitable primary response gene in 3T3 cells. J Cell Physiol 157:128–132

Gilbert RS, Herschman HR (1993b) Transforming growth factor β differentially modulates the inducible nitric oxide synthase gene in distinct cell types. Biochem Biophys Res Commun 195:380–384

Goetz RM, Morano I, Calovini T, Studer R, Holtz J (1994) Increased expression of endothelial constitutive nitric oxide synthase in rat aorta during pregnancy. Biochem Biophys Res Commun 205:905–910

Goureau O, Lepoivre M, Becquet F, Courtois Y (1993) Differential regulation of inducible nitric oxide synthase by fibroblast growth factors and transforming growth factor beta in bovine retinal pigmented epithelial cells: inverse correlation with cellular proliferation. Proc Natl Acad Sci U S A 90:4276–4280

Harrison DG (1997) Post transcriptional regulation of nitric oxide synthase. Jpn J Pharmacol 75:P-399

Hassall CJ, Saffrey MJ, Belai A, Hoyle CH, Moules EW, Moss J, Schmidt HH, Murad F, Förstermann U, Burnstock G (1992) Nitric oxide synthase immunoreactivity and NADPH-diaphorase activity in a subpopulation of intrinsic neurones of the guinea-pig heart. Neurosci Lett 143:65–68

Hecker M, Mülsch A, Busse R (1994) Subcellular localization and characterization of neuronal nitric oxide synthase. J Neurochem 62:1524–1529

Herdegen T, Brecht S, Mayer B, Leah J, Kummer W, Bravo R, Zimmermann M (1993) Long-lasting expression of JUN and KROX transcription factors and nitric oxide synthase in intrinsic neurons of the rat brain following axotomy. J Neurosci 13: 4130–4145

Hevel JM, White KA, Marletta MA (1991) Purification of the inducible murine macrophage nitric oxide synthase. Identification as a flavoprotein. J Biol Chem 266:22789–22791

Hortelano S, Genaro AM, Bosca L (1993) Phorbol esters induce nitric oxide synthase and increase arginine influx in cultured peritoneal macrophages. FEBS Lett 320:135–139

Kanno K, Hirata Y, Imai T, Marumo F (1993) Induction of nitric oxide synthase gene by interleukin in vascular smooth muscle cells. Hypertension 22:34–39

Kleemann R, Rothe H, Kolb-Bachofen V, Xie QW, Nathan C, Martin S, Kolb H (1993) Transcription and translation of inducible nitric oxide synthase in the pancreas of prediabetic BB rats. FEBS Lett 328:9–12

Kleinert H, Euchenhofer C, Ihrig-Biedert I, Förstermann U (1996) Glucocorticoids inhibit the induction of nitric oxide synthase II by downregulating cytokine-induced activity of transcription factor NF-κB. Mol Pharmacol 49:15–21

Kleinert H, Euchenhofer C, Fritz G, Ihrig-Biedert I, Förstermann U (1998a) Involvement of protein kinases in the induction of NO synthase II in human DLD-1 cells. Br J Pharmacol 123:1716–1722

Kleinert H, Wallerath T, Euchenhofer CE, Ihrig-Biedert I, Li H, Förstermann U (1998b) Estrogens increase transcription of the human endothelial NO synthase gene: analysis of the transcription factors involved. Hypertension 31:582–588

Kleinert H, Wallerath T, Fritz G, Ihrig-Biedert I, Rodriguez-Pascual F, Geller DA, Förstermann U (1998c) Cytokine induction of NO synthase II in human DLD-1 cells: roles of the JAK-STAT, AP-1 and NF-κB-signaling pathways. Br J Pharmacol 125:193–201

Kobzik L, Bredt DS, Lowenstein CJ, Drazen J, Gaston B, Sugarbaker D, Stamler JS (1993) Nitric oxide synthase in human and rat lung: immunocytochemical and histochemical localization. Am J Respir Cell Mol Biol 9:371–377

Kobzik L, Reid MB, Bredt DS, Stamler JS (1994) Nitric oxide in skeletal muscle. Nature 372:546–548

Koide M, Kawahara Y, Nakayama I, Tsuda T, Yokoyama M (1993) Cyclic AMP-elevating agents induce an inducible type of nitric oxide synthase in cultured vascular smooth muscle cells: synergism with the induction elicited by inflammatory cytokines. J Biol Chem 268:24959–24966

Komatsu T, Reiss CS (1997) IFN-gamma is not required in the IL-12 response to vesicular stomatitis virus infection of the olfactory bulb. J Immunol 159:3444–3452

Kornau H-C, Schenker LT, Kennedy MB, Seeburg PH (1995) Domain interaction between NMDA receptor subunits and the postsynaptic density protein PSD-95. Science 269:1737–1740

Lam HH, Hanley DF, Trapp BD, Saito S, Raja S, Dawson TM, Yamaguchi H (1996) Induction of spinal cord neuronal nitric oxide synthase (NOS) after formalin injection in the rat hind paw. Neurosci Lett 210:201–204

Lamas S, Michel T, Collins T, Brenner BM, Marsden PA (1992) Effects of interferon-gamma on nitric oxide synthase activity and endothelin-1 production by vascular endothelial cells. J Clin Invest 90:879–887

Lantin-Hermoso RL, Rosenfeld CR, Yuhanna IS, German Z, Chen Z, Shaul PW (1997) Estrogen acutely stimulates nitric oxide synthase activity in fetal pulmonary artery endothelium. Am J Physiol 273:L119–L126

Li H, Oehrlein SA, Wallerath T, Ihrig-Biedert I, Wohlfart P, Ulshöfer T, Jessen T, Herget T, Förstermann U, Kleinert H (1998) Activation of protein kinase Cα and/or ε enhances transcription of the human endothelial NO synthase gene. Mol Pharmacol 53:630–637

Liao JK, Zulueta JJ, Yu FS, Peng HB, Cote CG, Hassoun PM (1995) Regulation of bovine endothelial constitutive nitric oxide synthase by oxygen. J Clin Invest 96:2661–2666

Lin LH, Sandra A, Boutelle S, Talman WT (1997) Up-regulation of nitric oxide synthase and its mRNA in vagal motor nuclei following axotomy in rat. Neurosci Lett 221:97–100

Linn SC, Morelli PJ, Edry I, Cottongim SE, Szabo C, Salzman AL (1997) Transcriptional regulation of human inducible nitric oxide synthase gene in an intestinal epithelial cell line. Am J Physiol 272:G1499–G1508

Liu JW, Hughes TE, Sessa WC (1997) The first 35 amino acids and fatty acylation sites determine the molecular targeting of endothelial nitric oxide synthase into the golgi region of cells: a green fluorescent protein study. J Cell Biol 137:1525–1535

Liu SF, Adcock IM, Old RW, Barnes PJ, Evans TW (1996) Differential regulation of the constitutive and inducible nitric oxide synthase mRNA by lipopolysaccharide treatment in vivo in the rat. Crit Care Med 24:1219–1225

Lowenstein CJ, Alley EW, Raval P, Snowman AM, Snyder SH, Russell SW, Murphy WJ (1993) Macrophage nitric oxide synthase gene: two upstream regions mediate induction by interferon-gamma and lipopolysaccharide. Proc Natl Acad Sci U S A 90:9730–9734

Lumme A, Vanhatalo S, Sadeniemi M, Soinila S (1997) Expression of nitric oxide synthase in hypothalamic nuclei following axonal injury or colchicine treatment. Exp Neurol 144:248–257

Lyons CR, Orloff GJ, Cunningham JM (1992) Molecular cloning and functional expression of an inducible nitric oxide synthase from a murine macrophage cell line. J Biol Chem 267:6370–6374

Martin E, Nathan C, Xie QW (1994) Role of interferon regulatory factor 1 in induction of nitric oxide synthase. J Exp Med 180:977–984

Mayer B, John M, Böhme E (1990) Purification of a calcium/calmodulin-dependent nitric oxide synthase from porcine cerebellum. Cofactor role of tetrahydrobiopterin. FEBS Lett 277:215–219

McCall TB, Palmer RM, Moncada S (1992) Interleukin-8 inhibits the induction of nitric oxide synthase in rat peritoneal neutrophils. Biochem Biophys Res Commun 186:680–685

McQuillan LP, Leung GK, Marsden PA, Kostyk SK, Kourembanas S (1994) Hypoxia inhibits expression of eNOS via transcriptional and posttranscriptional mechanisms. Am J Physiol 267:H1921-H1927

Michel T, Li GK, Busconi L (1993) Phosphorylation and subcellular translocation of endothelial nitric oxide synthase. Proc Natl Acad Sci USA 90:6252–6256

Michel JB, Feron O, Sacks D, Michel T (1997) Reciprocal regulation of endothelial nitric-oxide synthase by Ca2+-calmodulin and caveolin. J Biol Chem 272:15583–15586

Mülsch A, B S-U, Mordvintcev PI, Hauschildt S, Busse R (1993) Diethyldithiocarbamate inhibits induction of macrophage NO synthase. FEBS Lett 321:215–218

Myatt L, Brockman DE, Eis AL, Pollock JS (1993) Immunohistochemical localization of nitric oxide synthase in the human placenta. Placenta 14:487–495

Nakane M, Mitchell J, Förstermann U, Murad F (1991) Phosphorylation by calcium calmodulin-dependent protein kinase II and protein kinase C modulates the activity of nitric oxide synthase. Biochem Biophys Res Commun 180:1396–1402

Nakane M, Schmidt HHHW, Pollock JS, Förstermann U, Murad F (1993) Cloned human brain nitric oxide synthase is highly expressed in skeletal muscle. FEBS Lett 316:175–180

Nathan CF, Hibbs JB (1991) Role of nitric oxide synthesis in macrophage antimicrobial activity. Curr Opin Immunol 3:65–70

Neugarten J, Ding Q, Friedman A, Lei J, Silbiger S (1997) Sex hormones and renal nitric oxide synthases. J Amer Soc Nephrol 8:1240–1246

Nishida K, Harrison DG, Navas JP, Fisher AA, Dockery SP, Uematsu M, Nerem RM, Alexander RW, Murphy TJ (1992) Molecular cloning and characterization of the constitutive bovine aortic endothelial cell nitric oxide synthase. J Clin Invest 90:2092–2096

Nunokawa Y, Oikawa S, Tanaka S (1996) Human inducible nitric oxide synthase gene is transcriptionally regulated by nuclear factor κB-dependent mechanism. Biochem Biophys Res Commun 223:347–352

Ohara Y, Sayegh HS, Yamin JJ, Harrison DG (1995) Regulation of endothelial constitutive nitric oxide synthase by protein kinase C. Hypertension 25:415–420

Phelan MW, Faller DV (1996) Hypoxia decreases constitutive nitric oxide synthase transcript and protein in cultured endothelial cells. J Cell Physiol 167:469–476

Pollock JS, Förstermann U, Mitchell JA, Warner TD, Schmidt HHHW, Nakane M, Murad F (1991) Purification and characterization of particulate endothelium-derived relaxing factor synthase from cultured and native bovine aortic endothelial cells. Proc Natl Acad Sci U S A 88:10480–10484

Pollock JS, Klinghofer V, Förstermann U, Murad F (1992) Endothelial nitric oxide synthase is myristylated. FEBS Lett 309:402–404

Pollock JS, Nakane M, Buttery LK, Martinez A, Springall D, Polak JM, Förstermann U, Murad F (1993) Characterization and localization of endothelial nitric oxide synthase using specific monoclonal antibodies. Am J Physiol 265:C1379–C1387

Ponting CP, Phillips C, Davies KE, Blake DJ (1997) PDZ domains: targeting signalling molecules to sub-membranous sites. Bioessays 19:469–479

Radomski MW, Palmer RM, Moncada S (1990) Glucocorticoids inhibit the expression of an inducible, but not the constitutive, nitric oxide synthase in vascular endothelial cells. Proc Natl Acad Sci USA 87:10043–10047

Reilly CM, Zamorano P, Stopper VS, Mills TM (1997) Androgenic regulation of NO availability in rat penile erection. J Androl 18:110–115

Reiser PJ, Kline WO, Vaghy PL (1997) Induction of neuronal type nitric oxide synthase in skeletal muscle by chronic electrical stimulation in vivo. J Appl Physiol 82:1250–1255

Robbins RA, Springall DR, Warren JB, Kwon OJ, Buttery LD, Wilson AJ, Adcock IM, Riveros-Moreno V, Moncada S, Polak J, Barnes PJ (1994) Inducible nitric oxide synthase is increased in murine lung epithelial cells by cytokine stimulation. Biochem Biophys Res Commun 198:835–843

Rodrigo J, Riveros-Moreno V, Bentura ML, Uttenthal LO, Higgs EA, Fernandez AP, Polak JM, Moncada S, Martinez-Murillo R (1997) Subcellular localization of nitric oxide synthase in the cerebral ventricular system, subfornical organ, area postrema, and blood vessels of the rat brain. J Comp Neurol 378:522–534

Rojas A, Delgado R, Glaria L, Palacios M (1993) Monocyte chemotactic protein-1 inhibits the induction of nitric oxide synthase in J774 cells. Biochem Biophys Res Commun 196:274–279

Saffrey MJ, Hassall CJ, Hoyle CH, Belai A, Moss J, Schmidt HH, Förstermann U, Murad F, Burnstock G (1992) Colocalization of nitric oxide synthase and NADPH-diaphorase in cultured myenteric neurones. Neuroreport 3:333–336

Schepens J, Cuppen E, Wieringa B, Hendriks W (1997) The neuronal nitric oxide synthase PDZ motif binds to -G(D,E)XV* carboxyterminal sequences. FEBS Lett 409:53–56

Schini VB, Durante W, Elizondo E, Scott BT, Junquero DC, Schafer AI, Vanhoutte PM (1992) The induction of nitric oxide synthase activity is inhibited by TGF-beta 1, PDGFAB and PDGFBB in vascular smooth muscle cells. Eur J Pharmacol 216:379–383

Schini VB, Catovsky S, Schray-Utz B, Busse R, Vanhoutte PM (1994) Insulin-like growth factor I inhibits induction of nitric oxide synthase in vascular smooth muscle cells. Circ Res 74:24–32

Schmidt HHHW, Pollock JS, Nakane M, Gorsky LD, Förstermann U, Murad F (1991) Purification of a soluble isoform of guanylyl cyclase-activating-factor synthase. Proc Natl Acad Sci USA 88:365–369

Schmidt HHHW, Gagne GD, Nakane M, Pollock JS, Miller MF, Murad F (1992a) Mapping of neural nitric oxide synthase in the rat suggests frequent co-localization with NADPH diaphorase but not with soluble guanylyl cyclase, and novel paraneural functions for nitrinergic signal transduction. J Histochem Cytochem 40:1439–1456

Schmidt HHHW, Warner TD, Ishii K, Sheng H, Murad F (1992b) Insulin secretion from pancreatic B cells caused by L-arginine-derived nitrogen oxides. Science 255:721–723

Schwarz PM, Gierten B, Boissel J-P, Förstermann U (1998) Expressional down-regulation of neuronal-type nitric oxide synthase I by glucocorticoids in N1E-115 neuroblastoma cells. Mol Pharmacol 54:258–263

Sessa WC, Harrison JK, Barber CM, Zeng D, Durieux ME, D'Angelo DD, Lynch KR, Peach MJ (1992) Molecular cloning and expression of a cDNA encoding endothelial cell nitric oxide synthase. J Biol Chem 267:15274–15276

Sessa WC, Barber CM, Lynch KR (1993) Mutation of N-myristoylation site converts endothelial cell nitric oxide synthase from a membrane to a cytosolic protein. Circ Res 72:921–924

Sessa WC, Pritchard K, Seyedi N, Wang J, Hintze TH (1994) Chronic exercise in dogs increases coronary vascular nitric oxide production and endothelial cell nitric oxide synthase gene expression. Circ Res 74:349–353

Sessa WC, Garcia-Cardena G, Liu J, Keh A, Pollock JS, Bradley J, Thiru S, Braverman IM, Desai KM (1995) The Golgi association of endothelial nitric oxide synthase is necessary for the efficient synthesis of nitric oxide. J Biol Chem 270:17641–17644

Sharma HS, Westman J, Alm P, Sjoquist PO, Cervos Navarro J, Nyberg F (1997) Involvement of nitric oxide in the pathophysiology of acute heat stress in the rat. Influence of a new antioxidant compound H-290/51. Ann N Y Acad Sci 813:581–590

Shaul PW, Smart EJ, Robinson LJ, German Z, Yuhanna IS, Ying Y, Anderson RG, Michel T (1996) Acylation targets endothelial nitric-oxide synthase to plasmalemmal caveolae. J Biol Chem 271:6518–6522

Sheng H, Schmidt HH, Nakane M, Mitchell JA, Pollock JS, Föstermann U, Murad F (1992) Characterization and localization of nitric oxide synthase in non-adrenergic non-cholinergic nerves from bovine retractor penis muscles. Br J Pharmacol 106:768–773

Sheng H, Gagne GD, Matsumoto T, Miller MF, Förstermann U, Murad F (1993) Nitric oxide synthase in bovine superior cervical ganglion. J Neurochem 61:1120–1126

Sherman MP, Aeberhard EE, Wong VZ, Griscavage JM, Ignarro LJ (1993a) Pyrrolidine dithiocarbamate inhibits induction of nitric oxide synthase activity in rat alveolar macrophages. Biochem Biophys Res Commun 191:1301–1308

Sherman PA, Laubach VE, Reep BR, Wood ER (1993b) Purification and cDNA sequence of an inducible nitric oxide synthase from a human tumor cell line. Biochemistry 32:11600–11605

Spitsin SV, Koprowski H, Michaels FH (1996) Characterization and functional analysis of the human inducible nitric oxide synthase gene promoter. Molecular Medicine 2:226–235

Stuehr DJ, Cho HJ, Kwon NS, Weise MF, Nathan CF (1991) Purification and characterization of the cytokine-induced macrophage nitric oxide synthase: an FAD- and FMN-containing flavoprotein. Proc Natl Acad Sci USA 88:7773–7777

Tascedda F, Molteni R, Racagni G, Riva MA (1996) Acute and chronic changes in K(+)-induced depolarization alter NMDA and nNOS gene expression in cultured cerebellar granule cells. Mol Brain Res 40:171–174

Taylor BS, deVera ME, Ganster RW, Wang Q, Shapiro RA, Morris SM, Billiar TR, Geller DA (1998) Multiple NF-κB enhancer elements regulate cytokine induction of the human inducible nitric oxide synthase gene. J Biol Chem 273:15148–15156

Tojo A, Gross SS, Zhang L, Tisher CC, Schmidt HH, Wilcox CS, Madsen KM (1994) Immunocytochemical localization of distinct isoforms of nitric oxide synthase in the juxtaglomerular apparatus of normal rat kidney. J Am Soc Nephrol 4:1438–1447

Tracey WR, Pollock JS, Murad F, Nakane M, Förstermann U (1994a) Identification of a type III (endothelial-like) particulate nitric oxide synthase in LLC-PK1 kidney tubular epithelial cells. Am J Physiol 266:C22–C26

Tracey WR, Xue C, Klinghofer V, Barlow J, Pollock JS, Förstermann U, Johns RA (1994b) Immunochemical detection of inducible NO synthase in human lung. Am J Physiol 266:L722–L727

Tsuchiya T, Kishimoto J, Nakayama Y (1996) Marked increases in neuronal nitric oxide synthase (nNOS) mRNA and NADPH-diaphorase histostaining in adrenal cortex after immobilization stress in rats. Psychoneuroendocrinology 21:287–293

Villar MJ, Ceccatelli S, Bedecs K, Bartfai T, Bredt D, Synder SH, Hokfelt T (1994) Upregulation of nitric oxide synthase and galanin message-associated peptide in hypothalamic magnocellular neurons after hypophysectomy. Immunohistochemical and in situ hybridization studies. Brain Res 650:219–228

Vizzard MA (1997) Increased expression of neuronal nitric oxide synthase in bladder afferent and spinal neurons following spinal cord injury. Dev Neurosci 19:232–246

Vodovotz Y, Bogdan C, Paik J, Xie QW, Nathan C (1993) Mechanisms of suppression of macrophage nitric oxide release by transforming growth factor β. J Exp Med 178:605–613

Weber CM, Eke BC, Maines MD (1994) Corticosterone regulates heme oxygenase-2 and NO synthase transcription and protein expression in rat brain. J Neurochem 63:953–962

Weiner CP, Knowles RG, Moncada S (1994a) Induction of nitric oxide synthases early in pregnancy. Am J Obstet Gynecol 171:838–843

Weiner CP, Lizasoain I, Baylis SA, Knowles RG, Charles IG, Moncada S (1994b) Induction of calcium-dependent nitric oxide synthases by sex hormones. Proc Natl Acad Sci U S A 91:5212–5216

Weisz A, Oguchi S, Cicatiello L, Esumi H (1994) Dual mechanism for the control of inducible-type NO synthase gene expression in macrophages during activation by interferon-gamma and bacterial lipopolysaccharide. Transcriptional and post-transcriptional regulation. J Biol Chem 269:8324–8333

Xie QW, Cho HJ, Calaycay J, Mumford RA, Swiderek KM, Lee TD, Ding A, Troso T, Nathan C (1992) Cloning and characterization of inducible nitric oxide synthase from mouse macrophages. Science 256:225–228

Xie QW, Whisnant R, Nathan C (1993) Promoter of the mouse gene encoding calcium-independent nitric oxide synthase confers inducibility by interferon gamma and bacterial lipopolysaccharide. J Exp Med 177:1779–1784

Xie QW, Kashiwabara Y, Nathan C (1994) Role of transcription factor NF-kappa b/Rel in induction of nitric oxide synthase. J Biol Chem 269:4705–4708

Xu DL, Martin PY, St John J, Tsai P, Summer SN, Ohara M, Kim JK, Schrier RW (1996) Upregulation of endothelial and neuronal constitutive nitric oxide synthase in pregnant rats. Am J Physiol 271:R1739–R1745

Xue C, Pollock J, Schmidt HHHW, Ward SM, Sanders KM (1994) Expression of nitric oxide synthase immunoreactivity by interstitial cells of the canine proximal colon. J Auton Nerv Syst 49:1–14

Yoshizumi M, Perrella MA, Burnett JC, Lee ME (1993) Tumor necrosis factor down-regulates an endothelial nitric oxide synthase messenger RNA by shortening its half-life. Circ Res 73:205–209

Yui Y, Hattori R, Kosuga K, Eizawa H, Hiki K, Kawai C (1991) Purification of nitric oxide synthase from rat macrophages. J Biol Chem 266:12544–12547

Zhang ZG, Chopp M, Zaloga C, Pollock JS, Förstermann U (1993) Cerebral endothelial nitric oxide synthase expression after focal cerebral ischemia in rats. Stroke 24:2016–2021

Zhang ZG, Chopp M, Gautam S, Zaloga C, Zhang RL, Schmidt HHHW, Pollock JS, Förstermann U (1994) Upregulation of neuronal nitric oxide synthase and mRNA, and selective sparing of nitric oxide synthase-containing neurons after focal cerebral ischemia in rat. Brain Res 654:85–95

Ziesche R, Petkov V, Williams J, Zakeri SM, Mosgoller W, Knofler M, Block LH (1996) Lipopolysaccharide and interleukin 1 augment the effects of hypoxia and inflammation in human pulmonary arterial tissue. Proc Natl Acad Sci USA 93:12478–12483

CHAPTER 4
Enzymology of Soluble Guanylyl Cyclase

D. Koesling and A. Friebe

A. Introduction

Soluble guanylyl cyclase (sGC) probably represents the most important receptor for nitric oxide (NO) as a signalling molecule (Fig. 1). The enzyme catalyses the conversion of guanosine triphosphate (GTP) into guanosine cyclic monophosphate (cGMP). Activation of sGC by NO leads to an enormous stimulation of cGMP production, yielding an up to 400-fold activation of the purified enzyme. Effector molecules for cGMP are the cGMP-dependent protein kinases, cGMP-gated ion channels and cGMP-regulated phosphodiesterases. Important cellular effects elicited by the NO-induced rise of the intracellular cGMP concentrations are smooth-muscle relaxation (Lincoln 1989), modulation of synaptic transmission (O'Dell et al. 1991; Zhuo and Hawkins 1995) and inhibition of platelet aggregation (Garthwaite et al. 1988; Walter 1989; Schmidt and Walter 1994; Moncada and Higgs 1995).

In addition to soluble guanylyl cyclases, another group of cGMP-synthesising enzymes exist, i.e. the membrane bound GCs (mGCs). These GCs are not stimulated by NO but are structurally related to the soluble enzymes (Fig. 2). These mGCs belong to the group of receptor-linked enzymes containing one membrane-spanning region (Wedel and Garbers 1997). The intracellular C-terminal domains possess the catalytic activity and, accordingly, are highly conserved, whereas the extracellular ligand-binding domains are rather diverse. Several isoforms of mGCs have been identified (GC-A to GC-F). Due to the diversity of the ligand-binding domains, the different isoforms of mGCs can be stimulated by various peptide hormones; GC-A, for example, is stimulated by the natriuretic peptides ANP and BNP, whereas GC-B exhibits the highest affinity for the C-type natriuretic peptide. Physiologically, GC-C is activated by the intestinal peptide hormone guanylin whereas, under pathophysiological conditions, stimulation of GC-C by the heat-stable enterotoxin of *Escherichia coli* results in diarrhoea. As expected, the catalytic domains of all cGMP-forming isoforms, i.e. both soluble and membrane-bound guanylyl cyclases, are related and are similar to the catalytic domains of the cyclic AMP (cAMP)-forming adenylyl cyclases (Fig. 2). The following article intends to give an overview on NO-sensitive sGC, with special emphasis on new modulators and on structural features of the enzyme.

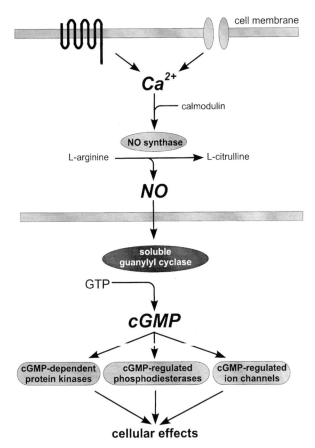

Fig. 1. Participants in the nitric oxide/cyclic guanosine monophosphate (*NO/cGMP*) signalling cascade. An increase in the intracellular calcium concentration leads to the activation of the endothelial or neuronal NO synthases and subsequent NO formation. NO can stimulate soluble guanylyl cyclase in the same or adjacent cells and, because of the increase in cGMP, the activity of the cGMP receptors (cGMP-dependent protein kinases, cGMP-regulated phosphodiesterases, cGMP-regulated ion channels) is altered, leading to changes in cellular function

B. Regulation of sGC

I. NO, the Physiological Activator of sGC

Although cGMP-forming activity of the enzyme sGC was first described in 1969 (HARDMAN and SUTHERLAND 1969; SCHULTZ et al. 1969; WHITE and AUERBACH 1969), it took until the late 1970s to find out that NO-containing compounds are potent activators of sGC (ARNOLD et al. 1977; BÖHME et al. 1978). Despite the stimulatory effect of these NO-containing compounds, the physiological significance of NO-induced activation of the enzyme did not

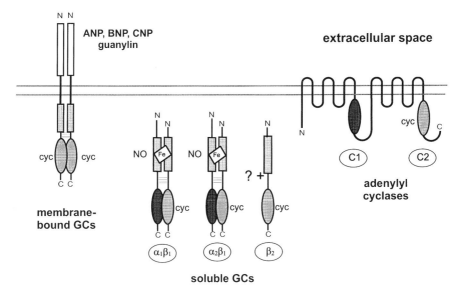

Fig. 2. Schematic representation of the overall structures of guanylyl cyclases (*GCs*) and adenylyl cyclases (ACs). Shown are the known heterodimers of soluble GC (sGC; $\alpha_1\beta_1$, $\alpha_2\beta_1$), a representative membrane-bound GC with different ligands at the respective receptor domain, and a representative AC. The dimerisation partner of the β_2 subunit of GC has not been identified. The catalytic domains (labelled *cyc*) are symbolised by *light grey ellipses* on the membrane-bound GC, the β subunit of sGC and, correspondingly, the C2 domain of AC. The *dark, shaded ellipses* correspond to the cyclase catalytic domains in the α subunits of sGC and the C1 domain of AC. *ANP* natriuretic peptide type A, *BNP* natriuretic peptide type B, *CNP* natriuretic peptide type C, *cyc* catalytic domains of GCs and ACs

become clear until the identification of endothelium-derived relaxing factor (EDRF) as NO (IGNARRO et al. 1987; PALMER et al. 1987). Formation of EDRF had been shown to occur in endothelial cells in response to vasodilatory agonists, such as acetylcholine, histamine or bradykinin, leading to vasodilation via the stimulation of sGC in smooth-muscle cells. After the discovery of NO in the vascular system, NO formation was reported to occur throughout the body (MONCADA and HIGGS 1995). The enzymes responsible for the synthesis of NO were identified and termed NO synthases. Three isoforms of NO synthases have been identified to date; these enzymes are discussed in more detail in other chapters of this book. The inducible isoform plays a role in the nonspecific immune response and exhibits direct toxic effects by the production of relatively high NO concentrations. The neuronal and endothelial NO synthases are constitutively expressed enzymes regulated by the intracellular calcium concentration. These isoforms produce relatively low NO concentrations (1–100nM). At these low concentrations, NO functions as a signalling molecule, and most of its effects are mediated via the activation of sGC.

II. Mechanism of Activation of sGC by NO

Among the three redox forms of NO (NO$^-$, NO$^•$, NO$^+$), only the uncharged NO radical (NO$^•$) has been shown to significantly activate the enzyme (DIERKS and BURSTYN 1996). sGC contains a prosthetic heme group that binds NO, resulting in up to 400-fold activation of the enzyme (HUMBERT et al. 1990; STONE and MARLETTA 1995).

The absorbance spectrum of the heme group of sGC exhibits a maximum at 431 nm. This peak indicates a five-coordinated ferrous heme (Fig. 3) with a histidine as the axial ligand at the fifth coordinating position (STONE and MARLETTA 1994). Using site-directed mutagenesis of all conserved histidines, histidine 105 of the β_1 subunit has been identified as the axial ligand (WEDEL et al. 1994). This study was confirmed recently (ZHOU and MARLETTA 1998). NO binds to the sixth coordination position of the heme iron and leads to the breakage of the histidine-to-iron bond, yielding a five-coordinated nitrosyl–heme complex with an absorbance maximum at 398 nm (Fig. 3). The opening of the histidine-to-iron bond is thought to initiate a conformational change resulting in the activation of the enzyme.

The finding that protoporphyrin IX, the iron-free precursor of heme, stimulates sGC independently of NO supports this mechanism of activation (IGNARRO et al. 1982a). The structure of protoporphyrin IX, a heme lacking iron, resembles that of the NO–heme complex, in which the iron is moved out of the plane of the porphyrin ring. Thus, in the protoporphyrin IX- and the NO-stimulated enzymes, the axial histidine is unbound. However, breakage of the histidine-to-iron bond is required but does not appear to be sufficient for activation, as a mutant without the proximal histidine (and, accordingly, without the histidine-to-iron bond) did not show an increased catalytic rate. These results indicate that the proximal histidine coordinates the heme and is required to mediate the stimulatory process (see "The Regulatory Heme-Binding Domain").

Like NO, carbon monoxide (CO) has a very high affinity for heme groups and, thus, also binds to the heme group of sGC. In contrast to NO, which activates sGC up to 400-fold, stimulation by CO only ranges between four- and sixfold. CO binding results in a six-coordinated heme, because CO does not lead to breakage of the heme-to-iron bond. This hexa-coordinated carbonyl–heme complex has an absorption maximum at 423 nm (Fig. 3). Therefore, the low stimulatory property of CO is in accordance with the assumption that the breakage of the histidine-to-iron bond is a prerequisite for sGC activation.

III. Termination of the NO-Induced Activation

It is widely accepted that the association of NO to a five-coordinate heme iron is nearly diffusion controlled, whereas the dissociation of NO is generally rather slow. Studies on the dissociation of NO from sGC in the absence of

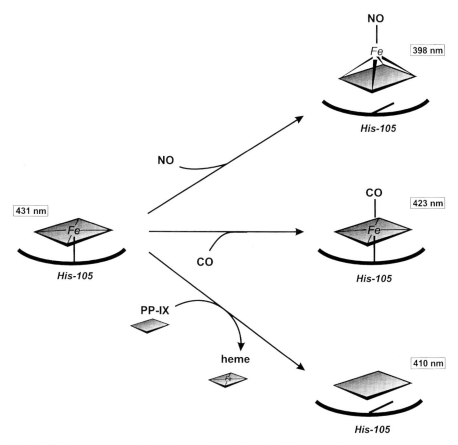

Fig. 3. The heme group of soluble guanylyl cyclase. In the non-activated state, the heme of sGC is five-coordinated, with histidine 105 of the β_1 subunit bound to the fifth coordination position of the heme iron (absorption maximum at 431 nm). After nitric oxide (*NO*) binds to the sixth coordination position, the heme is still five-coordinated, as NO binding leads to breakage of the histidine-to-iron bond so that the fifth coordination position is free (absorption maximum at 398 nm). Like NO, carbon monoxide (*CO*) binds to the sixth coordination position of the heme iron but, in contrast to NO, CO does not lead to the rupture of the histidine-to-iron bond; therefore, CO binding results in a six-coordinated heme (absorption maximum at 423 nm). Protoporphyrin IX (*PP-IX*) is able to substitute for the heme group and stimulates the enzyme (absorption maximum at 410 nm). For further explanation, see text

the substrate Mg–GTP revealed that the enzyme has a comparably comparatively high NO dissociation rate. This relatively high NO dissociation rate was increased 40-fold on addition of the substrate Mg–GTP, yielding a half-life of approximately 5 s at 37°C for the NO–sGC complex (KHARITONOV et al. 1997). This half-life appears to be fast enough to allow rapid deactivation of the enzyme in biological systems, although it has to be kept in mind that the presence of a NO scavenger is a prerequisite for such fast deactivation.

IV. CO: a Physiological Activator of sGC?

The role of CO as an activator of sGC has been a matter of debate during the last several years. Although CO was shown to induce effects similar to those evoked by NO, the gas has rather poor sGC-stimulating properties (see "Mechanism of Activation of sGC"). However, in the presence of the substance YC-1 (see "Modulators of sGC"), CO and NO were able to stimulate sGC to the same extent (FRIEBE et al. 1996b). Although the similarity in the abilities of CO and NO to stimulate sGC is striking, a physiological role for CO would require an endogenously occurring substance with YC-1-like properties and the rather high CO concentration required for sGC activation (RUSSWURM et al. 1998).

Nevertheless, there are reports of a role for CO in long-term potentiation, olfactory signal transduction, vasorelaxation and inhibition of platelet aggregation (MARKS et al. 1991; MAINES 1997). The most relevant enzymatic production of CO probably occurs through the heme oxygenases. These enzymes catalyse the degradation of heme to biliverdin and CO (MAINES 1988), although this appears to be a rather uneconomical way of producing a signalling molecule. In addition, with the exception of induction, no short-term regulation of the enzymes is known (MAINES 1997).

V. Redox Regulation of sGC?

Early research on sGC, especially before the identification of NO as the physiological activator, revealed a multitude of different activators and regulators of sGC, most of them exhibiting redox-active properties (WALDMAN and MURAD 1987). Substances that do not act by releasing NO ("non-NO activators") constitute a rather heterogeneous group; thus, several mechanisms of activation have been postulated and subsumed as "redox regulation" of sGC (BÖHME et al. 1984). Although many of the effects of redox-active compounds have been postulated to be mediated by the enzyme's cysteine residues (DERUBERTIS and CRAVEN 1977; BRANDWEIN et al. 1981; BÖHME et al. 1983), point mutation of 15 conserved cysteines did not impair catalytic activity or NO stimulation, indicating that the cysteines of sGC are unlikely to be the mediators of this redox regulation (FRIEBE et al. 1997).

Purified sGC was shown to be sensitive to the small amounts of NO present in the atmosphere (FRIEBE et al. 1996a). Studies with one of the reported non-NO activators of sGC, superoxide dismutase, revealed that the stimulatory effect is not caused by a direct effect on the enzyme. Rather, superoxide dismutase (which eliminates O_2^-) leads to an increase in the concentration of dissolved atmospheric NO in a solution by preventing the inactivation of NO to peroxynitrite by O_2^- (FRIEBE et al. 1998a). Other non-NO activators of sGC, such as the radical scavenger ascorbate, may also exert their action by decreasing the amount of O_2^- and thereby increasing the concentration of NO. Hence, the "redox regulation" of sGC may simply be an in vitro artefact due to the extreme NO sensitivity of sGC and the O_2^--mediated inactivation of the dissolved atmospheric NO.

VI. Modulators of sGC

1. ODQ: An Inhibitor of the Stimulated Activity of sGC

Methylene blue and LY-83583 have been used as inhibitors of sGC even though they show little specificity and, for example, interfere with NO formation and NO release from NO synthases (MAYER et al. 1993). In contrast, the quinoxalin derivative 1H-[1,2,4]oxadiazolo[4,3-a]-quinoxalin-1-one (ODQ) was shown to be a potent and selective inhibitor of sGC in brain slices (GARTHWAITE et al. 1995). ODQ did not lead to inhibition of the cyclic-nucleotide-forming membrane-bound guanylyl or adenylyl cyclases and did not interfere with the stimulation of NO synthase. Therefore, ODQ represents an important tool for discriminating cGMP-dependent and cGMP-independent effects of NO. Inhibition of sGC by ODQ has been demonstrated in a variety of cells and tissues (GARTHWAITE et al. 1995; BRUNNER et al. 1996; ABIBERGES et al. 1997).

Studies with purified sGC revealed that ODQ binds in an NO-competitive manner and leads to an apparently irreversible inhibition of the stimulated enzyme, leaving basal activity almost unchanged. Spectral analysis suggests oxidation of the heme iron as an underlying mechanism of the inhibitory effect of ODQ (SCHRAMMEL et al. 1996). The notion that the heme iron is the target of ODQ is confirmed by the finding that ODQ is not able to inhibit the stimulation of sGC induced by the iron-free protoporphyrin IX (KOESLING and FRIEBE 1999). A derivative of ODQ, NS-2028 (oxadiazolo (3,4-d)benz(b)(1,4)oxazin-1-one), was recently reported to have properties similar to those of ODQ (OLESEN et al. 1998).

2. YC-1: A Novel Activator of sGC

The compound YC-1, a benzylindazol derivative, was reported to be an inhibitor of platelet aggregation leading to an increase in the intracellular cGMP concentration (WU et al. 1995). YC-1 stimulated purified sGC approximately tenfold (FRIEBE et al. 1996b; MÜLSCH et al. 1997). This activation was independent of NO, as the NO scavenger oxyhaemoglobin was not able to prevent the stimulation. Despite this NO independence, YC-1 failed to stimulate the heme-free enzyme, showing that YC-1 stimulation required the presence of the prosthetic heme group. YC-1 exerted dramatic effects on the NO-stimulated enzyme, as it increased the maximal NO-induced catalytic rate by about 40% and potentiated the stimulatory effect of submaximally activating NO concentrations. The concentration–response curve of NO was shifted to the left by YC-1, indicating that YC-1 sensitised sGC to NO. Stimulation of sGC induced by the NO-independent activator protoporphyrin IX was also potentiated by YC-1; surprisingly, even the poor sGC activator CO became an effective stimulator (FRIEBE et al. 1996b).

YC-1 did not change the Soret absorption of basal or stimulated sGC. Together, our results suggest that YC-1 binds to an allosteric site of activated sGC molecules, thereby causing an increase in sGC, which makes a direct YC-1-heme interaction rather unlikely (FRIEBE and KOESLING 1998). In accordance with this idea, YC-1 was found to bind both to the non-stimulated and to the heme-depleted enzyme, suggesting an allosteric site on the enzyme. Dissociation of NO and sGC was slowed by YC-1, which suggests that YC-1 sensitised the enzyme to NO by decreasing the dissociation rate of the heme ligand (FRIEBE and KOESLING 1998). In addition, deactivation of the CO-activated enzyme by oxyhaemoglobin was slow in the presence of YC-1, suggesting that YC-1 either increases the association rate or decreases the dissociation rate of CO. This idea is contrasted by the results of STONE and MARLETTA (1998), who did not find a change in CO affinity in the presence of YC-1.

In conclusion, the effects of YC-1 on sGC suggest the existence of an unknown allosteric site on the enzyme; this site can modulate the catalytic rate and the responsiveness towards heme ligands. Ligands of this allosteric site may represent a novel class of drugs and may exert vasodilatory, anti-aggregatory and possibly other, unknown effects by sensitising sGC to its physiological activator, NO. In contrast to the NO donors commonly used in the treatment of coronary heart disease, YC-1-like substances would exert their action mainly at sites of endogenous NO production. In addition, it is tempting to speculate about the existence of a naturally occurring allosteric modulator of sGC that may alter the sensitivity of the enzyme to NO. When endogenously generated CO concentrations reach relevant levels, sGC may be significantly stimulated by CO in the presence of such an endogenous modulator.

YC-1 effects have been studied on purified sGC and have been demonstrated in a variety of cells and tissues. In vascular smooth-muscle cells, YC-1 was reported to increase cGMP levels and to induce a concentration-dependent relaxation of endothelium-free rat aortic rings precontracted with phenylephrine (MÜLSCH et al. 1997; WEGENER and NAWRATH 1997). Recently, we studied the possible synergistic effects of YC-1 and NO in intact platelets (FRIEBE et al. 1998b). YC-1 in combination with NO or CO led to complete inhibition of platelet aggregation at concentrations that were ineffective when tested without YC-1. The synergism of YC-1 and NO was emphasised by a drastic increase in intraplatelet cGMP levels. Although the application of NO or YC-1 at maximally effective concentrations only led to approximately 13-fold increases in cGMP concentration, both drugs together resulted in an over 1300-fold rise of the cGMP content. When using CO instead of NO, similar results were obtained.

This enormous cGMP increase was partly due to inhibition of cGMP degradation as, in addition to causing sGC stimulation, YC-1 was shown to inhibit cGMP-hydrolysing phosphodiesterases in platelet cytosol (FRIEBE et al. 1998b). Thus, YC-1 obviously has a dual effect on cGMP levels in intact platelets, as it stimulates cGMP formation through sGC and inhibits cGMP degradation through phosphodiesterases, thereby explaining the exceptional increase in the cGMP concentration in platelets. Because of this dual effect, YC-1 and YC-1-like substances are most interesting candidates as modulators of cGMP-mediated effects within the cardiovascular system.

C. Structure of sGC

I. Isoforms and Tissue Distribution

Several groups have succeeded in the purification of sGC. All isolation procedures yielded an enzyme with two subunits (termed α_1 and β_1) with molecular masses of 73 kDa and 70 kDa, respectively (GERZER et al. 1981a; HUMBERT et al. 1990; STONE and MARLETTA 1994). The purified enzyme was shown to contain heme as a prosthetic group (GERZER et al. 1981b) and was stimulated as much as 400-fold by NO. Specific activities of 25–40 μmol cGMP/min/mg have been reported under NO-stimulated conditions (STONE and MARLETTA 1995). sGC, similar to other nucleotide-converting enzymes, requires divalent metal ions as cofactors for catalysis. In addition to Mg^{2+}, the enzyme can also utilise Mn^{2+}, which leads to an approximately fourfold increase in the catalytic rate under non-stimulated conditions. However, stimulated enzyme activity is lower in the presence of Mn^{2+} than in the presence of Mg^{2+}.

Under non-stimulated conditions, sGC exhibits Michaelis-Menten kinetics in the presence of either Mg^{2+}–GTP or Mn^{2+}–GTP as substrate, with K_m values of approximately 100 μM and 10 μM, respectively. Under NO stimulated conditions, sGC exhibits K_m values of approximately 10 μM irrespective of the metal ion used. A detailed review of the kinetics of sGC can be found in WALDMAN and MURAD (1987).

Both the α_1 and β_1 subunits of the enzyme purified from lung have been cloned and sequenced. Expression experiments revealed that both subunits were required in order to form a catalytically active enzyme (HARTENECK et al. 1990; BUECHLER et al. 1991). In lung, brain and kidney, comparably high mRNA levels for the α_1 and β_1 subunits were found (NAKANE et al. 1990).

Two other sGC subunits, α_2 and β_2, were found by homology screening (YUEN et al. 1990; HARTENECK et al. 1991). The so-called α_3 and β_3 subunits of sGC (GIUILI et al. 1992) represent human variants of the α_1 and β_1 subunits rather than different isoforms, and changes in the reading frame probably account for the differences in amino acid sequences. Most laboratories have been unsuccessful in co-expressing the β_2 subunit with any of the known α or β subunits to form a cGMP-synthesising enzyme. However, there is a recent report on the formation of an active $\alpha_1\beta_2$ heterodimer (GUPTA et al. 1997). In contrast to β_2, the α_2 subunit is able to substitute for the α_1 subunit, as shown by the formation of a catalytically active, NO-sensitive heterodimer upon coexpression with the β_1 subunit (HARTENECK et al. 1991). In human placenta, the natural occurrence of the α_2 subunit has been demonstrated, and the β_1 subunit has been identified as its physiological dimerisation partner (RUSSWURM et al. 1999). Comparison of both heterodimers ($\alpha_2\beta_1$, $\alpha_2\beta_1$) did not reveal any differences between the isoforms with respect to the heme content, the sensitivity towards NO, the kinetic properties and the responsiveness towards modulators. Taking into consideration the fact that the N-terminal thirds of the α subunits share only 27% amino acids identity, this lack of differences between the isoforms is surprising. The physiological relevance of the two functionally similar isoforms cannot be explained.

II. Primary Structure and Homology among the Subunits of sGC

As outlined above, four subunits of sGC (α_1, α_2, β_1, β_2) have been identified (Figs. 1, 2). A comparison of the primary structures shows that the subunits can be divided into three domains: a C-terminal cyclase-catalytic domain, a central part and an N-terminal region.

The catalytic C-terminal domains of sGC reveal the highest degree of homology. These domains are also very similar to the respective regions in the mGCs and in the adenylyl cyclases (see "Catalytic Domain"). The C-terminus of the β_2 subunit has an 86-amino acid elongation that contains an isoprenylation consensus sequence (–CVVL). This suggests that the β_2 subunit may be attached to the membrane.

The central regions preceding the catalytic domains show considerable homology. In analogy to the membrane-bound enzymes (WILSON and CHINKERS 1995), these regions are probably involved in the dimerisation of the subunits.

The N-terminal regions of the subunits are relatively diverse. For example, only 27% amino acid identity is shared by the otherwise highly conserved α_1 and α_2 subunits. Despite these differences in primary structure, both α subunits reveal functional similarity (see "Isoforms and Tissue Distribution"), indicating that most of the regulatory features of sGC are determined by the β_1 subunit. Within the N-terminal regions, there is a stretch of about 100 amino acids that shows a higher degree of conservation among the α subunits or the β subunits than between these two groups. The structures of these regions may define the properties of α and β subunits. However, it should be remembered that histidine 105 of the β_1 subunit (see below), which acts as the proximal ligand of the heme group, is located within this region.

III. The Regulatory Heme-Binding Domain

The presence of a prosthetic heme group is mandatory for the stimulatory effect of NO (CRAVEN and DERUBERTIS 1978; IGNARRO et al. 1982b). Removal of the heme group abolishes NO-induced activation, which can be restored after reconstitution of sGC with the heme group (IGNARRO et al. 1986; FOERSTER et al. 1996). The heme content of the enzyme has been a matter of debate during the last several years. Whereas most groups found up to 1 mole of heme per mole of heterodimer (GERZER et al. 1981b; HUMBERT et al. 1990; TOMITA et al. 1997), the group of Marletta reported a heme content of 1.5 moles of heme per mole of heterodimer (STONE and MARLETTA 1995). However, recent results obtained by the same group are consistent with a heme-to-heterodimer ratio of 1:1 (STONE and MARLETTA 1998).

With the help of truncated enzymes, the N-termini of the α_1 and β_1 subunits have been shown to be required for proper heme binding (FOERSTER et al. 1996). However, the two subunits, α_1 and β_1, appear to contribute unequally. Obviously, the β_1 subunit plays a more important role in heme binding, although the N-terminus of the α_1 subunit is essential for a normal heme-binding capacity (FOERSTER et al. 1996). The N-terminus of the α_1 can also be substituted with the N-terminus of the β_1 subunit, as shown by ZHAO and MARLETTA (1997). They demonstrated that the N-terminal half of the β_1 subunit (amino acids 1–385) expressed in bacteria is itself able to form dimers. These homodimers were shown to be sufficient to bind heme in a manner similar to that displayed by the wild-type enzyme.

The heme group of sGC exhibits an absorption maximum at 431 nm, indicative of a histidine as the axial ligand. Using site-directed mutagenesis, histidine 105 of the β_1 subunit has been identified as the axial ligand of the heme group (WEDEL et al. 1995; ZHAO et al. 1998). Since the mutant lacking histidine 105 remained insensitive to NO even after heme reconstitution, histidine 105 may very likely be required to transduce NO-induced, heme-mediated activation within the protein.

IV. Catalytic Domain

The C-terminal catalytic domains of the subunits of sGC are also conserved in the mGCs and the adenylyl cyclases. Adenylyl cyclases are integral membrane proteins with two hydrophobic domains and two cytosolic domains (Fig. 2). Each of the hydrophobic domains supposedly contains six transmembrane helices. The two cytosolic domains, termed C1 and C2, are homologous both to each other and to the catalytic domains of both sGC and mGCs (SUNAHARA et al. 1996). Like sGC, adenylyl cyclases contain two different (though conserved) catalytic domains. However, in the adenylyl cyclases, both domains are located on one molecule whereas, in sGC, each of the two subunits contributes one catalytic domain. Comparison of the primary structures of the catalytic domains of adenylyl cyclases and sGC suggests that the catalytic domains of the α and β subunits correspond to the C1 and C2 domains of the adenylyl cyclases, respectively. It has to be remembered that dimer formation is a prerequisite for catalytic activity for all cyclases. Co-expression of an α and β subunit or of the truncated α_1 and β_1 subunits was required in order to yield a catalytically active sGC (WEDEL et al. 1995). The mGCs also consist of at least two subunits, and the truncated, catalytically active mGC has a homodimeric structure (THOMPSON and GARBERS 1995). Analogous results exist for the catalytic domains of the adenylyl cyclases, for which co-expression of the C1 and C2 domains yielded an enzymatically active enzyme (TANG and GILMAN et al. 1995; DESSAUER and GILMAN 1996). Although the C2 domain is able to form homodimers by itself, its cAMP-forming activity is negligible.

Recently, analysis of the crystal structure of the C1 and C2 domains of the adenylyl cyclases identified the ATP-binding catalytic site and suggested specific amino acid residues that may be responsible for nucleotide interaction (TESMER et al. 1997). Indeed, the exchange of the corresponding residues of sGC into the respective amino acids of the adenylyl cyclases resulted in a sGC that produced cAMP in an NO-sensitive manner (SUNAHARA et al. 1998). These data underline the structural similarity among the adenylyl cyclases and sGC.

The crystal structure revealed that the ATP-binding catalytic site of the adenylyl cyclase is formed by both the C1 and C2 domains. Due to the similarity of both domains, a counterpart of the ATP-binding catalytic centre that has lost its catalytic ability exists. Interestingly, forskolin, a stimulator of adenylyl cyclase, binds to this "catalytic-like" domain (SUNAHARA et al. 1997). It is tempting to speculate that, in analogy to forskolin, YC-1 binds to the "catalytic-like" domain of sGC, thereby stimulating the enzyme (Fig. 4). By binding to this site, YC-1 may slow NO dissociation (FRIEBE and KOESLING 1998), possibly counteracting the NO-dissociation-increasing effect of Mg–GTP (KHARITONOV et al. 1997). Further studies will show whether this

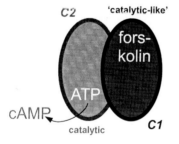
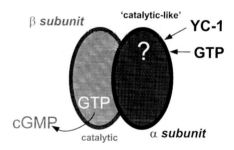

Fig. 4. Schematic diagram of the catalytic domains of adenylyl cyclase (AC) and soluble guanylyl cyclase (sGC). The C-terminal regions of heterodimeric AC and sGC are represented as *ellipses*; the C2 domain of AC and the corresponding β subunit of sGC are drawn in *light grey*; the C1 and α subunit are *dark grey*. Conversion of ATP to cAMP and of GTP to cGMP are shown for the respective cyclases. The "catalytic-like" domain of the AC binds the diterpene activator forskolin. The corresponding ligand for sGC has not been identified, although it is tempting to speculate that YC-1 is a possible ligand

"catalytic-like" domain of sGC (corresponding to the forskolin site of the adenylyl cyclase) indeed represents the YC-1-binding domain and whether GTP exerts its stimulatory effect on NO dissociation via this site.

D. Summary

Our knowledge about the NO receptor sGC has grown slowly but gradually during the last several years. We have gained insight into the regulation of this enzyme by NO; in particular, we have learned about the acceptor site for NO (i.e. the prosthetic heme group) and the mechanism of activation. The new compound YC-1 unveiled a novel mechanism of sensitisation of sGC towards NO, which may have broad pharmacological and even physiological implications. Endogenous ligands at the YC-1-binding site may exist; by adjusting the NO sensitivity of the enzyme, these ligands would be capable of modulating NO/cGMP signalling.

On the complementary DNA level, two α (α_1, α_2) and two β (β_1, β_2) subunits of sGC have been identified. However, on the protein level, the physiological occurrence of only two heterodimeric isoforms of sGC ($\alpha_1\beta_1$, $\alpha_2\beta_1$) has been demonstrated. The knowledge of the primary structure of the subunits revealed their close structural relationship to the mGCs and adenylyl cyclases. With the help of the recently published crystal structure of the catalytic adenylyl cyclase domain, it will be possible to develop a better spatial model of the catalytic centre of sGC.

Acknowledgements. The authors' research was supported by the Deutsche Forschungsgemeinschaft and Fonds der Chemischen Industrie.

References

Abiberges N, Hovemadsen L, Fischmeister R, Mery PF (1997) A comparative study of the effects of three guanylyl cyclase inhibitors on the L-type Ca^{2+} and muscarinic K^+ currents in frog. Br J Pharmacol 121:1369–1377

Arnold WP, Mittal CK, Murad F (1977) Nitric oxide activates guanylate cyclase and increases guanosine 3′:5′-cyclic monophosphate levels in various tissue preparations. Proc Natl Acad Sci USA 74:3203–3207

Böhme E, Graf H, Schultz G (1978) Effects of sodium nitroprusside and other smooth muscle relaxants on cyclic cGMP formation in smooth muscle and platelets. Adv Cyclic Nucleotide Res 9:131–143

Böhme E, Gerzer R, Grossmann G, Herz J, Mülsch A, Spies C, Schultz G (1983) Regulation of soluble guanylyl cyclase activity. In: Hormones and cell regulation Vol. 7 (Dumont JE, Nunez J, Denton RM, eds) Elsevier Biomedical Press, 147–161

Böhme E, Grossmann G, Herz J, Mülsch A, Spies C, Schultz G (1984) Regulation of cyclic GMP formation by soluble guanylate cyclase: stimulation by NO-containing compounds. Adv Cyclic Nucleotide Protein Phosphorylation Res 17:259–266

Brandwein HJ, Lewicki JA, Murad F (1981) Reversible inactivation of guanylate cyclase by mixed disulfide formation. J Biol Chem 256:2958–2962

Brunner F, Schmidt K, Nielsen EB, Mayer B (1996) Novel guanylyl cyclase inhibitor potently inhibits cyclic GMP accumulation in endothelial cells and relaxation of bovine pulmonary artery. J Pharmacol Exp Ther 277:48–53

Buechler WA, Nakane M, Murad F (1991) Expression of soluble guanylate cyclase activity requires both enzyme subunits. Biochem Biophys Res Commun 174: 351–357

Craven PA, DeRubertis FR (1978) Restoration of the responsiveness of purified guanylate cyclase to nitrosoguanidine, nitric oxide, and related activators by heme and heme proteins. Evidence for involvement of the paramagnetic nitrosyl–heme complex in enzyme activation. J Biol Chem 253:8433–8443

DeRubertis FR, Craven PA (1977) Activation of hepatic guanylate cyclase by N-methyl-N'-nitro-N-nitrosoguanidine. Effects of thiols, N-ethylmaleimide, and divalent cations. J Biol Chem 252:5804–5814

Dessauer CW, Gilman AG (1996) Purification and characterisation of a soluble form of mammalian adenylyl cyclase. J Biol Chem 271:16967–16974

Dierks EA, Burstyn JN (1996) Nitric oxide (NO), the only nitrogen monoxide redox form capable of activating soluble guanylyl cyclase. Biochem Pharmacol 51: 1593–1600

Foerster J, Harteneck C, Malkewitz J, Schultz G, Koesling D (1996) A functional heme-binding site of soluble guanylyl cyclase requires intact N-termini of α_1 and β_1 subunits. Eur J Biochem 240:380–386

Friebe A, Koesling D (1998) Mechanism of YC-1-induced activation of soluble guanylyl cyclase. Mol Pharmacol 53:123–127

Friebe A, Malkewitz J, Schultz G, Koesling D (1996a) Positive side-effects of pollution? Nature 382:120

Friebe A, Schultz G, Koesling D (1996b) Sensitizing soluble guanylyl cyclase to become a highly CO-sensitive enzyme. EMBO J 15:6863–6868

Friebe A, Wedel B, Foerster J, Harteneck C, Malkewitz J, Schultz G, Koesling D (1997) Function of conserved cysteine residues on soluble guanylyl cyclase. Biochemistry 36:1194–1198

Friebe A, Schultz G, Koesling D (1998a) Stimulation of soluble guanylyl cyclase by superoxide dismutase is mediated by NO. Biochem J 335:527–531

Friebe A, Müllershausen F, Smolenski A, Walter U, Schultz G, Koesling D (1998b) YC-1 potentiates NO- and CO-induced cGMP effects in human platelets. Mol Pharmacol 54:962–967

Garthwaite J, Charles SL, Chess-Williams R (1988) Endothelium-derived relaxing factor release on activation of NMDA receptors suggests role as intercellular messenger in the brain. Nature 336:385–388

Garthwaite J, Southam E, Boulton CL, Nielsen EB, Schmidt K, Mayer B (1995) Potent and selective inhibition of nitric oxide-sensitive guanylyl cyclase by 1H-[1,2,4]oxadiazolo[4,3-a]quinoxalin-1-one (ODQ). Mol Pharm 48:185–188

Gerzer R, Böhme E, Hofmann F, Schultz G (1981a) Soluble guanylate cyclase purified from bovine lung contains heme and copper. FEBS Lett 132:71–74

Gerzer R, Hofmann F, Schultz G (1981b) Purification of a soluble, sodium-nitroprusside-stimulated guanylate cyclase from bovine lung. Eur J Biochem 116:479–486

Giuili G, Scholl U, Bulle F, Guellaen (1992) Molecular cloning of the cDNAs coding for the two subunits of soluble guanylyl cyclase from human brain. FEBS Lett 304:83–88

Goldberg ND, O'Dea RF, Haddox MK (1973) Cyclic GMP. Adv Cycl Nucl Res 3:155–223

Gupta G, Azam M, Yang L, Danziger RS (1997) The β_2 subunit inhibits stimulation of the α_1/β_1 form of soluble guanylyl cyclase by nitric oxide. Potential relevance to regulation of blood pressure. J Clin Invest 100:1488–1492

Hardman JG, Sutherland EW (1969) Guanyl cyclase, an enzyme catalyzing the formation of guanosine $3',5'$-monophosphate from guanosine triphosphate. J Biol Chem 244:6363–6370

Harteneck C, Koesling D, Söling A, Schultz G, Böhme E (1990) Expression of soluble guanylate cyclase: catalytic activity requires two subunits. FEBS Lett 272:221–223

Harteneck C, Wedel B, Koesling D, Malkewitz J, Böhme E, Schultz G (1991) Molecular cloning and expression of a new α-subunit of soluble guanylyl cyclase. FEBS Lett 292:217–222

Hebeiss K, Kilbinger H (1998) Nitric oxide-sensitive guanylyl cyclase inhibits acetylcholine release and excitatory motor transmission in guinea-pig ileum. Neuroscience 82:623–629

Humbert P, Niroomand F, Fischer G, Mayer B, Koesling D, Hinsch KD, Gausepohl H, Frank R, Schultz G, Böhme E (1990) Purification of soluble guanylyl cyclase from bovine lung by a new immunoaffinity chromatographic method. Eur J Biochem 190:273–278

Ignarro LJ, Degnan JN, Baricos WH, Kadowitz PJ, Wolin MS (1982a) Activation of purified guanylate cyclase by nitric oxide requires heme. Comparison of heme-deficient, heme-reconstituted and heme-containing forms of soluble enzyme from bovine lung. Biochim Biophys Acta 718:49–59

Ignarro LJ, Wood KS, Wolin MS (1982b) Activation of purified soluble guanylate cyclase by protoporphyrin IX. Proc Natl Acad Sci USA 79:2870–2873

Ignarro LJ, Adams JB, Horwitz PM, Wood KS (1986) Activation of soluble guanylate cyclase by NO-hemoproteins involves NO-heme exchange: comparison of heme-containing and heme-deficient enzyme forms. J Biol Chem 261:4997–5002

Ignarro, LJ, Buga GM, Wood KS, Byrns RE, Chandhuri G (1987): Endothelium-derived relaxing factor produced and released from artery and vein is nitric oxide. Proc Natl Acad Sci USA 84:9265–9269

Kharitonov VG, Russwurm M, Magde D, Sharma VS, Koesling D (1997) Dissociation of nitric oxide from soluble guanylyl cyclase. Biochem Biophys Res Commun 239:284–286

Koesling D, Friebe A (1999) Soluble guanylyl cyclase: Structure and function. Rev Phys Biochem Pharmacol 135:41–65

Lincoln TM (1989) Cyclic GMP and mechanisms of vasodilation. Pharmacol Ther 41:479–502

Liu Y, Ruoho AE, Rao VD, Hurley JH (1997) Catalytic mechanism of the adenylyl and guanylyl cyclase: modeling and mutational analysis. Proc Natl Acad Sci USA 94: 13414–13419

Maines MD (1988) Heme oxygenase: function, multiplicity, regulatory mechanism, and clinical applications. FASEB J 2:2257–2568

Maines MD (1997) The heme oxygenase system: a regulator of second messenger gases. Annu Rev Pharmacol Toxicol 37: 517–554

Marks GS, Brien JF, Nakatsu KB and McLaughlin E (1991) Does carbon monoxide have a physiological function? Trends Pharmacol Sci 12:185–188

Mayer B, Brunner F, Schmidt K (1993) Inhibition of nitric oxide synthesis by methylene blue. Biochem Pharmacol 45:367–374

Moncada S, Higgs EA (1995) Molecular mechanisms and therapeutic strategies related to nitric oxide. FASEB J 9:1319–1330

Mülsch A, Bauersachs J, Schäfer A, Stasch JP, Kast R, Busse R (1997) Effect of YC-1, an NO-independent, superoxide-sensitive stimulator of soluble guanylyl cyclase, on smooth muscle responsiveness to nitrovasodilators. Br J Pharmacology 120: 681–689

Nakane M, Arai K, Saheki S, Kuno T, Buechler W, Murad F (1990) Molecular cloning and expression of cDNAs coding for soluble guanylyl cyclase from rat lung. J Biol Chem 265:16841–16845

O'Dell TJ, Hawkins RD, Kandel ER, Arancio O (1991) Tests of the roles of two diffusible substances in long-term potentiation: evidence for nitric oxide as a possible early retrograde messenger. Proc Natl Acad Sci USA 88:11285–11289

Olesen SP, Drejer J, Axelsson O, Moldt P, Bang L, Nielsen-Kudsk JE, Busse R, Mülsch A (1998) Characterization of NS 2028 as a specific inhibitor of soluble guanylyl cyclase. Br J Pharmacol 123:299–309

Palmer RMJ, Ferrige AG, Moncada S (1987) Nitric oxide release accounts for the biological activity of endothelium-derived relaxing factor. Nature 327:524–526

Russwurm M, Behrends S, Harteneck C, Koesling D (1998) Functional properties of a naturally occurring isoform of soluble guanylyl cyclase. Biochem J 335:125–130

Schmidt H, Walter U (1994) NO at work. Cell 78:919–925

Schrammel A, Behrends S, Schmidt K, Koesling D, Mayer B (1996) Characterisation of 1H-[1,2,4]oxadiazolo[4,3-a]quinoxalin-1-one (ODQ) as a heme site inhibitor of nitric oxide-sensitive guanylyl cyclase. Mol Pharmacol 50:1–5

Schultz G, Böhme E, Munske K (1969) Guanyl cyclase. Determination of enzyme activity. Life Sci 8:1323–1332

Stone JR, Marletta MA (1994) Soluble guanylyl cyclase from bovine lung: activation with nitric oxide and carbon monoxide and spectral characterisation of the ferrous and ferric states. Biochemistry 33:5636–5640

Stone JR, Marletta MA (1995) Heme stoichiometry of heterodimeric soluble guanylate cyclase. Biochemistry 34:14668–14674

Stone JR, Marletta MA (1998) Synergistic activation of soluble guanylate cyclase by YC-1 and carbon monoxide: implications for the role of cleavage of the iron-histidine bond during activation by nitric oxide. Chem Biol 5:255–261

Sunahara RK, Dessauer CW, Gilman AG (1996) Complexity and diversity of mammalian adenylyl cyclases. Annu Rev Pharmacol Toxicol 36:461–480

Sunahara RK, Beuve A, Tesmer JJG, Sprang SR, Garbers DL, Gilman AG (1998) Exchange of substrate and inhibitor specificities between adenylyl and guanylyl cyclases. J Biol Chem 273:16332–16338

Tang WJ, Gilman AG (1995) Construction of a soluble adenylyl cyclase activated by $Gs\alpha$ and forskolin. Science 268:1769–1772

Tang WJ, Stanzel M, Gilman AG (1995) Truncation and alanine-scanning mutants of type I adenylyl cyclase. Biochemistry 34:14563–14572

Tesmer JJG, Sunahara RK, Gilman AG, Sprang SR (1997) Crystal structure of the catalytic domains of adenylyl cyclase in a complex with $G_{s\alpha}GTP\gamma S$. Science 278:1907–1916

Thompson DK, Garbers DL (1995) Dominant negative mutations of the guanylyl cyclase-A receptor. J Biol Chem 270:425–430

Tomita T, Tsuyama S, Imai Y, Kitagawa T (1997) Purification of bovine soluble guanylate cyclase and ADP-ribosylation on its small subunit by bacterial toxins. J Biochemistry 122:531–536

Tucker CL, Hurley JH, Miller TR, Hurley JB (1998) Two amino acid substitutions convert a guanylyl cyclase, RetGC-1, into an adenylyl cyclase. Proc Natl Acad Sci USA 95:5993–5997

Waldman SA, Murad F (1987) Cyclic GMP synthesis and function. Pharmacol Rev 39:163–196

Walter U (1989) Physiological role of cGMP and cGMP-dependent protein kinase in the cardiovascular system. Rev Physiol Biochem Pharmacol 113:42–88

Wedel B, Humbert P, Harteneck C, Foerster J, Malkewitz J, Böhme E, Schultz G, Koesling D (1994) Mutation of His-105 of the β_1 subunit yields a nitric oxide-insensitive form of soluble guanylyl cyclase. Proc Natl Acad Sci USA 91:2592–2596

Wedel B, Harteneck C, Foerster J, Friebe A, Schultz G, Koesling D (1995) Functional domains of soluble guanylyl cyclase. J Biol Chem 270:24871–24875

Wedel BJ, Garbers DL (1997) New insights on the functions of the guanylyl cyclase receptors. FEBS Lett 410:29–33

Wegener JW, Nawrath H (1997) Differential effects of isoliquiritigenin and YC-1 in rat aortic smooth muscle. Eur J Pharmacol 3203:89–91

White AA, Aurbach GD (1969) Detection of guanyl cyclase in mammalian tissues. Biochim Biophys Acta 191:686–697

Wilson EM, Chinkers M (1995) Identification of sequences mediating guanylyl cyclase dimerisation. Biochemistry 34:4696–4701

Wu CC, Ko FN, Kuo SC, Lee FY, Teng CM (1995) YC-1 inhibited human platelet aggregation through NO-independent activation of soluble guanylate cyclase. Br J Pharmacol 116:1973–1978

Yuen PST, Potter LR, Garbers DL (1990) A new form of guanylyl cyclase is preferentially expressed in rat kidney. Biochemistry 29:10872–10878

Zhao Y, Marletta MA (1997) Localization of the heme binding region in soluble guanylate cyclase. Biochemistry 36:15959–15964

Zhao Y, Schelvis JP, Babcock GT, Marletta MA (1998) Identification of histidine 105 in the β_1 subunit of soluble guanylate cyclase as the heme proximal ligand. Biochemistry 37:4502–4509

Zhuo M, Hawkins RD (1995) Long-term depression: a learning-related type of synaptic plasticity in the mammalian central nervous system. Rev Neurosci 6:259–277

CHAPTER 5
Nitric Oxide Synthase Inhibitors I: Substrate Analogs and Heme Ligands

J.F. PARKINSON

A. Introduction

The purpose of the current review is to provide the interested investigator with an overview of the most recent advances in the identification, design and development of mechanism-based nitric-oxide synthase (NOS) inhibitors. A thorough review of the early work in this field was previously published by the author (PARKINSON and PHILLIPS 1997).

This review will focus on L-arginine-based substrate analogs and on those that directly ligate the heme at the NOS active site. NOS inhibitors based on the enzyme cofactor tetrahydrobiopterin (BH_4) are discussed in detail by Werner and Schmidt (Chap. 8). To put the field of NOS inhibitors into perspective, a brief overview of some therapeutic concepts for NOS inhibitors will be given, since these concepts have provided the impetus for the tremendous efforts that have been invested in finding isoform-selective NOS inhibitors. In addition, a brief discussion of results obtained with NOS-isoform knockout (–/–) mice is provided, since these animals can be regarded as the "ultimate test" of the effects of isoform-selective NOS inhibitors in vivo. As will be seen, the results have been encouraging in some cases but have been equivocal in others. The main thrust of the review will be on the description of the various classes of NOS inhibitors that are either commercially available or have been described by investigators in the pharmaceutical industry. Published examples of the in vitro and in vivo pharmacological uses of these compound classes are provided. Lastly, progress on NOS crystallography will be discussed. This is a dramatic development in the field, since tremendous progress was made in 1997 and 1998, and the era of rational design of NOS inhibitors is about to begin.

I. Therapeutic Concepts for NOS Inhibitors

The critical role played by the L-arginine/NO• pathway and the three major isoforms of NOS in regulating physiological and pathophysiological processes has led to significant efforts to discover and develop isoform-selective NOS inhibitors. There is tremendous therapeutic potential for NOS inhibitors in the treatment of human diseases in which excessive NO• production has been implicated in disease pathology. Neuronal NOS (NOS-I or nNOS) is implicated in the pathogenesis of neurodegenerative diseases, such as ischemic stroke (SAMDANI et al. 1997), amyotrophic lateral sclerosis (BECKMAN and KOPPENOL 1996) and Parkinson's disease (HANTRAYE et al. 1996; ARA et al. 1998). Inducible NOS (NOS-II or iNOS) is implicated in the pathogenesis of numerous inflammatory and autoimmune diseases. These include septic and hemorrhagic shock (MACMICKING et al. 1995; HIERHOLZER et al. 1998), rheumatoid arthritis and osteoarthritis (EVANS et al. 1995), psoriasis, cutaneous lupus erythematosus and other dermatoses (BRUCH-GERHARZ et al. 1998), multiple sclerosis (PARKINSON et al. 1997), inflammatory bowel diseases (RACHMILEWITZ et al. 1998) and insulin-dependent diabetes mellitus (McDANIEL et al. 1997).

Several lines of evidence support these therapeutic concepts.

1. NOS inhibitors have beneficial effects in vitro in animal and human tissue culture systems that serve as surrogates for processes of disease pathology
2. NOS inhibitors have shown efficacy in vivo in animal models of human disease
3. There is a correlation between the presence and activity of the NOS isoforms in active human disease and the downregulation of NOS activity by therapeutic interventions

II. NOS-Knockout Mice

Transgenic NOS-I$^{-/-}$, NOS-II$^{-/-}$ and endothelial NOS$^{-/-}$ (NOS-III$^{-/-}$) knockout mice have now been available for several years and have been used to test these therapeutic concepts. In the middle-cerebral-artery-occlusion (MCAO) model of cerebral ischemia, NOS-I$^{-/-}$ mice exhibit reduced infarct volume, less edema and improved neurological function compared with wild-type mice (HUANG et al. 1994; HARA et al 1996). NOS-I$^{-/-}$ mice show attenuated 1-methyl-4-phenylpyridinium-induced substantia nigra degeneration, supporting NOS-I involvement in Parkinson's disease (MATTHEWS et al. 1997). Thus, for NOS-I, the data from knockout mice are consistent with results obtained using NOS-I inhibitors in vivo (Sect. B). These data support the concept that selective NOS-I inhibitors may have applications in both acute and chronic diseases. It should be noted, however, that NOS-I has been implicated in the regulation of vascular and gastrointestinal muscle contractility. Indeed, NOS-I$^{-/-}$ mice exhibit stomach enlargement as a variably penetrant phenotype that is reminiscent of pyloric stenosis (HUANG et al. 1993). Since NOS inhibitors also modulate gastric muscle contractility in various in vitro and in vivo settings, the potential for gastrointestinal side effects of NOS-I inhibitors with chronic use should be considered.

The effect of the NOS-II$^{-/-}$ phenotype on inflammatory autoimmune and infectious diseases in mice has provided some consensus but has also generated much controversy (NATHAN 1997). In mice suffering from hemorrhagic shock and in NOS-II$^{-/-}$ mice, improvement in end-organ function is observed with a NOS-II inhibitor (HIERHOLZER et al. 1998). NOS-II involvement in severe hypotensive circulatory failure in endotoxic shock has been confirmed in NOS-II$^{-/-}$ mice (MACMICKING et al. 1995) and correlates with results recently obtained in endotoxemic rats treated with selective NOS-II inhibitors (WRAY et al. 1998). The ability of NOS-II inhibitors to improve survival in septic shock remains controversial. Contradictory results have been obtained in NOS-II$^{-/-}$ mice (compare MACMICKING et al. 1995 with WEI et al. 1995), and selective NOS-II inhibitors do not improve multiple organ failure in rats (WRAY et al. 1998; PARKINSON et al., unpublished observations). Similar controversy exists for the role of NOS-II in autoimmune pathology. The same laboratory that found that a NOS inhibitor blocked spontaneous inflammation in MRL-lpr/lpr mice found that the NOS-II$^{-/-}$ phenotype only had marginal effects when bred into the lpr/lpr background (WEINBERG et al. 1994; GILKESON et al. 1997). In experimental allergic encephalitis (EAE), the animal model of multiple sclerosis, two independent groups have reported disease exacerbation by the NOS-II$^{-/-}$ phenotype (FENYK-MELODY et al. 1998; SAHRBACHER et al. 1998). These results are contradictory to some studies with NOS inhibitors in the EAE model, which show disease inhibition (PARKINSON et al. 1997).

We are clearly not close to resolving the discrepancies in our understanding of the role of NOS-II in inflammatory disease. The results with the NOS-II$^{-/-}$ mice, however, do provide support for the hypothesis that NOS-II may be involved both as a non-specific effector of tissue damage in some of these animal models and in regulating the immune response. They also provide insights into potential side effects of NOS-II inhibitors. For example, NOS-II inhibitors render animals susceptible to infection by numerous intracellular pathogens, and the NOS-II$^{-/-}$ phenotype mimics these effects (NATHAN 1997). The NOS-II$^{-/-}$ phenotype also partially blocks the early, innate T-cell and NK-cell-mediated immune response to infection in mice by suppressing Th$_1$ cytokines (IL-12 and IFN-γ) and augmenting the Th$_2$ cytokine, TGF-β (DIEFENBACH et al. 1998). Further work with NOS-II$^{-/-}$ mice and pharmacological testing of NOS-II inhibitors in animal models will doubtless shed further light on the potential clinical utility of NOS-II inhibitors. Ultimately, of course, clinical testing in humans will be the true test of these therapeutic concepts.

For NOS-III, there seems to be general consensus that inhibition of endothelial NO$^{\bullet}$ generation would be a "bad thing", since this enzyme is critical to the normal homeostatic mechanisms of the vasculature (KNOWLES and MONCADA 1992; MONCADA and HIGGS 1993). As predicted, NOS-III$^{-/-}$ mice are hypertensive (SHESELY et al. 1996). They also exhibit other vascular pathologies, including abnormal remodeling in response to injury (RUDIC et al. 1998) and enhanced infarcts in stroke (HUANG et al. 1996). From these results and the study of endothelial NO$^{\bullet}$ function in many species (including humans), it is generally believed that NOS-I and NOS-II inhibitors should be highly selective so that NOS-III inhibition is minimized in vivo.

B. Mechanism-Based NOS Inhibitors

This section will review two classes of mechanism-based NOS inhibitors: L-arginine-based substrate analogs and heme ligands. For the former, much of the early work with "tool" compounds will be discussed only briefly, since these have previously been reviewed (PARKINSON and PHILLIPS 1997). The goals of this discussion will be to direct the reader to literature describing the pharmacological uses of these early compounds and to describe their limitations in relation to more recently developed compounds that have greater selectivity and potentially greater pharmacological utility. For ease of reference, Table 1 summarizes the key features of NOS inhibitors described in this section, in terms of potency, selectivity, advantages and disadvantages. All NOS isoforms catalyze the conversion of L-arginine into L-citrulline and NO$^{•}$ via N^{G}-OH-L-arginine, as shown in Scheme 1.

a l-arginine + O2 + NADPH → N^{G}-OH-L-arginine + NADP^{+} + H2O
b N^{G}-OH-L-arginine + O2 + 0.5NADPH → L-citrulline + NO$^{•}$ + 0.5NADP^{+} + H2O

Scheme 1. Stoichiometry of the nitric-oxide synthase reaction

The overall scheme for NOS catalysis and the role of heme in mediating both steps in this process have been reviewed extensively (MARLETTA 1993; STUEHR 1997). Oxidation of L-arginine proceeds via a mechanism that is analogous to the oxidation of substrates by cytochrome P_{450} (P450) and, like P450, is inhibited by the heme poison carbon monoxide. The purified NOS isoforms (either native or recombinant) all possess a prominent heme spectrum with a characteristic Soret peak that is modulated by the binding of heme ligands and substrates to the enzyme active site. NOS isoforms exhibit characteristic spectral changes very similar to those observed for P450. Type-I perturbation-difference spectra are obtained with L-arginine, N^{G}-OH-L-arginine and substrate-based inhibitors and are characterized by a peak at ~380 nm and a trough at ~420 nm. Type-II perturbation-difference spectra are obtained with classical direct-heme ligands, such as imidazole, and are characterized by the appearance of a peak at 430 nm and a trough at 395 nm (MCMILLAN and MASTERS 1993). These spectral changes thus demonstrate the close apposition of the L-arginine-binding site to the catalytic heme center and provide a simple basis for distinguishing the mode of binding of substrate analogs and heme ligands.

I. Substrate-Based NOS Inhibitors

1. Arginine Analogs

The compounds described in this section are shown in Fig. 1. The discovery that L-arginine was the substrate for NOS and that L-arginine oxidation took place on the gaunidino nitrogens led to the discovery of first-generation L-

Table 1. Summary profiles for selected substrate-based and heme-ligand nitric-oxide synthase (NOS) inhibitors

Compound name	Compound number[a]	Potency	Isoform selectivity	Advantages	Disadvantages	Other comments
L-NMMA	1	Low-micromolar range	Little to none in vitro or in vivo	Cheap; available; not toxic; orally available	Not selective	Tested in humans
NNA	2	Nanomolar range (NOS-I, NOS-III)	Selective for NOS-I and NOS-III	Cheap; available; not generally toxic; orally available; CNS available	Lack of selectivity for NOS-III may cause problems in vivo	Use methyl ester as prodrug
Aminoguanidine	3	Low-micromolar range	20-Fold for NOS-II	Cheap; available; not toxic; orally available	Affects other pathways; may be non-selective at high doses in vivo	Tested in humans
L-Thiocitrulline	4	Nanomolar range (NOS-I, NOS-III)	Selective for NOS-I and NOS-III	No advantages over NNA		Binds heme, gives type-II spectra
L-NIL	7	Low-micromolar range (NOS-II)	30-Fold for NOS-II	Available; not toxic; orally available	Weak selectivity for NOS-II	
Ethylisothiourea	9	Mid-nanomolar range	Not selective	No advantages over NNA		Some ITUs block NOS-II induction
PBITU	11	Mid-nanomolar range (NOS-II)	200-Fold for NOS-II at enzyme	None	Cell permeability, acute toxicity in vivo	
1400W	13	7 nM (NOS-II)	5000-Fold vs NOS-III; 500-fold vs NOS-I	Proven NOS-II selectivity in vivo; commercially available	Cleared rapidly by i.v. route; toxic as bolus; need for infusions	

Table 1. Continued

Compound name	Compound number[a]	Potency	Isoform selectivity	Advantages	Disadvantages	Other comments
α-Fluoro-N-(3-(aminomethyl)phenyl)acetamidine	15	40 nM (NOS-I)	100-Fold vs NOS-III; 50-fold vs NOS-II	Proven NOS-I selectivity in vivo; oral and CNS availability	Cleared rapidly	
ARL-17477	16	2 nM (NOS-I)	150-Fold vs NOS-III; 7000-fold vs NOS-II	Proven NOS-I selectivity in vivo; CNS availability		
(+)cis-4-Methyl-5-pentylpyrollidin-2-imine	21	250 nM (NOS-II)	900-Fold vs NOS-III; 13-fold vs NOS-I	Proven NOS-II selectivity in vivo; oral availability	NOS-I selectivity rather low	
7-NI	28	160 nM (NOS-I)	5- To 10-fold NOS-I selectivity	NOS-I selectivity in vivo	Also inhibits monoamine oxidase	
1-Phenylimidazole	24	Low-micromolar range (NOS-II)	20- To 100-fold selective vs NOS-I and NOS-III	No in vivo data available	P450 inhibition?	Binds heme, gives type-II spectra
Clotrimazole	25	Low-micromolar range (NOS-II) for dimerization	Selectivity versus other NOS isoforms not reported	Blocks NOS-II-dimer formation	No in vivo data available vs NOS-II	New mechanistic approach

1400W, *N*-(3-(aminomethyl)-benzyl)-acetamidine; *7-NI*, 7-nitroindazole; *ITU*, isothiourea; *L-NIL*, L-N^6-(1-iminoethyl)-lysine; *L-NMMA*, *N*-methyl-L-arginine; *NNA*, *N*-nitro-L-arginine; *PBITU*, *S*,*S*′-(1,3,phenylenebis(1,2-ethandiyl)-bisisothiourea.
[a] See text

Nitric Oxide Synthase Inhibitors I: Substrate Analogs and Heme Ligands

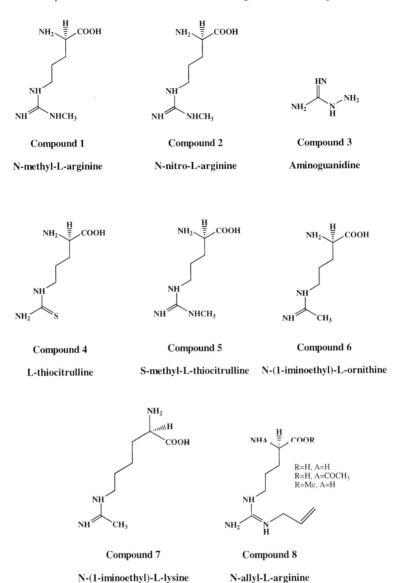

Fig. 1. Arginine analogs. The figure shows examples of simple substrate-based inhibitors of nitric-oxide synthase that have been commonly used for in vitro and/or in vivo pharmacology experiments

arginine-based NOS inhibitors (KERWIN et al. 1995; GRIFFITH and KILBOURN 1996). The tool compounds most commonly used for in vitro and in vivo pharmacological experiments were: N^G-methyl-L-arginine (L-NMMA; compound 1), N^G-nitro-L-arginine (L-NNA; compound 2) and aminoguanidine (compound 3). At the enzyme level, L-NMMA has little or no NOS isoform selectivity with IC_{50} against human enzymes in the 3.3–10 μM range. Aminoguanidine is a moderately potent ($IC_{50} = 6 \mu$M) and moderately selective inhibitor for human NOS-II (fivefold; FARACI et al. 1996) and murine NOS-II (30-fold; MISKO et al. 1993). L-NNA is selective for NOS-I and NOS-III versus NOS-II; the IC_{50}s against NOS-I, NOS-II and NOS-III are 0.5, 8 and 0.5 μM, respectively (MOORE et al. 1996). L-NNA is a slow-binding, irreversible inhibitor of NOS-I, and L-NMMA is a substrate for both NOS-I and NOS-II (OLKEN and MARLETTA 1993; KLATT et al. 1994a). In pharmacological experiments, L-NNA is commonly administered as the methyl ester (L-NAME), which is a prodrug that is activated by ester hydrolysis to L-NNA (SOUTHAN et al. 1995; PFEIFFER et al. 1996). Recently, N-propyl-L-arginine has been described as a potent and highly selective NOS-I inhibitor; the K_is for NOS-I, NOS-II and NOS-III are 0.057, 180 and 8.5 μM, respectively (ZHANG et al. 1997). This NOS-I selectivity is superior to that of L-NNA, but the compound has not been characterized in vivo.

L-Thiocitrulline (compound 4) and S-methyl-L-thiocitrulline (compound 5) are simple, substrate-based NOS inhibitors (NARAYANAN and GRIFFITH 1994). L-Thiocitrulline is similar in potency to L-NNA and is competitive with L-arginine, but it binds to NOS to give a type-II difference spectrum, indicating that it directly coordinates the sixth coordination site of the heme (FREY et al. 1994). This was the first example of a substrate-based NOS inhibitor that is also capable of heme coordination. Unlike the guanidine of the substrate, the thiourea in L-thiocitrulline is not basic and cannot form a salt bridge with an acidic residue, but it can form hydrogen bonds similar to those made by guanidine. Since the isosteric enzyme product L-citrulline is inactive as a NOS inhibitor, the sulfur in L-thiocitrulline is critical for binding. S-methyl-L-thiocitrulline is also competitive with L-arginine but binds NOS with a type-I difference spectrum, indicating that it does not coordinate the heme. S-methyl-L-thiocitrulline is isosteric with (and is less basic than) NMA but is significantly more potent (FURFINE et al. 1994). S-methyl-L-thiocitrulline has a higher potency than NMA; this implies an unspecified binding site for sulfur in the NOS active site.

N^G-(1-iminoethyl)-L-ornithine (compound 6; L-NIO) is a relatively potent but non-selective NOS inhibitor with an IC_{50} of approximately 3.3 μM (McCALL et al. 1991; MOORE et al. 1994). Replacement of the L-NIO chain nitrogen with sulfur or methylene results in inactive compounds (FURFINE et al. 1994; KERWIN et al. 1995). These findings indicate that the basic chain nitrogen (NH) is important for binding. Extension of the alkyl chain of L-NIO by one methylene group provides N-(1-imonoethyl)-L-lysine (compound 7; L-NIL), which has the same NOS-II potency as L-NIO but is 28-fold more selective than NOS-I (MOORE et al. 1994). Addition of a further methylene in the alkyl chain of L-NIL (to make homo-L-NIL) reduces the binding affinity dramatically. L-NIO and L-NIL are competitive with L-arginine and irreversibly inhibit NOS-II in a reduced nicotinamide adenine dinucleotide phosphate (NADPH)-dependent manner (GRANT et al. 1998). L-NIL has been used in vivo as a NOS-II-selective inhibitor (see below).

The interaction of the NOS active site with the primary amine and carboxyl groups of L-arginine were explored using analogs of N^G-allyl-L-arginine (compound 8; OLKEN and MARLETTA 1992; ROBERTSON et al. 1995). The ester and parent compound have similar potencies, but the acetamide of the amino acid nitrogen is less potent, indicating that the amine moiety is more important for binding than the acid function. Recent results suggest that des-amino- and des-carboxy-L-arginine analogs bind to NOS in a manner similar to the binding of L-arginine and that the amino function is required for catalysis (GRANT et al. 1998). While most groups have focused on chemical alterations in the guanidine moiety to enhance NOS-inhibitor potency or selectivity, one recent study suggests that modifications of the carboxyl group can be useful. Conversion of the carboxyl group of L-NIL to a vicinal diol causes a twofold decrease in potency for NOS-II, whereas selectivity versus NOS-III is enhanced 20-fold (HALLINAN et al. 1998).

The following are examples of the in vivo pharmacological uses of the simple L-arginine analogs summarized above. L-NMMA (KILBOURN et al. 1990; WRIGHT et al. 1992), L-NAME (GRAY et al 1991), aminoguanidine (WU et al. 1995) and L-thiocitrulline (NARAYANAN et at al. 1995) all reverse lipopolysaccharide (LPS)-induced hypotension in vivo. Prolonged hypotension induced by LPS in various species is primarily NOS-II mediated. With the exception of the effects caused by aminoguanidine, however, pressor effects due to non-selective NOS-III inhibition in normal animals are evidence of the lack of NOS-isoform selectivity of these compounds. Due to its apparent NOS-II selectivity, aminoguanidine has been used, frequently at very high doses, to inhibit NOS-II in chronic and acute inflammation models, such as EAE (ZHAO et al. 1996; BRENNER et al. 1997) and autoimmune diabetes (CORBETT and MCDANIEL 1996). The ability of aminoguanidine to affect other metabolic pathways, however, raises questions regarding the specificity of this compound when tested at the high doses used in these studies. L-NIL has been used by several groups to study the role of NOS-II in animal models of human diseases. L-NIL shows efficacy in a murine model of hemorrhagic shock (HIERHOLZER et al. 1998), in the dog cruciate ligament model of osteoarthritis (PELLETIER et al. 1998) and as a prophylactic but does not display therapeutic activity in rat adjuvant-induced arthritis (FLETCHER et al. 1998).

2. Amidine-Containing Inhibitors

The compounds described in this section are shown in Fig. 2. Several NOS-inhibitor series containing an amidine moiety ($C(=NH)NH_2$) have been

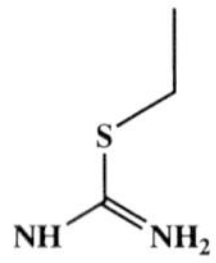

Compound 9
S-ethyl-isothiourea

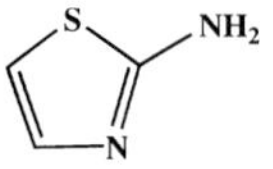

Compound 10
2-aminothiazoline

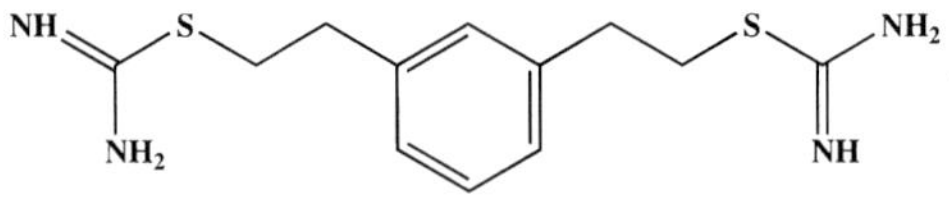

Compound 11
NOS-2-selective bis-isothiourea

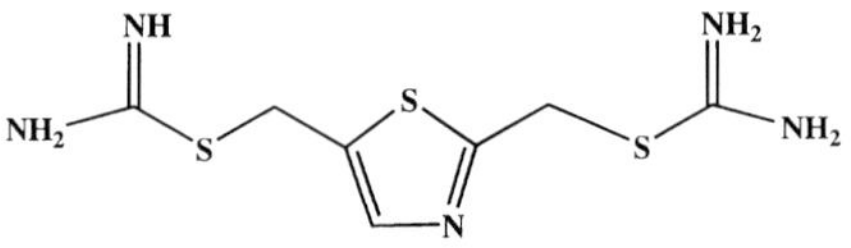

Compound 12
NOS-1-selective bis-isothiourea

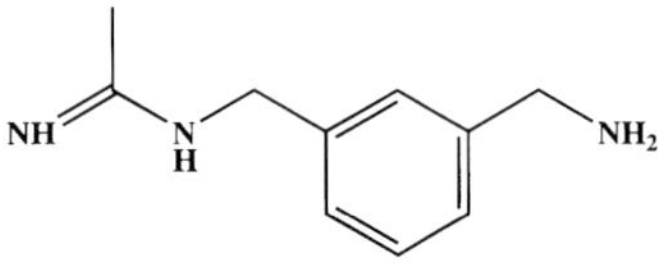

Compound 13
N-(3-(aminomethyl)-benzyl)-acetamidine (1400W)

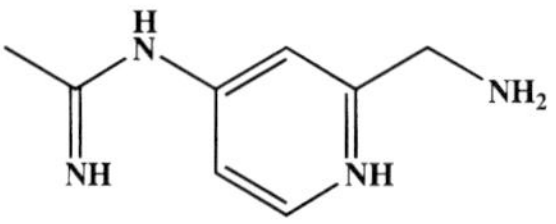

Compound 14
N-(3-(aminomethyl)-phenyl)-acetamidine

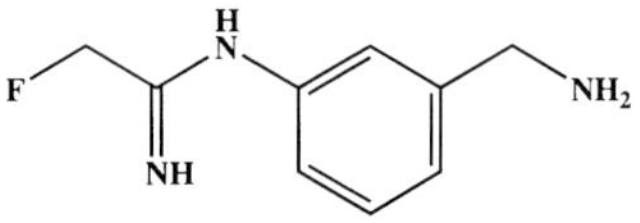

Compound 15
α-fluoro-N-(3-(aminomethyl)-phenyl)-acetamidine

Fig. 2A,B. Amidine-containing inhibitors. The figures show examples of amidine-containing nitric-oxide synthase (NOS) inhibitors from several classes: isothioureas (compound 9), cyclic isothioureas (compound 10), bis-isothioureas (compounds 11, 12), amidines (compounds 13–15), bis-amidines (compound 16), and cyclic amidines (compounds 17–22). Both NOS-I- and NOS-II-selective compounds have been identified in the amidine-based NOS-inhibitor class

described. Since this moiety is part of the guanidine structure of the substrate L-arginine, it is not surprising that these types of compounds are NOS inhibitors. Isothioureas (ITUs; ethyl-isothiourea; compound 9) are an example of this class and have been described by numerous investigators (GARVEY et al. 1994; NAKANE et al. 1995; MACDONALD 1996; STRATMAN et al. 1996). The ITUs are competitive with L-arginine and give type-I difference perturbation spectra. The mode of binding at the NOS active site of ITU has been proposed to result in: positioning of the SR group near the heme, localization of the R group in a hydrophobic pocket and binding of the nitrogens to the same site as that to which the non-reacting nitrogens of L-arginine bind (GARVEY et al. 1994). The size of the R group accommodated by the proposed hydrophobic

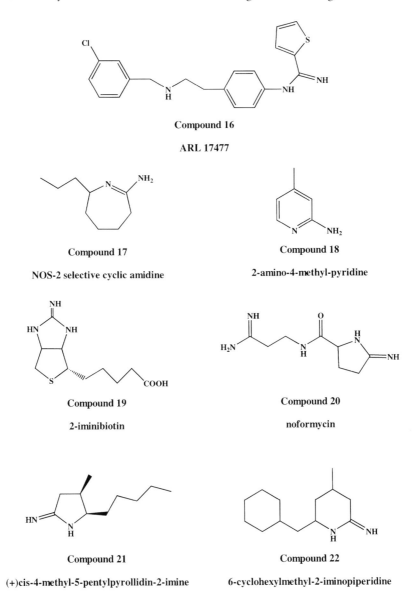

Fig. 2A,B. *Continued*

pocket has an upper limit, because there is a 2×log shift in potency when R is isopropyl compared with the potency when R is isobutyl. Some compounds in the ITU class are very potent, with IC_{50}s in the low- to mid-nanomolar range (compound 9 inhibits human NOS isoforms, with an IC_{50} of approximately 120–200 nM). Despite this potency, the selectivity of first-generation ITUs for the NOS isoforms can be described as moderate at best (HANDY et al. 1996) despite claims to the contrary (JANG et al. 1996). The NOS-II selectivity of aminoethylisothiourea derives (in part) from a moderate selectivity towards NOS-II at the enzyme level and from an unexpected ability of this compound to inhibit NOS-II induction both in vitro and in vivo (RUETTEN and THIEMERMANN 1996). Recently, a series of N-phenylisothioureas was described (SHEARER et al. 1997). The most effective compound had reasonable potency and selectivity for NOS-I versus NOS-II and NOS-III (K_is for NOS-I, NOS-II and NOS-III are 0.32, 37 and 9.4 μM, respectively). Although it is CNS permeable in vivo, the compound had poor pharmacokinetic properties and was subject to metabolism. Not surprisingly, cyclic ITUs have also been found to be NOS inhibitors that are competitive with L-arginine (GARVEY et al. 1994; NAKANE et al. 1995; CALAYCAY et al. 1996). The example shown is 2-aminothiazoline (compound 10; K_is for NOS-I, NOS-II and NOS-III are 0.41, 0.26 and 0.35 μM, respectively). Changing the heterocycle sulfur to oxygen dramatically decreases potency, indicating that, as with thiocitrulline (versus citrulline), the sulfur is important for binding.

Bis-ITUs have also been described as NOS inhibitors (GARVEY et al. 1994a). Compound 11 has moderate selectivity for human NOS-II (K_is for NOS-I, NOS-II and NOS-III are 0.25, 0.047 and 9.0 μM, respectively). Compound 12 is a potent and partially selective NOS-I inhibitor (K_is for NOS-I, NOS-II and NOS-III are 0.0013, 0.0087 and 0.090 μM, respectively). The symmetrical nature of bis-ITUs has led to the hypothesis that the compounds bind to both active sites in the dimeric NOS enzyme. This speculation remains unproven. Further optimization of a bis-ITU template led, via a bis-amidine intermediate, to the discovery of a very potent and selective NOS-II inhibitor, N-(3-(aminomethyl)benzyl)acetamidine (1400W; compound 13; GARVEY et al. 1997). 1400W is a slow-binding, irreversible/slowly reversible NOS-II inhibitor with a K_d of 7 nM; it inactivates NOS-II in a NADPH-dependent manner. In contrast, 1400W is a rapidly reversible and weak inhibitor of NOS-I (K_i = 50 μM) and NOS-III (K_i = 2 μM). The 5000-fold selectivity of 1400W for NOS-II versus NOS-III at the enzyme level is reflected in tissue-based assays for NOS-III and NOS-II activity in normal (endothelium-intact) and LPS-treated (endothelium-denuded) aortic rings, respectively. Thousand-fold selectivity is observed. 1400W is also active against NOS-II in vivo in an LPS-induced, delayed plasma leakage assay in the ileum with ED_{50} = 0.16 mg/kg and has superior selectivity compared with L-NMMA in vivo. Acute toxic effects of 1400W were observed at a 50 mg/kg bolus i.v. dose, but the compound was well tolerated as an i.v. infusion at doses greater than 100 mg/kg for 7 days. The impressive results obtained with 1400W, despite a poor pharmacokinetic profile, make it one of the most selective NOS-II inhibitors described to date for both in vitro and in vivo use. 1400W is commercially available for research purposes. 1400W has already been used in vivo to study the role of NOS-II in various pathological settings. These include coronary blood flow during ischemia-reperfusion injury (PARRINO et al. 1998), tumor growth

(THOMSEN et al. 1997), circulatory failure and multiple organ failure in endotoxic shock (WRAY et al. 1998) and cross-talk between the cyclooxygenase and NOS-II pathways during inflammation (HAMILTON and WARNER 1998).

The same group that discovered 1400W reported extensive structure–activity relationships for a series of N-phenylamidines that are potent and selective NOS-I inhibitors (COLLINS et al. 1998). N-(3-(aminomethyl)phenyl) acetamidine (compound 14) differs from 1400W only by removal of a methylene group but has remarkable potency and selectivity for NOS-I (K_is for NOS-I, NOS-II and NOS-III are 0.04, 2.0 and 6.2 μM, respectively). Unlike 1400W, which gains NOS-II selectivity partly by virtue of its slowly reversible mechanism of action, compound 14 is a rapidly reversible NOS-I inhibitor. In vitro and in vivo pharmacology data were reported for compound 14 and α-fluoro-N-(3-(aminomethyl)phenyl)acetamidine (compound 15). Compound 15 was slightly more potent in NOS-I enzyme inhibition than compound 14, with similar selectivity. Compound 15 was equivalent to NNA in NOS-I inhibition in brain slices in vitro, whereas compound 14 showed poor cell permeability in this assay. Compound 15 had a relatively rapid in vivo elimination rate ($t_{1/2} < 2$h) but was approximately 80% orally available and exhibited CNS-permeable NOS-I inhibition in vivo.

A similar approach to selective NOS-I inhibition was reported earlier by another group who used bis-amidines as a starting point (MACDONALD et al. 1994, 1996; MACDONALD 1995, 1996a, 1996b, 1996c). Starting with a weak, nonselective lead, this group discovered very potent and selective compounds. An example is ARL-17477 (compound 16; K_is for NOS-I, NOS-II and NOS-III are 0.002, 14.0 and 0.37 μM, respectively). ARL-17477 has been evaluated in vivo in the MCAO model of cerebral ischemia (ZHANG et al. 1996). ARL-17477 showed dose-dependent reductions in infarct volume and superior NOS I selectivity compared with L-NAME because of lower non-specific effects on regional cerebral blood flow and mean arterial pressure.

Cyclic amidines also inhibit NOS and have been described in several patents (HANSEN et al. 1995, 1996a, 1996b; GUTHIKONDA et al. 1996a, 1996b). One NOS-II-selective compound contained a seven-membered ring (compound 17; IC_{50}s for NOS-I and NOS-II are 7.8 and 0.081 μM, respectively). In a series of cyclic amidines with six-membered rings, differential placement of a methyl group had large effects on potency. Similar large effects of small steric changes have been noted in ITUs and cyclic ITUs (see above). 2-Aminopyridines (aromatized cyclic amidines) are relatively selective NOS-II inhibitors (FARACI et al. 1996). An example is 2-amino-4-methyl-pyridine (compound 18; IC_{50}s for NOS-I, NOS-II and NOS-III are 0.1, 0.04 and 0.09 μM, respectively). The methyl group enhances potency 30-fold compared with 2-aminopyridine, but further substitution at the 4 position decreases potency. Compound 18 is active in vivo against NOS-II in the LPS-induced plasma nitrite/nitrate assay but exhibits poor in vivo selectivity compared with NOS-III (measured as elevation in mean arterial pressure). This is one of the few studies in which the selectivity ratio for a NOS inhibitor has been directly determined in vivo. Two other cyclic amidines that are NOS inhibitors are 2-iminobiotin (compound 19; K_i for NOS-II is approximately $20\,\mu M$; SUP et al. 1994) and the natural product noformycin (compound 20; K_i for NOS-II is approximately $1.3\,\mu M$; GREEN et al. 1996).

The cyclic-amidine approach to selective NOS-II inhibition has borne further fruit. Initially, 2-iminopiperidine and 2-iminoazaheterocycles were found to be relatively weak and non-selective NOS inhibitors (MOORE et al. 1996). Two derivatives of this approach include a series of 2-iminopyrollidines and substituted 2-iminopiperidines that show both improved potency and selectivity (HAGEN et al. 1998; WEBBER et al. 1998). In the pyrollidine series, (+)cis-4-methyl-5-pentylpyrollidin-2-imine (compound 21) was relatively potent against NOS-II ($IC_{50} = 0.25\,\mu M$), with 900-fold and 13-fold selectivities versus NOS-III and NOS-I, respectively. In vivo, compound 21 showed dose-dependent inhibition of NOS-II in the LPS-induced plasma NOx assay, with an oral ED_{50} of approximately $3\,mg/kg$. The compound had minimal effects on NOS-III in vivo, as assessed by its lack of effect on mean arterial pressure at i.v. doses up to $10\,mg/kg$. In the iminopiperidine series, the most selective compound described was 6-cyclohexylmethyl-2-iminopiperidine (compound 22). Compounds in this series were not as selective as those in the pyrollidine series but were shown to be orally active in the LPS-induced plasma NOx assay for NOS-II in vivo.

3. Summary for Substrate Analogs

Despite the similarity in the active sites of the various NOS isoforms (see below), there has been substantial progress in the discovery and development of potent and isoform-selective substrate-based NOS inhibitors. The availability of these tools for basic in vitro and in vivo pharmacology research will lead to more specific answers regarding the roles of NOS-I, NOS-II and NOS-III in physiology and pathophysiology. From an analysis of the literature, it would appear that the current reagents of choice for in vivo studies are:

- ARL-17477 or compound 15 for NOS-I
- 1400W or compound 21 for NOS-II

The in vitro and in vivo pharmacology of these compounds suggests that their specificities are considerably higher than those of first-generation L-arginine analogs, such as L-NMMA, NNA, L-NIL and aminoguanidine.

II. Heme Ligands

The compounds described in this section are shown in Fig. 3. The presence of heme in the NOS active site and the critical function of heme in NOS catalysis led to the discovery and characterization of heme-binding compounds as NOS inhibitors. Imidazole and imidazole-containing compounds are known inhibitors of heme-containing enzymes, particularly P450 (COLE and

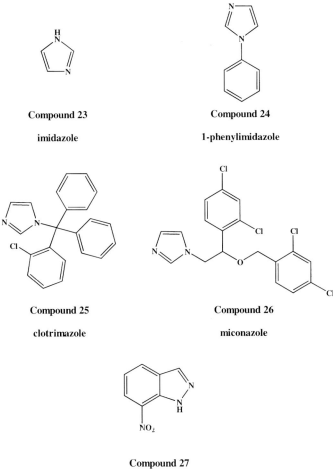

Fig. 3. Heme ligands. The figure shows examples of imidazole-based heme ligand nitric-oxide synthase (NOS) inhibitors. These include direct enzyme inhibitors (compounds 23, 24) and recently reported NOS-II dimerization inhibitors (compounds 25, 26). 7-Nitroindazole (compound 27) is only included in this figure for historical reasons, as it is no longer recognized as a heme ligand for NOS

ROBINSON 1990). Imidazole (compound 23) and 1-phenylimidazole (compound 24) are weak NOS inhibitors but are relatively selective for NOS-II (WOLFF et al. 1993a; MAYER et al. 1994; CHABIN et al. 1996). Imidazole is a reversible, arginine-competitive inhibitor of NOS-I and NOS-II (IC$_{50}$s for NOS-I, NOS-II and NOS-III are 175, 59 and 189 μM, respectively). Imidazole also competes with BH$_4$ for binding to NOS-II, and type-II perturbation-difference spectra indicate that it acts as the sixth ligand of the heme in NOS, as expected (CHABIN et al. 1996). Based on these observations, it was proposed that the substrate-binding site, catalytic heme center and BH_4-binding site in NOS are close to each other. The recent crystal-structure determinations for the catalytic center of NOS-II confirm this hypothesis (see below). 1-Phenylimidazole and some substituted variants of this compound retain the NOS-II selectivity of imidazole but only exhibit modest potency enhancements (CHABIN et al. 1996). Another series of arylimidazole-based NOS inhibitors has been patented (HOELSCHER et al. 1997). Details on specific modifications to achieve NOS-II potency and selectivity and the activity of these compounds in vivo will be reported in the near future (HOELSCHER et al., in preparation).

A range of P450 inhibitors have been tested for NOS inhibition (GRANT et al. 1997). Of 30 compounds tested, only one (chlorzoxazone, a P450-2E1 inhibitor) was found to have significant NOS inhibition, with $K_i = 3.3\,\mu M$. The compound was not NOS-isoform selective and caused type-I rather than type-II spectral shifts; thus, it is not a NOS heme ligand. These and other studies suggest that prototypical P450 heme ligands are not necessarily heme ligands for NOS and that the two enzyme classes have distinct structural and mechanistic properties. This is also supported by recent crystallographic data (see below).

Two P450 inhibitors that have been shown to inhibit NOS are the antifungal imidazoles clotrimazole and miconazole (compounds 25 and 26). These compounds were initially described as weak NOS inhibitors with mixed modes of enzyme inhibition via direct competition with arginine and interference with calmodulin-dependent NOS activation (WOLFF et al 1993b). Recent results have identified a new and potentially very interesting property of these compounds. By ligating the heme iron in both purified protein preparations and cell-culture systems, miconazole and clotrimazole prevent the assembly of murine NOS-II monomers into active dimers (SENNEQUIER and STUEHR 1998). The compound effects were weak (micromolar range) and are in contrast to the effects of imidazole and 1-phenylimidazole, which actually promoted NOS-II dimerization in vitro. Treatment of cells with clotrimazole led to accumulation of inactive NOS-II monomers. These results are consistent with observations that heme is critical for the assembly of NOS dimers from NOS monomers. The discovery that complex imidazoles can block NOS dimerization offers a new and unexpected approach to NOS inhibition. It remains to be determined whether this new mechanistic approach to NOS inhibition can be achieved with selective NOS-isoform inhibition. The effects of antifungal imidazoles on the dimerization of NOS-II in vivo have not been reported.

Various indazoles are NOS inhibitors, and 7-nitroindazole (compound 27) was reported to be a potent and weakly selective inhibitor of NOS-I (BABBEDGE et al. 1993; CHABIN et al. 1996). 7-Nitroindazole inhibits NOS-I competitively with respect to arginine and BH_4 (KLATT et al. 1994b). Although originally proposed to be a heme ligand (based on structural similarities with imidazole), recent spectral data suggests that 7-nitroindazole is not a heme ligand for NOS and functions as a competitive NOS inhibitor (TSAI et al. 1996; SENNEQUIER and STUEHR 1998). For reasons that may be related to selective cellular uptake rates, 7-nitrindazole appears to be a selective NOS-I inhibitor in vivo (SOUTHAN and SZABO 1996), although its in vivo selectivity versus NOS-III has recently been questioned (ZAGVAZDIN et al. 1996). Many groups studying the in vivo role of NOS-I in various physiological and pathophysiological settings have used 7-nitroindazole. For example, 7-nitroindazole has been shown to reduce infarct size in the MCAO model (YOSHIDA et al 1994) and to suppress 1-methyl-4-phenyl-1,2,3,6-tetrahydropyridine-mediated neurotoxicity in a model of Parkinson's disease (SCHULZ et al 1995). The ability of 7-nitroindazole to function as a monoamine-oxidase inhibitor, however, raises questions regarding its specificity and suitability for studying NOS-I in animal models of Parkinson's disease (CASTAGNOLI et al. 1997).

1. Summary for Heme Ligands

A number of potentially useful NOS inhibitors that function via direct heme ligation have been described. Usually, these have been characterized in vitro at the enzyme level, and very little information about them has been reported in vivo or using in vitro pharmacology experiments with whole tissue. For these reasons, compounds in this class cannot be recommended for in vivo use at this time. Nevertheless, for certain in vitro experiments and enzymology experiments, they are very useful reagents.

III. Towards Rational Design of NOS Inhibitors

The crystal structures of the catalytic domain of murine NOS-II with bound L-arginine, aminoguanidine, L-thiocitrulline and imidazole have been solved at high resolution (CRANE et al. 1997, 1998). The crystal structure of NOS-II is unique and is quite distinct from the structure of P450. Figure 4 shows details of the NOS-II active site with bound arginine and shows the relationship of the substrate to the heme prosthetic group and BH_4 cofactor. The key role played by the E371 glutamic acid side chain in stabilizing the guanidino nitrogens of L-arginine above the plane of the heme is shown and concurs with results of mutational analyses (GACCHUI et al. 1997). Residues from both NOS-II monomers form the active site in dimeric NOS-II, and heme, substrate and BH_4 are critical in stabilizing these structural elements (Chap. 3; CRANE et al 1998). The crystal structure for the NOS-III catalytic domain was also recently reported at an International Nitric Oxide conference (RAMAN et al. 1998). In addition, the crystal structures of rat NOS-I and human NOS-II catalytic domains have been solved in Dr. Poulos' laboratory (personal communication).

These major advances in the crystallography of NOS isoforms will lead to several new developments in the field. Medicinal chemists and NOS enzymologists will be able to obtain complexes of diverse NOS inhibitors with one

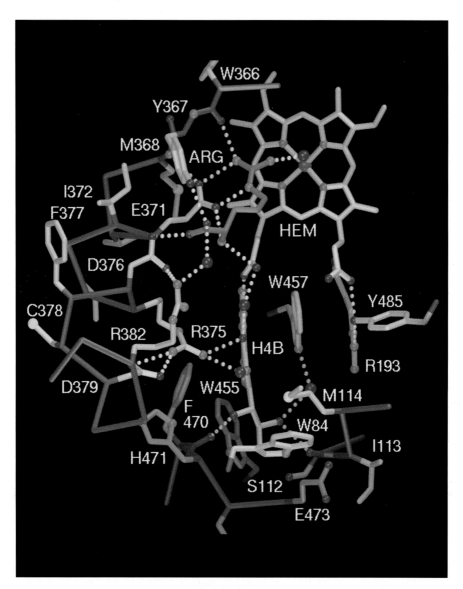

Fig. 4. The nitric-oxide synthase (NOS)-II active site. The figure, kindly provided by Dr. J.A. TAINER (Scripps Research Institute), shows details of the active site of the catalytic domain of murine NOS-II in a complex with the substrate L-arginine (CRANE et al. 1998). A glutamic acid (E371) side chain stabilizes the guanidino nitrogen atoms (*blue dots*) of L-arginine (ARG, *green*) so as to position the substrate in the proximity of the active-site heme (HEM, *gray*). The NOS cofactor tetrahydrobiopterin (H4B, *gray*) makes contact with the heme, the substrate and side chains of residues from both NOS-II monomers to form an integral part of the active site

or all NOS isoforms. Solving the bound conformations of these inhibitor complexes at high resolution will enable confirmation and extension of the interpretation of structure–activity relationships inferred from classical enzymology studies. These structures can then be used to assist in the rational design of new NOS inhibitors. Solutions of the structures of these rationally designed NOS inhibitors complexed with NOS isoforms can then be used (in an iterative fashion) to test and refine three-dimensional hypotheses that will lead to the optimization of second-generation NOS inhibitors. The era of isoform-selective NOS-inhibitor design is about to begin, and the output of this era is eagerly awaited, because it is likely to generate significant advances in addition to those that have already been made. In the near future, we can expect to have access to superior pharmaceuticals for testing the role of the various NOS isoforms in physiological and pathophysiological processes. We can also expect to have NOS inhibitors that have been optimized for target specificity and minimal side effects as therapeutics.

Acknowledgements. The author is very grateful to Dr. J.A. TAINER (Scripps Research Institute) for providing Fig. 4, to Dr. D.J. STUEHR (Cleveland Clinic) for his insights into NOS enzymology and to Dr. G. PHILLIPS (Berlex Biosciences) for his insights into NOS-inhibitor chemistry.

References

Ara J, Przedborski S, Naini AB, Jackson-Lewis V, Trifiletti RR, Horwitz J, Ischiropoulos H (1998) Inactivation of tyrosine hydroxylase by nitration following exposure to peroxynitrite and 1-methyl-4-phenyl-1,2,3,6-tetrahydropyridine. Proc Natl Acad Sci USA 95:7659–7663

Babbedge RC, Bland-Ward PA, Hart SL, Moore PK (1993) Inhibition of rat cerebellar nitric oxide synthase by 7-nitro indazole and related substituted indazoles. Br J Pharmacol 110:225–228

Beckman JS, Koppenol WH (1996) Nitric oxide, superoxide, and peroxynitrite: the good, the bad, and ugly. Am J Physiol 271:C1424–C1437

Brenner T, Brocke S, Szafer F, Sobel RA, Parkinson JF, Perez DH, Steinman L (1997) Inhibition of nitric oxide synthase for treatment of experimental autoimmune encephalomyelitis. J Immunol 158:2940–2946

Bruch-Gerharz D, Ruzicka T, Kolb-Bachofen V (1998) Nitric oxide in human skin: current status and future prospects. J Invest Dermatol 110:1–7

Calaycay JR, Kelly TM, MacNaul KL, McCauley ED, Qi H, Grant SK, Griffin PR, Klatt T, Raju SM, Nussler AK, Shah S, Weidner JR, Williams HR, Wolfe GC, Geller DA, Billiar TR, MacCoss M, Mumford RA, Tocci MJ, Schmidt JA, Wong KK, Hutchinson NI (1996) Expression and immunoaffinity purification of human inducible nitric oxide synthase. Inhibition studies with 2-amino-5,6-dihydro-4H-1,3-thiazine. J Biol Chem 271:28212–28219

Castagnoli K, Palmer S, Anderson A, Bueters T, Castagnoli N Jr (1997) The neuronal nitric oxide synthase inhibitor 7-nitroindazole also inhibits the monoamine oxidase-B-catalyzed oxidation of 1-methyl-4-phenyl-1,2,3,6-tetrahydropyridine. Chem Res Toxicol 10:364–8

Chabin, RM, McCauley E, Calaycay JR, Kelly TM, MacNaul KL, Wolfe GC, Hutchinson NI, Madhusudanaraju S, Schmidt JA, Kozarich JW, Wong KK (1996) Active-site structure analysis of recombinant human inducible nitric oxide synthase using imidazole. Biochemistry 35:9567–9575

Cole PA, Robinson CH (1990) Mechanism and inhibition of cytochrome P-450 aromatase. J Med Chem 33:2933–2942

Collins JL, Shearer BG, Oplinger JA, Lee S, Garvey EP, Salter M, Duffy C, Burnette TC, Furfine ES (1998) N-phenylamidines as selective inhibitors of human neuronal nitric oxide synthase: structure-activity studies and demonstration of in vivo activity. J Med Chem 41:2858–2871

Corbett JA, McDaniel ML (1996) The use of aminoguanidine, a selective iNOS inhibitor, to evaluate the role of nitric oxide in the development of autoimmune diabetes. Methods 10:21–30

Crane BR, Arvai AS, Gachhui R, Wu C, Ghosh DK, Getzoff ED, Stuehr DJ, Tainer JA (1997) The structure of nitric oxide synthase oxygenase domain and inhibitor complexes. Science 278:425–431

Crane BR, Arvai AS, Ghosh DK, Wu C, Getzoff ED, Stuehr DJ, Tainer JA (1998) Structure of nitric oxide synthase oxygenase dimer with pterin and substrate. Science 279:2121–2126

Diefenbach A, Schindler H, Donhauser N, Lorenz E, Laskay T, MacMicking J, Rollinghoff M, Gresser I, Bogdan C (1998) Type 1 interferon (IFNα/β) and type 2 nitric oxide synthase regulate the innate immune response to a protozoan parasite. Immunity 8:77–87

Evans CH, Stefanovic-Racic M, Lancaster J (1995) Nitric oxide and its role in orthopaedic disease. Clin Orthop 312:275–294

Faraci WS, Nagel AA, Verdries KA, Vincent LA, Xu H, Nichols LE, Labasi JM, Salter ED, Pettipher ER (1996) 2-Amino-4-methylpyridine as a potent inhibitor of inducible NO synthase activity in vitro and in vivo. Br J Pharmacol 119:1101–1108

Fenyk-Melody JE, Garrison AE, Brunnert SR, Weidner JR, Shen F, Shelton BA, Mudgett JS (1998) Experimental autoimmune encephalomyelitis is exacerbated in mice lacking the NOS2 gene. J Immunol 160:2940–2946

Fletcher DS, Widmer WR, Luell S, Christen A, Orevillo C, Shah S, Visco D (1998) Therapeutic administration of a selective inhibitor of nitric oxide synthase does not ameliorate the chronic inflammation and tissue damage associated with adjuvant-induced arthritis in rats. J Pharmacol Exp Ther 284:714–21

Frey C, Narayanan K, McMillan K, Spack L, Gross SS, Masters BS, Griffith OW (1994) L-Thiocitrulline. A stereospecific, heme-binding inhibitor of nitric-oxide synthases. J Biol Chem 269:26083–26091

Furfine ES, Harmon MF, Paith JE, Knowles RG, Salter M, Kiff RJ, Duffy C, Hazelwood R, Oplinger JA, Garvey EP (1994) Potent and selective inhibition of human nitric oxide synthases. Selective inhibition of neuronal nitric oxide synthase by S-methyl-L-thiocitrulline and S-ethyl-L-thiocitrulline. J Biol Chem 269:26677–26683

Gachhui R, Ghosh DK, Wu C, Parkinson J, Crane BR, Stuehr DJ (1997) Mutagenesis of acidic residues in the oxygenase domain of inducible nitric-oxide synthase identifies a glutamate involved in arginine binding. Biochemistry 36:5097–5103

Garvey EP, Oplinger JA, Tanoury GJ, Sherman PA, Fowler M, Marshall S, Harmon MF, Paith JE, Furfine ES (1994) Potent and selective inhibition of human nitric oxide synthases. Inhibition by non-amino acid isothioureas. J Biol Chem 269:26669–26676

Garvey EP, Oplinger JA, Furfine ES, Kiff RJ, Laszlo F, Whittle BJ, Knowles RG (1997) 1400W is a slow, tight binding, and highly selective inhibitor of inducible nitric-oxide synthase in vitro and in vivo, J Biol Chem 272:4959–4963

Gilkeson GS, Mudgett JS, Seldin MF, Ruiz P, Alexander AA, Misukonis MA, Pisetsky DS, Weinberg JB (1997) Clinical and serologic manifestations of autoimmune disease in MRL-lpr/lpr mice lacking nitric oxide synthase type 2. J Exp Med 186:365–373

Grant SK, Green BG, Wang R, Pacholok SG, Kozarich JW (1997) Characterization of inducible nitric-oxide synthase by cytochrome P-450 substrates and inhibitors. Inhibition by chlorzoxazone. J Biol Chem 272(2):977–83

Grant SK, Green BG, Stiffey-Wilusz J, Durette PL, Shah SK, Kozarich JW (1998) Structural requirements for human inducible nitric oxide synthase substrates and substrate analogue inhibitors. Biochemistry 37:4174–4180

Gray GA, Schott C, Julou-Schaeffer G, Fleming I, Parratt JR, Stoclet JC (1991) The effect of inhibitors of the L-arginine/nitric oxide pathway on endotoxin-induced loss of vascular responsiveness in anaesthetized rats. Br J Pharmacol 103:1218–1224

Green BG, Chabin R, Grant SK (1996) The natural product noformycin is an inhibitor of inducible-nitric oxide synthase. Biochem Biophys Res Commun 225:621–626

Griffith OW, Kilbourn RG (1996) Nitric oxide synthase inhibitors: amino acids. Methods in Enzymology, 268:375–392

Guthikonda RN, Hagmann WK, Maccoss M, et al. (1996a) Substituted 2-acylaminopyridines as inhibitors of nitric oxide synthase, WO patent 96/18617

Guthikonda RN, Grant SK, Maccoss M, et al. (1996b) Cyclic amidine analogs as inhibitors of nitric oxide synthase, WO patent 96/14844

Hagen TJ, Bergmanis AA, Kramer SW, Fok KF, Schmelzer AE, Pitzele BS, Swenton L, Jerome GM, Kornmeier CM, Moore WM, Branson LF, Connor JR, Manning PT, Currie MG, Hallinan EA (1998) 2-Iminopyrrolidines as potent and selective inhibitors of human inducible nitric oxide synthase. J Med Chem 41:3675–3683

Hallinan EA, Tsymbalov S, Finnegan PM, Moore WM, Jerome GM, Currie MG, Pitzele BS (1998) Acetamidine lysine derivative, N-(5(S)-amino-6,7-dihydroxyheptyl)-ethanimidamide dihydrochloride: a highly selective inhibitor of human inducible nitric oxide synthase. J Med Chem 41:775–777

Hamilton LC, Warner TD (1998) Interactions between inducible isoforms of nitric oxide synthase and cyclo-oxygenase in vivo: investigations using the selective inhibitors, 1400W and celecoxib. Br J Pharmacol 125:335–340

Handy RL, Wallace P, Moore PK (1996) Inhibition of nitric oxide synthase by isothioureas: cardiovascular and antinociceptive effects. Pharmacol Biochem 55:179–184

Hansen DW, Currie MG, Hallinan EA, et al. (1995) Amidino derivatives useful as nitric oxide synthase inhibitors, WO patent 95/11231

Hansen DW Jr, Hagen TJ, Kramer SW, et al. (1996a) Nitric oxide synthase inhibitors derived from cyclic amidines, WO patent 96/35677

Hansen DW, Jr, Hallinan EA, Hagen TJ, Kramer SW, et al. (1996b) Cyclic amidino agents useful as nitric oxide synthase inhibitors, WO patent 96/33175

Hantraye P, Brouillet E, Ferrante R, Palfi S, Dolan R, Matthews RT, Beal MF (1996) Inhibition of neuronal nitric oxide synthase prevents MPTP-induced Parkinsonism in baboons. Nat Med 2:1017–1021

Hara H, Huang PL, Panahian N, Fishman MC, Moskowitz MA (1996) Reduced brain edema and infarction volume in mice lacking the neuronal isoform of nitric oxide synthase after transient MCA occlusion. J Cereb Blood Flow Metab 16:605–611

Hierholzer C, Harbrecht B, Menezes JM, Kane J, MacMicking J, Nathan CF, Peitzman AB, Billiar TR, Tweardy DJ (1998) Essential role of induced nitric oxide in the initiation of the inflammatory response after hemorrhagic shock. J Exp Med 187:917–928

Hoelscher P, Rehwinkel H, Burton G, Phillips G, Parkinson JF (1997) New imidazole derivatives are nitric oxide synthase inhibitors. WO patent 9715555-A2

Huang PL, Dawson TM, Bredt DS, Snyder SH, Fishman MC (1993) Targeted disruption of the neuronal nitric oxide synthase gene. Cell 75:1273–1286

Huang Z, Huang PL, Panahian N, Dalkara T, Fishman MC, Moskowitz MA (1994) Effects of cerebral ischemia in mice deficient in neuronal nitric oxide synthase. Science 265:1883–1885

Huang Z, Huang PL, Ma J, Meng W, Ayata C, Fishman MC, Moskowitz MA (1996) Enlarged infarcts in endothelial nitric oxide synthase knockout mice are attenuated by nitro-L-arginine. J Cereb Blood Flow Metab 16(5):981–987

Jang D, Szabo C, Murrell GA (1996) S-substituted isothioureas are potent inhibitors of nitric oxide biosynthesis in cartilage. Eur J Pharmacol 312:341–347

Kerwin JF, Lancaster JR, Feldman PL (1995) Nitric oxide: a new paradigm for second messengers. J Med Chem 38:4343–4362

Kilbourn RG, Jubran A, Gross SS, Griffith OW, Levi R, Adams J, Lodato RF (1990) Reversal of endotoxin-mediated shock by N^G-methyl-L-arginine, an inhibitor of nitric oxide synthesis. Biochem Biophys Res Commun 172(3):1132–1138

Klatt P, Schmidt K, Brunner F, Mayer B (1994a) Inhibitors of brain nitric oxide synthase. Binding kinetics, metabolism, and enzyme inactivation. J Biol Chem 269:1674–1680

Klatt P, Schmid M, Leopold E, Schmidt K, Werner ER, Mayer B (1994b). The pteridine binding site of brain nitric oxide synthase. Tetrahydrobiopterin binding kinetics, specificity, and allosteric interaction with the substrate domain. J Biol 269:13861–13866

Knowles RG, Moncada S (1992) Nitric oxide as a signal in blood vessels. Trends Biochem Sci 17:399–402

Laszlo F, Whittle BJ (1997) Actions of isoform-selective and non-selective nitric oxide synthase inhibitors on endotoxin-induced vascular leakage in rat colon. Eur J Pharmacol 334:99–102

Macdonald JE (1995) Brain selective antagonists of nitric oxide synthase. Presentation at the 4th Annual International Business Communications Conference on Nitric Oxide; Philadelphia, PA; March 1995

Macdonald JE (1996a) Nitric oxide synthase inhibitors. In Annual Reports of Medicinal Chemistry, Bristol, J.A. ed., 221–230

Macdonald JE (1996b) Isothiourea derivatives as NO synthase inhibitors. WO patent 96/09286

Macdonald JE (1996c) Bicyclic isothiourea derivatives useful in therapy, WO patent 96/24588

Macdonald JE, Gentile RJ, Murray RJ (1994) Guanidine derivatives useful in therapy, WO patent 94/21621

Macdonald JE, Shakespeare WC, Murray RJ, Matz JR (1996) Bicyclic amidine derivatives as inhibitors of nitric oxide synthetase, WO patent 96/01817

MacMicking JD, Nathan C, Hom G, Chartrain N, Fletcher DS, Trumbauer M, Stevens K, Xie QW, Sokol K, Hutchinson N, Chen H, Mudgett JS (1995) Altered responses to bactcrial infection and endotoxic shock in mice lacking inducible nitric oxide synthase. Cell 81:641–650

Marletta MA (1993) Nitric oxide synthase structure and mechanism. J Biol Chem 268:12231–12234

Matthews RT, Beal MF, Fallon J, Fedorchak K, Huang PL, Fishman MC, Hyman BT (1997) MPP+ induced substantia nigra degeneration is attenuated in nNOS knockout mice. Neurobiol Dis 4:114–121

Mayer B, Klatt P, Werner ER, Schmidt K (1994) Identification of imidazole as L-arginine-competitive inhibitor of porcine brain nitric oxide synthase. FEBS Letters, 350:199–202

McCall TB, Feelisch M, Palmer RM, Moncada S (1991) Identification of N-iminoethyl-L-ornithine as an irreversible inhibitor of nitric oxide synthase in phagocytic cells. Br J Pharmacol 102:234–238

McDaniel ML, Corbett JA, Kwon G, Hill JR (1997). A role for nitric oxide and other inflammatory mediators in cytokine-induced pancreatic beta-cell dysfunction and destruction. Adv Exp Med Biol 426:313–319

McMillan K, Masters BS (1993) Optical difference spectrophotometry as a probe of rat brain nitric oxide synthase heme-substrate interaction. Biochemistry 32:9875–9880

Misko TP, Moore WM, Kasten TP, Nickols GA, Corbett JA, Tilton RG, McDaniel ML, Williamson JR, Currie MG (1993) Selective inhibition of the inducible nitric oxide synthase by aminoguanidine. Eur J Pharmacol 233:119–125

Moncada S, Higgs A (1993) The L-arginine-nitric oxide pathway. N Engl J Med 329: 2002–2012

Moore WM, Webber RK, Jerome GM, Tjoeng FS, Misko TP, Currie MG (1994) L-N6-(1-iminoethyl)lysine: a selective inhibitor of inducible nitric oxide synthase. J Med Chem 37:3886–3888

Moore WM, Webber RK, Fok KF, Jerome GM, Connor JR, Manning PT, Wyatt PS, Misko TP, Tjoeng FS, Currie MG (1996) 2-Iminopiperidine and other 2-iminoaza-heterocycles as potent inhibitors of human nitric oxide synthase isoforms. J Med Chem 39:669–672

Nakane M, Klinghofer V, Kuk JE, Donnelly JL, Budzik GP, Pollock JS, Basha F, Carter GW (1995) Novel potent and selective inhibitors of inducible nitric oxide synthase. Mol Pharm 47:831–834

Narayanan K, Griffith OW (1994) Synthesis of L-thiocitrulline, L-homothiocitrulline, and S-methyl-L-thiocitrulline: a new class of potent nitric oxide synthase inhibitors. J Med Chem 37:885–887

Narayanan K, Spack L, McMillan K, Kilbourn RG, Hayward MA, Masters BS, Griffith OW (1995) S-alkyl-L-thiocitrullines. Potent stereoselective inhibitors of nitric oxide synthase with strong pressor activity in vivo. J Biol Chem 270:11103–11110

Nathan C (1997) Inducible nitric oxide synthase: what difference does it make? J Clin Invest 100:2417–2423

Olken NM, Marletta MA (1992) N^G-allyl- and N^G-cyclopropyl-L-arginine: two novel inhibitors of macrophage nitric oxide synthase. J Med Chem 35:1137–1144

Olken NM, Marletta MA (1993) N^G-methyl-L-arginine functions as an alternate substrate and mechanism-based inhibitor of nitric oxide synthase. Biochemistry 32:9677–9685

Oplinger JA, Garvey EP, Furfine ES, et al. (1996), Acetamidine Derivatives and their use as inhibitors for the nitric oxide synthase, WO patent 96/19440

Parkinson JF, Phillips GB (1997) Nitric oxide synthases: enzymology and mechanism-based inhibitors. In: Rubanyi GM, Dzau VJ (eds), The endothelium in clinical practice: source and target of novel therapies. Marcel Dekker Inc., New York, pp 95–124

Parkinson JF, Mitrovic B, Merrill JE (1997) The role of nitric oxide in multiple sclerosis. J Mol Med 75:174–186

Parrino PE, Laubach VE, Gaughen JR Jr, Shockey KS, Wattsman TA, King RC, Tribble CG, Kron IL (1998) Inhibition of inducible nitric oxide synthase after myocardial ischemia increases coronary flow. Ann Thorac Surg 66:733–739

Pelletier JP, Jovanovic D, Fernandes JC, Manning P, Connor JR, Currie MG, Di Battista JA, Martel-Pelletier J (1998) Reduced progression of experimental osteoarthritis in vivo by selective inhibition of inducible nitric oxide synthase. Arthritis Rheum 41:1275–1286

Pfeiffer S, Leopold, Schmidt K, Brunner F, Mayer B (1996) Inhibition of nitric oxide synthesis by NG-nitro-L-arginine methyl ester (L-NAME): requirement for bioactivation to the free acid, NG-nitro-L-arginine. Br J Pharmacol 118:1433–1440

Rachmilewitz D, Eliakim R, Ackerman Z, Karmeli F (1998) Direct determination of colonic nitric oxide level–a sensitive marker of disease activity in ulcerative colitis. Am J Gastroenterol 93:409–412

Raman CS, Li H, Martasek P, Kral V, Masters BSS, Poulos TL (1998). Crystal structure of endothelial constitutive nitric oxide synthase: a paradigm for pterin function involving a novel metal center. Cell 95:939–950

Robertson JG, Bernatowicz MS, Dhalla AM, et al. (1995) Inhibition of bovine brain nitric oxide synthase by α-amino and α-carboxyl derivatives of N^G-allyl-L-arginine. Bioorg Chem 23:144–151

Rudic RD, Shesely EG, Maeda N, Smithies O, Segal SS, Sessa WC (1998) Direct evidence for the importance of endothelium-derived nitric oxide in vascular remodeling. J Clin Invest 101:731–736

Ruetten H, Thiemermann C (1996) Prevention of the expression of inducible nitric oxide synthase by aminoguanidine or aminoethyl–isothiourea in macrophages and in the rat. Biochem Biophys Res Commun 225:525–530

Sahrbacher UC, Lechner F, Eugster HP, Frei K, Lassmann H, Fontana A (1998) Mice with an inactivation of the inducible nitric oxide synthase gene are susceptible to experimental autoimmune encephalomyelitis. Eur J Immunol 28:1332–1338

Samdani AF, Dawson TM, Dawson VL (1997) Nitric oxide synthase in models of focal ischemia. Stroke 28:1283–1288

Schulz JB, Matthews RT, Muqit MM, Browne SE, Beal MF (1995) Inhibition of neuronal nitric oxide synthase by 7-nitroindazole protects against MPTP-induced neurotoxicity in mice. J Neurochem 64:936–939

Sennequier N, Stuehr DJ (1998) Antigungal imidazoles block assembly of inducible NO synthase into an active dimer. Nitric Oxide 2, abstract B12, p 92

Shearer BG, Lee S, Oplinger JA, Frick LW, Garvey EP, Furfine ES (1997) Substituted N-phenylisothioureas: potent inhibitors of human nitric oxide synthase with neuronal isoform selectivity. J Med Chem 40:1901–1905

Shesely EG, Maeda N, Kim HS, Desai KM, Krege JH, Laubach VE, Sherman PA, Sessa WC, Smithies O (1996) Elevated blood pressures in mice lacking endothelial nitric oxide synthase. Proc Natl Acad Sci USA 93:13176–13181

Southan GJ, Szabo C (1996) Selective pharmacological inhibition of distinct nitric oxide synthase isoforms, Biochem Pharmacol 51:383–394

Southan GJ, Gross SS, Vane JR (1995) Esters and amides of N^G-nitro-L-arginine act as prodrugs in their inhibition of inducible nitric oxide synthase. Enzymology and Biochemistry, 4–7

Stratman NC, Fiei GJ, Sethy VH (1996) U-19451A: a selective inducible nitric oxide synthase inhibitor, Life Sciences 59:945–951

Stuehr DJ (1997) Structure–function aspects in the nitric oxide synthases. Ann. Rev Pharmacol. Toxicol 37:339–359

Sup SJ, Green BG, Grant SK (1994) 2-Iminobiotin is an inhibitor of nitric oxide synthases. Biochem Biophys Res Commun 204:962–968

Thomsen LL, Scott JM, Topley P, Knowles RG, Keerie AJ, Frend AJ (1997) Selective inhibition of inducible nitric oxide synthase inhibits tumor growth in vivo: studies with 1400W, a novel inhibitor. Cancer Res 57:3300–3304

Tjocng FS, Fok KF, Webber RK (1995) Amidino derivatives useful as nitric oxide synthase inhibitors, WO patent 95/11014

Tsai AL, Berka V, Chen PF, Palmer G (1996) Characterization of endothelial nitric-oxide synthase and its reaction with ligand by electron paramagnetic resonance spectroscopy. J Biol Chem 1271:32563–32571

Webber RK, Metz S, Moore WM, Connor JR, Currie MG, Fok KF, Hagen TJ, Hansen DW Jr, Jerome GM, Manning PT, Pitzele BS, Toth MV, Trivedi M, Zupec ME, Tjoeng FS (1998) Substituted 2-iminopiperidines as inhibitors of human nitric oxide synthase isoforms. J Med Chem 41(1):96–101

Weinberg JB, Granger DL, Pisetsky DS, Seldin MF, Misukonis MA, Mason SN, Pippen AM, Ruiz P, Wood ER, Gilkeson GS (1994) The role of nitric oxide in the pathogenesis of spontaneous murine autoimmune disease: increased nitric oxide production and nitric oxide synthase expression in MRL-lpr/lpr mice, and reduction of spontaneous glomerulonephritis and arthritis by orally administered N^G-monomethyl-L-arginine. J Exp Med 179:651–660

Wolff DJ, Datto GA, Samatovicz RA, Tempsick RA (1993a) Calmodulin-dependent nitric-oxide synthase. Mechanism of inhibition by imidazole and phenylimidazoles. J Biol Chem 268:9425–9429

Wolff DJ, Datto GA, Samatovicz RA (1993b) The dual mode of inhibition of calmodulin-dependent nitric-oxide synthase by antifungal imidazole agents. J Biol Chem 1993 268(13):9430–9436

Wray GM, Millar CG, Hinds CJ, Thiemermann C (1998). Selective inhibition of the activity of inducible nitric oxide synthase prevents the circulatory failure, but not the organ injury/dysfunction, caused by endotoxin. Shock 9:329–335

Wright CE, Rees DD, Moncada S (1992) Protective and pathological roles of nitric oxide in endotoxin shock. Cardiovasc Res 26:48–57

Wu CC, Chen SJ, Szabo C, Thiemermann C, Vane JR (1995) Aminoguanidine attenuates the delayed circulatory failure and improves survival in rodent models of endotoxic shock. Br J Pharmacol 114:1666–1672

Yoshida T, Limmroth V, Irikura K, Moskowitz MA (1994) The NOS inhibitor, 7-nitroindazole, decreases focal infarct volume but not the response to topical acetylcholine in pial vessels. J Cereb Blood Flow Metab 14:924–929

Zagvazdin Y, Sancesario G, Wang YX, Share L, Fitzgerald ME, Reiner (1996) A Evidence from its cardiovascular effects that 7-nitroindazole may inhibit endothelial nitric oxide synthase in vivo. Eur J Pharmacol 303:61–69

Zhang HQ, Fast W, Marletta MA, Martasek P, Silverman RB (1997) Potent and selective inhibition of neuronal nitric oxide synthase by N^{ω}-propyl-L-arginine. J Med Chem 40:3869–3870

Zhang ZG, Reif D, Macdonald J, Tang WX, Kamp DK, Gentile RJ, Shakespeare WC, Murray RJ, Chopp M (1996) ARL-17477, a potent and selective neuronal NOS inhibitor decreases infarct volume after transient middle cerebral artery occlusion in rats. J Cereb Blood Flow Metab 16:599–604

Zhao W, Tilton RG, Corbett JA, McDaniel ML, Misko TP, Williamson JR, Cross AH, Hickey WF (1996) Experimental allergic encephalomyelitis in the rat is inhibited by aminoguanidine, an inhibitor of nitric oxide synthase. J Neuroimmunol 64: 123–133

CHAPTER 6
Nitric-Oxide-Synthase Inhibitors II – Pterin Antagonists/Anti-Pterins

E.R. WERNER and H.H.H.W. SCHMIDT

A. Introduction

In addition to heme, flavin adenine dinucleotide (FAD) and flavin mononucleotide (FMN), nitric oxide synthases (NOS) require tetrahydrobiopterin (H_4B). This was first demonstrated for homogenates of stimulated murine macrophages (KWON et al. 1989; TAYEH and MARLETTA 1989). Unlike other enzymes, e.g. aromatic-amino-acid hydroxylases, that also require H_4B, purified neuronal NOS (nNOS; MAYER et al. 1991; SCHMIDT et al. 1991) and endothelial NOS (eNOS; POLLOCK et al. 1993) co-purify with substoichiometric amounts of this cofactor bound to the enzyme. The main features of the H_4B-binding pocket, i.e. affinity and stereoselectivity, seem to be NOS-specific and not found in any other mammalian enzyme. Thus, the NOS H_4B-binding site provides a nearly ideal pharmacological target for drug development. First in vivo studies with the 4-amino analogue of H_4B (4-amino-H_4B; BAHRAMI et al. 1997) are, in fact, very encouraging. Thus, pterin antagonists of NOS may have the potential to yield new drugs with exciting applications, e.g. in ischemia-reperfusion injury and inflammation. This chapter will first describe the current knowledge on the function of the H_4B cofactor in the NOS reaction and then detail the present knowledge on biochemical properties of pterin antagonists, also termed anti-pterins, including first results on pharmacological actions of pterin antagonists in animals.

B. H_4B Dependence of the NOS Reaction

The role of H_4B in the mechanism of the NOS reaction is still a matter of debate (MAYER and WERNER 1995; BÖMMEL et al. 1998; CRANE et al. 1998; RAMAN et al. 1998; WERNER et al. 1998). Pterin antagonists of NOS were developed primarily to dissect various possible contributions of the pterin to the reaction mechanism. This work focussed primarily on the question whether or not H_4B contributes to the reaction solely by its binding in an allosteric way or by taking advantage of its remarkable redox properties to contribute stoichiometrically to electron transfer.

General aspects of the H_4B dependence of the NOS reaction have been recently reviewed (WERNER et al. 1998). We will, therefore, only briefly summarise the knowledge here. The reader is also referred to Chap. 3 for a discussion of the NOS reaction mechanism.

I. NOS-Associated H_4B

There is now consensus that all three isoforms of NOS require exogenous H_4B. In addition, endogenously bound H_4B has been detected in preparations of all three isoforms of NOS (MAYER et al. 1991; HEVEL and MARLETTA 1992; POLLOCK et al. 1993); the amount was frequently found to correlate with the basal activity observed in the absence of added H_4B (HEVEL and MARLETTA 1992). GIOVANELLI et al. 1991 described purified NOS to be initially H_4B-independent but did not consider NOS-bound H_4B. However, this picture is less clear for the recombinantly expressed forms of NOS (BÖMMEL et al. 1998). Recombinant NOS expressed in *Escherichia coli*, i.e. a system free of H_4B, shows no or nearly no activity but can be reconstituted by H_4B. Presumably due to the striking instability of NOS in the absence of H_4B, only about half of the activity that is observed when NOS is expressed in the presence of H_4B can be regained. While H_4B is by far the most potent and effective pterin cofactor for NOS, other tetrahydropterins can stimulate NOS when added in high enough concentrations (HEVEL and MARLETTA 1992; PRESTA et al. 1998).

II. Allosteric and Stabilising Effects

In addition to its effect on enzyme activity, H_4B exerts allosteric effects on NOS. Purified NOS is a homodimer (SCHMIDT et al. 1991), and H_4B seems to support the intracellular formation of the active dimeric conformation (BAEK et al. 1993; KLATT et al. 1995). In doing that, H_4B also shifts the heme low-vs-high-spin equilibrium towards the high-spin state (MCMILLAN and MASTERS 1995; RODRIGUEZ-CRESPO et al. 1996; WANG et al. 1995; GORREN et al. 1996), increases the affinity of NOS for arginine-based NOS inhibitors (KLATT et al. 1994) and lowers the apparent K_m for arginine during catalysis (BRAND et al. 1995; FRÖHLICH et al. 1998; KOTSONIS et al. submitted for publication).

Moreover, H_4B seems to stabilise the pure enzyme against loss of activity. GIOVANELLI et al. (1991) proposed that, to stay active, NOS-I requires exogenous H_4B only after some rounds of turnover. H_4B seems to stabilise the dimeric and active conformation during catalysis, possibly by reacting with various autodamaging products of coupled and uncoupled NOS catalysis (REIF et al. 1998; KOTSONIS et al. 1999).

III. Possible Electron-Transfer Role

Possible electronic contributions of H_4B to the NOS reaction mechanism are less clear. In the phenylalanine hydroxylase reaction, H_4B is converted to a 4α-hydroxy derivative, which is then recycled to H_4B by the action of two enzymes: carbinolamine dehydratase accelerates the dehydratation of the 4α-

hydroxy intermediate to 6,7[8H]-dihydrobiopterin (also called quinonoid dihydrobiopterin). This is then reduced by dihydropteridine reductase (at the expense of reduced nicotinamide adenine dinucleotide) to H₄B (KAUFMAN 1993). Compared with phenylalanine hydroxylase, stimulation of NOS by H₄B was found to be rather different in many aspects. A much lower concentration of H₄B was required for maximal stimulation of NOS. NOS was found to be much more selective for the side chain of H₄B, other tetrahydropterins acting with orders-of-magnitude-lower potency. Finally, no indications for the occurrence of a redox cycle in NOS (compared with the one found in phenylalanine hydroxylase) could be obtained experimentally (BÖMMEL et al. 1998; WERNER et al. 1998). However, H₄B accelerates the decay of the ferrous oxygen complex of the nNOS oxygenase domain by a factor of 70, while L-arginine

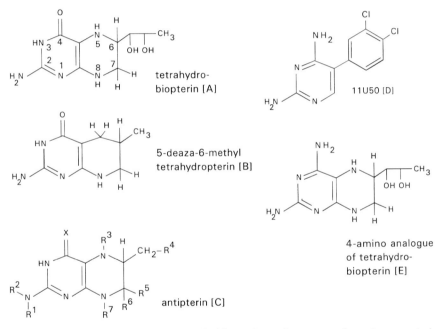

Fig. 1. Structural formulae of tetrahydrobiopterin and compounds acting as pterin anatagonists of nitric-oxide synthase. Tetrahydrobiopterin is a chiral compound; the configuration of the side chain in position 6 is L-erythro-1,2-dihydroxylpropyl. The formulae show the stereochemistry according to the Fischer's projections. Since C6 of the pterin ring is asymmetric and the side chain at C6 is chiral in most cases, most synthetic tetrahydropterins are mixtures of 6R and 6S diasteromers unless separated after the last step in the chemical synthesis, which usually is a catalytic hydrogenation converting the aromatic pterin to its tetrahydroform. The active tetrahydrobiopterin cofactor has the naturally occurring 6R configuration; the 6S-isomer is almost inactive. Notation of residues in the antipterins is as follows: X = O or NH; R1 = H, methyl, C1–5 alkanoyl; nicotinyl or (1-methyl-2-pyridino)-carbonyl; R2, R5, R6, and R7 = H or methyl; R3 = methyl, ethyl, C1–4 alkylthio or various other residues; R4 = H, C1–4 alkylthio, NH₂, OH, or various other residues

stabilised this complex (ABU-SOUD et al. 1997). Spectra of NOS reaction intermediates at low temperature suggest reductive activation of the ferrous oxygen complex of NOS by H_4B, for which a novel one-electron H_4B/H_3B redox cycle was suggested (BEC et al. 1998). Recently, a one-electron redox cycle between H_4B^+ and H_4B was proposed, based on the surprising observation that the pterin-binding site can also be occupied by the positively charged subtrate L-arginine (RAMAN et al. 1998). Most likely, the pterin antagonists/anti-pterins described in the present study will eventually also be instrumental in testing these two latter hypotheses.

IV. The Pterin-Binding Site

Binding studies using radiolabelled H_4B revealed both a complex behaviour of the association of H_4B to pterin-free nNOS and a monophasic dissociation curve (GORREN et al. 1996; ALDERTON et al. 1998). Binding of H_4B and L-arginine to NOS shows positive cooperativity (KLATT et al. 1994; GORREN et al. 1996; ALDERTON et al. 1998). Interpretation of the binding data yielded a model for two identical, highly anti-cooperative binding sites for H_4B and L-arginine (GORREN et al. 1996; ALDERTON et al. 1998). Thus, the first H_4B and L-arginine molecules bind to the dimer with extremely high affinity, creating a second binding site per dimer with moderate affinity ($K_d = 20-40\,\text{nM}$ for H_4B). In the recently elucidated crystal structure of the NOS-III/eNOS dimeric oxygenase domain, the pterin-free and H_4B-occupied pterin binding sites are identical (RAMAN et al. 1998).

C. Pterin-Based Inhibition of NOS

Based on the requirement of NOS for H_4B, two general approaches towards NOS activity modulation seem feasible, i.e. manipulation of intracellular H_4B levels and pterin-binding-site antagonists. This review will only briefly touch the former possibility and focus primarily on the development of pterin-antagonists or anti-pterins.

I. Manipulating Intracellular H_4B Levels

Manipulation of intracellular H_4B levels by drugs alters the amount of NO formed in cells (WERNER-FELMAYER et al. 1990, 1993; GROSS and LEVI 1992; SCHMIDT et al. 1992) and animals (KLEMM et al. 1995; BUNE et al. 1996). The first approach using a pterin derivative to study the requirement for the pterin cofactor in the NOS reaction in an intact cell was done by KWON et al. (1989). These authors found that, in macrophage homogenates, the anti-folate methotrexate could inhibit the stimulation of NOS by 7,8-dihydrobiopterin but not by 5,6,7,8-H_4B. The interpretation of these results is straightforward. Methotrexate has no effect on the NOS reaction itself but, by inhibiting dihydrofolate reductase, it blocks the conversion of 7,8-dihydrobiopterin to the active cofactor, 5,6,7,8-H$_4$B. However, inhibition of H$_4$B levels as a tool to inhibit NO formation in vivo suffers from the problem that the other H$_4$B-dependent enzymes (like phenylalanine-, tyrosine- and tryptophan hydroxylases, which are required for sympathetic neurotransmitter biosynthesis) have a much lower affinity for H$_4$B than NOS does. Thus, inhibition of H$_4$B biosynthesis will affect these pathways first and is expected to cause serious side effects before even acting on NOS.

Another line of research related to H$_4$B in the NOS reaction has focussed on restoring apparent cellular deficiencies in H$_4$B, such as those in arterial hypertension or atherosclerosis. Indeed, H$_4$B administration can, for example, restore endothelial function in patients suffering from hypercholesterolemia, with little effect on healthy control patients (STROES et al. 1997). This subject is beyond the scope of this article, and it remains to be seen in what way the observed effects of H$_4$B in this setting are related to the NOS reaction (WERNER et al. 1998). Interestingly, Harrisson and co-workers have recently postulated a mechanism by which increases in vascular superoxide levels lead to increases in H$_4$B breakdown (SOMERS et al. 1998).

II. Approaches to Pterin Antagonists

When examining the effects of 6-methyl analogues of tetrahydropterin, HEVEL and MARLETTA (1992) observed that the respective 5-deaza-analogue of 6-methyl-H$_4$ pterin (Fig. 1B) could not stimulate the reaction and even blocked the small stimulatory effect of 6-methyl tetrahydropterin. However, the effects of these these two 6-methyl derivatives are weak, possibly due to the lack of the 6 L erythro 1,2 dihydroxypropyl side chain, which is present in the natural H$_4$B cofactor (Fig. 1A) and seems to be required for tight binding (Werner et al. 1996). By combining the structural element required for tight binding, i.e. the 6-L-erythro-1,2-dihydroxypropyl side chain, with a structural element required for inhibition of dihydropteridine and folate reductase, respectively, i.e. the 4-amino substitution, WERNER et al. (1996) designed the 4-amino analogue of H$_4$B (Fig. 1E), yielding a potent, specific and well-characterised NOS inhibitor (K_i for binding = 13 nM) that is also effective in vivo. SCHENK et al. (1996 and 1997) identified several pteridine derivatives as effective inhibitors of NOS. These compounds were termed anti-pterins, in analogy to anti-folates, such as methotrexate (Fig. 1C). These compounds show very little effect on phenylalanine hydroxylase, an aromatic-amino-acid hydroxylase, and were subsequently further optimised for NOS inhibition (BÖMMEL et al. 1998; FRÖHLICH et al. 1999). While methotrexate does not affect NOS (see above), Russel found that 11U50 (Fig. 1D), another dihydrofolate-reductase inhibitor (DUCH et al. 1978), is also a pterin-site inhibitor of NOS (H$_4$B-binding K_i = 120 nM; ALDERTON et al.1998). In the following sections, the properties of 4-amino-H$_4$B (Werner at al. 1996), a series of other 4-amino pterins (FRÖHLICH et al., 1999) and novel 4-oxo pterins

(BÖMMEL et al. 1998; KOTSONIS et al., unpublished) will be discussed in more detail.

III. 4-Amino-H$_4$B

1. Effects of 4-Amino-H$_4$B on Purified Enzymes

When tested on purified NOS, the 4-amino analogue of H$_4$B has a similar effect on all three isoforms of NOS. In the presence of 10 μM H$_4$B, IC$_{50}$ values for 4-amino-H$_4$B are approximately 1 μM for rat nNOS (NOS-I; WERNER et al. 1996; PFEIFFER et al. 1997), 7 μM for the murine inducible isoform (iNOS, NOS-II; MAYER et al. 1997) and 14 μM for the bovine endothelial enzyme (NOS-III; SCHMIDT et al. 1999). The 4-amino analogue of H$_4$B does not inhibit the basal activity of NOS expressed and purified in the presence of H$_4$B but can potently inhibit the stimulation of these preparations by H$_4$B. Basal activity is thought to be due to substoichiometric of amounts enzyme-associated H$_4$B. However, when tested on H$_4$B-free enzyme preparations in presence of 10 μM exogenous H$_4$B, complete inhibition, with IC$_{50}$ values in the micromolar range, is observed (Fig. 2). These data indicate that the inhibitor can compete with the low-affinity binding site for H$_4$B but cannot displace H$_4$B that is already bound to the high-affinity site. The complete inhibition of the H$_4$B-free enzyme, however, was reversible when diluting the enzyme in the presence of H$_4$B (MAYER et al. 1997; PFEIFFER et al. 1997). The 4-amino analogue of H$_4$B inhibited both steps of the NOS reaction: the formation of N'-hydroxy-L-arginine from L-arginine and the formation of NO and citrulline from N'-hydroxy-L-arginine (PFEIFFER et al. 1997). When compared with methotrexate, 4-amino-H$_4$B requires 60-fold higher concentrations to inhibit the target enzyme of methotrexate dihydrofolate reductase and has a similar potency with respect to dihydropteridine reductase. Strikingly, when compared with NOS inhibition, the concentrations of 4-amino-H$_4$B required to inhibit phenylalanine hydroxylase, a H$_4$B-dependent aromatic-amino-acid hydroxylase, are three orders of magnitude higher (Table 1).

Table 1. Inhibition of pure enzymes by the 4-amino analogue of tetrahydrobiopterin (4-Amino-H$_4$B) and by methotrexate (Schmidt et al. 1999)

Enzyme	IC$_{50}$ (μM) 4-Amino-H$_4$B	Methotrexate	Ratio
Rat neuronal nitric-oxide synthase	1.1	>300	>273
Chicken liver dihydrofolate reductase	0.63	0.01	0.016
Sheep liver dihydropteridine reductase	20	100	5
Phenylalanine hydroxylase	970		Not determined

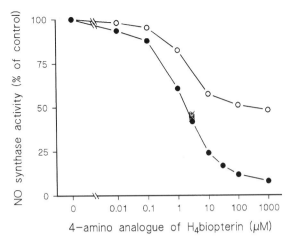

Fig. 2. Inhibitory potential of the 4-amino analogue of tetrahydrobiopterin for nitric-oxide (NO) synthases with or without endogenously bound tetrahydrobiopterin. Recombinant rat neuronal NO synthase was purified from an insect cell-expression system with or without inhibition of tetrahydrobiopterin biosynthesis, as described by LIST et al. (1996). The tetrahydrobiopterin-free enzyme (*full circles*) and the tetrahydrobiopterin-containing enzyme (*open circles*) were incubated in presence of $10\,\mu$M tetrahydrobiopterin and the indicated concentrations of the 4-amino analogue of tetrahydrobiopterin. Activity without addition of inhibitor was set to 100%. Presumably due to the lower stability of the enzyme in absence of tetrahydrobiopterin, the specific activity of the pure, tetrahydrobiopterin-free NO synthase reached only about 50% of the specific activity of the enzyme isolated in presence of tetrahydrobiopterin. NO synthase activity was assayed by the formation of ^{3}H-citrulline from ^{3}H-L-arginine

Surprisingly, 4-amino-H$_4$B has allosteric effects on both nNOS (PFEIFFER et al. 1997) and iNOS (MAYER et al. 1997; NOS-I and NOS-II, respectively) that are similar to the allosteric effects of the H$_4$B cofactor. These data indicate that the allosteric actions of H$_4$B alone are not sufficient to explain its role in the NOS reaction. What could be the mechanism of inhibition of NOS by the 4-amino analogue of H$_4$B? To address this question, a comparison with the mechanism of inhibition of dihydrofolate reductase by methotrexate may be helpful. Like the 4-amino analogue of H$_4$B, methotrexate derives from the natural substrate folic acid by a substitution of the 4-oxo function with an amino group. In addition to that, it also carries an additional methyl group at N10 of folate, but this substitution is not essential for the inhibitory action. Crystal structures of dihydrofolate reductase with methotrexate bound revealed that, due to the 4-amino substitution, the orientation of the molecule in the crystal structure is changed so that the hydride can no longer be transferred to N5 (BOLIN et al. 1982). To date, no crystal structure of a full-length NOS has been published. However, a crystal structure of a part of the iNOS oxygenase domain with substrate and pterin has been resolved at 2.6-Å resolution. In this crystal, the 4-oxo group of H_4B interacts with the guanidino group of arginine 375 (CRANE et al. 1998), which corresponds to arginine 367 in the endothelial oxygenase domain (RAMAN et al. 1998). The 4-amino substitution dramatically alters the charge distribution in the H_4B molecule (Reibnegger, personal communication). In contrast to the 4-oxo function of H_4B, the 4-amino group in the analogue may, therefore, be repelled rather than attracted by the guanidino group of arginine 375. Thus, it is likely that the 4-amino analogue binds to the active centre with distorted geometry. This may prohibit a possible electron-transfer reaction that may physiologically involve H_4B, e.g. from the reductase domain to the heme. However, it has been questioned whether H_4B is in close enough proximity to the substrate to allow for direct electronic interactions (CRANE et al. 1998).

2. Effects of 4-Amino-H_4B on Cultured Cells

Examination of the action of the 4-amino analogue of H_4B on cultured cells revealed that it was as active as N^G-methyl-L-arginine in inhibiting NO formation by cytokine-stimulated murine fibroblasts ($IC_{50} = 15\,\mu M$; SCHMIDT et al. 1999). Treatment of cells with sepiapterin, which is enzymatically converted to H_4B, attenuates the action of the pterin antagonist but not of the arginine analogue. In porcine endothelial cells, in contrast, the compound shows almost no NOS-inhibitory activity ($IC_{50} = 420\,\mu M$; SCHMIDT et al. 1999). However, the 4-amino analogue of H_4B enters both cell types to reach similar intracellular concentrations (170 pmol/mg protein in fibroblasts; 224 pmol/mg protein in endothelial cells) and does not affect the expression of iNOS messenger RNA in murine fibroblasts. However, the IC_{50} for purified iNOS and eNOS (NOS-II and NOS-III) are in a similar range. How can the preferential inhibition of the inducible isoform of NOS by the 4-amino analogue of H_4B in intact cells be explained? One possible explanation may be the greatly different turnover rates of the two NOS proteins. Whereas the half-life of the iNOS is 2–4h (WALKER et al. 1997), the endothelial isoform is turned over at a much lower rate ($\sim$20h; LIU et al. 1995). Since the 4-amino analogue of H_4B inhibits only the H_4B-free state of NOS (Fig. 2), eNOS cannot be inhibited, because a major part of it is already expressed, with H_4B bound to the high-affinity binding site. In case of a cell expressing the inducible isoform, however, there is a continuous de novo synthesis and degradation of the NOS protein. The 4-amino analogue of H_4B thus continuously comes in contact with the newly synthesised, presumably H_4B-free enzyme and can, therefore, efficiently inhibit its activity (SCHMIDT et al. 1999).

3. Effects of 4-Amino-H_4B in Animals

Several studies of the effects of the 4-amino analogue of H_4B in animals are currently under way; however, none of these have been completed yet. Results of a pilot study in a rat model of septic shock showed that a single intravenous injection of 10 mg/kg of the 4-amino analogue of H_4B can protect the animals from the lethal effects of endotoxin (BAHRAMI et al. 1997). The compound is well tolerated by the animals. The tolerated maximal daily dose is currently estimated to be approximately 0.5 g/kg, which is 50 times the therapeutic dose applied by BAHRAMI et al. (1997). The plasma half-life of the 4-amino analogue of H_4B in mice is about 1 h, and almost all of it is excreted renally within the first 4 h following intramuscular injection (BRANDACHER et al., unpublished observations). When applied by continuous infusion, the compound is effective as an inhibitor of iNOS in a rat model of septic shock in vivo (BAHRAMI et al. 1997).

IV. Further 4-Aminopteridines

In order to establish a structure–activity relationship (SAR) of the NOS pterin/anti-pterin binding site, the pterin substitution pattern and functionalities within a 4-amino substituted-pteridine derivative as a scaffold were systematically varied at positions 2 (Table 3), 4, 5, 6 and 7 (Tables 3, 4) and tested on H_4B-free and native nNOS (NOS-I; FRÖHLICH et al. 1999). This approach yielded several new and potent derivatives that inhibited NOS-I activity, provided further insight into the NOS pterin-binding pocket and suggested a pharmacologically relevant adjacent exosite.

1. The 4-Amino Function

Increasing the basicity of the 4-amino function of the aromatic 4-amino pterin derivative resulted in potent and complete inhibitors of NOS-I activity, with IC_{50} values in the low micromolar range. For example, further dialkylation of the 4-amino function with a dimethyl group (2a, Table 2), a diethyl group (2b), a morpholine- (2c) or piperidine-ring system (2d) or $(CH_2C_6H_5)_2$ (2e) yielded effective and potent (2d, 2e) NOS inhibitors. This inhibition is consistent with an optimised hydrophilic interaction with hydrogen-providing groups of the NOS pterin-binding site, and agrees with the recently published crystal structure of the dimeric murine NOS-II (iNOS) and NOS-III (eNOS) oxygenase domain (CRANE et al. 1998; RAMAN et al. 1998). Also in this class falls the type-II anti-pterin PHS-330 (Fig. 3), which inhibits human, porcine and rat NOS.

2. The 2, 5 and 7 Positions

In contrast, varying the substitution pattern in the 2, 5 and 7 positions has no significant effect on the inhibition of NOS-I activity. Varying the substitution pattern at the 2 position leads to no significant change, suggesting that this functionality is not important in the activation of NOS-I enzyme. Similar conclusions can be drawn for the 7 position, since chemical modifications of this position alone also showed no effect on the inhibitor profile of these anti-pterins. CRANE et al. (1998) and RAMAN et al. (1998) observed a hydrophilic interaction in the catalytic centre between the 4-oxo group of the natural H_4B, the propionic-acid side chain of the heme and the free amino group of the guanidino moiety of Arg^{375}. Given the high degree of sequence homology, one

Table 2. Effects of 4-amino modifications of aromatic pteridines on nitric-oxide synthase (NOS)-I activity. Inhibition of total NOS activity at an inhibitor concentration of 100 µM in the presence of 2 µM tetrahydrobiopterin (H_4B). IC_{50} values were only determined for those compounds that, at 100 µM, inhibited total NOS activity by at least 50% of the control activity. Adapted from FRÖHLICH et al. (1999)

Number	R_2	R_4	R_6	NOS-activity	IC_{50} (µM)
2a	H	$(CH_3)_2$	C_6H_5	92 ± 11	
2b	H	$(C_2H_5)_2$	C_6H_5	75 ± 3	
2c	H	$(C_2H_4)_2O$	C_6H_5	41 ± 8	82
2d	H	$(CH_2)_5$	C_6H_5	0 ± 0.05	62
2e	H	$(CH_2 C_6H_5)_2$	C_6H_5	0 ± 0.05	3

Table 3. Effects of 4-amino 5,6,7,8-tetrahydropterin derivatives on nitric-oxide synthase (NOS)-I activity. Inhibition of total NOS activity at an inhibitor concentration of 100 µM in the presence of 2 µM tetrahydrobiopterin (H_4B). IC_{50} values were only determined for those compounds that, at 100 µM, inhibited total NOS activity by at least 50% of the control activity. Adapted from FRÖHLICH et al. (1999)

Number	R_2	R_5	R_6	R_7	NOS-activity	IC_{50} (µM)
	H	H	H	H	102 ± 8	
3a	CH_3	H	H	H	108 ± 23	
3b	CH_3	H	C_6H_5	C_6H_5	86 ± 23	
3c	CH_3	H	CH_3	CH_3	57 ± 5	
3d	H	H	C_6H_5	H	17 ± 1	6
3e	H	H	$CH_2OCH_2CH_3$	H	3 ± 1	30
3f	H	H	$CHOHCHOHCH_3$	H	38 ± 2	6
3g	H	H	H	C_6H_5	39 ± 1	24

Table 4. Effects of 6- and 7-substituted heteroaromatic 4-amino pterin derivatives on nitric-oxide synthase (NOS)-I activity. Inhibition of total NOS activity at an inhibitor concentration of 100 μM in the presence of 2 μM H$_4$B. IC$_{50}$ values were only determined for those compounds that, at 100 μM, inhibited total NOS activity by at least 50% of the control activity. Adapted from FRÖHLICH et al. (1999)

Number	R$_6$	R$_7$	NOS-activity	IC$_{50}$ (μM)
4a	C$_6$H$_5$	H	85 ± 24	
4b	H	CH$_2$C$_6$H$_5$	89 ± 6	
4c	C$_6$H$_5$	C$_6$H$_5$	62 ± 4	
4d	CH$_2$OCO(CH$_2$)$_2$C$_6$H$_4$COC$_6$H$_4$	H	29 ± 6	48
4e	CH$_2$O(CH$_2$)$_9$CH$_3$	H	16 ± 1	30

would expect that this interaction would also be applicable to native NOS-I and our tested anti-pterins. By increasing the basicity of the 4 position (and, therefore, the ability to form more effective bonding interactions with the hydrogen-donating heme and the structural arginine), the inhibitor potency may increase, unless a different charge distribution and wrong insertion would prevent this (see above).

3. The C6 Side Chain and Pterin Exosite

Furthermore, CRANE et al. (1998) demonstrated a hydrophilic interaction between the 9,10-dihydroxylated propyl side chain of the natural H$_4$B and the NOS amino acid residues Phe[470] and Ser[112] of the enzyme. In contrast, this chemical interaction appears to be functionally insignificant for the 6 position of aromatic pterin derivatives. This difference may be due to the planar nature of the pyrazin ring of the aromatic pterin derivatives, which probably fails to effectively insert in the NOS pterin-binding pocket, although an increased π-electron stacking with Trp[457] should be favoured when compared with conformations of the 6-substituted 4-amino- and 5,6,7,8-tetrahydro-pterin derivatives (Table 4). Conversely, altering the side chain to a 6-hydroxymethyl (PFEIFFER et al., unpublished) or ethoxymethyl (3e) compound leads to a more or less dramatic loss in the inhibitory potential of 4-amino-H$_4$B (3f). Thus, like the 4-oxo parent compounds (H$_4$B, tetrahydrodictyopterin, tetrahydro-6-methylpterin (WERNER et al. 1996), some tetrahydrated inhibitors may be bind more tightly and may, therefore, be more active when containing the L-erythro-

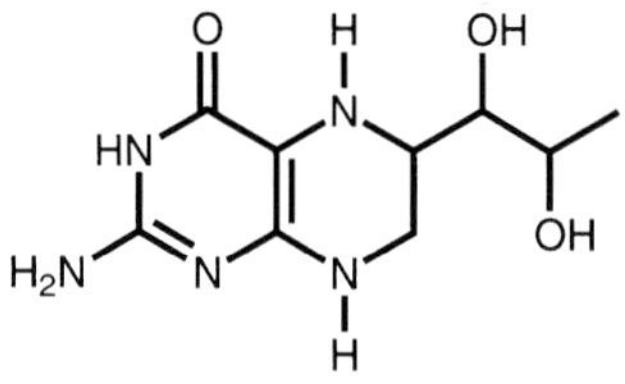

Fig. 3. Structural formulae of tetrahydrobiopterin (*H₄Bip*) and PHS compounds with prototypic type-I (*PHS-32*) and type-II (*PHS-72* and *PHS-330*) anti-pterin profiles. For details, refer to the text

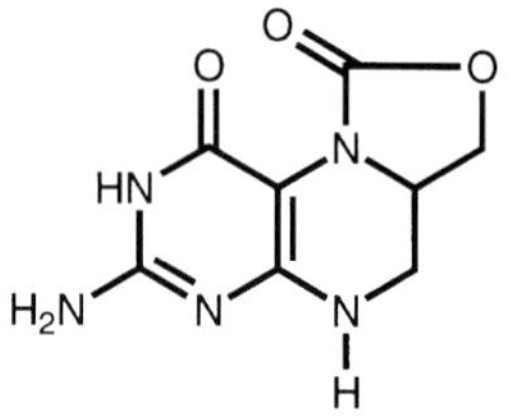

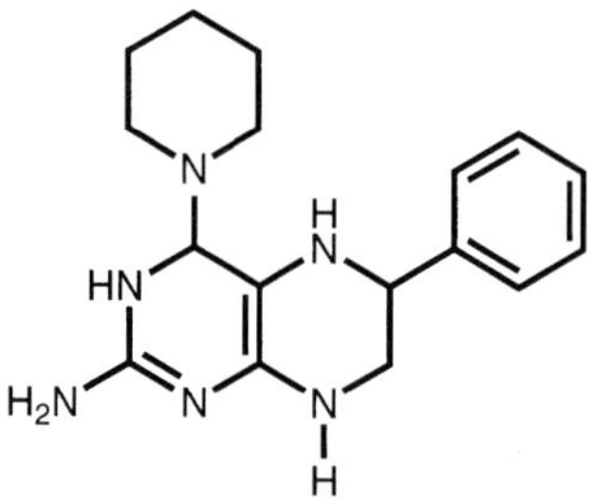

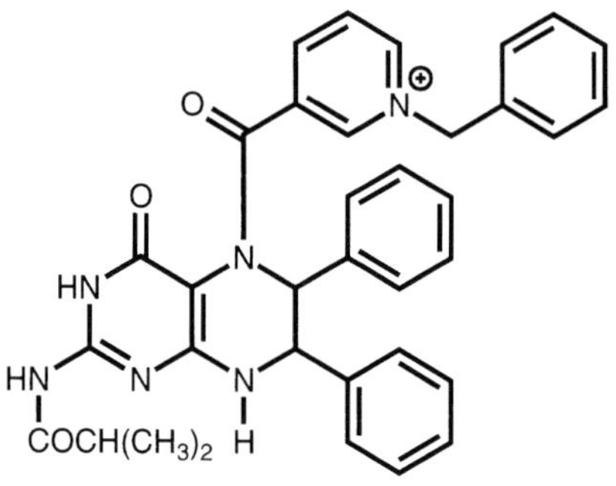

1,2-dihydroxypropyl side chain at position 6 of the pterin ring. In contrast, more bulky substituents in that 6-position, like phenyl- or mesityl groups, markedly affected the inhibitory potency of these compounds in both directions. Inhibition by 6-substituted aromatic compounds may be due to additional hydrophobic interactions with the aromatic moiety of Phe[470]. Interestingly, substitution of the 6 position with either a phenyl or mesityl group of the aromatic 4-amino pterin derivative alone has no effect but, when combined with functionalities that increase the basicity of the 4 position, a mesityl group in the 6 position results in potent inhibitors of NOS-I (FRÖHLICH et al., 1999).

Methyl substitution at the 2-amino group of the 4-amino-H₄B derivative (3a, Table 3) was without effect on NOS-I activity. In contrast, additional 6,7-diphenyl (3b) or 6,7-dimethyl substitution (3c) began to convey NOS-inhibitory activity. The effectiveness of 4-amino-H₄B is enhanced by substituting the dihydroxypropyl side chain in 6 position with either a single phenyl group (3d) or an ether group (3e). This results in some of the most effective and potent NOS inhibitors based on a reduced 4-amino-5,6,7,8-

tetrahydropteridine scaffold. Most noteworthy, mono-phenyl substitution at the 7 position (3g) results in weaker NOS inhibitors. In contrast to the reduced pterin backbone, 6-mono (4a), 7-mono (4b) and 6,7-disubstitution (4c) of 4-aminopteridines in the heteroaromatic series (Table 4) yielded only weak NOS-I inhibitors. This prompted us to change further the chemical functionalities of the substituents at the 6 position. Introduction of an ester function containing a long side chain (4d) or a long-chain aliphatic substituent in the 6 position, such as the decylmethylether (4e), finally led to aromatic compounds with significant NOS-I inhibition. Such compounds may later be favourable towards reduced anti-pterins because of lower redox sensitivity. We have no evidence for any relevant hydrophilic interactions between a substituent in the 6 position of the aromatic 4-amino pteridine nucleus and the pterin-binding pocket (FRÖHLICH et al. 1999). Thus, hydrophobic interactions, as conveyed by long aliphatic side chains (4d, 4e), are most likely favourable in order to augment NOS-I inhibition at this position.

4. Conclusion

In conclusion, the inhibitory effect of the aromatic 4-amino substituted anti-pterins appears to be due to an increased hydrophilic interaction of the 4 position or displacement of water within the binding pocket of NOS when compared with that in the naturally occurring H_4B cofactor. In addition, a hydrophobic interaction of bulky aromatic substituents in the 6 position of aromatic 4-amino pterin derivatives with Phe^{470} from NOS also appear to be of relevance. This seems to be part of a pterin exosite that is relevant for anti-pterin binding, but not for H_4B binding. An intimate localisation of the pterin adjacent to the heme- and arginine-binding sites in the crystal structures of the NOS-II and III oxygenase domains is consistent with our observations using the 4-amino pterin derivatives. The resulting SAR suggests that the upper portion of the pterin molecule (positions 4,5,6 of the H_4B cofactor) is responsible for efficient insertion into the pterin-binding pocket of NOS. In addition, the present SAR agrees with a comparative molecular field analysis (CoMFA) of a data set of selected NOS-I inhibitors that resulted in a statistically significant and predictive three-dimensional quantitative-SAR model of the pterin-binding-site interactions (MATTER et al., unpublished). Some of these interactions, in particular in the exosite, may be highly species- and/or NOS-isoform specific. For example, we found that the inhibitory effect of selected compounds on human and NOS I–III activity was slightly lower than in the case with the native porcine NOS I and was even absent, in some cases (FREY et al., unpublished). Thus, following the anti-pterin approach may lead to specific, isozyme-selective inhibitors of NOS.

V. 4-Oxopteridines as Inhibitors of NOS

In addition to 4-amino derivatives of H_4B, 4-oxo pteridines (Table 5, Fig. 3) also inhibit NOS-I activity (SCHENK et al. 1996) regardless of whether the

Table 5. Chemical structure of the six most potent 4-oxo-tetrahydrobiopterin (H_4B)-based anti-pterins, differentiated into type-I and type-II anti-pterins. Pterin-derived nitric-oxide-synthase (NOS) inhibitors were chemically synthesised. With respect to the classification of the anti-pterins into types I and II, refer to the text. Activity of purified porcine NOS-I was determined in the absence (control) or presence of the anti-pterin (100 μM). The results shown represent mean values ± the standard errors of the mean of 3–6 experiments, each performed in triplicate. From Bömmel et al. (1998)

1, H_4Pte; *2*, H_2Pte; *3*, Pte

Anti-pterin (100 μM)	Pterin scaffold and substituents						NOS activity with 2 μM H_4B (% of control)	IC_{50} (μM)
	Type	Scaffold	R_1	R_2	R_3	R_4	X	
H_2B	I	(2)	H	H	CH(OH)CH(OH)CH₃		O	18.0 ± 1.2	25
PHS-32	I	(1)	H	H	CH₂–O–CO (cyclic R_3/R_4)		O	31.4 ± 5.5	35
PHS-52	II	(1)	H	H	H	CSNHPh	O	10.1 ± 3.5	50
PHS-72	II	(1)	COCH(CH₃)₂	Ph	Ph	(N+-benzyl pyridinium-CO)	O	0.1 ± 0.9	16
PHS-176	II	(3)	H	H	(ethyl 3-(4-benzoylphenyl)propanoate)		O	2.3 ± 1.2	50

(1), H_4Pte; *(2)*, 5,6,7,8-tetrahydro, H_2Pte; *(3)*, dihydro or Pte aromatic pterin; R_1, R_2, R_3, R_4, chemical substituents of the pterin-backbone.

oxygenation state of the pterin backbone is tetrahydro (H_4Pte in Table 5), dihydro (H_2Pte) or aromatic (Pte; BÖMMEL et al. 1998). PHS-32 and PHS-72 (Fig. 3) are prototypic inhibitors within this 4-oxo class.

1. Specificity and the Anti-Pterin-Binding Domain

While the 4-oxo anti-pterin PHS-32 lowers the V_{max} value of NOS, the EC_{50} value for L-arginine is not significantly changed, and that for reduced nicotinamide adenine dinucleotide phosphate (NADPH) is even slightly increased. In addition, there is no anti-pterin-induced loss of enzyme-bound flavins (FAD, FMN). These data make a direct interaction of PHS-32 with the L-arginine-, NADPH- and flavin-binding sites of NOS unlikely and suggest pterin-specific interactions. Using a photolabile, tritiated anti-pterin, [^{3}H]PHS-176, anti-pterin binding occurs preferentially in the oxygenase/dimerisation domain of different NOS constructs (SCHENK et al. 1997; BÖMMEL et al. 1998). The *C*-terminal reductase domain of NOS-I is not labelled, while the *N*-terminus of NOS-I lowers anti-pterin binding to the oxygenase/dimerisation domain. This may differentiate the NOS-I pterin interaction from those of the other two NOS isoforms, which have shorter *N*-termini. However, no structural data on NOS-I are available yet.

2. Type-I and -II Anti-Pterins

Enzyme-kinetic experiments using 4-oxo anti-pterins revealed two distinct inhibitor profiles (type-I and -II anti-pterins, Fig. 4). This finding has meanwhile also been extended to the 4-amino compounds and is, thus, a general phenomenon of anti-pterins. Both types of anti-pterins are able to compete with exogenous, radiolabelled [^{3}H]H_4B. Type I anti-pterins antagonise the concentration-dependent stimulation of purified NOS-I by H_4B in a classical, competitive manner. In the presence of 100 μM PHS-32, the concentration–response curve for H_4B is shifted to the right, towards higher concentrations. The inhibitory effect of PHS-32 on NOS is markedly (although not fully) reversed by excess H_4B, due to the fact that H_4B at concentrations above 100 μM has an inhibitory effect on NOS-I. In contrast, inhibition by type-II anti-pterins is not fully competitive with respect to H_4B, though both anti-pterins compete with exogenous [^{3}H]H_4B. This may be related to an interaction between the more complex chemical substituents of type-II anti-pterins at position 3 or 4 and the hydrophobic exosite next to the NOS pterin-binding site (Fig. 5). Importantly, the inhibition of NOS-I by the prototypic anti-pterins PHS-32 and PHS-72 was fully reversible within 1 min when H_4B was added and the inhibitor was diluted to a sub-threshold concentration. Typically, inhibition of NOS-I by type-I anti-pterins is submaximal, as these compounds have no effect on basal enzyme activity, i.e. the activity that is observed without addition of H_4B. Consistent with this, H_4B-stimulated enzyme activity can only be inhibited to the level of basal activity. Surprisingly, the type-I anti-pterin PHS-32 (under different conditions) effectively displaces endogenous pterin

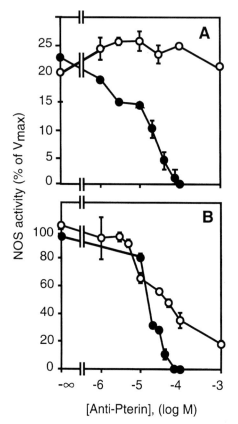

Fig. 4A,B. Enzyme-kinetic analysis of type-I (PHS-32) and type-II (PHS-72) anti-pterins. Purified porcine cerebellum nitric-oxide synthase 1 was assayed either in the absence (**A**) or presence (**B**) of 2 µM tetrahydrobiopterin and increasing concentrations of PHS-32 (*open symbols*) or PHS-72 (*closed symbols*). Data represent means ± standard error of the mean. Adapted from BÖMMEL et al. (1998)

(>80%) although it has no effect on basal enzyme activity and shows no partial agonism. One possible explanation for this finding is that the basal activity of NOS might be H_4B-independent and requires H_4B mainly for enzyme stability (BÖMMEL et al. 1998; REIF et al. 1999; KOTSONIS et al., submitted). A similar proposal had been made earlier by GIOVANELLI et al. (1991) although, at that time, the fact that NOS co-purifies with substoichiometric amounts of H_4B was not known.

In contrast, type-II anti-pterins abolish both basal and H_4B-stimulated NOS activities. The full inhibition of basal activity by the type-II anti-pterins is not due to loss of endogenous H_4B from NOS. Possibly, type-II anti-pterins can bind only to the unoccupied pterin binding-site of the NOS dimer. Nevertheless, inhibition by PHS-72 was fully reversible and partially competitive with respect to H_4B. The exact mechanism by which type-II anti-pterins inhibit NOS activity remains to be elucidated. However, recent co-crystallisation studies confirm their binding site within the pterin-binding site of the NOS oxygenase domain (RAMAN et al., unpublished).

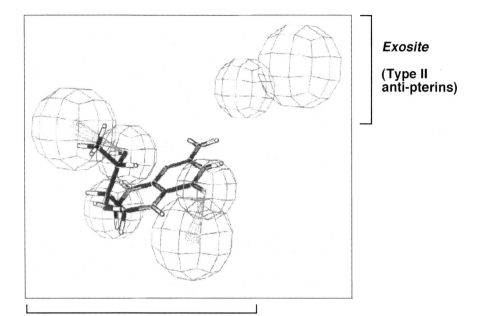

Pterin binding pocket

(H₄Bip and anti-pterins)

Fig. 5. Structure–activity relationship of 4-oxo anti-pterins, as modeled by using the CATALYST software. *Green*, hydrophilic interaction; *blue*, hydrophobic interaction. The structural formula shows tetrahydrobiopterin. The predicted interaction in the far right upper corner depicts the exosite, characteristic of type-II anti-pterins (recently confirmed by co-crystallisation analysis; RAMAN et al., unpublished)

3. 4-Oxo Anti-Pterins in Intact Cells

Anti-pterins of the 4-oxo type not only inhibit purified NOS-I but are also effective inhibitors of NO release from intact cells, as has been noted in the NOS-I expressing cell line N1E-115. Ten minutes after the addition of inhibitors, intracellular NOS was activated by adding the Ca^{2+} ionophore A23187. Ca^{2+}-induced NO formation and release, as determined by nitrite and nitrate accumulation over time, was markedly inhibited by both types of antipterins without affecting cell viability. The efficacy of anti-pterins was similar to that of N'-methyl-L-arginine, a classical, arginine-based NOS inhibitor (BÖMMEL et al. 1998).

4. Conclusions

In conclusion, not only 4-amino but also 4-oxo compounds yield excellent pterin-based NOS inhibitors. The pharmacological activity seems to be independent of the pteridin backbone (reduced or aromatic) and has very char-

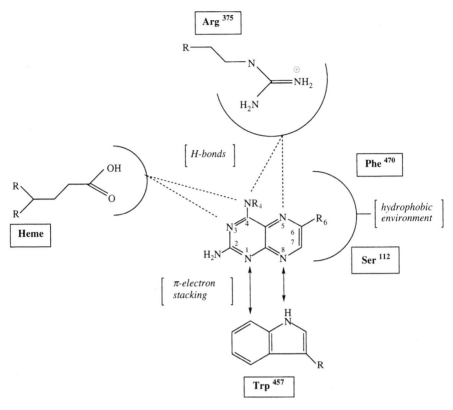

Fig. 6. Possible anti-pterin and nitric-oxide-synthase (NOS)-I interactions within the pterin-binding site predicted from a published crystal structure of the dimeric NOS oxygenase domains (CRANE et al. 1998; RAMAN et al. 1998). Adapted from FRÖHLICH et al. 1999.

acteristic steric requirements (particularly at position 6) to lead to NOS inhibition. A detailed analysis of the SAR of this class of anti-pterin compounds (Fig. 6; FRÖHLICH et al. 1999) is in good agreement with a recently performed CoMFA analysis of the 4-amino anti-pterin family (MATTER et al., in preparation).

D. Outlook

Compared with other pharmacological approaches to develop NOS inhibitors, the H_4B antagonists/anti-pterins seem to be very promising. The NOS pterin-binding site appears to be unique amongst all known H_4B-dependent proteins and enzymes with respect to function and SAR. Thus, unwanted side effects, e.g. on adrenergic neurotransmitter biosynthesis or by anti-folate activity of anti-pterins, are not likely to occur. This is different from the case of the NOS arginine-binding site, which leaves very little space for modifications to increase affinity or specificity of arginine-based inhibitors versus other arginine-binding or metabolising proteins and enzymes. Future work will clearly be aimed at developing anti-pterins that are active in the nanomolar range and are membrane permeable. 4-Amino-H_4B is clearly an important milestone along this road. Moreover, preliminary data indicate that these compounds have the highest potential for providing pharmacophores that selectively affect only one of the three NOS isoforms.

Acknowledgements. The original work of the authors cited in this article was supported by the Austrian Research Funds "zur Förderung der Wissenschaftlichen Forschung" (P13793MoB; ERW), the Bundesministerium für Bildung, Wissenschaft, Forschung und Technologie (BMBF; HHHWS), the Deutsche Forschungsgemeinschaft (SFB 355/C7; HHHWS) and Vasopharm Biotech (HHHWS). We thank Lothar G. Fröhlich and Peter Kotsonis for critically reading the manuscript and making valuable suggestions.

References

Abu-Soud HM, Gachhui R, Raushel FM, Stuehr DJ (1997) The ferrous–dioxy complex of neuronal nitric oxide synthase: divergent effects of L-arginine and tetrahydrobiopterin on its stability. J Biol Chem 272:17349–17353

Alderton WK, Boyhan A, Lowe PN (1998) Nitroarginine and tetrahydrobiopterin binding to haem domain of neuronal nitric oxide synthase using a scintillation proximity assay. Biochem J 332:195–201

Baek KJ, Thiel BA, Lucas G, Stuehr DJ (1993) Macrophage nitric oxide synthase subunits. Purification, characterization, and role of prosthetic groups and substrate in regulating their association into a dimeric enzyme. J Biol Chem 268:21120–21129

Bahrami S, Strohmaier W, Gasser H, Peichl G, Fürst W, Fitzal F, Werner ER, Schlag G (1997) 2,4-Diamino-5,6,7,8-tetrahydro-6-(L-erythro-1,2-dihydroxypropyl)pteridine (4-ABH_4) reduces nitric oxide formation and improves survival rate in rat experimental shock. Shock 8 (Suppl):159 (abstr)

Bec N, Gorren ACF, Voelker C, Mayer B, Lange R (1998) Reaction of neuronal nitricoxide synthase with oxygen at low temperature. Evidence for reductive activation of the oxy–ferrous complex by tetrahydrobiopterin. J Biol Chem 273:13502–13508

Bolin JT, Filman DJ, Matthews DA, Hamlin RC, Kraut J (1982) Crystal structures of *Escherichia coli* and *Lactobacillus casei* dihydrofolate reductase refined at 1.7-A resolution. I. General features and binding of methotrexate. J Biol Chem 257:13650–13662

Bömmel HM, Reif A, Fröhlich LG, Frey A, Hofmann H, Marecak DM, Groehn V, Kotsonis P, La M, Köster S, Meinecke M, Bernhardt M, Weeger M, Ghisla S, Prestwich GD, Pfleiderer W, Schmidt HHHW (1998) Anti-pterins as tools to characterize the function of tetrahydrobiopterin in NO synthase. J. Biol. Chem. 273:33142–33149

Brand MP, Heales SJ, Land JM, Clark JB (1995) Tetrahydrobiopterin deficiency and brain nitric oxide synthase in the hph1 mouse. J Inherit Metab Dis 18:33–39

Bune AJ, Brand MP, Heales SJ, Shergill JK, Cammack R, Cook HT (1996) Inhibition of tetrahydrobiopterin synthesis reduces in vivo nitric oxide production in experimental endotoxic shock. Biochem Biophys Res Commun 220:13–19

Crane BR, Arvai AS, Ghosh DK, Wu CQ, Getzoff ED, Stuehr DJ, Tainer JA (1998) Structure of nitric-oxide synthase–oxygenase dimer with pterin and substrate. Science 279:2121–2126

Duch DS, Bowers SW, Nichol CA (1978) Elevation of brain histamine levels by diaminopyrimidine inhibitors of histamine *N*-methyl transferase. Biochem Pharmacol 27:1507–1509

Fröhlich L, Kotsonis P, Reif A, Frey A, Pfleiderer W, Schmidt HHHW (1998) Allosteric pterin and L-arginine-binding site interactions of neuronal NO synthase. Pteridines 9:148 (abstr)

Fröhlich LG, Kotsonis P, Traub H, Taghavi-Moghadam S, Almasoudi N, Hofman H, Strobel H, Matter H, Pfleiderer W, Schmidt HHHW (1999) Inhibition of neuronal nitric-oxide synthase by 4-amino pteridine derivatives: structure–activity relationship of antagonists of (6R)-5,6,7,8-tetrahydrobiopterin cofactor. J Med Chem 42:4108–4121

Giovanelli J, Campos KL, Kaufman S (1991) Tetrahydrobiopterin, a cofactor for rat cerebellar nitric-oxide synthase, does not function as a reactant in the oxygenation of arginine. Proc Natl Acad Sci U S A 88:7091–7095

Gorren AC, List BM, Schrammel A, Pitters E, Hemmens B, Werner ER, Schmidt K, Mayer B (1996) Tetrahydrobiopterin-free neuronal nitric-oxide synthase: evidence for two identical highly anti-cooperative pteridine-binding sites. Biochemistry 35: 16735–16745

Gross SS, Levi R (1992) Tetrahydrobiopterin synthesis: an absolute requirement for cytokine-induced nitric-oxide generation by vascular smooth muscle. J Biol Chem 267:25722–25729

Hevel JM, Marletta MA (1992) Macrophage nitric oxide synthase: relationship between enzyme-bound tetrahydrobiopterin and synthase activity. Biochemistry 31:7160–7165

Kaufman S (1993) New tetrahydrobiopterin-dependent systems. Annu Rev Nutr 13:261–286

Klatt P, Schmid M, Leopold E, Schmidt K, Werner ER, Mayer B (1994) The pteridine site of brain nitric oxide synthase. Tetrahydrobiopterin-binding kinetics, specificity, and allosteric interaction with the substrate domain. J Biol Chem 269: 13861–13866

Klatt P, Schmidt K, Lehner D, Glatter O, Bachinger HP, Mayer B (1995) Structural analysis of porcine brain nitric-oxide synthase reveals a role for tetrahydrobiopterin and L-arginine in the formation of a SDS-resistant dimer. EMBO J 14:3687–3695

Klemm P, Hecker M, Stockhausen H, Wu CC, Thiemermann C (1995) Inhibition by N-acetyl-5-hydroxytryptamine of nitric-oxide synthase expression in cultured cells and in the anaesthetized rat. Br J Pharmacol 115:1175–1181

Kotsonis P, Frey A, Fröhlich LG, Hofmann H, Reif A, Wink D, Feelisch M, Schmidt HHHW (1999) Autoinhibition of neuronal nitric oxide synthase: distinct effects of reactive nitrogen and oxygen species on enzyme activity. Biochem. J 340:745–752

Kwon NS, Nathan CF, Stuehr DJ (1989) Reduced biopterin as a cofactor in the generation of nitrogen oxides by murine macrophages. J Biol Chem 264:20496–20501

List BM, Klatt P, Werner ER, Schmidt K, Mayer B (1996) Overexpression of neuronal nitric-oxide synthase in insect cells reveals requirement of haem for tetrahydrobiopterin binding. Biochem J 315:57–63

Liu JW, Garciagardena G, Sessa WC (1995) Biosynthesis and palmitoylationof endothelial nitric-oxide synthase: mutagenesis of palmytoylation sites, cysteines 15 and/or 26, argues against depalmitoylation-induced translocation of the enzyme. Biochemistry 34:12333–12340

Mayer B, Werner ER (1995) In search of a function for tetrahydrobiopterin in the biosynthesis of nitric oxide. Naunyn Schmiedebergs Arch Pharmacol 351:453–463

Mayer B, John M, Bohme E (1990) Purification of a Ca2+/calmodulin-dependent nitric-oxide synthase from porcine cerebellum. Cofactor role of tetrahydrobiopterin. FEBS Lett 277:215–219

Mayer B, John M, Heinzel B, Werner ER, Wachter H, Schultz G, Bohme E (1991) Brain nitric-oxide synthase is a biopterin- and flavin-containing multi-functional oxido-reductase. FEBS Lett 288:187–191

Mayer B, Wu C, Gorren ACF, Pfeiffer S, Schmidt K, Clark P, Stuehr DJ, Werner ER (1997) Tetrahydrobiopterin binding to macrophage inducible nitric-oxide synthase expressed in *Escherichia coli*. Heme spin shift and dimer stabilization by the potent pterin antagonist 4-amino tetrahydrobiopterin. Biochemistry 36:8422–8427

McMillan K, Masters BSS (1995) Prokaryotic expression of the heme- and flavin-binding domains of rat neuronal nitric-oxide synthase as distinct polypeptides: identification of the heme-binding proximal thiolate ligand as cysteine 415. Biochemistry 34:3686–3693

Pfeiffer S, Gorren ACF, Pitters E, Schmidt K, Werner ER, Mayer B (1997) Allosteric modulation of rat brain nitric-oxide synthase by the pterin-site enzyme inhibitor 4-amino tetrahydrobiopterin. Biochem J 328:349–352

Pollock JS, Werner ER, Mitchell JA, Förstermann U (1993) Particulate endothelial nitric-oxide synthase: requirement and content of tetrahydrobiopterin, FAD, and FMN. Endothelium 1:147–152

Presta A, Siddhanta U, Wu CQ, Sennequier N, Huang LX, Abusoud HM, Erzurum S, Stuehr DJ (1998) Comparative functioning of dihydro- and tetrahydropterins in supporting electron transfer, catalysis, and subunit dimerization in inducible nitric-oxide synthase. Biochemistry 37:298–310

Raman CS, Li H, Martasek P, Kral V, Masters BS, Poulos T (1998). Crystal structure of constitutive endothelial NO synthase: a paradigm for pterin function involving a novel metal center. Cell 95:939–950

Reif A, Fröhlich L, Frey A, Kotsonis P, Bömmel H, Pfleiderer W, Schmidt HHHW (1998) Tetrahydrobiopterin stabilizes the NO-synthase quaternary structure and reacts with a catalysis product but has no catalytic role. Pteridines 9:162 (abstr)

Reif A, Fröhlich LG, Kotsonis P, Frey A, Bömmel H, Wink DA, Pfleiderer W, Schmidt HHHW (1999) Tetrahydrobiopterin inhibits monomerization and is consumed during catalysis in neuronal NO synthase. J Biol Chem 274:24921–24926

Rodriguez Crespo I, Gerber NC, Ortiz de Montellano PR (1996) Endothelial nitric-oxide synthase. Expression in *Escherichia coli*, spectroscopic characterization, and role of tetrahydrobiopterin in dimer formation. J Biol Chem 271:11462–11467

Schenk H, Hofmann H, Sennefelder H, Groehn V, Pfleiderer W, Schmidt HHHW (1996) Tetrahydrobiopterin antagonists: novel inhibitors of NO synthase. Pteridines 7:60 (abstr)

Schenk H, Hofman H, Fröhlich L, Meinecke M, Weeger M, Weinberg R, Marecak D, Prestwich G, Groehn V, Pfleiderer W, Schmidt HHHW (1997) Identification and characterization of the NO synthase pterin-binding domain and function. Pteridines 8:46–47 (abstr)

Schmidt HHHW, Pollock JS, Nakane M, Gorsky LD, Förstermann U and Murad F (1991) Purification of a soluble isoform of guanylyl cyclase-activating-factor synthase. Proc Natl Acad Sci USA 88: 365–369

Schmidt HHHW, Smith RM, Nakane M and Murad F (1992) Ca^{2+}/calmodulin-dependent NO synthase type I: a biopteroflavoprotein with Ca^{2+}/calmodulin-independent diaphorase and reductase activities. Biochemistry 31: 3243–3249

Schmidt K, Werner ER, Mayer B, Wachter H, Kukovetz WR (1992) Tetrahydrobiopterin-dependent formation of endothelium-derived relaxing factor (nitric oxide) in aortic endothelial cells. Biochem J 281:297–300

Schmidt K, Werner–Felmayer G, Mayer B, Werner ER (1999) Preferential inhibition of inducible nitric-oxide synthase in intact cells by the 4-amino analogue of tetrahydrobiopterin. Eur J Biochem 259:25–31

Somers MJ, Falgui BT, Laursen JB, Harrison DG (1998) Tetrahydrobiopterin and peroxynitrite interactions as a mechanism of endothelial dysfunction. Free Rad Biol Med. 25:S68 (abstract)

Stroes E, Kastelein J, Cosentino F, Erkelens W, Wever R, Koomans H, Luscher T, Rabelink T (1997) Tetrahydrobiopterin restores endothelial function in hypercholesterolemia. J Clin Invest 99:41–46

Tayeh MA, Marletta MA (1989) Macrophage oxidation of L-arginine to nitric oxide, nitrite, and nitrate. Tetrahydrobiopterin is required as a cofactor. J Biol Chem 264:19654–19658

Walker G, Pfeilschifter J, Kunz D (1997) Mechanisms of suppression of inducible nitric-oxide synthase (iNOS) expression in interferon (IFN)-γ stimulated RAW 264.7 cells by dexamethasone – evidence for glucocorticoid-induced degradation of iNOS protein by calpain as a key step in post-transcriptional regulation. J Biol Chem 272:16679–16687

Wang J, Stuehr DJ, Rousseau DL (1995) Tetrahydrobiopterin-deficient nitric oxide synthase has a modified heme environment and forms a cytochrome P-420 analogue. Biochemistry 34:7080–7087

Werner ER, Pitters E, Schmidt K, Wachter H, Werner-Felmayer G, Mayer B (1996) Identification of the 4-amino analogue of tetrahydrobiopterin as a dihydropteridine-reductase inhibitor and a potent pteridine antagonist of rat neuronal nitric-oxide synthase. Biochem J 320:193–196

Werner ER, Werner-Felmayer G, Mayer B (1998) Minireview: tetrahydrobiopterin, cytokines and nitric-oxide synthesis. Proc Soc Exp Biol Med 219:172–182

Werner-Felmayer G, Werner ER, Fuchs D, Hausen A, Mayer B, Reibnegger G, Weiss G, Wachter H (1993) Ca2+/calmodulin-dependent nitric-oxide synthase activity in the human cervix carcinoma cell line ME–180. Biochem J 289:357–361

Werner-Felmayer G, Werner ER, Fuchs D, Hausen A, Reibnegger G, Wachter H (1990) Tetrahydrobiopterin-dependent formation of nitrite and nitrate in murine fibroblasts. J Exp Med 172:1599–1607

CHAPTER 7
Mechanisms of Cellular Resistance Against Nitric Oxide

B. Brüne, A. von Knethen, and K. Sandau

A. Introduction
I. Cell Death: Apoptosis Versus Necrosis

Cell death is often defined by morphological criteria and is believed to occur by either necrosis or apoptosis. Necrotic death comprises cell and organelle swelling, ultimately followed by cell dissolution. Intraorganelle and cytoplasmic contents leak out into the extracellular space as a result of membrane damage and cause an inflammatory reaction. DNA is exposed to lysosomal nucleases, causing DNA degradation, with fragments displaying a continuous spectrum of sizes.

In 1972, Kerr, Wyllie, and Currie marshaled morphological evidence to draw a clear distinction between cell deaths that occur in both animal development and tissue homeostasis and in some pathological states, and pathological necrotic cell deaths (Kerr et al. 1972). Although the term "apoptosis" was originally coined to describe morphological alterations, it now generally refers to the evolutionary conserved pathway of biochemical and molecular events leading to cell demise (Hale et al. 1996; McConkey and Orrenius 1997; Peter et al. 1997; Leist and Nicotera 1998). The term "programmed cell death", adopted from developmental biology, is now used as a synonym for apoptosis to appreciate genetic programs that regulate cell death (Vaux and Strasser 1996). It is accepted that animal cells have a built-in suicide program, a fact which largely became established through genetic studies in the nematode *Caenorhabditis elegans* (Jacobson et al. 1997). Elements of a core program controlling the executive phase of apoptosis seem to be constitutively expressed in virtually all cells. External triggers of apoptosis are multiple and may be influenced by the appearance or disappearance of hormones, growth factors, or cytokines or may be subject to modulation by the intracellular milieu (Ruoslathi and Reed 1994).

Apoptotic cells usually shrink and condense, display surface alterations (i.e. phosphatidylserine exposure that normally is confined to the cell interior), and cleave DNA into large and often small oligonucleosomal-sized (200bp and multiples) fragments, while organelles and the plasma membrane retain their integrity. Dead cells or their fragments (apoptotic bodies) are rapidly phagocytosed by neighboring cells or macrophages before lysis occurs.

There is accumulating evidence that, during apoptosis, multiple signaling pathways intersect, a process known as crosstalk; therefore, a cell response to a given stimulus may vary significantly among different cells within one cell population. It is conceivable that apoptotic pathways converge to one, or very few, common final executive steps (GOLDSTEIN 1997; PETER et al. 1997). Regulators, such as the tumor-suppressor p53, caspases, or the regulatory role of Bcl-2-family members are consistent with a convergence of positive and negative apoptotic-signal-transducing steps.

The tumor-suppressor gene p53 is considered a master guardian of the genome and a member of the DNA-damage-response pathway (ALMOG and ROTTER 1997; OREN 1997; WANG and HARRIS 1997), with the inherent ability to induce growth arrest or apoptosis. Induction of growth arrest requires sequence-specific DNA binding and transcriptional activation of p53 target genes. Induction of apoptosis is less well understood and may involve trans-activation-dependent or -independent mechanisms. p53 activation is marked by increased protein levels, probably due to an increased protein half-life or decreased protein degradation.

Caspases are proteases that specifically cleave a growing number of cellular substrates during the progression of apoptosis (NICHOLSON and THORNBERRY 1997; SALVESEN and DIXIT 1997). Currently, the caspase family consists of 12 members, many of which have a proven role in inflammation or apoptosis (HUMKE et al. 1998). All members share a number of amino acid residues crucial for substrate binding and catalysis. The trivial name "caspase" reflects a cysteine protease ("c") that cleaves after aspartic acid ("aspase") (cysteine ASPartASE). The role of individual caspases, their crosstalk, and the role of substrate cleavage are currently being investigated (COHEN 1997). For apoptotic signal propagation, activation of downstream caspases (i.e. caspase-3 or -7) seems important and is considered the point of no return in the cell-destruction cascade. This has recently been highlighted by the fact that an active caspase promotes DNase activation and, thus, DNA cleavage (ENARI et al. 1998).

Apoptosis is controlled, in part, by an interplay between regulatory proteins. The prototype negative regulator of mammalian cell death is the proto-oncogene *bcl-2* (REED 1997). The *bcl-2* gene was first discovered because of its involvement in t(14:18) chromosomal translocations found in follicular B-cell lymphomas. Constitutive overexpression of Bcl-2 or related proteins, such as Bcl-x_L, has proven that these family members protect many cell types from apoptosis induced by a wide variety of stimuli. Although questions regarding their biochemical mechanism of action remain unanswered, it is imperative to conclude that Bcl-2 can prevent apoptosis by interacting with caspase activators, thereby preventing amplification of pro-apoptotic signals (HENGARTNER 1998). Among the inducers of apoptosis, the molecule nitric oxide (NO•) has recently been recognized.

II. NO•: Formation and Signaling

The diffusible messenger NO• has unexpectedly come to occupy a central role in mammalian pathology/physiology. NO• is catalytically produced by NO• synthase (NOS) isoforms from the terminal guanido nitrogen of L-arginine (MAYER and HEMMENS 1997). NOSs are broadly distinguished as constitutively versus inducibly expressed and are referred to as nNOS (neuronal NOS or NOS-I), eNOS (endothelial NOS or NOS-III), and iNOS (inducible NOS or NOS-II). Although it is appreciated that constitutive NOS-enzymes may be induced and, conversely, that iNOS may function as a "constitutive" enzyme, an accepted nomenclature still describes nNOS, iNOS, and eNOS isoforms, reflecting the original tissue of the protein and complementary DNA isolates (NATHAN 1992; MICHEL and FERON 1997). Once activated, NOS not only produces NO•, the primary reaction product, but also those species resulting from oxidation, reduction, or adduction of NO• in physiological milieus, thereby producing various NO•s, S-nitrosothiols, peroxynitrite (ONOO⁻), and transition-metal adducts (STAMLER et al. 1992, STAMLER 1994). Signaling refers to the ability to diffuse between cells and to initiate signaling pathways due to the formation of distinctive chemical species that differ with respect to target interactions. Moreover, the determinant of isoenzyme activity allows one to approximate a low versus high output system for NO• and a rough correspondence between toxic and homeostatic functions of the molecule. NOS inhibitors, such as N^G-monomethyl-L-arginine (L-NMMA), intervene pharmacologically, thus allowing us to describe the involvement of NO• formation (MONCADA et al. 1991). In contrast, to study NO•-signal transduction irrespective of NOS involvement, NO•-releasing compounds are valuable tools. NO•-releasing compounds, termed nitrovasodilators or NO• donors, preserve NO• or a related species in their molecular structure and evoke activity after decomposition. Examples are 3-morpholinosydnonimine, S-nitrosothiols (i.e. S-nitrosoglutathione, GSNO), or compounds containing the N(O)NO⁻ functional group [i.e. diethylamine–nitric oxide complex or spermine-NO].

Biological activity of NO• is classified into cyclic guanosine monophosphate (cGMP)-dependent and -independent pathways, both attributed to physiology and pathology (STAMLER 1994; SCHMIDT and WALTER 1994). Activation of soluble guanylyl cyclase, formation of cGMP, and concomitant protein phosphorylation is considered the main physiological signaling system for NO•. However, for cytostatic or cytotoxic signal-conveying pathways, cGMP-independent reactions appear to be of greater importance.

B. Cytotoxicity of Nitric Oxide

I. NO•-Mediated Cytotoxicity/Apoptosis

NO•, generated in high quantities by activated macrophages, is part of the inflammatory response against bacteria, viruses, and tumor cells (KOLB and KOLB-BACHOFEN 1992). The ability of macrophages to kill tumor cells and bacteria in an L-arginine-dependent fashion was originally noticed by Hibbs and colleagues (HIBBS et al. 1987). Despite these initial observations (STUEHR and NATHAN 1989), the precise mechanisms causing cell injury are still unclear. It is evident that NO•-mediated toxicity is complex, varies considerably between cell types, and occurs as a result of either necrosis or apoptosis (BONFOCO et al. 1995).

Initial reports documented cell lysis as a result of an active NOS or in response to NO donors (KRÖNKE et al. 1991; BERGMANN et al. 1992; DUERKSEN-HUGHES et al. 1992). Later, NO•-mediated apoptosis was noticed (ALBINA et al. 1993; SARIH et al. 1993; XIE et al. 1993) in association with accumulation of the tumor suppressor p53 (MESSMER et al. 1994). It was reported that, mechanistically, NO• targets naked DNA (WINK et al. 1991; NGUYEN et al. 1992) and induces DNA damage in activated macrophages (DE ROJAS-WALKER et al. 1995). Still, it is believed that a NO•-damaged DNA elicits a rapid stress response in mammalian cells and subsequent attachment of poly(adenosine diphosphateribose) polymerase (PARP) to strand breaks (DE MURCIA and MENESSIER-DE MURCIA 1994). Conceptually, PARP activation leads to nicotinamide adenine dinucleotide ion (NAD$^+$) depletion. In an effort to resynthesize NAD$^+$, adenosine triphosphate becomes depleted, which ultimately leads to cell death due to energy deprivation. This scenario may account for neurotoxicity (ZHANG et al. 1994) and islet cell death (HELLER et al. 1995) but is unlikely to represent a general pathway leading to NO•-elicited death, because (1) apoptotic cell death is an energy-requiring process (EGUCHI et al. 1997; LEIST et al. 1997a) and (2) PARP seems fully dispensable for apoptosis (LEIST et al. 1997b; WANG et al. 1997).

NO•-mediated apoptosis is assured by chromatin condensation and internucleosomal DNA fragmentation, i.e. DNA laddering. A cause–effect relationship was indirectly established in L-arginine-restricted medium and by using L-NMMA to block adverse effects. In the meantime, numerous reports confirmed the ability of NO• to initiate apoptosis. This holds for macrophage cell lines, β cells, or corresponding cell lines, such as the RINm5F cells, thymocytes, chondrocytes, mesangial cells, neurons, mast cells, vascular endothelial cells, smooth muscle cells, various tumor cells, and several more (BRÜNE et al. 1998a, 1998b).

II. Apoptotic-Signal Transduction: p53 Accumulation and Caspase Activation

p53 accumulation and NO•-mediated cell death were linked in macrophages or β cells (RINm5F) (MESSMER et al. 1994, MESSMER and BRÜNE 1996b). Activation of iNOS or the use of NO donors caused p53 accumulation, clearly preceding late apoptotic features, i.e. DNA fragmentation. The use of L-NMMA established a cause–effect relation between iNOS activation and the p53 response. Apparently, apoptotic cell death correlated with the degree of p53 accumulation (MESSMER and BRÜNE 1996a, 1996b). In addition, RAW cells stably transfected with plasmids encoding p53 antisense RNA (MESSMER and BRÜNE 1996b) exhibited reduced p53 levels in response to GSNO and showed reduced DNA fragmentation. The ability of NO• to promote a functional p53 response has been confirmed in murine and human systems (Ho et al. 1996; CALMELS et al. 1997; ZHAO et al. 1997). However, experiments in p53-negative cells (U937 cells) implied that p53-independent pathways are also operative during NO•-mediated apoptosis (BROCKHAUS and BRÜNE 1998).

We found activation of caspases in close association with NO•-evoked apoptosis, as indicated by the cleavage of PARP, an established caspase substrate (MESSMER et al. 1996d, MOHR et al. 1998). Caspase activation was initiated by endogenously produced or exogenously supplied NO•. A time-dependent analysis of the apoptotic program established p53 accumulation prior to caspase activation. Caspase activation by NO• donors has been confirmed in human leukemia cells (YABUKI et al. 1997), mesangial cells (SANDAU et al. 1998) and human neoplastic lymphoid cells (CHLICHLIA et al. 1998), and neuronal excitotoxicity has been triggered by NO• (LEIST et al. 1997c). Although PARP cleavage in response to NO• intoxication is widely accepted, a cause–effect relationship between PARP degradation and propagation of the apoptotic signal must be questioned based on data from PARP knock-out animals (LEIST et al. 1997b; WANG et al. 1997). We also noticed PARP cleavage under non-apoptotic conditions in U937 cells and macrophages (BROCKHAUS and BRÜNE 1998; MESSMER et al. 1998), which implies that PARP cleavage and apoptosis can be separated. Under conditions that allow NO•-mediated apoptosis, activation of caspases seems to be an inherent step in the death pathway, and pharmacological interference at this point attenuates cell injury (LEIST et al. 1997c; BROCKHAUS and BRÜNE 1998; MESSMER et al. 1998). Figure 1 shows accumulation of p53 and caspase activation i.e. PARP cleavage specifically for a pro-apoptotic signaling response of S-nitrosoglutathione.

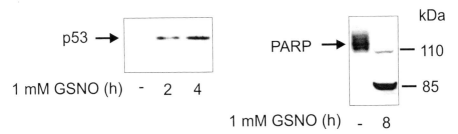

Fig. 1. Accumulation of p53 and poly(adenosine diphosphate–ribose) polymerase (*PARP*) cleavage in response to *S*-nitrosoglutathione (*GSNO*). Typical apoptotic-associated events, such as p53 accumulation and caspase activation, i.e. PARP cleavage (116 kDa holoenzyme; 85 kDa cleaved product), were examined by Western-blot analysis after RAW 264.7 macrophages had been exposed to GSNO

C. Resistance Against NO•-Mediated Toxicity

I. Antagonism by Bcl-2-Family Members

Initiation of apoptosis is often accompanied by an altered expression of Bcl-2-family members, although a general cause–effect relation remains to be established (HAENDELER et al. 1996; BROCKHAUS and BRÜNE 1998). In macrophages and mesangial cells, pro-apoptotic proteins, such as Bax, are up-regulated in response to NO• (MESSMER et al. 1996c; SANDAU et al. 1997b). In line with a general assumption that Bcl-2 members regulate apoptosis, it might be concluded that the balance between anti- and pro-apoptotic proteins determines the choice between life and death under conditions of NO• formation.

To address this question, we stably transfected macrophages with Bcl-2. Transfected clones were resistant towards endogenously generated or exogenously supplied NO•, although the NO•-evoked p53 response remained unchanged (MESSMER et al. 1996c). Conclusively, Bcl-2 acts downstream of p53. In contrast, caspase activation and PARP cleavage were attenuated, thus indicating an interference of Bcl-2 upstream of caspases (MESSMER et al. 1996d). Protection against NO•-mediated cytotoxicity by Bcl-2 was confirmed by others (ALBINA et al. 1996; BONFOCO et al. 1997; MELKOVA et al. 1997) and further supported by overexpression of the Bcl-2-family member Bcl-x_L (OKADA et al. 1998).

II. Protection by NO• and O_2^- Co-Generation

During our studies with rat mesangial cells, we unexpectedly noticed that NO•-mediated apoptotic cell death was antagonized by the simultaneous formation of superoxide (O_2^-). The production of O_2^- from reduced nicotinamide adenine dinucleotide (phosphate)-like oxidases, mitochondrial respiration, and the xanthine oxidase is a naturally occurring process and is likely to occur simultaneously with NO• production. Despite the notion that a massive formation of oxygen radicals is cell damaging and life threatening, O_2^- also operates as a physiological signal mediator (LANDER 1997). Part of the signal transmission of both NO• and O_2^- may stem from the diffusion-controlled NO•/O_2^- interaction that results in the formation of peroxynitrite (ONOO⁻) (PRYOR and SQUADRITO 1995). ONOO⁻ causes oxidation and nitration of proteins and is considered to account in part for NO• toxicity (ESTEVEZ et al. 1995; SZABO et al. 1996). Often, ONOO⁻ is supplied at a relatively high bolus concentration due to the rapid decomposition. In our studies, we addressed the NO•/O_2^- interaction by exposing cells to NO donors and O_2^--generating systems, such as the redox cycler DMNQ (2,3-dimethoxy-1,4-naphtoquinone) or the hypoxanthine/xanthine-oxidase system. This allowed the continuous formation of both radicals over an extended period. The balanced and simultaneous generation of both radicals turned out to be protective for mesangial cells, whereas the unopposed generation of either NO•- or O_2^--elicited apoptosis and, in higher concentrations, necrotic cell death (SANDAU et al. 1997a, 1997b). While apoptosis was accompanied by increased p53 and Bax expression, active caspases, and DNA fragmentation, these alterations were attenuated under conditions of $NO^{\bullet}/O_2^-$ co-administration. Protection was reproduced with different NO donors in combination with two O_2^--generating systems, i.e. DMNQ versus the xanthine/xanthine-oxidase system.

To cause protection, the release of $NO^{\bullet}$ must match the production of O_2^- with respect to time and concentration. Of particular importance is the simultaneous generation of both radicals. If the generation of either $NO^{\bullet}$ or O_2^- is offset, protection is less efficient. We conclude that signaling mechanisms as a consequence of the $NO^{\bullet}/O_2^-$ interaction redirect apoptotic-initiating signals to harmless pathways. Although transducing mechanisms are uncharacterized so far, protection demands reduced glutathione (GSH) (SANDAU and BRÜNE, unpublished observations). Depletion of GSH by preincubation with buthionine–sulfoximine abrogated protection and promoted necrotic cell destruction. The assumption that the radical interaction increased oxidative stress was substantiated by exaggerated GSSG (oxidized glutathione) formation under conditions of $NO^{\bullet}/O_2^-$ co-generation, as compared with GSH oxidation as a result of either $NO^{\bullet}$ or O_2^- donation. Severe oxidative stress as a result of the $NO^{\bullet}/O_2^-$ interaction is in line with in vitro experiments performed by Wink and colleagues (WINK et al. 1993, 1997). They observed GSSG formation by incubating NO donors, O_2^-, and GSH and concluded that $NO^{\bullet}$-evoked nitrosative reactions are quenched by the resultant oxidative stress. In analogy to our system, protection was achieved by a radical–radical interaction that acted as a chain-breaking system for apoptosis as long as GSH is available, most likely by detoxifying the $NO^{\bullet}/O_2^-$-interaction products. In extension to these findings, we observed the upregulation of protective proteins, such as Bcl-2 or heme oxygenase-1, in close correlation with oxidative stress conditions, i.e. GSSG formation (SANDAU et al. 1998). Although no cause–effect relation has been established so far, one might speculate that oxidative conditions resulting from the $NO^{\bullet}/O_2^-$-interaction promote protective gene induction to reduce the risk of $NO^{\bullet}$-mediated apoptosis.

In analogy to our observation, $NO^{\bullet}$ attenuated O_2^--mediated toxicity in chondrocytes (BLANCO et al. 1995) or stretch-induced programmed myocyte cell death that resulted from O_2^- formation (CHENG et al. 1995). Again, the formation of $NO^{\bullet}$ signals protection in the presence of O_2^-, while the interaction product $ONOO^-$ appears to be non-destructive. In support of this are reports that $NO^{\bullet}$ donors inhibit the toxicity of oxidized low-density lipoprotein to endothelial cells, whereby $NO^{\bullet}$ may play an anti-atherogenic role and/or act as an inhibitor of lipid peroxidation (STRUCK et al. 1995). Figure 2 schematically describes the action of $NO^{\bullet}$ and O_2^- in causing apoptosis and promoting protection.

As a general concept, it appears that, in some systems, the balanced formation and interaction of physiologically relevant radicals resembles a protective principle, thereby eliminating harmful reactions that operate as a

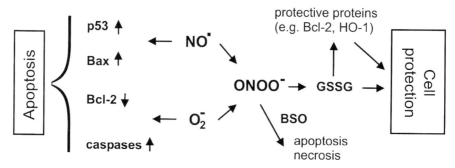

Fig. 2. The role of NO• and O_2^- in promoting cell damage and protection. Proposed signaling events during NO•- and O_2^--initiated apoptosis versus their action in promoting cell protection in mesangial cells. Importantly, *arrows* do not imply a direct cause–effect relation. For details, see text

consequence of unopposed radical generation. These results and hypothetical considerations are in some agreement with studies in organ systems or animal experiments where NO• functions as a protective signal during ischemia–reperfusion, peroxide-induced toxicity, lipid peroxidation, or myocardial injury (OURY et al. 1992; WINK et al. 1993; RUBBO et al. 1994).

III. Protective Protein Expression

It is known that lipopolysaccharide (LPS) given at sublethal dosages can induce a significant enhancement of non-specific resistance to pathogenic infections, inhibition of tumor growth, and a state of resistance to the effects of LPS itself. In analogy, analyzing macrophage programmed cell death in more detail, we realized that desensitization to NO•-elicited apoptosis occurs upon pre-activation with a combination of LPS and interferon γ (IFN-γ) under conditions of blocked NOS or results form pre-stimulation with a low, non-destructive dose of NO donors (GSNO) (BRÜNE et al. 1996). We determined induction of cyclooxygenase-2 (COX-2) to represent a critical regulator of macrophage apoptosis (VON KNETHEN and BRÜNE 1997; VON KNETHEN et al. 1998). Non-activated macrophages do not express COX-2, whereas LPS/IFN-γ/L-NMMA caused protein expression within 6–12 h. In analogy, a low and, thus, nontoxic dose of GSNO promoted COX-2 up-regulation and protection. A functional role of COX-2 during protection was assured by stably transfecting a rat COX-2 expression vector into RAW cells. COX-2-overexpressing macrophages with an activated phospholipase A_2 caused protection against NO•. Protection achieved by LPS/IFN-γ/L-NMMA pre-stimulation was antagonized by the COX-2-selective inhibitor NS398 or by a stably transfected antisense COX-2 expression vector. Activation of the nuclear transcription factor nuclear factor κB (NF-κB) appears to be obligatory for protection and/or expression of COX-2 (VON KNETHEN and BRÜNE, unpublished observations).

NF-κB supershift analysis implied an active p50/p65 heterodimer following NO• or LPS/IFN-γ/L-NMMA addition. Degradation of I-κBα and activation of a luciferase reporter construct containing four copies of the NF-κB site derived from the murine COX-2 promoter confirmed NF-κB activation. Furthermore, a NF-κB-decoy approach attenuated not only COX-2 expression and inducible protection but also restored DNA fragmentation and p53 accumulation in response to high doses of GSNO. These examinations provided evidence for an anti-apoptotic role for NO, which is transmitted by NF-κB activation. It is interesting to note that NF-κB activation seems to represent a more general pathway to eliminate adverse, pro-apoptotic effects of diverse agonists (HAUNSTETTER and IZUMO 1998).

Experiments performed in microglial cells (BV-2 line), where LPS pretreatment protects against NO• toxicity, appear to be analogous to our examinations (SUGAYA et al. 1997). Further support for NO•-inducible protection comes from chondrocytes, where low-dose GSNO enhances the tolerance to a second high-dose GSNO exposure (TURPAEV et al. 1997) or from studies where peroxynitrite pretreatment protects macrophages from cytokine-induced cytotoxicity (SCIVITTARO et al. 1997).

Alternative studies further substantiate a functional role of NO• in attenuating cell death. Low doses of NO• induced autoprotection in association with suppression of NO•-mediated hepatocyte necrotic cell death (KIM et al. 1995). Based on inhibitor studies with tin–protoporphyrin IX, a suggested heme oxygenase 1 (HO-1)-blocking agent, it is argued that up-regulation of HO-1 resembles a protective principle. Two other reports focused on endorsed expression of the heat-shock protein 70 (HSP70), which protects against NO• damage (BELLMANN et al. 1996; KIM et al. 1997a). The study of KIM and coworkers describes HSP70 expression in response to exogenous or endogenous NO• in close correlation to rat hepatocyte protection from tumor-necrosis-factor-α-induced apoptosis. A cause–effect relation of HSP70 expression was confirmed by antisense oligonucleotides directed against HSP70. Consistent with these and our studies, which show a protective role of NO•, is the notion that cell survival is a result of protective associated protein expression. It remains to be established how these protective proteins use NO• to circumvent cell death. It might be speculated that interaction with specific death-inducing pathways allows abrogation of apoptosis. However, this assumption only makes sense if we consider that cell death as a result of NO• intoxication uses signaling pathways to initiate cell destruction rather than causing necrotic cell death in association with unspecific DNA damage.

IV. cGMP Formation and Protein Thiol Modification

Cellular protection as a result of NO• formation is also noted in association with activation of soluble guanylyl cyclase. Experimental support for the involvement of cGMP in mediating protection came from the use of lipophilic cGMP analogues or from intervention studies at the level of guanylyl cyclase, thereby blocking cGMP formation. cGMP-mediated protection is reported for endothelial cells exposed to tumor necrosis factor-α (POLTE et al. 1997). In the case of B lymphocytes, cGMP-dependent (GENARO et al. 1995) and cGMP-independent observations are claimed as apoptosis-antagonizing pathways (MANNICK et al. 1994). In T lymphocytes NO•, via the formation of cGMP, protects against CD95/Fas-evoked apoptosis at multiple sites located upstream and downstream of ceramide generation (SCIORATI et al. 1997). Additionally, NO• prevents Fas-mediated apoptosis in human eosinophils (HEBESTREIT et al. 1998). In this case, NO• disrupts Fas signaling at the level of, or proximal to, Jun kinase but distal to ceramide generation, with the implication that an interleukin-1β-converting-enzyme-like protease (caspase-1) is not blocked by NO•. Moreover, in hepatocytes, NO• partially prevents tumor necrosis factor-α- or Fas-mediated apoptosis via a cGMP-dependent mechanism (KIM et al. 1997b). Although an attenuating role of cyclic GMP seems established, molecular targets and mechanistic insights into antagonistic pathways mostly remain elusive. It will be imperative to know whether cGMP exerts its anti-apoptotic role via immediate phosphorylation-associated alterations or as a result of cGMP-initiated gene activation.

Caspases are emerging as a new potential target of NO•. Caspase activation is inherent to the final executive phase of apoptosis, and inhibition at this point appears to be a rational pharmacological approach. Indeed, pharmacological inhibition of caspases blocks NO•-mediated apoptosis (LEIST et al. 1997c; BROCKHAUS and BRÜNE 1998). Inhibition of caspases, which contain a catalytically reactive cysteine moiety, by NO•-mediated S-nitrosation or oxidation emerges as a rational approach, and has now been experimentally proven by several groups (DIMMELER et al. 1997; KIM et al. 1997b; LI et al. 1997; MELINO et al. 1997; MOHR et al. 1997; TENNETI et al. 1997). However, most of these studies have been performed in cell extracts or with purified proteins, and it appears that inhibition is largely reversed under stringent reducing conditions. This makes it difficult to extrapolate the mentioned results to cellular conditions and the question of whether endogenously produced NO• inhibits apoptosis due to a direct interaction with caspases remains unanswered (with the exception of the hepatocyte system, where NO• inhibited caspase activity by roughly 50% under cellular conditions) (KIM et al. 1997b).

One has to consider that any interference of NO• with the apoptotic-signaling cascade upstream of caspase activation would result in attenuated caspase activity and would not necessarily reflect direct enzyme inhibition. Moreover, from the numerous studies where NO• initiated apoptosis and caused activation of caspases, it is not immediately apparent how and under which circumstances NO• actually blocks progression of programmed cell death. The easiest explanation would be the concentration of NO• and the respective time that NO• is present. It appears that NO• initiates signal transduction that later on progresses to caspase activation. If NO• only intervenes with an active caspase, it can be speculated that NO•, at these later time points, attenuates apoptosis via caspase inhibition. However, it should be noted that NO•-initiated apoptosis in human neopastic lymphoid cells requires activation of caspases, in particular Fas-associated-death-domain-protein-like interleukin-1-converting enzyme (caspase-8), the most CD95-receptor-proximal caspase (CHLICHLIA et al. 1998). The modulatory role of NO•, with regard to CD95/Fas-signaling, is not uniform. It is noticed that NO• up-regulates the Fas system and thereby contributes to apoptosis in pancreatic β cells or lymphoid cells (STASSI et al. 1997; CHLICHLIA et al. 1998) while it inhibits Fas-induced apoptosis in human leukocytes and T cell clones via a cGMP-independent mechanism (MANNICK et al. 1997). Again, the pro- and anti-apoptotic action of NO• can only be explained by the concurrent crosstalk of other signaling components. Assuming autocrine or paracrine generation of NO• as an early signaling event in pathways that regulate apoptosis allows one to predict modulation of immune and inflammatory responses by NO•.

D. Conclusions

The toxicity of NO• is not constant (Fig. 3). It is influenced by the existing biological milieu. Relative rates of NO• formation, its oxidation and reduction, the combination with oxygen, superoxide, and other biomolecules determines agonistic and antagonistic signaling pathways of NO•. In some systems, activation of iNOS generates sufficient amounts of NO• to promote cell death, as defined by morphological and biochemical apoptotic features. Apoptosis in response to NO donors or endogenous NO• generation is accompanied by

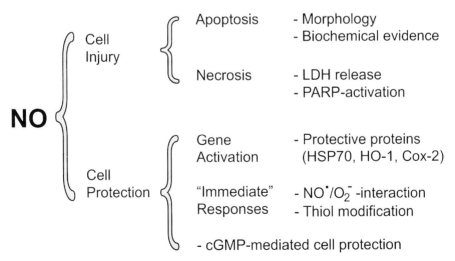

Fig. 3. NO•: a pro-apoptotic and anti-apoptotic molecule. Signaling pathways that initiate or inhibit cell death/apoptosis as a result of NO• formation are indicated. For details, see text

caspase activation, alterations in the expression of Bcl-2-family members, DNA fragmentation, and, in some cases, by accumulation of the tumor-suppressor protein p53. Bcl-2-gene transfer attenuated apoptotic alterations distal to p53 accumulation and proximal to caspase activation.

Not all systems that up-regulate iNOS enter the death pathway (for example, rat mesangial cells) (SANDAU et al. 1997a). Antagonistic and/or protective principles exist. An antagonistic principle arises in the presence of a balanced rate of O_2^- production, which redirects cell destruction to cell protection. In addition to the adverse effects of NO•, the molecule also signals cell protection. NO•-derived cell protection is rationalized by up-regulating protective proteins, such as heat-shock proteins or COX-2 and by signaling pathways that demand cGMP formation or thiol modification, i.e. caspase inactivation (Fig. 3).

It will be essential to define the versatility of NO•-signaling mechanisms in relation to their apoptosis-inducing ability and to explore how NO•-responsive targets serve both sensory and regulatory roles in transducing a signal. The switch from physiology to pathophysiology, the action of potentially protective and destructive NO• species, and the molecular recognition of these balances will be central to the understanding of the pro- and anti-apoptotic actions of NO•.

Acknowledgements. We apologize to investigators whose primary papers (which form the basis for our current knowledge in this active field of investigation) could only be cited indirectly by reference to more recent reviews. During recent years, we received support from the Deutsche Forschungsgemeinschaft, Deutsche Krebshilfe, and the European Community, which made these studies possible.

Abbreviations

cGMP	cyclic guanosine monophosphate
COX-2	cyclooxygenase 2
DMNQ	2,3-dimethoxy-1,4-naphtoquinone
eNOS	endothelial nitric oxide synthase
GSH	glutathione (reduced form)
GSNO	S-nitrosoglutathione
GSSG	oxidized glutathione
HO-1	heme oxygenase 1
HSP	heat-shock protein 70
IFN-γ	interferon γ
iNOS	inducible nitric oxide synthase
LPS	lipopolysaccharide
NAD$^+$	nicotinamide adenine dinucleotide ion
NF-κB	nuclear factor κB
L-NMMA	N^G-monomethyl-L-arginine

nNOS   neuronal nitric oxide synthase
NOS   nitric oxide synthase
PARP   poly(adenosine diphosphate–ribose) polymerase

References

Albina JE, Cui S, Mateo RB, Reichner JS (1993) Nitric oxide-mediated apoptosis in murine peritoneal macrophages. J Immunol 150:5080–5085

Albina JE, Martin BA, Henry WL, Louis CA, Reichner JS (1996) B cell lymphoma-2 transfected P815 cells resist reactive nitrogen intermediate-mediated macrophage-dependent cytotoxicity. J Immunol 157:279–283

Almog N, Rotter V (1997) Involvement of p53 in cell differentiation and development. Biochim Biophys Acta 1333:F1–F27

Bellmann K, Jäättelä M, Wissing D, Burkart V, Kolb H (1996) Heat shock protein hsp70 overexpression confers resistance against nitric oxide. FEBS Lett 391:185–188

Bergmann L, Kröncke KD, Suschek D, Kolb-Bachofen V (1992) Cytotoxic action of IL-1β against pancreatic islets is mediated via nitric oxide formation and is inhibited by N^G-monomethyl-L-arginine. FEBS Lett 299:103–106

Blanco FJ, Ochs RL, Schwarz H, Lotz M (1995) Chondrocyte apoptosis induced by nitric oxide. Am J Pathol 146:75–85

Bonfoco E, Krainc D, Ankarcrona M, Nicotera P, Lipton SA (1995) Apoptosis and necrosis: two distinct events induced, respectively, by mild and intense insults with N-methyl-D-aspartate or nitric oxide/superoxide in cortical cultures. Proc Natl Acad of Sci USA 92:7162–7166

Bonfoco E, Zhivotovsky B, Rossi AD, Santelises MA, Orrenius S, Lipton SA, Nicotera P (1997) Bcl-2 delays apoptosis and PARP cleavage induced by NO donors in GT1-7 cells. Neuroreport 8:272–276

Brockhaus F, Brüne B (1998) U937 apoptotic cell death by nitric oxide: Bcl-2 down-regulation and caspase activation. Exp Cell Res 238:33–41

Brüne B, Gölkel C, Von Knethen A (1996) Cytokine and low-level nitric oxide prestimulation blocked p53 accumulation and apoptosis of RAW 264.7 macrophages. Biochem Biophys Res Com 229:396–401

Brüne B, Sandau K, von Knethen A (1998a) Apoptotic cell death and nitric oxide: activating and antagonistic transducing pathways. Biochem (Moscow) 63:817–825

Brüne B, Sandau K, von Knethen A (1998b) Nitric oxide and its role in apoptosis. Eur J Pharmacol 351:261–272

Calmels S, Hainaut P, Ohshima H (1997) Nitric oxide induces conformational and functional modifications of wild-type p53 tumor suppressor protein. Cancer Res 57:3365–3369

Cheng W, Li B, Kajstura J, Wolin MS, Sonnenblick EH, Hintze TH, Olivetti G, Anversa P (1995) Stretch-induced programmed myocyte cell death. J Clin Invest 96:2247–2259

Chlichlia K, Peter ME, Rocha M, Scaffidi C, Bucur M, Krammer PH, Schirrmacher V, Umansky V (1998) Caspase activation is required for nitric oxide-mediated CD95 (APO-1/Fas)-dependent and independent apoptosis in human neoplastic lymphoid cells. Blood 91:4311–4320

Cohen GM (1997) Caspases: the executioners of apoptosis. Biochem J 326:1–16

De Murcia G, Menessier-De Murcia J (1994) Poly(ADP–ribose) polymerase: a molecular nick-sensor. Trends in Biol Sci 19:172–176

De Rojas-Walker T, Tamir S, Ji H, Wishnok JS, Tannenbaum SR (1995) Nitric oxide induces oxidative damage in addition to deamination in macrophage DNA. Chem Res Toxicol 8:473–477

Dimmeler S, Haendeler J, Nehls M, Zeiher AM (1997) Suppression of apoptosis by nitric oxide via inhibition of interleukin-1β-converting enzyme (ICE)-like and cysteine protease protein (CPP)-32-like protease. J Exp Med 185:601–607

Duerksen-Hughes PJ, Day DB, Laster SM, Zachariades NA, Aquino L, Gooding LR (1992) Both tumor necrosis factor and nitric oxide participate in lysis of Siam virus 40-transformed cells by activated macrophages. J Immunol 149:2114–2122

Eguchi Y, Shimizu S, Tsujimoto Y (1997) Intracellular ATP levels determine cell death fate by apoptosis or necrosis. Cancer Res 57:1835–1840

Enari M, Sakahira H, Yokoyama H, Okawa K, Iwamatsu A, Nagata S (1998) A caspase-activated DNase that degrades DNA during apoptosis, and its inhibitor ICAD. Nature 391:43–50

Estevez AG, Radi R, Barbeito L, Shin JT, Thompson JA, Beckman JS (1995) Peroxynitrite-induced cytotoxicity in PC12 cells: evidence for an apoptotic mechanism differentially modulated by neurotropic factors. J Neurochem 65:1543–1550

Genaro AM, Hortelano S, Alvarez A, Martinez-A C, Bosca L (1995) Splenic B lymphocyte programmed cell death is prevented by nitric oxide release through mechanisms involving sustained Bcl-2 levels. J Clin Invest 95 1884–1890

Goldstein P (1997) Controlling cell death. Science 275:1081–1082

Haendeler J, Messmer UK, Brüne B, Neugebauer E, Dimmeler S (1996) Endotoxic shock leads to apoptosis in vivo and reduces Bcl-2. Shock 6:405–409

Hale AJ, Smith CA, Sutherland LC, Stoneman VEA, Longthorne VL, Culhane AC, Williams GT (1996) Apoptosis: molecular regulation of cell death. Eur J Biochem 236:1–26

Haunstetter A, Izumo S (1998) Basic mechanisms and implications for cardiovascular disease. Circ Res 82:1111–1129

Hebestreit H, Dibbert B, Balatti I, Braun D, Schapowal A, Blaser K, Simon HU (1998) Disruption of Fas receptor signaling by nitric oxide in eosinophils. J Exp Med 187:415–425

Heller B, Wang Z, Wahner EF, Radons J, Bürkle A, Fehsel K, Burkart V, Kolb H (1995) Inactivation of the poly(ADP–ribose) polymerase gene affects oxygen radical and nitric oxide toxicity in islet cells. J Biol Chem 270:11176–11180

Hengartner MO (1998) Death cycle and Swiss army knives. Nature 391:441–442

Hibbs JB Jr, Taintor RR, Vavrin Z (1987) Macrophage cytotoxicity: role of L-arginine deiminase and imino nitrogen oxidation to nitrite. Science 235:473–476

Ho YS, Wang YJ, Lin JK (1996) Induction of p53 and p21/WAF1/CIP1 expression by nitric oxide and their association with apoptosis in human cancer cells. Mol Carcinogen 16:20–31

Humke EW, Ni J, Dixit VM (1998) ERICE, a novel FLICE-activatable caspase. J Biol Chem 273:15702–15707

Jacobson MD, Weil M, Raff MC (1997) Programmed cell death in animal development. Cell 88:347–354

Kerr JFR, Wyllie AH, Currie AR (1972) Apoptosis: a basic biological phenomenon with wide-ranging implications in tissue kinetics. Br J Cancer 26:239–257

Kim Y-M, Bergonia H, Lancaster JR Jr (1995) Nitrogen oxide-induced autoprotection in isolated rat hepatocytes. FEBS Lett 374:228–232

Kim Y-M, De Vera ME, Watkins SC, Billiar TR (1997a) Nitric oxide protects cultured rat hepatocytes from tumor necrosis factor-α-induced apoptosis by inducing heat shock protein 70 expression. J Biol Chem 272:1402–1411

Kim Y-K, Talanian RV, Billiar TR (1997b) Nitric oxide inhibits apoptosis by preventing increases in caspase-3-like activity via two distinct mechanisms. J Biol Chem 272:31138–31148

Kolb H, Kolb-Bachofen V (1992) Nitric oxide: a pathogenetic factor in autoimmunity. Immunol Today 13:157–160

Krönke KD, Kolb-Bachofen V, Berschick B, Burkart V, Kolb H (1991) Activated macrophages kill pancreatic islet cells via arginine-dependent nitric oxide generation. Biochem Biophys Res Com 175:752–758

Lander HM (1997) An essential role for free radicals and derived species in signal transduction. FASEB J 11:118–124

Leist M, Nicotera P (1998) Apoptosis, excitotoxicity, and neuropathology. Exp Cell Res 239:183–201
Leist M, Single B, Castoldi AF, Kühnle S, Nicotera P (1997a) Intracellular ATP concentration: a switch deciding between apoptosis and necrosis. J Exp Med 185:1481–1486
Leist M, Single B, Künstle G, Volbracht C, Hentze H, Nicotera P (1997b) Apoptosis in the absence of poly-(ADP–ribose) polymerase. Biochem Biophys Res Com 233:518–522
Leist M, Volbracht C, Kühnle S, Fava E, Ferrando-May E, Nicotera P (1997c) Caspase-mediated apoptosis in neuronal excitotoxicity triggered by nitric oxide. Mol Med 3:750–764
Li J, Billiar TR, Talanian RV, Kim YM (1997) Nitric oxide reversibly inhibits seven members of the caspase family via S-nitrosylation. Biochem Biophys Res Com 240:419–424
Mannick JB, Asano K, Izumi K, Kieff E, Stamler JS (1994) Nitric oxide produced by human B lymphocytes inhibits apoptosis and Epstein-Barr virus reactivation. Cell 79:1137–1146
Mannick JB, Miao XQ, Stamler JS (1997) Nitric oxide inhibits Fas-induced apoptosis. J Biol Chem 272:24125–24128
Mayer B, Hemmens B (1997) Biosynthesis and action of nitric oxide in mammalian cells. Trends Biol Sci 22:477–481
McConkey DJ, Orrenius S (1997) The role of calcium in the regulation of apoptosis. Biochem Biophys Res Com 239:357–366
Melino G, Bernassola F, Knight RA, Corasaniti MT, Nistico G, Finazzi-Agro A (1997) S-nitrosylation regulates apoptosis. Nature 388:432–433
Melkova Z, Lee SB, Rodriguez D, Esteban M (1997) Bcl-2 prevents nitric oxide-mediated apoptosis and poly(ADP–ribose)polymerase cleavage. FEBS Lett 403:273–278
Messmer UK, Brüne B (1996a) Nitric oxide (NO) in apoptotic versus necrotic cell RAW 264.7 macrophage cell death: the role of NO-donor exposure, NAD^+ content, and p53 accumulation. Arch Biochem Biophys 327:1–10
Messmer UK, Brüne B (1996b) Nitric oxide-induced apoptosis: p53-dependent and p53-independnet signaling pathways. Biochem J 319:299–305
Messmer UK, Ankarcrona M, Nicotera P, Brüne B (1994) p53 expression in nitric oxide-induced apoptosis. FEBS Lett 355:23–26
Messmer UK, Reed JC, Brüne B (1996a) Bcl-2 protects macrophages from nitric oxide-induced apoptosis. J Biol Chem 271:20192–20197
Messmer UK, Reimer DM, Reed JC, Brüne B (1996b) Nitric oxide induced poly(ADP–ribose)polymerase cleavage in RAW 264.7 macrophage apoptosis is blocked by Bcl-2. FEBS Lett 384:162–166
Messmer UK, Reimer DM, Brüne B (1998) Protease activation during nitric oxide-induced apoptosis: comparison between poly(ADP–ribose)polymerase and U1–70 kDa cleavage. Eur J Pharmacol 349:333–343
Michel TM, Feron O (1997) Nitric oxide synthases: which, where, how, and why? J Clin Invest 100:2146–2152
Mohr S, Zech B, Lapetina EG, Brüne B (1997) Inhibition of caspase-3 by S-nitrosation and oxidation caused by nitric oxide. Biochem Biophys Res Com 238:387–391
Mohr S, McCormick TS, Lapetina EG (1998) Macrophages resistant to endogenously generated nitric oxide-mediated apoptosis are hypersensitive to exogenously added nitric oxide donors: dichotomous apoptotic response independent of caspase 3 and reversal by the mitogen-activated protein kinase kinase (MEK) inhibitor DP098059. Proc Natl Acad Sci USA 95:5045–5050
Moncada S, Palmer RMJ, Higgs EA (1991) Nitric oxide: physiology, pathophysiology, and pharmacology. Pharmacol Rev 43:109–142
Nathan C (1992) Nitric oxide a secretory product of mammalian cells. FASEB J 6:3051–3064

Nguyen T, Brunson D, Crespi CL, Penman BW, Wishnok JS, Tannenbaum SR (1992) DNA damage and mutation in human cells exposed to nitric oxide in vitro. Proc Natl Acad Sci USA 89:3030–3034

Nicholson DW, Thornberry NA (1997) Caspases: killer proteases. Trends Biol Sci 22:299–306

Okada S, Zhang H, Hatano M, Tokuhisa T (1998) A physiologic role of Bcl-x_L induced in activated macrophages. J Immunol 160:2590–2596

Oren M (1997) Lonely no more: p53 finds its kin in a tumor suppressor haven. Cell 90:829–832

Oury TD, Ho Y-S, Piantadosi CA, Crapo JD (1992) Extracellular superoxide dismutase, nitric oxide, and central nervous system O_2 toxicity. Proc Natl Acad Sci USA 89:9715–9719

Peter ME, Heufelder AE, Hengartner MO (1997) Advances in apoptosis research. Proc Natl Acad Sci USA 94:12736–12737

Polte T, Oberle S, Schröder H (1997) Nitric oxide protects endothelial cells from tumor necrosis factor-α-induced cytotoxicity: possible involvement of cyclic GMP. FEBS Lett 409:46–48

Pryor WA, Squadrito GL (1995) The chemistry of peroxynitrite: a product from the reaction of nitric oxide with superoxide. Am J Physiol 268:L699–L722

Reed JC (1997) Double identity for proteins of the Bcl-2 family. Nature 387:773–776

Rubbo H, Radi R, Freeman BA (1994) Nitric oxide regulation of superoxide and peroxynitrite-dependent lipid peroxidation. J Biol Chem 269:26066–26075

Ruoslahti E, Reed JC (1994) Anchorage dependence, integrins, and apoptosis. Cell 77:477–478

Salvesen GS, Dixit VM (1997) Caspases: intracellular signaling by proteolysis. Cell 91 443–446

Sandau K, Pfeilschifter J, Brüne B (1997a) The balance between nitric oxide and superoxide determines apoptotic and necrotic cell death of rat mesangial cells. J Immunol 158:4938–4946

Sandau K, Pfeilschifter J, Brüne B (1997b) Nitric oxide and superoxide induced p53 and Bax accumulation during mesangial cell apoptosis. Kidney Intern 52:378–386

Sandau K, Pfeilschifter J, Brüne B (1998) Nitrosative and oxidative stress induced heme oxygenase-1 accumulation in rat mesangial cells. Eur J Pharmacol 342:77–84

Sarih M, Souvannavong V, Adam A (1993) Nitric oxide induces macrophage death by apoptosis. Biochem Biophys Res Com 191:503–508

Schmidt HHHW, Walter U (1994). NO at work. Cell 78:919–925

Sciorati C, Rovere P, Ferrarini M, Heltai S, Manfredi AA, Clementi E (1997) Autocrine nitric oxide modulates CD95-induced apoptosis in $\gamma\delta$ T lymphocytes. J Biol Chem 272:23211–23215

Scivittaro V. Boggs S, Mohr S, Lapetina EG (1997) Peroxynitrite protects RAW 264.7 macrophages from lipopolysaccharide/interferon-γ-induced cell death. Biochem Biophys Res Com 241:37–42

Stamler JS (1994) Redox signaling: nitrosylation and related target interactions of nitric oxide. Cell 78:931–936

Stamler JS, Singel DJ, Loscalzo J (1992) Biochemistry of nitric oxide and its redox-activated forms. Science 258:1898–1902

Stassi G, de Maria R, Trucco G, Rudert W, Testi R, Galluzzo A, Giordano C, Trucco M, (1997) Nitric oxide primes pancreatic β cells for Fas-mediated destruction in insulin-dependent diabetes mellitus. J Exp Med 8:1193–1200

Struck AT, Hogg N, Thomas JP, Kalyanaraman B (1995) Nitric oxide donor compounds inhibit the toxicity of oxidized low-density lipoprotein to endothelial cells. FEBS Lett 261:291–294

Stuehr DJ, Nathan C (1989) Nitric oxide: a macrophage product responsible for cytostasis and respiratory inhibition in tumor target cells. J Exp Med 169:1543–1545

Sugaya K, Chouinard M, McKinney M (1997) Immunostimulation protects microglial cells from nitric oxide-mediated apoptosis. Neurorep 8:2241–2245

Szabo C, Zingarelli B, O'Connor M, Salzman AL (1996) DNA strand breakage, activation of poly(ADP–ribose)polymerase synthetase, and cellular energy depletion are involved in the cytotoxicity in macrophages and smooth muscle cells exposed to peroxynitrite. Proc Natl Acad Sci USA 93:1753–1758

Tenneti L, Emilia DM, Lipton SA (1997) Suppression of neuronal apoptosis by S-nitrosylation of caspases. Neurosci Letter 263:139–142

Turpaev KT, Amchenkova AM, Narovlyansky AN (1997) Two pathways of the nitric oxide-induced cytocydal action. Biochem Mol Biol Interact 41:1025–1033

Vaux DL, Strasser A (1996). The molecular biology of apoptosis. Proc Natl Acad Sci USA 93:2239–2244

Von Knethen A, Brüne B (1997) Cyclooxygenase-2: an essential regulator of NO-mediated apoptosis. FASEB J 11:887–895

Von Knethen A, Lotero A, Brüne B (1998) Etoposide and cisplatin induced apoptosis in activated RAW 264.7 macrophages is attenuated by cAMP-induced gene expression. Oncogene 17:387–394

Wang XW, Harris CC (1997) p53 tumor suppressor gene: clues to molecular carcinogenesis. J Cell Physiol 173:247–255

Wang Z, Stingl L, Morrison C, Jantsch M, Los M, Schulze-Osthoff K, Wagner EF (1997) PARP is important for genomic stability but dispensable in apoptosis. Genes and Develop 11:2347–2358

Wink DA, Kasprzak KS, Maragos CM, Elespuru RK, Misra M, Dunams TM, Cebula TA, Koch WH, Andrews AW, Allen JS, Keefer LK (1991) DNA deaminating ability and genotoxicity of nitric oxide and its progenitors. Science 254:1001–1003

Wink DA, Hanbauer I, Krishna MC, DeGraff W, Gamson J, Mitchell JB (1993) Nitric oxide protects against cellular damage and cytotoxicity from reactive oxygen species. Proc Natl Acad Sci USA 90:9813–9817

Wink DA, Cook JA, Kim SY, Vodovotz Y, Pacelli R, Krishna MC, Russo A, Mitchell JB, Jourd'heuil D, Miles AM, Grisham MB (1997) Superoxide modulates the oxidation and nitrosation of thiols by nitric oxide-derived reaction intermediates. J Biol Chem 272:11147–11151

Xie K, Huang S, Dong Z, Fidler IJ (1993) Cytokine-induced apoptosis in transformed murine fibroblasts involves synthesis of endogenous nitric oxide. Internat J Oncol 3:1043–1048

Yabuki M, Kariya S, Inai Y, Hamazaki K, Yoshioka T, Yasuda T, Horton AA, Utsumi K (1997) Molecular mechanism of apoptosis in HL-60 cells induced by a nitric oxide-releasing compound. Free Rad Res 27:325–335

Zhang J, Dawson V, Dawson TM, Snyder SH (1994) Nitric oxide activation of poly(ADP–ribose) synthase in neurotoxicity. Science 263:687–689

Zhao Z, Francis CE, Welch G, Loscalzo J, Ravid K (1997) Reduced glutathione prevents nitric oxide-induced apoptosis in vascular smooth muscle cells. Biochim Biophys Acta 1359:143–152

Section III
Physiological Functions of NO

CHAPTER 8
Nitric Oxide and Regulation of Vascular Tone

R. BUSSE and I. FLEMING

A. Regulation of Vascular Tone

In all higher organisms, the cardiovascular system represents an elaborated transport network, which is essential for the maintenance of vital functions by supplying oxygen and nutrients to tissue and removing by-products of metabolism. In order to adapt to the varying demands of the tissues, the circulatory system has evolved central and local control mechanisms that act in concert to maintain an adequate blood flow. At a given blood pressure, the blood flow to each organ is determined by the peripheral vascular resistance of this organ, which is adjusted by a variety of local mechanisms affecting the tone of the smooth muscle cells in the so-called resistance vessels, i.e. small terminal arteries and large and small arterioles. In the last two decades, a large body of experimental and clinical evidence has been accumulated, demonstrating that nitric oxide (NO) released from endothelial cells is a crucial regulator of arterial conductance and, in this way, plays an indispensable role in the adequate adjustment of tissue perfusion.

The blood flowing through a vessel exerts a frictional force on the luminal surface of the endothelium. This fluid shear stress represents the major stimulus for a continuous production of NO in vivo and is a highly effective and sensitive system to counteract myo- or neurogenically-induced vascular contraction. Although there are a number of additional endothelium-derived vasodilator and vasoconstrictor autacoids [endothelin-1, prostacyclin (PGI_2), prostaglandin H_2, the superoxide anion (O_2^-), and the endothelium-derived hyperpolarizing factor (EDHF)], none of these autacoids play such a central role in the regulation of vascular tone and homeostasis as the free radical NO.

More than 60 years ago, SCHRETZENMAYER (1933) provided the first experimental evidence for flow-induced dilation. In the hind leg of anaesthetised cats he showed that, whenever blood flow to the leg was increased, there was a concomitant increase in the diameter of the feeding femoral artery. The author concluded that this flow-dependent dilator response, which improves conductivity of feeding vessels, was due to a tissue-derived signal transmitted along the vascular tree. Further studies led to the concept of an "ascending dilation" in conduit arteries under conditions of high tissue oxygen demand. Since distal trans-section did not impede the dilation of conduit arteries in response to flow (LIE et al. 1970), it became evident that this dilator response was a locally generated phenomenon of the vascular wall instead of being elicited by an ascending signal. The potential role of endothelial cells in this flow-dependent dilation gained considerable interest after the pioneering observation by FURCHGOTT and ZAWADZKI (1980) that the endothelium can actively induce changes of vascular tone by the release of a labile relaxing factor. In fact, an obligatory role of endothelial cells in sensing flow signals and transducing them into vasodilator responses was demonstrated in large conduit arteries as well as resistance arteries in vitro and in vivo. Furthermore, clinical studies revealed that the flow-dependent dilation, which occurs in different vascular beds in humans, is reduced or even abolished by hypercholesterolaemia and arteriosclerosis.

B. Endothelial Nitric Oxide Synthase

The nitric oxide synthase (NOS) present in vascular endothelial cells is a multidomain enzyme consisting of an N-terminal oxygenase domain (amino acids 1–491) that contains binding sites for heme, L-arginine (CHEN et al. 1997) and tetrahydrobiopterin (H_4B), and a reductase domain (amino acids 492–1205) containing binding sites for flavin mononucleotide (FMN), flavin adenine dinucleotide (FAD), reduced nicotinamide adenine dinucleotide phosphate (NADPH) and calmodulin (CaM) (SESSA et al. 1992). The functional endothelial NOS (eNOS), like the other isoforms, is a dimer comprised of two identical subunits, both of which are myristoylated and palmitoylated. Only the dimer retains the ability to bind substrate and cofactor and the presence of H_4B is critical for dimer formation (CRANE et al. 1998). Maintenance of the integrity of the H_4B-binding site on the eNOS oxygenase domain appears to involve a zinc tetrathiolate or $Zn[S-cysteine]_4$ positioned equidistant from each heme and H_4B (RAMAN et al. 1998). The identification of ZnS_4 in eNOS may well be of physiological relevance and it is feasible that increasing nitrosative stress in vivo may result in the release of the zinc from the oxygenase domain, destabilise the H_4B and/or deplete eNOS of this essential cofactor and favour the generation of O_2^-.

During the synthesis of NO, NADPH-derived electrons pass into the reductase domain flavins, and then must be transferred to the heme located in the oxygenase domain so that the heme iron can bind O_2 and catalyse stepwise NO synthesis from L-arginine (PRESTA et al. 1997; LIST et al. 1997; CHEN et al. 1994; MARTASEK et al. 1996; ABU-SOUD et al. 1997). The binding of CaM to its binding site (amino acids 493–512) is generally accepted to activate NO synthesis by enabling the reductase domain to transfer electrons to the oxygenase domain (ABU-SOUD and STUEHR 1993; PRESTA et al. 1997).

In addition to NO, all of the NOS isoforms can generate O_2^- in a Ca^{2+}/CaM-dependent manner, especially in the absence of L-arginine and H_4B (POU et al. 1992; COSENTINO and KATUSIC 1995; WEVER et al. 1997; VASQUEZ-

VIVAR et al. 1998; XIA et al. 1998). While some investigators initially attributed the ability of eNOS to generate O_2^- to an artefact associated with endothelial cell culture, i.e. H_4B depletion in passaged cells, evidence is continuously accumulating to suggest that H_4B depletion, and thus an eNOS-dependent generation of O_2^-, could be implicated in hypertension and atherosclerosis. Indeed, H_4B can be oxidised by peroxynitrite ($ONOO^-$), to dihydrobiopterin, such that $ONOO^-$ may attenuate eNOS activity by essentially depleting endothelial cells of H_4B. Under these conditions, eNOS Ca^{2+}-dependently produces O_2^- rather than NO (XIA et al. 1998) and amplifies intracellular oxidative stress.

I. Ca^{2+}-Dependent eNOS Activation

The intracellular Ca^{2+} level under resting conditions is sufficient to allow the NOS to "tick over" and produce low amounts of NO. This basal NO synthesis can be enhanced by an increase in $[Ca^{2+}]_i$ following cell stimulation with receptor-dependent stimuli, such as bradykinin, and receptor-independent stimuli, such as Ca^{2+} ionophores (NEWBY and HENDERSON 1990), whereas the removal of extracellular Ca^{2+} abolishes both agonist-induced NO formation and vasodilatation (SINGER and PEACH 1982; LÜCKHOFF et al. 1988). The identification of a CaM-binding domain in the primary structure of eNOS (NISHIDA et al. 1992; BREDT et al. 1991; MARSDEN et al. 1992) together with the finding that CaM binding proteins inhibited enzyme activity (BUSSE and MÜLSCH 1990) suggested that the binding of a Ca^{2+}/CaM complex is essential to activate this NOS isoform.

1. The Interaction of eNOS with CaM

Alterations in the CaM-binding domain determine the Ca^{2+} sensitivity of the various NOS isoforms, and substitution of eNOS and iNOS CaM-binding domains in eNOS/iNOS chimeric proteins produces major alterations in the Ca^{2+}/CaM dependence of the intact enzymes (VENEMA et al. 1996). One distinctive difference between the Ca^{2+}-regulated and Ca^{2+}-independent NOS isoforms is the existence of a unique polypeptide (45 amino acid) insert in the FMN binding domains of the Ca^{2+}-dependent enzymes which is not shared by iNOS or other related flavoproteins (SALERNO et al. 1997). Three-dimensional molecular modelling suggests that the insert originates from a site immediately adjacent to the CaM-binding sequence and synthetic peptides derived from the 45 amino acid insert were found to potently inhibit the binding of CaM to eNOS as well as enzyme activity (SALERNO et al. 1997). Based on these observations, it was suggested that the polypeptide insert is an auto-inhibitory control element, docking with a site on eNOS which physically impedes CaM binding and thus enzyme activation. Such a control mechanism would imply that CaM must displace the insert on binding to eNOS. The insert peptide is also a potential site for phosphorylation as 12 of the 45 amino acids are either serine or threonine (SALERNO et al. 1997); thus, phosphorylation and/or dephosphorylation may influence the affinity of insert peptides for binding and thus sensitivity and/or affinity of eNOS/CaM binding.

2. The Interaction of eNOS with Caveolin-1

Caveolae are small plasmalemmal invaginations, which can function as dynamic vesicular carriers and may also be organised cell surface signal transduction centres (LISANTI et al. 1994; ANDERSON 1993; LIU et al. 1997b). The chief structural components of caveolae are cholesterol and caveolar-specific structural proteins, such as the caveolins (caveolin-1, -2 and -3). Caveolin-1 has a hairpin structure and is abundant in endothelial cells. A component of the C-terminal, membrane-proximal, segment of caveolin-1, termed the scaffolding domain, is responsible for association of a number of lipid-anchored signalling molecules with the caveolae, including G-protein subunits, phosphatidylinositol 3-kinase and the Src family protein tyrosine kinases (ROTHBERG et al. 1992; OKAMOTO et al. 1998).

The subcellular localisation of eNOS and alterations in its cellular compartmentalisation following cell stimulation are controversial. Previous reports have assigned eNOS to the Golgi apparatus (SESSA et al. 1995; O'BRIEN et al. 1995; LIU et al. 1997a), while others have localised eNOS in plasma membranes (HECKER et al. 1994a) or partially/exclusively in plasmalemmal caveolae (SHAUL et al. 1996; FERON et al. 1996; LIU et al. 1996; GARCÍA-CARDENA et al. 1996a). The truth lies somewhere in between, as immunostaining of porcine coronary arteries for eNOS reveals an association with the plasma membrane and the Golgi apparatus. Moreover, co-staining of endothelial cells with antibodies raised against eNOS and caveolin-1 shows that not all eNOS is co-localised with caveolin-1. In which fraction eNOS is active in unstimulated cells, and can account for the basal production of NO, is also controversial, since the eNOS in caveolae is thought to be mostly inactive, and disruption of the Golgi apparatus in rabbit carotid arteries failed to affect NO production (BAUERSACHS et al. 1997).

Interaction between both the N- and C-terminal domains of caveolin-1 and the oxygenase domain of eNOS have been described (GARC'A-CARDENA et al. 1997) (JU et al. 1997) but it was recently demonstrated that caveolin-1 must bind to the reductase domain of eNOS in order to inhibit enzyme activity (GHOSH et al. 1998). The interaction of caveolin-1 with the reductase domain was independent of the caveolin binding motif and reversed by CaM. It would therefore appear that the binding of caveolin-1 to the reductase domain of eNOS compromises its ability to bind CaM and to donate electrons to the heme subunit, thereby inhibiting NO synthesis (GHOSH et al. 1998). This suggests that eNOS activity may be determined by the relative proportion of eNOS–Ca^{2+}/CaM to eNOS–caveolin-1 binding (MICHEL et al. 1997).

Although there is overwhelming evidence for the existence of such a reversible protein–protein interaction, the exact mechanism in vivo is still uncertain. Indeed, although a shear stress-induced dissociation of caveolin from eNOS and a concomitant eNOS association with calmodulin has been described in caveolae isolated from native rat lung endothelial cells (Rizzo et al. 1999), shear stress-induced activation of eNOS is apparently Ca^{2+} independent and insensitive to calmodulin antagonists (see below).

3. Other Modulators of eNOS Activity

There is a clear disparity in the temporal relationship between agonist-induced increases in endothelial $[Ca^{2+}]_i$ and the activation of eNOS, the duration of the Ca^{2+}-response being significantly shorter than that of the subsequent NO production. Indeed, apart from changes in intracellular levels of Ca^{2+}, a number of post-translational mechanisms have been proposed to regulate eNOS activity; including the interaction of eNOS with associated proteins, or membrane phospholipids, and phosphorylation. In addition, relatively small pH changes in the physiological range markedly alter the activity of the eNOS derived from two different species (Fleming et al. 1994).

a) Endothelial NOS-Associated Protein-1

Endothelial NOS-associated protein-1 (ENAP-1) (Venema et al. 1996) is a tyrosine-phosphorylated, 90-kDa protein observed to interact with eNOS immunoprecipitated from cultured bovine aortic endothelial cells. The Ca^{2+}-elevating, receptor-dependent agonist bradykinin, which enhances NO production, is reported to stimulate cycles of tyrosine phosphorylation/dephosphorylation of ENAP-1. The functional significance of the temporal waves of tyrosine phosphorylation remain to be elucidated (Venema et al. 1996).

b) Hsp90

The molecular chaperone Hsp90, has recently been identified as an eNOS-associated protein, and its binding to the enzyme increases catalytic activity (García-Cardena et al. 1998). A certain amount of Hsp90 appears to complex with eNOS in unstimulated endothelial cells as immunoprecipitation of Hsp90 results in the recovery of eNOS, and vice versa. In response to cell stimulation with histamine or vascular endothelial growth factor (VEGF), Hsp90 is rapidly recruited to eNOS and exposure of endothelial cells to fluid shear stress stimulates the association of both proteins, albeit with a slower time course. In all cases, the association of Hsp90 with eNOS increased NO production and was prevented by pretreatment with the Hsp90-binding protein, geldanamycin (García-Cardena et al. 1998). The eNOS-associated Hsp90 may also serve as a scaffolding protein, facilitating the organisation of additional associated regulatory proteins.

Given the similarity in the molecular weight of ENAP and Hsp90, it is possible that these are one and the same protein. Although Hsp90 has been described as a serine/threonine phosphorylated protein (Mimnaugh et al.

1995), co-immunoprecipitation experiments showed that genistein inhibited ligand-induced release of Hsp90 from the glucocorticoid receptor. Thus, the interaction of proteins with Hsp90 may also be regulated by a tyrosine-kinase-dependent pathway (GRADIN et al. 1998).

c) Phosphorylation

The role of protein kinases in the regulation of endothelial NO production is a topic of intense current investigation, since several consensus sequence sites for phosphorylation by protein kinase (PK) A, PKB (Akt), PKC and CaM kinase II are found in eNOS. Although eNOS was initially reported to be basally phosphorylated solely on serine residues (MICHEL et al. 1993; HIRATA et al. 1995; ROBINSON et al. 1995; CORSON et al. 1996) evidence that eNOS is also threonine and tyrosine phosphorylated has recently been provided (GARCÍA-CARDENA et al. 1996b; FLEMING et al. 1998a). In cultured endothelial cells, bradykinin was initially described to enhance the serine phosphorylation of eNOS, an effect which was maximal after 5min and was maintained for at least 20min (MICHEL et al. 1993). This bradykinin-induced phosphorylation of eNOS appears to be a Ca^{2+}-dependent phenomenon and is inhibited either by a CaM antagonist or the removal of extracellular Ca^{2+}. A rough comparison of the time course of NO production and eNOS activation in response to bradykinin tends to suggest that at least this serine phosphorylation of eNOS may be an inactivating mechanism and agrees well with the observations that PKC [which phosphorylates eNOS in vitro (HIRATA et al. 1995)] negatively regulates endothelial NO production (TSUKAHARA et al. 1993; DAVADA et al. 1994; AYAJIKI et al. 1996; HIRATA et al. 1995). In addition, the exposure of pulmonary artery endothelial cells to sodium nitroprusside (SNP), enhances the serine phosphorylation of eNOS in a PKC-dependent manner and attenuates eNOS activity (SHEEHY et al. 1998). Fluid shear stress, which elicits the maintained Ca^{2+}-independent production of NO (AYAJIKI et al. 1996), also induces the serine and tyrosine phosphorylation of eNOS(FLEMING et al. 1998b). While definitive proof of an enhanced serine or tyrosine phosphorylation of eNOS under Ca^{2+}-free conditions is at the moment lacking, it would seem that, under these experimental conditions, the increase in phosphorylation is unlikely to be an inhibitory signal. It is most probable that the effect of phosphorylation on eNOS activity very much depends on the specific residue phosphorylated, as phosphorylation of Ser741 in the CaM-binding domain of nNOS prevents the binding of CaM and attenuates enzyme activity (ZOCHE et al. 1997), while the PKB-mediated phosphorylation of eNOS Ser1177 increases enzyme activity (DIMMELER et al. 1999).

The AMP-activated protein kinase (AMPK) co-precipitates with eNOS from cardiac homogenates and this kinase, like the PKB in endothelial cells (DIMMELER et al. 1999), phosphorylates eNOS on Ser1177 and increases enzyme activity (CHEN et al. 1999). The phosphorylation of Ser1177 by AMPK unlike that of PKB activates eNOS only in the presence of Ca^{2+}/CaM (CHEN et al. 1999). Currently, it is unclear whether the phosphorylation of eNOS by AMPK occurs in endothelial cells or only in cardiac myocytes. There is a precedent for a differential effect of AMP on NOS activity depending on the cell type studied. Forskolin, for example, does not induce the phosphorylation of eNOS in endothelial cells (MICHEL et al. 1993), but does phosphorylate the nonpalmitoylated 150-kDa eNOS isoform in cardiac myocytes and inhibits both the post-translational processing of eNOS and its translocation from internal membranes to the sarcolemma (BELHASSEN et al. 1997). As this 150-kDa eNOS is not detectable in endothelial cells, it is possible that AMPK may only phosphorylate eNOS and modulate its activity in cardiac myocytes.

II. Ca^{2+}-Independent eNOS Activation

A basal eNOS activity was originally reported at Ca^{2+} concentrations as low as 10nmol/l in lysates prepared from native endothelial cells, indicating that a significant portion of the NO produced by unstimulated endothelial cells may be formed via a Ca^{2+}-independent pathway (MÜLSCH et al. 1989). Little physiological relevance was attributed to this phenomenon, especially after the identification of a CaM-binding domain in the primary structure of eNOS. More recent biochemical studies have, however, reinforced the original observation that eNOS may produce NO in an apparently Ca^{2+}-independent manner.

eNOS activation in response to the application of shear stress to endothelial cells differs from that activated by receptor-dependent agonists in that it is maintained (over hours) and can be observed in the absence of extracellular Ca^{2+}, and is not inhibited by the CaM antagonist calmidazolium, which abrogates the agonist-induced vasodilatation to acetylcholine (AYAJIKI et al. 1996; KUCHAN and FRANGOS 1994). In native endothelial cells, eNOS exists as part of a multi-molecular complex and its Ca^{2+}-independent activation seems to be linked to eNOS phosphorylation, and association of the enzyme with other proteins, as well as the redistribution of the eNOS complex to a TritonX-100-insoluble/cytoskeletal cell fraction (FLEMING et al. 1998a). As a change in detergent solubility is frequently indicative of the formation of a protein complex, it is tempting to speculate that fluid shear stress and tyrosine phosphatase inhibitors may alter the conformation and/or protein coupling of eNOS, facilitating its interaction with specific phospholipids, proteins and/or protein kinases that enhance/maintain its activation. Indeed, the time course of the association of Hsp90 with eNOS (GARCÍA-CARDENA et al. 1998) parallels the changes in the detergent solubility of eNOS induced by stimuli that elicit its Ca^{2+}-independent activation (FLEMING et al. 1998a). Moreover, Hsp90-binding proteins inhibit both the shear stress-induced increase in NO production and redistribution of eNOS to a Triton X-100-insoluble (cytoskeletal) cell fraction (FLEMING et al. 1998a). However, since the formation of an Hsp90/eNOS complex also rapidly occurs following endothelial cell stimulation with Ca^{2+}-elevating agonists, it is unlikely that the binding of Hsp90 to eNOS alone is sufficient to render its activation independent of a maintained increase in $[Ca^{2+}]_i$.

A pharmacologically identical activation of eNOS can be induced by shear stress and protein tyrosine phosphatase inhibitors suggesting that the tyrosine phosphorylation of eNOS or an associated regulatory protein might be relevant for its Ca^{2+}-independent activation (FLEMING et al. 1998a). However, ultimately, it is the phosphorylation of eNOS on Ser1177 that renders its activation independent of Ca^{2+} in response to fluid shear stress (DIMMELER et al. 1999).

The hypothesis that eNOS can be activated by at least two independent signalling pathways is supported by the observation that disruption of the cytoskeleton attenuated the flow-induced release of NO in native endothelial cells but failed to affect the agonist-induced production of NO (HUTCHESON and GRIFFITH 1996). In addition, RGD-containing peptides have been reported to attenuate the flow-induced dilation of isolated coronary arterioles without affecting the response to substance P (MULLER et al. 1997).

III. The Link Between Fluid Shear Stress and NO Production

The endothelial cell can be viewed as a membrane stretched over a frame composed of microtubules, intermediate filaments and actin fibres, which transverse the cells and end in characteristic adhesion complexes as well as in the vicinity of caveolae. Even under non-stimulated conditions the entire endothelial cytoskeleton is maintained under tension and, in response to an externally applied stimulus, intracellular tension is redistributed over the cytoskeletal network. This tensegrity architecture within cells permits forces to be directly transmitted from the cell surface, through the cytoskeleton, across physically interconnecting filaments to the nucleus [for review see INGBER (1997)]. Thus extracellular forces are superimposed upon pre-existing forces within cells attached to the extracellular matrix at focal adhesion points, and to each other at cell–cell contacts. Generally signalling molecules are clustered around and inherent to these contact sites, as well as to plasmalemmal caveolae, such that it is conceivable that the application of a stress signal that is transmitted through the entire cell by the actin cytoskeleton activates signal transduction cascades without the need of a specific shear stress or stretch receptor. Recently, molecular connections between integrins, cytoskeletal filaments and nuclear scaffolds have been proposed to provide a pathway for signal transfer, thus raising the possibility that mechanical stimuli may be passed on to the nucleus in the absence of/or simultaneously with mechano-chemical signalling processes (MANIOTIS et al. 1997). The caveolar protein, caveolin, has also been proposed as a mechanoreceptor as this protein oligomerizes to form the structural frame for the caveolae and create the characteristic invaginated form (LISANTI et al. 1994; ROTHBERG et al. 1992; MONIER et al. 1995; LEE and SCHMID-SCHONBEIN 1995). This geometric configuration may act as a coiled spring acutely responsive to changes in haemodynamic forces experienced at the cell surface. Haemodynamic forces are known to induce a strain on caveolae which can distort them (LEE and SCHMID-SCHONBEIN 1995) and may, for instance, modulate caveolin conformation sufficiently to permit local activation and translocation events.

The localisation of eNOS, in the vicinity of the cell–cell contact (ANDRIES et al. 1998) and within caveolin-rich membrane domains (SHAUL et al. 1996; FERON et al. 1996), both subcompartments of the plasma membrane in which key signal transducing complexes are concentrated, is likely to have profound repercussions on enzyme activity as well as on its sensitivity to activation by signal transduction cascades, other than those resulting in an increase in $[Ca^{2+}]_i$.

C. Mechanisms of Action of NO on Vascular Smooth Muscle

I. Effects of NO on $[Ca^{2+}]_i$

Generally NO is described to induce relaxation by activating the soluble guanylyl cyclase to increase intracellular concentrations of cyclic GMP, which in turn activates the G kinase and subsequently reduces smooth muscle $[Ca^{2+}]_i$. NO has been shown to affect most of the processes regulating $[Ca^{2+}]_i$ in smooth muscle cells, but not all of these effects are dependent on an increase in cyclic GMP.

NO can attenuate the release of Ca^{2+} from intracellular stores by activating the G kinase, which in turn phosphorylates and inactivates inositol trisphosphate (IP_3) receptors (KOMALAVILAS and LINCOLN 1996). The importance of this mechanism in the control of $[Ca^{2+}]_i$ seems to vary with the cells studied, a fact perhaps related to the heterogeneity of IP_3 receptors in different smooth muscle cells. Other mechanisms for lowering smooth muscle $[Ca^{2+}]_i$ include inhibition of voltage-gated Ca^{2+} channels, activation of the Na^+/Ca^{2+} exchanger, inhibition of phospholipase C and IP_3 formation and the stimulation of K^+_{Ca} channels.

NO has been shown to induce hyperpolarisation of the resting membrane potential of smooth muscle cells, as well as repolarisation (i.e. reversal of contractile agonist-induced depolarisation) (TARE et al. 1990; GARLAND and MCPHERSON 1992; PLANE et al. 1995; COHEN et al. 1997). In addition, cyclic GMP and NO (either directly or by increasing cyclic GMP), can activate delayed rectifier K^+ channels (LI et al. 1998a) and charybdotoxin-sensitive K^+_{Ca} channels in freshly isolated smooth muscle cells (ROBERTSON et al. 1993; BOLOTINA et al. 1994; LI et al. 1997; GEORGE and SHIBATA 1995; LI et al. 1998a). Activation of K^+_{Ca} channels in smooth muscle leads to hyperpolarisation, and the subsequent inactivation of voltage-gated Ca^{2+} channels causes a decrease in $[Ca^{2+}]_i$ and thus promotes relaxation.

Experimental data have documented a tight correlation between NO release, smooth muscle repolarisation, and relaxation. Three distinct mechanisms are implicated in the NO-mediated relaxation; a cyclic GMP-dependent voltage-independent pathway, a cyclic GMP-mediated smooth muscle cell repolarisation and a cyclic GMP-independent, charybdotoxin-sensitive smooth muscle repolarisation (PLANE et al. 1998). It is interesting to note that relaxation to authentic and endothelium-derived NO is mediated by parallel cyclic GMP-dependent and -independent pathways, while the relaxation mediated by NO donors can be attributed solely to cyclic GMP-dependent mechanisms (PLANE et al. 1998).

In situations in which smooth muscle contraction occurs without the direct involvement of L-type Ca^{2+} channels (e.g. in angiotensin or phenylephrine-contracted rabbit aorta), NO initiates its response by accelerating sequestration of Ca^{2+} into intracellular stores via the sarcoplasmic/endoplasmic reticulum Ca^{2+}-ATPase (SERCA), thereby rapidly decreasing $[Ca^{2+}]_i$. The resulting refilling of Ca^{2+} stores directly inhibits store-operated Ca^{2+} influx, thereby further decreasing $[Ca^{2+}]_i$ and facilitating smooth muscle cell relaxation (COHEN et al. 1999). Since NO induces the cyclic GMP-dependent phosphorylation of phospholamban in vascular smooth muscle (CORNWELL et al. 1991; KARCZEWSKI et al. 1998), dissociation of phosphorylated phospholamban from SERCA could be the molecular mechanism by which NO regulates SERCA activity and thus capacitative Ca^{2+} entry (LINCOLN et al. 1995). Inconsistent with this model at present is the observation that the activation of SERCA by NO is not affected by inhibitors of the soluble guanylyl cyclase (WEISBROD et al. 1998; TREPAKOVA et al. 1999). Cyclic GMP-independent effects of NO on SERCA, via redox signalling and probably mediated by O_2^- are possible, as well as signal transduction pathways involving tyrosine kinases and phosphatases (Fig. 1).

A further possibility is that NO could be able to induce changes in the intracellular concentration of cyclic ADP-ribose (cADPR), which affects Ca^{2+}-signalling in various cell types by virtue of its ability to activate ryanodine receptors (GALIONE et al. 1993; GRAIER et al. 1992). Since NO is known to modify various NAD-utilising enzymes either through direct nitrosation or stimulation of auto-ADP-ribosylation, it has been proposed that NO could increase cADPR hydrolase activity and/or decrease the ADP-ribosylcyclase activity and thus the Ca^{2+}-depressing effects of NO (LEE 1994). Although the relationship between NO and the generation of cADPR has not been intensively investigated in vascular smooth muscle cells, both cADPR and ADPR are formed and can modulate smooth muscle $[Ca^{2+}]_i$ and K^+_{Ca} channel activity (KANNAN et al. 1996; de TOLEDO et al. 1997; LI et al. 1998b).

There have been a number of reports suggesting that NO generated upon endothelial cell stimulation with receptor-dependent agonists is able to attenuate eNOS activity by a feedback inhibitory mechanism since, NO and NO donors were found to non-competitively inhibit endothelial NO production in response to receptor-dependent agonists and increased flow (ASSREUY et al. 1993). An alteration in Ca^{2+} signalling was proposed to underlie this effect, as inhibition of NOS activity has been reported to enhance the Ca^{2+}-response to

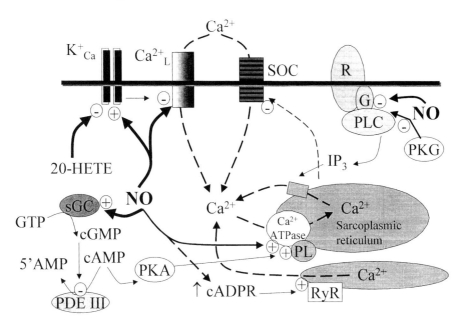

Fig. 1. Schematic representation of the effects of nitric oxide (NO) on mechanisms regulating $[Ca^{2+}]_i$ in vascular smooth muscle cells. *20-HETE* 20-hydroxyeicosatetraenoic acid; K^+Ca Ca^{2+}-dependent K^+ channels; Ca^{2+}_L L-type Ca^{2+} channels; *cADPR* cyclic ADP-ribose; *IP₃* inositol 1,4,5-trisphosphate; *PKA* protein kinase A; *PKG* protein kinase G; *PL* phospholamban; *PLC* phospholipase C; *RyR* ryanodine receptor; *sGC* soluble guanylyl cyclase; *SOC* store-operated Ca^{2+} channel

receptor-dependent agonists such as ATP and bradykinin (SHIN et al. 1992). NO and NO donors, on the other hand, were found to attenuate the agonist-induced Ca^{2+} response to various stimuli in endothelial cells (MORGAN and NEWBY 1989; GARG and HASSID 1991; HOYT et al. 1992; BAUERSACHS et al. 1996), effects which may be related to the finding that NO can interfere with G-protein signalling and attenuate IP₃ production in endothelial cells (LANG and LEWIS 1989). While these effects can certainly be observed using high concentrations of NO, there is no evidence to suggest that lower, more physiological concentrations, of NO are involved in the fine regulation of endothelial cell Ca^{2+} signalling.

II. Effects of NO on Cyclic Nucleotide Phosphodiesterase III

Phosphodiesterase (PDE) III is a cyclic GMP-inhibited, cyclic AMP phosphodiesterase, which is expressed in native vascular smooth muscle cells (KOMAS et al. 1991). Inhibition of PDE III leads to increases in cyclic AMP and a subsequent relaxation in arterial ring preparations, implying that the cyclic GMP-dependent inhibition of PDE III contributes to NO-induced relaxation (KOMAS et al. 1991). Such a mechanism could account for the observation that soluble guanylyl cyclase, but not G-kinase inhibitors are able to attenuate responses to NO donors in some vascular beds (KURTZ et al. 1998; SANDER et al. 1999).

III. Effects of NO on Other Systems Involved in the Control of Vascular Tone

NO is also able to affect the production of other endothelium-derived autacoids. For example, NO increases prostacyclin synthesis both in vitro (SALVEMINI et al. 1993; SALVEMINI et al. 1994; DAVIDGE et al. 1995) and in vivo (SAUTEBIN et al. 1995) by enhancing prostaglandin H synthase (DAVIDGE et al. 1995) and/or cyclo-oxygenase activity (SALVEMINI et al. 1993) via a cyclic GMP-independent mechanism. In contrast, the release of the endothelium-derived hyperpolarising factor, pharmacologically characterised as a cytochrome P_{450}-derived metabolite of arachidonic acid (HECKER et al. 1994b; FULTON et al. 1994), appears to be suppressed at physiological concentrations of NO (BAUERSACHS et al. 1996).

1. Endothelin-1

The circulating levels of the potent vasoconstrictor peptide endothelin-1 (ET-1) in vivo are very low (ANDO et al. 1989), suggesting that little ET-1 is produced under normal conditions. Therefore, there is not much support for the hypothesis that ET-1-induced vascular tone plays a role in the control of blood pressure under physiological conditions. In cultured endothelial cells, however, the mRNA encoding pre-proendothelin is constitutively expressed and ET-1 can be detected in the extracellular medium (YANAGISAWA et al. 1988). One reason for the low expression of ET-1 in vivo, is that NO is thought to intrinsically inhibit ET-1 expression. This hypothesis is supported by the observation that endothelial production of ET-1 can be attenuated by the endothelium-derived vasodilatory autacoids. NO, nitrovasodilators, atrial natriuretic peptide (ANP) and cyclic GMP analogues prevent the thrombin-stimulated release of ET-1, whereas inhibitors of NOS and soluble guanylyl cyclase augment the thrombin-induced production of ET-1 (BOULANGER and LÜSCHER 1990; SAIJONMAA et al. 1990). PGI_2 via the formation of cyclic AMP can also inhibit the production of ET-1 (LÜSCHER et al. 1992). Since ET-1 stimulates the production of both NO and PGI_2, the endothelium appears to possess cyclic GMP- and cyclic AMP-dependent mechanisms which exert a negative feedback on the production of ET-1.

2. Noradrenaline

Noradrenaline release is under the pre-junctional inhibitory control of endogenous NO produced in stellate ganglion cells and in neurones of the heart. In the Langendorff-perfused rat heart, NOS inhibitors have been shown to evoke the release of noradrenaline in a concentration dependent manner, while NO donors, added in presence of the NOS inhibitors, restored the suppression of noradrenaline release (SCHWARZ et al. 1995). More recent evidence obtained with PC12 cells suggests that S-nitroso-cysteine, but not other NO donors, inhibits some common processes occurring during noradrenaline release and neither NO radicals, peroxynitrite, nor cyclic GMP mediate the inhibitory effects of S-nitroso-cysteine (NAGANUMA et al. 1998).

Porcine cerebral arteries, however, were reported to be innervated by NOS-containing nerves that liberate NO on excitation as a neurotransmitter to produce muscular relaxation. The function of these nerves is protected from O_2^-, by endogenously expressed superoxide dismutase. The relaxation elicited by the activation of such nitroxergic nerves is inhibited by acetylcholine released from cholinergic nerves, possibly because of an impaired production or release of NO (TANAKA et al. 1999).

3. NO and Iron-Containing Proteins

Intracellular iron is a primary target for NO, and NO binds to the heme moiety in oxygen carrying haemoproteins, such as the soluble guanylyl cyclase, haemoglobin and myoglobin. In the case of the soluble guanylyl cyclase, NO alters the conformation of the heme moiety of the enzyme, causing its activation and stimulating the production of cyclic GMP. There is evidence to suggest that NO interacts with the prosthetic heme group of cyclo-oxygenase to either activate or inhibit enzymatic activity depending on cellular conditions. Other heme-containing proteins, including NOS itself, and various cytochrome P_{450} (CYP) enzymes can be inhibited by NO, and the interaction of NO with haemoglobin and myoglobin to form iron-nitrosyl heme complexes may represent a mechanism by which the biological actions of NO are attenuated (for review see KIM et al. 1997). Indeed, the interaction of NO with oxyhaemoglobin resulting in the formation of S-nitrosohemoglobin, which relinquishes bound NO upon deoxygenation of hemoglobin within the capillary circulation, is an alternative mechanism by which local oxygen concentrations may modulate vascular function in an NO-dependent manner (JIA et al. 1996). NO, can of course, also interact with enzymes containing non-heme iron, such as mitochondrial *cis*-aconitase and the cytosolic aconitase (also known as iron regulatory protein or IRP). Such reactions are, however, more likely to be associated with cytotoxic effects of large amounts of NO produced by activated macrophages rather than with the relatively modest amounts produced by the endothelium.

4. NO and Mitochondrial Respiration

NO also has a multitude of effects on cellular metabolism and is reported to reversibly inhibit mitochondrial respiration by increasing the apparent K_m of cytochrome-C oxidase for oxygen, an effect which could account for the observed increase in respiration in certain cell systems following NOS inhibition. Thus, NO may act as a physiological regulator of the affinity of mitochondrial respiration for oxygen, enabling mitochondria to act as oxygen sensors over the physiological range. Recently it was reported that low concentrations of NO inhibit complex IV in a reversible manner, whereas prolonged exposure to NO results in the gradual and persistent S-nitrosation and inhibition of complex I which may eventually lead to cell damage (CLEMENTI et al. 1998). NO may play a role in the regulation of mitochondrial respiration and tissue O_2 consumption under physiological conditions (SHEN et al. 1994; BERNSTEIN et al. 1996). Recent observations indicate that, of the three NOS isoforms expressed in vascular tissue, it is eNOS which plays a pivotal role in the control of O_2 consumption in the myocardium (LOKE et al. 1999).

IV. Dinitrosyl Iron Complexes, Nitrosothiol-Containing Proteins and Vascular Tone

Storage forms of NO may exist in vascular tissue and protein-bound dinitrosyl-iron complexes detected at high concentrations in certain tissues can provide a reservoir of S-nitrosating species, such as low molecular dinitrosyl-iron complexes (BOESE et al. 1995). In the aortae from rats treated with bacterial lipopolysaccharide, NO generated by the inducible NOS can be stored as protein-bound dinitrosyl-iron complexes within the vascular media and/or adventitia. NO, in the form of low-molecular-mass dinitrosyl-iron complexes, can be released from these storage sites following the application of N-acetyl-L-cysteine and induce vascular relaxation (MULLER et al. 1996). Although exogenously applied dinitrosyl-iron complexes can activate soluble guanylyl cyclase in vascular preparations, the release of dinitrosyl-iron complexes induced by N-acetyl-L-cysteine elicited a cyclic GMP-independent relaxation, which reportedly involves the activation of K^+_{Ca} channels (MULLER et al. 1998). Storage forms of NO exert biological actions distinct from NO. Thus, glutathione reductase is irreversibly inhibited by dinitrosyl-iron-di-L-cysteine and dinitrosyl-iron-di-glutathione in a concentration- and time-dependent manner. This inhibition was not accompanied by formation of a protein-bound dinitrosyl-iron complex and/or S-nitrosation of active site thiols, but could be attributed to the similarity of the NO carrier to S-nitroso glutathione and the enhanced transfer of NO^+ from the NO-carrier to the glutathione reductase (BOESE et al. 1997).

Important cellular targets for NO are sulphydryl (SH)-containing proteins. NO derivatives, such as the nitrosonium ion (NO^+), interact with these SH-groups to produce biologically active S-nitrosoproteins (PAWSON 1994). These S-nitrosylated proteins, like NO, are potent vasodilators and inhibitors of platelet aggregation, effects mediated via the activation of soluble guanylyl cyclase in the target cells. Nitrosothiols have a significantly longer half life in vivo (>30min) than NO (PAWSON 1994; LEE and GILMAN 1994) and, therefore, may serve as a storage form of NO and thus prolong its activity and/or facilitate its biological action.

D. NO and the Control of Blood Flow

Selective inhibition of NO synthesis causes an increase in mean arterial blood pressure and a reduction in blood flow, underlining the significance of shear stress-dependent endothelium-derived NO in global cardiovascular homeostasis. Knocking out the gene encoding eNOS in mice results in a maintained hypertension, and endothelium-intact aortic rings removed ex vivo from these animals display no relaxation to acetylcholine and are unaffected by treatment with NOS inhibitors (HUANG et al. 1995; SHESELY et al. 1996). Surprisingly, the systemic administration of NOS inhibitors to mice deficient in eNOS resulted in a decrease in mean arterial blood pressure which was prevented by L-arginine. Thus, more than one isoform of NOS, e.g. nNOS which is present both in vasomotor centres of the central nervous system and in peripheral nerves as well as certain vascular smooth muscle cells (BOULANGER et al. 1998), is likely to play a role in the global regulation of blood pressure.

Most vascular beds, except the lungs, tend to maintain their blood flow constant over a wide range of perfusion pressure. This phenomenon is commonly referred to as autoregulation of blood flow and involves complex interactions between three basic mechanisms: myogenic contraction, flow-induced dilation and the wash-out of vasodilator metabolites. Thus, the extent of autoregulation reflects the delicate balance between constrictor and dilator mechanisms, which varies from one vascular bed to another. For example the cutaneous circulation that normally exhibits only a very weak autoregulation (due to the low metabolic activity of the skin), autoregulates following NOS inhibition. However, in organs with high metabolic activity, NO exerts less influence on autoregulation. For example, in the heart, there is a predominant metabolite wash-out-mediated autoregulation. Thus, NOS inhibition does not markedly change the autoregulatory range but shifts this to a lower flow level.

Myogenic contraction occurs in response to instantaneous increases in transmural pressure, a response which forms the basis of the autoregulatory properties of any given vascular bed. Normally this response is functionally antagonised by the shear stress-induced release of NO; however, following inhibition of eNOS or a decrease in the availability of NO, this feedback mechanism is lost and the myogenic contraction of arteries/arterioles is unopposed. In many physiological situations in vivo, for example during exercise, intravascular pressure and flow are simultaneously elevated. In such situations, the net effect on vascular tone is the result of the dynamic interaction between pressure-induced myogenic tone and flow-induced dilation. The role of NO in this reaction is best demonstrated by studying the effects of a step increase in pres-

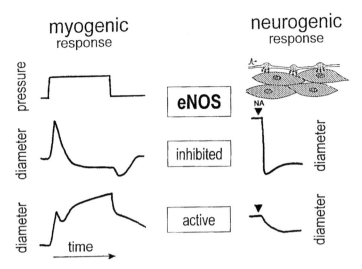

Fig. 2. Schematic representation of the diameter changes of small, in situ perfused arteries in response to rapid changes in transmural pressure or nerve stimulation, with and without a nitric oxide synthase (NOS) inhibitor. *NA* noradrenaline

sure and flow on the diameter of two arteries in an arterial bed, one treated with a NOS inhibitor and the other not. Following NOS inhibition, an increase in pressure results in a typical myogenic response, i.e. an initial transient vasodilatation followed by a prolonged vasoconstriction, such that the resulting arterial diameter is smaller than that measured prior to the pressure step. In the absence of a NOS inhibitor, the simultaneous increase in intraluminal pressure and flow converts this response to a maintained NO-mediated vasodilatation (Fig. 2). A similar behaviour can be demonstrated in a perfused arterial segment in response to a neurogenically-induced contraction. The contraction of this segment at constant flow leads to an increase in shear stress at the endothelial surface and thus elicits the shear stress-dependent production of NO which, in turn, functionally antagonises contraction and decreases shear stress. In the presence of a NOS inhibitor, this mechanism cannot function and arterial contraction is exaggerated. Thus each and every vasoconstriction of a perfused vessel increases the shear stress acting on endothelial cells, which respond by increasing the production of NO and this in turn attenuates the vasoconstriction.

The modulation of myogenic constriction by NO is more pronounced in larger than in small arterioles. In small (terminal) arterioles, the products of energy metabolism are effective dilators, but the accumulation of metabolites alone has relatively little effect on vascular resistance. The reason for this is that, in order to achieve optimal vascular conductance, both small arterioles and the larger arterioles feeding them must dilate in concert. This phenomenon of "conducted dilation" is thought to be determined by the electric propagation of local changes in membrane potential as well as by flow-dependent dilation of larger arterioles (de WIT et al. 1998). The flow-induced dilation is the consequence of an increase in endothelial NO production elicited by an enhanced fluid shear stress that results from the metabolite-induced dilation of downstream arterioles.

At this point it is necessary to note that not all vessels within one vascular bed respond in the same way to an accumulation of vascular metabolites or to mechanical stimuli (transmural pressure and shear stress). Metabolic control exerts a dominant influence on the smallest arterioles ($<20\,\mu$m), but is a generally less important stimulus in more upstream vessels. The pressure-induced myogenic response dominates in middle-sized arteries (20–30 μm), whereas its influence wanes upstream in relatively large arterioles (100–200 μm) in which fluid shear stress predominantly governs tone (Fig. 3).

I. Interaction Between NO and O_2^-

The regulation of vascular tone is complicated by the fact that the vascular endothelium can also favour constriction rather than dilation, as stimuli such as shear stress and circumferential stretch elicit the simultaneous production of both NO and O_2^-/H_2O_2 (HOWARD et al. 1997). A rapid increase in transmural pressure has been reported to trigger an active increase in tone in endothelium-intact canine coronary arteries as well as in endothelium-denuded coronary artery rings situated downstream from the donor vessel; both effects were dependent on the presence of the endothelium in the donor (RUBANYI 1988). Increased transmural pressure has also been shown to decrease the relaxant response to acetylcholine suggesting that the stretch may either attenuate the production of endothelium-derived relaxing factors or enhance that of endothelium-derived contracting factors from the donor segment (RUBANYI 1988; HUTCHESON and GRIFFITH 1991). This inhibitory effect on agonist-induced, NO-mediated dilator responses is even more pronounced in rhythmically stretched arterial segments (RYAN et al. 1995). Under such conditions, superoxide dismutase was able to prevent the impairment of relaxation to acetylcholine, suggesting an increase in the production of O_2^- (RYAN et al. 1995). Potential enzymatic sources of the increased endothelial O_2^- production are eNOS (PRITCHARD, JR. et al. 1995), PLA_2 (GINSBURG et al. 1989), COX (HOLLAND et al. 1988; KATUSIC and VANHOUTTE 1989), the NADH oxidoreductase (MOHAZZAB-H et al. 1994), xanthine oxidase (BULKLEY 1993) and cytochrome P_{450} enzymes (PRITCHARD, JR. et al. 1990). Which of these enzymes is activated by increased transmural pressure and is responsible for the increase in endothelium-derived O_2^- is unclear as N^Gnitro-L-arginine methyl ester (HUTCHESON and GRIFFITH 1991), indomethacin (RYAN et al. 1995) and the O_2^- generator/soluble guanylyl cyclase inhibitor, methylene blue (RUBANYI 1988), have been reported to inhibit the pressure-induced contraction and attenuation of the response to acetylcholine.

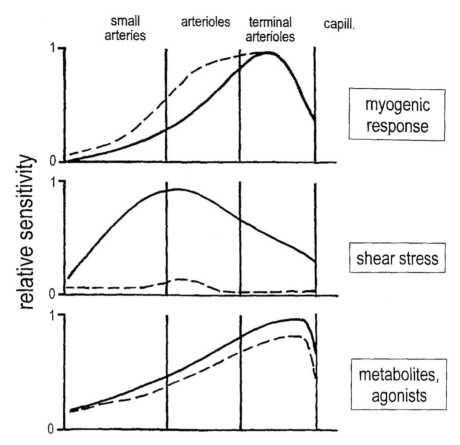

Fig. 3. Relative sensitivity of small arteries and arterioles to myogenic responses, fluid shear stress and metabolite accumulation. The *broken line* represents the sensitivity observed following nitric oxide synthase (NOS) inhibition

It may be possible to alter the balance between vascular NO and O_2^- production and to improve endothelium-dependent vasodilator responses. Regular exercise over prolonged periods is, for example, a method by which NO production and endothelium-dependent dilation can be improved, and myogenic contraction attenuated in humans and animals (JUNGERSTEN et al. 1997; JONSDOTTIR et al. 1998; DE KEULENAER et al. 1998). This beneficial effect is attributed to the increase in shear stress exerted on the endothelium which, in addition to eliciting the acute production of NO, stimulates the upregulation of "vascular protective" or antioxidant genes such as eNOS (SUN et al. 1994; WANG et al. 1993; SESSA et al. 1994; WOODMAN et al. 1997) and superoxide dismutase (DE KEULENAER et al. 1998; WOODMAN et al. 1999) and possibly the decrease in the expression of the O_2^- producing NADPH oxidase (DE KEULENAER et al. 1998).

II. NO and 20-HETE

20-Hydroxyeicosatetraenoic acid (20-HETE), an ω-hydroxylation product of arachidonic acid catalysed by cytochrome $P_{450}4A$, is produced by vascular smooth muscle cells and has been proposed to mediate the development of myogenic tone (HARDER et al. 1997). Initially, the effects of 20-HETE on vascular tone were described as endothelium-dependent and attributed to the metabolism of 20-HETE to an endoperoxide (ESCALANTE et al. 1989; SCHWARTZMAN et al. 1989) or PGI_2 by COX (CARROLL et al. 1996). However, 20-HETE is now known to be endogenously produced by smooth muscle cells from small renal and cerebral vessels in response to stretch or increases in transmural pressure (MA et al. 1997; HARDER et al. 1997; CARROLL et al. 1996) and is not affected by COX inhibition and endoperoxide/thromboxane receptor antagonism (MA et al. 1997). $P_{450}4A$ mRNA and protein are expressed in vascular smooth muscle cells (HARDER et al. 1994) and, once formed, 20-HETE acts as an intracellular second messenger and increases smooth muscle tone by inhibiting large conductance K^+_{Ca} channels, inducing depolarization and increasing $[Ca^{2+}]_i$, probably by activating L-type Ca^{2+} channels (MA et al. 1997; ZOU et al. 1996; IMIG et al. 1996). NO, which also increases in response to pulsatile flow, may also modulate the formation of 20-HETE (ALONSO-GALICIA et al. 1997) by binding and inactivating the P_{450} heme moiety in much the same way that it inhibits the EDHF synthase (BAUERSACHS et al. 1996). This effect may account for the cyclic GMP-independent vasodilator effects of NO in certain arterial preparations (TAKEUCHI et al. 1996) and may also explain the observation that NO donors inhibit L-type channel activity (SCHOBERSBERGER et al. 1997). Indeed, selective inhibition of 20-HETE production markedly decreased the vasodilator response to SNP in rat renal arterioles and attenuated the SNP-induced decrease in renal vascular resistance in anaesthetised rats (ALONSO-GALICIA et al. 1997).

20-HETE is not only involved in the regulation of myogenic tone, since cytochrome $P_{450}4A$ has recently been proposed to act as an O_2 sensor in the hamster skeletal muscle microcirculatory bed, such that the vasoconstrictor response observed in response to an increase in tissue O_2 delivery can be attributed to the enhanced production of 20-HETE (LOMBARD et al. 1999). Taken together, it is tempting to speculate that the open probability of large conductance K^+_{Ca} channels in arterial segments is determined by the balance in the vascular production of 20-HETE and EDHF/NO, and that EDHF and NO affect vascular tone by counteracting the 20-HETE-induced inhibition of K^+_{Ca} channels.

Acknowledgements. Experiments performed in the authors laboratory were supported by the Deutsche Forschungsgemeinschaft (SFB 553, B1 and B5).

References

Abu-Soud HM, Gachhui R, Raushel FM, Stuehr DJ (1997) The ferrous-dioxy complex of neuronal nitric oxide synthase. Divergent effects of L-arginine and tetrahydrobiopterin on its stability. J Biol Chem 272:17349–17353

Abu-Soud HM, Stuehr DJ (1993) Nitric oxide synthases reveal a role for calmodulin in controlling electron transfer. Proc Natl Acad Sci USA 90:10769–10772

Alonso-Galicia M, Drummond HA, Reddy KK, Falck JR, Roman RJ (1997) Inhibition of 20-HETE production contributes to the vascular responses to nitric oxide. Hypertension 29:320–325

Anderson RG (1993) Caveolae: where incoming and outgoing messengers meet. Proc Natl Acad Sci USA 90:10909–10913

Ando K, Hirata Y, Schichiri M, Emori T, Maruno F (1989) Presence of immunoreactive endothelin in human plasma. FEBS Lett 245:164–166

Andries LJ, Brutsaert DL, Sys SU (1998) Nonuniformity of endothelial constitutive nitric oxide synthase distribution in cardiac endothelium. Circ Res 82:195–203

Assreuy J, Cunha FQ, Liew FY, Moncada S (1993) Feedback inhibition of nitric oxide synthase activity by nitric oxide. Br J Pharmacol 108:833–837

Ayajiki K, Kindermann M, Hecker M, Fleming I, Busse R (1996) Intracellular pH and tyrosine phosphorylation but not calcium determine shear stress-induced nitric oxide production in native endothelial cells. Circ Res 78:750–758

Bauersachs J, Fleming I, Scholz D, Popp R, Busse R (1997) Endothelium-derived hyperpolarizing factor but not nitric oxide is reversibly inhibited by brefeldin A. Hypertension 30:1598–1605

Bauersachs J, Popp R, Hecker M, Sauer E, Fleming I, Busse R (1996) Nitric oxide attenuates the release of endothelium-derived hyperpolarizing factor. Circulation 94:3341–3347

Belhassen L, Feron O, Kaye DM, Michel T, Kelly RA (1997) Regulation by cAMP of post-translational processing and subcellular targeting of endothelial nitric-oxide synthase (type 3) in cardiac myocytes. J Biol Chem 272:11198–11204

Bernstein RD, Ochoa FY, Xu X, Forfia P, Shen W, Thompson CI, Hintze TH (1996) Function and production of nitric oxide in the coronary circulation of the conscious dog during exercise. Circ Res 79:840–848

Boese M, Keese MA, Becker K, Busse R, Mülsch A (1997) Inhibition of glutathione reductase by dinitrosyl-iron- dithiolate complex. J Biol Chem 272:21767–21773

Boese M, Mordvintcev PI, Vanin AF, Busse R, Mülsch A (1995) S-nitrosation of serum albumin by dinitrosyl-iron complex. J Biol Chem 270:29244–29249

Bolotina VM, Najibi S, Palacino JJ, Pagano PJ, Cohen RA (1994) Nitric oxide directly activates calcium-dependent potassium channels in vascular smooth muscle. Nature 368:850–853

Boulanger CM, Lüscher TF (1990) Release of endothelin from the porcine aorta: inhibition by endothelium-derived nitric oxide. J Clin Invest 85:587–590

Boulanger, CM, Heymes C, Benessiano J, Geske RS, Levy BI, Vanhoutte PM (1998) Neuronal nitric oxide synthase is expressed in rat vascular smooth muscle cells: activation by angiotensin II in hypertension. Circ Res 83:1271–1278

Bredt DS, Huang PM, Glatt CE, Lowenstein C, Reed RR, Snyder SH (1991) Cloned and expressed nitric oxide synthase structurally resembles cytochrome P-450 reductase. Nature 351:714–718

Bulkley GB (1993) Endothelial xanthine oxidase: a radical transducer of inflammatory signals for reticuloendothelial activation. Br J Surg 80:684–686

Busse R, Mülsch A (1990) Calcium-dependent nitric oxide synthesis in endothelial cytosol is mediated by calmodulin. FEBS Lett 265:133–136

Carroll MA, Balazy M, Margiotta P, Huang DD, Falck JR, McGiff JC (1996) Cytochrome *P-450*-dependent HETEs: Profile of biological activity and stimulation by vasoactive peptides. Am J Physiol 271:R863–R869

Chen P-F, Tsai AL, Wu KK (1994) Mutation of Glu-361 in human endothelial nitric-oxide synthase selectively abolishes L-arginine binding without perturbing the behavior of heme and other redox centers. J Biol Chem 269:25062–25066

Chen PF, Tsai AL, Berka V, Wu KK (1997) Mutation of Glu-361 in human endothelial nitric-oxide synthase selectively abolishes L-arginine binding without perturbing the behaviour of heme and other redox centers. J Biol Chem 272:6114–6118

Chen Z-P, Mitchelhill KI, Michell BJ, Stapelton D, Rodriguez-Crespo I, Witters LA, Power DA, Ortiz de Montellano PR, Kemp BK (1999) AMP-activated protein kinase phosphorylation of endothleial NO synthase. FEBS Lett 443:285–289

Clementi E, Brown GC, Feelisch M, Moncada S (1998) persistent inhibition of cell respiration by nitric oxide: Crucial role of S-nitrosylation of mitochondrial complex I and protective action of glutathione. Proc Natl Acad Sci USA 95:7631–7636

Cohen RA, Plane F, Najibi S, Huk I, Malinski T, Garland CJ (1997) Nitric oxide is the mediator of both endothelium-dependent relaxation and hyperpolarization of the rabbit caroid artery. Proc Natl Acad Sci USA 94:4193–4198

Cohen RA, Weisbrod R, Gerecke M, Yaghoubi M, Bierl C, Bolotina V (1999) Mechanism of nitric oxide induced vasodilatation. Refilling of intracellular stores by sarcoplasmic reticulun Ca^{2+} ATPase and inhibition of store-operated Ca^{2+} influx. Circ Res 84:210–219

Cornwell TL, Pryzwansky KB, Wyatt TA, Lincoln TM (1991) Regulation of sarcoplasmic reticulum protein phosphorylation by localized cyclic GMP dependent kinase in vascular smooth muscle cells. Mol Pharmacol 40:923–931

Corson MA, James NL, Latta SE, Nerem RM, Berk BC, Harrison DG (1996) Phosphorylation of endothelial nitric oxide synthase in response to fluid shear stress. Circ Res 79:984–991

Cosentino F, Katusic ZS (1995) Tetrahydrobiopterin and dysfunction of endothelial nitric oxide synthase in coronary arteries. Circulation 91:139–144

Crane BR, Arvai AS, Ghosh DK, Wu CQ, Getzoff ED, Stuehr DJ, Tainer JA (1998) Structure of nitric oxide synthase oxygenase dimer with pterin and substrate. Science 279:2121–2126

Davada RK, Chandler LJ, Guzman NJ (1994) Protein kinase C modulates receptor-independent activation of endothelial nitric oxide synthase. Eur J Pharmacol 266:237–244

Davidge ST, Baker PN, McLaughlin MK, Roberts JM (1995) Nitric oxide produced by endothelial cells increases production of eicosanoids through activation of prostaglandin H synthase. Circ Res 77:274–283

De Keulenaer GW, Chappell DC, Alexander RW, Nerem RM, Griendling KK (1998) Oscillatory and steady laminar shear stress differentially affect human endothelial redox state: role of a superoxide-producing NADH oxidase. Circ Res 82:1094–1101

de Toledo FG, Cheng J, Dousa TP (1997) Retinoic acid and triiodothyronine stimulate ADP-ribosyl cyclase activity in rat vascular smooth muscle cells. Biochem Biophys Res Commun 238:847–850

de Wit C, Jahrbeck B, Schäfer C, Bolz SS, Pohl U (1998) Nitric oxide opposes myogenic pressure responses predominantly in large arterioles in vivo. Hypertension 31:787–794

Dimmeler S, Fleming I, Fisslthaler B, Hermann C, Busse R, Zeiher AM (1999) Activation of nitric oxide synthase in endothelial cells by Akt-dependent phosphorylation. Nature 399:601–605

Escalante B, Sessa WC, Falck JR, Yadagiri P, Schwartzman M (1989) Vasoactivity of 20-hydroxyeicosatetraenoic acid is dependent on metabolism by cyclooxygenase. J Pharmacol Exp Ther 248:229–232

Feron O, Belhassen L, Kobzik L, Smith TW, Kelly RA, Michel T (1996) Endothelial nitric oxide synthase targeting to caveolae – Specific interactions with caveolin isoforms in cardiac myocytes and endothelial cells. J Biol Chem 271:22810–22814

Fleming I, Bauersachs J, Fisslthaler B, Busse R (1998a) Calcium-independent activation of the endothelial nitric oxide synthase in response to tyrosine phosphatase inhibitors and fluid shear stress. Circ Res 82:686–695

Fleming I, Fisslthaler B, Busse R (1998b) Shear stress-induced phosphorylation of the endothelial NO synthase. Nitric Oxide 2:76(Abstract)

Fleming I, Hecker M, Busse R (1994) Intracellular alkalinization induced by bradykinin sustains activation of the constitutive nitric oxide synthase in endothelial cells. Circ Res 74:1220–1226

Fulton D, McGiff JC, Quilley J (1994) Role of K^+ channels in the vasodilator response to bradykinin in the rat heart. Br J Pharmacol 113:954–958

Furchgott RF, Zawadzki JV (1980) The obligatory role of endothelial cells in the relaxation of arterial smooth muscle by acetylcholine. Nature 288:373–376

Galione A, White A, Willmott N, Turner M, Potter BVL, Watson SP (1993) cGMP mobilizes intracellular Ca^{2+} in sea urchin eggs by stimulating ADP-ribose synthesis. Nature 365:456–459

García-Cardena G, Fan G, Stern DF, Liu J, Sessa WC (1996b) Endothelial nitric oxide synthase is regulated by tyrosine phosphorylation and interacts with caveolin-1. J Biol Chem 271:27237–27240

García-Cardena G, Fan R, Shah V, Sorrentino R, Cirino G, Papapetropoulos A, Sessa WC (1998) Dynamic activation of endothelial nitric oxide synthase by Hsp90. Nature 292:821–824

García-Cardena G, Martasek P, Masters BS, Skidd PM, Couet J, Li SW, Lisanti MP, Sessa WC (1997) Dissecting the interaction between nitric oxide synthase (NOS) and caveolin – Functional significance of the NOS caveolin binding domain in vivo. J Biol Chem 272:25437–25440

García-Cardena G, Oh P, Liu J, Schnitzer JE, Sessa WC (1996a) Targeting of nitric oxide synthase to endothelial caveolae via palmitoylation: implications for nitric oxide signaling. Proc Natl Acad Sci USA 93:6448–6453

Garg UC, Hassid A (1991) Nitric oxide decreases cytosolic free calcium in Balb/c 3T3 fibroblasts by a cyclic GMP-independent mechanism. J Biol Chem 266:9–12

Garland CJ, McPherson GA (1992) Evidence that nitric oxide does not mediate the hyperpolarization and relaxation to acetylcholine in the rat small mesenteric artery. Br J Pharmacol 105:429–435

George MJ, Shibata EF (1995) Regulation of calcium-activated potassium channels by S-nitrosothiol compounds and cyclic guanosine monophosphate in rabbit aoronary artery myocytes. J Invest Med 43:451–458

Ghosh S, Gachhui R, Crooks C, Wu CQ, Lisanti MP, Stuehr DJ (1998) Interaction between caveolin-1 and the reductase domain of endothelial nitric-oxide synthase – Consequences for catalysis. J Biol Chem 273:22267–22271

Ginsburg I, Gibbs DF, Schuger L, Johnson KJ, Ryan US, Ward PA, Varani J (1989) Vascular endothelial cell killing by combinations of membrane-active agents and hydrogen peroxide. Free Radic Biol Med 7:369–376

Gradin K, Whitelaw ML, Toftgard R, Poellinger L, Berghard A (1998) A tyrosine kinase-dependent pathway regulates ligand-dependent activation of the dioxin receptor in human keratinocytes. J Biol Chem 269:23800–23807

Graier WF, Groschner K, Schmidt K, Kukovetz WR (1992) Increases in endothelial cyclic AMP levels amplify agonist-induced formation of endothelium-derived relaxing factor (EDRF). Biochem J 288:345–349

Harder DR, Gebremedhin D, Narayanan J, Jefcoat C, Falck JR, Campbell WB, Roman R (1994) Formation and action of a P-450 4A metabolite of arachidonic acid in cat cerebral microvessels. Am J Physiol 266:H2098–H2107

Harder DR, Lange AR, Gebremedhin D, Birks EK, Roman RJ (1997) Cytochrome P450 metabolites of arachidonic acid as intracellular signaling molecules in vascular tissue. J Vasc Res 34:237–243

Hecker M, Bara AT, Bauersachs J, Busse R (1994b) Characterization of endothelium-derived hyperpolarizing factor as a cytochrome P450-derived arachidonic acid metabolite in mammals. J Physiol (Lond.) 481:407–414

Hecker M, Mülsch A, Bassenge E, Förstermann U, Busse R (1994a) Subcellular localization and characterization of nitric oxide synthase(s) in endothelial cells: physiological implications. Biochem J 299:247–252

Hirata K, Kuroda R, Sakoda T, Katayama M, Inoue N, Suematsu M, Kawashima S, Yokoyama M (1995) Inhibition of endothelial nitric oxide synthase activity by protein kinase C. Hypertension 25:180–185

Holland JA, Pritchard KA Jr, Rogers NJ, Stemerman MB (1988) Preturbation of cultured human endothelial cells by atherogenic levels of low density lipoprotein. Am J Pathol 132:474–478

Howard AB, Alexander RW, Nerem RM, Griendling KK, Taylor WR (1997) Cyclic strain induces an oxidative stress in endothelial cells. Am J Physiol 41:C421–C427

Hoyt KR, Tang L-H, Aizenman E, Reynolds IJ (1992) Nitric oxide modulates NMDA-induced increases in intracellular Ca^{2+} in cultured rat forebrain neurones. Brain Res 592:310–316

Huang PL, Huang Z, Mashimo H, Bloch KD, Moskowitz MA, Bevan JA, Fishman MC (1995) Hypertension in mice lacking the gene for endothelial nitric oxide synthase. Nature 377:239–242

Hutcheson IR, Griffith TM (1991) Release of endothelium-derived relaxing factor is modulated both by frequency and amplitude of pulsatile flow. Am J Physiol 261:H257–H262

Hutcheson IR, Griffith TM (1996) Mechanotransduction through the endothelial cytoskeleton: mediation of flow- but not agonist-induced EDRF release. Br J Pharmacol 118:720–726

Imig JD, Zou AP, Stec DE, Harder DR, Falck JR, Roman RJ (1996) Formation and actions of 20-hydroxyeicosatetraenoic acid in rat renal arterioles. Am J Physiol 270:R217–R227

Ingber DE (1997) Tensegrity: The architectural basis of cellular mechanotransduction. Annu Rev Physiol 59:575–599

Jia L, Bonaventura C, Bonaventura J, Stamler JS (1996) S-nitrosohaemoglobin: a dynamic activity of blood involved in vascular control. Nature 380:221–226

Jonsdottir IH, Jungersten L, Johansson C, Wennmalm A, Thoren P, Hoffmann P (1998) Increase in nitric oxide formation after chronic voluntary exercise in spontaneously hypertensive rats. Acta Physiol Scand 162:149–153

Ju H, Zou R, Venema VJ, Venema RC (1997) Direct interaction of endothelial nitric-oxide synthase and caveolin-1 inhibits synthase activity. J Biol Chem 272:18522–18525

Jungersten L, Ambring A, Wall B, Wennmalm A (1997) Both physical fitness and acute exercise regulate nitric oxide formation in healthy humans. J Appl Physiol 82:760–764

Kannan MS, Fenton AM, Prakash YS, Sieck GC (1996) Cyclic ADP-ribose stimulates sarcoplasmic reticulum calaium release in porcine coronary artery smooth muscle. Am J Physiol 270:H801–H806

Karczewski P, hendrischke T, Wolf-Peter W, Morano I, Bartel S, Schrader J (1998) Phosphorylation of phospholamban correlates with relaxation of coronary artery induced by nitric oxide, adenosine, and prostacyclin in the pig. J Cell Biochem 70:49–59

Katusic ZS, Vanhoutte PM (1989) Superoxide anion is an endothelium-derived contracting factor. Am J Physiol 257:H33–H37

Kim YM, Tzeng E, Billiar TR (1997) Role of NO and nitrogen intermediates in regulation of cell functions. In: *Nitric oxide and the kidney*, edited by Goligorsky, M, Gross, S.S. Chapman & Hall, New York, pp 22–51

Komalavilas P, Lincoln TM (1996) Phosphorylation of the inositol 1,4,5-trisphosphate receptor – Cyclic GMP-dependent protein kinase mediates cAMP and cGMP dependent phosphorylation in the intact rat aorta. J Biol Chem 271:21933–21938

Komas N, Lugnier C, Stoclet J-C (1991) Endothelium-dependent and independent relaxation of the rat aorta by cyclic nucleotide phosphodiesterase inhibitors. Br J Pharmacol 104:495–503

Kuchan MJ, Frangos JA (1994) Role of calcium and calmodulin in flow-induced nitric oxide production in endothelial cells. Am J Physiol 266:C628–C636

Kurtz A, Götz KH, Hamann M, Kieninger M, Wagner C (1998) Stimulation of renin secretion by NO donors is related to the cAMP pathway. Am J Physiol 274:F709–F712

Lang D, Lewis MJ (1989) Endothelium-derived relaxing factor inhibits the formation of inositol trisphosphate by rabbit aorta. J Physiol (Lond.) 411:45–52

Lee G, Gilman M (1994) Dual modes of control of c-*fos* mRNA induction by intracellular calcium in T cells. Mol Cell Biol 14:4579–4587

Lee HC (1994) A signaling pathway involving cyclic ADP-ribose, cGMP, and nitric oxide. News Physiol Sci 9:134–137

Lee J, Schmid-Schonbein GW (1995) Biomechanics of skeletal muscle capillaries: hemodynamic resistance, endothelial distensibility, and pseudopod formation. Ann Biomed Eng 23:226–246

Li PL, Jin MW, Campbell WB (1998a) Effect of selective inhibition of soluble guanylyl cyclase on the K- Ca channel activity in coronary artery smooth muscle. Hypertension 31:303–308

Li PL, Zou AP, Campbell WB (1997) Regulation of potassium channels in coronary arterial smooth muscle by endothelium-derived vasodilators. Hypertension 29:262–267

Li PL, Zou AP, Campbell WB (1998b) Regulation of KCa-channel activity by cyclic ADP-ribose and ADP-ribose in coronary arterial smooth muscle. Am J Physiol 275:H1002–H1010

Lie M, Sejersted OM, Kiil F (1970) Local regulation of vascular cross section during changes in femoral arterial blood flow in dogs. Circ Res 27:727–737

Lincoln TM, Komalavilas P, Boerth NJ, MacMillan-Crow LA, Cornwell TL (1995) cGMP signaling through cAMP- and cGMP-dependent protein kinases. Adv Pharmacol 34:305–322

Lisanti MP, Scherer PE, Tang Z, Sargiacomo M (1994) Caveolae, caveolin and caveolin-rich membrane domains: a signalling hypothesis. Trends Cell Biol 4:231–235

List BM, Klosch B, Volker C, Gorren ACF, Sessa WC, Werner ER, Kukovetz WR, Schmidt K, Mayer B (1997) Characterization of bovine endothelial nitric oxide synthase as a homodimer with down-regulated uncoupled NADPH oxidase activity: Tetrahydrobiopterin binding kinetics and role of heme in dimerization. Biochem J 323:159–165

Liu J, García-Cardena G, Sessa WC (1996) Palmitoylation of endothelial nitric oxide synthase is necessary for optimal stimulated release of nitric oxide: implications for caveolae localization. Biochemistry 35:13277–13281

Liu J, Hughes TE, Sessa WC (1997a) The first 35 amino acids and fatty acylation sites determine the molecular targeting of endothelial nitric oxide synthase into the Golgi region of cells: a green fluorescent protein study. J Cell Biol 137:1525–1535

Liu J, Oh P, Horner T, Rogers R, Schnitzer JE (1997b) Organized endothelial cell surface signal transduction in caveolae distinct from glycosylphosphatidylinositol-anchored protein microdomains. J Biol Chem 272:7211–7222

Loke KE, McConnell PI, Tuzman JM, Shesely EG, Smith CJ, Stackpole CJ, Thompson CI, Kaley G, Wolin MS, Hintze TH (1999) Endogenous endothelial nitric oxide synthase-derived nitric oxide is a physiological regulator of myocardial oxygen consumption. Circ Res 84:840–845

Lombard JH, Kunert MP, Roman RJ, Falck JR, Harder DR, Jackson WF (1999) Cytochrome P-450 omega-hydroxylase senses O2 in hamster muscle, but not cheek pouch epithelium, microcirculation. Am J Physiol 276:H503–H508

Lückhoff A, Pohl U, Mülsch A, Busse R (1988) Differential role of extra- and intracellular calcium in the release of EDRF and prostacyclin from cultured endothelial cells. Br J Pharmacol 95:189–196

Lüscher TF, Boulanger CM, Dohi Y, Yang Z (1992) Endothelium-derived contracting factors. Hypertension 19:117–130

Mülsch A, Bassenge E, Busse R (1989) Nitric oxide synthesis in endothelial cytosol: evidence for a calcium-dependent and a calcium-independent mechanism. Naunyn Schmiedebergs Arch Pharmacol 340:767–770

Ma Y-H, Gebremedhin D, Schwartzman M, Falck JR, Masters BS, Harder DR, Roman RJ (1997) 20-Hydroxyeicosatetraenoic acid is an endogenous vasoconstrictor of canine renal arcuate arteries. Circ Res 72:136

Maniotis AJ, Chen CS, Ingber DE (1997) Demonstration of mechanical connections between integrins cytoskeletal filaments, and nucleoplasm that stabilize nuclear structure. Proc Natl Acad Sci USA 94:849–854

Marsden PA, Schappert KT, Chen HS, Flowers M, Sundell CL, Wilcox JN, Lamas S, Michel T (1992) Molecular cloning and characterization of human endothelial nitric oxide synthase. FEBS Lett 307:287–293

Martasek P, Liu Q, Liu J, Roman LJ, Gross SS, Sessa WC, Masters BS (1996) Characterization of bovine endothelial nitric oxide synthase expressed in E. coli. Biochem Biophys Res Commun 219:359–365

Michel JB, Feron O, Sacks D, Michel T (1997) Reciprocal regulation of endothelial nitric-oxide synthase by Ca^{2+}-calmodulin and caveolin. J Biol Chem 272:15583–15586

Michel T, Li GK, Busconi L (1993) Phosphorylation and subcellular translocation of endothelial nitric oxide synthase. Proc Natl Acad Sci USA 90:6252–6256

Mimnaugh EG, Worland PJ, Whitesell L, Neckers LM (1995) Possible role for serine/threonine phosphorylation in the regulation of the heteroprotein complex between the hsp90 stress protein and the pp60v-src tyrosine kinase. J Biol Chem 270:28654–28659

Mohazzab-H KM, Kaminski PM, Wolin MS (1994) NADH oxidoreductase is a major source of superoxide anion in bovine coronary artery endothelium. Am J Physiol 266:H2568–H2572

Monier S, Parton RG, Vogel F, Behlke J, Henske A, Kurzchalia TV (1995) VIP21-caveolin, a membrane protein constituent of the caveolar coat, oligomerizes in vivo and in vitro. Mol Biol Cell 6:911–927

Morgan RO, Newby AC (1989) Nitroprusside differentially inhibits ADP-stimulated calcium influx and mobilization in human platelets. Biochem J 258:447–454

Muller B, Kleschyov AL, Malblanc S, Stoclet J-C (1998) Nitric oxide-related cyclic GMP-independent relaxing effect of N-acetylcysteine in lipopolysaccharide-treated rat aorta. Br J Pharmacol 123:1221–1229

Muller B, Kleschyov AL, Stoclet J-C (1996) Evidence for N-acetylcysteine-sensitive nitric oxide storage as dinitrosyl-iron complexes in lipopolysaccharide-treated rat aorta. Br J Pharmacol 119:1281–1285

Muller JM, Chilian WM, Davis MJ (1997) Integrin signaling transduces shear stress-dependent vasodilation of coronary arterioles. Circ Res 80:320–326

Naganuma T, Miyakoshi M, Murayama T, Nomura Y (1998) Regulation of noradrenaline release by S-nitroso-cysteine: inhibition in PC12 cells in a cyclic GMP-independent manner. Eur J Pharmacol 361:277–283

Newby AC, Henderson AH (1990) Stimulus-secretion coupling in vascular endothelial cells. Ann Rev Physiol 52:661–674

Nishida K, Harrison DG, Navas JP, Fisher AA, Dockery SP, Uematsu M, Nerem RM, Alexander RW, Murphy TJ (1992) Molecular cloning and characterization of the constitutive bovine aortic endothelial cell nitric oxide synthase. J Clin Invest 90:2092–2096

O'Brien AJ, Young HM, Povey JM, Furness JB (1995) Nitric oxide synthase is localized predominantly in the Golgi apparatus and cytoplasmic vesicles of vascular endothelial cells. Histochemistry 103:221–225

Okamoto T, Schlegel A, Scherer PE, Lisanti MP (1998) Caveolins, a family of scaffolding proteins for organizing "preassembled signaling complexes" at the plasma membrane'. J Biol Chem 273:5419–5422

Pawson T (1994) Regulation of the Ras signalling pathway by protein-tyrosine kinases. Biochem Soc Transact 22:455–460

Plane F, Pearson T, Garland CJ (1995) Multiple pathways underlying endothelium-dependent relaxation in the rabbit isolated femoral artery. Br J Pharmacol 115:31–38

Plane F, Wiley KE, Jeremy JY, Cohen RA, Garland CJ (1998) Evidence that different mechanisms underlie smooth muscle relaxation to nitric oxide and nitric oxide donors in the rabbit isolated carotid artery. Br J Pharmacol 123:1351–1358

Pou S, Pou WS, Bredt DS, Snyder SH, Rosen GM (1992) Generation of superoxide by purified brain nitric oxide synthase. J Biol Chem 267:24173–24176

Presta A, Liu J, Sessa WC, Stuehr DJ (1997) Substrate binding and calmodulin binding to endothelial nitric oxide synthase coregulate its enzymatic activity. Nitric Oxide 1:74–87

Pritchard KA Jr, Groszek L, Smalley DM, Sessa WC, Wu M, Villalon P, Wolin MS, Stemerman MB (1995) Native low-density lipoprotein increases endothelial cell nitric oxide synthase generation of superoxide anion. Circ Res 77:510–518

Pritchard KA Jr, Wong P, Stemerman MB (1990) Atherogenic concentrations of low density lipoprotein enhance endothelial cell generation of epoxyeicosatrienoic acid products. Am J Pathol 136:1381–1391

Raman CS, Li H, Martásek P, Král V, Masters BS, Poulos TL (1998) Crystal structure of constitutive endothelial nitric oxide synthase: a paradigm for pterin function involving a novel metal center. Cell 95:939–950

Rizzo V, McIntosh DP, Oh P, Schnitzer JE (1999) *In situ* flow activates endothleial nitric oxide synhtase in luminal caveolae of endothelium with rapid caveolin dissociation and calmodulin association. J Biol Chem 273:34724–34729

Robertson BE, Schubert R, Hescheler J, Nelson MT (1993) cGMP-dependent protein kinase activates Ca-activated K channels in cerebral artery smooth muscle cells. Am J Physiol 265:C299–C303

Robinson LJ, Busconi L, Michel T (1995) Agonist-modulated palmitoylation of endothelial nitric oxide synthase. J Biol Chem 270:995–998

Rothberg KG, Heuser JE, Donzell WC, Ying Y-S, Glenney JR, Anderson RGW (1992) Caveolin, a protein component of caveolae membrane coats. Cell 68:673–682

Rubanyi GM (1988) Endothelium-dependent pressure-induced contraction of isolated canine coronary arteries. Am J Physiol 255:H783–H788

Ryan SM, Waack BJ, Weno BL, Heistad DD (1995) Increases in pulse pressure impair acetylcholine-induced vascular relaxation. Am J Physiol 268:H359–H363

Saijonmaa O, Ristimäki A, Fyhrquist F (1990) Atrial natriuretic peptide, nitroglycerine, and nitroprusside reduce basal and stimulated endothelin production from cultured endothelial cells. Biochem Biophys Res Commun 173:514–520

Salerno JC, Harris DE, Irizarry K, Patel B, Morales AJ, Smith SM, Martasek P, Roman LJ, Masters BS, Jones CL, Weissman BA, Lane P, Liu Q, Gross SS (1997) An autoinhibitory control element defines calcium- regulated isoforms of nitric oxide synthase. J Biol Chem 272:29769–29777

Salvemini D, Misko TP, Masferrer JL, Siebert K, Currie MG, Needleman P (1993) Nitric oxide activates cyclo-oxygenase enzymes. Proc Natl Acad Sci USA 90:7240–7244

Salvemini D, Seibert K, Masferrer JL, Misko TP, Currie MG, Needleman P (1994) Endogenous nitric oxide enhances prostaglandin production in a model of renal inflammation. J Clin Invest 93:1940–1947

Sander P, Kornfeld M, Ruan X, Arendshorst WJ, Kurtz A (1999) Nitric oxide/ cAMP interactions in the control of rat renal vascular resistance. Circ Res 84:186–192

Sautebin L, Ialenti A, Ianaro A, Di Rosa M (1995) Modulation by nitric oxide of prostaglandin biosynthesis in the rat. Br J Pharmacol 114:323–328

Schobersberger W, Friedrich F, Hoffmann G, Volkl H, Dietl P (1997) Nitric oxide donors inhibit spontaneous depolarizations by L-type Ca^{2+} currents in alveolar epithelial cells. Am J Physiol 272:L1092–L1097

Schretzenmayr A (1933) über kreislaufregulatorische Vorgänge an den großen Arterien bei der Muskelarbeit. Pfluger's Arch Ges Physiol 232:743–748

Schwartzman ML, Flack JR, Yadagiri P, Escalante B (1989) Metabolism of 20-hydroxyeicosatetraenoic acid by cyclooxygenase: formation and identification of a novel endothelium-dependent vasoconstrictor metabolite. J Biol Chem 264: 11658–11662

Schwarz P, Diem R, Dun NJ, Förstermann U (1995) Endogenous and exogenous nitric oxide inhibits norepinephrine release from rat heart sympathetic nerves. Circ Res 77:841–848

Sessa WC, García-Cardena G, Liu J, Keh A, Pollock JS, Bradley J, Thiru S, Braverman IM, Desai KM (1995) The Golgi association of endothelial nitric oxide synthase is necessary for the efficient synthesis of nitric oxide. J Biol Chem 270:17641–17644

Sessa WC, Harrison JK, Barber CM, Zeng D, Durieux ME, D'Angelo DD, Lynch KR, Peach MJ (1992) Molecular cloning and expression of a cDNA encoding endothelial cell nitric oxide synthase. J Biol Chem 267:15274–15276

Sessa WC, Pritchard KA, Seyedi N, Wang J, Hintze TH (1994) Chronic exercise in dogs increases coronary vascular nitric oxide production and endothelial cell nitric oxide synthase gene expression. Circ Res 74:349–353

Shaul PW, Smart EJ, Robinson LJ, German Z, Yuhanna IS, Ying YS, Anderson RGW, Michel T (1996) Acylation targets endothelial nitric oxide synthase to plasmalemmal caveolae. J Biol Chem 271:6518–6522

Sheehy AM, Burson MA, Black SM (1998) Nitric oxide exposure inhibits endothelial NOS activity but not gene expression: a role for superoxide. Am J Physiol 247: L833–L841

Shen W, Xu X, Ochoa M, Zhao G, Wolin MS, Hintze TH (1994) Role of nitric oxide in the regulation of oxygen consumption in concious dogs. Circ Res 75:1086–1095

Shesely EG, Maeda N, Kim HS, Desai KM, Krege JH, Laubach VE, Sherman PA, Sessa WC, Smithies O (1996) Elevated blood pressures in mice lacking endothelial nitric oxide synthase. Proc Natl Acad Sci USA 93:13176–13181

Shin WS, Sasaki T, Kato M, Hara K, Seko A, Yang W-D, Shimamoto N, Sugimoto T, Toyo-oka T (1992) Autocrine and paracrine effects of endothelium-derived relaxing factor on intracellular Ca^{2+} of endothelial cells and vascular smooth muscle cells. J Biol Chem 267:20377–20382

Singer AH, Peach MJ (1982) Calcium- and endothelial-mediated vascular smooth muscle relaxation in rabbit aorta. Hypertension 4:II-19–II-25

Sun D, Huang A, Koller A, Kaley G (1994) Short-term daily exercise activity enhances endothelial NO synthesis in skeletal muscle arterioles of rats. J Appl Physiol 76:2241–2247

Takeuchi T, Kishi M, Ishii T, Nishio H, Hata F (1996) Nitric oxide-mediated relaxation without concomitant changes in cyclic GMP content of rat proximal colon. Br J Pharmacol 117:1204–1208

Tanaka T, Okamura T, Handa J, Toda N (1999) Neurogenic vasodilation mediated by nitric oxide in porcine cerebral arteries. J Cardiovasc Pharmacol 33:56–64

Tare M, Parkington HC, Coleman HA, Neild TO, Dusting GJ (1990) Hyperpolarisation and relaxation of arterial smooth muscle caused by nitric oxide derived from the endothelium. Nature 346:69–71

Trepakova ES, Cohen RA, Bolotina V (1999) Nitric oxide inhibits capacitative cation influx in human platelets by promoting sarcoplasmic/endoplasmic reticulum Ca^{2+}-ATPase-dependent refilling of Ca^{2+} stores. Circ Res 84:201–209

Tsukahara H, Gordienko DV, Goligorsky M (1993) Continuous monitoring of nitric oxide release from human umbilical vein endothelial cells. Biochem Biophys Res Commun 193:722–729

Vasquez-Vivar J, Kalyanaraman B, Martasek P, Hogg N, Masters BS, Karoui H, Tordo P, Pritchard KA (1998) Superoxide generation by endothelial nitric oxide synthase: The influence of cofactors. Proc Natl Acad Sci USA 95:9220–9225

Venema RC, Sayegh HS, Kent JD, Harrison DG (1996) Identification, characterization, and comparison of the calmodulin-binding domains of the endothelial and inducible nitric oxide synthases. J Biol Chem 271:6435–6440

Venema VJ, Marrero MB, Venema RC (1996) Bradykinin-stimulated protein tyrosine phosphorylation promotes endothelial nitric oxide synthase translocation to the cytoskeleton. Biochem Biophys Res Commun 226:703–710

Wang J, Wolin MS, Hintze TH (1993) Chronic exercise enhances endothelium-mediated dilation of epicardial coronary artery in conscious dogs. Circ Res 73: 829–838

Weisbrod RM, Griswold MC, Yaghoubi M, Komalavilas P, Lincoln TM, Cohen RA (1998) Evidence that additional mechanisms to cyclic GMP mediate the decrease in intracellular calcium and relaxation of rabbit aortic smooth muslce cells to nitric oxide. Br J Pharmacol 125:1695–1707

Wever RM, Vandam T, Vanrijn HJ, Degroot F, Rabelink TJ (1997) Tetrahydrobiopterin regulates superoxide and nitric oxide generation by recombinant endothelial nitric oxide synthase. Biochem Biophys Res Commun 237:340–344

Woodman CR, Muller JM, Laughlin MH, Price EM (1997) Induction of nitric oxide synthase mRNA in coronary resistance arteries isolated from exercise-trained pigs. Amer J Physiol -Heart Circ Phy 42:H2575–H2579

Woodman CR, Muller JM, Rush JWE, Laughlin MH, Price EM (1999) Flow regulation of ecNOS and Cu/Zu SOD mRNA expression in porcine coronary arteries. Am J Physiol 276:H1058–H1063

Xia Y, Tsai AL, Berka V, Zweier JL (1998) Superoxide generation from endothelial nitric oxide synthase. J Biol Chem 273:25804–25808

Yanagisawa M, Kurihara H, Klimura S, Tomobe Y, Kobayashi M, Mitsui Y, Yazaki Y, Goto, K, Masaki, T (1988) A novel potent vasoconstrictor peptide produced by vascular endothelial cells. Nature 332:411–415

Zoche M, Beyermann M, Koch KW (1997) Introduction of a phosphate at serine(741) of the calmodulin-binding domain of the neuronal nitric oxide synthase (NOS-I) prevents binding of calmodulin. Biol Chem 378:851–857

Zou AP, Fleming JT, Falck JR, Jacobs ER, Gebremedhin D, Harder DR, Roman RJ (1996) 20-HETE is an endogenous inhibitor of the large-conductance Ca^{2+}-activated K^+ channel in renal arterioles. Am J Physiol 270:R228–R237

CHAPTER 9
Regulation of Cardiac Function by Nitric Oxide

J.-L. BALLIGAND

A. Introduction

There is now ample evidence for a regulatory role of nitric oxide (NO) in various aspects of cardiac myocyte biology, with important resulting effects on heart function. After the initial cloning of the three main isoforms of nitric oxide synthase (NOS), the discovery of their expression in various cell types comprising cardiac muscle emphasised the physiological importance of NO (produced autocrinally or paracrinally) on the modulation of cardiomyocyte contraction. The discovery also put in context many previously published results on the functional effect of drugs acting through the release of exogenous NO. After an overview of the aspects of NOS regulation relevant to the cardiomyocyte biology, we examine the evidence supporting the functional effect of NO on cardiac function and review the intracellular mechanisms mediating its contractile effects.

B. Specifics on Cardiac NOS Biology

Despite the apparent promiscuity of having all three NOS isoforms within cardiac muscle, a tight regulation of both the expression and the activity of each NOS isoform within a specific cell type ensures the coordinate physiological regulation of cardiac muscle contraction by the NO produced by each of these isoforms. Accordingly, the predominant cellular source of the NO that is produced in the heart may vary according to specific transcriptional and post-transcriptional stimuli differentially affecting each cell type and/or isoform [as will subsequently be illustrated for endothelial NOS (eNOS)].

I. Which Isoform(s)?

Despite the lack of detectable NOS-I in cardiac myocytes, either in its canonical form (BALLIGAND et al. 1995a; BELHASSEN et al. 1996) or muscle-specific isoform [neuronal NOS-μ (nNOS-μ)] (SILVAGNO et al. 1996; BELHASSEN et al. 1997) in the rat species, NOS-I protein has been identified in both cholinergic and non-adrenergic, non-cholinergic nerve terminals, in specialised conduction tissue and in sympathetic nerve terminals of the guinea-pig heart (SCHMIDT et al. 1992; TANAKA et al. 1993). A protein immunoreactive for anti-nNOS antibodies, but of slightly higher molecular weight, was also detected in the sarcoplasmic reticulum of rabbit cardiomyocytes (XU et al. 1999). Immunohistochemical co-localisation of NOS-I with tyrosine hydroxylase in cardiac neurones definitively demonstrated that orthosympathetic nerve terminals express NOS-I in the rodent heart (SCHWARTZ et al. 1995), where they were shown to regulate the release of catecholamines during electrical sympathetic nerve stimulation (SCHWARZ et al. 1995). Similar observations in PC-12 cells in vitro (KAYE et al. 1997) supported this conclusion.

NOS-III (eNOS) is expressed in endothelial cells from the endocardium and from arterial capillaries and veins in rodents and a variety of species (BALLIGAND 1995a; BALLIGAND and CANNON 1997; ANDRIES et al. 1998), including man (BALLIGAND and SMITH 1997; GAUTHIER et al. 1998a). The abundance of the immunostained proteins seems to be heterogenous along the coronary vascular bed (ANDRIES et al. 1998). In addition to endothelial cells, there is unequivocal evidence that eNOS is expressed in cardiomyocytes from atria, atrioventricular nodal and ventricular tissue in several species (BALLIGAND et al. 1995a; HAN et al. 1996; SEKI et al. 1996; WEI et al. 1996; GAUTHIER et al. 1998a). Several independent experimental approaches were used to support this conclusion, i.e. single-cell contractility or ionic current measurements, in situ hybridisation, and immunohistochemistry. Moreover, Feron and colleagues recently showed that, as in endothelial cells, palmitoylated and myristoylated eNOS in cardiac myocytes is localised to detergent-insoluble glycosphingolipid-rich microdomains in the plasmalemma, termed caveolae (FERON et al. 1996). These authors showed that, in extracts of rat cardiac myocytes, eNOS is co-immunoprecipitated with an isoform of caveolin, caveolin-3, which is specifically expressed in myocytes but not endothelial cells, thereby definitively demonstrating myocyte-specific expression of eNOS (FERON et al. 1996). The identification of a caveolin-binding domain within the eNOS protein is consistent with a direct protein–protein interaction between eNOS and caveolin-3 in cardiomyocytes, such as had been previously demonstrated between eNOS and caveolin-1 (MICHEL and FERON 1997; GARCIA-CARDENA et al. 1997). However, additional post-translational lipidations (e.g. palmitoylation) are necessary for the correct targeting of eNOS to caveolae (FERON et al. 1998a). Interestingly, it had been suggested for years that the T-tubular system developed from the coalescence of clusters of caveolae at the sarcolemmal membrane (ISHIKAWA 1968), and Parton and colleagues (PARTON et al. 1997) recently showed that caveolin-3 is also expressed in the T-tubular system in skeletal and cardiac muscle. Together, these pieces of evidence emphasise the potential role of eNOS in the regulation of excitation–contraction coupling in the heart, as will be reviewed below.

Cardiac microvascular endothelial cells and cardiac myocytes are among the many cell types within heart muscle known to express inducible NOS (iNOS) upon appropriate stimulation with specific combinations of inflammatory cytokines (BALLIGAND and CANNON 1997; KELLY et al. 1996). Even though many cardiovascular diseases resulting in heart failure are associated with increased expression and production of inflammatory cytokines [as directly demonstrated for tumor necrosis factor α (TNF-α); HABIB et al. 1996; TORRE-AMIONE et al. 1996], which are known to increase iNOS gene transcription in vitro, the relative abundance of iNOS expression within the different cell types in pathophysiological situations in vivo may be substantially different from the results observed after exposure of the cells to recombinant cytokines in vitro. In human allograft rejection, for example, there is a clear expression of iNOS within macrophages and myocytes, which (along with endothelial cells) account for most of the iNOS immunoreactivity (SZABOLCS et al. 1998). The amount of iNOS expression, as detected by immunostaining, appears to be more heterogeneous in the myocardium of patients with heart failure (HABIB et al. 1996; HAYWOOD et al. 1996; LEWIS et al. 1996). In a recent study in patients with end-stage heart failure due to ischaemic cardiomyopathy, sepsis and myocarditis (FUKUCHI et al. 1998), iNOS immunostaining was present in cardiac myocytes within infarcted and noninfarcted regions and was present in infiltrating macrophages. iNOS activity was variable and correlated significantly with the density of infiltrating macrophages. The results mentioned above suggest that the positivity of iNOS immunostaining varies according to the aetiology of the cardiomyopathy, with some studies reporting increase in iNOS expression restricted to idiopathic cardiomyopathy (HABIB et al. 1996) and others reporting increased expression only in myocardial dysfunction associated with sepsis (THOENES et al. 1996) and in patients with heart failure secondary to ischaemic coronary artery disease (FUKUCHI et al. 1998). As a potential explanation for this apparent inconsistency, one should keep in mind that iNOS protein expression is discontinuous over time, and iNOS expression in different cells in the myocardium may vary depending on the time interval of sampling and the progression of the disease in the tissue samples that have been examined.

II. How Are They Regulated?

We will focus on the expressional control and regulation of activity of the two isoforms known to be expressed in cardiomyocytes, namely eNOS and iNOS.

1. Endothelial Nitric Oxide Synthase

a) Expressional Control

In cultured ventricular cardiac myocytes, as in macrovascular endothelial cells (YOSHIZUMI et al. 1995), the abundance of eNOS mRNA is downregulated by inflammatory cytokines, such as TNF-α, interleukin 1β (IL-1β) and interferon γ(IFN-γ) (BALLIGAND et al. 1995a). The expression of eNOS is similarly downregulated in animal models of heart failure, such as the chronically paced dog, at least in the vasculature (SMITH et al. 1996). Studies of NO- and endothelial-dependent relaxation in the vasculature of patients with heart failure have led to similar observations of eNOS downregulation, which may explain, in part, the increased vascular resistance in heart failure. The question of whether eNOS is similarly downregulated in cardiac myocytes, however, is still unresolved despite some preliminary evidence in a rat model of cardiac hypertrophy (BAYRAKTUTAN et al. 1998). Importantly, a previous study (BELHASSEN et al. 1996) had demonstrated a differential regulation of eNOS expressed in cardiac myocytes versus microvascular endothelial cells in rat ventricular muscle. Increases in cyclic adenosine monophosphate (cAMP) following treatment of cultured cells with β-adrenergic agonists in vitro or treatment of rats with milrinone in vivo were associated with selective downregulation of both eNOS mRNA and protein in cardiac myocytes but not in microvascular endothelial cells. Such treatment in vitro was also associated with the appearance of a higher-molecular-weight form of eNOS (150kD) and its translocation to cytosolic fractions (BELHASSEN et al. 1996, 1997). Whether this shift towards a higher-molecular-weight form is associated with any alteration of eNOS activity is unknown. There is also some evidence that eNOS activity and protein levels are influenced by pertussis toxin in the rat heart through mechanisms that are presently uncharacterised (HARE et al. 1998b).

b) Acute Regulation of Activity

α. Mechanical Forces

Even though the constitutive isoforms of NOS are classically activated by acute increases in intracellular calcium, promoting the binding of calmodulin and consecutive allosteric activation of the electron flow for the oxidation of L-arginine (NATHAN et al. 1994), mechanical forces like flow-dependent shear stress stimulate the basal (i.e. in the absence of agonists) release of NO in the vasculature (HARRISON 1997) through calcium-independent mechanisms. This receptor-independent activation of eNOS is thought to involve changes in the phosphorylation state of eNOS-associated cytoskeletal proteins in endothelial cells (FLEMING et al. 1998). Similar mechanical forces may be acting on endothelial or endocardial cells within cardiac muscle. Using a NO-specific porphyrinic electrode implanted within different layers of beating ventricular myocardium, Pinsky and colleagues demonstrated that the NO signals were modulated by changes in preload and afterload. Since de-endothelisation of these preparations abrogated the NO signals, the authors concluded that endothelial cells were the predominant source of NO production under their experimental conditions (PINSKY et al. 1997).

β. Beating Rate

The K_a for calcium and calmodulin activation of eNOS activity is about $1\,\mu M$; therefore, it is within the range of intracellular calcium concentrations observed after the binding of agonists known to elevate intracellular calcium concentrations. It is also close to intracellular calcium levels measured during systole in cardiac myocytes, which prompted experiments testing the hypothesis that eNOS activation may be proportional to cardiac beating rate. Using electrically stimulated cultures of adult rat ventricular myocytes, Kaye and colleagues showed that myocyte NO production is activated by increasing the frequency of contraction and can be inhibited by NOS inhibitors (KAYE et al. 1996). Similarly, the force–frequency relationship was shown to be modulated by endogenously produced NO in guinea-pig papillary muscle through a N^G-monomethyl-L-arginine (L-NMMA)-sensitive mechanism (FINKEL et al. 1995).

γ. β-Adrenergic Agonists

Catecholamines acting on β1 and β2 receptors are well known to increase intracellular calcium concentration in cardiac myocytes through cyclic AMP elevations and protein kinase A (PKA)-dependent upregulation of L-type calcium current. Accordingly, exposure of cultured neonatal or adult rat ventricular myocytes to β1 or β2 (but not α) adrenergic agonists evoked L-NMMA-inhibitable increases in NO release (as measured with a porphyrinic microelectrode; KANAI et al. 1997) that are consistent with the previous demonstration of eNOS expression within cardiac myocytes, as mentioned above. Cyclic AMP analogues, guanosine triphosphates (GTPs) and the calcium channel activating dihydropyridine (BAYK 8644) all enhanced the NO signal, suggesting eNOS activation through the classical G-protein-coupled β1 or β2 adrenergic pathway. More recently, however, we observed that activation of the β3 adrenoceptor subtype in human ventricular muscle, which was previously shown to produce a striking negative inotropic effect (GAUTHIER et al. 1996), is coupled to eNOS activation and induces a L-NMMA-inhibitable increase in the intracellular concentration of cyclic guanosine monophosphate (cGMP). These observations were reproduced upon exposure of ventricular tissue both to specific adrenoceptor agonists of the β3 subtype and norepinephrine in the presence of full α1-, β1- and β2-adrenoceptor blockade. The fact that both the increases in cyclic GMP and the negative inotropic effect were fully inhibited upon incubation with pertussis toxin strongly suggests that β3 adrenoceptors in human heart are coupled to NOS activation through $G\alpha_{i/o}$ proteins (GAUTHIER et al. 1998a).

δ. Muscarinic Cholinergic Agonists

M2 receptors, which are predominantly expressed on cardiomyocytes, mediate acetylcholine's negative actions on heart rate and force of contraction through $G\alpha_i$-mediated inhibition of adenylyl cyclase or $G\alpha_q$ coupling to specific potassium currents (I_{K-Ach}). The specific coupling of any muscarinic receptor subtype to eNOS activation is poorly characterised but could involve muscarinic-receptor-mediated release of intracellular calcium, including in heart cells (FELDER 1995). Recently, FERON et al. demonstrated that muscarinic-receptor stimulation in isolated ventricular myocytes from rat hearts promotes the translocation of the receptors to caveolae (FERON et al. 1997), where components of the phosphoinositide pathway (including phospholipase C), sarcoplasmic reticulum calcium release channels and eNOS (FERON et al. 1996) are co-localised, thus rendering their functional interaction likely. However, concurrent evidence from several laboratories supports a "caveolin/eNOS regulatory cycle" in endothelial cells, where the formation of a heteroduplex between eNOS and caveolin inhibits the enzyme activity, and this protein–protein interaction is in equilibrium with the calcium-dependent binding of calmodulin to eNOS (MICHEL and FERON 1997).

According to this scheme, agonist-dependent increases in intracellular calcium and calmodulin binding would promote eNOS activation by displacing caveolin, thereby relieving its tonic inhibition of eNOS. As a corollary, upon increases in the abundance of caveolin, the efficient coupling of muscarinic receptors to eNOS activation could theoretically be hampered by the displacement of the binding equilibrium towards the formation of the inhibitory heteroduplex between eNOS and caveolin. Indeed, this was directly demonstrated in experiments in which the muscarinic cholinergic-induced increase in intracellular cyclic GMP and the decrease in beating rate (which were previously shown to be eNOS-dependent; BALLIGAND et al. 1993a) were abolished upon overexpression of caveolin-3 in transfected neonatal rat myocytes (FERON et al. 1998b). Similarly, reversible permeabilisation of the myocytes with peptides containing the caveolin scaffolding sequence (which mimicked the formation of the inhibitory heteroduplex with eNOS) uncoupled the muscarinic receptor, whereas control peptides with a scrambled sequence were inactive (FERON et al. 1998b). In addition, the physical co-localisation of all the necessary signaling proteins within the same subcellular compartment may be required for the efficient coupling of any membrane receptor to eNOS activation. This was suggested by the observation that the cholinergic muscarinic-dependent increase in cyclic GMP and decrease in beating rate, which are lost in cultured neonatal myocytes from mice deficient in the eNOS gene (eNOS$^{-/-}$), can be restored upon transfection of the myocytes with the wild-type eNOS but not the eNOS mutant (G2AeNOS) deficient for the miristoylation site, which cannot be targeted to the plasma membrane (FERON et al. 1998b). In addition to facilitating receptor-dependent activation of the enzyme, targeting of eNOS to the plasma membrane may also promote its interaction with other proteins (from the cytoskeleton) that could participate in mechanical (i.e. through increases in stretch, shear) or other potentially calcium-independent mechanisms of eNOS activation (FLEMING et al. 1997).

ε. Acute Effect of Cytokines

Aside from their effect on the transcriptional regulation of the iNOS gene, cytokines and inflammatory mediators may also increase NO production in cardiovascular tissue through direct activation of the constitutively expressed eNOS. This was suggested by the early observations of the acute release of a

NO-like factor from aortic endothelial cells induced by endotoxin within a time course that would exclude de novo gene expression (SALVEMINI et al. 1990). A similar acute effect independent of increased gene transcription or protein expression was identified in the study by Finkel and colleagues, who demonstrated that exposure of hamster papillary muscles to TNF-α, IL-2 or IL-6 decreased the amplitude of the muscles' contraction within minutes (FINKEL et al. 1992). The inhibition of this negative inotropic effect by NOS inhibitors, such as L-NMMA, strongly suggested the involvement of a constitutively expressed isoform of NOS. In support of the above conclusion, Goldhaber and colleagues subsequently showed that recombinant TNF-α decreased the contractile amplitude of shortening of isolated rabbit ventricular myocytes through L-NMMA-sensitive mechanisms (GOLDHABER et al. 1996).

2. Inducible Nitric Oxide Synthase

a) Expressional Control

A comprehensive enumeration of the transcriptional and post-transcriptional regulations of iNOS expression in cardiovascular cells is beyond the scope of this review but can be found elsewhere (KELLY et al. 1996; BALLIGAND and CANNON, 1997). Several studies have examined the effects of many cytokines, including IL-1β, TNF-α, IFN-γ and transforming growth factor β, on the expression of iNOS in the specific context of cardiac myocytes (ROBERTS et al. 1992; BALLIGAND et al. 1994; PINSKY et al. 1995a; UNGUREANU-LONGROIS et al. 1995a; SINGH et al. 1996). Using neonatal rat ventricular myocytes transfected with the iNOS promoter, key transcriptional regulators of iNOS gene expression in cardiomyocytes were identified; these include nuclear factor κB, IFN regulatory factor 1 and cyclic-AMP-responsive element-binding protein (CREB; KINUGAWA et al. 1997). Interactions with CREB may explain previous observations of potentiation of iNOS induction following angiotensin-II or phenylephrine stimulation by increases in intracellular cyclic AMP (IKEDA et al. 1995, 1996). Additional experiments, however, suggested that the effect of cyclic AMP was mediated through enhanced iNOS mRNA stability (ODDIS et al. 1995).

b) Acute Regulation of Activity

Contrary to the situation for constitutive isoforms, the K_a for calcium/calmodulin binding to iNOS is well below the cytoplasmic calcium concentration in most cells, so once iNOS protein is expressed, it retains calmodulin bound to its consensus sequence regardless of acute physiological fluctuations of intracellular calcium. Therefore, the enzyme activity is not regulated through receptor- (or beating-) dependent increases in calcium. Once expressed, it produces large amounts of NO for longer periods of time, provided both substrate and cofactors are not critically limited (NATHAN et al. 1994). Accordingly, the activity of iNOS is regulated post-translationally by increases in the availability of its substrate, L-arginine, following cytokine induction of L-arginine transporters in cardiac myocytes and microvascular endothelial cells (SIMMONS et al. 1996a). The same cytokines also induce the expression of GTP cyclohydrolase, the rate-limiting enzyme for de novo synthesis of tetrahydrobiopterin, a key cofactor for NOS activity (BALLIGAND et al. 1994; SIMMONS et al. 1996b) in rat hearts. A similar effect was shown to impact the cytokine-induced production of NO in rat cardiac myocytes (KASAI et al. 1997).

C. Intracellular Mechanisms of Action of NO in Cardiac Muscle Cells

The effects of NO are classically distinguished between those that are dependent of cGMP formation following the activation of guanylyl cyclase (including through the formation of S-nitrosoglutathione; MAYER et al. 1998) and those that are independent of cyclic GMP (Fig. 1). This classification, however, does not preclude from simultaneous actions of NO through both mechanisms, depending on such parameters as the amount of NO produced, the intracellular versus extracellular source, intracellular compartmentation and local redox conditions, all of which are likely to impact NO reactivity with its intracellular targets. In addition, there are important variations in the expression and distribution of effectors of NO among species and regions of the heart considered, which may explain some of the apparent discrepancies in the results of previous studies on the effect of NO or cyclic GMP on cardiac contraction. We will systematically review those cGMP-dependent and -independent mechanisms that likely result either in an increase or a decrease in cardiomyocyte contractile function, as listed in Table 1.

I. Cyclic GMP-Dependent Mechanisms

1. Contraction-Enhancing Mechanisms

In the early 90s, Galione and colleagues provided evidence that intracellular cyclic GMP could activate intracellular calcium release from the ryanodine channel in sea-urchin eggs through activation of adenosine diphosphate (ADP) ribosyl cyclase and subsequent increases in cyclic ADP ribose (GALIONE et al. 1991). The same group subsequently demonstrated that the ryanodine calcium channel could be activated by exogenous NO not only in reconstituted lipid bilayers but also in intact sea-urchin eggs; this prompted the search for a putative similar action of NO to regulate calcium-induced calcium release in intact cardiac myocytes. Despite early negative data (apparently ruling out such a pathway in intact guinea-pig cardiac myocytes; GUO et al. 1996), Galione and colleagues more recently demonstrated that, at physiological temperatures, exogenous NO was able to increase calcium

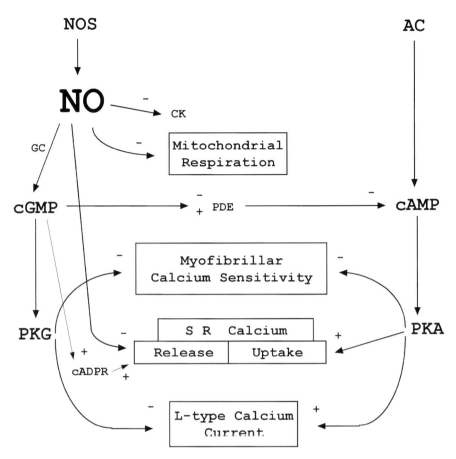

Fig. 1. Multiple targets for the effect of nitric oxide (*NO*) on cardiomyocyte contraction. *Right side*: Cyclic adenosine monophosphate (*cAMP*), produced by the membrane-bound enzyme adenylyl cyclase (*AC*), activates protein kinase A, which mediates most of the classic inotropic effects of adrenergic stimulation of heart muscle. (i) Positive inotropic effect through increase in L-type calcium current. (ii) Positive lusitropic effect through increase in sarcoplasmic reticulum (*SR*) calcium uptake and phosphorylation of troponin I to decrease myofilament calcium sensitivity. *Left side*: NO produced by endothelial nitric oxide synthase (*eNOS*) or inducible nitric oxide synthase in cardiac myocytes (or neighboring cells) may oppose or potentiate the above-mentioned effects through mechanisms that are dependent or independent of cyclic guanosine monophosphate (*cGMP*), formed by NO's activation of guanylyl cyclase (*GC*). Through allosteric interactions, cGMP stimulates or inhibits specific isoforms of phosphodiesterases that hydrolyze cyclic AMP. cGMP activates protein kinase G, which either phosphorylates troponin-I to decrease myofilament calcium sensitivity, or acts on the L-type calcium channel (or an intermediate protein) to decrease calcium current. cGMP also stimulates the formation of cyclic adenosine diphosphate ribose, which activates the release of calcium through the ryanodine-sensitive channel. NO binds non-heme iron and inactivates enzymes (complex I, II, IV) in the mitochondrial chain to decrease mitochondrial respiration and oxygen consumption. Through *S*-nitrosation of critical thiol residues, NO inhibits creatine kinase and can reversibly (or irreversibly upon further oxidation) activate the ryanodine channel; alternatively, NO produced by eNOS may downregulate the ryanodine receptor and decrease calcium-induced calcium release

Table 1. Cyclic guanosine monophosphate (cGMP)-dependent and -independent mechanisms of the regulation of cardiac myocyte contraction by nitric oxide

Effector	Target	Mechanism	Effect	Reference
cGMP-dependent				
cGMP	cGMP PDE-II	Allosteric interaction	Activation of cAMP hydrolysis inhibition	Mery et al. 1993; Han et al. 1998
cADPR	cGMP PDE-III	Sensitization of RyR to CICR	Increased SR calcium release	Ono et al. 1991
	RyR			Ino et al. 1997
PKG	Troponin-I	Phosphorylation	Decreased myofilament calcium sensitivity	Pfitzer et al. 1982; Robertson et al. 1982
	L-type calcium channel (or regulatory protein)	Phosphorylation	Decrease in calcium current	Mery et al. 1991
cGMP-independent				
NO (or derivative)	complex I, II, IV	Binding to non-heme iron	Inhibition of mitochondrial respiration	Shen et al. 1994; Oddis et al. 1995
S-nitrosothiol	Creatine kinase	S-nitrosation of thiol residue	Enzyme inhibition	Gross et al. 1996; Arstall et al. 1998
S-nitrosothiol	L-type calcium channel	S-nitrosation?	Decrease in calcium current	Hu et al. 1997
Peroxinitrite	L-type calcium channel	Oxidation	Increase in calcium current	Campbell et al. 1996
S-nitrosothiol	RyR	Reversible poly-S nitrosation	Increase in SR calcium release	Xu et al. 1998
Peroxinitrite	RyR	Oxidation	Deregulated calcium release	Ibid.
NO or derivative produced by eNOS	RyR	?	Decrease in SR calcium release and decreased CICR	Zahradnikova et al. 1997

cADPR, cyclic adenosine diphosphate ribose; *CICR*, calcium-induced calcium release; *PDE*, phosphodiesterase; *PKG*, protein kinase G (cyclic GMP-dependent protein kinase); *RyR*, ryanodine receptor; *SR*, sarcoplasmic reticulum.

release from the sarcoplasmic reticulum in intact guinea-pig myocytes (INO et al. 1997).

Cardiomyocytes from most mammalian species express isoforms of phosphodiesterases, such as PDE-III, which is allosterically inhibited by cyclic GMP. PDE-III was suggested to mediate NO-induced increases in cyclic AMP with subsequent activation of L-type calcium current and inotropy in isolated cardiac myocytes from rodents and human atria (MERY et al. 1993; KIRSTEIN et al. 1995). Low concentrations of NO donors also induced a moderate, positively inotropic effect in adult rat ventricular myocytes (KOJDA et al. 1996) and in open-chest dog hearts (PRECKEL et al. 1997). In feline cardiac myocytes, the same pathway may mediate the rebound increase in L-type calcium current, calcium transient and contraction after the acute removal of acetylcholine (WANG et al. 1998).

2. Contraction-Decreasing Mechanisms

Cardiac myocytes from amphibian, mammalian and human species also express PDE-2 that is activated by cGMP. Cyclic GMP-dependent activation of PDE-2 produces effects opposite to those of PDE-III, with decreases in intracellular cyclic AMP and its downstream effectors (including L-type calcium current), thereby decreasing myocyte contraction. This pathway was shown to be operative in the muscarinic cholinergic-dependent "accentuated antagonism" (i.e. following initial adrenergic stimulation and cyclic AMP increases) that involves eNOS activation in sino-atrial, atrio-ventricular and ventricular myocytes from rabbits and rats (HAN et al. 1995, 1996, 1998). Evidence provided from several groups has led to the suggestion that low concentrations of exogenous NO potentiate inotropic parameters, whereas higher concentrations do the opposite (ONO and TRAUTWEIN 1991; KOJDA et al. 1996). In cardiac myocytes in which iNOS was induced with cytokines, endogenous NO and cyclic GMP attenuate both the increase of isoproterenol-stimulated cyclic AMP as well as the shortening of adult rat cardiac myocytes in culture (JOE et al. 1998). The fact that the attenuation in cyclic AMP generation was partly abrogated upon treatment of the cells with 3-isobutyl-1-methylxanthine clearly implicated phosphodiesterase stimulation as one mechanism of the attenuation of cardiac myocyte contraction by endogenously produced NO. In addition to activating PDE-2, cyclic GMP can attenuate cardiomyocyte contraction by activating cyclic GMP-dependent protein kinase (PKG). PKG may, in turn, downregulate the L-type calcium current stimulated by PKA (MERY et al. 1991; WAHLER and DOLLINGER 1995) or the contractile responses of contractile proteins, independently of changes in calcium transients (GOLDHABER et al. 1996; YASUDA and LEW 1997). The latter mechanism could operate through the previously described phosphorylation of troponin-I, which desensitised cardiac myofilaments to calcium (PFITZER et al. 1982; ROBERTSON et al. 1982). However, the demonstration that such mechanism contributes to the contractile response to endogenously produced NO is still lacking.

II. Cyclic GMP-Independent Mechanisms

1. Contraction-Enhancing Mechanisms

Accumulating evidence indicates that the oxidation of critical thiol residues on regulatory proteins by NO or peroxynitrite may account for effects of NO on various parameters of cardiac contraction. Xu and coworkers recently reported that poly-S-nitrosation of the calcium-release channel (ryanodine receptor) resulted in its activation in lipid bilayers and that this effect was reversed by denitrosation. More profound oxidation of the channel resulted in its deregulated activation (Xu et al. 1998). Similarly, thiol oxidants were shown to stimulate L-type calcium currents in intact ferret ventricular myocytes in a redox-dependent manner (Campbell et al. 1996).

2. Contraction-Decreasing Mechanisms

Conversely, heterologously expressed cardiac L-type calcium channels were shown to be directly inhibited by S-nitrothiols (Hu et al. 1997). In addition, experiments from Meszaros and colleagues recently demonstrated in membrane preparations that the cardiac-specific ryanodine receptor channel was inhibited by NO produced endogenously by the eNOS protein (Zahradnikova et al. 1997). These observations emphasise the need to integrate the various, often antagonistic effects of NO on excitation–contraction coupling and the need to systematically verify that any effect produced with pharmacological concentrations of NO donors is representative of the action of NO as produced by eNOS targeted to specific membrane compartments. In addition to the effects on excitation–contraction coupling and calcium transients discussed above, NO may also potentially regulate key glycolytic enzymes and mitochondrial respiration, thereby affecting oxygen consumption and adenosine triphosphate generation within heart muscle (Shen et al. 1995; Kelly et al. 1996; Balligand and Cannon 1997; Kelm et al. 1997). S-nitrosation of creatinine kinase was shown to decrease its activity in intact isolated guinea-pig hearts and isolated rat ventricular myocytes in vitro (Gross et al. 1996; Arstall et al. 1998). This effect was reproduced by endogenously produced NO following iNOS induction in endothelial cells co-cultured with cardiomyocytes (Arstall et al. 1998). In studies using slices of canine myocardium, NO reduced myocardial oxygen consumption, presumably by inhibiting mitochondrial electron transport. This effect was reproduced in experiments where endogenous NO production by iNOS was stimulated in neonatal myocytes treated with IL-1β (Shen et al. 1994,1995; Oddis and Finkel 1995). Peroxynitrite, a strong oxidant that is produced from the interaction of NO and superoxide at high concentrations, can produce nitration of tyrosines of myocardial proteins, potentially altering their function (Beckman and Koppenol 1996). DNA damage induced by peroxynitrite was also shown to activate poly-ADP-ribose synthase and result in mitochondrial inhibition and depletion of cellular energy stores in cultures of cardiac myoblasts (Gilad et al. 1997). These effects of peroxynitrite probably account for some of the myocardial dysfunction in circumstances, such as septic shock, where high amounts of NO and superoxide anions are likely to be produced by the high-output iNOS under cytokine stimulation. High levels of oxidative stress may result from the additional inhibition of glutathione peroxidase by NO, as was shown in cultured cardiac myocytes following iNOS induction, with a resultant increase in the susceptibility of the cells to oxidant-mediated injury (IGARASHI et al. 1998).

D. Regulation of Cardiac Function by eNOS
I. Basal Systolic and Diastolic Function

A positive lusitropic effect of NO, produced endogenously by endocardial eNOS, was shown in ferret papillary muscle, in which application of substance P shortened the duration of contraction by inducing earlier onset of relaxation, with little effect on the peak force of contraction (SMITH et al. 1991). Such a paracrine effect of NO was also observed in co-cultures of isolated guinea-pig cardiomyocytes and endothelial cells from macrovessels, in which paracrinally produced NO following stimulation of endothelial cells with bradykinin reduced the amplitude of shortening of the co-cultured myocytes (BRADY et al. 1993). Similarly, stimulation of endothelial NOS with bradykinin or substance P in isolated ejecting guinea-pig hearts induced an earlier onset and acceleration of the diastolic left ventricular (LV) pressure fall, with no significant alteration of systolic parameters (GROCOTT-MASON et al. 1994). More recently, the same group reported that inhibition of eNOS decreased the Frank-Starling responses of isolated, perfused hearts by a mechanism that was apparently independent of changes in coronary vasomotor tone. The data were interpreted to support a role of constitutively expressed eNOS in modulating appropriate diastolic relaxation in response to increases in preload (PRENDERGAST et al. 1997). In contrast to these results, inhibition of eNOS did not change any parameter of systolic contraction or diastolic relaxation in electrically stimulated isolated rat cardiomyocytes or human endomyocardial biopsy specimens (which contain a mixture of endocardial, microvascular endothelial cells and myocytes) in the absence of agonist stimulation (BALLIGAND et al. 1993a; GAUTHIER et al. 1998a). Qualitatively and quantitatively similar results were also recently obtained in isolated ventricular myocytes from homozygous mice in which the eNOS gene had been disrupted; the myocytes from the eNOS-deficient mice exhibited baseline contractility parameters which were not significantly different from those of the wild-type animals (HAN et al. et al. 1998).

Substance P, which is known to increase the release of NO from the vasculature when infused intracoronarily, produced a fall in peak LV systolic pressure and an earlier onset of LV relaxation in the absence of any change in instantaneous LV pressure or LV end diastolic (ED) volume in patients with normal cardiac function and in transplant recipients free of rejection or vasculopathy. The earlier onset of relaxation in the absence of changes in ED volume or pressure was consistent with an increase in distensibility induced by paracrinally produced NO (PAULUS et al. 1995). Others have also shown that inhibition of endogenous eNOS with NOS inhibitors decreases myocardial oxygen consumption, an effect that has been correlated with the small positive inotropic effect of low doses of NO (SHERMAN et al. 1997). Conversely, increases in endogenous NO production with angiotensin-converting enzyme inhibitor treatment depress basal myocardial oxygen consumption (LAURSEN and HARRISON 1997; ZHANG et al. 1997). The latter observation is consistent with an effect of NO to modulate mitochondrial respiration, as outlined above.

II. Regulation of β-Adrenergic Response

Experiments with NO-donor drugs have identified either a potentiation or a downregualtion of the contractile response to adrenergic agonists in isolated hearts or cardiomyocytes, depending mainly on the concentration of NO, as outlined in Sect. C. Most studies evaluating the contractile effect of NO endogenously produced by NOS concluded that it produced a negative inotropic effect attenuating the effect of catecholamines on heart muscle, as reviewed below.

Accordingly, we initially showed that inhibition of eNOS potentiates the positive inotropic response to isoproterenol stimulation in isolated rat ventricular myocytes (BALLIGAND et al. 1993a). The same observation was later reproduced in rat isolated atria (STERIN-BORDA et al. 1998). Inhibition of NOS in human atrial strips in vitro also potentiated the inotropic responsiveness to isoproterenol (GAUTHIER et al. 1998b). In the same preparation, NOS inhibitors enhanced the arrhythmogenic effects of sub-maximal concentrations of isoproterenol (GAUTHIER et al. 1998b). The latter observation suggests that, in addition to modulation of the inotropic state, endogenous production of NO in the myocardium may also elevate the threshold for adrenergically induced ventricular arrhythmia. This was recently shown in open-chest dogs with acute coronary-artery occlusion, in which intrapericardial perfusion with L-arginine to increase NOS activity (as reflected by an increased NO effluent in the coronary sinus) also decreased the occurrence of ventricular fibrillation (FEI et al. 1997). Altogether, these results point to a role of endogenously produced NO in attenuating the contractile responsiveness to β-adrenergic stimulation. Such a role has been clearly demonstrated for the contractile response to isoproterenol in anesthetised dogs (KEANEY et al. 1996) and in preliminary experiments in transgenic mice deficient for eNOS that exhibited an enhanced contractile responsiveness to positive inotropic agents in comparison to wild-type controls (GYURKO et al. 1997; GODECKE et al. 1998).

To evaluate the physiological significance of these findings in humans, Hare and colleagues administered an intracoronary infusion of L-NMMA to patients and examined the inotropic response following intravenous infusion of dobutamine (a peripheral infusion to avoid confounding effects of a higher amine concentration in a vasoconstricted coronary vascular bed). In patients with heart failure, they observed a significant potentiation of the positive inotropic effect of dobutamine, an effect which was not observed in patients with normal heart function (HARE et al. 1995a, 1998a). A potential explanation for this discrepancy may be the increased sensitivity of the failing myocardium to changes in the amount and consecutive regulatory influence of endogenously produced NO. Consistent with this interpretation, the inotropic response to dobutamine was markedly blunted when cardiac eNOS was activated by intracoronary substance P infusion in transplant patients and patients with dilated non-ischaemic cardiomyopathy (BARTUNEK et al. 1997).

III. Regulation of Muscarinic Cholinergic Response

The paradigm of the obligatory role of endogenously expressed eNOS in mediating the negative chronotropic effect of muscarinic cholinergic stimulation was first demonstrated in cultures of rat neonatal ventricular myocytes (BALLIGAND et al. 1993a). This paradigm was later verified in isolated rabbit sino-atrial node cells (HAN et al. 1994, 1995), rabbit atrioventricular cells (HAN et al. 1996) and rat ventricular myocytes (BALLIGAND et al. 1995a), where eNOS was shown to mediate the parasympathetic accentuated antagonism. Despite earlier negative data (KILTER et al. 1995), the same paradigm was also verified in atrial strips from human hearts, in which the muscarinic cholinergic attenuation of isoproterenol-stimulated contraction was also found to be significantly blunted on NOS inhibition with L-NMMA (GAUTHIER et al. 1998b). Earlier negative data obtained in frog atrial myocytes (MERY et al. 1996) and guinea-pig cardiomyocytes (STEIN et al. 1993) apparently received some confirmation from the observation, by the same group, of the unperturbed parasympathetic responsiveness of isolated myocytes or atria from mice genetically deficient for eNOS (VANDECASTEELE et al. 1999). However, the presence of significant hypertrophy in these older (3–6 months) eNOS$^{-/-}$ mice, a phenotype not independently accounted for in the control group, may invalidate these conclusions (HARE and STAMLER 1999). By contrast, Han et al demonstrated that the accentuated antagonism is completely lost in younger (<3 months) mice homozygously deficient for eNOS. In the latter study, the mice were exempt from significant hypertrophy and also had been backcrossed into C57B/6 X 129SvEv strains to obtain control animals with the appropriate genetic background (HAN et al. 1998). Others have since demonstrated a similar involvement of a constitutive NOS pathway in mediating the accentuated antagonism of the L-type calcium current in guinea-pig cardiomyocytes (GALLO et al. 1998) and in the purinergic attenuation of L-type calcium current in rabbit pacemaker and atrioventricular nodal cells (MARTYNYUK et al. 1996, 1997; SHIMONI et al. 1996).

From earlier and recent reports on this subject it appears that both the experimental protocol and species variation critically influences the functional effects observed with eNOS inhibition. A critical parameter is probably the temperature at which the functional role of eNOS is examined. It is noteworthy that most of the negative studies were performed at room temperature, or below 37°C, which would impede full catalytic activity of eNOS and detection of functional effects of endogenous NO. Other apparent discrepancies may be resolved upon careful consideration of the protocols used. For example, when the role of endogenous NO is examined in the absence of pre-stimulation with β-adrenergic agonists in cat cardiomyocytes, it is found that eNOS inhibition does not affect the attenuating effect of acetylcholine on L-type calcium current and contractile shortening but clearly blunts the rebound increase in these parameters following abrupt removal of the muscarinic cholinergic agonist (WANG et al. 1998). The latter effect probably involves NO- and cyclic GMP-dependent-inhibition of PDE-II, resulting in increases in intracellular cyclic AMP, as mentioned above. Finally, electrophysiological experiments in single isolated cardiomyocytes in transgenic eNOS-deficient mice apparently exclude any modulating effect of eNOS on ionic currents other than the L-type calcium current, such as chloride currents (ZAKHAROV et al. 1996) or muscarinic cholinergic-activated potassium currents (I_{K-Ach}; HAN et al. 1998).

To gain more insight into the physiological significance of the above-mentioned paradigm, it was important to demonstrate that NO mediates some of the effects of vagal stimulation on the heart in vivo. Using a model of open-chest, anaesthetised dogs, Hare and colleagues observed that inhibition of myocardial NOS on intracoronary infusion of L-NMMA attenuated the effect of vagal stimulation on the contractile response to β-adrenergic agonists. NOS inhibition also significantly attenuated the blunting of the positive inotropic effect of dobutamine upon stimulation of the vagus nerve in these dogs (thereby reproducing the classic accentuated antagonism; HARE et al. 1995b). In a different set of experiments in dogs, Zipes and colleagues found that selective infusion of the NOS inhibitor L-NMMA into the sino-atrial nodal artery modulated sinus discharge rate and atrioventricular conduction (ELVAN et al. 1997).

E. Regulation of Cardiac Function by iNOS

Large concentrations of NO produced by the high-output iNOS are likely to produce a myocardial depression by one or several of the mechanisms reviewed above, especially under the oxidative conditions that prevail in inflammatory reactions. Despite supporting evidence from in vitro experiments with isolated hearts or cardiomyocytes, the picture is much less clear in vivo. This may be due to redundant inflammatory pathways leading to myocardial dysfunction despite iNOS inhibition, side-effects of NOS inhibitors when used in vivo, or both.

I. Basal Contractile Function

NOS inhibitors were shown to ameliorate the contractile dysfunction of isolated perfused hearts treated with IL-1β and TNF-α (SCHULZ et al. 1995), of isolated cardiomyocytes from lipopolysaccharide (LPS)-treated guinea pigs (BRADY et al. 1992) or guinea-pigs cardiocytes treated in vitro with cytokines or LPS (MCKENNA et al. 1995; YASUDA and LEW 1997). Similarly, the relatively selective iNOS inhibitor aminoguanidine (which also has antioxidant and other properties) was shown to reverse the decrease in basal shortening of cardiomyocytes taken from heterotransplanted rat hearts (ZIOLO et al. 1998) and to improve rat cardiac allograft survival and developed tension in papillary muscles from the allografted hearts in another study (WORRALL et al. 1995).

The development of transgenic mice deficient for the iNOS gene (iNOS$^{-/-}$) provided a valuable model with which to study the impact of iNOS in sepsis in vivo (LAUBACH et al. 1995; MACMICKING et al. 1995; WEI et al. 1995). Of the three studies on the effect of LPS injection in three different iNOS$^{-/-}$ lines of mice, two reported that iNOS$^{-/-}$ mice exibited significantly less mortality in comparison to wild-type controls. In one of these two, however, the benefit in terms of mortality was lost when mice were co-injected with *Propionibacterium acnes*, a model of infection that may mimic the clinical situation more closely (MACMICKING et al. 1995). None of these studies specifically examined the potential participation of myocardial dysfunction in hypotensive shock.

In cardiac transplant recipients, the expression of iNOS mRNA in endomyocardial biopsies was reported to be inversely related to LV performance i.e. ejection fraction (LEWIS et al. 1996). iNOS enzyme activity and iNOS protein measured by immunohistochemistry in cardiac muscle of patients with idiopathic dilated cardiomyopathy were increased and correlated with impaired contractile performance (HAYWOOD et al. 1996; LEWIS et al. 1996; DE BELDER et. al. 1997). Even though increases in iNOS activity or protein have been found in many disease states with cardiac dysfunction, such as myocardial infarction, immune cardiomyopathy, post-partum myopathy and idiopathic dilated cardiomyopathy (Sect. I; FUKUCHI et al. 1998), this circumstantial evidence is insufficient to demonstrate a direct causal role of NO in the pathogenesis of these diseases. Moreover, results using NOS inhibitors in vivo are discordant at best, perhaps due to indiscriminate inhibition of all NOS isoforms, resulting in myocardial ischaemia (AVONTUUR et. al, 1995). This could result from direct vasoconstriction or to an increase in the adhesion of platelets or neutrophils through the inhibition of NOS in endothelial cells (BALLIGAND et al. 1995b; UNGUREANU-LANGROIS et al. 1995a). NOS inhibition may also reverse the NO-mediated downregulation of reduced nicotinamide adenine dinucleotide phosphate oxidase activity in neutrophils that control oxygen-radical formation, thereby enhancing oxidative damage and the permeability of the endothelium to macromolecules (KUBES 1991, 1995; FUJII et al. 1997). In addition, other cardiodepressant factors are probably involved in complete septic-shock syndrome (UNGUREANU-LANGROIS et al. 1995a) and other cardiac diseases. For example, infusion of LPS was reported to be followed by myocardial depression in the absence of measurable increases in NOS activity in one study (DECKING et al. 1995).

One of these factors could be TNF-α, an inflammatory cytokine involved in iNOS induction which is also known to be elevated in sepsis and in heart failure (LEVINE et al. 1990). TNF-α has recently received much attention because of the report of beneficial effects of soluble TNF receptors in heart-failure patients (DESWAL et al. 1997). Despite the clear demonstration of co-expression and localisation of both TNF protein and iNOS in myocardium from failing hearts (HABIB et al. 1996; DE BELDER et al. 1997), a specific role for iNOS in the beneficial effect of TNF-α suppression is only speculative. TNF-α at pathophysiologically relevant concentrations has been shown to induce cardiomyocyte dysfunction through alternative mechanisms, at least in vitro (MÜLLER-WERDAN et al. 1997). These may include activation of the sphingomyelinase pathway (ORAL 1997) and induction of myocyte apoptosis (ANVERSA 1998; ANVERSA and KAJSTURA 1998).

II. Regulation of β-Adrenergic Response

In the early 90s, experiments in neonatal cardiac myocytes had shown that cytokine treatment resulted in a decreased contractile responsiveness to adrenergic agonists, which was associated with an attenuation of the normal increase of intracellular cyclic AMP (GULICK et al. 1989; CHUNG et al. 1990). A similar decreased responsiveness to β-adrenergic agonists was later observed in ventricular myocytes isolated from rejecting cardiac allografts, although the specific involvement of the NO pathway was not determined in these experiments (PYO and WAHLER 1995). Using specific combinations of recombinant cytokines or a mixture of inflammatory cytokines obtained from the culture supernatants of rat macrophages stimulated with LPS, we found that the reduced contractile response of isolated myocytes exposed to sub-maximal concentrations of isoproterenol was fully reversed upon co-treatment of the cells with NOS inhibitors, thereby clearly implicating NO production in the contractile dysfunction, at least under these experimental conditions (BALLIGAND et al. 1993b). We subsequently made similar findings using cultures of rat ventricular myocytes in co-culture with iNOS-expressing endothelial cells (UNGUREANU-LANGROIS et al. 1995b), as did many other groups using either isolated contracting cardiomyocytes or papillary muscles exposed to LPS alone or in combination with other cytokines (BALLIGAND and CANNON 1997; YASUDA et al. 1997; SUN et al. 1998).

III. iNOS and Cardiomyocyte Biology

The role of iNOS on cardiac myocyte biology is not restricted to alterations in contractile function. Consistent with the role of this isoform as part of the early, innate immune response, iNOS expression in the myocardium was shown to modulate the degree of myocardial injury after viral infection. Experiments in a murine model of viral myocarditis induced by coxsackie-B3 have suggested a deleterious role for the NO produced, since low doses of the NOS inhibitor L-nitro-arginine methyl ester decreased myocardial injury and mortality (MIKAMI et al. 1997), although high doses of the NOS inhibitor had the opposite effect. A protective role for iNOS, however, was reported by others using a similar model. In these experiments, where iNOS induction was demonstrated in macrophages infiltrating the myocardium, treatment of mice with NOS inhibitors increased mortality and induced higher viral titers (LOWENSTEIN et al. 1996). The results regarding the role of iNOS in modulating the oxidative stress are equally discordant in several models. Studies with a porphyrinic electrode in a rat transplantation model of ischaemia reperfusion indicated that low levels of NO quenched oxygen-derived radicals (PINSKY et al. 1994). However, induction of iNOS with recombinant cytokines was shown to increase oxidative damage through inhibition of glutathione synthesis in cultured myocytes (IGARASHI et al. 1998). The latter finding would be consistent with prior observations that iNOS expression is associated with myocardial damage and cell death in models of cardiac allograft rejection (YANG et al. 1994; WORRALL et al. 1995). Induction of apoptosis of cardiac myocytes by NO and/or peroxynitrite is one of the pathways by which iNOS may promote cell death (SZABOLCS et al. 1996, 1998).

F. Conclusion and Perspectives

It is clear that the NO-dependent contractile regulation of cardiac muscle integrates NO produced from many cellular sources, where the expression of each isoform of NOS is controlled in a cell-specific manner. In addition, the biological effect of NO is likely influenced by the microenvironment where it is produced, most importantly by the redox conditions, which may differentially impact the effect of either NO produced paracrinally from neighbouring cells or NO produced endogenously by the enzyme targeted to subcellular compartments, where specific effectors are co-localised. In cardiac myocytes, such co-localisation of eNOS with receptors, signalling proteins or calcium-release channels both adds another level of complexity and opens exciting new perspectives in the understanding of the enzyme's physiological role, e.g. in excitation–contraction coupling. In addition to spatial constraints, temporal restriction of iNOS activation within the heart may be key for the delineation between its protective role in a physiological immune response and the deleterious effect of extensive NO (and/or peroxinitrite) production, leading to generalised myocardial dysfunction. As the biology of NO and molecular regulation of NOS become more thoroughly characterised, the idea that exogenous NO-donor drugs faithfully mimic the functional effect of NOS activation in cardiomyocytes appears increasingly naive. Further characterisation of the molecular regulation of NOS in cardiac cells should help resolve some of the controversies that remain after many pioneering studies with classic pharmacological approaches over the last 20 years, and should also hopefully help design new therapeutic strategies for the improvement of heart function in patients.

References

Andries LJ, Brutsaert DL, Sys SU (1998) Nonuniformity of endothelial constitutive nitric oxide synthase distribution in cardiac endothelium. Cir Res 82:195–203

Anversa P (1998) Myocyte apoptosis and heart failure. Eur Heart J 19:359–360

Anversa P, Kajstura J (1998) Myocyte cell death in the diseased heart. Circ Res 82:1231–1233

Arstall MA, Bailey C, Gross WL, Bak M, Balligand JL, Kelly RA (1998) Reversible S-nitrosation of creatinine kinase by nitric oxide in adult rat ventricular myocytes. J Mol Cell Cardiol 30:979–988

Avontuur JA, Bruining HA, Ince C (1995) Inhibition of nitric oxide synthesis causes myocardial ischemia in endotoxemic rats. Circ Res 76:418–425

Balligand JL, Cannon PJ (1997) Nitric oxide synthases and cardiac muscle: autocrine and paracrine influences. Arteriosclerosis Thromb Vasc Biol 17:1846–185

Balligand JL, Smith TW (1997) Molecular regulation of NO synthase in the heart. In: Shah AM, Lewis MJ (eds) Endothelial modulation of cardiac contraction. Harwood Academic, pp 53–71

Balligand JL, Kelly RA, Marsden PA, Smith TW, Michel T (1993a) Control of cardiac muscle cell function by an endogenous nitric oxide signaling system. Proc Natl Acad Sci U S A 90:347–351

Balligand JL, Ungureanu D, Kelly RA, Kobzik L, Pimental D, Michel T, Smith TW (1993b) Abnormal contractile function due to induction of nitric oxide synthesis in rat cardiac myocytes follows exposure to activated macrophage-conditioned medium J Clin Invest 91: 2314–2319

Balligand JL, Ungureanu-Longrois D, Simmons WW, Pimental D, Malinski TA, Kapturczak M, Taha Z, Lowenstein CJ, Davidoff AJ, Kelly RA, Smith TW, Michel T (1994) Cytokine-inducible nitric oxide synthase (iNOS) expression in cardiac myocytes. Characterization and regulation of iNOS expression and detection of iNOS activity in single cardiac myocytes in vitro. J Biol Chem 269:27580–27588

Balligand JL, Kobzik L, Han X, Kaye DM, Belhassen L, O'Hara DS, Kelly RA, Smith TW, Michel T (1995a) Nitric oxide-dependent parasympathetic signaling is due to activation of constitutive endothelial (type III) nitric oxide synthase in cardiac myocytes. J Biol Chem 270:14582–14586

Balligand JL, Ungureanu-Longrois D, Simmons WW, Kobzik L, Lowenstein CJ, Lamas S, Kelly RA, Smith TW, Michel T (1995b) Induction of NO synthase in rat cardiac microvascular endothelial cells by IL-1β and IFN-γ. Am J Physiol 268:H1293–1301

Bartunek J, Shah AM, Vanderheyden M, Paulus WJ (1997) Dobutamine enhances cardiodepressant effects of receptor-mediated coronary endothelial stimulation. Circulation 95:90–96

Bayraktutan U, Yang Z-K, Shah AM (1998) Selective dysregulation of nitric oxide synthase type 3 in cardiac myocytes but not coronary microvascular endothelial cells of spontaneously hypertensive rat. Cardiovasc Res 38:719–726

Beckman JS, Koppenol WH (1996) Nitric oxide, superoxide and peroxinitrite: the good, the bad and the ugly. Am J Physiol 271:C1424–C1437

Belhassen L, Kelly RA, Smith TW, Balligand JL (1996) Nitric oxide synthase (NOS3) and contractile responsiveness to adrenergic and cholinergic agonists in the heart: regulation of NOS3 transcription in vitro and in vivo by cAMP in rat cardiac myocytes. J Clin Invest 97:1908–1915

Belhassen L, Feron O, Kaye DM, Michel T, Kelly RA (1997) Regulation by cAMP of post-translational processing and subcellular targeting of endothelial nitric-oxide synthase (type 3) in cardiac myocytes. J Biol Chem 272:11198–11204

Brady AJ, Poole-Wilson PA, Harding SE, Warren JB (1992) Nitric oxide production within cardiac myocytes reduces their contractility in endotoxemia. Am J Physiol 263:H1963–H1966

Brady AJ, Warren JB, Poole-Wilson PA, Williams TJ, Harding SE (1993) Nitric oxide attenuates cardiac myocyte contraction. Am J Physiol 265:H176–H182

Campbell DL, Stamler JS, Strauss HC (1996) Redox modulation of L-type calcium channel in ferret ventricular myocytes. Dual mechanism regulation by nitric oxide and S-nitrosothiols. J Gen Physiol 108:277–293

Chung MK, Gulick TS, Rotondo RE, Schreiner GF, Lange LG (1990) Mechanism of cytokine inhibition of β-adrenergic agonist stimulation of cyclic AMP in rat cardiac myocytes. Impairment of signal transduction. Circ Res 67:753–763

De Belder A, Robinson N, Richardson P, Martin J, Moncada S (1997) Expression of inducible nitric oxide synthase in human heart failure. Circulation 95:1672–1673

Decking UK, Flesche CW, Goedecke A, Schrader J (1995) Endotoxin-induced contractile dysfunction in guinea pig hearts is not mediated by nitric oxide. Am J Physiol 268:H2460–H2465

Deswal A, Seta Y, Blosch CM, Mann DL (1997) A phase-I trial of tumor necrosis factor receptor (p75) fusion protein (TNR:FC) in patients with advanced heart failure. Circulation 96:I-323,1802

Elvan A, Rubart M, Zipes DP (1997) NO modulates autonomic effects on sinus discharge rate and AV nodal conduction in open-chest dogs. Heart Circ Physiol 41:H263–H271

Fei L, Baron AD, Henry DP, Zipes DP (1997) Intrapericardial delivery of L-arginine reduces the increased severity of ventricular arrhythmias during sympathetic stimulation in dogs with acute coronary occlusion: nitric oxide modulates sympathetic effects on ventricular electrophysiological properties. Circulation 96:4044–4049

Felder CC (1995) Muscarinic acetylcholine receptors: signal transduction through mutiple effectors. FASEB J 9:619–625

Feron O, Belhassen L, Kobzik L, Smith TW, Kelly RA, Michel T (1996) Endothelial nitric oxide synthase targeting to caveolae: specific interaction with caveolin isoforms in cardiac myocytes and endothelial cells. J Biol Chem 271:22810–22814

Feron O, Smith TW, Michel T, Kelly RA (1997) Dynamic targeting of the agonist-stimulated m2 muscarinic acetylcholine receptor to caveolae in cardiac myocytes. J Biol Chem 272:17744–17748

Feron O, Saldana JB, Michel JB, Michel T (1998a) The eNOS-caveolin regulatory cycle J Biol Chem 273:3125–3128

Feron O, Dessy C, Opel DJ, Arstall MA, Kelly RA, Michel T (1998b) Modulation of the endothelial nitric oxide synthase–caveolin interaction in cardiac myocytes: implications for the autonomic regulation of heart rate. J Biol Chem 273:30249–30253

Finkel MS, Oddis CV, Jacob TD, Watkin SC, Hattler BG, Simmons RL (1992) Negative inotropic effects of cytokines on the heart mediated by nitric oxide. Science 257:387–389

Finkel MS, Oddis CV, Mayer OH, Hattler BG, Simmons RL (1995) Nitric oxide synthase inhibitor alters papillary muscle force-frequency relationship. J Pharmacol Exp Ter 272:945–952

Fleming I, Bauersachs J, Busse R (1997) Calcium-dependent and calcium-independent activation of the endothelial NO synthase. J Vasc Res 34(3):165–174

Fleming I, Bauersachs J, Fisslthaler B, Busse R (1998) Ca2+-independent activation of the endothelial nitric oxide synthase in response to tyrosine phosphatase inhibitors and fluid shear stress. Circ Res 82:686–695

Fujii H, Ichimori K, Hoshiai K, Nakazawa H (1997) Nitric oxide inactivates NADPH oxidase in pig neutrophils by inhibiting its assembling process. J Biol Chem 272: 32773–32778

Fukuchi M, Hussain SNA, Giaid A (1998) Heterogeneous expression and activity of endothelial and inducible nitric oxide synthases in end-stage human heart failure: their relation to lesion site and β-adrenergic receptor therapy. Circulation, 98: 132–139

Galione A, Lee HC, Busa WB (1991) Ca(2+)-induced Ca2+ release in sea urchin egg homogenates: modulation by cyclic ADP-ribose. Science 252:1143–1146

Gallo MP, Ghigo D, Bosia A, Allaotti G, Costamagna C, Penna C, Levi RC (1998) Modulation of guinea-pig cardiac L-type calcium current by nitric oxide synthase inhibitors. J Physiol (Lond) 506:639–651

Garcia-Cardena G, Martasek P, Siler Masters BS, Skidd PM, Couet J, Li S, Lisanti MP, Sessa WC (1997). Dissecting the interaction between nitric oxide synthase and caveolin. J Biol Chem 272: 25437–25440

Gauthier C, Tavernier G, Charpentier F, Langin D, Le Marec H (1996) Functional b3-adrenoceptor in the human heart. J Clin Invest 98: 556–562

Gauthier C, Leblais V, Trochu JN, Balligand JL, Le Marec H (1998a) The negative inotropic effect of β-3-adrenoceptor stimulation is mediated by activation of a nitric oxide synthase pathway in human ventricle. J Clin Invest 102:1377–1384

Gauthier C, Erfanian M, Baron O, Balligand JL (1998b) Control of contractile and rhythmic properties of human atrial tissue by a NO pathway. Circulation 98:I70, 3838

Gilad E, Zingarelli B, Salzman AL, Szabo C (1997) Protection by inhibition of poly (ADP-ribose) synthetase against oxidant injury in cardiac myoblasts in vitro. J Mol Cell Cardiol 29:2585–2597

Godecke A, Heinicke T, Decking UKM, Stumpe T, Schrader J (1998) β-adrenergic stimulation of eNOS-deficient mice hearts. Circulation 98:I 70, 353

Goldhaber JI, Kim KH, Natterson PD, Lawrence T, Yang P, Weiss JN (1996) Effects of TNF-α on $[Ca^{2+}]_i$ and contractility in isolated adult rabbit ventricular myocytes. Am J Physiol 271:H1449–H1455

Grocott-Mason R, Fort S, Lewis MJ, Shah AM (1994) Myocardial relaxant effect of exogenous nitric oxide in isolated ejecting hearts. Am J Physiol 266:H1699–H1705

Gross WL, Bak MI, Ingwall JS, Arstall MA, Smith TW, Balligand JL, Kelly RA (1996) Nitric oxide inhibits creatine kinase and regulates rat heart contractile reserve. Proc Natl Acad Sci U S A 93:5604–5609

Gulick T, Chung MK, Pieper SJ, Lange LG, Schreiner GF (1989) Interleukin 1 and tumor necrosis factor inhibit cardiac myocyte β-adrenergic responsiveness. Proc Natl Acad Sci U S A 86:6753–6757

Guo X, Laflamme A, Becker PL (1996) Cyclic ADP-ribose does not regulate sarcoplasmic reticulum Ca2+ release in intact cardiac myocytes. Circ Res 79:147–151

Gyurko R, Fishman MC, Huang PL (1997) Enhanced systolic contractility and preserved diastolic relaxation in mice deficient in endothelial nitric oxide synthase. Presented at the 4th International Meeting on the Biology of Nitric Oxide, Kyoto, September 1997

Habib FM, Springall DR, Davies GJ, Oakley CM, Yacoub MH, Polak JM (1996) Tumor necrosis factor and inducible nitric oxide synthase in dilated cardiomyopathy. Lancet 347:1151–1155

Han X, Shimoni Y, Giles WR (1994) An obligatory role of nitric oxide in autonomic control of mammalian heart rate. J Physiol 476:309–314

Han X, Shimoni Y, Giles WR (1995) A cellular mechanism for nitric oxide-mediated cholinergic control of mammalian heart rate. J Gen Physiol 106:45–65

Han X, Kobzik L, Balligand JL, Kelly RA, Smith TW (1996) Nitric oxide synthase (NOS3)-mediated cholinergic modulation of Ca2+ current in adult rabbit atrioventricular nodal cells. Circ Res 78:998–1008

Han X, Kubota I, Feron O, Opel DJ, Arstall MA, Zhao YY, Huang P, Fishman MC, Michel T, Kelly RA (1998) Muscarinic cholinergic regulation of cardiac myocyte ICa-L is absent in mice with targeted disruption of endothelial nitric oxide synthase. Proc Natl Acad Sci USA 95:6510–6515

Hare JM, Stamler JS (1999) NOS: modulator, not mediator of cardiac performance. Nature Med 5:273–274

Hare JM, Loh E, Creager MA, Colucci WS (1995a) Nitric oxide inhibits the positive inotropic response to β-adrenergic stimulation in humans with left ventricular dysfunction. Circulation 92:2198–2203

Hare JM, Keaney JF, Balligand JL, Loscalzo J, Smith TW, Colucci WS (1995b) Role of nitric oxide in parasympathetic modulation of β-adrenegic myocardial contractility in normal dogs. J Clin Invest 95:360–366

Hare JM, Gibertz MM, Creager MA, Colucci WS (1998a) Increased sensitivy to nitric oxide synthase inhibition in patients with heart failure: potentiation of β-adrenergic inotropic responsiveness. Circulation 97:161–166

Hare JM, Kim B, Flavahan NA, Ricker KM, Peng X, Colman L, Weiss RG, Kass DA (1998b) Pertussis-toxin-sensitive g proteins influence nitric oxide synthase-III activity and protein levels in rat heart. J Clin Invest 1901:1424–1431

Harrison DG (1997) Cellular and molecular mechanisms of endothelial cell dysfunction. J Clin Invest 100:2153–2157

Hausladen A, Fridovich K (1994) Superoxide and peroxinitrite inactivate aconitases, but nitric oxide does not. J Biol Chem 269:29405–29408

Haywood GA, Tsao PS, von der Leyen HE, Mann MS, Keeling PJ, Trindade PT, Lewis NP, Byrne CD, Rickenbacher PR, Bishopric NH, Cooke JP, McKenna WS, Fowler MB (1996) Expression of inducible nitric oxide synthase in human heart failure. Circulation 93:1087–1094

Hu H, Chiamvimonvat N, Yamagishi T, Marban E (1997) Direct inhibition of expressed cardiac L-type Ca^{2+} channels by S-nitrosothiol nitric oxide donors. Circ Res 81: 742–752

Igarashi J, Nishida M, Hoshida S, Yamashita N, Kosaka H, Hori M, Kuzuya T, Tada M (1998) Inducible nitric oxide synthase augments injury elicited by oxidative stress in rat cardiac myocytes. Am J Physiol 274:C245–C252

Ino S, Cui Y, Galione A, Terrar DA (1997) Actions of cADP-ribose and its antagonists on contraction in guinea pig isolated ventricular myocytes. Influence of temperature. Circ Res 81:879–884

Ikeda U, Maeda Y, Kawahara Y, Yokoyama M, Shimada K (1995) Angiotensin-II augments cytokine-stimulated nitric oxide synthesis in rat cardiac myocytes. Circulation 92:2683–2689

Ikeda U, Murakami Y, Kanbe T, Shimada K (1996) A-adrenergic stimulation enhances inducible nitric oxide synthase expression in rat cardiac myocytes. J Mol Cell Cardiol 28:1539–1545

Ishikawa H (1968) Formation of elaborate networks of T-system tubules in cultured skeletal muscle with special reference to the T-system formation. J Cell Biol 38:51–66

Joe EK, Schussheim AE, Ungureanu-Longrois D, Mäki T, Kelly RA, Smith TW, Balligand JL (1998) Regulation of cardiac myocyte contractile function by inducible nitric oxide synthase (iNOS): mechanisms of contractile depression by nitric oxide. J Mol Cell Cardiol 30:303–315

Kanai AJ, Mesaros S, Finkel MS, Oddis CV, Birder LA, Malinski T (1997) B-adrenergic regulation of constitutive nitric oxide synthase in cardiac myocytes. Am J Physiol 273:C1371–C1377

Kasai K, Hattori Y, Banba N, Hattori S, Motohashi S, Shimoda S, Nakanishi N, Gross SS (1997) Induction of tetrahydrobiopterin synthesis in rat cardiac myocytes: impact on cytokine-induced NO generation. Am J Physiol 273:H665–H672

Kaye DM, Wiviott SD, Balligand JL, Simmons WW, Smith TW, Kelly RA (1996) Frequency-dependent activation of a constitutive nitric oxide synthase and regulation of contractile function in adult rat ventricular myocytes Circ Res 78:217–224

Kaye DM, Wiviott D, Kobzik L, Kelly RA, Smith TW (1997) S-nitrisothiols inhibit neuronal norepinephrine transport. Am J Physiol 272:H875–H883

Keaney JF, Hare JM, Balligand JL, Loscalzo J, Smith TW, Colucci WS (1996) Inhibition of nitric oxide synthase augments myocardial contractile responses to β-adrenergic stimulation. Am J Physiol 40:H2646–H2652

Kelly RA, Balligand JL, Smith T (1996). Nitric oxide and cardiac function. Circ Res 79:363–380

Kelm M, Schafer S, Dahmann R, Dolu B, Perings S, Decking UK, Schrader J, Strauer BE (1997) Nitric oxide induced contractile dysfunction is related to a reduction in myocardial energy generation. Cardiovasc Res 36:185–194

Kilter H, Lenz O, La Rosée K, Flesch M, Schwinger RHG, Mädge M, Kuhn-Regnier F, Böhm M (1995) Evidence against a role of nitric oxide in the indirect negative inotropic effect of M-cholinoreceptor stimulation in human ventricular myocardium. Arch Pharmacol 352:308–312

Kinugawa K, Schimizu T, Yao A, Kohmoto O, Serizawa T, Takahashi T (1997) Transcriptional regulation of inducible nitric oxide synthase in cultured neonatal rat cardiac myocytes. Circ Res 81:911–921

Kirstein M, Rivet-Bastide M, Hatem S, Benardeau A, Mercadier JJ, Fischmeister R (1995) Nitric oxide regulates the calcium current in isolated human atrial myocytes. J Clin Invest 95:794–802

Kojda G, Kottenberg K, Nix P, Schluter KD, Piper HM, Noack (1996) Low increase of cGMP induced by organic nitrates and nitrovasodilators improves contractile response of rat ventricular myocytes. Circ Res 78:91–101

Kubes P (1995) Nitric oxide affects microvascular permeability in the intact and inflamed vasculature. Microcirculation 2:235–244

Kubes P, Grisham MB, Barrowman JA, Gaginella T, Granger DN (1991) Leukocyte-induced vascular protein leakage in cat mesentery. Am J Physiol 261:H1872–H1879

Laubach VE, Sheseley EG, Smithies O, Sherman PA (1995) Mice lacking inducible nitric oxide synthase are not resistant to lipopolysaccharide-induced death (1995) Proc Natl Acad Sci USA 92: 10688–10692

Laursen JB, Harrisson DG (1997) Modulation of myocardial oxygen consumption through ACE inhibitors. No effect? Circulation 95:14–16

Levine B, Kalman J, Mayer L, Fillit HM, Packer M. Elevated levels of tumor necrosis factor in severe chronic heart failure (1990) N Engl J Med 323: 236–241

Lewis NP, Tsao PS, Rickenbacher PR, Xue C, Johns RA, Haywood GA, von der Leyen H, Trindade PT, Cooke JP, Hunt SA, Billingham ME, Valantine HA, Fowler MB (1996) Induction of nitric oxide synthase in the human cardiac allograft is associated with contractile dysfunction of the left ventricle. Circulation 93:720–729

Lowenstein CJ, Hill SL, Lafond-Walker A, Wu J, Allen G, Landavere M, Rose NR, Herskowitz A (1996) Nitric oxide inhibits viral replication in murine myocarditis. J Clin Invest 97:1837–1843

MacMicking JD, Nathan C, Hom G, Chartrain N, Fletcher DS, Trumbauer M, Stevens K, Xie QW, Sokol K, Hutchinson N, Chen H, Mudgett JS (1995) Altered responses to bacterial infection and endotoxic shock in mice lacking inducible nitric oxide synthase. Cell 81:641–650

Martynyuk AE, Kane KA, Cobbe SM, Rankin AC (1996) Nitric oxide mediates the anti-adrenergic effect of adenosine on calcium current in isolated rabbit atrioventricular nodal cells. Pflugers Arch 431:452–457

Martynyuk AE, Kane KA, Cobbe SM, Rankin AC (1997) Role of nitric oxide, cyclic GMP and superoxide in inhibition by adenosine of calcium current in rabbit atrioventricular nodal cells. Cardiovasc Res 34:360–367

Mayer B, Pfeiffer S, Schrammel A, Koesling D, Schmidt K, Brunner F (1998) A new pathway of nitric oxide/cyclic GMP signaling involving S-nitrosoglutathione. J Biol Chem 273:3263–270

McKenna TM, Li S, Tao S (1995) PKC mediates LPS- and phorbol-induced cardiac cell nitric oxide synthase activity and hypocontractility. Am J Physiol 269:H1891–H1898

Mery PF, Lohmann SM, Walter U, Fischmeister R (1991) Ca^{2+} current is regulated by cyclic GMP-dependent protein kinase in mammalian cardiac myocytes. Proc Natl Acad Sci U S A 88:1197–1201

Mery PF, Pavoine C, Belhassen L, Pecker F, Fischmeister R (1993) Nitric oxide regulates cardiac Ca2+ current. Involvement of cGMP-inhibited and cGMP-stimulated phosphodiesterases through guanylyl cyclase activation. J Biol Chem 268:26286–26295

Mery-PF, Hove-Madsen L, Chesnais JM, Hartzell HC, Fischmeister R (1996) Nitric oxide synthase does not participate in negative inotropic effect of acetylcholine in frog heart. Am J Physiol 270:H1178–H1188

Michel T, Feron F (1997) Nitric oxide synthases: which, where, how and why? J Clin Invest 100:2146–2152

Mikami S, Kawashima S, Kanazawa K, Hirata K, Hotta H, Hayashi Y, Itoh H, Yokoyama M (1997) Low-dose N^{ω}-nitro-L-arginine methyl ester treatment improves survival rate and decreases myocardial injury in a murine model of viral myocarditis induced by coxackie virus B3. Circ Res 81:504–511

Muller-Werdan U, Schumann H, Fuchs R, Reithmann C, Loppnow H, Koch S, Zimny-Arndt U, He C, Darmer D, Jungblut P, Stadler J, Holz J, Werdan K (1997) TNFα is cardiodepressant in pathophysiologically relevant concentration without inducing iNOS or triggering cytotoxicity. J Mol Cell Cardiol 29:2915–2923

Nathan C, Xie QW (1994) Regulation of biosynthesis of nitric oxide. J Biol Chem 269:13725–13728

Oddis CV, Finkel MS (1995) Cytokine-stimulated nitric oxide production inhibits mitochondrail activity in cardiac myocytes. Biochem Biophys Res Com 213:1002–1009

Oddis CV, Simmons RL, Hattler BG, Finkel MS (1995) cAMP enhances inducible nitric oxide synthase mRNA stability in cardiac myocytes. Am J Physiol 269:H2044–H2050

Ono K, Trautwein W (1991) Potentiation by cyclic GMP of β-adrenergic effect on Ca^{2+} current in guinea-pig ventricular cells. J Physiol (Lond) 443:387–404

Oral H, Dorn GW, Mann DL (1997) Sphingosine mediates the immediate negative inotropic effects of tumor necrosis factor-α in the adult mammalian cardiac myocyte. J Biol Chem 272:4836–4842

Parton RG, Way M, Zorzi N, Stang E (1997) Caveolin-3 associates with developing T-tubules during muscle differentiation. J Cell Biol 136:137–154

Paulus WJ, Vantrimpont PJ, Shah AM (1995) Paracrine coronary endothelial control of left ventricular function in humans. Circulation 92:2119–2126

Pfitzer G, Ruegg JC, Flockerzi V, Hofmann F (1982) cGMP-dependent protein kinase decreases calcium sensitivity of skinned cardiac fibers. FEBS Lett 149:171–175

Pinsky DJ, Oz MC, Koga S, Taha Z, Broekman MJ, Marcus AJ, Cannon PJ, Nowygrod R, Malinsky T, Stern DM (1994) Cardiac preservation is enhanced in a heterotopic rat transplant model by supplementing the nitric oxide pathway J Clin Invest 93:2291–2297

Pinsky DJ, Cai B, Yang X, Rodriguez C, Sciacca RR, Cannon PJ (1995) The lethal effects of cytokine-induced nitric oxide on cardiac myocytes are blocked by nitric oxide synthase antagonism or transforming growth factor β. J Clin Invest 95:677–685

Pinsky DJ, Patton S, Mesaros S, Brovkovych V, Kubaszewski E, Grunfeld S, Malinski T (1997) Mechanical transduction of nitric oxide synthesis in the beating heart. Circ Res 81:372–379

Preckel B, Kojda G, Schlack W, Ebel D, Kottenberg K, Noack E, Thamer V (1997) Inotropic effects of glyceryltrinitrate and spontaneous NO donors in the dog heart. Circulation 96:3675–2682

Prendergast BD, Sagach VF, Shah AM (1997) Basal release of nitric oxide augments the Frank-Starling response in the isolated heart. Circulation 96:1320–1329

Pyo RT, Wahler GM (1995) Ventricular myocytes isolated from rejecting cardiac allografts exhibit a reduced β-adrenergic contractile response J Mol Cell Cardiol 27:773–776

Roberts AB, Vodovotz Y, Roche NS, Sporn MB, Nathan CF (1992) Role of nitric oxide in antagonistic effects of transforming growth factor-β and interleukin-1β on the beating rate of cultured cardiac myocytes. Mol Endocrinol 6:1921–1930

Robertson SP, Johnson JD, Holroyde MJ, Kranias EG, Potter JD, Solaro RJ (1982) The effect of troponin I phosphorylation on the Ca2+-binding properties of the Ca^{2+}-regulatory site of bovine cardiac troponin. J Biol Chem 257.260–263

Salvemini D, Korbut R, Anggard E, Vane J (1990) Immediate release of a nitric oxide-like factor from bovine aortic endothelial cells by *Escherichia coli* lipopolysaccharide. Proc Natl Acad Sci USA 87:2593–2597

Schmidt HHHW, Gagne GD, Nakane M, Pollock JS, Miller MF, Murad F (1992) Mapping of neural nitric oxide synthase in the rat suggests frequent co-localization with NADPH diaphorase but not with soluble guanylyl cyclase and novel paraneural functions for nitrinergic signal transduction. J Histochem Cytochem 40:1439–1456

Schulz R, Panas DL, Catena R, Moncada S, Olley PM, Lopaschuk GD (1995) The role of nitric oxide in cardiac depression induced by interleukin 1 and tumor necrosis factor-α. Br J Pharmacol 114:27–34

Schwarz P, Diem R, Dun NJ, Försterman U (1995) Endogenous and exogenous nitric oxide inhibits norepinephrine release from rat heart sympathetic nerves. Circ Res 77:841–848

Seki T, Hagiwara H, Naruse K, Kadowaki M, Kashiwagi M, Demura H, Hirose S, Naruse M (1996) In situ identification of messenger RNA of endothelial type nitric oxide synthase in rat cardiac myocytes. Biochem Biophys Res Commun 218:601–605

Shen W, Xu X, Ochoa M, Zhao G, Wolin MS, Hintze TH (1994) Role of nitric oxide in the regulation of oxygen consumption in conscious dogs. Circ Res 75:1086–1095

Shen W, Hintze TH, Wolin MS (1995) Nitric oxide: an important signaling mechanism between vascular endothelium and parenchymal cells in the regulation of oxygen consumption. Circulation 92:3505–3512

Sherman AJ, Davis CA 3rd, Klocke FJ, Harris KR, Srinivasan G, Yacoub AS, Quinn DA, Ahlin KA, Jang JJ (1997) Blockade of nitric oxide synthesis reduces myocardial oxygen consumption in vivo. Circulation 95:1328–1334

Shimoni Y, Han X, Severson D, Giles WR (1996) Mediation by nitric oxide of the indirect effects of adenosine on calcium current in rabbit heart pacemaker cells. 119:1463–1469

Silvagno F, Xia HH, Bredt DS (1996) Neuronal nitric oxide synthase MU, an alternatively spliced isoform expressed in differentiated skeletal muscle. J Biol Chem 271:11204–11208

Simmons WW, Closs EI, Cunningham JM, Smith TW, Kelly RA (1996a) Cytokines and insulin induce cationic amino acid transporter (CAT) expression in cardiac myocytes. Regulation of L-arginine transport and NO production by CAT-1, CAT-2A, and CAT-2B. J Biol Chem 271:11694–11702

Simmons WW, Ungureanu-Longrois D, Smith GK, Smith TW, Kelly RA (1996b) Glucocorticoïds regulate inducible nitric oxide synthase by inhibiting tetrahydrobiopterin synthesis and L-arginine transport. J Biol Chem 271:23928–23937

Singh K, Balligand JL, Fischer TA, Smith TW, Kelly RA (1996) Regulation of cytokine-inducible nitric oxide synthase (NOS2) in cardiac myocytes and microvascular endothelial cells: role of ERK1/ERK2 (p44/p42) mitogen-activated protein kinases and STAT1 α. J Biol Chem 271:1111–1117

Smith CJ, Sun D, Hoegler C, Roth BS, Zhang X, Zhao G, Xu XB, Kobari Y, Pritchard K Jr, Sessa WC, Hintze TH (1996) Reduced gene expression of vascular endothelial NO synthase and cyclooxygenase-1 in heart failure. Circ Res 78:58–64

Smith JA, Shah AM, Lewis MJ (1991) Factors released from endocardium of the ferret and pig modulate myocardial contraction. J Physiol Lond 439:1–14

Stein B, Drogmuller A, Mulsch A, Schmitz W, Scholz H (1993) Ca (++)-dependent constitutive nitric oxide synthase is not involved in the cyclic GMP-increasing effects of carbachol in ventricular cardiomyocytes. J Pharmacol Exp Ther 266:919–925

Sterin-Borda L, Genaro A, Perez Leiros C, Cremaschi G, Echague AV, Borda E (1998) Role of nitric oxide in cardiac β-adrenoceptor-inotropic response. Cell Signal 10: 253–257

Sun X, Delbridge LMD, Dusting GY (1998) Cardiodepressant effects of interferon-γ and endotoxin reversed by inhibition of NO synthase 2 in rat myocardium. J Mol Cell Cardiol 30:989–997

Szabolcs M, Michler RE, Yang X, Aji W, Royd, Athan E, Sciacca RR, Minanov OP, Cannon PJ (1996) Apoptosis of cardiac myocytes during cardiac allograft rejection. Relation to induction of nitric oxide synthase. Circulation 94:1665–1673

Szabolcs MJ, Ravalli S, Minanov O, Sciacca RR, Michler RE, Cannon PJ (1998) Apoptosis and increased expression of inducible nitric oxide synthase in human allograft rejection. Transplantation 65:804–812

Tanaka K, Hassall CJ, Burnstock G (1993) Distribution of intracardiac neurons and nerve terminals that contain a marker for nitric oxide, NADPH-diaphorase, in the guinea-pig heart. Cell Tissue Res 273:293–300

Thoenes M, Förstermann U, Tracey WR, Bleese NM, Nüssler AK, Scholz H, Stein B (1996) Expression of inducible nitric oxide synthase in failing and non-failing human heart. J Mol Cell Cardiol 28:165–169

Torre-Amione G, Kapadia S, Benedict C, Oral H, Young JB, Mann DL (1996) Proinflammatory cytokine levels in patients with depressed left ventricular ejection fraction: a report from the Studies of Left Ventricular Dysfunction (SOLVD). J Am Coll Cardiol 27:1201 1206

Ungureanu-Longrois D, Balligand JL, Kelly RA, Smith TW (1995a) Myocardial contractile dysfunction in the systemic inflammatory response syndrome: role of a cytokine-inducible nitric oxide synthase in cardiac myocytes. J Mol Cell Cardiol 27:155–167

Ungureanu-Longrois D, Balligand JL, Okada I, Simmons WW, Kobzik L, Lowenstein CJ, Kunkel S, Michel T, Kelly RA, Smith TW (1995b) Contractile responsiveness of ventricular myocytes to isoproterenol is regulated by induction of nitric oxide synthase activity in cardiac microvascular endothelial cells in heterotypic primary culture. Circ Res 77:486–493

Vandecasteele G, Eschenhagen T, Scholz H, Stein B, Verde I, Fischmeister R (1999) Muscarinic and β-adrenergic regulation of heart rate, force of contraction and calcium current is preserved in mice lacking endothelial nitric oxide synthase. Nature Med 5:331–334

Wahler GM, Dollinger SJ (1995) Nitric oxide donor SIN-1 inhibits mammalian cardiac calcium current through cGMP-dependent protein kinase. Am J Physiol 268:C45–C54

Wang YG, Rechenmecher CE, Lipsius SL (1998) Nitric oxide signaling mediates stimulation of L-type Ca^{2+} current elicited by withdrawal of acetylcholine in cat atrial myocytes. J Gen Physiol 111:113–125

Wei CM, Jiang SW, Lust JA, Daly RC, MacGregor CGA (1996) Genetic expression of endothelial nitric oxide synthase in human atrial myocardium. Mayo Clin Proc 71:346–350

Wei XQ, Charles IG, Smith A, Ure J, Feng GJ, Huang FP, Xu D, Muller W, Moncada S, Liew FY (1995) Altered immune responses in mice lacking inducible nitric oxide synthase. Nature 375: 408–411

Worrall NK, Lazenby WD, Misko TP, Lin TS, Rodi CP, Manning PT, Tilton RG, Williamson JR, Ferguson TB Jr (1995) Modulation of in vivo alloreactivity by inhibition of inducible nitric oxide synthase. J Exp Med 181:63–70

Xu KY, Huso DL, Dawson TM, Bredt DS, Becker LC (1999) Nitric oxide synthase in cardiac sarcoplasmic reticulum. Proc Natl Acad Sci USA 96:657–662

Xu L, Elu JP, Meissner G, Stamler JS (1998) Activation of the cardiac calcium release channel (ryanodine receptor) by poly-S-nitrosylation. Science 279:234–237

Yang X, Chowdhury N, Cai B, Brett J, Marboe C, Sciacca RR, Cannon PJ (1994) Induction of myocardial nitric oxide synthase by cardiac allograft rejection. J Clin Invest 94:714–721

Yasuda S, Lew WY (1997) Lipopolysaccharide depresses cardiac contractility and β-adrenergic contractile response by decreasing myofilament response to Ca^{2+} in cardiac myocytes. Circ Res 85:1011–1020

Yoshizumi M, Perrella MA, Burnett JC Jr, Lee HE (1995) Tumor necrosis factor down-regulates an endothelial nitric oxide synthase mRNA by shortening its half-life. Circ Res 73:205–209

Zahradnikova A, Minarovic I, Venema RC, Meszaros LG (1997) Inactivation of the cardiac ryanodine receptor calcium release channel by nitric oxide. Cell Calcium 22:447–454

Zakharov SI, Pieramici S, Kumar GK, Prabhakar NR, Harvey RD (1996) Nitric oxide synthase activity in guinea-pig ventricular myocytes is not involved in muscarinic inhibition of cAMP-regulated ion channels. Circ Res 78:925–935

Zhang X, Xie Y-W, Nasjletti A, Xu X, Wolin MS, Hintze TH (1997) ACE inhibitors promote nitric oxide accumulation to modulate myocardial oxygen consumption. Circulation 95:176–182

Ziolo MT, Dollinger SJ, Wahler GM (1998) Myocytes isolated from rejecting transplanted rat hearts exhibit reduced basal shortening which is reversible by aminoguanidine. J Mol Cell Cardiol 30:1009–1017

CHAPTER 10
Regulation of Platelet Function

L. McNaughton, A. Radomski, G. Sawicki, and Marek W. Radomski

A. Introduction

Blood platelets are small (2μm in diameter) blood elements that play a vital role in *vascular haemostasis*, i.e. the ability of the vasculature to seal accidental rents in the vessel wall, thus containing blood loss. Platelets *adhere* to the damaged portions of the vessel wall then *aggregate*, forming the haemostatic plug in order to seal the vessel and contain blood loss.

This ability of the vasculature must be carefully balanced against the necessity to maintain the fluid state of blood in order to provide oxygen and nutrients to the cells and tissues. The failure to maintain blood in its fluid state leads to *thrombosis*, one of the most serious manifestations of vascular dysfunction.

I. Platelet Rheology

Although freshly drawn blood samples give no appearance of cellular arrangement and order, this is not the case in blood in vivo subjected to the shear stress resulting from the interactions between flowing liquid and the vascular wall. Under conditions of laminar flow, more numerous and larger red cells are found close to the central longitudinal axis of the vessel, forcing platelets to assume positions in the proximal vicinity of the endothelial cells (Radomski and Radomski 1999). Knowledge of platelet rheology is crucial to the understanding of the regulatory effects of agents controlling haemostasis, in particular prostacyclin and nitric oxide (NO).

II. Platelet Control

In order to maintain tight control over the process of platelet activation, often referred to as the *haemostatic–thrombotic balance*, regulatory mechanisms operate in the interactive system of reciprocal agents that stimulate or inhibit the process of platelet activation. While thromboxane A_2, adenosine diphosphate (ADP) and matrix metalloproteinase-2 mediate the activator platelet reactions (Sawicki et al. 1997), prostacyclin, ADPase and NO provide the platelet-inhibitor pathways of regulation (Radomski and Radomski 1999).

B. Nitric Oxide

I. NO in Platelets: the Quest

The appreciation of the presence of endogenous inhibitors of platelet activation was stimulated by the discovery of prostacyclin (MONCADA et al. 1976), a major platelet-regulatory prostaglandin. However, it became apparent that the generation and release of this eicosanoid can only account in part for non-thrombogenic properties of vascular endothelium. Indeed, the endothelial cells express the anti-aggregatory activity even under conditions of complete inhibition of prostaglandin generation (RADOMSKI et al. 1987a).

The discovery of endothelium-dependent relaxation (FURCHGOTT and ZAWADZKI 1980) and the fact that endothelium-derived relaxing factor (EDRF) appeared not to be a metabolite of arachidonic acid have added new impetus to this search. The study by Azuma and colleagues (1986) suggested that EDRF might be acting as an inhibitor of platelet aggregation.

Between 1986 and 1987, we analysed the non-eicosanoid vasodilator and platelet-inhibitory properties of endothelial cells and found that these could be accounted for by the release of EDRF. Moreover, comparison of pharmacological properties of EDRF with those of NO gas and the measurement of NO metabolites (nitrite and nitrate) released during these reactions clearly showed that generation of NO could account for the platelet-inhibitory activities of NO on platelet adhesion and aggregation (PALMER et al. 1987; RADOMSKI et al. 1987a, 1987b, 1987c, 1987d). Interestingly, endogenous NO proved to be identical in its anti-platelet spectrum with exogenous NO generated during cigarette smoking and by pharmacologically related vasodilators (MELLION et al. 1981).

The description of the L-arginine-to-NO pathway in the endothelial cells and NO synthase enzyme (NOS) (PALMER et al. 1988) stimulated the search for the presence of NOS in platelets. The rationale for this research was provided by the observation that platelet activation is associated with increased generation of cyclic GMP (cGMP) that mediates some of the biological effects of NO (GOLDBERG et al. 1975). Between 1990 and 1993, we provided both biochemical and pharmacological characteristics of platelet NOS (RADOMSKI et al. 1990a, 1990b) and measured the release of NO from stimulated platelets using a selective porphyrinic microsensor (MALINSKI et al. 1993). During the past four years, several groups have described molecular characteristics of platelet NOS (MURUGANANDAM and MUTUS 1994; CHEN and MEHTA 1996; BERKELS et al. 1997; WALLERATH et al. 1997).

II. Molecular Biology of Platelet NOS

As platelets are anucleate blood elements containing tracing amounts of DNA and small amounts of RNA, the identification of complementary DNA fragments coding NOS proteins requires application of reverse-transcriptase polymerase chain reaction. With RNA isolated from platelets, this technique amplified DNA fragments consistent with endothelial NOS (eNOS) but not inducible NOS (iNOS) or neuronal NOS (MEHTA et al. 1995; SASE and MICHEL 1995; WALLERATH et al. 1997).

There is general consensus for the presence of eNOS mRNA in platelets (MEHTA et al. 1995; SASE and MICHEL 1995); however, the identification of iNOS mRNA proved to be controversial. CHEN and MEHTA (1996), but not Wallerath and colleagues (1997), found mRNA for iNOS in platelets. It is possible that a short-half-life (6h) iNOS mRNA (GENG and LOTZ 1995) accounts for this discrepancy.

Muruganandam and MUTUS (1994) first purified platelet NOS protein to homogeneity. The isolated protein was cytosolic, and the molecular weight of the monomer was found to be approximately 80kDa.

Western blot, the immunocytochemistry and reduced nicotinamide adenine dinucleotide phosphate (NADPH) diaphorase staining have now been used to probe NOS isoforms in platelets. The immunological methods detected the presence of eNOS and iNOS immunoreactivity in normal human and porcine platelets (CHEN and MEHTA 1995; BERKELS et al. 1997; WALLERATH et al. 1997). In contrast to the work by MURUGANANDAM and MUTUS (1994) and MEHTA et al. (1995), some studies found platelet eNOS, similar to endothelial eNOS, to be associated with the particulate fraction of the platelets (BERKELS et al. 1997; WALLERATH et al. 1997). It is possible that platelet eNOS undergoes intracellular translocation and activation during platelet activation (RADOMSKI et al. 1990a, 1990b; BERKELS et al. 1997).

The significance of iNOS presence in normal platelets is uncertain. Platelets have a limited capacity to synthesise protein de novo and acquire most of their proteins from megakaryocytes. Our group and others have shown that cytokine-stimulated megakaryoblasts and bone-marrow megakaryocytes express iNOS (LELCHUK et al. 1992; WALLERATH et al. 1997). As megakaryocytopoiesis and thrombopoiesis are cytokine dependent and regulated by the stimulatory and inhibitory cytokines (BROWN and MARTIN 1994), the presence or absence of iNOS in platelets may reflect interactions between these hematopoietic factors.

III. Regulation of NO Generation in Platelet Microenvironment

Nitric oxide available for platelet regulation is generated both by endothelium- and platelet-derived NOS.

1. Cell Activation

Endothelial and platelet activation plays an important role in the generation of NO. Tonic release of NO from the endothelial cells is likely to be shear stress mediated (BUSSE and FLEMING 1998).

Resting platelets generate small amounts of NO (MALINSKI et al. 1993; ZHOU et al. 1995). Platelet adhesion and aggregation stimulate platelet NOS, leading to the release of NO (Radomski et al. 1990a; Malinski et al. 1993; Polanowska-Grabowska and Gear 1994; Lantoine et al. 1995).

2. Role of Substrate

Both hypoxia and hyperoxia are known to influence the activity of eNOS (Radomski et al. 1998 and references therein). As L-arginine is abundant both in plasma and intracellular compartments in quantities several fold higher than the reported k_m for NOS, the pharmacological effects of this amino acid are difficult to explain. L-Arginine is readily converted to NO by platelet NOS when the enzyme is activated (Radomski et al. 1990a; Bode-Böger et al. 1998).

3. Role of Co-Factors

Both platelet- and endothelium-derived eNOS are NADPH, biopterin and flavin dependent (Berkels et al. 1997b, Salas et al. 1997). The availability of calcium appears to be crucial, as its removal abolishes the enzyme activity (Radomski et al. 1990a). Interestingly, there is an apparent lack of correlation between the changes in the intraplatelet calcium levels and the activation of platelet-derived eNOS (Malinski et al. 1993; Lantoine et al. 1995). The reasons for this discrepancy remain unknown.

4. Rheology

It is very important to point out that the unique position of platelets in the proximal vicinity of the endothelial lining facilitates regulation of platelet function by endothelium-derived NO (Radomski and Radomski 1999). Indeed, under physiological conditions red cells containing hemoglobin, a NO scavenger, are localised in the centre of blood flow. In this context, it is interesting to point out that turbulent blood flow (as encountered at vessel branching) may facilitate cell–vessel-wall interactions and the development of vascular disorders.

IV. Physiological Effects of NO on Platelets

1. Effects of NO on Platelet Function In Vitro

Cultured and fresh endothelial cells, when stimulated with bradykinin, release NO in quantities sufficient to inhibit platelet adhesion (Radomski et al. 1987b, 1987c; Sneddon and Vane 1988). Moreover, the coronary and pulmonary vasculatures generate NO to inhibit platelet adhesion under constant flow conditions (Pohl and Busse 1989; Venturini et al. 1989).

In vitro, platelet aggregation induced by a variety of agonists and by shear stress is inhibited by NO released from fresh or cultured endothelial cells (Busse et al. 1987; Furlong et al. 1987; Radomski et al. 1987a, 1987d; Macdonald et al. 1988; Alheid et al. 1989; Houston et al. 1990; Broekman et al. 1991). This NO also causes disaggregation of pre-formed platelet aggregates (RADOMSKI et al. 1987d) and inhibits platelet recruitment to the aggregate (FREEDMAN et al. 1997).

2. Effects of NO on Platelet Function In Vivo

Basal or stimulated release of NO results in both inhibition of platelet aggregation induced by some aggregating agents or endothelial injury and an increase in bleeding time (ROSENBLUM et al. 1987; BHARDWAJ et al. 1988; HOGAN et al. 1988; HUMPHRIES et al. 1990; HERBACZYNSKA-CEDRO et al. 1991; MAY et al. 1991; GOLINO et al. 1992; YAO et al. 1992; HOUSTON and BUCHANAN 1994). In addition, there is luminal release of NO from human vasculature, causing increases in intraplatelet cGMP levels (ANDREWS et al. 1994a). Finally, the administration of NOS inhibitor N-monomethyl-L-arginine into healthy volunteers increased platelet aggregation granule release (BODZENTA-LUKASZYK et al. 1994) and shortened bleeding time (SIMON et al. 1995), whereas L-arginine leads to the inhibition of platelet activation (CAREN and CORBO 1973; ADAMS et al. 1995; BODE-BÖGER et al. 1998). Both the vasodilator (HOUSTON and BUCHANAN 1994) and platelet-inhibitory (GOLINO et al. 1992; YAO et al. 1992; BODZENTA-LUKASZYK et al. 1994; ADAMS et al. 1995; BODE-BÖGER et al. 1998) components contribute to the haemostatic action of NO.

3. NO in Synergistic Regulation of Platelet Function

The studies performed between 1980 and 1989 (RADOMSKI and MONCADA 1991 for references) showed that platelet activation and thrombus formation occur as a result of synergistic interactions between pro-aggregating agents. In 1987, we found that the synergistic interactions also take place between agents that inhibit platelet aggregation. Indeed, the endothelial cells generate and release NO and prostacyclin that act jointly to maximise the extent of platelet inhibition and vessel-wall protection (RADOMSKI et al. 1987d). Similar synergistic interactions have now been found to occur between NO and other inhibitors of platelet activation (SALAS et al. 1997b for references).

V. The Mechanisms of NO Action on Platelets

NO activates the soluble guanylyl cyclase (GC-S) by binding to the heme moiety of the enzyme (CRAVEN and DERUBERTIS 1978). Indeed, platelet fractions containing GC-S readily sequester NO gas (LIU et al. 1993). The interaction of NO with GC-S results in the conversion of magnesium guanosine 5′-triphosphate to guanosine 3′,5′-monophosphate (cGMP) (MELLION et al. 1981).

Three proteins mediate the cellular actions of cGMP: cGMP-dependent protein kinase, cGMP-binding cyclic adenosine monophosphate (cAMP) phosphodiesterase and cGMP-regulated ion channels (WALTER 1989). Stimulation of cGMP-dependent protein kinase results in protein phosphorylation

(WALTER 1989), including 46/50-kDa vasodilator-stimulated phosphoprotein (VASP, HAFFNER et al. 1995). In adhering platelets, VASP is associated with actin filaments and focal contact areas, i.e. transmembrane junctions between microfilaments and the extracellular matrix (REINHARD et al. 1992). The binding of VASP with the platelet cytoskeleton may mediate its inhibitory effect on the fibrinogen receptor (HORSTRUP et al. 1994). cGMP-induced protein phosphorylation may be also involved in the uptake of serotonin by platelets (LAUNAY et al. 1994).

cGMP decreases basal and stimulated concentrations of intracellular Ca^{2+} (NAKASHIMA et al. 1986; JOHANSSON and HAYNES 1992). A number of Ca^{2+}-handling systems have been identified in platelets, including receptor-operated channels, passive leak, Ca^{2+}–adenosine triphosphatase (ATPase) extrusion pump, the Na^+/Ca^{2+} exchanger, the Ca^{2+}-accumulating ATPase pump of the dense tubular membrane (an intraplatelet membrane Ca^{2+} store) and passive leakage and receptor-operated Ca^{2+} channels in the dense tubular membrane. In principle, all these processes could be affected by cGMP. It has been shown that cGMP increases the activity of Ca^{2+}–ATPase extrusion pump and leakage across the plasma membrane (JOHANSSON and HAYNES 1992). In addition, cGMP causes inhibition of Ca^{2+} mobilisation from intraplatelet stores, including the dense tubular membrane (NAKASHIMA et al. 1986). Metabolism of membrane phospholipids may be also a target for the action of cGMP. Indeed, the inhibition of both phospholipase C and A_2 has been implicated in the mechanism of this action on platelets (NAKASHIMA et al. 1986; SANE et al. 1989).

Cyclic GMP down-regulates the function of some platelet receptors, including the fibrinogen receptor IIb/IIIa and protein kinase C-induced expression of P-selectin and the release of lysosomal protein CD63 (MENDELSOHN et al. 1990; SALAS et al. 1994; MUROHARA et al. 1995; MICHELSON et al. 1996). Interestingly, von Willebrand factor and fibronectin receptors appear not to be regulated by cGMP (SHAHBAZI et al. 1994; MICHELSON et al. 1996).

Phosphorylation of an cGMP-inhibited cAMP phosphodiesterase by the cAMP-dependent protein kinase increases phosphodiesterase activity, and this may represent a negative-feedback mechanism of cellular cAMP levels. It has been shown that cGMP, by inhibiting cGMP-inhibited cAMP phosphodiesterase, may delay the hydrolysis of cAMP and enhance the biological effects of the latter nucleotide (MAURICE and HASLAM 1990). However, the physiological and pharmacological relevance of this "cross-talk" between cAMP and cGMP pathways is unclear (RADOMSKI et al. 1992).

Some authors suggested that, in addition to cGMP-dependent effects, some actions of NO on platelets might be independent of the generation of this cyclic nucleotide. These could be due to direct actions of NO on calcium flux (MENSHIKOV et al. 1993), metabolism via inhibition of ADP ribosylation (BRÜNE and LAPETINA, 1989) and inhibition of 12-lipoxygenase (NAKATSUKA and OSAWA, 1994).

Studies attempting to investigate the role of GC-S in the mediation of the physiological effects of NO on platelets have been handicapped for a long time by the lack of potent and selective inhibitors of this enzyme. Both methylene blue and LY83583, which were widely used to inhibit GC-S, lack selectivity and interact with a number of molecular targets in addition to GC-S (MORO et al. 1996 for references).

Recently characterised 1H-[1,2,4]oxadiazolo[4,3-a]quinoxalin-1-one (ODQ) was characterised as a potent and selective inhibitor of GC-S (GARTH-WAITE et al, 1995). Using this compound, we have shown that both the anti-aggregatory and adhesion-inhibitory effects of NO in vitro are sensitive to ODQ and, thus, are cGMP-dependent (MORO et al 1996, MARINEZ-CUESTA and RADOMSKI 1998).

Interestingly, some NO donors, such as *S*-nitroso-glutathione may inhibit platelet function using both cGMP-dependent and -independent mechanisms (GORDGE et al. 1998). The actions, independent of cGMP formation, may require the presence of copper and membrane thiols. The biological actions of cGMP are terminated by cGMP phosphodiesterase and by its efflux from platelets and may also depend on the activity of protein phosphatases (WALTER 1989).

C. The Role of NO in the Pathogenesis of Vascular Disorders Associated with Platelet Activation

I. Pathomechanism

The vasodilator and platelet-regulatory functions of endothelium are impaired during the course of vascular disorders, including atherosclerosis, coronary artery disease, essential hypertension, diabetes mellitus and pre-eclampsia (DEBELDER and RADOMSKI, 1994); however, the reasons underlying this impairment are not clear. A number of researchers correlated changes in endothelial function with the generation of NO. The endothelial dysfunction was ascribed to both decreased and enhanced generation of NO. To explain this discrepancy, it was proposed that these changes in NO generation are often accompanied by *reduced bioactivity of NO* (RADOMSKI and SALAS 1995). It is likely that the metabolism of NO and its interactions with reactive oxygen species account for the reduced bioactivity of NO. In 1990, Beckman and associates first found that the reaction of NO with superoxide could take place under physiological conditions and lead to formation of peroxynitrite (ONOO$^-$) (BECKMAN and TSAI 1994 for references). Peroxynitrite is a highly reactive oxidant that can oxidise various biomolecules in the cellular microenvironment. In 1994, we found that ONOO$^-$ can decrease the vasodilator and platelet-inhibitory activities of NO and prostacyclin (MORO et al. 1994; VILLA et al. 1994). Thiols and glucose (MORO et al. 1994, 1995) attenuated these detrimental effects of ONOO$^-$. The reaction of ONOO$^-$ with thiols in cell membranes and glucose in the extracellular fluid results in synthesis of NO donors that counteract the vasoconstrictor and platelet-aggregatory activities of the parent oxidant (MORO et al. 1994, BROWN et al. 1998). Interestingly, there is now evidence that small amounts of ONOO⁻ may be generated during aggregation of normal platelets (R. BRUCKDORFER, personal communication). Thus, it is likely that, when generated by platelets under physiological conditions, ONOO⁻ is rapidly detoxified and converted to NO donors following reactions with platelet membrane thiols (BROWN et al. 1998). The oxidising stress could decrease the efficiency of this regulating mechanism and precipitate platelet dysfunction and damage.

II. Atherosclerosis, Thrombosis and Hypertension

Atherogenesis is associated with profound changes in the oxidative status of the vascular wall. Since oxidative modification of low-density lipoproteins (LDL) plays a key role in atherogenesis, a number of studies (RADOMSKI and SALAS 1995) have examined the effects of native and oxidised LDL on NO-mediated vascular functions. In most of these studies, lipoproteins decreased the bioactivity of NO (COOKE and TSAO 1992 and references therein; LÜSCHER et al. 1993). The decreased bioactivity of NO in atherosclerosis could also result from changes in the metabolism of this molecule and generation of ONOO⁻ from superoxide and inducible NO (BECKMAN and TSAI 1994). In addition, LDL may inhibit L-arginine uptake into platelets and, through this mechanism, decrease NOS activity and promote thrombosis (CHEN and MEHTA 1994), an effect reversible by the administration of L-arginine in the diet (TSAI et al. 1994). In contrast to LDL, high-density lipoproteins (HDL) decreased platelet activation and thrombosis by increasing NOS activity in platelets (CHEN and MEHTA 1994). Moreover, human apolipoprotein E, which mediates hepatic clearance of lipoproteins, exerts a significant inhibitory effect on platelets through stimulation of platelet NOS (RIDDELL et al. 1997). The ischemic heart disorder and myocardial infarction are common manifestations of coronary atherosclerosis. The endogenous NO inhibited microthromboembolism in the ischemic heart, protected the myocardium against intracoronary thrombosis and decreased platelet deposition due to the carotid endarterectomy (KOMAMURA et al. 1994; OLSEN et al. 1996). Moreover, decreased generation of NO by platelets predicts the presence of acute coronary syndromes in patients with coronary atherosclerosis (FREEDMAN et al. 1998). Finally, an impaired NO generation or action may also underlie the pathomechanism of vasospastic and thrombotic changes of essential hypertension (CALVER et al. 1992; CADWGAN and BENJAMIN 1993).

III. Diabetes Mellitus and Stress

There are indications that changes in the bioactivity and metabolism of NO are involved in the pathogenesis of diabetes mellitus. Insulin, at physiological concentrations, inhibits platelet activation via stimulation of platelet NOS and the resultant increase in cGMP and possibly cAMP levels (TROVATI et al. 1996, 1997). This suggests that a reduced generation of NO in insulin-deficient states could contribute to platelet hyperactivity and diabetic angiopathy. Recently, we examined the formation of NO in the vasculature and platelets of JCR-LA-cp rat, a model of insulin-resistant states associated with obesity and complicated by atherosclerosis. We have found that generation of NO is crucial for the preservation of vascular homeostasis under these conditions (McKENDRICK et al. 1998).

Generation of vascular NO in response to stress protects the vasculature from the vasoconstrictor and platelet-activator effects of stress hormones (LEZA et al. 1998). However, the capacity of the vasculature to offset the detrimental effects of stress appears to be limited, and longer-lasting stress exposure leads to decreased generation of platelet NO, thus facilitating platelet activation and thrombosis (LEZA et al. 1998).

IV. Pre-Eclampsia

Vasoconstriction and increased platelet activation are also characteristic for pre-eclampsia, a severe disease that may complicate normal pregnancy. We have measured the activity of NOS and GC-S in pre-eclamptic women and compared it with the levels in non-pregnant and healthy pregnant subjects. Increased generation of NO in the platelets of pre-eclamptic women was associated with increased platelet activation (SALAS et al. 1997a). However, despite increased formation of NO, the activity of the GC-S was reduced in pre-eclamptic women, suggesting that this NO was not bioactive, i.e. it failed to inhibit platelet activation. We hypothesised that enhanced generation of $ONOO^-$ could contribute to platelet dysfunction and damage in pre-eclampsia.

V. Septicaemia

The invasion of gram-negative bacteria and the exposure of cells to bacterial toxins and cytokines (DEBELDER and RADOMSKI 1994 for references; RADOMSKI 1995) lead to a life-threatening syndrome often referred to as septic shock. Bacterial toxins and cytokines cause induction of a number of enzymatic systems, including iNOS. The expression of iNOS is likely to have complex repercussions for vascular hemostasis. Inducible NO could attenuate the detrimental effects on hemostasis brought about by stimulation of clotting cascade (disseminated intravascular coagulation) by bacterial toxins. Indeed, inhibition of NO generation by inhibitors potentiated cytokine-stimulated platelet adhesion to cultured human endothelial cells (RADOMSKI et al. 1993), precipitated renal glomerular thrombosis (SCHULTZ and RAIJ 1992) and exacerbated sepsis-induced renal hypoperfusion (SPAIN et al. 1994) in vivo. However, the exposure of endothelial cells to cytokine-induced NO may result in cell toxicity and destruction (PALMER et al. 1992), and it has been reported that the inhibition of NOS may be beneficial in the treatment of septic shock (KILBOURN et al. 1990; WRIGHT et al. 1992). A partial explanation for this discrepancy may be that the currently available inhibitors of iNOS are not selective and inhibit the activities of other NOS isoenzymes. Indeed, an intact generation of constitutive NO may be important to maintain the integrity of the microvasculature during sepsis.

VI. Uraemia

Platelet diathesis associated with bleeding is a classic complication of uraemia due to the suppression of platelet function by the disease process. As early as in 1970, it was found that L-arginine and some other metabolites of the urea cycle are accumulated in uraemia (Horowitz et al. 1970). More recent work showed that uraemia leads to an increase in cytokine levels (Noris et al. 1993). Platelets obtained from uremic patients generate more NO than controls, so that increased expression and/or activity of NOS may play a role in the platelet dysfunction observed in this condition (Noris et al. 1993).

VII. Cancer

Platelets contribute to cytotoxic cell effector system controlling neoplasia (Okada et al. 1996 for references). A part of this cytotoxic mechanism of platelets could be NO dependent (Okada et al. 1996).

Platelets also play a role in the pathogenesis of tumour metastasis by increasing the formation of tumour cell–platelet aggregates, thus facilitating cancer cell arrest in the microvasculature. Tumour cell-induced platelet aggregation in vitro is modulated by the ability of tumour cells to generate NO, and this correlates with their propensity for metastasis (Radomski et al. 1991). Indeed, human colon carcinoma cells isolated from metastases exhibited lower NO activity than cells isolated from the primary tumour. Moreover, the expression of iNOS by murine melanoma cells inversely correlated with their ability to form metastases in vivo (Dong et al. 1994). These data suggest that differential synthesis of NO may distinguish between cells of low and high metastatic potential. Interestingly, NOS has been found in some human gynaecological malignancies and the highest NOS activities detected in poorly differentiated tumours (Thomson et al. 1994). Thus, further work is needed to unravel the biological significance of NO for the growth of tumour, tumour metastasis and platelet–tumour cell interactions.

D. Pharmacological Modulation of Formation and Action of NO on Platelets

I. L-Arginine

The amounts of endogenous L-arginine in the platelet microenvironment are high (mM), suggesting that the availability of substrate is unlikely to constitute a rate-limiting factor for activation of platelet eNOS. Indeed, pharmacological stimulation of resting platelets with L-arginine does not result in generation of NO (RADOMSKI et al. 1990a). Platelet activation is a potent stimulus for stimulation of NOS and, under these conditions, exogenous L-arginine is promptly converted by platelet eNOS to NO (RADOMSKI et al. 1990a; MALINSKI et al. 1993; BODE-BÖGER et al. 1998). The presence of extracellular calcium appears to be crucial for the activation of platelet eNOS by L-arginine, as the platelet-inhibitory activity of L-arginine in the citrated platelet-rich plasma (low calcium levels) is lower than in whole blood anti-coagulated with hirudin, which preserves physiological calcium levels (RADOMSKI et al. 1990a; BODE-BÖGER et al. 1998).

In contrast to eNOS, the expression of iNOS appears to provide an appropriate stimulus for the conversion of pharmacologically administered L-arginine to NO. Under physiological conditions platelets have been shown to express both the message and protein for iNOS (WALLERATH et al. 1997), although evidence for the expression of enzyme activity under these conditions is lacking. In contrast, atherosclerosis and pre-eclampsia lead to a clear-cut increase in the activity of Ca^{2+}-independent NOS, implying that iNOS is activated during these pathologies (DUBE et al. 1998; SALAS et al. 1998). Interestingly, Cooke and colleagues (TSAO et al. 1994) provided convincing evidence for platelet-inhibitory actions of L-arginine when administered to animals and humans with atherosclerosis. Moreover, pharmacological administration of L-arginine to patients suffering from coronary artery disease and peripheral arterial obstructive disease alleviated the symptoms of arterial insufficiency (SLAWINSKI et al. 1996; CEREMUZYNSKI et al. 1997). Thus, L-arginine may be a useful pharmacological agent in the treatment of vascular disorders associated with platelet activation.

II. Stimulators of NOS

Some pharmacological agents including relaxin, a uterine hormone (BANI et al. 1995) and trilinolein, a triacylglycerol obtained from medicinal herb *Panax pseudoginseng*) inhibit platelet function via stimulation of the activity of platelet NOS (SHEN and HONG 1995)

III. Inhibitors of NOS and NO Scavengers

Inhibitors of NOS (RADOMSKI et al. 1990a) oppose the platelet-inhibitory effects of L-arginine. These compounds shorten the bleeding time in humans (SIMON et al. 1995) and may precipitate thrombosis under conditions of vascular stress exemplified by septicemia (SCHULZ and RAIJ 1992). Interestingly, long-term smoking impairs the activity of platelet NOS (ICHIKI et al. 1996), although the mechanism responsible for this impairment remains to be elucidated.

Purified human haemoglobins, cross-linked to prevent renal damage, are currently undergoing clinical trials as oxygen-carrying agents. In addition to binding oxygen, haemoglobin has a very high affinity for NO. OLSEN and colleagues (1996) showed that cross-linked haemoglobin enhances platelet deposition in a rat carotid endarterectomy model, an effect prevented by administration of L-arginine. Thus, cross-linked haemoglobins may stimulate platelet activation through a mechanism involving NO scavenging.

IV. NO Gas

Studies using NO gas in vitro showed that the molecule was a potent but short-duration (the biological half-life less than 4min) inhibitor of platelet adhesion and aggregation and a stimulator of platelet disaggregation (RADOMSKI et al. 1987a, 1987b, 1987c, 1987d).

Because of its short-lasting pharmacological effects, inhaled NO is used increasingly as a selective pulmonary vasodilator to treat critically ill adults and infants (CHEUNG et al. 1997). Some investigators studied the effects of inhaled NO on platelet function in health and disease. Interestingly, there are very limited effects of inhaled NO on the platelets of healthy subjects (ALBERT et al. 1996). However, both in adults with adult respiratory distress syndrome (SAMAMA et al. 1995) and critically ill neonates (CHEUNG et al. 1998), NO treatment causes inhibition of platelet function. Under some conditions, this effect is unexpectedly long lasting and may enhance the risk of the intracranial bleeding. Therefore, caution should be taken during administration of NO gas to critically ill patients.

Interestingly, animal experiments showed that inhaled NO might inhibit the development of coronary thrombosis (ADRIE et al. 1996). The pharmacological significance of these findings remains to be investigated.

V. NO Donors

The mechanism of pharmacological activity of NO donors is related to the release of NO (KATSUKI et al. 1977; FEELISCH and NOACK 1987). The mechanism of NO release from these compounds is spontaneous or catalyst- and enzyme-dependent. In contrast to the relatively short-lived pharmacological effects of NO gas, those of NO donors are longer lasting.

Organic nitrates are poor spontaneous releasers of NO and require the presence of a thiol co-factor for acceleration of this liberation (FEELISCH 1991). In vivo, however, the release of NO from organic nitrates is greatly enhanced by thiols and enzyme(s), including plasma glutathione-S-transferases (CHEN et al. 1996). Whether or not these or similar enzymes are present in platelets remains controversial (GERZER et al. 1988; WEBER et al. 1996). Nitrate-induced inhibition of platelet aggregation in vitro can be greatly potentiated in the presence of thiols or cultured vascular cells (LOSCALZO 1985; BENJAMIN et al. 1991; FEELISCH 1991). This indicates that the conversion of organic nitrates by the vascular tissue in vivo can result in the release of sufficient amounts of NO for inhibition of platelet function. Indeed, in experimental animals, as well as in healthy volunteers, oral and intravenous administrations of glyceryl trinitrate and isosorbide mononitrates resulted in inhibition of platelet aggregation ex vivo (PLOTKINE et al. 1991; WERNS et al. 1994; SALAS et al. 1997b). The effectiveness of organic nitrates as antithrombotics increases with the extent of vascular injury (LAM et al. 1988). Furthermore, short- and long-lasting administration of nitroglycerin and isosorbide dinitrate to patients suffering from coronary artery disease and acute myocardial infarction resulted in a significant inhibition of platelet adhesion and aggregation (DIODATI et al. 1990; GEBALSKA 1990; SINZINGER et al. 1992).

What is the position of organic nitrates among "classical" inhibitors of platelet function? Actelylsalicylic acid (aspirin) is by far the most widely used anti-platelet drug in clinical practice, and its benefits (in terms of decreasing mortality due to re-infarction) have been unequivocally demonstrated (International Study of Infarct Survival Collaborative Group 1988, 1992; PATRONO 1989), while those of organic nitrates have not yet been established. A meta-analysis found significant reduction in mortality when intravenous glyceryl trinitrate or nitroprusside were used during acute course of myocardial infarction (YUSUF et al. 1988). Moreover, when combined with N-acetylcysteine, glyceryl trinitrate substantially reduced myocardial infarction in unstable angina, an effect compatible with an anti-platelet effect of glyceryl trinitrate (HOROWITZ et al. 1988). Surprisingly, GISSI III (Gruppo Italiano per lo Studio della Sopravvivenza nell'infarto MIOCARDICO 1994) and ISIS-4 (International Study of Infarct Survival Collaborative Group 1993) studies failed to show clinically beneficial effects of organic nitrates on mortality after myocardial infarction. However, further analysis of GISSI III suggests that the apparent additive effect of glyceryl trinitrate and lisinopril could be attributed to anti-platelet effects of this NO donor (ANDREWS et al. 1994b). In addition, it is possible that nitrates may act by reducing the infarct size in small rather than large infarcts, so the neutral results of GISSI-III and ISIS-4 may be explained by the heterogeneity of the effect. Interestingly aspirin, a cyclooxygenase inhibitor, blocks only thromboxane-mediated platelet aggregation (PATRONO 1989 for references), leaving the remaining pathways of adhesion and aggregation unopposed. In contrast, NO inhibits the activation cascade of mediators generated by all known pathways of platelet aggregation (SALAS et al. 1997b), including recently described matrix metalloproteinase-2-dependent platelet aggregation (SAWICKI et al. 1997) and some pathways of platelet adhesion to sub-endothelium (SHAHBAZI et al. 1994).

Whether or not platelets, similar to vessel-wall cells, become tolerant to the platelet-inhibitory effects of organic nitrates is again controversial (BOOTH et al. 1996; WEBER et al. 1996). There have been many attempts to synthesise tolerance-free NO donors. One of more promising groups of drugs is cysteine-containing nitrates. The incorporation of a cellular thiol cysteine into the structures of organic nitrates resulted in a high effectiveness of these compounds as inhibitors of platelet and leukocyte functions both in vitro and in vivo (LEFER et al. 1993).

The phenomenon of tolerance is of lesser pharmacological significance for other NO donors, including sodium nitroprusside, molsidomine and SIN-1. Because of its powerful vasodilator action, sodium nitroprusside is often used to treat vascular emergencies associated with hypertensive crisis. Since this compound shows some anti-platelet activity both in vitro and in vivo (LEVIN et al. 1982; HINES and BARASH 1989), its acute clinical effects may also be mediated, in part, through inhibition of platelet function. Recently, sodium nitroprusside was administered intrapericardially to treat experimentally induced coronary thrombosis in dogs (WILLERSON et al. 1996). As this route of administration of sodium nitroprusside produced less vasodilatation than systemic one, localised administration of this drug may offer new therapeutic possibilities for the treatment of the coronary thrombosis.

Molsidomine and its active metabolite SIN-1 inhibit experimental thrombosis and platelet aggregation in healthy volunteers and in patients suffering from acute myocardial infarction (WAUTIER et al. 1989). Interestingly, SIN-1 generates superoxide and $ONOO^-$ in addition to NO (HOGG et al. 1993). Since $ONOO^-$ causes platelet aggregation and counteracts the platelet-inhibitory activity of NO (MORO et al. 1994), the formation of this radical may offset the anti-platelet activity of NO released from SIN-1.

VI. Novel NO Donors

The platelet-inhibitory actions of organic nitrates cannot be separated from their effects on vascular wall. The concept of platelet-selective NO donors has arisen from our experiments with *S*-nitrosoglutathione (GSNO) (RADOMSKI et al. 1992). GSNO is a tripeptide *S*-nitrosothiol that is formed by *S*-nitrosylation of glutathione, the most abundant intracellular thiol. We have found that the intravenous administration of GSNO into conscious rats inhibits platelet aggregation at doses that have only a small effect on the blood pressure (RADOMSKI et al. 1992). Moreover, similar platelet/vascular differentiation is detected following intra-arterial administration of GSNO into the circulation of the human forearm (DEBELDER et al. 1994). Finally, we have infused GSNO into patients undergoing balloon angioplasty and found that this NO donor effectively protects platelets from activation at the site of angioplastic injury without altering blood pressure (LANGFORD et al. 1994). Interestingly, the exposure of human neutrophils to NO led to depletion of glutathione stores, activation of hexose monophosphate shunt, synthesis of endogenous GSNO and inhibition of superoxide generation by neutrophils. Synthetic GSNO resulted in similar effects (CLANCY et al. 1994). Moreover, the administration of GSNO inhibited leukocyte activation, expression of iNOS and bypass-induced myocardial lesion in dogs (MAYERS et al. 1998). These observations show that GSNO is a potent regulator of platelet and neutrophil functions, and it may be a prototype for the development of blood cell-selective NO donors.

Recently, nitro derivatives of aspirin and RGDS peptide have been synthesised in order to capitalise on the synergy of NO with cyclooxygenase inhibitors (aspirin) and the antagonists of the fibrinogen receptor (RGDS) (LECHI et al. 1996; GUREVICH et al. 1997). The pharmacological and clinical potentials of such hybrid compounds remain to be investigated.

VII. NO-Independent Activators of GC-S

Recently, benzyl indazole derivatives, including YC-1, have been synthesised and shown to be effective inhibitors of platelet adhesion and aggregation (WU et al. 1995). The pharmacological potential of direct stimulators of GC-S as anti-thrombotics remains to be studied.

E. Conclusions

NO plays an important role in regulation of vessel-wall haemostasis. Pathological derangement in its generation, action or metabolism contributes to the pathogenesis of occlusive vascular disorders. Exploring the potential of NO donors as anti-thrombotics holds promise for pharmacological development and therapy of vascular disorders associated with platelet activation and damage.

Acknowledgements. This work was supported by a grant from Medical Research Council of Canada. MWR is an Alberta Heritage Foundation for Medical Research Scholar.

References

Adams MR, Forsyth CJ, Jessup W, Robinson J, Celermajer DS (1995) Oral L-arginine inhibits platelet aggregation but does not enhance endothelium-dependent dilation in healthy young men. J Am Coll Cardiol 26:1054–1061
Adrie C, Bloch KD, Moreno PR, et al. (1996) Inhaled nitric oxide increases coronary artery patency after thrombolysis. Circulation 94:1919–1926
Albert J, Wallen NH, Broijersen A, Frostell C, Hjemdahl P (1996) Effects of inhaled nitric oxide compared with aspirin on platelet function in vivo in healthy volunteers. Clin Sci 91:225–231
Alheid U, Reichwehr I, Förstermann U (1989) Human endothelial cells inhibit platelet aggregation by separately stimulating platelet cyclic AMP and cyclic GMP. Eur J Pharmacol 164:103–110
Andrews NP, Dakak N, Schenke WH, Quyyumi AA (1994a) Platelet-endothelium interactions in humans: changes in platelet cyclic guanosine monophosphate content in patients with endothelial dysfunction. Circulation 90:I-397
Andrews R, May JA, Vickers J, Heptinstall S (1994b) Inhibition of platelet aggregation by transdermal glyceryl trinitrate. Br Heart J 72:575–579
Azuma H, Ishikawa M, Sekizaki S (1986) Endothelium-dependent inhibition of platelet aggregation. Br J Pharmacol 88:411–415
Bani D, Bigazzi M, Massini E, Bani G, Sacchi TB (1995) Relaxin depresses platelet aggregation: in vitro studies on isolated rabbit and human platelets. Lab Invest 73:709–716
Beckman J, Tsai JH (1994) Reactions and diffusion of nitric oxide and peroxynitrite. The Biochemist 16:8–10

Benjamin N, Dutton JAE, Ritter JM (1991) Human vascular smooth muscle cells inhibit platelet aggregation when incubated with glyceryl trinitrate: evidence for generation of nitric oxide. Br J Pharmacol 102:847–850

Berkels R, Stockklauser K, Rosen P, Rosen R (1997) Current status of platelet NO synthases. Thromb Res 87:51–55

Bhardwaj R, Page CP, May GR, Moore PK (1988) Endothelium-derived relaxing factor inhibits platelet aggregation in human whole blood in vitro and in the rat in vivo. Eur J Pharmacol 157:83–91

Bode-Böger SM, Boger RH, Galland A, Frölich JC (1998) Differential inhibition of human platelet aggregation and thromboxane A_2 formation by L-arginine in vivo and in vitro. Arch Pharmacol 357:143–150

Bodzenta-Lukaszyk A, Gabryelewicz A, Lukaszyk A, et al. (1994) Nitric oxide synthase inhibition and platelet function. Thromb Res 75:667–672

Booth BP, Jacob S, Bauer JA, Fung HL (1996) Sustained antiplatelet properties of nitroglycerin during hemodynamic tolerance in rats. J Cardiovasc Pharmacol 28:432–438

Broekman MJ, Eiroa AM, Marcus AJ (1991) Inhibition of human platelet reactivity by endothelium-derived relaxing factor from human umbilical vein endothelial cells in suspension: blockade of aggregation and secretion by an aspirin-insensitive mechanism. Blood 78:1033–1040

Brown AS, Martin JF (1994) The megakaryocyte platelet system and vascular disease. Eur J Clin Invest 24(S1):9–15

Brown AS, Moro MA, Masse JM, Cramer EM, Radomski MW, Darley-Usmar V (1998) Nitric oxide-dependent and -independent effects on human platelet treated with peroxynitrite. Cardiovasc Res 40:380–388

Brüne B, Lapetina EG (1989) Activation of a cytosolic ADP-ribosyltransferase by nitric oxide-generating agents. J Biol Chem 264:8455–8458

Busse R, Fleming I (1998) Pulsatile stretch and shear stress: physical stimuli determining the production of endothelium-derived relaxing factors. J Vasc Res 35:73–84

Busse R, Lückhoff A, Bassenge E (1987) Endothelium-derived relaxing factor inhibits platelet activation. Arch Pharmacol 336:566–571

Cadwgan TM, Benjamin N (1993) Evidence for altered platelet nitric oxide synthesis in essential hypertension. J Hypertens 11:417–420

Calver A, Collier J, Moncada S, Vallance P (1992) Effect of local infusion of N^G-monomethyl-L-arginine in patients with hypertension. The nitric oxide dilator mechanism appears abnormal. J Hypertens 10:1025–1031

Caren R, Corbo L (1973) Response of plasma lipids and platelet aggregation to intravenous arginine. Proc Soc Exp Biol Med 143:1067–1071

Ceremuzynski L, Chamiec T, Herbaczynska-Cedro K (1997) Effect of supplemental oral L-arginine on exercise capacity in patients with stable angina pectoris. Am J Cardiol 80:331–333

Chen LY, Mehta JL (1994) Inhibitory effect of high-density lipoprotein on platelet function is mediated by increase in nitric oxide synthase activity in platelets. Life Sci 23:1815–1821

Chen LY, Mehta JL (1996) Further evidence of the presence of constitutive and inducible nitric oxide synthase isoforms in human platelets. J Cardiovasc Pharmacol 27:154–158

Chen LY, Mehta P, Mehta JL (1996) Platelet inhibitory effect of nitroglycerin in platelet-rich plasma: relevance of glutathione-S-transferase in plasma. J Invest Med 44:5611–565

Cheung P-Y, Salas E, Schulz R, Radomski MW (1997) Nitric oxide and platelet function: implications for neonatology. Sem Perinatol 21:409–417

Cheung P-Y, Salas E, Etches PC, Phillipos E, Schulz R, Radomski MW (1998) Inhaled nitric oxide and inhibition of platelet aggregation in critically ill neonates. Lancet 351:1181–1182

Clancy RM, Levartovsky D, Leszczynska-Piziak J, Yegudin J, Abramson SB (1994) Nitric oxide reacts with intracellular glutathione and activates the hexose monophosphate shunt in human neutrophils: evidence for S-nitrosoglutathione as a bioactive intermediary. Proc Natl Acad Sci USA 91:3680–3684

Cooke JP, Tsao P (1992) Cellular mechanisms of atherogenesis and the effects of nitric oxide. Curr Opin Cardiol 7:799–804

Craven PA, DeRubertis FR (1978) Restoration of the responsiveness of purified guanylate cyclase to nitrosoguanidine, nitric oxide, and related activators by heme and heme proteins. Evidence for involvement of the paramagnetic nitrosyl–heme complex in enzyme activation. J Biol Chem 253:8433–8443

DeBelder AJ, Radomski MW (1994) Nitric oxide in the clinical arena. J Hypertens 12:617–624

DeBelder AJ, MacAllister R, Radomski MW, Moncada S, Vallance PJ (1994) Effects of S-nitrosoglutathione in the human forearm circulation. Evidence for selective inhibition of platelet activation. Cardiovasc Res 28:691–694

Diodati J, Theroux P, Latour JG, et al. (1990) Effects of nitroglycerin at therapeutic doses on platelet aggregation in unstable angina pectoris and acute myocardial infarction. Am J Cardiol 66:683–688

Dong Z, Staroselsky AH, Qi X, Xie K, Fiedler I (1994) Inverse correlation between expression of inducible nitric oxide synthase activity and production of metastasis in K-1735 murine melanoma cells. Cancer Res 54:789–793

Dube G, Salas E, Kurtz W, Christie R, Radomski MW (1998). Effects of estrogen and an estrogen receptor modulator on atherogenesis in rabbits: role of NO. In: The biology of nitric oxide. Part 6. Moncada S, Toda N, Maeda H, Higgs EA eds. Portland, London, p 146

Feelisch M (1991) The action and metabolism of organic nitrates and their similarity with endothelium-derived relaxing factor (EDRF). In: Clinical relevance of nitric oxide in the cardiovascular system Moncada S, Higgs EA, Berrazueta JR eds, Edicomplet, Madrid, pp 29–43

Feelisch M, Noack EA (1987) Correlation between nitric oxide formation during degradation of organic nitrates and activation of guanylyl cyclase. Eur J Pharmacol 139:19–30

Freedman JE, Loscalzo J, Barnard MR, Alpert C, Keaney JF Jr, Michelson AD (1997) Nitric oxide released from activated platelets inhibits platelet recruitment J Clin Invest 100:350–356

Freedman JE, Ting B, Hankin B, Loscalzo J, Keaney JF, Vita JA (1998) Impaired platelet production of nitric oxide predicts presence of acute coronary syndromes. Circulation 98:1481–1486

Furchgott RF, Zawadzki JV (1980) The obligatory role of endothelial cells in the relaxation of arterial smooth muscle by acetylcholine. Nature 288:373–376

Furlong B, Henderson AH, Lewis MJ, Smith JA (1987) Endothelium-derived relaxing factor inhibits in vitro platelet aggregation Br J Pharmacol 90:687–692

Garthwaite J, Southam E, Boulton CL, Nielsen EB, Schmidt K, Mayer B (1995) Potent and selective inhibition of nitric oxide-sensitive guanylyl cyclase by 1H-[1,2,4]oxadiazolo[4,3]quinoxalin-1-one. Mol Pharmacol 48 (2):184–188.

Gebalska J (1990) Platelet adhesion and aggregation in relation to clinical course of acute myocardial infarction. MD thesis Warsaw, in Polish

Geng Y, Lotz M (1995) Increased intracellular Ca2+ selectively suppresses IL-1-induced NO production by reducing iNOS mRNA stability. J Cell Biol 129:1651–1657

Gerzer R, Karrenbrock B, Siess W, Heim JM (1988) Direct comparison of the effects of nitroprusside, SIN-1 and various nitrates on platelet aggregation and soluble guanylyl cyclase activity. Thromb Res 52:11–21

Goldberg ND, Haddox MK, Nicol SE, et al. (1975) Biological regulation through opposing influences of cyclic GMP and cyclic AMP: the yin-yang hypothesis. Adv Cyc Nucleotide Res 5:307–330

Golino P, Capelli-Bigazzi M, Ambrosio G, et al. (1992) Endothelium-derived relaxing factor modulates platelet aggregation in an in vivo model of recurrent platelet activation Circ Res 71:1447–1456

Gordge MP, Hothersall JS, Noronha-Dutra AA (1998) Evidence for a cyclic GMP-independent mechanism in the antiplatelet action of S-nitroso-glutathione. Br J Pharmacol 124:141–148

Gruppo Italiano per lo Studio della Sopravvivenza nell'infarto Miocardico (1994) GISSI-3 Effects of lisinopril and transdermal glyceryl trinitrate singly and together on 6-week mortality and ventricular function after acute myocardial infarction. Lancet 343:1115–1122

Gurevich VS, Lominadze DG, Adeagbo ASO, et al. (1997) Antithrombotic and vasorelaxant properties of a novel synthetic RGD peptide containing nitric oxide. Pharmacology 55:1–9

Haffner C, Jarchau T, Reinhard M, et al. (1995) Molecular cloning, structural analysis and functional expression of the proline-rich focal adhesion and microfilament-associated protein VASP. EMBO J 14:19–27

Herbaczynska-Cedro K, Lembowicz K, Pytel B (1991) N^G-monomethyl-L-arginine increases platelet deposition on damaged endothelium in vivo. A scanning electron microscopy study Thromb Res 64:1–9

Hines R, Barash PG (1989) Infusion of sodium nitroprusside induces platelet dysfunction in vitro. Anesthesiology 71:805–806

Hogan JC, Lewis MJ, Henderson AH (1988) In vivo EDRF activity influences platelet function. Br J Pharmacol 94:1020–1022

Hogg N, Darley-Usmar VM, Wilson MT, Moncada S (1993) Oxidation of alpha-tocopherol in human low density lipoprotein by the simultaneous generation of superoxide and nitric oxide. FEBS Lett 326:199–203

Horowitz HI, Stein IM, Cohen BD, White JG (1970) Further studies on the platelet-inhibitory effect of guanidinosuccinic acid and its role in uremic bleeding. Am J Med 49:336–345

Horowitz JD, Henry CA, Syrjanen ML, et al. (1988) Combined use of nitroglycerin and N-acetylcysteine in the management of unstable angina pectoris. Circulation 77:787–794

Horstrup K, Jablonka B, Hönig-Liedl P, et al. (1994) Phosphorylation of focal adhesion vasodilator-stimulated phosphoprotein at Ser157 in intact human platelets correlates with fibrinogen receptor inhibition. Eur J Biochem 225:21–27

Houston DS, Buchanan MR (1994) Influence of endothelium-derived relaxing factor on platelet function and hemostasis in vivo. Thromb Res 74:25–37

Houston DS, Robinson P, Gerrard JM (1990) Inhibition of intravascular platelet aggregation by endothelium-derived relaxing factor: reversal by red blood cells. Blood 76:953–958

Humphries RG, Tomlinson W, O'Connor SE, Leff P (1990) Inhibition of collagen- and ADP-induced platelet aggregation by substance P in vivo: involvement of endothelium-derived relaxing factor. J Cardiovasc Pharmacol 16:292–297

Ichiki K, Ikeda H, Haramaki N, Ueno T, Imaizumi T (1996) Long-term smoking impairs platelet-derived nitric oxide release. Circulation 94:3109–3114

International Study of Infarct Survival Collaborative Group (1988) Randomized trial of intravenous streptokinase, oral aspirin, both or neither among 17187 cases of suspected acute myocardial infarction: ISIS-2. Lancet 2:349–360

International Study of Infarct Survival Collaborative Group (1992) A randomized comparison of streptokinase vs tissue plasminogen activator vs anistreplase and of aspirin plus heparin vs aspirin alone among 41299 cases of suspected acute myocardial infarction. Lancet 339:753–770

International Study of Infarct Survival Collaborative Group (1993) Randomised study of oral isosorbide mononitrate in over 50000 patients with suspected acute myocardial infarction. Circulation 88:I-394

Johansson JS, Haynes DH (1992) Cyclic GMP increases the rate of the calcium extrusion pump in intact platelets but has no direct effect on the dense tubular calcium accumulation system. Biochim Biophys Acta 1105:40–50

Katsuki S, Arnold W, Mittal C, Murad F (1977) Stimulation of guanylate cyclase by sodium nitroprusside, nitroglycerin and nitric oxide in various tissue preparations and comparison to the effects of sodium azide and hydroxylamine. J Cyc Nucl Res 3:23–35

Kilbourn RG, Gross SS, Adams J, et al. (1990) N^G-methyl-L-arginine inhibits tumor necrosis factor-induced hypotension: implications for the involvement of nitric oxide. Proc Natl Acad Sci USA 87:3029–3032

Komamura K, Node K, Kosaka H, Inoue M (1994) Endogenous nitric oxide inhibits microthromboembolism in the ischemic heart. Circulation 90:I-345

Lam JYT, Chesebro JH, Fuster V (1988) Platelets, vasoconstriction, and nitroglycerin during arterial wall injury. A new antithrombotic role for an old drug. Circulation 78:712–716

Langford EJ, Brown AS, Wainwright RJ, et al. (1994) Inhibition of platelet activity by S-nitrosoglutathione during coronary angioplasty. Lancet 344:1458–1460

Lantoine F, Brunet A, Bedioui F, Devynck J, Devynck MA (1995) Direct measurement of nitric oxide production in platelets: relationship with cytosolic Ca^{2+} concentration. Biochem Bioph Res Commun 215:842–848

Launay JM, Bondoux D, Oset-Gasque MJ, et al. (1994) Increase of human platelet serotonin uptake by atypical histamine receptors. Am J Physiol 266:R526–R536

Lechi C, Gaino S, Tommasoli V, et al. (1996) In vitro study of the anti-aggregating activity of two nitroderivatives of acetylsalicylic acid. Blood Coagulation Thromb 7:206–209

Lefer DJ, Nakanishi K, Vinten-Johansen J (1993) Endothelial and myocardial cell protection by a cysteine-containing nitric oxide donor after myocardial ischemia and reperfusion. J Cardiovasc Pharmacol 22(Suppl 7):S34–S43

Lelchuk R, Radomski MW, Martin JF, Moncada S (1992) Constitutive and inducible nitric oxide synthases in human megakaryoblastic cells. J Pharmacol Exp Ther 262:1220–1224

Levin RL, Weksler BB, Jaffe EA (1982) The interaction of sodium nitroprusside with human endothelial cells and platelets: nitroprusside and prostacyclin synergistically inhibit platelet function. Circulation 66:1299–1307

Leza JC, Salas E, Sawicki G, Russell JCR, Radomski MW (1998) The effects of stress on homeostasis in JCR-LA-cp rats: the role of nitric oxide. J Pharmacol Exp Ther 286:1397–1403

Liu Z, Nakatsu K, Brien JF, et al. (1993) Selective sequestration of nitric oxide by subcellular components of vascular smooth muscle and platelets: relationship to nitric oxide stimulation of the soluble guanylyl cyclase. Can J Physiol Pharmacol 71: 938–945.

Loscalzo J (1985) N-acetylcysteine potentiates inhibition of platelet aggregation nitroglycerin. J Clin Invest 76:703–738

Lüscher TF, Tanner FC, Tschudi MR, Noll G (1993) Endothelial dysfunction in coronary artery disease. Ann Rev Med 44:395–418

MacDonald PS, Read MA, Dusting GJ (1988) Synergistic inhibition of platelet aggregation by endothelium-derived relaxing factor and prostacyclin Thromb Res 49: 437–449

Malinski T, Radomski MW, Taha Z, Moncada S (1993) Direct electrochemical measurement of nitric oxide released from human platelets Biochem Biophys Res Commun 194:960–965

Martinez-Cuesta MA, Radomski MW (1998) Cyclic GMP mediates the inhibitory effect of nitric oxide on platelet adhesion. In: The biology of nitric oxide. Part 6. Moncada S, Toda N, Maeda H, Higgs EA eds. Portland, London, p 141

Maurice DH, Haslam RJ (1990) Molecular basis of the synergistic inhibition of platelet function by nitrovasodilators and activators of adenylate cyclase: inhibition of cyclic AMP breakdown by cyclic GMP. Mol Pharmacol 37:671–681

May GR, Crook P, Moore PK, Page CP (1991) The role of nitric oxide as an endogenous regulator of platelet and neutrophil activation within the pulmonary circulation. Br J Pharmacol 102:759–763

Mayers I, Johnson D, Salas E, Hurst T, Radomski MW (1998) Myocardial activity is increased following canine cardiopulmonary bypass. In: The biology of nitric oxide. Part 6. Moncada S, Toda N, Maeda H, Higgs EA Portland, London, p 181

McKendrick JD, Salas E, Dube GP, Murat J, Russell JC, Radomski MW (1998) Inhibition of nitric oxide generation unmasks vascular dysfunction in insulin-resistant, obese JCR:LA-cp rats. Br J Pharmacol 124:361–369

Mehta JL, Chen LY, Kone CB, Mehta P, Turner P (1995) Identification of constitutive and inducible forms of nitric oxide synthase in human platelets. J Lab Clin Med 125:370–377

Mellion BT, Ignarro LJ, Ohlstein EH, et al. (1981) Evidence for the inhibitory role of guanosine 3′-5′-monophosphate in ADP-induced human platelet aggregation in the presence of nitric oxide and related nitrovasodilators. Blood 57:946–955

Mendelsohn ME, O'Neill S, George D, Loscalzo J (1990) Inhibition of fibrinogen binding to human platelets by S-nitroso-N-acetylcysteine. J Biol Chem 265:19028–19034

Menshikov MY, Ivanova K, Schaefer M, Drummer C, Gerzer R (1993) Influence of the cGMP analogue 8-PCPT-cGMP on agonist-induced increases in ionized calcium and on aggregation of human platelets. Eur J Pharmacol 245:281–284

Michelson AD, Benoit SE, Furman MI, et al. (1996) Effects of nitric oxide/EDRF on platelet surface glycoproteins. Am J Physiol 39:H1640–H1648

Moncada S, Gryglewski RJ, Bunting S, Vane JR (1976) An enzyme isolated from arteries transforms prostaglandin endoperoxides to an unstable substance that inhibits platelet aggregation. Nature 263:663–665

Moro MA, Darley-Usmar VM, Goodwin DA, et al. (1994) Paradoxical fate and biological action of peroxynitrite on human platelets. Proc Natl Acad Sci USA 91:6702–6706

Moro MA, Darley-Usmar VM, Lizasoain I, et al. (1995) The formation of nitric oxide donors from peroxynitrite. Br J Pharmacol 116:1999–2004

Moro MA, Russell RJ, Cellek S, et al. (1996) cGMP mediates the vascular and platelet actions of nitric oxide. Confirmation using an inhibitor of soluble guanylyl cyclase. Proc Natl Acad Sci U S A 93:1480–1485

Murohara T, Parkinson SJ, Waldman SA, Lefer AM (1995) Inhibition of nitric oxide biosynthesis provides P-selectin expression in platelets. Role of protein kinase C. Arter Thromb Vasc Biol 15:2068 2075

Muruganandam A, Mutus B (1994) Isolation of nitric oxide synthase from human platelets. Biochim Biophys Acta 1200:1–6

Nakashima S, Tohmatsu T, Hattori H, Okano Y, Nozawa Y (1986) Inhibitory action of cyclic GMP on secretion, phosphoinositide hydrolysis and calcium mobilization in thrombin-stimulated human platelets. Biochem Biophys Res Commun 135:1099–1104

Nakatsuka M, Osawa Y (1994) Selective inhibition of the 12-lipoxygenase pathway of arachidonic acid metabolism by L-arginine or sodium nitroprusside in intact human platelets. Biochem Biophys Res Commun 200:1630–1634

Noris M, Benigni A, Boccardo P, et al. (1993) Enhanced nitric oxide synthesis in uremia: implications for platelet dysfunction and dialysis hypotension. Kidney Int 44:445–450

Okada M, Sagawa T, Tominaga A, Kodama T, Hitsumoto Y (1996) Two mechanisms of platelet-mediated killing of tumour cells: one cyclo-oxygenase dependent and the other nitric oxide dependent. Immunology 89:158–164

Olsen SB, Tang DB, Jackson MR, et al. (1996) Enhancement of platelet deposition by cross-linked hemoglobin in a rat carotid endarterectomy model. Circulation 93:327–332

Palmer RMJ, Ferrige AG, Moncada S (1987) Nitric oxide release accounts for the biological activity of endothelium-derived relaxing factor. Nature 327:524–526

Palmer RMJ, Ashton DS, Moncada S (1988) Vascular endothelial cells synthesize nitric oxide from L-arginine. Nature 333:664–666

Palmer RMJ, Bridge L, Foxwell NA, Moncada S (1992) The role of nitric oxide in endothelial cell damage and its inhibition by glucocorticoids. Br J Pharmacol 105:11–12

Patrono C (1989) Aspirin and human platelets: from clinical trials to acetylation of cyclooxygenase and back. Trends Pharmacol Sci 10:453–458

Plotkine M. Allix M, Guillou J, Boulu R (1991) Oral administration of isosorbide dinitrate inhibits arterial thrombosis in rats. Eur J Pharmacol 201:115–116

Pohl U, Busse R (1989) EDRF increases cyclic GMP in platelets during passage through the coronary vascular bed. Circ Res 65:1798–1803

Polanowska-Grabowska R, Gear ARL (1994) Role of cyclic nucleotides in rapid platelet adhesion to collagen. Blood 83:2508–2515

Radomski A, Sawicki G, Olson DM, Radomski MW (1998) The role of nitric oxide and metalloproteinases in the pathogenesis of hyperoxia-induced lung injury in newborn rats. Br J Pharmacol 125:1455–1462

Radomski MW (1995) Nitric oxide-biological mediator, modulator and effector molecule. Ann Med 27:321–330

Radomski MW, Moncada S (1991) Role of nitric oxide in endothelial cell-platelet interactions In: Antithrombotics. Herman AG ed. Kluwer Academic Dordrecht, London, pp 25–48

Radomski MW, Radomski AS (1999) Regulation of blood cell function by the endothelial cells. In: Vascular endothelium in human physiology and pathophysiology, eds. Vallance P, Webb D. Kluwer Academic, in press

Radomski MW, Salas E (1995). Nitric oxide-biological mediator, modulator and factor of injury: its role in the pathogenesis of atherosclerosis. Atherosclerosis 118(Suppl):S69–S80

Radomski MW, Palmer RMJ, Moncada S (1987a) Comparative pharmacology of endothelium-derived relaxing factor, nitric oxide and prostacyclin in platelets. Br J Pharmacol 92:181–187

Radomski MW, Palmer RMJ, Moncada S (1987b) Endogenous nitric oxide inhibits human platelet adhesion to vascular endothelium. Lancet 2:1057–1058

Radomski MW, Palmer RMJ, Moncada S (1987c) The role of nitric oxide and cGMP in platelet adhesion to vascular endothelium. Biochem Biophys Res Commun 148:1482–1489

Radomski MW, Palmer RMJ, Moncada S (1987d) The anti-aggregating properties of vascular endothelium: interactions between prostacyclin and nitric oxide. Br J Pharmacol 92:639–646

Radomski MW, Palmer RMJ, Moncada S (1990a). An L-arginine/nitric oxide pathway present in human platelets regulates aggregation. Proc Nat Acad Sci USA 87:5193–5197

Radomski MW, Palmer RMJ, Moncada S (1990b). Characterization of the L-arginine: nitric oxide pathway in human platelets Br J Pharmacol 101:325–328

Radomski MW, Jenkins DC, Holmes L, Moncada S (1991) Human colorectal adenocarcinoma cells: differential nitric oxide synthesis determines their ability to aggregate platelets. Cancer Res 51:6073–6078

Radomski MW, Rees DD, Dutra A, Moncada S (1992) S-Nitrosoglutathione inhibits platelet activation in vitro and in vivo. Br J Pharmacol 107:745–749

Radomski MW, Vallance P, Whitley G, Foxwell N, Moncada S (1993) Platelet adhesion to human vascular endothelium is modulated by constitutive and cytokine induced nitric oxide. Cardiovasc Res 27:1380–1382

Reinhard M, Halbrügge M, Scheer U, et al. (1992) The 46/50 kDa phosphoprotein VASP purified from human platelets is a novel protein associated with actin filaments and focal contacts. EMBO J 11:2063–2070

Riddell DR, Graham A, Owen JS (1997) Apolipoprotein E inhibits platelet aggregation through the L-arginine:nitric oxide pathway. Implications for vascular disease J Biol Chem 272:89–95

Rosenblum WI, Nelson GH, Povlishock JT (1987) Laser-induced endothelial damage inhibits endothelium-dependent relaxation in the cerebral microcirculation of the mouse. Circ Res 60:169–176

Salas E, Busletta A, Radomski MW (1997a) Generation of NO, cGMP and platelet function in normal pregnancy and preeclampsia. The Fifth International Meeting on Biology of Nitric Oxide, 15–19 September 1997, Kyoto, Japan. Jpn J Pharmacol 75(Suppl 1):116P

Salas E, Miszta-Lane H, Radomski MW (1997b) Regulation of platelet function by nitric oxide and other nitrogen- and -oxygen-derived species. In: Handbook of experimental pharmacology vol. 126. Platelets and their factors. Eds. Von Bruchhausen F, Walter U. Springer, Berlin Heidelberg New York, pp 371–397

Salas E, Moro MA, Askew S, et al. (1994) Comparative pharmacology of analogues of S-nitroso-N-acetyl-DL-penicillamine in platelets. Br J Pharmacol 112:1071–1076

Samama CM, Diaby M, Fellahi JL, et al. (1995) Inhibition of platelet aggregation by inhaled nitric oxide in patients with acute adult respiratory distress syndrome. Anesthesiology 83:56–65

Sane DC, Bielawska A, Greenberg CS, Hannun YA (1989) Cyclic GMP analogs inhibit gamma thrombin-induced arachidonic acid release in human platelets. Biochem Biophys Res Commun 165:708–714

Sase K, Michel T (1995) Expression of constitutive, endothelial nitric oxide synthase in human blood platelets. Life Sci 57:2049–2055

Sawicki G, Salas, E, Murat J, Miszta-Lane H, Radomski MW (1997) Release of gelatinase A from human platelets mediates aggregation. Nature 386:616–619

Schultz PJ, Raij L (1992) Endogenously synthesized nitric oxide prevents-endotoxin-induced glomerular thrombosis. J Clin Invest 90:1718–1725

Shahbazi T, Jones N, Radomski MW, Moro MA, Gingell D (1994) Nitric oxide donors inhibit platelet spreading on surfaces coated with fibrinogen but not fibronectin. Thromb Res 75:631–642

Shen YC, Hing CY (1995) Effect of trilinolein on cyclic nucleotide formation in human platelets: relationship with its antiplatelet effect and nitric oxide synthesis. Br J Pharmacol 116:1644–1648

Simon DI, Stamler JS, Loh E, Loscalzo J, Francis SA, Craeger MA (1995) Effect of nitric oxide synthase inhibition on bleeding time in humans. J Cardiovasc Pharmacol 26:339–342

Sinzinger H, Fitscha P, O'Grady J, et al. (1990) Synergistic effect of prostaglandin E_1 and isosorbide dinitrate in peripheral vascular disease. Lancet 335:627–628

Slawinski M, Grodzinska L, Kostka-Trabka E, Goszcz A, Gryglewski RJ (1996) L-Arginine, substrate for NO synthesis, its beneficial effects in therapy of patients with peripheral arterial disease: comparison with placebo-preliminary results. Acta Physiol Hungarica 84:457–458

Sneddon JM, Vane JR (1988) Endothelium-derived relaxing factor reduces platelet adhesion to bovine endothelial cells. Proc Natl Acad Sci U S A 85:2800–2804

Spain DA, Wilson MA, Garrison RN (1994) Nitric oxide synthase inhibition exacerbates sepsis-induced renal hypoperfusion. Surgery 116:322–331

Thomson L, Lawton FG, Knowles RG, et al. (1994) Nitric oxide synthase activity in human gynecological cancer. Cancer Res 54:1352–1354

Trovati M, Massucco P, Mattiello L, et al. (1996) The insulin-induced increase of guanosine-3',5'-cyclic monophosphate in human platelets is mediated by nitric oxide. Diabetes 45:768–770

Trovati M Anfossi G, Massucco P, et al. (1997) Insulin stimulates nitric oxide synthesis in human platelets and, through nitric oxide, increases platelet concentrations of both guanosine-3',5'-cyclic monophosphate and adenosine-3',5'-cyclic monophosphate. Diabetes 46:742–749

Tsao PS, Theilmeier G, Singer AH, Leung LLK, Cooke JP (1994) L-Arginine attenuates platelet reactivity in hypercholesterolemic rabbits. Arterioscler Thromb 14:1529–1533

Venturini CM, Del Vecchio PJ, Kaplan JE (1989) Thrombin-induced platelet adhesion to endothelium is modified by endothelial derived relaxing factor (EDRF). Biochem Biophys Res Commun 159:349–354

Villa LM, Salas E, Darley-Usmar VM, Radomski MW, Moncada S (1994) Peroxynitrite induces both vasodilatation and impaired vascular relaxation in the rat isolated perfused heart. Proc Natl Acad Sci USA 91:12383–12387

Wallerath T, Gath I, Aulitzky WE, Pollock JS, Kleinert H, Förstermann U (1997) Identification of NO synthase isoforms expressed in human neutrophil granulocytes, megakaryocytes and platelets. Thromb Haemostas 77:163–167

Walter U (1989) Physiological role of cGMP and cGMP-dependent protein kinase in the cardiovascular system. Rev Physiol Biochem Pharmacol 113:41–88

Wautier JL, Weill D, Kadeva H, Maclouf J, Soria C (1989) Modulation of platelet function by SIN-1A. J Cardiovasc Pharmacol 14:S111–S114

Weber AA, Neuhaus T, Seul C, et al. (1996) Biotransformation of glyceryl trinitrate by blood platelets as compared vascular smooth muscle cells. Eur J Pharmacol 309:209–213

Werns SW, Rote WE, Davis JH, Guevara T, Lucchesi BR (1994) Nitroglycerin inhibits experimental thrombosis and reocclusion after thrombolysis. Am Heart J 127:727–737

Willerson JT, Igo SR, Yao SK, Ober JC, Macris MP, Ferguson JJ (1996) Localised administration of sodium nitroprusside enhances its protection against platelet aggregation in stenosed and injured coronary arteries. Texas Heart Inst J 23:1–8

Wright CE, Rees DD, Moncada S (1992) Protective and pathological role of nitric oxide in endotoxin shock. Cardiovasc Res 26:48–57

Wu C-C, Ko F-N, Kuo S-C, Lee F-Y, Teng C-M (1995) YC-1 inhibited platelet aggregation through NO-independent activation of soluble guanylate cyclase. Br J Pharmacol 116:1973–1978

Yao SK, Ober JC, Krishnaswami A, et al. (1992) Endogenous nitric oxide protects against platelet aggregation and cyclic flow variations in stenosed and endothelium-injured arteries. Circulation 86:1302–1309

Yusuf S, MacMahon S, Collins R, Peto R (1988) Effect of intravenous nitrates on mortality in acute myocardial infarction: an overview of the randomised trials. Lancet 1:1088–1092

Zhou Q, Hellermann GR, Solomonson LP (1995) Nitric oxide release from resting platelets. Thromb Res 77:87–96

CHAPTER 11
The Physiological Roles of Nitric Oxide in the Central Nervous System

J. GARTHWAITE

A. Introduction

The CNS has an especially rich capacity to synthesise NO. By virtue of its high rate of diffusion in both lipid and aqueous environments, a single source of NO is likely to influence neural function over a large tissue volume, perhaps equivalent to one containing as many as 1–2 million synapses (WOOD and GARTHWAITE 1994). This large sphere of influence, enclosing a large number of heterogeneous targets (dendrites, axons, presynaptic nerve terminals, glial cells, blood vessels), indicates that the principles governing the action of NO are likely different from those applying to locally-acting, fast neurotransmitters, whose influence is exerted chiefly within single-synapse domains (RUSAKOV et al. 1999). Monoamine neurotransmitters, such as 5-hydroxytryptamine, which is probably able to influence structures within a range of tens of micrometers, might be analogous (BUNIN and WIGHTMAN 1999). However, for the monoamines, neural cells are differentially equipped with different classes of surface receptors coupled to different transduction pathways, allowing a selectivity of action. The major receptor for NO, soluble guanylyl cyclase (sGC), however, appears to display limited molecular heterogeneity and yet has a wide distribution complementary to that of NO synthase (NOS; FURUYAMA et al. 1993; SOUTHAM and GARTHWAITE 1993). How NO achieves selectivity of action in this scenario is one of the important questions still to be answered satisfactorily.

A common trigger for NO formation in the CNS is activation of neuronal receptors for the major excitatory neurotransmitter, glutamate. Glutamate typically mediates fast (millisecond) synaptic transmission through the family of α-amino-3-hydroxy-5-methyl-4-isoxazolepropionic acid (AMPA) receptors and less fast (tens of milliseconds) excitation through N-methyl-D-aspartate (NMDA) receptors. The latter receptors usually participate under conditions of a depolarised membrane potential. Glutamate can also act on postsynaptically located G-protein-coupled (metabotropic) receptors to generate a slow (on the order of seconds) form of synaptic excitation (BATCHELOR and GARTHWAITE 1997). Of particular significance for NO production are NMDA receptors, whose associated ion channel permits substantial entry of Ca^{2+} (GARTHWAITE et al. 1988), resulting in the formation of a Ca^{2+}–calmodulin complex that activates NOS (BREDT and SNYDER 1990). Other receptors

(AMPA receptors and other neurotransmitter receptors) and channels (voltage-dependent Ca^{2+} channels) can also be coupled to NOS activation. The predominant synthesising enzyme in the CNS under normal conditions is the neuronal NOS isoform (nNOS; HUANG et al. 1993), and this enzyme is expressed to a greater or lesser extent in most brain regions (BREDT et al. 1990; VINCENT and HOPE 1992; SOUTHAM and GARTHWAITE 1993). It has been reported (from immunohistochemical studies) that certain neurones express endothelial NOS (eNOS; DINERMAN et al. 1994), but the failure to detect this isoform in neurones by in situ hybridisation casts doubt on this conclusion (ELIASSON et al. 1999).

Given the widespread distribution of nNOS, it is anticipated that NO will participate in many different brain functions and that interference with the NO pathway will have a variety of behavioural repercussions. This appears to be the case. Much evidence from histochemical, anatomical and physiological studies has indicated that, in vertebrates, NO regulates neural networks associated with: the circadian light–dark cycle and the sleep–waking cycle (PAPE 1995; DZOLJIC et al. 1997); feeding and drinking (NELSON et al. 1997); male and female sexual behaviour (MELIS and ARGIOLAS 1997; NELSON et al. 1997); learning and memory (HAWKINS 1996; HOLSCHER 1997); motor coordination (MORENO-LOPEZ et al. 1996, 1998; KRIEGSFELD et al. 1999); the activity of the sympathetic nervous system (OWLYA et al. 1997); and pain (MELLER and GEBHART 1993; DICKENSON 1995; THOMSEN and OLESEN 1998) and other sensory pathways (BREER and SHEPHERD 1993; BROILLET and FIRESTEIN 1996; CUDEIRO and RIVADULLA 1999). NO signalling is also important in the nervous systems of invertebrates, in which its main functions appear to be in olfaction, feeding, vision, memory formation and development (JACKLET 1997; MÿLLER 1997; BICKER 1998).

This article attempts to review the physiological aspects of NO signalling in the CNS. Extensive literature dealing with various effects of *exogenous* NO (or, usually more correctly, of chemicals able to release NO) on neural cells exists, but often the physiological (or even pathophysiological) relevance of the findings is difficult to gauge. This is particularly the case when agents (e.g. sodium nitroprusside) having many non-specific actions are used. Therefore, the focus will be on (1) the types of function for which there is convincing evidence for a participation of *endogenous* NO and (2) the possible cellular and molecular mechanisms through which endogenous NO modifies neural function. In this regard, evidence accrued during the last few years favours activation of sGC, leading to cyclic guanosine monophosphate (cGMP) accumulation as the predominant NO signal-transduction pathway in the CNS. As with all other signalling mechanisms, the identification of more reliable pharmacological tools for studying this receptor enzyme (GARTHWAITE et al. 1995; KOESLING 1998) has been of importance to progress.

B. Acute Actions of NO

Frequently, manipulating the NO and/or cGMP pathways in brain tissue as one would in investigating a neurotransmitter pathway has no obvious electrophysiological effects on the way neurones or synapses operate, suggesting that much of what NO is doing is covert at this level of investigation. There are, however, several exceptions.

I. Synaptic Transmission

To date, there is no good evidence that NO performs a neurotransmitter-like function in the vertebrate CNS; that is, there is no evidence that it directly mediates excitation or inhibition at synapses analogous to the manner in which it has been proposed to function at peripheral neuroeffector junctions (RAND 1992; BENNETT 1997). Examples of such an action, however, have been found in invertebrates. In synapses associated with the feeding circuits of two types of snail, *Aplysia* and *Lymnea*, slow, excitatory, post-synaptic potentials (EPSPs) are elicited by brief, high-frequency stimulation of the pre-synaptic neurone. These EPSPs can be inhibited by NOS inhibitors or manoeuvres that inactivate NO; the EPSPs are mimicked by exogenous NO application (JACKLET 1995; PARK et al. 1998). In the *Aplysia* synapse, the excitation appears to involve sGC, as it is reduced by sGC inhibition and is mimicked by application of a cGMP derivative (KOH and JACKLET 1999). In these situations, NO is assumed to act in an orthograde manner, being released from the presynaptic nerve terminals and acting on the postsynaptic cell. This may also happen elsewhere, as (in other *Aplysia* neurones) NO, through cGMP, is able to inhibit postsynaptic potassium currents or activate sodium currents (SAWADA et al 1995, 1996).

At other synapses in *Aplysia*, NO affects synaptic transmission differently, by acting on presynaptic terminals to influence the release of the neurotransmitter acetylcholine. Interestingly, the direction of the change in acetylcholine release varies according to the synapse investigated. In one synapse (in the abdominal ganglion), where acetylcholine causes excitation, NO enhances EPSP amplitudes; in another (in the buccal ganglion), where acetylcholine is inhibitory, NO reduces the amplitude of the inhibitory postsynaptic potentials. These actions appear to involve sGC and cGMP and to be relevant to what occurs naturally at these synapses when the frequency of synaptic transmission increases (MOTHET et al. 1996).

In vertebrates, acute modifications of neuronal function that are attributable to endogenous NO have been observed rather rarely under physiological conditions. The best example is in the cat visual system in vivo, at the levels of both the thalamus and the visual cortex. The thalamic dorsal–lateral geniculate nucleus is a major target for retinal outputs, and a natural light stimulus to the eye causes an increased rate of firing of the thalamic neurones. Local administration of NOS inhibitors inhibits this light-evoked response to about 50% of the control value; moreover, the same inhibitors reduce spontaneous activity and selectively block excitation mediated by NMDA receptors, whereas NO-donors selectively increase NMDA-induced excitation (CUDEIRO et al. 1996). In this structure, NOS is located in cholinergic nerve terminals, and it is suggested that, through NO, this input serves to amplify visual signals by acting postsynaptically to selectively increase NMDA-receptor function (not necessarily directly) in the retino-geniculate neurones.

In the visual cortex, the situation is somewhat more complicated because of the existence of different populations of NO-responsive neurones (CUDEIRO et al. 1997; KARA and FRIEDLANDER 1999). The major population responds to exogenous NO with an increase in action potential firing in response to light. Importantly, endogenous NO appears to contribute significantly to the normal, light-evoked firing response, because it is reduced when NOS is inhibited and is enhanced by the NOS substrate, L-arginine. In one of the studies (CUDEIRO et al. 1997), NOS inhibitors also reduced the excitation produced by NMDA, AMPA and acetylcholine, indicating that NO is acting to increase postsynaptic neuronal excitability generally. Administration of 8-bromo-cGMP had effects similar to that of NO, consistent with mediation by sGC and cGMP. While the effector mechanism and the relevant source of NO are unclear, by increasing neuronal signal-to-noise characteristics, NO serves to enhance the signal-detection capabilities of these visual cortex neurones. In the other, minor population of responsive neurones in this brain region, NO has the opposite effect, in that NOS inhibition enhances visual responses but exogenous NO depresses them.

An involvement of NO in the processing of sensory signals in the thalamus appears to extend to non-visual signals and other thalamic nuclei because, in the rat ventrobasal thalamus, neuronal firing in response to an air-jet directed at the animal's whiskers was markedly reduced by NOS inhibition, as were the responses to locally applied NMDA and AMPA (Do et al. 1994). One effect of NO and cGMP in the thalamus is to modify the behaviour of a class of potassium channels (PAPE and MAGER 1992), but the question of whether this mechanism is responsible for the effects observed in vivo remains to be investigated.

These examples indicate that endogenous NO can participate acutely in synaptic transmission, at least in some brain regions. The electrophysiological effects (neuronal excitation or inhibition) of exogenously applied NO and cGMP in other areas of the CNS – such as the noradrenergic neurones of the locus coeruleus (PINEDA et al. 1996), the spinal cord (PEHL and SCHMID 1997) and neurones of the paraventricular nucleus of the hypothalamus (BAINS and FERGUSON 1997) – suggest that this mode of action may be more widespread than previously appreciated. If so, it is probable that, as in the above examples, the net effect of NO on synaptic transmission will not be predictable but will vary from synapse to synapse.

Details of how cGMP acts to modify neuronal excitability are unclear. A variety of ion channels (K^+ and Ca^{2+} channels) may become affected following activation of cGMP-dependent protein kinase (PKG) or modulation of cyclic adenosine monophosphate levels as a result of cGMP-mediated changes in phosphodiesterase activity (GARTHWAITE 1991; FAGNI and BOCKAERT 1996). Alternatively, as in sensory receptor cells (photoreceptor and olfactory cells), cGMP may directly operate ion channels in the CNS (GARTHWAITE and BOULTON 1995; KINGSTON et al. 1996; BRADLEY et al. 1997).

Similarly, clear details of how cGMP affects neurotransmitter release are missing, although recent evidence implicates PKG (YAWO 1999) or cGMP-operated ion channels (RIEKE and SCHWARTZ 1994; SAVCHENKO et al. 1997). It should be noted that there are many reports of exogenous NO donors influencing neurotransmitter release independently of cGMP. Almost without exception, however, no checks were made to ensure that the observed effects were not simply the result of mitochondrial dysfunction caused by high NO concentrations (McNAUGHT and BROWN 1998).

II. Gap Junctions

Gap junctions are complex intercellular channels made up of proteins (connexins) donated by both participating cells. More than a dozen different connexins conferring heterogeneous properties to the final assemblies have been identified. The channels provide a means for the diffusion of relatively low-molecular-weight substances from one cell to another (sometimes unidirectionally) and for the synchronisation of the electrical activity of populations of cells. The channels are not inert conduits, however, as abundant evidence suggests that their functional properties can be modified by diverse stimuli, including the NO–cGMP pathway.

In the retina, there is abundant cell–cell coupling both between cells of the same class and between different cell types. Modifications of the coupling are considered important in transmitting and shaping visual signals (BALDRIDGE et al. 1998). In the horizontal cells, which transmit lateral inhibitory signals that serve to sharpen receptive fields, gap-junctional conductance (studied in fish retinae) is inhibited by various NO donors and by cGMP (DEVRIES and SCHWARTZ 1989; MIYACHI et al. 1990; LU and McMAHON 1997). The effect is caused by a reduction in the frequency of channel opening and appears to involve PKG (LU and McMAHON 1997). Retinal gap junctions between amacrine cells and cone bipolar cells (but not those between amacrine cells themselves) are also inhibited by the NO–cGMP pathway, an effect that could contribute to light adaptation (MILLS and MASSEY 1995). These are suggestive findings, but there is no evidence yet that endogenous NO performs this function physiologically in response to light.

In the brain, NO and cGMP can inhibit coupling between neurones in the cerebral cortex early in development (RORIG and SUTOR 1996); again, however, the physiological significance of this observation has not yet been investigated.

The opposite effect, an increase in gap-junction permeability in response to NO, has also been observed. Certain interneurones of the developing hippocampal formation, for example, are extensively linked by gap junctions, and the conductance of the junctions appears to be maintained by endogenous NO, since inhibition of NOS abolishes the electrophysiological correlate (depolarisation-induced hyperpolarisation) of electrical coupling (STRATA et al. 1998). In response to stimulation of afferent fibres originating in the cerebral cortex, endogenous NO appears to mediate increases in gap-junction permeability of inhibitory neurones in the striatum of adult animals (O'DONNELL and GRACE 1997), indicating that NO may serve to synchronise the activity of these neurones and contribute to the motor functions of this brain region. In the supraoptic nucleus of the hypothalamus, increased gap-junctional coupling between neurones is associated with physiological conditions of increased demand for peptide-hormone release (dehydration or lactation). Recordings from these neurones in vitro have indicated that the NO–cGMP pathway could be a key regulator of this response, since NO donors markedly increase gap-junctional coupling in a manner that is blocked by sGC inhibition and is mimicked by a membrane-permeant cGMP analogue. The net effect is likely to be increased release of the hormone vasopressin (HATTON and YANG 1996; YANG and HATTON 1999).

III. Local Cerebral Blood Flow

Several years ago, it was suggested that neuronally derived NO, formed locally in response to increased synaptic activity, could access nearby blood vessels and cause them to relax, thereby increasing delivery of blood to that region (GALLY et al. 1990). It soon became clear that cerebral circulation is regulated by NO (at least that formed by eNOS in the vessels themselves and by nNOS present in cranial autonomic nerves; IADECOLA et al. 1994). Testing the role of parenchymal NO in blood-flow regulation is more problematic. Consistent with such a role is the demonstration that NMDA-induced dilation of cerebral pial arterioles can be blocked by NOS inhibitors, including those deemed to act selectively on nNOS (FARACI and BREESE 1993; FARACI and BRIAN 1995). More direct evidence has been obtained in studies of the cerebellum, where alterations in blood flow in response to electrical stimulation of glutamatergic pathways can be observed. Several reports have now shown that the blood-flow response to stimulation of one such pathway, the parallel fibres, is reduced not only by NOS inhibition but also by procedures that inhibit synaptic transmission generally and by blockers of glutamate receptors (AKGOREN et al. 1994; IADECOLA et al. 1996). The major synaptic targets of the parallel fibres are the Purkinje cells, which lack NMDA receptors and NOS, and inhibitory interneurones, which express NMDA receptors and NOS. Interestingly, it is activation of the AMPA receptors (rather than NMDA receptors) that is responsible for much of the increased blood flow. Purkinje cells appear to contribute to an NO-independent mechanism (possibly involving adenosine), whereas the inhibitory interneurones putatively mediate the NO-dependent component (YANG et al. 1998). Stimulation of the other input to the cerebellar cortex, the climbing fibres, also causes blood-flow changes that are mediated by AMPA receptors (but not NMDA receptors) and, in large part, by NO formation (YANG and IADECOLA 1998).

These investigations provide good in vivo evidence in support of the hypothesis that blood vessels are targets for neuronally derived NO. In such studies, however, it is difficult to investigate the underlying mechanism in detail and to exclude the possibility of indirect effects resulting from the activation of neuronal networks. The finding that NMDA application results in NO-dependent vasodilatation of microvessels in hippocampal slices maintained in vitro adds strong evidence in favour of a local mechanism (LOVICK et al. 1999). Somewhat unexpectedly, the NMDA response in the slices was sensitive to tetrodotoxin, as has been found in vivo (FARACI and BREESE 1993), implying that the NO causing the vasodilation is not produced as a direct result of NMDA-receptor activation but through a secondary, action-potential-dependent mechanism. This mechanism might involve depolarisation (directly or indirectly) of the intrinsic, hippocampal, NOS-containing-fibre network followed by action-potential generation in those fibres and a subsequent liberation of NO at sites close to the microvessels (LOVICK et al. 1999).

IV. Glial Cells

In addition to neuronal and vascular elements, NO can also act on sGC in glial cells. Glial cells have intimate functional and anatomical relationships with neurones and blood vessels, but relatively little has been done to examine the functional consequences of the NO–cGMP pathway in these cell types. In the developing cerebellum maintained in slice culture, inhibition of NOS reduced the migration of granule cells (which occurs along the radial fibres of the Bergmann glia) and also reduced the length of the Bergmann glial fibres (TANAKA et al. 1994), implying a role for endogenous NO in the differentiation of these cells. The involvement of cGMP, whose levels in Bergmann glia are powerfully regulated by NO (de VENTE et al. 1990), was not examined. In human Muller glial cells of the retina, NO and cGMP activate a Ca^{2+}-permeable, non-selective cation current, an effect that may be mediated by a cyclic-nucleotide-gated channel and (perhaps secondarily) a Ca^{2+}-activated potassium current (KUSAKA et al. 1996). In cultured cortical glia, apparently by utilising cGMP-dependent mechanisms, NO can inhibit the release of the neurotrophin nerve-growth factor (XIONG et al. 1999) and stimulate the Na^+–Ca^{2+} exchanger, reducing agonist-induced increases in intracellular Ca^{2+} concentrations (TAKUMA et al. 1996). These studies indicate that functional changes are likely to be produced in glia in response to NO, but the significance of this pathway for information processing, ion homeostasis or other aspects of neural behaviour remains unclear.

C. NO and Synaptic Plasticity

Synaptic plasticity refers to the ability of synapses to alter their strength, either upwards or downwards, over different time scales (seconds–years) following brief periods of altered activity. Normally, this is viewed as a physiological event that serves to encode information in the CNS and, hence, is a cellular correlate of learning behaviour (BLISS and COLLINGRIDGE 1993). It may also be relevant to pathophysiology, because chronic pain, epilepsy and neurodegeneration following transient brain ischaemia have all been associated with long-lasting increases in synaptic strength in the relevant neuronal circuits. Hence, understanding the underlying mechanisms has both physiological and pathological implications. The most-studied examples of synaptic plasticity are long-term potentiation (LTP) and depression (LTD), both typically studied over periods of an hour or more but which may (in vivo) last for much longer. Shorter-term (seconds–minutes) changes also occur.

A key factor in the induction of most types of synaptic plasticity is a rise in intracellular Ca^{2+} concentration, which engages second-messenger pathways, ultimately leading to enduring changes in postsynaptic responsiveness to the neurotransmitter and/or its release. The direction of the change (potentiation or depression) is likely to be determined by the precise temporal and spatial patterns of the Ca^{2+} signals and by their amplitudes (YANG et al. 1999). Given the requirement for NMDA-receptor activation in many (but not all) forms of synaptic plasticity, the link between NMDA receptors and NO formation and the putative involvement of an intercellular signalling molecule (BLISS and COLLINGRIDGE 1993), a role for NO in these phenomena is an attractive possibility.

I. Short-Term Plasticity

NOS is strongly expressed in sympathetic, preganglionic neurones in the spinal cord, and experiments carried out in vitro in young rat spinal cord indicate that NO functions to transiently alter the activities of excitatory and inhibitory synapses. Potentiation of inhibitory currents is observed after a brief (10-s) burst of depolarisation (30Hz) applied to the neurones. The effect was abolished by blockers of the NO–cGMP pathway and was reproduced by L-arginine or a permeant cGMP analogue. Since the postsynaptic properties of the neurones (including their sensitivity to glycine, the presumed inhibitory transmitter) were unaltered, it was suggested that NO is released from the postsynaptic neurones; it then acts retrogradely on the nerve terminals and enhances transmitter release (WU and DUN 1996). Similar experiments indicated that the NO–cGMP pathway can also elicit a transient (usually less than 1-h) potentiation of excitatory (glutamatergic) currents (WU et al. 1997b), though the physiological conditions under which such an effect might occur were not explored. It was suggested that the potentiating effects observed are relevant to the role of NO as a central modulator of the discharge of sympathetic nerves. Transient potentiations lasting about 20min can also be elicited in the visual cortex by pairing intracellular depolarising pulses with weak afferent-fibre stimulation (HARSANYI and FRIEDLANDER 1997). The plastic changes were dependent on NMDA-receptor activation, but NO appeared to have unpredictable effects, since NOS inhibition increased the potentiation in some cells and decreased it in others.

II. Long-Term Potentiation

The possible role of NO as an effector molecule in enduring forms of synaptic plasticity has been investigated in several brain regions. Evidence in favour of such a role, based on manoeuvres that inhibit NO formation or inactivate NO, has been obtained for LTP in the hippocampal CA1 region, the dentate gyrus, the cerebral cortex and the amygdala (GARTHWAITE and BOULTON 1995; HOLSCHER 1997). Some of the findings, however, have not been replicated in other laboratories, and it has become clear that the experimental conditions, notably the strength of stimulation, represent one of the factors that govern the relative expression of NO-dependent and -independent forms of LTP. A further complication is added by the finding that, when evident, the NO-dependent form of hippocampal LTP is typically seen together with NO-independent forms (BOULTON et al. 1995; SON et al. 1996), necessitating careful dissection of the two. Furthermore, in the CA1 region of the hippocampus and in the cerebral cortex, the requirement for NO varies according to the exact synapses under investigation, even in the same postsynaptic neurone (HALEY et al. 1996; SON et al. 1996; WAKATSUKI et al. 1998).

An important next step has been to try to unravel the mechanisms through which NO causes long-term changes. Tests of the obvious possibility, activation of sGC and cGMP formation, have generated controversial findings. Several earlier studies using sGC inhibitors, cGMP analogues and PKG inhibitors supported the involvement of this pathway (HALEY et al. 1992; ZHUO et al. 1994; ARANCIO et al. 1995; BLITZER et al. 1995; BOULTON et al. 1995). Other groups, however, failed to replicate some of these results (SCHUMAN et al. 1994; SELIG et al. 1996). SELIG et al. combined the results of three independent laboratories that aimed to test the prediction that application of 8-bromo-cGMP will restore tetanus-induced LTP under conditions of NMDA-receptor blockade. SCHUMAN et al. conducted similar tests using dibutyryl cGMP. The findings were all negative. A recent re-examination of this issue by SON et al. (1999), however, reconciles the disparate results. It transpires that, for 8-bromo-cGMP (plus tetanic stimulation) to elicit a potentiation, a brief (5-min) application is successful whereas a longer one (10min) is not, implying some sort of adaptation to prolonged administration. The duration of pre-incubation with the NMDA antagonist is also relevant, apparently because the magnitude of the effect of cGMP is dependent on the degree to which NMDA receptors are blocked. In other words, cGMP plus some NMDA-receptor activity is required for optimal LTP. We have found analogous results when examining the mechanism of potentiation of hippocampal synaptic transmission by exogenous NO (Bon and Garthwaite, unpublished results). This may reflect the requirement for activation of both pre- and postsynaptic mechanisms to generate full LTP (Son et al. 1999).

III. Long-Term Depression

The phenomenon of LTD has been studied most extensively in the cerebellum, specifically at synapses between parallel fibres and Purkinje cells, where it is induced by brief conjoint activity in the two glutamatergic, afferent pathways: the parallel fibres and the climbing fibres. Here, LTD has been linked to motor learning behaviour, and abundant evidence exists that the NO–cGMP cascade participates (Garthwaite and Boulton 1995; Daniel et al. 1998). The mechanism appears to involve a release of NO from the activated parallel fibres; NO then acts on Purkinje cell dendrites to elevate cGMP levels. Under conditions where the Ca^{2+} concentration in Purkinje cells is simultaneously increased (normally brought about by climbing-fibre activity), LTD ensues through a mechanism involving PKG. Downstream, one putative target of the kinase in this scenario is a protein (known as the G-substrate) which, when phosphorylated, acts as a phosphatase inhibitor (Endo et al. 1999). LTD is considered to be caused by a lasting decrease in the sensitivity of postsynaptic AMPA receptors located specifically at the activated parallel-fibre synapses, but the link between phosphatase inhibition and changes in AMPA-receptor sensitivity remains to be identified.

LTD can be induced in many other central synapses, and there is evidence for an involvement of NO in LTD in the hippocampus (Izumi and Zorumski 1993), dentate gyrus (Wu et al. 1997a) and, most recently, the striatum (Calabresi et al. 1999). In this latter study, LTD in striatal spiny neurones was induced by brief, high-frequency stimulation of the glutamatergic fibres in the cerebral cortex. A comprehensive series of tests (including the use of NOS inhibitors, exogenous sources of NO, a sGC inhibitor, a phosphodiesterase inhibitor, cGMP analogues and PKG inhibitors) produced a convincing case that the phenomenon, like that in the cerebellum, requires the NO–cGMP pathway and a downstream alteration in protein phosphorylation in the postsynaptic neurones. In this instance, the source of NO does not appear to be the afferent fibres but is a subpopulation of NOS-containing striatal interneurones that are presumed to be synaptically co-activated by the corticostriatal fibres.

D. NO and Developmental Plasticity

At embryonic stages of development in the rat brain, NOS is expressed prominently (but transiently) in the cerebral cortex (Bredt and Snyder 1994), though the significance of this is unclear. Elsewhere, NOS appears in neurones once they have started to undergo differentiation. In both the developing cerebellum (SOUTHAM et al. 1991) and hippocampus (EAST and GARTHWAITE 1991), NMDA-receptor stimulation is coupled to a robust, NO-dependent cGMP accumulation. These findings indicate that the NOS–sGC pathway is present and active in the developing CNS, consistent with a role for this pathway in synapse formation. Several aspects of the development of neuronal connections may be affected by this pathway, including the regulation of growth-cone behaviour (SONG et al. 1998), neuronal survival (ESTEVEZ et al. 1998), neuronal migration, glial cell differentiation (TANAKA et al. 1994) and dendritic development (INGLIS et al. 1998).

A role in the refinement of synaptic connections is indicated by studies of the developing visual system. Thus, in the ferret, NO (together with NMDA receptors) plays an important role in segregating axonal connections from the retina, sending each to appropriate locations in the lateral geniculate nucleus of the thalamus (CRAMER et al. 1996). In chicks, NO (and NMDA receptors) are required for the efficient removal of transient retinotectal connections (WU et al. 1994; ERNST et al. 1999). An interesting further example is found in *Drosophila* (GIBBS and TRUMAN 1998). During development, the photoreceptor cells and their axons respond to NO with an increase in cGMP while axons are growing and making initial contact with their synaptic partners in the optic lobes. At the same time, the target neurones in the optic lobes express NOS. From investigations in a tissue-culture system, it was shown that inhibition of NOS or sGC resulted in a major disorganisation of the retinal projections and a failure of the photoreceptor axons to stop growing when they reached their normal targets. The disruptive effect of NOS inhibition was overcome by low concentrations of a permeant cGMP analogue. From these results, it appears (1) that the formation of NO in optic-lobe neurones provides an arrest signal to incoming photoreceptor axons prior to the formation of permanent synaptic connections and (2) that it does so by stimulating axonal cGMP accumulation. This suggestion is consistent with one made earlier from observations on mouse neuroblastoma cells, where it was found that intracellular injection of cGMP results in a rapid and long-lasting arrest of growth-cone motility, frequently followed by retraction of motile processes (BOLSOVER et al. 1992).

The way in which cGMP acts on growth cones is not yet known; one possibility is that it acts on cyclic nucleotide-gated channels, which are expressed on the growth cones of cultured embryonic hippocampal neurones prior to synapse formation (BRADLEY et al. 1997). This channel class is also found (at least at the mRNA level) to be particularly expressed in the developing rat brain (ROY and BARNSTABLE 1999). In this respect, it may be of relevance that, in *Caenorhabditis elegans*, mutation of the cyclic-nucleotide-channel genes, *Tax-2* and *Tax-4*, resulted in abnormalities in axon outgrowth and in predicted disturbances in sensory function (COBURN and BARGMANN 1996). Specifically, certain sensory axons invaded regions from which they are normally excluded, consistent with the idea that cGMP acts through these channels to provide an inhibitory signal to axon growth (COBURN et al. 1998), presumably initiated by an influx of Ca^{2+} (BOLSOVER et al. 1992; FRINGS et al. 1995).

E. Concluding Remarks

It has taken many years of research to begin to understand how NO contributes to information processing in the CNS under physiological conditions. This was not altogether surprising, given the potential influence of NO on virtually every aspect of brain function, the unique manner in which NO operates, the lack of clarity about NO signal-transduction mechanisms and the lack of good research tools. During recent years, however, evidence from different neuroscientific disciplines applied to different animals has begun to provide a more cohesive picture. At the system level, NO participates in the processing of sensory and motor signals, in the regulation of neuroendocrine function and sexual behaviour, in the regulation of local blood flow and in certain forms of learning and memory formation. Several of these functions appear to have been preserved across millions of years of evolution. However, it is important to stress that, in all these functions, NO is just one of many participants, and the way in which NO interacts with other signalling pathways is not well understood. At the cellular level (and more than 20 years after this action of NO was first discovered), increasing evidence implicates sGC as the principal physiological receptor for NO in the CNS. The way in which sGC functions and is regulated within cells and the way the product of this enzyme, cGMP, acts at the subcellular and molecular levels to induce changes in neural function on time scales varying from milliseconds to years are fascinating topics for future research.

References

Akgoren N, Fabricius M, Lauritzen M (1994) Importance of nitric oxide for local increases of blood flow in rat cerebellar cortex during electrical stimulation. Proc Natl Acad Sci USA 91:5903–5907

Arancio O, Kandel ER, Hawkins RD (1995) Activity-dependent long-term enhancement of transmitter release by presynaptic 3′,5′-cyclic GMP in cultured hippocampal neurons. Nature 376:74–80

Bains JS, Ferguson AV (1997) Nitric oxide depolarizes type II paraventricular nucleus neurons in vitro. Neuroscience 79:149–159

Baldridge WH, Vaney DI, Weiler R (1998) The modulation of intercellular coupling in the retina. Semin Cell Dev Biol 9:311–318

Batchelor AM, Garthwaite J (1997) Frequency detection and temporally dispersed synaptic signal association through a metabotropic glutamate receptor pathway. Nature 385:74–77

Bennett MR (1997) Non-adrenergic non-cholinergic (NANC) transmission to smooth muscle: 35 years on. Prog Neurobiol 52:159–195

Bicker G (1998) NO news from insect brains. Trends Neurosci 21:349–355

Bliss TV, Collingridge GL (1993) A synaptic model of memory: long-term potentiation in the hippocampus. Nature 361:31–39

Blitzer RD, Wong T, Nouranifar R, Iyengar R, Landau EM (1995) Postsynaptic cAMP pathway gates early LTP in hippocampal CA1 region. Neuron 15:1403–1414

Bolsover SR, Gilbert SH, Spector I (1992) Intracellular cyclic AMP produces effects opposite to those of cyclic GMP and calcium on shape and motility of neuroblastoma cells. Cell Motil Cytoskeleton 22:99–116

Boulton CL, Southam E, Garthwaite J (1995) Nitric oxide-dependent long-term potentiation is blocked by a specific inhibitor of soluble guanylyl cyclase. Neuroscience 69:699–703

Bradley J, Zhang Y, Bakin R, Lester HA, Ronnett GV, Zinn K (1997) Functional expression of the heteromeric "olfactory" cyclic nucleotide-gated channel in the hippocampus: a potential effector of synaptic plasticity in brain neurons. J Neurosci 17:1993–2005

Bredt DS, Snyder SH (1990) Isolation of nitric oxide synthetase, a calmodulin-requiring enzyme. Proc Natl Acad Sci USA 87:682–685

Bredt DS, Snyder SH (1994) Transient nitric oxide synthase neurons in embryonic cerebral cortical plate, sensory ganglia, and olfactory epithelium. Neuron 13:301–313

Bredt DS, Hwang PM, Snyder SH (1990) Localization of nitric oxide synthase indicating a neural role for nitric oxide. Nature 347:768–770

Breer H, Shepherd GM (1993) Implications of the NO/cGMP system for olfaction. Trends Neurosci 16:5–9

Broillet MC, Firestein S (1996) Gaseous second messengers in vertebrate olfaction. J Neurobiol 30:49–57

Bunin MA, Wightman RM (1999) Paracrine neurotransmission in the CNS: involvement of 5-HT. Trends Neurosci 22:377–382

Calabresi P, Gubellini P, Centonze D, Sancesario G, Morello M, Giorgi M, Pisani A, Bernardi G (1999) A critical role of the nitric oxide/cGMP pathway in corticostriatal long-term depression. J Neurosci 19:2489–2499

Coburn CM, Bargmann CI (1996) A putative cyclic nucleotide-gated channel is required for sensory development and function in *C. elegans*. Neuron 17:695–706

Coburn CM, Mori I, Ohshima Y, Bargmann CI (1998) A cyclic nucleotide-gated channel inhibits sensory axon outgrowth in larval and adult *Caenorhabditis elegans*: a distinct pathway for maintenance of sensory axon structure. Development 125:249–258

Cramer KS, Angelucci A, Hahm JO, Bogdanov MB, Sur M (1996) A role for nitric oxide in the development of the ferret retinogeniculate projection. J Neurosci 16:7995–8004

Cudeiro J, Rivadulla C (1999) Sight and insight – on the physiological role of nitric oxide in the visual system. Trends Neurosci 22:109–116

Cudeiro J, Rivadulla C, Rodriguez R, Grieve KL, Martinez-Conde S, Acuna C (1997) Actions of compounds manipulating the nitric oxide system in the cat primary visual cortex. J Physiol (Lond) 504:467–478

Cudeiro J, Rivadulla C, Rodriguez R, Martinez-Conde S, Martinez L, Grieve KL, Acuna C (1996) Further observations on the role of nitric oxide in the feline lateral geniculate nucleus. Eur J Neurosci 8:144–152

Daniel H, Levenes C, Crepel F (1998) Cellular mechanisms of cerebellar LTD. Trends Neurosci 21:401–407

de Vente J, Bol JGJM, Berkelmans HS, Schipper J, Steinbusch (1990) Immunocytochemistry of cGMP in the cerebellum of the immature, adult, and aged rat: the involvement of nitric oxide. A micropharmacological study. Eur J Neurosci 2:845–862

DeVries SH, Schwartz EA (1989) Modulation of an electrical synapse between solitary pairs of catfish horizontal cells by dopamine and second messengers. J Physiol (Lond) 414:351–375

Dickenson AH (1995) Central acute pain mechanisms. Ann Med 27:223–227

Dinerman JL, Dawson TM, Schell MJ, Snowman A, Snyder SH (1994) Endothelial nitric oxide synthase localized to hippocampal pyramidal cells: implications for synaptic plasticity. Proc Natl Acad Sci USA 91:4214–4218

Do KQ, Binns KE, Salt TE (1994) Release of the nitric oxide precursor, arginine, from the thalamus upon sensory afferent stimulation, and its effect on thalamic neurons in vivo. Neuroscience 60:581–586

Dzoljic E, van Leeuwen R, de Vries R, Dzoljic MR (1997) Vigilance and EEG power in rats: effects of potent inhibitors of the neuronal nitric oxide synthase. Naunyn Schmiedebergs Arch Pharmacol 356:56–61

East SJ, Garthwaite J (1991) NMDA receptor activation in rat hippocampus induces cyclic GMP formation through the L-arginine-nitric oxide pathway. Neurosci Lett 123:17–19

Eliasson MJL, Huang Z, Ferrante RJ, Sasamata M, Molliverf ME, Snyder SH, Moskowitz MA (1999) Neuronal nitric oxide synthase activation and peroxynitrite formation in ischemic stroke linked to neural damage. J Neurosci 19:5910–5918

Endo S, Suzuki M, Sumi M, Nairn AC, Morita R, Yamakawa K, Greengard P, Ito M (1999) Molecular identification of human G-substrate, a possible downstream component of the cGMP-dependent protein kinase cascade in cerebellar Purkinje cells. Proc Natl Acad Sci USA 96:2467–2472

Ernst AF, Wu HH, El Fakahany EE, McLoon SC (1999) NMDA receptor-mediated refinement of a transient retinotectal projection during development requires nitric oxide. J Neurosci 19:229–235

Estevez AG, Spear N, Thompson JA, Cornwell TL, Radi R, Barbeito L, Beckman JS (1998) Nitric oxide-dependent production of cGMP supports the survival of rat embryonic motor neurons cultured with brain-derived neurotrophic factor. J Neurosci 18:3708–3714

Fagni L, Bockaert J (1996) Effects of nitric oxide on glutamate-gated channels and other ionic channels. J Chem Neuroanat 10:231–240

Faraci FM, Breese KR (1993) Nitric oxide mediates vasodilatation in response to activation of N-methyl-D-aspartate receptors in brain. Circ Res 72:476–480

Faraci FM, Brian JE Jr (1995) 7-Nitroindazole inhibits brain nitric oxide synthase and cerebral vasodilatation in response to N-methyl-D-aspartate. Stroke 26:2172–2175

Frings S, Seifert R, Godde M, Kaupp UB (1995) Profoundly different calcium permeation and blockage determine the specific function of distinct cyclic nucleotide-gated channels. Neuron 15:169–179

Furuyama T, Inagaki S, Takagi H (1993) Localizations of $\alpha 1$ and $\beta 1$ subunits of soluble guanylate cyclase in the rat brain. Brain Res Mol Brain Res 20:335–344

Gally JA, Montague PR, Reeke GN, Jr., Edelman GM (1990) The NO hypothesis: possible effects of a short-lived, rapidly diffusible signal in the development and function of the nervous system. Proc Natl Acad Sci USA 87:3547–3551

Garthwaite J (1991) Glutamate, nitric oxide and cell-cell signalling in the nervous system. Trends Neurosci 14:60–67

Garthwaite J, Boulton CL (1995) Nitric oxide signaling in the central nervous system. Annu Rev Physiol 57:683–706

Garthwaite J, Charles SL, Chess Williams R (1988) Endothelium-derived relaxing factor release on activation of NMDA receptors suggests role as intercellular messenger in the brain. Nature 336:385–388

Garthwaite J, Southam E, Boulton CL, Nielsen EB, Schmidt K, Mayer B (1995) Potent and selective inhibition of nitric oxide-sensitive guanylyl cyclase by 1H-[1,2,4]oxadiazolo[4,3-a]quinoxalin-1-one. Mol Pharmacol 48:184–188

Gibbs SM, Truman JW (1998) Nitric oxide and cyclic GMP regulate retinal patterning in the optic lobe of *Drosophila*. Neuron 20:83–93

Haley JE, Schaible E, Paulidis P, Murdock A, Madison DV (1996) Basal and apical synapses of CA1 pyramidal cells employ different LTP induction mechanisms. Learning and Memory 3:289–295

Haley JE, Wilcox GL, Chapman PF (1992) The role of nitric oxide in hippocampal long-term potentiation. Neuron 8:211–216

Harsanyi K, Friedlander MJ (1997) Transient synaptic potentiation in the visual cortex. I. Cellular mechanisms. J Neurophysiol 77:1269–1283

Hatton GI, Yang QZ (1996) Synaptically released histamine increases dye coupling among vasopressinergic neurons of the supraoptic nucleus: mediation by H1 receptors and cyclic nucleotides. J Neurosci 16:123–129

Hawkins RD (1996) NO honey, I don't remember. Neuron 16:465–467

Holscher C (1997) Nitric oxide, the enigmatic neuronal messenger: its role in synaptic plasticity. Trends Neurosci 20:298–303

Huang PL, Dawson TM, Bredt DS, Snyder SH, Fishman MC (1993) Targeted disruption of the neuronal nitric oxide synthase gene. Cell 75:1273–1286

Iadecola C, Li J, Xu S, Yang G (1996) Neural mechanisms of blood flow regulation during synaptic activity in cerebellar cortex. J Neurophysiol 75:940–950

Iadecola C, Pelligrino DA, Moskowitz MA, Lassen NA (1994) Nitric oxide synthase inhibition and cerebrovascular regulation. J Cereb Blood Flow Metab 14:175–192

Inglis FM, Furia F, Zuckerman KE, Strittmatter SM, Kalb RG (1998) The role of nitric oxide and NMDA receptors in the development of motor neuron dendrites. J Neurosci 18:10493–10501

Izumi Y, Zorumski CF (1993) Nitric oxide and long-term synaptic depression in the rat hippocampus. Neuroreport 4:1131–1134

Jacklet JW (1995) Nitric oxide is used as an orthograde cotransmitter at identified histaminergic synapses. J Neurophysiol 74:891–895

Jacklet JW (1997) Nitric oxide signaling in invertebrates. Invert Neurosci 3:1–14

Kara P, Friedlander MJ (1999) Arginine analogs modify signal detection by neurons in the visual cortex. J Neurosci 19:5528–5548

Kingston PA, Zufall F, Barnstable CJ (1996) Rat hippocampal neurons express genes for both rod retinal and olfactory cyclic nucleotide-gated channels: novel targets for cAMP/cGMP function. Proc Natl Acad Sci USA 93:10440–10445

Koesling D (1998) Modulators of soluble guanylyl cyclase. Naunyn Schmiedebergs Arch Pharmacol 358:123–126

Koh H, Jacklet JW (1999) Nitric oxide stimulates cGMP production and mimics synaptic responses in metacerebral neurons of *Aplysia*. J Neurosci 19:3818–3826

Kriegsfeld LJ, Eliasson MJ, Demas GE, Blackshaw S, Dawson TM, Nelson RJ, Snyder SH (1999) Nocturnal motor coordination deficits in neuronal nitric oxide synthase knock-out mice. Neuroscience 89:311–315

Kusaka S, Dabin I, Barnstable CJ, Puro DG (1996) cGMP-mediated effects on the physiology of bovine and human retinal Muller (glial) cells. J Physiol (Lond) 497:813–824

Lovick TA, Brown LA, Key BJ (1999) Neurovascular relationships in hippocampal slices: physiological and anatomical studies of mechanisms underlying flow-metabolism coupling in intraparenchymal microvessels. Neuroscience 92:47–60

Lu C, McMahon DG (1997) Modulation of hybrid bass retinal gap junctional channel gating by nitric oxide. J Physiol (Lond) 499:689–699

McNaught KS, Brown GC (1998) Nitric oxide causes glutamate release from brain synaptosomes. J Neurochem 70:1541–1546

Melis MR, Argiolas A (1997) Role of central nitric oxide in the control of penile erection and yawning. Prog Neuropsychopharmacol Biol Psychiatry 21:899–922

Meller ST, Gebhart GF (1993) Nitric oxide (NO) and nociceptive processing in the spinal cord. Pain 52:127–136

Mills SL, Massey SC (1995) Differential properties of two gap junctional pathways made by AII amacrine cells. Nature 377:734–737

Miyachi E, Murakami M, Nakaki T (1990) Arginine blocks gap junctions between retinal horizontal cells. Neuroreport 1:107–110

Moreno-Lopez B, Escudero M, Delgado Garcia JM, Estrada C (1996) Nitric oxide production by brain stem neurons is required for normal performance of eye movements in alert animals. Neuron 17:739–745

Moreno-Lopez B, Estrada C, Escudero M (1998) Mechanisms of action and targets of nitric oxide in the oculomotor system. J Neurosci 18:10672–10679

Mothet JP, Fossier P, Tauc L, Baux G (1996) Opposite actions of nitric oxide on cholinergic synapses: which pathways? Proc Natl Acad Sci USA 93:8721–8726

MŸller U (1997) The nitric oxide system in insects. Prog Neurobiol 51:363–381

Nelson RJ, Kriegsfeld LJ, Dawson VL, Dawson TM (1997) Effects of nitric oxide on neuroendocrine function and behavior. Front Neuroendocrinol 18:463–491

O'Donnell P, Grace AA (1997) Cortical afferents modulate striatal gap junction permeability via nitric oxide. Neuroscience 76:1–5

Owlya R, Vollenweider, L, Trueb L, Sartori C, Lepori M, Nicod P, Scherrer U (1997) Cardiovascular and sympathetic effects of nitric oxide inhibition at rest and during static exercise in humans. Circulation 96:3897–3903

Pape H (1995) Nitric oxide: an adequate modulatory link between biological oscillators and control systems in the mammalian brain. Semin Neurosci 7:329–340

Pape HC, Mager R (1992) Nitric oxide controls oscillatory activity in thalamocortical neurons. Neuron 9:441–448

Park JH, Straub VA, O'Shea M (1998) Anterograde signaling by nitric oxide: characterization and in vitro reconstitution of an identified nitrergic synapse. J Neurosci 18:5463–5476

Pehl U, Schmid HA (1997) Electrophysiological responses of neurons in the rat spinal cord to nitric oxide. Neuroscience 77:563–573

Pineda J, Kogan JH, Aghajanian GK (1996) Nitric oxide and carbon monoxide activate locus coeruleus neurons through a cGMP-dependent protein kinase: involvement of a nonselective cationic channel. J Neurosci 16:1389–1399

Rand MJ (1992) Nitrergic transmission: nitric oxide as a mediator of non-adrenergic, non-cholinergic neuro-effector transmission. Clin Exp Pharmacol Physiol 19:147–169

Rieke F, Schwartz EA (1994) A cGMP-gated current can control exocytosis at cone synapses. Neuron 13:863–873

Rorig B, Sutor B (1996) Nitric oxide-stimulated increase in intracellular cGMP modulates gap junction coupling in rat neocortex. Neuroreport 7:569–572

Roy DRS, Barnstable CJ (1999) Temporal and spatial pattern of expression of cyclic nucleotide-gated channels in developing rat visual cortex. Cerebral Cortex 9:340–347

Rusakov DA, Kullman DM, Stewart MG (1999) Hippocampal synapses: do they talk to their neighbours? Trends Neurosci 22:382–388

Savchenko A, Barnes S, Kramer RH (1997) Cyclic-nucleotide-gated channels mediate synaptic feedback by nitric oxide. Nature 390:694–698

Sawada M, Ichinose M, Hara N (1995) Nitric oxide induces an increased Na^+ conductance in identified neurons of Aplysia. Brain Res 670:248–256

Sawada M, Ichinose M, Stefano GB (1996) Inhibition of the Met-enkephalin-induced K^+ current in B-cluster neurons of Aplysia by nitric oxide donor. Brain Res 740:124–130

Schuman EM, Meffert MK, Schulman H, Madison DV (1994) An ADP-ribosyltransferase as a potential target for nitric oxide action in hippocampal long-term potentiation. Proc Natl Acad Sci USA 91:11958–11962

Selig DK, Segal MR, Liao D, MALENKA RC, Malinow R, Nicoll RA, Lisman JE (1996) Examination of the role of cGMP in long-term potentiation in the CA1 region of the hippocampus. Learning and Memory 3:42–48

Son H, Hawkins RD, Martin K, Kiebler M, Huang PL, Fishman MC, Kandel ER (1996) Long-term potentiation is reduced in mice that are doubly mutant in endothelial and neuronal nitric oxide synthase. Cell 87:1015–1023

Son H, Lu Y, Zhuo M, Arancio O, Kandel ER, Hawkins RD (1999) The specific role of cGMP in hippocampal LTP. Learning and Memory 5:231–245

Song H, Ming G, He Z, Lehmann M, Tessier-Lavigne M, Poo M (1998) Conversion of neuronal growth cone responses from repulsion to attraction by cyclic nucleotides. Science 281:1515–1518

Southam E, East SJ, Garthwaite J (1991) Excitatory amino acid receptors coupled to the nitric oxide/cyclic GMP pathway in rat cerebellum during development. J Neurochem 56:2072–2081

Southam E, Garthwaite J (1993) The nitric oxide-cyclic GMP signalling pathway in rat brain. Neuropharmacology 32:1267–1277

Strata F, Atzori M, Molnar M, Ugolini G, Berretta N, Cherubini E (1998) Nitric oxide sensitive depolarization-induced hyperpolarization: a possible role for gap junctions during development. Eur J Neurosci 10:397–403

Takuma K, Matsuda T, Hashimoto H, Kitanaka J, Asano S, Kishida Y, Baba A (1996) Role of Na^+-Ca^{2+} exchanger in agonist-induced Ca^{2+} signaling in cultured rat astrocytes. J Neurochem 67:1840–1845

Tanaka M, Yoshida S, Yano M, Hanaoka F (1994) Roles of endogenous nitric oxide in cerebellar cortical development in slice cultures. Neuroreport 5:2049–2052

Thomsen LL, Olesen J (1998) Nitric oxide theory of migraine. Clin Neurosci 5:28–33

Vincent SR, Hope BT (1992) Neurons that say NO. Trends Neurosci 15:108–113

Wakatsuki H, Gomi H, Kudoh M, Kimura S, Takahashi K, Takeda M, Shibuki K (1998) Layer-specific NO dependence of long-term potentiation and biased NO release in layer V in the rat auditory cortex. J Physiol (Lond) 513:71–81

Wood J, Garthwaite J (1994) Models of the diffusional spread of nitric oxide: implications for neural nitric oxide signalling and its pharmacological properties. Neuropharmacology 33:1235–1244

Wu HH, Williams CV, McLoon SC (1994) Involvement of nitric oxide in the elimination of a transient retinotectal projection in development. Science 265:1593–1596

Wu J, Wang Y, Rowan MJ, Anwyl R (1997a) Evidence for involvement of the neuronal isoform of nitric oxide synthase during induction of long-term potentiation and long-term depression in the rat dentate gyrus in vitro. Neuroscience 78:393–398

Wu SY, Dun NJ (1996) Potentiation of IPSCs by nitric oxide in immature rat sympathetic preganglionic neurones in vitro. J Physiol (Lond) 495:479–490

Wu SY, Dun SL, Forstermann U, Dun NJ (1997b) Nitric oxide and excitatory postsynaptic currents in immature rat sympathetic preganglionic neurons in vitro. Neuroscience 79:237–245

Xiong H, Yamada K, Jourdi H, Kawamura M, Takei N, Han D, Nabeshima T, Nawa H (1999) Regulation of nerve growth factor release by nitric oxide through cyclic GMP pathway in cortical glial cells. Mol Pharmacol 56:339–347

Yang G, Feddersen RM, Zhang F, Clark HB, Beitz AJ, Iadecola C (1998) Cerebellar vascular and synaptic responses in normal mice and in transgenics with Purkinje cell dysfunction. Am J Physiol 274:R529–40

Yang G, Iadecola C (1998) Activation of cerebellar climbing fibers increases cerebellar blood flow: role of glutamate receptors, nitric oxide, and cGMP. Stroke 29:499–507

Yang QZ, Hatton GI (1999) Nitric oxide via cGMP-dependent mechanisms increases dye coupling and excitability of rat supraoptic nucleus neurons. J Neurosci 19:4270–4279

Yang S, Tang Y, Zucker RS (1999) Selective induction of LTP and LTD by postsynaptic $[Ca^{2+}]_i$ elevation. J Neurophysiol 81:781–787

Yawo H (1999) Involvement of cGMP-dependent protein kinase in adrenergic potentiation of transmitter release from the calyx-type presynaptic terminal. J Neurosci 19:5293–5300

Zhuo M, Hu Y, Schultz C, Kandel ER, Hawkins RD (1994) Role of guanylyl cyclase and cGMP-dependent protein kinase in long-term potentiation. Nature 368:635–639

CHAPTER 12

The Role of Nitric Oxide in the Peripheral Nervous System

W. Martin

A. Introduction

I. Nomenclature

The term "nitrergic", initially coined by Rand (1992) to describe nerves whose transmission process utilises the L-arginine–nitric oxide (NO) pathway, has now gained acceptance by the International Union of Pharmacology Committee on Nomenclature of the Pharmacology of Nitric Oxide (Moncada et al. 1997). As was the case when Dale (1933) proposed the terms adrenergic and cholinergic, nitrergic was introduced to facilitate classification and discussion and not to indicate the precise chemical nature of the neurotransmitter involved. Indeed, as will be discussed, intense debate has raged as to whether the nitrergic transmitter is NO per se or a NO-like molecule. Most of the major objections to acceptance of NO, however, now seem to have been largely overcome. The term nitrergic is, however, far from ideal. Although seemingly appropriate in circumstances where a particular transmission process is wholly explained by the L-arginine–NO pathway, it appears less so in situations of co-transmission both in the periphery and central nervous system (CNS), where this pathway is not the sole or even the dominant mechanism for transmission. Nevertheless, in the absence of a more suitable alternative, the term nitrergic is employed here to include any nerve that utilises the L-arginine–NO pathway, irrespective of the co-involvement of additional neurotransmitters.

II. Historical Perspective

Only a brief account can be given here of the historical landmarks leading to our current acceptance of the presence of nerves that utilise the L-arginine–NO pathway for neurotransmission in the peripheral autonomic nervous system. The reader is directed to more comprehensive insights published elsewhere (Gillespie et al. 1990; Martin and Gillespie 1991; Gibson et al. 1995; Rand and Li 1995b).

III. The Concept of Non-Adrenergic, Non-Cholinergic Neurotransmission

The earliest descriptions of processes which we now recognise as nitrergic were provided by nineteenth-century physiologists and pre-dated the very concept of chemical neurotransmission (LANGLEY and ANDERSON 1895). In these, atropine, which proved highly effective in blocking the effects of parasympathetic nerve stimulation in the heart, had no effect whatever on parasympathetic stimulation at other sites, particularly in the pelvic viscera. Following the now classic demonstrations of chemical neurotransmission and the involvement of acetylcholine and noradrenaline as neurotransmitters at post-ganglionic parasympathetic and sympathetic sites, respectively, the problem of "atropine resistance" was explained by two erroneous trains of thought. The first, proposed by DALE and GADDUM (1930), envisaged atropine establishing a "barrier" in tissues; where acetylcholine was released outside this barrier, parasympathetic stimulation was blocked whereas, when released inside this barrier, its effects remained untouched. The second was the assumption that certain atropine-resistant parasympathetic responses might be explained by the presence of "rogue" sympathetic fibres in parasympathetic nerves (LANGLEY 1898; HARRISON and MCSWINNEY 1936). This arose largely through the absence of an effective drug to block sympathetic responses. Ergotoxine was available to early physiologists, but its actions were variable, producing blockade of sympathetic responses at certain sites, including the vasculature, but paradoxical enhancement at others, such as the heart (JANG 1940). We can now explain these observations by the selective ability of ergotoxine to block post-junctional α_1- but not β-adrenoceptors and enhance the release noradrenaline through inhibition of the negative-feedback mechanism mediated through prejunctional α_2-adrenoceptors. At the time, these observations presented such a confusing picture that a unifying hypothesis failed to appear. The advent of the adrenergic neurone-blocking agents, such as guanethidine and bretylium, however, finally brought clarity. Their use in the guinea-pig taenia coli provided unequivocal evidence that, in the presence of effective adrenergic blockade, relaxant responses and associated hyperpolarisation of smooth muscle could still be elicited by nerve stimulation (BURNSTOCK et al. 1963a, 1963b). Subsequent use of the combination of atropine and an adrenergic neurone blocker revealed the widespread distribution of peripheral non-adrenergic, non-cholinergic (NANC) autonomic nerves in the gastrointestinal, respiratory and urogenital tracts and in certain blood vessels. Although the focus of this article is on nitrergic nerves that relax smooth muscle, it should be borne in mind that numerous examples of excitatory NANC transmission have been reported (BURNSTOCK 1972; WHITE 1991).

Numerous reviews have been compiled describing the presence and functions of NANC nerves; these demonstrate that no single nerve type can explain all these manifestations of anomalous neurotransmission. Rather, it is now accepted that a wide range of substances, including adenosine triphosphate

(ATP) (BURNSTOCK 1972; WHITE 1991), various neuropeptides (FAHRENKRUG 1991; POTTER 1998) and either NO or an NO-like material (GILLESPIE et al. 1990; MARTIN and GILLESPIE 1991; GIBSON et al. 1995; RAND and LI 1995b), are employed as neurotransmitters by distinct nerves that together constitute the peripheral NANC system. Recent evidence also supports the existence of inhibitory nerves, analogous to the nitrergic system, but which utilise carbon monoxide as their neurotransmitter (RATTAN and CHAKDER 1993; NY et al. 1996).

IV. The Concept of Nitrergic Nerves

Despite the widespread distribution of such nerves, it was work on a restricted number of smooth muscle preparations, including the anococcygeus and retractor penis muscles of a variety of species, that laid the foundations for development of the concept of nitrergic nerves. These muscles are characterised by the presence of a conventional adrenergic motor innervation arising via the lumbar sympathetic supply and an inhibitory innervation via the sacral parasympathetic nerves (LUDUENA and GRIGAS 1966; GILLESPIE 1972; KLINGE and SJOSTRAND 1974; GIBSON and WEDMORE 1981). Some, including the bovine retractor penis (BRP) (KLINGE and SJOSTRAND 1974), also contain cholinergic fibres, but these provide a prejunctional inhibitory system controlling the adrenergic supply and do not affect smooth-muscle function directly. This simple pattern of innervation permitted the properties of their NANC nerves to be studied with greater ease than was possible with much more complex structures, such as the gut.

Another vital development that transformed the field was the discovery, in 1980, of endothelium-derived relaxing factor (EDRF) in the vasculature (FURCHGOTT and ZAWADZKI 1980). Indeed, from that point on, developments in our understanding of the neurotransmission process in the anococcygeus and retractor penis muscles and the phenomenon of endothelium-dependent relaxation were to proceed in parallel. The key events began with attempts to extract the NANC transmitter from the BRP muscle. Today, we know the nitrergic transmitter is not stored in nerve fibres but, at the time, this seemed a valid approach, since all established transmitters could be isolated by such means. Fortuitously, the extraction process using dilute acid led to the isolation of a novel, labile activity termed inhibitory factor (AMBACHE et al. 1975; GILLESPIE and MARTIN 1980; GILLESPIE et al. 1981). Although its identity as a mixture of NO-releasing S-nitrosothiols (YUI et al. 1989; KERR et al. 1993), formed from the acidification of nitrite in the presence of tissue thiols was not established until some years later, it was clear from the outset that this novel activity was an excellent mimic of the relaxation seen on stimulation of the NANC nerves. The ability to extract inhibitory factor from tissues lacking an NANC innervation (GILLESPIE and MARTIN 1980; BOWMAN et al. 1981) did, however, cast some early doubt on its role as a neurotransmitter. Nevertheless, a striking link between inhibitory factor, the NANC nerves and EDRF was to come following the resolution of an anomalous finding that inhibitory factor had no effect on the blood pressure of the anaesthetised rat yet relaxed every isolated blood vessel and vascular bed to which it was added (BOWMAN et al. 1979, 1981). Blood explained this blockade of inhibitory factor activity in vivo, and further analysis localised the action to the erythrocytes and subsequently to oxyhaemoglobin (BOWMAN et al. 1981, 1982; BOWMAN and GILLESPIE 1982). Although a seemingly unlikely tool, oxyhaemoglobin, was to prove highly selective; it had no effect on relaxations to prostaglandin E_1 or the phosphodiesterase (PDE) inhibitor 3-isobutyl-1-methylxanthine but blocked those produced by electrical field stimulation (EFS) in the BRP and rat anococcygeus. Later, it was shown to block the actions of EDRF in rabbit aorta (MARTIN et al. 1985a, 1985b), thus providing the first of many steps in the unification of two separate strands of biological research. Many other observations served to strengthen these links, including the findings that the relaxant effects of EDRF, inhibitory factor and the NANC nerves in the BRP were all associated with a rise in smooth-muscle cyclic guanosine monophosphate (GMP) content (RAPOPORT and MURAD 1983; BOWMAN and DRUMMOND 1984), blocked by an inhibitor of soluble guanylate cyclase, methylene blue (MARTIN et al. 1985b; BOWMAN et al. 1986), and potentiated by a selective inhibitor of PDE V, zaprinast (M&B 22, 948) (BOWMAN and DRUMMOND 1984; MARTIN et al. 1986).

Definitive proof of the link between NANC nerves and EDRF came following elucidation of the L-arginine-NO pathway in the vascular endothelium. The demonstration that EDRF was NO, synthesised from the precursor L-arginine by the action of nitric oxide synthase (NOS) (PALMER et al. 1987, 1988), and the development of analogues of L-arginine which inhibit this process competitively provided vital new analytical tools (REES et al. 1990). The first inhibitor of NOS available was N^G-monomethyl-L-arginine (L-NMMA) (REES et al. 1989), and a number of research groups demonstrated almost simultaneously the ability of this agent to inhibit NANC relaxation in the rat anococcygeus muscle (GILLESPIE et al. 1989; LI and RAND 1989; RAMAGOPAL and LEIGHTON 1989), with evidence of blockade in the mouse anococcygeus following soon thereafter (GIBSON et al. 1990). As in the vasculature, this blocking action was both stereoselective and competitive, with L- but not D-NMMA being active, and reversal was obtained with excess L- but not D-arginine. Surprisingly, L-NMMA did not block NANC relaxation in the BRP, but the subsequent development of the more potent and effective blocker N^G-nitro-L-arginine (HOBBS and GIBSON 1990; MOORE et al. 1990) confirmed the role of the L-arginine–NO pathway as the means of NANC neurotransmission in this tissue also (LIU et al. 1991). Thus, demonstration of an L-arginine-reversible blockade by an inhibitor of NOS became the primary means with which to define the role of the L-arginine–NO pathway in NANC transmission. This, together with the localisation of neuronal NOS (nNOS, NOS-I) either by immunocytochemistry (BREDT et al. 1990; SHENG et al. 1992; BRAVE et al. 1993b) or reduced nicotinamide adenine dinucleotide phosphate (NADPH)-diaphorase staining (DAIL et al. 1993), the assessment of NO release either by bioassay (BULT et al. 1990), chemiluminescence analysis (CHAKDER and RATTAN 1993; VIALS et al. 1997) or visualisation (WIKLUND et al. 1997), and the use of the nNOS knockout mouse (HUANG et al. 1993) have all contributed to our present understanding of the widespread distribution of nitrergic nerves in the peripheral autonomic nervous system.

B. Properties of Nitrergic Nerves

I. Properties of nNOS

Much of our understanding of the properties of nNOS (NOS-I) has come from studies on brain tissue. Indeed, the enzyme was first purified from rat and porcine cerebellum (BREDT and SNYDER 1990; MAYER et al. 1990). Like endothelial NOS (eNOS, NOS-III) (MITCHELL et al. 1991a), nNOS is constitutively expressed (SCHMIDT et al. 1991) but, unlike eNOS, it exists mainly as a cytosolic rather than a particulate enzyme. The complementary DNAs encoding nNOS have been cloned from rat and human brain, and the deduced amino acid sequences predict molecular masses of 160kDa and 161kDa, respectively, which are in accord with those obtained for the purified enzymes. In humans, the gene for nNOS has been localised, by in situ hybridisation, to chromosome 12 (XU et al. 1993). Furthermore, the enzyme is highly conserved across species, with 93% amino acid homology between rat and human forms. All isoforms of the enzyme use L-arginine and molecular oxygen as substrates, require the cofactors NADPH, tetrahydrobiopterin, flavin adenine dinucleotide (FAD) and flavin mononucleotide (FMN), and contain a haem prosthetic group. In common with eNOS, the activity of nNOS is regulated by the calcium/calmodulin complex, with a sensitivity to calcium within the physiological range for excitable cells; threshold activation is around 80nM calcium, and maximal activity is around 500nM (KNOWLES et al. 1989; SCHMIDT et al. 1991). The isoforms of nNOS purified from nitrergic nerves in the anococcygeus of the rat (MITCHELL et al. 1991b) and mouse (GIBSON et al. 1992) and the BRP (SHENG et al. 1992) exhibit the same substrate and cofactor requirements and calcium/calmodulin dependence as those from brain. Furthermore, the demonstration of cross-reactivity of antisera to rat cerebellum nNOS with that from BRP observed by immunoprecipitation (SHENG et al. 1992), the widespread utility of antisera to the brain-derived isoform for purposes of immunohistochemical localisation in peripheral nerves (BREDT et al. 1990; SHENG et al. 1992) and the absence of nitrergic activity in peripheral tissues from the nNOS knockout mouse (MASHIMO et al. 1996) strongly suggest that the nNOS of nitrergic nerves is homologous with that of brain.

II. Localisation of nNOS in Nitrergic Nerves

The most reliable means of localising nNOS in tissues is the use of immunohistochemistry, and a number of mono- and polyclonal preparations are available commercially for this purpose. An alternative is to employ NADPH-diaphorase histochemistry (DAIL et al. 1993). This technique, which is based on the reduction of nitro blue tetrazolium to insoluble formazan dye, must be used with caution since, although all three isoforms of NOS possess NADPH-diaphorase activity, these represent only a fraction of the total cellular activity (TRACEY et al. 1993). The use of immunohistochemistry has successfully localised nNOS activity to nerves in the periphery (STEINBERG et al. 1990; BRAVE et al. 1993b), spinal cord (DUN et al. 1992; LIN and BENNETT 1994) and CNS (BREDT et al. 1990). Somewhat surprisingly, nNOS has also been found in non-nervous tissue, including epithelial cells in the uterus, lung and stomach, cells of the macula densa (SCHMIDT et al. 1992a), pancreatic islet cells, where it mediates insulin release (SCHMIDT et al. 1992b), and fast-twitch skeletal muscle fibres (NAKANE et al. 1993), where it opposes force development (KOBZIC et al. 1994).

The pattern of staining of nitrergic nerves revealed by immunohistochemistry and NADPH-diaphorase histochemistry is essentially identical, with nNOS located throughout the nerve cell body, axon and varicose fibres. Such a pattern suggests the possibility of release of the transmitter along the entire length of the nerve fibre rather than exclusively at varicose regions, as with all other autonomic fibres. Indeed, by employing a chemiluminescence system involving luminol and hydrogen peroxide, NO release was observed along the entire length of the nitrergic nerves in the guinea-pig hypogastric-nerve–myenteric-plexus preparation (WIKLUND et al. 1997). Consequently, nitrergic nerves may operate a unique mode of transmission in the nervous system, involving synaptic and non-synaptic components.

III. Anatomical Distribution and Physiological Functions of Nitrergic Nerves

In keeping with functional studies, the most common location of nNOS is in post-ganglionic nerves of the parasympathetic division of the autonomic nervous system (STEINBERG et al. 1990; BRAVE et al. 1993b; LUMME et al. 1996), but nNOS and NADPH-diaphorase staining has also been found in certain pre-ganglionic (ANDERSON 1992; LUMME et al. 1996) and post-ganglionic (SHENG et al. 1993; JEN et al. 1996) sympathetic fibres. In fact, ganglionic transmission at both sympathetic (SOUTHAM et al. 1996) and parasympathetic (SCOTT and BENNETT 1993) sites appears to be enhanced by NO via a pre-synaptic enhancement of transmitter output (LIN and BENNETT 1994; SOUTHAM et al. 1996), analogous to the mechanisms underlying long-term potentiation in the hippocampus.

The physiological roles of nitrergic nerves at their major anatomical locations, including the urogenital (GILLESPIE et al. 1990; MARTIN et al. 1991a; ANDERSSON 1993; ANDERSSON and PERSSON 1993; ANDERSSON and WAGNER 1995), gastrointestinal (STARK and SZURSZEWSKI 1992; LEFEBVRE 1997) and respiratory (BARNES and BELVISI 1993; BELVISI et al. 1995) tracts, have been reviewed. Reviews on the role of nitrergic nerves in cerebral (TODA 1995) and penile (KLINGE and SJOSTRAND 1974) vessels have also been published, but there is increasing awareness that these nerves are more generally distributed in the vasculature at other sites, including the kidney (OKAMURA et al. 1995), lung (LIU et al. 1992), uterus (TODA et al. 1994, 1995a), eye (TODA et al. 1997) and arteriovenous anastomoses (FUNK et al. 1994). At each of these sites, the nerves elicit smooth-muscle relaxation. Nitrergic fibres are also present within the heart (TANAKA et al. 1993), although their functions are as yet unknown.

IV. Unitary Transmission, Dual Transmission and Co-Transmission

Although the introduction of L-NMMA provided definitive evidence of the role of the L-arginine–NO system in NANC transmission in the anococcygeus of the rat and mouse (GILLESPIE et al. 1989; LI and RAND 1989; RAMAGOPAL and LEIGHTON 1989; GIBSON et al. 1990), this agent typically produced a maximum blockade of around 50%. The introduction of the more potent and effective analogues, N-nitro-L-arginine (L-NNA) and N-nitro-L-arginine methyl ester (L-NAME), showed, however, that complete blockade could be established in the anococcygeus and BRP muscles, suggesting that NANC transmission was mediated solely by nitrergic nerves (HOBBS and GIBSON 1990; MARTIN et al. 1991b). The presence of a unitary transmission process in the anococcygeus of the rat has, however, recently been challenged on the basis that the almost universal use of guanethidine generates such powerful tone and membrane depolarisation in this tissue that an additional NANC transmission process is masked (SELEMIDIS and COCKS 1997). These authors propose that if lower levels of tone are generated using phenylephrine, EFS elicits relaxation that is resistant to blockade by L-NNA but inhibited by nifedipine. They further suggest that this relaxation is mediated by a nerve-derived hyperpolarising factor (NDHF) analogous to the putative vascular endothelium-derived hyperpolarising factor (EDHF). It remains to be determined if others can confirm these observations, but experiments conducted under similar conditions failed to demonstrate the presence of nerves other than nitrergic nerves in the mouse anococcygeus muscle (FONSECA et al. 1998).

Nevertheless, numerous examples have been reported where NANC transmission processes involve NO acting in concert with additional transmitters. These include: NO with vasoactive intestinal peptide (VIP) in the gastric fundus of the rat, cat and pig (LI and RAND 1990; BARBIER and LEFEBVRE 1993; LEFEBVRE et al. 1995), colon of the dog (KEEF et al. 1994) and trachea of the guinea pig (LI and RAND 1991); NO with ATP in the proximal urethra of the hamster (PINNA et al. 1998) and colon of the guinea pig (ZAGORODNYUK and MAGGI 1994); and NO with VIP and ATP in the colon of the mouse (MASHIMO et al. 1996). In many situations, it has not been possible to determine if the mixed-transmission process results from release of transmitters from different nerves (dual transmission) or from the same nerve (co-transmission). In such circumstances, immunohistochemistry has proved vital in discriminating between these two possible modes of transmission. Using this approach, examples of co-transmission have been highlighted by the co-localisation of nNOS and VIP in nerves throughout the enteric nervous system (FURNESS et al. 1992a, 1992b; LEFEBVRE et al. 1995) of many species, including humans (GUO et al. 1997), in bovine cerebral arteries (GONZALEZ et al. 1997) and in the porcine lower urinary tract (PERSSON et al. 1995). Furthermore, histochemical evidence has been obtained for co-transmission involving NO with noradrenaline in the human male vas deferens, seminal vesicles, prostate and bladder neck (JEN et al. 1996), and with neuropeptide Y and acetylcholine in the porcine lower urinary tract (PERSSON et al. 1995). A comprehensive review of the nature and sites of co-transmission processes has been published (LUNDBERG 1996).

C. Nature of the Nitrergic Neurotransmitter

Although there is consensus that the L-arginine–NO pathway mediates nitrergic transmission, the precise nature of the neurotransmitter involved has been the subject of intense debate. This has arisen because many agents that block the actions of authentic NO either fail to block or have only a minor effect on nitrergic transmission. Such a discrimination is seen with superoxide anion-generating agents, including pyrogallol in the BRP (GILLESPIE and SHENG 1990; MARTIN et al. 1994), rat anococcygeus (LIU et al. 1997) and opossum lower oesophageal sphincter (KNUDSEN et al. 1992), LY 83583 in the rat gastric fundus (BARBIER and LEFEBVRE 1992) and BRP (MARTIN et al. 1994), 7-ethoxyresorufin in the rat anococcygeus (LI and RAND 1996) and duroquinone in the mouse anococcygeus (LILLEY and GIBSON 1995). Hydroquinone, which, depending on the redox milieu of the tissue concerned, can act either as a superoxide anion generator (PAISLEY and MARTIN 1996) or radical scavenger (HOBBS et al. 1991), also discriminates between NO and the nitrergic transmitter in the BRP (PAISLEY and MARTIN 1996), mouse corpus cavernosum (GOCMEN et al. 1997), guinea-pig trachea, mouse anococcygeus and rat gastric fundus (HOBBS et al. 1991; LEFEBVRE 1996). Furthermore, several agents that scavenge NO directly, including haemoglobin, 2-phenyl-(4-carboxyphenyl)-4,4,5,5-tetramethylimidazoline-1-oxyl-3-oxide potassium salt (carboxy-PTIO) and hydroxocobalamin, exhibit either quantitative or qualitative differences in their effects on relaxation induced by nitrergic nerves and authentic NO. Haemoglobin binds NO with high affinity (GIBSON and ROUGHTON 1957), resulting in the rapid formation of methaemoglobin and nitrate (FEELISCH and NOACK 1987; KELM and SCHRADER 1990), and proved a vital early tool in establishing the concept of nitrergic transmission. Nevertheless, its potency in blocking nitrergic transmission in the rat anococcygeus (BOWMAN and GILLESPIE 1982; BOWMAN et al. 1982) is some 100-fold lower than its ability to inhibit relaxation induced by NO itself (LA et al. 1996). Indeed, in the rat gastric fundus, it failed to block nitrergic transmission

(JENKINSON et al. 1995). Carboxy-PTIO, which rapidly oxidises NO to nitrite and nitrate (AKAIKE et al. 1993), produces blockade of nitrergic transmission in the BRP (PAISLEY and MARTIN 1996) but not in the rat anococcygeus or gastric fundus (RAND and LI 1995a). Hydroxocobalamin inactivates NO by forming nitrosocobalamin (KACZKA et al. 1951; BROUWER et al. 1996) and blocks nitrergic transmission in the mouse anococcygeus (LILLEY and GIBSON 1996) and corpus cavernosum (GOCMEN et al. 1997) but not in the rat gastric fundus (JENKINSON et al. 1995). In the rat anococcygeus and BRP, it produces effective blockade under light but not dark conditions (RAJANAYAGAM et al. 1993; PAISLEY and MARTIN 1996; LA et al. 1997).

A number of explanations have been proposed to explain the differential effects of the above agents on authentic NO and nitrergic transmission. These include the proposals that:

1. The transmitter is NO, but differences in the effects of drugs on nerve-derived and bath-applied NO should be expected
2. The transmitter is not NO per se but an NO-like or NO-releasing molecule
3. The transmitter is NO and is protected from inactivation within the tissue

I. Predicted Differences in the Effects of Drugs on Nerve-Derived and Bath-Applied NO

A detailed mathematical model describing the kinetic and concentration profiles of NO generated from single or multiple sources has been used to account for the differential effects of drugs on responses to authentic NO and to nitrergic nerve stimulation (WOOD and GARTHWAITE 1994). On the basis that the path length which bath-applied NO must traverse is many orders of magnitude greater than the distance across the neuroeffector junction, this model predicts that drugs will have more opportunity to interfere in the former than in the latter situation. This difference becomes more acute for high-molecular-weight, poorly diffusable substances, such as haemoglobin, which have difficulty entering the interstitial space (RATTAN et al. 1995).

II. Evidence that the Nitrergic Neurotransmitter is a NO-Like or NO-Releasing Molecule

The ability of the agents described above to discriminate between authentic NO and the nitrergic transmitter was taken by many to indicate that the two are not identical. Consequently, a widespread search began for NO-like or NO-releasing compounds that have a pharmacological profile more closely matching that of the nitrergic transmitter. Prime candidates for this role were the NO-releasing S-nitrosothiols. Since L-cysteine and glutathione represent the two most abundant tissue thiols, S-nitrosocysteine (CYSNO) and S-nitrosoglutathione (GSNO) were regarded as the two most likely candidates, but a wide range of other synthetic analogues, including S-nitroso-N-acetylpenicillamine (SNAP), S-nitroso-N-acetylcysteine (SNAC) and S-nitrosocoenzyme A (CoASNO) also produced relaxation on all nitrergically innervated tissues tested (Gibson et al. 1992; Liu et al. 1994; Rand and Li 1994a). Unfortunately, direct attempts to investigate the role of this class of compound in nitrergic transmission by preventing their formation through the use of compounds that either oxidise or covalently modify thiols proved unsatisfactory. Although ethacrynic acid blocked nitrergic transmission in the rat anococcygeus (Li et al. 1994) and diamide and N-ethylmaleimide had a similar effect in the BRP (Liu et al. 1994), responses to isoprenaline and, to a lesser extent, papaverine were also blocked. It was evident that the thiol-inactivating agents produced non-selective effects, blocking the effector pathways for these diverse relaxants, and any additional action in preventing formation of S-nitrosothiols by the nitrergic nerves was impossible to discern.

Other blocking agents appeared to discriminate between S-nitrosothiols and the nitrergic transmitter and only served to highlight as many obstacles for acceptance of the candidacy of this class of NO donor as for NO itself. Evidence against CYSNO was considerable: methylene blue potentiated its relaxant actions in the opossum lower oesophageal sphincter but blocked relaxation produced by the nitrergic nerves (Knudsen et al. 1992); hydroquinone blocked its relaxant actions in the mouse anococcygeus but had no effect on relaxation to GSNO, SNAP, CoASNO or the nitrergic nerves (Gibson et al. 1992); and LY 83583 blocked its relaxant actions in rat gastric fundus but had no effect on relaxation to GSNO, SNAP, SNAC or the nitrergic nerves. Evidence against GSNO was also strong: hydroxocobalamin enhanced its relaxant actions in the rat anococcygeus (Rand and Li 1994a) and gastric fundus (Jenkinson et al. 1995) but failed to affect nitrergic transmission in both tissues. The most compelling evidence against the involvement of S-nitrosothiols in nitrergic transmission has come, however, from studies investigating the actions of copper and certain antioxidants. Specifically, copper is known to promote the release of NO from S-nitrosothiols (Askew et al. 1995; Dicks et al. 1996) and, indeed, the addition of $CuSO_4$ to rat gastric fundus enhanced the relaxant effects of CYSNO, GSNO and SNAP (De Man et al. 1996a). This enhancement took place, however, without any effect on nitrergic relaxation. Furthermore, the copper chelators bathocuproine and ethylene diamine tetraacetic acid inhibited relaxations to CYSNO, GSNO and SNAP in rat gastric fundus but had no effect on that produced by the nitrergic nerves (De Man et al. 1996a, 1998). The antioxidants ascorbate and α-tocopherol also enhanced relaxation of rat gastric fundus induced by CYSNO, GSNO and SNAP, presumably through augmented release of NO, but failed to affect that produced by the nitrergic nerves (De Man et al. 1998). On the basis of this evidence, it seems unlikely that S-nitrosothiols participate in nitrergic transmission.

The possibility that the alternative redox forms of NO, i.e. nitroxyl (NO^-) and nitrosonium (NO^+) (Stamler et al. 1992), might mediate nitrergic transmission rather than NO radical has been considered (GIBSON et al. 1995). Indeed, one report suggests that the primary product of nNOS is NO⁻ as opposed to the NO radical (SCHMIDT et al. 1996), and this species too appears to possess smooth-muscle relaxant properties (PINO and FEELISCH 1994; ZAMORA et al. 1995). At present, however, no direct evidence linking either NO⁻ or NO⁺ to nitrergic transmission has been published.

III. Evidence that NO is the Nitrergic Neurotransmitter and is Protected from Inactivation

Many of the agents that discriminate between authentic NO and the nitrergic transmitter in tissue-bath experiments are superoxide-generating agents. Consequently, the possibility was considered that high levels of superoxide dismutase (SOD) activity within the tissues might provide an explanation for these differential actions (MARTIN et al. 1994). According to this scheme, SOD within the tissue would protect NO produced by nitrergic nerves, whereas NO added exogenously would lack such protection and so be vulnerable to destruction by superoxide anions present within the bathing fluid. Three distinct isoforms of SOD exist in mammalian cells: two are Cu/Zn-containing isoforms, one extracellularly (MARKLUND 1984) and one intracellularly in the cytosol and nucleus (CRAPO et al. 1992), and the third is a Mn-containing isoform located mainly within the mitochondrial matrix (WESIGER and FRIDOVICH 1973). Since the two Cu/Zn-containing isoforms are ideally placed to protect nerve-derived NO, the effects of inhibiting these irreversibly with the copper chelator, diethyldithiocarbamate (DETCA) (Cocco et al. 1981; KELNER et al. 1989), were examined on strips of BRP muscle (MARTIN et al. 1994; PAISLEY and MARTIN 1996; MOK et al. 1998). This treatment reduced the total SOD activity (from 73.1 ± 15.7 U/mg protein to 8.2 U/mg protein) but, more importantly, resulted in the normally resistant nitrergic transmission process becoming sensitive to inhibition by all superoxide anion generators tested, including pyrogallol, LY 83583 and hypoxanthine/xanthine oxidase. In each case, the blockade was clearly due to destruction of the nitrergic transmitter by superoxide anion, since restoration of transmission was achieved by the addition of exogenous SOD or certain SOD mimetics. Clearly, therefore, endogenous Cu/Zn SOD protected the nitrergic transmitter from destruction in the BRP and, thus, provided an explanation for the ability of superoxide-anion generators to discriminate between nitrergic nerve-derived and bath-applied NO.

Subsequent investigations using DETCA provided evidence that Cu/Zn SOD exerted a similar protective role in other tissues, since nitrergic transmission became susceptible to inhibition by duroquinone in mouse anococcygeus (LILLEY and GIBSON, 1995), by pyrogallol in rat anococcygeus (LIU et al. 1997), by duroquinone, pyrogallol and LY 83583 in rat gastric fundus

(DE MAN et al. 1996b; LEFEBVRE 1996) and by xanthine/xanthine oxidase in opossum oesophagus (THOMAS et al. 1996). A complication arose, however, with the use of hydroquinone: in the mouse anococcygeus and rat gastric fundus, where it acts as a radical scavenger, treatment with DETCA failed to enhance blockade (LEFEBVRE 1996; LILLEY and GIBSON 1995) whereas, in the BRP, where the redox milieu results in the sequential generation of the semi-quinone radical and superoxide anion, treatment with DETCA powerfully potentiated blockade (PAISLEY and MARTIN 1996). In keeping with this protective role, immunohistochemical evidence shows co-localisation of the intracellular isoform of Cu/Zn SOD with nNOS in nitrergic nerves in the rat anococcygeus muscle (LIU et al. 1997). Further investigations will be required to determine if protection of transmission by Cu/Zn SOD represents a universal property of tissues possessing a nitrergic innervation. The possibility that the third isoform of SOD, i.e. Mn SOD, might provide additional protection of the nitrergic transmitter has not been investigated, largely because of lack of availability of a selective inhibitor analogous to DETCA. Such a role might be suggested, however, by the findings that unusually high levels of this isoform are co-localised with nNOS in enteric neurones in the opossum oesophagus (THOMAS et al. 1996) and with NADPH-diaphorase throughout the rat gut (FANG and CHRISTENSEN 1995).

The isoforms of SOD may not, however, represent the only means by which the nitrergic transmitter is protected within tissues. Recent findings suggest an additional process, mediated by low-molecular-weight antioxidants may also be involved (LILLEY and GIBSON 1996). In keeping with its ability to scavenge superoxide anion (SOM et al. 1983; GOTOH and NIKI 1992), ascorbic acid mimicked the ability of SOD to protect exogenous NO from destruction by the superoxide generators, duroquinone and xanthine/xanthine oxidase in the mouse anococcygeus. In contrast to SOD, however, this protective effect was found to extend also to the NO scavengers, hydroquinone and carboxy-PTIO. A more limited spectrum of protection was also seen with α-tocopherol and glutathione: each failed to inhibit blockade by the free-radical generators, but the former blocked the effects of carboxy-PTIO and the latter blocked those of hydroquinone. Moreover, further investigation by the same workers provided strong evidence that ascorbate was released from nitrergic nerves following their activation in the rat anococcygeus (GIBSON and LILLEY 1997; LILLEY and GIBSON 1997).

On the basis of the above, it appears that the presence of antioxidant mechanisms involving Cu/Zn SOD, ascorbate and perhaps other low-molecular-weight molecules within the tissues protects nitrergic nerve-derived NO from inactivation by a variety of scavenging agents. Quite apart from exerting a vital physiological protective function safeguarding nitrergic transmission, the existence of these mechanisms provides a satisfactory explanation for the ability of many drugs to discriminate between authentic NO and the nitrergic transmitter. Accordingly, the major objections to accepting NO per se as the nitrergic transmitter have been largely overcome.

D. Pre-Junctional Mechanisms

I. Activation of Nitrergic Nerves

A wide variety of methods are employed to activate nitrergic nerves. That which most closely mimics the physiological processes of activation is through stimulation of reflex arcs; this has even been achieved in isolated preparations through distension-activated receptive relaxation in the guinea-pig stomach (DESAI et al. 1991) and peristalsis in the guinea-pig ileum (SUZUKI et al. 1994). The most commonly adopted mode of activation, however, is by EFS. This technically simple but extremely versatile technique can be employed for the activation of extrinsic nerve trunks both in vivo and in vitro and of nerve fibres running deep within isolated tissues. In common with other nerve types, activation of nitrergic nerves by this technique results from generation of action potentials (GILLESPIE 1972; KLINGE and SJOSTRAND 1974), since it is readily abolished by tetrodotoxin (TTX), an inhibitor of voltage-gated Na^+ channels. Other procedures which lead to TTX-sensitive activation of nitrergic nerves include the use of depolarising solutions of K^+ (GIBSON and JAMES 1977; WERKSTROM et al. 1997), scorpion venom and its purified components (GWEE et al. 1996; GONG et al. 1997; TEIXEIRA et al. 1998), α-latrotoxin from the black widow spider (WERKSTROM et al. 1997), and γ-aminobutyric acid type-A-receptor stimulation (MAGGI et al. 1984; BOECKXSTAENS et al. 1990). There has also been the suggestion that VIP activates nitrergic nerves in the gastrointestinal tract (GRIDER et al. 1992), but this has been disputed (KEEF et al. 1994). Activation of pre-junctional nicotinic receptors by high concentrations of acetylcholine (RAND and LI 1993b), dimethylphenyl piperazinium (DMPP) (BOECKXSTAENS et al. 1993a; BORJESSON et al. 1997) or nicotine itself (TODA 1975, 1981) also leads to activation of nitrergic nerves. Such activation does not, however, appear to involve the generation of action potentials, since it is resistant to blockade by TTX (TODA 1981) and may arise simply from depolarisation of the nitrergic nerve terminals.

II. Role of Ca^{2+} in Activation of Nitrergic Nerves

Ca^{2+} has a well-established role in stimulating the exocytotic secretion of stored neurotransmitters (AUGUSTINE et al. 1987). Experiments employing solutions low in Ca^{2+} or high in Mg^{2+}, which antagonises Ca^{2+}, have also provided evidence for the involvement of this cation in the release of NO from nitrergic nerves in the canine small intestine (STARK et al. 1991), ileocolonic junction (BOECKXSTAENS et al. 1993a) and cerebral artery (TODA et al. 1995b). The role of Ca^{2+} in these nerves is unlikely to be for the exocytotic release of NO although, at sites of co-transmission, stored co-transmitters will be released by this mechanism. Rather, the function of Ca^{2+} in this case is to activate nNOS, a Ca^{2+}/calmodulin-dependent enzyme (KNOWLES et al. 1989; SCHMIDT et al. 1991).

The mode of entry of Ca^{2+} into nitrergic nerves following activation has been investigated using an extensive array of inhibitors of the different types of voltage-operated channel (VOC). Treatment with the L-type blockers nimodipine, verapamil and nifedipine in the rabbit urethra and detrusor (ZYGMUNT et al. 1993) and canine ileocolonic junction (BOECKXSTAENS et al. 1993a), with the P-type blocker ω-agatoxin in the rabbit and pig urethra (WERKSTRÖM et al. 1995; ZYGMUNT et al. 1995) and with the T-type blocker tetramethrin in the rabbit urethra and detrusor (ZYGMUNT et al. 1993) all failed to inhibit nitrergic transmission. N-type VOCs do, however, appear to be involved, since ω-conotoxin GVIA blocks nitrergic transmission in the rat gastric fundus and anococcygeus (DE LUCA et al. 1990), canine ileocolonic junction (BOECKXSTAENS et al. 1993a) and cerebral artery (TODA et al. 1995b) and pig urethra (WERKSTRÖM et al. 1995). It has generally been found that ω-conotoxin GVIA blocks responses to low- but not high-frequency nitrergic stimulation. Moreover, the blockade is often less effective than that obtained against adrenergic, cholinergic or purinergic transmission in the same tissues (ZYGMUNT et al. 1993; ZAGORODNYUK and MAGGI 1994). Indeed, nitrergic relaxation to EFS in the sheep middle cerebral artery is unaffected by ω-conotoxin GVIA (MATTHEW and WADSWORTH 1997). Nitrergic relaxations elicited by DMPP in canine ileocolonic junction (BOECKXSTAENS et al. 1993a) and nicotine in cerebral artery (TODA et al. 1995b) are also unaffected by ω-conotoxin GVIA, although those elicited by EFS in these tissues are blocked effectively. The ability of ω-conotoxin MVIIC to augment the blockade of nitrergic transmission in the rabbit urethra produced by ω-conotoxin GVIA (ZYGMUNT et al. 1995) suggests an additional entry of Ca^{2+} through Q-type voltage-operated channels in this tissue. Whether a component of Ca^{2+} entry through Q-type VOCs explains the inability of ω-conotoxin GVIA to block completely nitrergic transmission at other sites remains to be determined.

III. Pre-Junctional Augmentation of Nitrergic Transmission

A universal feature of the nerve action potential is its termination by closure of voltage-operated Na^+ channels and opening of delayed rectifier K^+ channels. Accordingly, it is well established that blockade of the delayed rectifier prolongs action-potential duration, thus augmenting Ca^{2+} entry and consequent transmitter output (AUGUSTINE 1990). Nitrergic nerves also exhibit this property, since the non-selective K^+ channel blockers tetraethylammonium (TEA) and 4-aminopyridine and the large-conductance, Ca^{2+}-activated K^+-channel blocker charybdotoxin each augment NO release from nitrergic nerves in the canine ileocolonic junction (DE MAN et al. 1994). TEA has also been shown to enhance nitrergic relaxation of the rat anococcygeus muscle (GILLESPIE and TILMISANAY 1976).

Ginsenosides have been shown to augment nitrergic transmission in the rabbit corpus cavernosum (CHEN and LEE 1995). This occurs by a pre-junctional mechanism that is poorly understood but may underlie the aphrodisiac properties of these agents.

IV. Blockade of Nitrergic Transmission by Inhibition of NOS

The introduction of N^G-substituted analogues of L-arginine (L-NMMA, L-NNA and L-NAME) that act as competitive inhibitors of NOS and their importance in the development of the concept of nitrergic transmission has been discussed already (Sect. A.IV). Subsequent to these early studies, other structural analogues of L-arginine, including N^G-nitro-L-arginine-*p*-nitroanilide (MATTHEW and WADSWORTH 1997) and N^G-iminoethyl-L-ornithine (VIALS et al. 1997) have also been shown to block nitrergic transmission. The most striking feature of all these inhibitors is their rapidity of action, typically blocking nitrergic relaxation to EFS within a few minutes. This, coupled with the high diffusability of NO, which precludes its storage in secretory granules, indicates that, unlike all other known neurotransmitters, the nitrergic transmitter is not stored within the nerve in vesicles and released in quanta. Rather, it is synthesised and released on demand, most likely via graded production regulated by the intracellular levels of free Ca^{2+} achieved.

As already discussed, L-NMMA is a less potent and effective inhibitor of nitrergic transmission than the other inhibitors. Indeed, in certain tissues, including the BRP and bovine penile artery (LIU et al. 1991; MARTIN et al. 1993) and the rabbit anococcygeus (CELLEK and MONCADA 1997a), it fails to block nitrergic transmission. Surprisingly, in these particular tissues, L-NMMA behaves likes the endogenous substrate L-arginine and actually protects nitrergic transmission from blockade by L-NNA. In fact, L-NMMA is more potent in doing so than L-arginine, ruling out the possibility that protection stems from its metabolic conversion to L-arginine. It is more likely that L-NMMA is preferred to L-arginine as a substrate for the nNOS in these tissues.

Recovery from blockade of nitrergic transmission by NOS inhibitors has also been reported in the mouse anococcygeus following treatment with N^G-hydroxy-L-arginine (GIBSON et al. 1992), and this is in keeping with its formation as an intermediate in the biosynthesis of NO. L-Citrulline, the co-product of NO synthesis by NOS, has also been shown to reverse the blockade of nitrergic transmission produced by NOS inhibitors in the porcine cerebral artery (CHEN and LEE 1995; LEE et al. 1996), mouse anococcygeus (GIBSON et al. 1990) and colon (SHUTTLEWORTH et al. 1997) and opossum internal anal sphincter (RATTAN and CHAKDER 1997). These findings suggest the existence, in nitrergic nerves, of a metabolic pathway for the recycling of L-citrulline to L-arginine by arginosuccinate synthetase and arginosuccinate lyase. The first intermediate formed, arginosuccinate, is an inhibitor of nitrergic transmission in the rat anococcygeus muscle (RAND and LI 1992), but its actions are short lived, probably as a consequence of its rapid conversion to L-arginine in the tissue.

Treatment with L-arginine itself is reported to have different effects on nitrergic transmission in different tissues; in the mouse (GIBSON et al. 1990) and rabbit anococcygeus (KASAKOV et al. 1995), a small but significant enhancement was seen, whereas no effect was observed in the BRP and bovine penile artery (LIU et al. 1991). In the rat anococcygeus, both no effect (LIU et al. 1991) and a small enhancement (LI and RAND 1989) have been reported. It is likely, therefore, that in some tissues the availability of L-arginine for formation of NO is limiting. In this regard, it is interesting to note that humans fed large quantities of L-arginine exhibited augmented relaxation of the lower oesophageal sphincter and gall bladder (LIUKING et al. 1998), actions attributed to enhancement of nitrergic transmission.

The inhibitors discussed above block each of the isoforms of NOS, with no particular selectivity for nNOS. Others have been introduced, however, which are reported to be nNOS-specific. Of these, 7-nitroindazole (MOORE et al. 1993a, 1993b) and 1-(2-trifluoromethylphenyl)imidazole (TRIM) (HANDY et al. 1996) have proven effective in blocking nNOS within the CNS and spinal cord. Disappointingly, however, they fail to inhibit nitrergic transmission in the rat and mouse anococcygeus muscles (P.K. Moore, personal communication). Moreover, TRIM also failed to inhibit nitrergic relaxation of the rabbit corpus cavernosum despite powerful blockade by L-NAME (TEIXEIRA et al. 1998). An explanation for the lack of effect of these nNOS-specific inhibitors on nitrergic transmission is awaited.

It is not only structural analogues of L-arginine that have the ability to inhibit nNOS; in keeping with the requirement for FAD and FMN, the flavoprotein inhibitor diphenylene iodonium blocks nitrergic transmission in the rat anococcygeus (RAND and LI 1993a). Furthermore, KN62, an inhibitor of Ca^{2+}/calmodulin-dependent protein kinase II, blocks nitrergic transmission in the canine cerebral artery (TODA et al. 1997a). This last finding is somewhat surprising, given that phosphorylation of nNOS from rat brain by this protein kinase has been shown to inhibit its activity (NAKANE et al. 1991), but may reflect a species or site difference.

Targeted disruption of the nNOS gene represents an alternative, extremely powerful means of preventing NO release from nitrergic nerves (HUANG et al. 1993). These knockout mice have grossly enlarged stomachs with hypertrophy of the pyloric sphincter (HUANG et al. 1993), resembling the human condition infantile pyloric stenosis, perhaps suggesting a role for nNOS in the embryological development of this tissue. Inhibitory-junction potentials (IJPs) recorded in the circular muscle layer of the gastric fundus of these animals exhibit only a fast, transient, purinergic component and lack the slow nitrergic component seen in wild-type mice (MASHIMO et al. 1996). Surprisingly, despite the emerging roles played by nNOS in the central and peripheral nervous systems, these animals are viable and fertile and lack any major histopathological abnormalities in the CNS. Compensatory mechanisms involving converging or normally redundant pathways may, therefore, mask the true roles of nNOS in wild-type mice.

E. Nerve–Nerve Interactions

Since EFS activates all nerve fibres in isolated smooth-muscle preparations, the resultant response represents the net sum of the effects of the different transmitters released. Consequently, use of this technique has highlighted many interactions in which nitrergic transmission modulates or is modulated by other autonomic nerve fibres at both the pre- and post-junctional levels.

I. Nitrergic–Adrenergic Interactions

Blockade of nitrergic transmission by hypoxia or treatment with haemoglobin, L-NMMA, L-NNA, L-NAME, diphenylene iodonium, N-methyl-hydroxylamine or 1-H-[1,2,4]oxadiazolo-[4,3-a]quinoxalin-1-one (ODQ) results in marked augmentation of adrenergic contractions in the BRP (BOWMAN and MCGRATH 1985), anococcygeus of the rat (BRAVE et al. 1993a; RAND and LI 1993a; KASAKOV et al. 1994; MUDUMBI et al. 1996; HOYO et al. 1997), mouse (GIBSON et al. 1990) and rabbit (LUDUENA and GRIGAS 1966) and the rabbit renal artery (VIALS et al. 1997). Adrenergic contractions are thus inhibited by NO released simultaneously from nitrergic nerves. Indeed, in the human and rabbit corpus cavernosum, nitrergic activity dominates such that adrenergic contractions are seen only following cessation of EFS. Blockade of the synthesis or actions of NO, however, results in adrenergic contractions appearing immediately upon commencement of stimulation, and these are augmented in magnitude (CELLEK and MONCADA 1997b). Although some have proposed that NO may inhibit noradrenaline release by a pre-junctional mechanism (ADDICKS et al. 1994; MAEKAWA et al. 1996), the lack of effect of inhibitors of NOS on release of $[^3H]$-noradrenaline and on contractions to noradrenaline (BRAVE et al. 1993a; RAND and LI 1993a; KASAKOV et al. 1994; HOYO et al. 1997) suggest that the modulation results from post-junctional physiological antagonism.

Adrenergic modulation of nitrergic transmission has also been reported in many tissues. BHT-920, UK 14,304 and clonidine, which act as selective agonists at the α_2-adrenoceptor, inhibit nitrergic transmission in the guinea-pig colon (KOJIMA et al. 1988), canine ileocolonic junction (BOECKXSTAENS et al. 1993b), rat gastric fundus (LEFEBVRE and SMITS 1992) and horse penile resistance arteries (SIMONSEN et al. 1997) but not in the canine cerebral artery (TODA et al. 1995b). These inhibitory effects were reversed by the α_2-adrenoceptor antagonists yohimbine and rauwolscine. Responses to NO or to NO donors were, however, unaffected by the α_2-adrenoceptor agonists, suggesting a pre-junctional site of action. Thus, nitrergic nerves at many sites possess pre-junctional α_2-adrenoceptors that inhibit the output of NO.

II. Nitrergic–Cholinergic Interactions

Blockade of nitrergic transmission following treatment with L-NMMA, L-NNA, L-NAME and methylene blue enhances cholinergic excitatory-junction potentials in the guinea-pig gastric fundus (OHNO et al. 1996) and contractions in the guinea-pig trachea (BELVISI et al. 1993), rat bronchi (MISAWA and SATO 1997) and trachea (SEKIZAWA et al. 1993). In the opossum, lower oesophageal sphincter nitrergic transmission dominates such that cholinergic contractions are seen only upon cessation of EFS (CELLEK and MONCADA 1997c). Following blockade of NO synthesis, however, cholinergic contractions appear at the onset of EFS, and these are augmented in magnitude. Although a large component of these inhibitory actions is likely to have resulted from physiological antagonism between cholinergic contraction and nitrergic relaxation, an additional pre-junctional mechanism by which NO inhibits acetylcholine release, measured directly by high-performance liquid chromatography or as overflow of [^{3}H]-choline, has been uncovered in the rat trachea and bronchi (SEKIZAWA et al. 1993; MISAWA and SATO 1997). No evidence for pre-junctional control of acetylcholine release by nitrergic nerves was obtained, however, in experiments measuring [^{3}H]-choline overflow in guinea-pig ileum (WIKLUND et al. 1993) or taenia coli (WARD et al. 1996).

Conversely, treatment with acetylcholine has been shown to inhibit nitrergic relaxation in canine (TODA et al. 1995b) and monkey (TODA et al. 1997b) cerebral artery by an action at pre-junctional M_2 muscarinic receptors. The canine artery lacks cholinergic nerves, but in the monkey, where they are present, release of endogenous acetylcholine by EFS activates this inhibitory mechanism on nitrergic transmission.

III. Nitrergic–NANC Interactions

As previously discussed (Sect. B.IV), NO and VIP are often released as co-transmitters, usually with each augmenting the action of the other (GRIDER et al. 1992; DANIEL et al. 1994; MASHIMO et al. 1996). In the guinea-pig ileum (DANIEL et al. 1994) and stomach (GRIDER et al. 1992), reduced overflow of VIP is seen following inhibition of NOS with L-NAME and L-NNA, respectively, suggesting pre-junctional augmentation of release of the peptide by NO. The controversy regarding the proposed ability of VIP to stimulate NO release has already been addressed (Sect. C.I).

Following blockade of cholinergic contractions with atropine, EFS of guinea-pig (WIKLUND et al. 1993; YUNKER and GALLIGAN 1996) and canine ileum (FOXTHRELKELD et al. 1997) evokes an NANC contraction, probably involving release of substance P. These NANC contractions are enhanced following blockade of nitrergic transmission by L-NNA or haemoglobin. Since contractions to substance P or its methyl ester were unaffected, these authors concluded that the enhancement resulted from loss of a NO-mediated pre-junctional inhibition of substance-P release rather than a post-junctional physiological antagonism by NO. Such a conclusion must be viewed with caution, however, since contractions to substance P and its methyl ester were assessed in the absence of EFS, so any potential post-junctional physiological antagonism by nitrergic nerve-derived NO would not have been operative.

Met-enkephalin inhibits nitrergic relaxation, presumably by a prejunctional mechanism, in the guinea-pig ileum (IVANCHEVA and RADOMIROV 1996) and canine lower oesophageal sphincter (BARNETTE et al. 1990). Met-enkephalin and [D-Ala^2,D-Leu^5]enkephalin, a δ-opioid agonist, also inhibit nitrergic IJPs in the canine pyloric sphincter (BAYGUINOV and SANDERS 1993a). In this last study, use of the opioid antagonist naloxone provided evidence that endogenous opioids participate in this inhibitory mechanism during long periods of EFS but not during brief ones. Prejunctional inhibitory opiate receptors are not universally found on nitrergic nerves, however, since the δ-opioid agonist [D-Pen^2,D-Pen^5]enkephalin had no effect on the slow nitrergic component of the IJP in guinea-pig colon, although it did inhibit the fast purinergic component (ZAGORODNYUK and MAGGI 1994). Furthermore, morphine and naloxone had no effect on nitrergic relaxation of the canine cerebral artery (TODA et al. 1995b).

High concentrations of NO are known to inhibit NOS present in macrophages (ASSREUY et al. 1993), endothelium (RENGASAMAY and JOHNS 1998) and cerebellar neurones (ROGERS and IGNARRO 1992). The possibility that NO might regulate its own release from nitrergic nerves by an autocrine mechanism has also been suggested by the finding that prolonged treatment with the NO donors SIN-1 and glyceryl trinitrate inhibits relaxations to EFS but not to authentic NO in the rat gastric fundus (DE MAN et al. 1995). Similar experiments conducted on pig gastric fundus also resulted in a reduction in nitrergic relaxation, but here the reduction was attributed to postjunctional tolerance to NO and not to inhibition of NO release (LEFEBVRE and VANDEKERCKHOVE 1998).

F. Junctional and Post-Junctional Mechanisms

I. Scavengers of NO

The ability of haemoglobin, carboxy-PTIO, hydroxocobalamin and generators of superoxide anion to inhibit nitrergic transmission has already been addressed (Sect. C.). Ethanol and other aliphatic alcohols also inhibit nitrergic transmission in the BRP and rat and rabbit anococcygeus (GILLESPIE et al. 1982; RAND and LI 1994b), perhaps as a result of destruction of NO and the formation of the corresponding nitroso alcohol.

II. Blockade of Soluble Guanylate Cyclase

Many agents are available that inhibit soluble guanylate cyclase. The first to be introduced was N-methylhydroxylamine (DEGUCHI et al. 1978), and this blocks nitrergic relaxation and the associated rise in cyclic GMP in the BRP (BOWMAN and DRUMMOND 1984) and mouse (GIBSON and MIRZAZADEH 1989; GIBSON et al. 1992) and rat (MUDUMBI et al. 1996) anococcygeus. Its potency is low, however, being active only at millimolar concentrations. Methylene blue is more potent and blocks nitrergic transmission in many tissues, including the BRP (BOWMAN et al. 1986), human (PICKARD et al. 1991) and rabbit (IGNARRO et al. 1990) corpus cavernosum and canine colon (HUIZINGA et al. 1992). It is now known, however, to have additional actions, including an ability to inhibit nNOS (LUO et al. 1995) and generate superoxide anion (WOLIN et al. 1990; MARCZIN et al. 1992), so its use is in decline. LY 83583 was also introduced as a selective inhibitor of soluble guanylate cyclase (SCHMIDT et al. 1985; MÜLSCH et al. 1988), and its effects on nitrergic transmission have already been discussed. Like methylene blue, however, this agent inhibits nNOS (LUO et al. 1995) and generates superoxide anion (MÜLSCH et al. 1988; MARTIN et al. 1994), so it should be used with caution. The most recent inhibitor of soluble guanylate cyclase to be introduced is ODQ (GARTHWAITE et al. 1995), and it has rapidly become the drug of choice. It is highly selective, blocking relaxations in the rabbit anococcygeus produced by nitrergic nerves but not those due to atrial natriuretic peptide, which arise thorough activation of particulate guanylate cyclase (CELLEK et al. 1996). It is also the most potent inhibitor available, producing complete blockade of nitrergic transmission in many tissues in the low micromolar range. Other tissues in which its has proven effective in blocking nitrergic transmission include the guinea-pig colon (OLGART et al. 1998) and basilar artery (JIANG et al. 1998), canine gastric fundus (BAYGUINOV and SANDERS 1998) and proximal colon (FRANCK et al. 1997) and horse, rabbit and human corpus cavernosum (CELLEK and MONCADA 1997b; RECIO et al. 1998).

III. Post-Junctional Potentiation of Nitrergic Transmission

The actions of the intracellular messenger cyclic GMP are terminated upon hydrolysis by PDE types I and V (BEAVO et al. 1994). Inhibition of isoform V by zaprinast (M&B 22, 948) potentiates the relaxation, both in magnitude and duration, as well as the rise in cyclic-GMP content resulting from stimulation of the nitrergic nerves in the BRP (BOWMAN and DRUMMOND 1984) and mouse (GIBSON and MIRZAZADEH 1989) and rat (MIRZAZADEH et al. 1991) anococcygeus. The membrane hyperpolarisations resulting from activation of nitrergic nerves in the canine proximal colon (WARD et al. 1992) and pyloric sphincter (BAYGUINOV and SANDERS 1993b) are also potentiated. Although it was a vital early tool, the low potency and relatively poor isoform selectivity of zaprinast precluded its use in the clinic. In contrast, sildenafil (Viagra R), the most recently introduced inhibitor of PDE V, shows great clinical utility. This compound is some 100–240-fold more potent than zaprinast, inhibiting PDE V from human corpus cavernosum with an IC_{50} value in the nanomolar range (BALLARD et al. 1998; MORELAND et al. 1998). By itself, it produces relaxation and elevation of the cyclic-GMP content of isolated strips of rabbit and human corpus cavernosum (JEREMY et al. 1997; STIEF et al. 1998), and it potentiates the effects of nitrergic nerve stimulation in the rabbit corpus cavernosum (BALLARD et al. 1998). It also potentiates the increase in intracavernosal pressure resulting from stimulation of the pelvic nerve in the dog in vivo (CARTER et al. 1998). Unlike established treatments for male impotence, which require intracavernosal injection or transurethral delivery, sildenafil is active when taken orally. This, combined with its ability to produce erection in individuals with impotence resulting from a variety of psychological and organic causes (BOOLELL et al. 1996; GOLDSTEIN et al. 1998), makes it likely that sildenafil will revolutionise the treatment of this condition.

G. Post-Junctional Transduction Pathway
I. Role of Cyclic GMP

The ability of inhibitors of soluble guanylate cyclase and of PDE V to block or potentiate nitrergic relaxation, respectively, highlights a central role for cyclic GMP in signal transduction. Indeed, increased levels of this cyclic nucleotide are associated with nitrergic relaxation in the BRP (BOWMAN and DRUMMOND 1984), rat (MIRZAZADEH et al. 1991) and rabbit (CELLEK et al. 1996) anococcygeus, rabbit (ZYGMUNT et al. 1995) and pig (WERKSTRÖM et al. 1995) urethra, rabbit (IGNARRO et al. 1990; CHEN and LEE 1995) and human (PICKARD et al. 1991) corpus cavernosum and canine proximal colon (WARD et al. 1992). Furthermore, two membrane-permeant analogues of cyclic GMP, 8-bromo cyclic GMP and 8-p-chlorophenylthio-cyclic GMP, mimic the relaxation produced by nitrergic nerves in the BRP (BOWMAN and DRUMMOND 1984), mouse (GIBSON et al. 1994b) and rabbit (CELLEK et al. 1996) anococcygeus and guinea-pig basilar artery (JIANG et al. 1998). The membrane hyperpolarisation produced by stimulation of nitrergic nerves was also mimicked by these analogues in the canine (WARD et al. 1992; FRANCK et al. 1997) and guinea-pig (WATSON et al. 1996a) proximal colon, canine pyloric sphincter (BAYGUINOV and SANDERS 1993b) and rat gastric fundus (KITAMURA et al. 1993).

II. Inhibition of Calcium Mobilisation

The multiple mechanisms by which the NO/cyclic-GMP pathway inhibits calcium mobilisation in vascular smooth muscle through reduction of inositol 1,4,5-trisphosphate (IP_3) production, calcium entry through voltage-operated (VOC) and voltage-independent channels, calcium sensitivity of the contractile proteins and stimulation of calcium extrusion and intracellular sequestration have been reviewed (LINCOLN 1989; LINCOLN and CORNWELL 1993). It is likely that these mechanisms also underlie relaxation in blood vessels receiving a nitrergic innervation. In non-vascular smooth muscle, including the rat anococcygeus (RAMAGOPAL and LEIGHTON 1989; RAYMOND and WENDT 1996) and canine gastric fundus (BAYGUINOV and SANDERS 1998), relaxation produced by nitrergic nerves or NO donors is also associated with falls in the intracellular level of calcium. As is often the case in vascular smooth muscle, these falls do not completely account for the relaxations seen, suggesting a reduction in the sensitivity of the contractile proteins to calcium. Unlike the situation in vascular smooth muscle, however, nitrergic relaxation of the rat anococcygeus does not inhibit production of IP_3 (GIBSON et al. 1994a). It does, however, inhibit the capacitative entry of calcium into cells of the mouse anococcygeus resulting from the IP_3-stimulated discharge of the calcium store in the sarcoplasmic reticulum (WAYMAN et al. 1996, 1998). Capacitative entry of calcium rather than entry via L-type VOCs sustains active tone in this tissue.

III. Role of Membrane Hyperpolarisation

It is well established that membrane hyperpolarisation blocks calcium entry through L-type VOCs in smooth muscle (CLAPP and GURNEY 1991). The occurrence of membrane hyperpolarisation following nitrergic nerve stimulation does, however, vary considerably between tissues. Powerful hyperpolarisation is seen in the BRP (BYRNE and MUIR 1984), rabbit anococcygeus (CREED and GILLESPIE 1977), rat gastric fundus (KITAMURA et al. 1993), opossum oesophagus (CAYABYAB and DANIEL 1995) and guinea-pig (WATSON et al. 1996a) and dog (WARD et al. 1992) colon, but little or none is seen in the rat anococcygeus (CREED and GILLESPIE 1977) and rabbit urethra (WALDECK et al. 1998). It is, therefore, likely that the magnitude of the hyperpolarisations observed reflects the contribution of closure of L-type VOCs to nitrergic relaxation in the different tissues. Nevertheless, hyperpolarisation was shown to play no role in the lowering of intracellular calcium produced by nitrergic nerve stimulation in the canine gastric fundus (BAYGUINOV and SANDERS 1998).

The hyperpolarisations produced are believed to result from the opening of a number of distinct potassium channels by NO (KOH et al. 1995). Although there are suggestions that NO can directly open potassium channels in smooth muscle (BOLOTINA et al. 1994; KOH et al. 1995), the ability of ODQ to abolish nitrergic nerve-induced hyperpolarisation (FRANCK et al. 1997) strongly suggests a role for cyclic GMP in the opening of these channels. The effects of drugs that block different potassium channels have been examined on nitrergic relaxation or hyperpolarisation in order to investigate the nature of the channels involved. The non-selective blockers of Ca^{2+}-activated K^+ channels tetraethylammonium and 4-aminopyridine have no effect on the opossum oesophagus (CAYABYAB and DANIEL 1995) or guinea-pig colon (WATSON et al. 1996b) but produce a partial block in guinea-pig basilar artery (JIANG et al. 1998). A role for small-conductance Ca^{2+}-activated K^+ channels can be excluded following the lack of effect of apamin in the BRP (BYRNE and MUIR 1984), horse corpus cavernosum (RECIO et al. 1998), pig ureter (HERNANDEZ et al. 1997) and guinea-pig basilar artery (JIANG et al. 1998). An apamin-sensitive component observed in response to EFS in the opossum oesophagus (CAYABYAB and DANIEL 1995), guinea-pig colon (WATSON et al. 1996b) and rat gastric fundus (KITAMURA et al. 1993) is likely to result from release of an additional transmitter along with NO, probably ATP. The ability of glibenclamide to block nitrergic relaxation in the pig ureter (HERNANDEZ et al. 1997) suggests a role for ATP-sensitive K$^+$ channels in this tissue, but this agent has no effect in the opossum oesophagus (CAYABYAB and DANIEL 1995), horse corpus cavernosum (RECIO et al. 1998) or guinea-pig basilar artery (JIANG et al. 1998). On the basis of the actions of charybdotoxin and iberiotoxin, large-conductance Ca^{2+}-activated K$^+$ channels appear to contribute to nitrergic responses in the guinea-pig basilar artery (JIANG et al. 1998) but not in the opossum oesophagus (CAYABYAB and DANIEL 1995), pig ureter (HERNANDEZ et al. 1997) or horse corpus cavernosum (RECIO et al. 1998).

H. Concluding Remarks

It is now widely accepted that the L-arginine–NO system can account for the NANC transmission process at many anatomical sites. The neurotransmitter released from these nitrergic nerves is almost certainly NO per se, although some still dispute this. The transmission process differs in many respects from that of all other established nerves; the transmitter is not stored in vesicles or released in quanta but is synthesised on demand. Moreover, release of NO is not confined to the varicose regions but occurs throughout the entire length of the nerve fibre. Accordingly, certain of the criteria laid down by ECCLES (1964) for acceptance of a substance as a neurotransmitter are not satisfied by NO. These criteria must, therefore, be revised.

References

Addicks K, Bloch W, Feelisch M (1994) Nitric oxide modulates sympathetic neurotransmission at the prejunctional level. Microscopy Res Technique 29:161–168

Akaike T, Yoshida M, Miyamoto Y, et al. (1993) Antagonistic action of imidazolineoxyl N-oxides against endothelium-derived relaxing factor/NO through a radical reaction. Biochemistry 32:827–832

Ambache N, Killick SW, Zar MA (1975) Extraction from ox retractor penis of an inhibitory substance which mimics its atropine-resistant neurogenic relaxation. Br J Pharmacol 54:409–410

Anderson CR (1992) NADPH diaphorase-positive neurons in the rat spinal cord include a subpopulation of autonomic preganglionic neurons. Neurosci Lett 139:280–284

Andersson K-E (1993) Pharmacology of lower urinary tract smooth muscles and penile erectile tissues. Pharmacol Rev 45:253–308

Andersson K-E, Persson K (1993) The L-arginine/nitric oxide pathway and non-adrenergic, non-cholinergic relaxation of the lower urinary tract. Gen Pharmacol 24:833–839

Andersson K-E, Wagner G (1995) Physiology of penile erection. Physiol Rev 75:191–236

Askew SC, Barnett DJ, McAninly J, Williams DHL (1995) Catalysis by Cu^{2+} of nitric oxide release from S-nitrosothiols (RSNO). J Chem Soc Perkin Trans 2:741–745

Assreuy J, Cunha FQ, Liew FY, Moncada S (1993) Feedback inhibition of nitric oxide synthase by nitric oxide. Br J Pharmacol 108:833–837

Augustine GJ (1990) Regulation of transmitter release at the giant squid synapse by presynaptic delayed rectifier potassium current. J Physiol 431:343–364

Augustine GJ, Charlton MP, Smith SJ (1987) Calcium action in synaptic transmitter release. Ann Rev Neurosci 10:633–693

Ballard SA, Gingell CJ, Tang K, Turner LA, Price ME, Naylor AM (1998) Effects of sildenafil on the relaxation of human corpus cavernosum tissue in vitro and on activities of cyclic nucleotide phosphodiesterase isoenzymes. J Urol 159:2164–2171

Barbier AJM, Lefebvre RA (1992) Effect of LY 83583 on relaxation induced by non-adrenergic, non-cholinergic nerve stimulation and exogenous nitric oxide in the rat gastric fundus. Eur J Pharmacol 219:331–334

Barbier AJM, Lefebvre RA (1993) Involvement of the L-arginine: nitric oxide pathway in nonadrenergic noncholinergic relaxation of the cat gastric fundus. J Pharmacol Exp Ther 266:172–178

Barnes PJ, Belvisi MG (1993) Nitric oxide and lung disease. Thorax 48:1034–1043

Barnette MS, Grous M, Manning CD, Callahan JF, Barone FC (1990) Inhibition of neuronally induced relaxation of canine esophageal sphincter by opioid peptides. Eur J Pharmacol 182:363–368

Bayguinov O, Sanders KM (1993a) Regulation of neural responses in the canine pyloric sphincter by opioids. Br J Pharmacol 108:1024–1030

Bayguinov O, Sanders KM (1993b) Role of nitric oxide as an inhibitory neurotransmitter in the canine pyloric sphincter. Am J Physiol 246:G975-G983

Bayguinov O, Sanders KM (1998) Dissociation between electrical and mechanical responses to nitrergic stimulation in the canine gastric fundus. J Physiol 509:437–448

Beavo JA, Conti M, Heaslip RJ (1994) Multiple cyclic nucleotide phosphodiesterases. Mol Pharmacol 46:399–405

Belvisi MG, Miura M, Stretton D, Barnes PJ (1993) Endogenous vasoactive intestinal peptide and nitric oxide modulate cholinergic neurotransmission in guinea-pig trachea. Eur J Pharmacol 231:97–102

Belvisi MG, Ward JK, Mitchell JA, Barnes PJ (1995) Nitric oxide as a neurotransmitter in human airways. Arch Int Pharmacodyn Therap 329:97–110

Boeckxstaens GE, Pelckmans PA, Rampart M, et al. (1990) $GABA_A$ receptor-mediated stimulation of non-adrenergic non-cholinergic neurones in the dog ileocolonic junction. Br J Pharmacol 101:460–464

Boeckxstaens GE, De Man JG, Pelckmans PA, Cromheeke KM, Herman AG, Van Maercke YM (1993a) Ca^{2+} dependency of the release of nitric oxide from non-adrenergic non-cholinergic nerves. Br J Pharmacol 110:1329–1334

Boeckxstaens GE, De Man JG, Pelckmans PA, Herman AG, Van Maercke YM (1993b) α_2-Adrenoceptor-mediated modulation of the nitrergic innervation of the canine isolated ileocolonic junction. Br J Pharmacol 109:1079–1084

Bolotina VM, Najibi S, Palacino PJ, Cohen RA (1994) Nitric oxide directly activates calcium-dependent potassium channels in vascular smooth muscle. Nature 368:850–853

Boolell M, Gepiattee S, Gingell JC, Allen MJ (1996) Sildenafil, a novel effective oral therapy for male erectile dysfunction. Br J Urol 78:257–261

Borjesson L, Nordgren S, Delbro DS (1997) DMPP causes relaxation of rat distal colon by a purinergic and nitrergic mechanism. Eur J Pharmacol 334:223–231

Bowman A, Drummond AH (1984) Cyclic GMP mediates neurogenic relaxation in the bovine retractor penis muscle. Br J Pharmacol 81:665–674

Bowman A, Gillespie JS (1982) Block of some non-adrenergic inhibitory responses of smooth muscle by a substance from haemolysed erythrocytes. J Physiol 328:11–25

Bowman A, McGrath JC (1985) The effect of hypoxia on neuroeffector transmission in the bovine retractor penis and rat anococcygeus muscles. Br J Pharmacol 85:869–875

Bowman A, Gillespie JS, Martin W (1979) The inhibitory material in extracts from the bovine retractor penis muscle is not an adenine nucleotide. Br J Pharmacol 67:327–328

Bowman A, Gillespie JS, Martin W (1981) Actions on the cardiovascular system of an inhibitory material extracted from the bovine retractor penis. Br J Pharmacol 72:365–372

Bowman A, Gillespie JS, Pollock D (1982) Oxyhaemoglobin blocks non-adrenergic inhibition in the bovine retractor penis muscle. Eur J Pharmacol 85:221–224

Bowman A, Gillespie JS, Soares-De-Silva P (1986) A comparison of the actions of endothelium-derived relaxant factor and the inhibitory factor from the bovine retractor penis on rabbit aortic smooth muscle. Br J Pharmacol 87:175–181

Brave SR, Bhat S, Hobbs AJ, Tucker JF, Gibson A (1993a) The influence of L-N^G-nitro-arginine on sympathetic nerve-induced contractions and noradrenaline release in the rat isolated anococcygeus muscle. J Autonomic Pharmacol 13:219–225

Brave SR, Tucker JF, Gibson A, et al. (1993b) Localisation of nitric oxide synthase within nonadrenergic, noncholinergic nerves in the mouse anococcygeus. Neuroscience Lett 161:93–96

Bredt DS, Snyder SH (1990) Isolation of nitric oxide synthetase, a calcium-requiring enzyme. Proc Natl Acad Sci USA 87:682–685

Bredt DS, Hwang PM, Snyder SH (1990) Localisation of nitric oxide synthase indicating a neural role for nitric oxide. Nature 347:768–770

Brouwer M, Chamulitrat W, Ferruzzi G, Sauls DL, Weinberg JB (1996) Nitric oxide interactions with cobalamins: biochemical and functional consequences. Blood 88:1857–1864

Bult H, Boeckxstaens GE, Pelckmans PA, Jordaens FH, Van Maercke YM, Herman AG (1990) Nitric oxide as an inhibitory non-adrenergic, non-cholinergic neurotransmitter. Nature 345:346–347

Burnstock G (1972) Purinergic nerves. Pharmacol Rev 24:509–581

Burnstock G, Campbell G, Bennett M, Holman ME (1963a) Inhibition of the smooth muscle of the taenia coli. Nature 200:581–582

Burnstock G, Campbell G, Bennett M, Holman ME (1963b) The effects of drugs on transmission of inhibition from autonomic nerves to the smooth muscle of the guinea pig taenia coli. Biochem Pharmacol 12:134

Byrne NG, Muir TC (1984) Electrical and mechanical responses of the bovine retractor penis to nerve stimulation and to drugs. J Autonom Pharmacol 4:261–271

Carter AJ, Ballard SA, Naylor AM (1998) Effect of the selective phosphodiesterase type 5 inhibitor sildenafil on erectile function in the anaesthetised dog. J Urol 160:242–246

Cayabyab FS, Daniel EE (1995) K^+ channel opening mediates hyperpolarizations by nitric oxide donors and IJPs in opossum oesophagus. Am J Physiol 31:G831-G842

Cellek S, Moncada S (1997a) Modulation of noradrenergic responses by nitric oxide from inducible nitric oxide synthase. Nitric Oxide Biology and Chemistry 1:204–210

Cellek S, Moncada S (1997b) Nitrergic control of peripheral sympathetic responses in the human corpus cavernosum: a comparison with other species. Proc Natl Acad Sci U S A 15:8226–8231

Cellek S, Moncada S (1997c) Nitrergic modulation of cholinergic responses in the opossum lower oesophageal sphincter. Br J Pharmacol 122:1043–1046

Cellek S, Kasakov L, Moncada S (1996) Inhibition of nitrergic relaxations by a selective inhibitor of the soluble guanylate cyclase. Br J Pharmacol 118:137–140

Chakder S, Rattan S (1993) Release of nitric oxide by activation of nonadrenergic noncholinergic neurons of internal anal sphincter. Am J Physiol 264:G7-G12

Chen F, Lee TJ-F (1995a) Arginine synthesis from citrulline in perivascular nerves of cerebral artery. J Pharmacol Exp Ther 273:895–901

Chen X, Lee TJ-F (1995b) Ginsenoside-induced nitric oxide-mediated relaxation of the rabbit corpus cavernosum. Br J Pharmacol 115:15–18

Clapp LH, Gurney AM (1991) Modulation of calcium movements by nitroprusside in isolated vascular smooth muscle cells. Pfluger's Archiv 418:462–470

Cocco D, Calabrese L, Rigo A, Argese E, Rotilo G (1981) Re-examination of the reaction of diethyldithiocarbamate with the copper of superoxide dismutase. J Biol Chem 256:8983–8986

Crapo JD, Oury TD, Rabouillec C (1992) Copper zinc superoxide dismutase is primarily a cytosolic protein in human cells. Proc Natl Acad Sci U S A 89:10405–10409

Creed KE, Gillespie JS (1977) Some electrical properties of the rabbit anococcygeus muscle and a comparison of the effects of inhibitory nerve stimulation in the rat and rabbit. J Physiol 273:137–153

Dail WG, Galloway B, Bordegaray J (1993) NADPH diaphorase innervation of the rat anococcygeus and retractor penis muscles. Neuroscience Lett 160:17–20

Dale HH (1933) Nomenclature of fibres in the autonomic system and their effects. J Physiol 80:10–11

Dale HH, Gaddum JH (1930) Reactions of denervated voluntary muscle and their bearing on the mode of action of parasympathetic and related nerves. J Physiol 70:109–144

Daniel EE, Haugh C, Woskowska Z, Cipris S, Jury J, Foxthrelkeld JET (1994) Role of nitric oxide-related inhibition in intestine function – relation to vasoactive intestinal polypeptide. Am J Physiol 266:G31-G39

De Luca A, Li CG, Rand MJ, Reid JJ, Thaina P, Wong-Dusting HK (1990) Effects of w-conotoxin GVIA on autonomic neuroeffector transmission in various tissues. Br J Pharmacol 101:437–447

De Man JG, Boeckxstaens GE, Herman AG, Pelckmans PA (1994) Effect of potassium channel blockade and α_2-adrenoceptor activation on the release of nitric oxide from non-adrenergic non-cholinergic nerves. Br J Pharmacol 112:341–345

De Man JG, Boeckxstaens GE, De Winter BY, Moreels TG, Herman AG, Pelckmans PA (1995) Inhibition of nonadrenergic noncholinergic relaxations by nitric oxide donors. Eur J Pharmacol 285:269–274

De Man JG, De Winter BY, Boeckxstaens GE, Herman AG, Pelckmans PA (1996a) Effect of Cu^{2+} on relaxations to the nitrergic neurotransmitter, NO and S-nitrosothiols in the rat gastric fundus. Br J Pharmacol 119:990–996

De Man JG, De Winter BY, Boeckxstaens GE, Herman AG, Pelckmans PA (1996b) Effects of thiol modulators and Cu/Zn superoxide dismutase inhibition on nitrergic relaxations in the rat gastric fundus. Br J Pharmacol 119:1022–1028

De Man JG, De Winter BY, Moreels TG, Herman AG, Pelckmans PA (1998) S-nitrosothiols and the nitrergic neurotransmitter in the rat gastric fundus: effect of antioxidants and metal chelation. Br J Pharmacol 123:1039–1046

Deguchi T, Saito M, Kono M (1978) Blockade by N-methylhydroxylamine of activation of guanylate cyclase and elevations of guanosine 3′,5′-monophosphate levels in nervous tissues. Biochim Biophys Acta 544:8–19

Desai KM, Sessa WC, Vane JR (1991) Involvement of nitric oxide in the reflex relaxation of the stomach to accommodate food or fluid. Nature 351:477–479

Dicks AP, Swift HR, Williams DHL, Butler AR, Al-Sa'doni HH, Cox BG (1996) Identification of Cu^+ as the effective agent in nitric oxide formation from S-nitrosothiols (RSNO). J Chem Soc Perkin Trans 3:481–487

Dun NL, Dun SL, Förstermann U, Tseng LF (1992) Nitric oxide synthase immunoreactivity in the rat spinal cord. Neurosci Lett 147:217–220

Eccles JC (1964) The physiology of synapses. Springer, Berlin

Fahrenkrug J: (1991) Vasoactive intestinal peptide (VIP) and automomic neurotransmission. In: Bell C (ed) Novel peripheral neurotransmitters. Pergamon, New York, pp 113–134

Fang S, Christensen J (1995) Manganese superoxide dismutase and reduced nicotinamide adenine dinucleotide diaphorase colocalize in the rat gut. Gastroenterology 109:1429–1436

Feelisch M, Noack EA (1987) Correlation between nitric oxide formation during degradation of organic nitrates and activation of guanylate cyclase. Eur J Pharmacol 139:19–30

Fonseca M, Uddin N, Gibson A (1998) No evidence for a significant non-nitrergic, hyperpolarising factor contribution to field stimulation-induced relaxation of the muose anococcygeus. Br J Pharmacol 124:524–528

Foxthrelkeld JET, Woskowska Z, Daniel EE (1997) Sites of nitric oxide (NO) actions in control of circular muscle motility of the perfused isolated canine ileum. Can J Physiol Pharmacol 75:1340–1349

Franck H, Sweeney KM, Sanders KM, Shuttleworth CWR (1997) Effects of a novel guanylate cyclase inhibitor on nitric oxide-dependent inhibitory transmission in canine proximal colon. Br J Pharmacol 122:1223–1229

Funk RHW, Mayer B, Worl J (1994) Nitrergic innervation and nitrergic cells in arteriovenous anastomoses. Cell Tiss Res 277:477–484

Furchgott RF, Zawadzki JV (1980) The obligatory role of endothelial cells in the relaxation of arterial smooth muscle by acetylcholine. Nature 288:373–376

Furness JB, Bornstein JC, Murpht R, Pompolo S (1992a) Roles of peptides in transmission in the enteric nervous system. Trends Neurosci 15:66–71

Furness JB, Pompolo S, Shuttleworth CWR, Burleigh DE (1992b) Light- and electron-microscopic immunohistochemical analysis of nerve fibre types innervating the taenia coli of the gunea pig caecum. Cell Tissue Res 270:125–137

Garthwaite J, Southam E, Boulton CL, Nielsen EB, Schmidt K, Mayer B (1995) Potent and selective inhibition of nitric oxide-sensitive guanylyl cyclase by 1H-[1,2,4]oxadiazolo[4,3-a]quinoxalin-1-one. Mol Pharmacol 48:184–188

Gibson A, James TA (1977) The nature of potassium chloride-induced relaxations of the rat anococcygeus muscle. Br J Pharmacol 60:141–145

Gibson A, Lilley E (1997) Superoxide anions, free radical scavengers, and nitrergic neurotransmission. Gen Pharmacol 28:489–493

Gibson A, Mirzazadeh S (1989) N-methyl-hydroxylamine inhibits and MandB 22948 potentiates relaxations of the mouse anococcygeus muscle to non-adrenergic, non-cholinergic field stimulation and nitrovasodilator drugs. Br J Pharmacol 96:637–644

Gibson A, Wedmore CV (1981) Responses of the isolated anococcygeus muscle of the mouse to drugs and to field stimulation. J Autonomic Nervous System 1:225–233

Gibson A, Mirzazadeh S, Hobbs AJ, Moore PK (1990) L-N^G-monomethyl arginine and L N^G nitro arginine inhibit non-adrenergic, non-cholinergic relaxation of the mouse anococcygeus muscle. Br J Pharmacol 99:602–606

Gibson A, Babbedge RC, Brave SR, et al. (1992) An investigation of some S-nitrosothiols, and of hydroxy-arginine, on the mouse anococcygeus. Br J Pharmacol 107:715–721

Gibson A, Brave SR, McFadzean I, Mirzazadeh S, Tucker JF, Wayman C (1994a) Nitrergic stimulation does not inhibit carbachol-induced inositol phosphate generation in the rat anococcygeus. Neurosci Lett 178:35–38

Gibson A, McFadzean I, Tucker JF, Wayman C (1994b) Variable potency of nitrergic-nitrovasodilator relaxations of the mouse anococcygeus against different forms of induced tone. Br J Pharmacol 113:1494–1500

Gibson A, Brave SR, McFadzean I, Tucker JF, Wayman C (1995) The nitrergic transmitter of the anococcygeus – NO or not? Arch Int Pharmacodyn Therap 329:39–51

Gibson QH, Roughton FJW (1957) The kinetics and equilibria of the reactions of nitric oxide with sheep haemoglobin. J Physiol 136:507–526

Gillespie JS (1972) The rat anococcygeus muscle and its response to nerve stimulation and to some drugs. Br J Pharmacol 45:404–416

Gillespie JS, Martin W (1980) A smooth muscle inhibitory material from the bovine retractor penis and rat anococcygeus muscles. J Physiol 309:55–64

Gillespie JS, Sheng H (1990) The effects of pyrogallol and hydroquinone on the response to NANC nerve stimulation in the rat anococcygeus and bovine retractor penis muscles. Br J Pharmacol 99:194–196

Gillespie JS, Tilmisanay AK (1976) The action of tetraethylammonium chloride on the reponse of the rat anococcygeus muscle to motor and inhibitory nerve stimulation and to some drugs. Br J Pharmacol 58:47–55

Gillespie JS, Hunter JC, Martin W (1981) Some chemical and physical properties of the smooth muscle inhibitory factor in extracts of the bovine retractor penis muscle. J Physiol 315:111–125

Gillespie JS, Hunter JC, McKnight AT (1982) The effects of ethanol on inhibitory and motor responses in the rat and rabbit anococcygeus and the bovine retractor penis muscles. Br J Pharmacol 75:189–198

Gillespie JS, Liu X, Martin W (1989) The effects of L-arginine and N^G-monomethyl L-arginine on the response of the rat anococcygeus to NANC nerve stimulation. Br J Pharmacol 98:1080–1082

Gillespie JS, Liu X, Martin W (1990) The neurotransmitter of the non-adrenergic, non-cholinergic inhibitory nerves to smooth muscle of the genital system. In: Moncada S, Higgs EA (eds) Nitric oxide from L-arginine: a bioregulatory system. Elsevier, Amsterdam, pp 147–164

Gocmen C, Ucar P, Singirik E, Dikmen A, Baysal F (1997) An in vitro study of nonadrenergic-noncholinergic activity on the cavernous tissue of mouse. Urological Res 25:269–275

Goldstein I, Lue TF, PadmaNathan H, Rosen RC, Steers WD (1998) Oral sildenafil in the treatment of erectile dysfunction. New Eng J Med 338:1397–1404

Gong JP, Gwee MCE, Gopalakrishnakone P, Kini RM, Chung MCM (1997) Adrenergic and nitrergic responses of the rat isolated anococcygeus muscle to a new toxin (makatoxin I) from the venom of the scorpion *Buthus martensi Karsch*. J Autonom Pharmacol 17:129–135

Gonzalez C, Barroso C, Martin C, Gulbenkian S, Estrada C (1997) Neuronal nitric oxide synthase activation by vasoactive intestinal peptide in bovine cerebral arteries. J Cereb Blood Flow Metab 17:977–984

Gotoh N, Niki E (1992) Rates of interactions of superoxide with vitamin E, vitamin C, and related compounds measured by chemiluminescence. Biochim Biophys Acta 1115:201–207

Grider JR, Murthy KS, Jin J-G, Makhlouf GM (1992) Stimulation of nitric oxide from muscle cells by VIP: prejunctional enhancement of VIP release. Am J Physiol 262:G774–G778

Guo RS, Nada O, Suita S, Taguchi T, Masumoto K (1997) The distribution and co-localisation of nitric oxide synthase and vasoactive intestinal polypeptide in nerves of the colons with Hirschsprung's disease. Virchows Archiv Int J Pathol 430:53–61

Gwee MCE, Cheah LS, Gopalakrishnakone P, Wong PTH, Gong JP, Kini RM (1996) Studies on venoms from the black scorpion *Heterometrus longimanus* and some other scorpion species. J Toxicol Toxin Rev 15:37–57

Handy RLC, Harb HL, Wallace P, Gaffen Z, Whitehead KJ, Moore PK (1996) Inhibition of nitric oxide synthase by 1-(2-trifluoromethylphenyl) imidazole (TRIM) in vitro: antinociceptive and cardiovascular effects. Br J Pharmacol 119:423–431

Harrison JS, McSwinney BA (1936) The chemical transmitter of motor impulses to the stomach. J Physiol 87:79–86

Hernandez M, Prieto D, Orensanz FM, et al. (1997) Involvement of a glibenclamide-sensitive mechanism in the nitrergic neurotransmission of the pig intravesical ureter. Br J Pharmacol 120:609–616

Hobbs AJ, Gibson A (1990) L-N^G-nitro-L-arginine and its methyl ester are potent inhibitors of non-adrenergic, non-cholinergic transmission in the rat anococcygeus. Br J Pharmacol 100:749–752

Hobbs AJ, Tucker JF, Gibson A (1991) Differentiation by hydroquinone of relaxations induced by exogenous and endogenous nitrates in non-vascular smooth muscle. Br J Pharmacol 104:645–650

Hoyo Y, Giraldo J, Vila E (1997) Effects of L-N^G-nitro-arginine on noradrenaline induced contraction in the rat anococcygeus muscle. Br J Pharmacol 120: 1035–1038

Huang PL, Dawson TM, Bredt DS, Snyder SH, Fishman MC (1993) Targeted disruption of the neuronal nitric oxide synthase gene. Cell 75:1273–1286

Huizinga JD, Tomlinson J, Pinto-Quezada J (1992) Involvement of nitric oxide in nerve-mediated inhibition and action of vasoactive intestinal peptide in colonic smooth muscle. J Pharmacol Exp Ther 260:803–808

Ignarro LJ, Bush PA, Buga GM, Wood KS, Fukuto JM, Rafer J (1990) Nitric oxide and cyclic GMP formation upon electrical field stimulation cause relaxation of corpus cavernosum smooth muscle. Biochem Biophys Res Commun 170:843–850

Ivancheva C, Radomirov R (1996) Met-enkephalin-dependent nitrergically mediated relaxation in the guinea-pig ileum. Methods and Findings in Experimental and Clinical Pharmacology 18:521–525

Jang CS (1940) The potentiation and paralysis of adrenergic effects by ergotoxine and other substances. J Pharmacol Exp Ther 71:87–94

Jen PHY, Dixon JS, Gearhart JP, Gosling JA (1996) Nitric oxide synthase and tyrosine hydroxylase are colocalised in nerves supplying the postnatal human male genitourinary organs. J Urol 155:1117–1121

Jenkinson KM, Reid JJ, Rand MJ (1995) Hydroxocobalamin and hemoglobin differentiate between exogenous and neuronal nitric oxide in the rat gastric fundus. Eur J Pharmacol 275:145–152

Jeremy JY, Ballard SA, Naylor AM, Miller MAW, Angelini GD (1997) Effects of sildenafil, a type 5 cGMP phosphodiesterase inhibitor, and papaverine on cyclic GMP and cyclic GMP levels in the rabbit corpus cavernosum in vitro. Br J Urol 79: 958–963

Jiang F, Li CG, Rand MJ (1998) Role of potassium channels in the nitrergic nerve stimulation-induced vasodilatation in the guinea-pig isolated basilar artery. Br J Pharmacol 123:106–102

Kaczka EA, Wolf DE, Kuehl FAJ, Folkers K (1951) Vitamin B_{12}. Modifications of cyano-cobalamin. J Am Chem Soc 73:3569–3572

Kasakov L, Belai A, Vlaskovska M, Burnstock G (1994) Noradrenergic-nitrergic interactions in the rat anococcygeus muscle – evidence for post-junctional modulation by nitric oxide. Br J Pharmacol 112:403–410

Kasakov L, Cellek S, Moncada S (1995) Characterisation of nitrergic neurotransmission during short-term and long-term electrical stimulation of the rabbit anococcygeus muscle. Br J Pharmacol 115:1149–1154

Keef KD, Shuttleworth CWR, Xue C, Bayguinov O, Publicover NG, Sanders KM (1994) Relationship between nitric oxide and vasoactive intestinal polypeptide in enteric inhibitory neurotransmission. Neuropharmacol 33:1303–1314

Kelm M, Schrader J (1990) Control of coronary vascular tone by nitric oxide. Circ Res 66:1561–1575

Kelner MJ, Bagnell R, Hale B, Alexander NM (1989) Inactivation of intracellular copper–zinc superoxide dismutase by copper chelating agents without glutathione depletion and met-hemoglobin formation. Free Radical Biol Med 6:355–360

Kerr SW, Buchanan LV, Bunting S, Mathews WR (1993) Evidence that S-nitrosothiols are responsible for the smooth muscle relaxant activity of the bovine retractor penis inhibitory factor. J Pharmacol Exp Ther 263:285–292

Kitamura K, Lian Q, Carl A, Kuriyama H (1993) S-nitrosocysteine, but not sodium nitroprusside, produces apamin-sensitive hyperpolarisation in rat gastric fundus. Br J Pharmacol 109:415–425

Klinge E, Sjostrand NO (1974) Contraction and relaxation of the retractor penis muscle and penile artery of the bull. Acta Physiol Scand Suppl 420:1–88

Knowles RG, Palacios M, Palmer RMJ, Moncada S (1989) Formation of nitric oxide from L-arginine in the central nervous system: A transduction system for the stimulation of soluble guanylate cyclase. Proc Natl Acad Sci USA 86:5159–5162

Knudsen MA, Svane D, Totrup A (1992) Action profiles of nitric oxide, S-nitroso-cysteine, SNP, and NANC responses in opossum lower esophageal sphincter. Am J Physiol 262:G840-G846

Kobzic L, Reid MB, Bredt DS, Stamler JS (1994) Nitric oxide in skeletal muscle. Nature 372:546–548

Koh SD, Campbell JD, Carl A, Sanders KM (1995) Nitric oxide activates multiple potassium channels in canine colonic smooth muscle. J Physiol 489:735–743

Kojima S, Sakato M, Shimo Y (1988) An α_2-adrenoceptor-mediated inhibition of non-adrenergic non-cholinergic responses of the isolated proximal colon of the guinea-pig. Asia Pac J Pharmacol 3:69–75

La M, Li CG, Rand MJ (1996) Comparison of the effects of hydroxocobalamin and oxyhaemoglobin on responses to NO, EDRF and the nitrergic transmitter. Br J Pharmacol 117:805–810

La M, Paisley K, Martin W, Rand MJ (1997) Effects of hydroxocobalamin on nitrergic transmission in rat anococcygeus and bovine retractor penis muscles: sensitivity to light. Eur J Pharmacol 321:R5-R6

Langley JN (1898) On the inhibitory fibres in the vagus to the end of the oesophagus and stomach. J Physiol 23:407–414

Langley JN, Anderson HK (1895) The innervation of the pelvic and adjoining viscera Part III. J Physiol 19:85–121

Lee TJ-F, Sarwinski S, Ishine T, Lai CC, Chen FY (1996) Inhibition of cerebral neurogenic vasodilatation by L-glutamine and nitric oxide synthase inhibitors and its reversal by L-citrulline. J Pharmacol Exp Ther 276:353–358

Lefebvre RA (1996) Influence of superoxide dismutase inhibition on the discrimination between NO and the nitrergic neurotransmitter in the rat gastric fundus. Br J Pharmacol 118:2171–2177

Lefebvre RA (1997) Nitric oxide as a non-adrenergic non-cholinergic neurotransmitter in gastrointestinal motility. Periodicum Biologorum 99:455–459

Lefebvre RA, Smits GJM (1992) Modulation of non-adrenergic non-cholinergic inhibitory transmission in rat gastric fundus by the α_2-adrenoceptor agonist, UK 14,304. Br J Pharmacol 107:256–261

Lefebvre RA, Vandekerckhove K (1998) Effect of nitroglycerin and long-term electrical stimulation on nitrergic relaxation in the pig gastric fundus. Br J Pharmacol 123:143–149

Lefebvre RA, Smits GJM, Timmermans J-P (1995) Study of NO and VIP as non-adrenercic non-cholinergic neurotransmitters in the pig gastric fundus. Br J Pharmacol 116:2017–2026

Li CG, Rand MJ (1989) Evidence for a role of nitric oxide in the neurotransmitter system mediating relaxation of the rat anococcygeus muscle. Clin Exp Pharmacol Physiol 16:933–938

Li CG, Rand MJ (1990) Nitric oxide and vasoactive intestinal polypeptide mediate non-adrenergic, non-cholinergic inhibitory transmission to smooth muscle of the rat gastric fundus. Eur J Pharmacol 191:303–309

Li CG, Rand MJ (1991) Evidence that part of the NANC relaxant response of guinea-pig trachea to electrical field stimulation is mediated by nitric oxide. Br J Pharmacol 102:91–94

Li CG, Rand MJ (1996) Inhibition of NO-mediated responses by 7-ethoxyresorufin, a substrate and competitive inhibitor of cytochrome P_{450}. Br J Pharmacol 118:57–62

Li CG, Brosch SF, Rand MJ (1994) Inhibition by ethacrynic acid of NO-mediated relaxations of the rat anococcygeus muscle. Clin Exp Pharmacol Physiol 21:293–299

Lilley E, Gibson A (1995) Inhibition of relaxations to nitrergic stimulation of the mouse anococcygeus by duroquinone. Br J Pharmacol 116:3231–3236

Lilley E, Gibson A (1996) Antioxidant protection of NO-induced relaxations of the mouse anococcygeus against inhibition by superoxide anions, hydroquinone and carboxy-PTIO. Br J Pharmacol 119:432–438

Lilley E, Gibson A (1997) Release of the antioxidants ascorbate and urate from a nitrergically-innervated smooth muscle. Br J Pharmacol 122:1746–1752

Lin Y-Q, Bennett MR (1994) Nitric oxide modulation of quantal secretion in chick ciliary ganglia. J Physiol 481:385–394

Lincoln TM (1989) Cyclic GMP and mechanisms of relaxation. Pharmacol Ther 41:479–502

Lincoln TM, Cornwell TL (1993) Intracellular cyclic GMP receptor proteins. FASEB J 7:328–338

Liu SF, Crawley DE, Rohde JAL, Evans TW, Barnes PJ (1992) Role of nitric oxide and guanosine 3',5'-cyclic monophosphate in mediating nonadrenergic, noncholinergic relaxation in guinea-pig pulmonary arteries. Br J Pharmacol 107:861–866

Liu X, Gillespie JS, Gibson IF, Martin W (1991) Effects of N^G-substituted analogues of L-arginine on NANC relaxation of the rat anococcygeus and bovine retractor penis muscles and the bovine penile artery. Br J Pharmacol 104:53–58

Liu X, Gillespie JS, Martin W (1994) Non-adrenergic, non-cholinergic relaxation of the bovine retractor penis muscle: role of S-nitrosothiols. Br J Pharmacol 111:1287–1295

Liu X, Miller SM, Szurszewski JH (1997) Protection of nitrergic neurotransmission by and colocalisation of neural nitric oxide synthase with copper zinc superoxide dismutase. J Autonomic Nervous System 62:126–133

Liuking YC, Weusten BLAM, Portincasa P, Van Der Meer R, Smout AJPM, Akkermans LMA (1998) Effects of long-term oral L-arginine on esophageal motility and gallbladder dynamics in healthy humans. Am J Physiol 37:G984-G991

Luduena FP, Grigas EO (1966) Pharmacological study of autonomic innervation of dog retractor penis. Am J Physiol 210:435–444

Lumme A, Vanhatalo S, Soinila S (1996) Axonal transport of nitric oxide synthase in autonomic nerves. J Autonomic Nervous System 56:207–214

Lundberg JM (1996) Pharmacology of cotransmission in the autonomic nervous system – integrative aspects on amines, neuropeptides, adenosine triphosphate, amino acids and nitric oxide. Pharmacol Rev 48:113–178

Luo D, Das S, Vincent SR (1995) Effects of methylene blue and LY83583 on neuronal nitric oxide synthase and NADPH-diaphorase. Eur J Pharmacol 290:247–251

Maekawa H, Matsumura Y, Matsuo G, Morimoto S (1996) Effect of sodium nitroprusside on norepinephrine overflow and antidiuresis induced by stimulation of renal nerves in anesthetised dogs. J Cardiovasc Pharmacol 27:211–217

Maggi CA, Manzini S, Meli A (1984) Evidence that $GABA_A$ receptors mediate relaxation of rat duodenum by activating intramural nonadrenergic noncholinergic neurons. J Autonom Pharmacol 4:77–81

Marczin N, Ryan US, Catravas JD (1992) Methylene blue inhibits nitrovasodilator- and endothelium-derived relaxing factor-induced cyclic GMP accumulation in cultured pulmonary arterial smooth muscle cells via generation of superoxide anion. J Pharmacol Exp Ther 263:170–179

Marklund SL (1984) Extracellular superoxide dismutase in human tissues and cell lines. J Clin Invest 74:1398–1403

Martin W, Gillespie JS (1991) L-arginine-derived nitric oxide: the basis of inhibitory transmission in the anococcygeus and retractor penis muscles. In: Bell C (ed) Novel peripheral neurotransmitters. Pergamon, New York, pp 65–79

Martin W, Villani GM, Jothianandan D, Furchgott RF (1985a) Blockade of endothelium-dependent and glyceryl trinitrate-induced relaxation of rabbit aorta by certain ferrous hemoproteins. J Pharmacol Exp Ther 233:679–685

Martin W, Villani GM, Jothianandan D, Furchgott RF (1985b) Selective blockade of endothelium-dependent and glyceryl trinitrate-induced relaxation by hemoglobin and by methylene blue in the rabbit aorta. J Pharmacol Exp Ther 232:708–716

Martin W, Furchgott RF, Villani GM, Jothianandan D (1986) Phosphodiesterase inhibitors induce endothelium-dependent relaxation of rat and rabbit aorta by potentiating the effects of spontaneously released endothelium-derived relaxing factor. J Pharmacol Exp Ther 237:539–547

Martin W, Gibson IF, Gillespie JS, Liu X (1991a) Nitric oxide as a neurotransmitter in smooth muscle. In: Stone TW (ed) Aspects of synaptic transmission. Taylor and Francis, London, pp 258–282

Martin W, Gillespie JS, Liu X, Gibson IF (1991b) Effects of N^G-substituted analogues of L-arginine on NANC relaxation of the anococcygeus, retractor penis and penile artery. Br J Pharmacol 102:83P

Martin W, Gillespie JS, Gibson IF (1993) Actions and interactions of N^G-substituted analogues of L-arginine on NANC neurotransmission in the bovine retractor penis and rat anococcygeus muscles. Br J Pharmacol 108:242–247

Martin W, McAllister HM, Paisley K (1994) NANC neurotransmission in the bovine retractor penis muscle is blocked by superoxide anion following inhibition of superoxide dismutase with diethyldithiocarbamate. Neuropharmacology 33:1293–1301

Mashimo H, Xue DH, Huang PL, Fishman MC, Goyal RK (1996) Neuronal constitutive nitric oxide synthase is involved in murine enteric inhibitory neurotransmission. J Clin Invest 98:8–13

Matthew JD, Wadsworth RM (1997) The role of nitric oxide in inhibitory neurotransmission in the middle cerebral artery of the sheep. Gen Pharmacol 28:393–399

Mayer B, John M, Bohme E (1990) Purification of a Ca^{2+}/calmodulin-dependent nitric oxide synthase from porcine cerebellum. FEBS Lett 277:215–219

Mirzazadeh S, Hobbs AJ, Tucker JF, Gibson A (1991) Cyclic nucleotide content of the rat anococcygeus during relaxations induced by drugs or by non-adrenergic, non-cholinergic field stimulation. J Pharm Pharmacol 43:247–257

Misawa M, Sato J (1997) Abnormal modulation of cholinergic neurotransmission by endogenous nitric oxide in the bronchus of rats with hyper-responsiveness induced by allergen challenge. Jpn J Pharmacol 73:125–132

Mitchell JA, Förstermann U, Warner TD, et al. (1991a) Endothelial cells have a particulate enzyme system responsible for EDRF formation: measurement by vascular relaxation. Biochem Biophys Res Commun 176:1417–1423

Mitchell JA, Sheng H, Förstermann U, Murad F (1991b) Characterisation of nitric oxide synthases in non-adrenergic, non-cholinergic nerve containing tissue from the rat anococcygeus muscle. Br J Pharmacol 104:289–291

Mok JSL, Paisley K, Martin W (1998) Inhibition of nitrergic neurotransmission in the bovine retractor penis muscle by an oxidant stress: effects of superoxide dismutase mimetics. Br J Pharmacol 124:111–118

Moncada S, Higgs EA, Furchgott RF (1997) International Union of Pharmacology nomenclature in nitric oxide research. Pharmacol Rev 49:137–142

Moore PK, al-Swayeh OA, Chong NWS, Evans RA, Gibson A (1990) L-N^G-nitroarginine (L-NOARG), a novel, L-arginine-reversible inhibitor of endothelium-dependent relaxation in vitro. Br J Pharmacol 99:408–412

Moore PK, Babbedge RC, Wallace P, Gaffen ZA, Hart SL (1993a) 7-Nitroindazole, an inhibitor of nitric oxide synthase, exhibits anti-nociceptive activity in the mouse without increasing blood pressure. Br J Pharmacol 108:296–297

Moore PK, Wallace P, Gaffen Z, Hart SL, Babbedge RC (1993b) Characterization of the novel nitric oxide synthase inhibitor 7-nitro indazole and related indazoles: antinociceptive and cardiovascular effects. Br J Pharmacol 110:219–224

Moreland RB, Goldstein I, Traish A (1998) Sildenafil, a novel inhibitor of phosphodiesterase type 5 in human corpus cavernosum smooth muscle cells. Life Sci 62:309–318

Mudumbi RV, Parmeter LL, McIntyre MS, Leighton HJ (1996) Interactions between neurotransmitters and exogenous norepinephrine in isolated rat anococcygeus muscle. Gen Pharmacol 27:193–197

Mülsch A, Busse R, Liebau S, Förstermann U (1988) LY 83583 interferes with the release of the endothelium-derived relaxing factor and inhibits soluble guanylate cyclase. J Pharmacol Exp Ther 247:283–288

Nakane M, Mitchell JA, Förstermann U, Murad F (1991) Phosphorylation by calcium calmodulin-dependent protein kinase II and protein kinase C modulates the activity of nitric oxide synthase. Biochem Biophys Res Commun 180:1396–1407

Nakane M, Schmidt HHHW, Pollock JS, Förstermann U, Murad F (1993) Cloned human brain nitric oxide synthase is highly expressed in skeletal muscle. FEBS Lett 316:175–180

Ny L, Alm P, Ekström P, Larsson B, Grundemar L, Andersson K-E (1996) Localization and activity of haem oxygenase and functional effects of carbon monoxide in the feline lower oesophageal sphincter. Br J Pharmacol 118:392–399

Ohno N, Xue L, Yamamoto Y, Suzuki H (1996) Properties of the inhibitory junction potential in smooth muscle of the guinea-pig gastric fundus. Br J Pharmacol 117:974–978

Okamura T, Yoshida K, Toda N (1995) Nitroxidergic innervation in dog and monkey renal arteries. Hypertension 25:1090–1095

Olgart C, Hallen K, Wiklund NP, Iversen HH, Gustafsson LE (1998) Blockade of nitrergic neuroeffector transmission in guinea-pig colon by a selective inhibitor of soluble guanylate cyclase. Acta Physiol Scand 162:89–95

Paisley K, Martin W (1996) Blockade of nitrergic transmission by hydroquinone, hydroxocobalamin and carboxy-PTIO in bovine retractor penis: role of superoxide anion. Br J Pharmacol 117:1633–1638

Palmer RMJ, Ferrige AG, Moncada S (1987) Nitric oxide release accounts for the biological activity of endothelium-derived relaxing factor. Nature 327:524–526

Palmer RMJ, Ashton DS, Moncada S (1988) Vascular endothelial cells synthesise nitric oxide from L-arginine. Nature 333:664–666

Persson K, Alm P, Johansson K, Larsson B, Andersson K-E (1995) Co-existence of nitrergic, peptidergic and acetylcholine esterase-positive nerves in the pig lower urinary tract. J Autonomic Nervous System 52:225–236

Pickard RS, Powell PH, Zar MA (1991) The effect of inhibitors of nitric oxide biosynthesis and cyclic GMP formation on nerve evoked relaxation of human cavernosal smooth muscle. Br J Pharmacol 104:755–759

Pinna C, Puglisi L, Burnstock G (1998) ATP and vasoactive intestinal polypeptide relaxant responses in hamster isolated proximal urethra. Br J Pharmacol 124:1069–1074

Pino RZ, Feelisch M (1994) Bioassay discrimination between nitric oxide ($NO^.$) and nitroxyl (NO^-) using L-cysteine. Biochem Biophys Res Commun 201:54–62

Potter E: (1998) Neuropeptide Y as an autonomic transmitter. In: Bell C (ed) Novel peripheral neurotransmitters. Pergamon, New York, pp 81–112

Rajanayagam MAS, Li CG, Rand MJ (1993) Differential effects of hydroxocobalamin on NO-mediated relaxations in rat aorta and anococcygeus muscle. Br J Pharmacol 108:3–5

Ramagopal MV, Leighton HJ (1989) Effect of N^G-monomethyl-L-arginine on field stimulation-induced decreases in cytosolic Ca^{2+} levels and relaxation in the rat anococcygeus muscle. Eur J Pharmacol 174:297–299

Rand MJ (1992) Nitrergic transmission: nitric oxide as a mediator of non-adrenergic, non-cholinergic neuroeffector transmission. Clin Exp Pharmacol 19:147–169

Rand MJ, Li CG (1992) Effects of arginosuccinic acid on nitric oxide mediated relaxations in rat aorta and anococcygeus muscle. Clin Exp Pharmacol Physiol 19:331–334

Rand MJ, Li CG (1993a) The inhibition of nitric oxide-mediated relaxations in rat aorta and anococcygeus muscle by diphenylene iodonium. Clin Exp Pharmacol Physiol 20:141–148

Rand MJ, Li CG (1993b) Modulation of acetylcholine-induced contractions of the rat anococcygeus muscle by activation of nitrergic nerves. Br J Pharmacol 110:1479–1482

Rand MJ, Li CG (1994a) Differential effects of hydroxocobalamin on relaxations induced by nitrosothiols in rat aorta and anococcygeus muscles. Eur J Pharmacol 241:294–255

Rand MJ, Li CG (1994b) Effects of ethanol and other aliphatic alcohols on NO-mediated relaxations in rat anococcygeus muscles and gastric fundus. Br J Pharmacol 111:1089–1094

Rand MJ, Li CG (1995a) Discrimination by the NO-trapping agent, carboxy-PTIO, between NO and the nitrergic transmitter but not between NO and EDRF. Br J Pharmacol 116:1906–1910

Rand MJ, Li CG (1995b) Nitric oxide as a neurotransmitter in peripheral nerves: nature of transmitter and mechanism of transmission. Ann Rev Physiol 57:659–682

Rapoport RM, Murad F (1983) Agonist-induced endothelium-dependent relaxation may be mediated through cyclic GMP. Circ Res 52:352–357

Rattan S, Rosenthal GJ, Chakder S (1995) Human recombinant hemoglobin (rHb1.1) inhibits nonadrenergic noncholinergic (NANC) nerve-mediated relaxation of internal anal sphincter. J Pharmacol Exp Ther 272:1211–1216

Rattan S, Chakder S (1993) Inhibitory effects of CO on internal anal sphincter: heme oxygenase inhibitor inhibits NANC relaxation. Am J Physiol 265:G799-G804

Rattan S, Chakder S (1997) L-citrulline recycling in the opossum internal anal sphincter: relaxation by nonadrenergic, noncholinergic nerve stimulation. Gastroenterology 112:1250–1259

Raymond GL, Wendt IR (1996) Force and intracellular Ca^{2+} during cyclic nucleotide-mediated relaxation of rat anococcygeus muscle and the effects of cyclopiazonic acid. Br J Pharmacol 119:1029–1037

Recio P, Lopez PG, Hernandez M, Prieto D, Contreras J (1998) Nitrergic relaxation of the horse corpus cavernosum. Role of cGMP. Eur J Pharmacol 351:85–94

Rees DD, Palmer RMJ, Hodson HF, Moncada S (1989) A specific inhibitor of nitric oxide formation form L-arginine attenuates endothelium-dependent relaxations. Br J Pharmacol 96:418–424

Rees DD, Palmer RMJ, Schulz R, Hodson HF, Moncada S (1990) Characterisation of three inhibitors of endothelial nitric oxide synthase in vitro and in vivo. Br J Pharmacol 101:746–752

Rengasamay A, Johns RA (1998) Regulation of nitric oxide synthase by nitric oxide. Mol Pharmacol 44:124–128

Rogers NE, Ignarro LJ (1992) Constitutive nitric oxide synthase from cerebellum is reversibly inhibited by nitric oxide formed from L-arginine. Biochem Biophys Res Commun 189:242–249

Schmidt HHHW, Pollock JS, Nakane M, Gorsky LD, Förstermann U, Murad F (1991) Purification of a soluble isoform of guanylyl cyclase-activating factor synthase. Proc Natl Acad Sci U S A 88:365–369

Schmidt HHHW, Gagne GD, Nakane M, Pollock JS, Miller MF, Murad F (1992a) Mapping of neural nitric oxide synthase in the rat suggests frequent co-localisation with NADPH diaphorase but not with soluble guanylate cyclase, and novel paraneural functions for nitrinergic signal transduction. J Histochem Cytochem 40:1439–1456

Schmidt HHHW, Warner TD, Ishii K, Sheng H, Murad F (1992b) Insulin secretion from pancreatic β cells caused by L-arginine-derived nitrogen oxides. Science 255:721–723

Schmidt HHHW, Hofmann H, Schindler U, Shutenko ZS, Cunningham DD, Feelisch M (1996) No NO from NO synthase. Proc Natl Acad Sci USA 93:14492–14497

Schmidt MJ, Sawyer BD, Truex LL, Marshall WS, Fleisch JH (1985) LY 83583: An agent that lowers intracellular levels of cyclic guanosine 3',5'-monophosphate. J Pharmacol Exp Ther 232:764–769

Scott TRD, Bennett MR (1993) The effect of nitric oxide on the efficacy of synaptic transmission through the chick ciliary ganglion. Br J Pharmacol 110:627–632

Sekizawa K, Fukushima T, Ikarashi Y, Maruyama Y, Sasaki H (1993) The role of nitric oxide in cholinergic neurotransmission in rat trachea. Br J Pharmacol 110:816–820

Selemidis S, Cocks TM (1997) Evidence that both nitric oxide (NO) and a non-NO hyperpolarising factor elicit NANC nerve-mediated relaxation in the rat isolated anococcygeus. Br J Pharmacol 120:662–666

Sheng H, Schmidt HHHW, Nakane M, et al. (1992) Characterisation and localisation of nitric oxide synthase in non-adrenergic non-cholinergic nerves from bovine retractor penis muscles. Br J Pharmacol 106:768–773

Sheng H, Gagne GD, Matsumoto T, Miller MF, Förstermann U, Murad F (1993) Nitric oxide synthase in bovine superior cervical ganglion. J Neurochem 61:1120–1126

Shuttleworth CWR, Conlon SB, Sanders KM (1997) Regulation of citrulline recycling in nitric oxide-dependent neurotransmission in the murine colon. Br J Pharmacol 120:707–713

Simonsen U, Prieto D, Hernandez M, de Tejada IS, Garcia-Sacristan A (1997) Prejunctional α_2-adrenoceptors inhibit nitrergic neurotransmission in horse penile resistance arteries. J Urol 157:2356–2360

Som S, Raha C, Chatterjee IB (1983) Ascorbic acid: a scavenger of superoxide radical. Acta Vitaminol Enzymol 5:243–250

Southam E, Charles SL, Garthwaite J (1996) The nitric oxide cyclic GMP pathway and synaptic plasticity in the rat superior cervical ganglion. Br J Pharmacol 119:527–532

Stamler JS, Singel DJ, Loscalzo J (1992) Biochemistry of nitric oxide and its redox-activated forms. Science 258:1898–1902

Stark ME, Szurszewski JH (1992) Role of nitric oxide in gastrointestinal and hepatic function and disease. Gastroenterology 103:1928–1948

Stark ME, Bauer AJ, Szurszewski JH (1991) Effect of nitric oxide on circular muscle of the canine small intestine. J Physiol 444:743–761

Steinberg C, Aiska K, Gross SS, Griffith OW, Levi P (1990) Vasopressor effects of N^G-substituted arginine analogs in the anaesthetised guinea-pig. Eur J Pharmacol 183:165

Stief CG, Uckert S, Becker AJ, Truss MC, Jonas U (1998) The effects of specific phosphodiesterase (PDE) inhibitors on human and rabbit cavernous tissue in vitro and in vivo. J Urol 159:1390–1399

Suzuki N, Mizuno K, Gomi Y (1994) Role of nitric oxide in the peristalsis of isolated guinea-pig ileum. Eur J Pharmacol 251:221–227

Tanaka K, Hassall CJS, Burnstock G (1993) Distribution of intracardiac neurons and nerve terminals that contain a marker for nitric oxide, NADPH-diaphorase, in the guinea-pig heart. Cell Tiss Res 273:293–300

Teixeira CE, Bento AC, Lopes–Martins RAB, et al. (1998) Effects of *Tityus serrulatus* scorpion venom on the rabbit isolated corpus cavernosum and the involvement of NANC nitrergic nerve fibres. Br J Pharmacol 123:435–442

Thomas RM, Fang S, Leichus LS, et al. (1996) Antioxidant enzymes in intramural nerves of the opossum esophagus. Am J Physiol 270:G136–G142

Toda N (1975) Nicotine-induced relaxation in isolated canine cerebral arteries. J Pharmacol Exp Ther 193:376–384

Toda N (1981) Non-adrenergic, non-cholinergic innervation in monkey and human cerebral arteries. Br J Pharmacol 72:281–283

Toda N (1995) Nitric oxide and the regulation of cerebral arterial tone. In: Vincent S (ed) Nitric oxide in the nervous system. Academic, London, pp 207–225

Toda N, Kimura T, Yoshida K, et al. (1994) Human uterine arterial relaxation induced by nitroxidergic nerve stimulation. Am J Physiol 266:H1446–H1450

Toda N, Kimura T, Okamura T (1995a) Nitroxidergic nerve stimulation relaxes human uterine vein. J Autonomic Nervous System 55:198–192

Toda N, Uchiyama M, Okamura T (1995b) Prejunctional modulation of nitroxidergic nerve function in canine cerebral arteries. Brain Res 700:213–218

Toda N, Ayajiki K, Okamura T (1997a) Effects of Ca^{2+}/calmodulin-dependent protein kinase II inhibitors on the neurogenic cerebroarterial relaxation. Eur J Pharmacol 340:59–65

Toda N, Ayajiki K, Okamura T (1997b) Inhibition of nitroxidergic nerve function by neurogenic acetylcholine in monkey cerebral arteries. J Physiol 498:435–461

Toda M, Okamura T, Azuma I, Toda N (1997c) Modulation by neurogenic acetylcholine of nitrodoxidergic nerve function in porcine ciliary arteries. Invest Opthalmol Vis Sci 38:2261–2269

Tracey WR, Nakane M, Pollock JS, Förstermann U (1993) Nitric oxide synthases in neuronal cells, macrophages and endothelium are NADPH diaphorases, but represent only a fraction of total cellular NADPH diaphorase activity. Biochem Biophys Res Commun 195:1035–1040

Vials AJ, Crowe R, Burnstock G (1997) A neuromodulatory role for neuronal nitric oxide in the rabbit renal artery. Br J Pharmacol 121:213–220

Waldeck K, Ny L, Persson K, Andersson K-E (1998) Mediators and mechanisms of relaxation in rabbit urethral smooth muscle. Br J Pharmacol 123:617–624

Ward SM, Dalziel HH, Bradley ME, et al. (1992) Involvement of cyclic GMP in non-adrenergic, non-cholinergic inhibitory neurotransmission in dog proximal colon. Br J Pharmacol 107:1075–1082

Ward SM, Dalziel HH, Khoyi MA, Westfall A, Sanders KM, Westfall DP (1996) Hyperpolarisation and inhibition of contraction mediated by nitric oxide released from enteric inhibitory neurons in guinea-pig taenia coli. Br J Pharmacol 118:49–56

Watson MJ, Bywater RAR, Taylor GS, Lang RJ (1996a) Effects of nitric oxide (NO) and NO donors on the membrane conductance of circular smooth muscle cells of the guinea-pig proximal colon. Br J Pharmacol 118:1605–1614

Watson MJ, Lang RJ, Bywater RAR, Taylor GS (1996b) Characterisation of the membrane conductance changes underlying the apamin-resistant NANC inhibitory junction potential in the guinea-pig proximal and distal colon. J Autonomic Nervous System 60:31–42

Wayman C, McFadzean I, Gibson A, Tucker JF (1996) Inhibition by sodium nitroprusside of a calcium store depletion-activated non-selective cation current in smooth muscle cells of the mouse anococcygeus. Br J Pharmacol 118:2001–2008

Wayman C, Gibson A, McFadzean I (1998) Depletion of either ryanodine- or IP_3-sensitive calcium stores activates capacitative calcium entry in mouse anococcygeus. Pfluger's Archiv Eur J Physiol 435:231–239

Werkström V, Persson K, Ny L, Bridgewater M, Brading AF, Andersson K-E (1995) Factors involved in the relaxation of female pig urethra evoked by electrical field stimulation. Br J Pharmacol 116:1599–1604

Werkström V, Ny L, Persson K, Andersson K-E (1997) Neurotransmitter release evoked by α-latrotoxin in the smooth muscle of the female pig urethra. Naunyn-Schmiedeberg's Arch Pharmacol 356:151–158

Wesiger RA, Fridovich I (1973) Superoxide dismutase: organelle specificity. J Biol Chem 248:3583–3592

White TD: (1991) Role of ATP and adenosine in the autonomic nervous system. In: Bell C (ed) Novel peripheral neurotransmitters. Pergamon, New York, pp 9–64

Wiklund CU, Olgart C, Wiklund NP, Gustafsson LE (1993) Modulation of cholinergic and substance P-like neurotransmission by nitric oxide in the guinea-pig ileum. Br J Pharmacol 110:833–839

Wiklund NP, Cellek S, Leone AM, et al. (1997) Visualisation of nitric oxide released by nerve stimulation. J Neuroscience Res 47:224–232

Wolin MS, Cherry PD, Rodenburg JM, Messina EJ, Kaley G (1990) Methylene blue inhibits vasodilatation of skeletal muscle arterioles to acetylcholine and nitric oxide via the extracellular generation of superoxide anions. J Pharmacol Exp Ther 254:872–876

Wood J, Garthwaite J (1994) Models of the diffusional spread of nitric oxide (NO): implications for neural NO signalling and its pharmacological properties. Neuropharmacology 33:1235–1244

Xu WM, Gorman P, Sheer D, et al. (1993) Regional localization of the gene coding for human brain nitric oxide synthase (NOS1) to 12q24.2–24.31 by fluorescent in situ hybridisation. Cytogenet Cell Genet 64:62–63

Yui Y, Ohkawa S, Ohnishi K, et al. (1989) Mechanism for the generation of an active smooth muscle inhibitory factor (IF) from bovine retractor penis muscle (BRP). Biochem Biophys Res Commun 164:544–594

Yunker AMR, Galligan JJ (1996) Endogenous NO inhibits NANC but not cholinergic neurotransmission to circular muscle of guinea-pig ileum. Am J Physiol 34:904–912

Zagorodnyuk V, Maggi CA (1994) Electrophysiological evidence for different release mechanisms of ATP and NO as inhibitory NANC transmitters in guinea-pig colon. Br J Pharmacol 112:1077–1082

Zamora R, Grzesiok A, Weber H, Feelisch M (1995) Oxidative release of nitric oxide accounts for guanylyl cyclase stimulating, vasodilator and antiplatelet activity of Piloty's acid – a comparison with Angeli's salt. Biochem J 312:333–339

Zygmunt PM, Zygmunt PKE, Högestätt ED, Andersson K-E (1993) Effects of w-conotoxin on adrenergic, cholinergic and NANC neurotransmission in the rabbit urethra and detrusor. Br J Pharmacol 110:1285–1290

Zygmunt PKE, Zygmunt PM, Högestätt ED, Andersson K-E (1995) NANC neurotransmission in lamina propria of the rabbit urethra: regulation by different subsets of calcium channels. Br J Pharmacol 115:1020–1026

CHAPTER 13
Nitric Oxide and Neuroendocrine Function

P. NAVARRA, A. COSTA, and A. GROSSMAN

A. Introduction

Nitric oxide (NO) is a short-lived, unstable gas which is synthesized by a number of isoforms of the enzyme nitric oxide synthase (NOS), all of which are related to the cytochrome P_{450} family but which have diverged from a speculated common precursor in the distant past (MONCADA et al. 1991). Three NOS isoforms which are differentially localized and regulated have been described so far. Two isoforms appear to depend on calcium for their activation: the endothelial isoform (eNOS), a vital component in the regulation of blood pressure (PALMER et al. 1987), and the neuronal isoform (nNOS), which is mainly neural in location and is functionally related to the endothelial form in terms of its calcium dependence. However, there are now numerous situations where nNOS can be induced, although not usually by inflammatory cytokines. Neuronal NO acts in large part by activating the haemoproteins cytoplasmic guanylate cyclase and cyclo-oxygenase, thus utilizing cyclic guanosine monophosphate (GMP) and prostanoids, respectively, as second messengers.

The third isoform, inducible NOS (iNOS), is active in a manner independent of the prevailing calcium milieu, principally because calmodulin is bound with high affinity even in the presence of a low calcium concentration. This isoform is synthesized by a variety of cell types, particularly resident macrophages (microglia and astrocytes in the central nervous system) and, most importantly, is induced by inflammatory mediators, such as interleukin-1 (IL-1). The function of this enzyme appears to be the generation of large amounts of local NO as part of an anti-microbial defence process. At high levels, NO becomes cytocidal, partially through an interaction with superoxides, which produce peroxynitrite; this, in turn, can disable a number of important haem-containing enzymes, such as bacterial cytochromes (FANG 1997), and in part by inhibiting ribonucleotide diphosphate reductase, thus reducing DNA biosynthesis (NOCENTINI 1996). Each of these isoforms is transcribed from separate genetic loci on different chromosomes.

B. NO Biosynthesis in the Hypothalamus: Relationship Between Localization and Function

Within the hypothalamus, all three isoforms can be readily identified. It is not known whether the endothelial isoform has a specific intra-cerebral function, but it may act, at least in part, by coordinating local blood supply with neuronal metabolism (CECCATELLI et al. 1992). There is very little iNOS present under basal conditions (BHAT et al. 1996) but, after in vivo stimulation with endotoxin, we and others have recently shown an increase in message transcription (WONG et al. 1996; JACOBS et al. 1997; SATTA et al. 1998). It has been postulated that vascular and perivascular induction of iNOS by inflammatory cytokines – IL-1 in particular – represents a mechanism for signalling across the blood–brain barrier (BBB) by circulating inflammatory mediators (LICINIO and WONG 1997). A role for NO in mediating signalling through the BBB is also suggested by the finding of nNOS immunoreactivity localized within circumventricular organs (which are outside the BBB), including the organum vasculosum of the lamina terminalis, the subfornical organ, the median eminence and the choroid plexus in the rat (ALM et al. 1997).

nNOS is the major NOS isoform within the hypothalamus; it is localized to discrete cell clusters and groups. The majority of nNOS mRNA, as revealed by in situ hybridization, is located within the paraventricular (PVN) and supraoptic nuclei (SON), with scattered cells distributed in the anterior periventricular region (BREDT et al. 1990; GROSSMAN et al. 1994). However, there is an additional, relatively intense location in the anterior hypothalamus, principally in the diagonal band of Broca and the islands of Calleja (GROSSMAN et al. 1994). Immunohistochemical staining for NOS and reduced nicotinamide adenine dinucleotide phosphate (NADPH)-diaphorase, a tinctorial characteristic of NOS-containing neurons (DAWSON et al. 1991), shows a broadly similar distribution (VINCENT and KIMURA 1992).

The PVN and SON are the origin of the hypophysiotropic peptides corticotrophin-releasing hormone (CRH), arginine vasopressin (AVP) and oxytocin and, in general, there is evidence favouring co-storage with each of these neuropeptides. According to one report, most of the NOS immunoreactivity is co-localised to oxytocin-containing cells (TORRES et al. 1993), but at least some of the NOS is also found in CRHergic and vasopressinergic neurons. HATAKEYAMA et al. (1996) found that the distribution of NOS was segregated from that of CRH-containing parvicellular neurons in the posterior PVN but overlapped with that of magnocellular neurons. These authors found that about 70% of oxytocin and 30% of AVP neurons were also NADPH-diaphorase-positive; this was about one half the total number of NOS neurons. About 40–50% of the latter displayed increased Fos immunoreactivity after injection of lipopolysaccharide or hypertonic saline, but only 10–15% did so in response to restraint stress or pain. Only NADPH-diaphorase immunoreactivity in AVP-containing magnocellular neurons (but not that in parvicellular neurons) increased to some extent after adrenalectomy (SANCHEZ et al. 1996).

Apart from stress hormones, NOS co-storage appears to be a widespread phenomenon in the hypothalamus: NOS-like immunoreactivity has been shown to coexist with substance P in the medial preoptic area, the ventromedial and dorsomedial nuclei and the mamillary region; with cholecystokinin in the SON and mamillary region; with galanin in the SON and dorsomedial nuclei; with enkephalin in the PVN, the arcuate, ventromedial and dorsomedial nuclei, and the mamillary region; with somatostatin in the ventromedial nucleus; and with vasoactive intestinal polypeptide in the suprachiasmatic nucleus (REUSS et al. 1995; YAMADA et al. 1996).

It was originally noted that the distribution of NOS mRNA within the anterior hypothalamus was strikingly similar to that of gonadotrophin-releasing hormone (GnRH), suggesting co-storage of NOS in GnRH neurons. However, more detailed analysis using double-staining techniques revealed that the NOS-containing cells were adjacent to, but not identical with, the GnRH neurons (GROSSMAN et al. 1994). A recent elegant and detailed study has confirmed that the GnRH neurons are surrounded and almost encapsulated by the NOS containing interneurons (HERBISON et al. 1996), which have also been shown to bear receptors for excitatory amino acids, such as the N-methyl-D-asparate (NMDA) receptor and the oestrogen receptor.

The analysis of NOS localization within the hypothalamus suggests two possible modes of action for NO: first, the gas might act in a *paracrine* fashion, signalling through the BBB or regulating hypothalamic peptides, such as GnRH. In the latter case, NO has been proposed to function as a synchronizing agent for GnRH neurons, which represent a scattered population of cells within the anterior hypothalamus and require an external diffusible factor to synchronize their pulsatile secretory pattern (GROSSMAN et al. 1997; LOPEZ et al. 1997). Second, NO might act as an *autocrine* agent to control the release and perhaps the synthesis of several peptides in specific neuronal populations where co-storage with NOS has been demonstrated (see above). However, diffusibility of the gas implies that both the autocrine and paracrine modes of action may act simultaneously in or near the sites of NO generation.

C. Physiology of Hypothalamic NO

I. Vasopressin and Oxytocin

There is considerable evidence that the generation of NO inhibits the secretion of AVP and oxytocin. The stimulated release of both peptides is inhibited by NO precursors and donors from hypothalamic explants in vitro (YASIN et al. 1993). The in vivo secretion of oxytocin (but not AVP) induced by osmotic stimuli was enhanced by blockade of NO generation (SUMMY-LONG et al. 1993).

The neurohypophyseal content of NOS (SAGAR and FERRIERO 1987; POW 1992; CALKA and BLOCK 1993) and its mRNA (KADOWAKI et al. 1994) was increased during dehydration. An increase in NOS gene expression and activity was also seen in the SON and PVN after water deprivation (O'SHEA and GUNDLACH 1996). Up-regulation of NOS mRNA in oxytocin neurons (but not in AVP neurons) was also reported in another model of osmotic stimulation, i.e. lactation (LUCKMAN et al. 1997). At variance with the osmotic stimuli, food deprivation produced conflicting data, with one group showing increased (O'SHEA and GUNDLACH 1996) and another decreased (UETA et al. 1995a) NOS mRNA expression in the SON and PVN of the rat. Thus, NO generated within the SON and PVN may act as a negative-feedback inhibitor of neurohypophyseal peptide release, particularly oxytocin, thereby reducing the sensitivity of the hypothalamus to osmotic stimuli. An additional control mechanism over oxytocin and AVP release may include NO modulation of γ-aminobutyric acid (GABA) levels within the hypothalamus (CHIODERA et al. 1996).

Other hypothalamic functions also appear to be regulated by an interplay between NO and oxytocin neurons; these include the central control of penile erection and yawning. These behaviours are induced by the activation of oxytocin as well as dopaminergic and excitatory amino acid transmission within the PVN. Pharmacological evidence has shown that NO has a facilitatory effect on these behaviours, since locally injected NOS inhibitors prevent and NO donors induce penile erection and yawning indistinguishable from those elicited by oxytocin or dopamine agonists (MELIS and ARGIOLAS 1997). An interplay between NO and oxytocin also seems to take place in the regulation of GnRH secretion (see below).

II. Corticotrophin-Releasing Hormone and the Hypothalamo–Pituitary–Adrenal Axis

Conflicting data exist regarding the modulation of CRH release in vitro; our own findings have quite clearly shown that generation of NO markedly suppresses the stimulated release of CRH (COSTA et al. 1993), while a number of other groups have shown *stimulation* of CRH by NO (BRUNETTI et al. 1993; KARANTH et al. 1993; RABER and BLOOM 1994; SANDI and GUAZA 1995). Such discrepancies might be due to the fact that the experimental models used are not strictly comparable. It is possible that NO may activate a number of different enzymes, including guanylate cyclase and cyclooxygenase, which have opposing effects; the relative activation of two (or more) pathways would then determine the net secretion of CRH and might well be highly preparation dependent. We were able to show activation of guanylate cyclase, but not cyclooxygenase, by NO donors (MANCUSO et al. 1998), but at least one group has claimed that increasing NO will generate prostaglandins, specifically prostaglandin E_2 (RETTORI et al. 1992). In a broadly comparable system – CRH release from human placenta and cultured human trophoblasts – NO has been shown to inhibit CRH secretion (but not synthesis) through a mechanism involving soluble guanylate cyclase and cyclic GMP production (ROE et al. 1996; NI et al. 1997).

In vivo studies indicate that NO may play a dual role in regulating the hypothalamo–pituitary–adrenal axis – either inhibition or stimulation, depending on the nature of the stressors. In fact, antagonists of NOS further enhance the activation of the hypothalamo–pituitary–adrenal axis following systemic endotoxaemia or local inflammation in the rat (RIVIER and SHEN 1994; RIVIER 1995; TURNBULL and RIVIER 1996). On the contrary, NO appears to *attenuate* the response to a noxious stressor (RIVIER 1994). These findings have been basically confirmed in BALB/c mice (GIORDANO et al. 1996). In the human, NOS inhibitors (given at doses that do not affect blood pressure) enhance adrenocorticotrophic hormone rises in response to hypoglycaemia but not to angiotensin II (VOLPI et al. 1996). Evidence from one group showed that the inhibitory effect of NO on hypothalamo–pituitary–adrenal axis function occurs through an action at the level of the adrenal cortex, where increased NO generation would suppress glucocorticoid secretion under stress conditions (TSUCHIYA et al. 1997).

It may be possible to reconcile these differing responses of NO on the release of CRH and the activity of the hypothalamo–pituitary–adrenal axis if one considers that NO can also modulate hypothalamic IL-1β. Activation of NO pathways will increase basal (Fig. 1) and stimulated (Fig. 2) IL-1β release in vitro through a guanylyl cyclase pathway (Fig. 3; MANCUSO et al. 1998), and this may, in turn, lead to a secondary release of CRH, as IL-1 is a CRH secretagogue (TSAGARAKIS et al. 1992) acting through a prostanoid second messenger (NAVARRA et al. 1992). Thus, the *net* effect of of NO on CRH may depend on the relative activation of the IL-1-dependent (CRH-stimulatory) and IL-1-independent (CRH-inhibitory) pathways. Alternatively, it has been suggested that NO may stimulate via the PVN and inhibit via the median eminence (KIM and RIVIER 1998). Of course, it is possible that these theories may

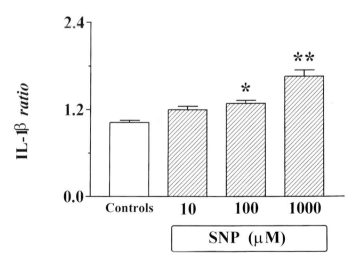

Fig. 1. The basal release of interleukin 1 from rat hypothalamic explants induced by the nitric oxide donor sodium nitroprusside (*SNP*). The data in this and subsequent figures are taken from MANCUSO et al. (1998). *$P < 0.05$ compared to control incubation; **$P < 0.01$ compared to control incubation

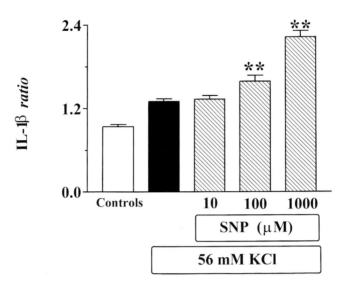

Fig. 2. The effect of the nitric oxide donor sodium nitroprusside (*SNP*) on the potassium-stimulated release of interleukin 1 from rat hypothalamic explants. **$P < 0.01$ compared to control incubation with potassium chloride

be combined; immune or other stressful signals could activate intrahypothalamic NO to increase IL-1, which in turn would stimulate CRH via an eicosanoid second messenger. A second direct route independent of IL-1 may operate at the level of the median eminence (Fig. 4). It is also of interest that the inhibitory route may involve eNOS as the generating enzyme (KIM and RIVIER 1998). Such complexities clearly render this a difficult area in which to predict the outcome to experimental interventions but should allow for precise assessment of the theory.

III. Hypothalamo–Pituitary–Gonadal Axis

There is consensus that both in vivo and in vitro generation of NO activates the hypothalamo–pituitary–gonadal axis by stimulating the secretion of GnRH (BONAVERA et al. 1993; MORETTO et al. 1993; RETTORI et al. 1994; CHIODERA et al. 1995). In vitro evidence also shows that NO mediates the increase in GnRH release induced by oxytocin (RETTORI et al. 1997) or leptin (YU et al. 1997).

A complex interplay between NO, NMDA-related pathways and oestrogens appears to regulate the oestrus surge in circulating lutenising hormone (LH) mediated by the release of anterior hypothalamic GnRH. As far as oestrogens are concerned, anterior hypothalamic NOS content (OKAMURA et al. 1994) and mRNA (CECCATELLI et al. 1996) are up-regulated by oestradiol, while the oestrogen-induced surge in LH release is enhanced by augmenta-

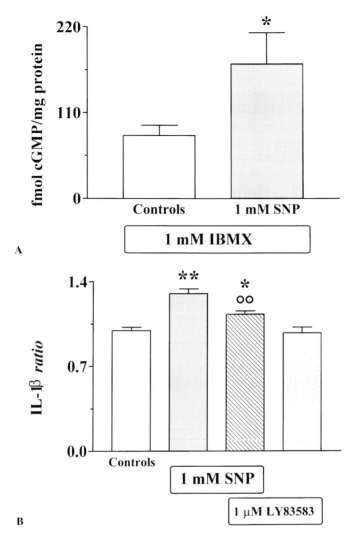

Fig. 3. A The effect of sodium nitroprusside (*SNP*) on the release of cyclic guanosine monophosphate (cGMP) in the presence of the phosphodiesterase inhibitor 3-isobutyl-1-methylxanthine (*IBMX*). *$P < 0.05$ compared to control incubation. **B** SNP stimulation of interleukin 1 is antagonised by the inhibitor of cGMP generation, *LY 83583*. *$P < 0.05$ compared to control incubation; **$P < 0.01$ compared to control incubation; ∞$P < 0.001$ compared to SNP-stimulated release

tion of NO. This probably involves NO-induced stimulation of neuropeptide Y, which in turn stimulates GnRH release (BONAVERA et al. 1996). Conversely, the oestradiol-induced LH surge in ovariectomized rats can be antagonized by nNOS and eNOS (but not iNOS) mRNA antisense oligonucleotides (AGUAN et al. 1996). However, involvement of excitatory amino acids is sug-

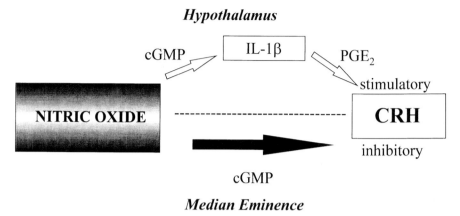

Fig. 4. Suggested scheme whereby nitric oxide can both stimulate and inhibit the secretion of corticotropin-releasing hormone (*CRH*)

gested by the fact that, in immortalized GT1-1 and GT1-7 cells, NOS is contained within the GnRH cells themselves (SORTINO et al. 1994; LOPEZ et al. 1997) and mediates direct stimulation of GnRH release by NMDA (MAHACHOKLERTWATTANA et al. 1994). As noted above, in normal rats, the NOS is found adjacent to but not actually within GnRH-containing neurons but may still be involved in NMDA-induced secretion, as NMDA-induced LH release in vivo can be antagonized by NOS inhibitors (PU et al. 1996). At least one other neurosecretory action of NMDA, stimulation of α-melanocyte-stimulating hormone release from the rat hypothalamus, is also mediated by NO (WAYMAN et al. 1994).

IV. Other Hormonal Systems

There is some evidence for a functional interplay between NO and the hypothalamo–pituitary–thyroid axis. NOS gene expression in the PVN and SON of the rat is under the control of thyroid hormones, since experimental hypothyroidism reduces and hyperthyroidism up-regulates such expression (UETA et al. 1995b). There is also evidence for the involvement of NO in thyrotropin-releasing hormone-induced thyroid-stimulating hormone secretion in the human (COIRO et al. 1995).

Early studies showed that the administration of the NO precursor, L-arginine, via the NOS pathway is followed by an increase in circulating levels of growth hormone (GH) (MERIMEE et al. 1969). This effect was thought to occur *via* the suppression of somatostatin secretion (ALBA-ROTH et al. 1966). However, the notion that L-arginine effects on GH release are mediated by NO has recently been questioned, since the effect of L-arginine could not be mimicked using NO donors (KORBONITS et al. 1996). Moreover, NO seems to increase rather than inhibit the gene expression and release of somatostatin induced in the rat by such stimuli as GH-releasing factor (AGUILA 1994). While it is unlikely that L-arginine acts on GH release via the generation of NO, the latter seems to exert a permissive role on GH release stimulated by GH-releasing hexapeptide and N-methyl-D-aspartate in vivo in the rat (TENA-SEMPERE et al. 1996).

It is not clear whether NO is involved in the control of prolactin (PRL) secretion. In vitro studies on isolated pituitaries show that increased NO levels are usually associated with reduced PRL release, although one group found NO-induced increase in PRL secretion under basal conditions or after IL-1β stimulation (BRUNETTI et al. 1995). NO seems to mediate the inhibition of PRL release induced by interferon-γ, and it might also be involved in inhibition caused by dopamine or atrial natriuretic factor (DUVILANSKI et al. 1995; VANKELECOM et al. 1997). The inhibitory effect of NO on PRL release might be mediated by increases in GABA generation induced by the gas (DUVILANSKI et al. 1996). Interpretation of in vivo studies is difficult; in one report, intracerebroventricular administrations of NO donors elicited dose-dependent increases in PRL circulating levels, but both L-arginine and NOS inhibitors failed to affect PRL secretion (GONZALES et al. 1996). Another study showed potentiation and inhibition by N^G-nitro-L-arginine methyl ester of morphine- and stress-induced PRL rises, respectively (MATTON et al. 1997). We were unable to demonstrate any role for NO in the induction of PRL release by L-arginine in human studies (KORBONITS et al. 1996).

A complex interplay seems to occur between NO and melatonin, since the latter reduces NOS activity in rat hypothalamus (BETTAHI et al. 1996) whereas NO donors inhibit melatonin synthesis in rat and bovine pineal cells (MARONDE et al. 1995). NO has also been shown to mediate certain effects of melatonin in the suprachiasmatic nuclei in vitro (STARKEY 1996).

References

Aguan K, Mahesh VB, Ping L, Bhat G, Brann DW (1996) Evidence for a physiological role for nitric oxide in the regulation of the LH surge: effect of central administration of antisense oligonucleotides to nitric oxide synthase. Neuroendocrinology 64:449–455

Aguila MC (1994) Growth hormone-releasing factor increases somatostatin release and mRNA levels in the rat periventricular nucleus via nitric oxide by activation of guanylate cyclase, Proc Natl Acad Sci USA 91:782–786

Alba-Roth J, Muller GA, Schophol J, Werder K (1966) Arginine stimulates growth hormone secretion by suppressing endogenous somatostatin secretion. J Clin Endocrinol Metab 67:1186–1189

Alm P, Skagerberg G, Nylen A, Larsson B, Andersson KE (1997) Nitric oxide synthase and vasopressin in rat circumventricular organs. An immunohistochemical study. Exp Brain Res 117:59–66

Bettahi I, Pozo D, Osuna C, Reiter RJ, Acuna-Castrviejo D, Guerrero JM (1996) Melatonin reduces nitric oxide synthase activity in rat hypothalamus. J Pineal Res 20:205–210

Bhat G, Mahesh VB, Aguan K, Brann DW (1996) Evidence that brain nitric oxide synthase is the major nitric oxide synthase isoform in the hypothalamus of the adult female rat and that nitric oxide potently regulates hypothalamic cGMP levels. Neuroendocrinology 64:93–102

Bonavera JJ, Sahu A, Kalra PS, Kalra SP (1993) Evidence that nitric oxide may mediate the ovarian steroid-induced luteinizing hormone surge: involvement of excitatory amino acids. Endocrinology 133:2481–2487

Bonavera JJ, Kalra PS, Kalra SP (1996) L-arginine/nitric oxide amplifies the magnitude and duration of the luteinizing hormone surge induced by estrogn: involvement of neuropeptide Y. Endocrinology 137:1956–1962

Bredt DS, Hwang PM, Snyder SH (1990) Localization of nitric oxide synthase indicating a neural role for nitric oxide. Nature 347:768–770

Brunetti L, Preziosi P, Ragazzoni E, Vacca M (1993) Involvement of nitric oxide in basal and interleukin-1β-induced CRH and ACTH release in vitro. Life Sci 53:PL219–222

Brunetti L, Ragazzoni E, Preziosi P, Vacca M (1995) A possible role for nitric oxide but not for prostaglandin E2 in basal and interleukin-1-beta-induced PRL release in vitro. Life Sci 56:PL277–PL283

Calka J, Block CH (1993) Relationship of vasopressin with NADPH-diaphorase in the hypothalamo-neurohypophysial system, Brain Res Bull 32:207–210

Ceccatelli S, Lundberg JM, Fahrenkrug J, Bredt DS, Snyder SH and Hockfelt T (1992) Evidence for involvement of nitric oxide in the regulation of hypothalamic portal blood flow. Neuroscience 51:769–772

Ceccatelli S, Grandison L, Scott REM, Pfaff DW, Kow L-M (1996) Estradiol regulation of nitric oxide synthase mRNAs in rat hypothalamus. Neuroendocrinology 64:357–363

Chiodera P, Volpi R, Caffarri G, Capretti L, Magotti MG, Coiro V (1995) Mediation by nitric oxide of LH-RH-stimulated gonadotropin secretios in human subjects. Neuropeptides 29:321–324

Chiodera P, Volpi R, Capretti L, Coiro V (1996) Gamma-aminobutyric acid mediation of the inhibitory effect of nitric oxide on the arginine vasopressin and oxytocin responses to insulin-induced hypoglycemia, Regul Pept 67:21–25

Coiro V, Volpi R, Chiodera P (1995) Mediation by nitric oxide of TRH-, but not metoclopramide-stimulated TSH secretion in humans. Neuroreports 6:1174–1176

Costa A, Trainer P, Besser M, Grossman A (1993) Nitric oxide modulates the release of corticotrophin-releasing hormone from the rat hypothalamus in vitro. Brain Res 605:187–192

Dawson TM, Bredt DS, Fotuhi M, Hwang PM, Snyder SH (1991) Nitric oxide synthase and neuronal NADPH-diaphorase are identical in brain and peripheral tissues. Proc Natl Acad Sci USA 88:7797–7801

Duvilanski BH, Zambruno C, Seilicovich A, Pisera D, Lasaga M, Diaz MC, Belova N, Rettori V, McCann SM (1995) Role of nitric oxide in control of prolactin release by the adenohypophysis. Proc Natl Acad Sci USA 92:170–174

Duvilanski BH, Zambruno C, Lasaga M, Pisera D, Seilicovich A (1996) Role of nitric oxide/cyclic GMP pathway in the inhibitory effect of GABA and dopamine on prolactin release. J Neuroendocrinol 8:909–913

Fang FC (1997) Mechanisms of nitric oxide-related antimicrobial activity. J Clin Invest 99:2818–2825

Giordano M, Vermeulen M, Trevani AS, Dran G, Andonegui G, Geffner JR (1996) Nitric oxide synthase inhibitors enhance plasma levels of corticosterone and ACTH. Acta Physiol Scand 157:259–264

Gonzalez MC, Linares JD, Santos M, Llorente E (1996) Effects of nitric oxide donors sodium nitroprusside and 3-morpholino-sydnonimine on prolactin secretion in conscious rats. Neurosci Lett 203:167–170

Grossman A, Rossmanith WG, Kabigting EB, Cadd G, Clifton D, Steiner RA (1994) The distribution of hypothalamic nitric oxide synthase mRNA in relation to gonadotropin-releasing hormone neurons. J Endocrinol 140:R5–R8

Grossman A, Costa A, Forsling ML, Jacobs R, Kostoglou-Athanassiou I, Nappi G, Navarra P, Satta MA (1997) Gaseous neurotransmitters in the hypothalamus. The role of nitric oxide and carbon monoxide in neuroendocrinology. Horm Metabol Res 29:477–482

Hatakeyama S, Kawai Y, Ueyama T, Semba E (1996) Nitric oxide synthase-containing magnocellular neurons of the rat hypothalamus synthesize oxytocin and vasopressin and express FPS following stress stimuli. J Chem Neuroanatom 11:243–256

Herbison AE, Simonian SX, Norris PJ, Emson PC (1996) Relationship of neuronal nitric oxide synthase immunoreactivity to GnRH neurons in the ovariectomized and intact female rat, J Neuroendocrinol 8:73–82

Jacobs RA, Dahia PLM, Satta MA, Chew SL, Grossman AB (1997) Induction of nitric oxide synthase and interleukin-1β, but not heme oxygenase, messenger RNA in rat brain following peripheral administration of endotoxin. Mol Brain Res 49: 238–246

Kadowaki K, Kishimoto J, Leng G, Emson PC (1994) Up-regulation of NOS gene expression together with NOS activity in the rat hypothalamo-hypophysial system after chronic salt loading: evidence of a neuromodulatory role of nitric oxide on arginine vasopressin and oxytocin secretion. Endocrinology 134:1011–1017

Karanth S, Lyson K, McCann SM (1993) Role of nitric oxide in interleukin-2-induced corticotropin-releasing factor release from incubated hypothalami. Proc Natl Acad Sci USA 90:3383–3387

Kim CK, Rivier C (1998) Influence of nitric oxide synthase inhibitors on the ACTH and cytokine responses to peripheral immune signals. J Neuroendocrinol 10:353–362

Korbonits M, Trainer PJ, Fanciulli G, Oliva O, Pala A, Dettori A, Besser M, Delitala G, Grossman AB (1996) L-arginine is unlikely to exert neuroendocrine effects in humans via the generation of nitric oxide. Eur J Endocrinol 135:543–547

Licinio J and Wong M (1997) Pathways and mechanisms for cytokine signaling of the central nervous system. J Clin Invest 100:2941–47

Lopez FJ, Moretto M, Merheethalei I, Negro-Vllar A (1997) Nitric oxide is involved in the genesis of pulsatile LHRH secretion from immortalized LHRH neurons. J Neuroendocrinol 9:647–654

Luckman SM, Huckett L, Bicknell RJ, Voisin DL, Herbison AE (1997) Up-regulation of nitric oxide synthase messenger RNA in an integrated forebrain circuit involved in oxytocin secretion. Neuroscience 77:37–48

Mahachoklertwattana P, Black SM, Kaplan SL, Bristow JD, Grumbach MM (1994) Nitric oxide synthesized by gonadotrophin-releasing hormone neurons is a mediator of N-methyl-D-asparate (NMDA)-induced GnRH secretion. Endocrinology 135:1709–1712

Mancuso C, Tringali G, Grossman A, Preziosi P, Navarra P (1998) The generation of nitric oxide and carbon monoxide produces opposite effects on the release immunoreactive interleukin-1β from the rat hypothalamus in vitro: evidence for the involvement of different signaling pathways. Endocrinology 139:1031–1037

Maronde E, Middlendorff R, Mayer B, Olcese J (1995) The effect of NO-donors in bovine and rat pineal cells: stimulation of cGMP and cGMP-independent inhibition of melatonin synthesis. J Neuroendocrinol 7:207–214

Matton A, Bollengier F, Finne E, Vanhaelst L (1997) Effect of N^{ω}-nitro-L-arginine methyl ester, a nitric oxide synthesis inhibitor, on stress- and morphine-induced prolactin release in male rats, Br J Pharmacol 120:268–272

Melis MR, Argiolas A (1997) Role of central nitric oxide in the control of penile erection and yawning. Prog Neuropsychoparmacol Biol Psychiatry 21:899–822

Merimee TJ, Rabinowitz Y, Fineberg SE (1969) Arginine initiated release of human growth hormone: factors modifying the release in normal men. New Engl J Med 280:1434–1438

Moncada S, Palmer RMJ, Higgs EA (1991) Nitric oxide: physiology, pathophysiology, and pharmacology. Pharmacol Rev 43:109–142

Moretto M, Lopez FJ, Negro-Vilar J (1993) Nitric oxide regulates luteinizing hormone-releasing hormone secretion. Endocrinology 133:2399–2402

Navarra A, Pozzoli G, Brunetti L, Ragazzoni E, Besser GM, Grossman A 1992 Interleukins-1β and -6 specifically increase the release of prostaglandin E2 from rat hypothalamic explants in vitro. Neuroendocrinology 56:61–68

Ni X, Chan EC, Fitter JT, Smith R (1997) Nitric oxide inhibits corticotrophin-releasing hormone exocytosis but not synthesis by cultured human trophoblasts. J Clin Endocrinol Metab 82:4171–4175

Nocentini G (1996) Ribonucleotide reductase inhibitors: new strategies for cancer chemotherapy, Crit Rev Oncol Hematol 22:89–126

Okamura H, Yokosuka M, McEwen BS, Hayashi S (1994) Colocalization of NADPH-diaphorase and estrogen receptor immunoreactivity in the rat ventromedial hypothalamic nucleus: stimulatory effect of estrogen on NADPH-diaphorase activity. Endocrinology 135:1705–1708

O'Shea RD and Gundlach AL (1996) Food or water deprivation modulate nitric oxide synthase (NOS) activity and gene expression in rat hypothalamic neurones: correlation with neurosecretory activity? J Neuroendocrinol 8:417–425

Palmer RMJ, Ferridge AG, Moncada S (1987) Nitric oxide release accounts for the biological activity of endothelium-derived relaxing factor. Nature 327:524–526

Pow DV (1992) NADPH-diaphorase (nitric oxide synthase) staining in the rat supraoptic nucleus is activity-dependent: possible functional implications. J Neuroendocrinol 4:377–380

Pu S, Xu B, Kalra SP, Kalra PS (1996) Evidence that gonadal steroids modulate nitric oxide influx in the medial preoptic area: effects of N-methyl-S-disparate and correlation with luteinizing hormone secretion. Endocrinology 137:1949–1955

Raber J, Bloom FE (1994) IL-2 induces vasopressin release from the hypothalamus and the amygdala: role of nitric oxide-mediated signalling. J Neurosci 14:6187–6195

Rettori V, Gimeno M, Lyson K, McCann S (1992) Nitric oxide mediates norepinephrine-induced prostaglandin E2 release from the hypothalamus. Proc Natl Acad Sci USA 89:11543–11546

Rettori V, Belova N, Kamat A, Lyson K, McCann SM (1994) Blockade by interleukin-1β of nitricoxidergic control of luteinizing-hormone-releasing hormone release in vivo and in vitro. Neuroimmunomodulation 1:86–91

Rettori V, Canteros G, Renoso R, Gimeno M, McCann SM (1997) Oxytocin stimulates the release of luteinizing hormone-releasing hormone from medial basal hypothalamic explants by releasing nitric oxide. Proc Natl Acad Sci USA 94:2741–2744

Reuss S, Decker K, Rosseler L, Layes E, Schollmayer A, Spessert R (1995) Nitric oxide synthase in the hypothalamic suprachiasmatic nucleus of rat: evidence from histochemistry, immunohistochemistry and western blot; and colocalization with VIP. Brain Res 695:257–262

Rivier C (1994) Endogenous nitric oxide participates in the activation of the hypothalamic–pituitary–adrenal axis by noxious stimuli. Endocr J 2:367–373

Rivier C (1995) Blockade of nitric oxide formation augments adrenocorticotropin released by blood-borne interleukin-1β: role of vasopressin, prostaglandins, and α1-adrenergic receptors, Endocrinology 136:3597–3603

Rivier C, Shen GH (1994) In the rat, nitric oxide modulates the response of the hypothalamic-pituitary-adrenal axis to interleukin-1β, vasopressin, and oxytocin. J Neurosci 14:1985–1993

Roe CM, Leitch IM, Boura AL, Smith R (1996) Nitric oxide regulation of corticotrophin-releasinh hormone release from the human perfused placenta in vitro. J Clin Endocrinol Metab 81:763–769

Sagar SM, Ferriero DM (1987) NADPH diaphorase activity in the posterior pituitary: relation to neuronal function. Brain Res 400:348–352

Sanchez F, Rubio M, Hernandez V, Santos M, Carretero J, Vasquez RJ, Vasquez R (1996) NADPH-diaphorase activity and vasopressin in the paraventricular nucleus of the hypothalamus following adrenalectomy. Neuropeptides 30:515–520

Sandi C, Guaza C (1995) Evidence for a role of nitric oxide in the corticotrophin-releasing factor release induced by interleukin-1β. Eur J Pharmacol 274:17–23

Satta M A, Jacobs RA, Kaltsas GA, Grossman AB (1998) Endotoxin induces interleukin-1β and nitric oxide synthase mRNA in rat hypothalamus and pituitary. Neuroendocrinology 67:109–116

Sortino MA, Aleppo G, Scapagnini U, Canonico PL (1994) Involvement of nitric oxide in the regulation of gonadotropin-releasing hormone release from the GT1–1 cell line. Endocrinology 134:1782–1787

Starkey SJ (1996) Melatonin and 5-hydroxytryptamine phase-advance the rat circadian clock by activation of nitric oxide synthesis. Neurosci Lett 211:199–202

Summy-Long JY, Bui V, Mantz S, Koehler E, Weisz J, Kadekaro M (1993) Central inhibition of nitric oxide synthase preferentially augments release of oxytocin during dehydration. Neurosci Lett 152:190–193

Tena-Sempere M, Pinilla L, Gonzalez D, Aguilar E (1996) Involvement of endogenous nitric oxide in the control of pituitary responsiveness to different elicitors of growth hormone release in prepubertal rats. Neuroendocrinology 64:146–152

Torres G, Lee S, Rivier C (1993) Ontogeny of the rat nitric oxide synthase and colocalization with neuropeptides. Mol Cell Neurosci 4:155–163

Tsagarakis S, Gillies G, Rees LH, Besser GM, Grossman A (1989) Interleukin–1 directly stimulates the release of corticotrophin releasing factor from rat hypothalamus. Neuroendocrinology 49:98–101

Tsuchiya T, Kishimoto J, Koyama J, Ozawa T (1997) Modulatory effect of L-NAME, a specific nitric oxide synthase (NOS) inhibitor, on stress-induced changes in plasma adrenocorticotropic hormone (ACTH) and corticosterone levels in rats: physiological significance of stress-induced NOS activation in hypothalamic-pituitary-adrenal axis. Brain Res 776:68–74

Turnbull AV, Rivier C (1996) Corticotopin-releasing factor, vasopressin, and prostaglandins mediate, and nitric oxide restrains, the hypothalamic-pituitary-adrenal response to acute local inflammation in the rat. Endocrinology 137:455–463

Ueta Y, Levy A, Chowdrey HS, Lightman SL (1995a) Inhibition of hypothalamic nitric oxide synthase gene expression in the rat paraventricular nucleus by food deprivation is independent of serotonin depletion. J Neuroendocrinol 7:861–865

Ueta Y, Levy A, Chowdrey HS, Lightman SL (1995b) Hypothalamic nitric oxide synthase gene expression is regulated by thyroid hormones. Endocrinology 136:4182–4187

Vankelecom H, Matthys P, Denef C (1997) Involvement of nitric oxide in the interferon-gamma-induced inhibition of growth hormone and prolactin secretion in anterior pituitary cell cultures. Mol Cell Endocrinol 129:157–167

Vincent SR, Kimura H (1992) Histochemical mapping of nitric oxide synthase in rat brain. Neuroscience 46:755–784

Volpi R, Chiodera P, Caffarri G, Vescovi PP, Capretti L, Gatti C, Coiro V (1996) Influence of nitric oxide on hypoglycemia- or angiotensin II-stimulated ACTH and GH secretion in normal men. Neuropeptides 30:528–532

Wayman CP, Pike NV, Wilson JF (1994) N-methyl-D-aspartate (NMDA) stimulates release of α-MSH from the rat hypothalamus through release of nitric oxide. Brain Res 666:201–206

Wong ML, Rettori V, Al-Sheklee A, Bongiorno PB, Canteros G, McCann SM, Gold PWE, Licinio J (1996) Inducible nitric oxide synthase gene expression in the brain during systemic inflammation. Nature Medicine 2:581–584

Yamada K, Emson P, Hokfelt T (1996) Immunohostochemical mapping of nitric oxide synthase in the rat hypothalamus and colocalization with neuropeptides. J Chem Neuroanat 10:295–316

Yasin S, Costa A, Trainer P, Windle R, Forsling ML, Grossman A (1993) Nitric oxide modulates the release of vasopressin from rat hypothalamic explants. Endocrinology 133:1466–1469

Yu WH, Walczewska A, Karanth S, McCann SM (1997) Nitric oxide mediates leptin-induced luteinizing hormone-releasing hormone (LHRH) and LHRH and leptin-induced LH release from the pituitary gland. Endocrinology 138:5505–5048

CHAPTER 14
The Role of Nitric Oxide in Kidney Function

E. HACKENTHAL

A. Introduction

There is increasing evidence that nitric oxide (NO) is one of the most important paracrine modulators and mediators in the control of renal functions, such as overall and regional renal blood flow (RBF), renal autoregulation, glomerular filtration, renin secretion and salt excretion. NO also plays an important role in the pathogenesis of several renal disease states, such as diabetic nephropathy, inflammatory glomerular disease, acute renal failure in septic shock, chronic renal failure and nephrotoxicity of drugs, conveying both beneficial effects via its hemodynamic functions and detrimental effects via its cytotoxicity when produced in large amounts by inducible nitric oxide synthase (iNOS). The extensive literature on these pathophysiological functions of NO will not be covered here, and the reader is referred to the pertinent chapters (Chaps. 18, 21, 22, 23) for discussion.

In the kidney, an unusually large diversity of cells of different types is integrated in a delicate and complex architecture, forming various types of functional units. The complexity of the spatial arrangement of cells in these functional units, for example in the glomerulus and the juxtaglomerular apparatus (JGA), is not only a crucial prerequisite for proper functioning but also a serious obstacle to the analysis of these functions. This problem also extends to the analysis of the intrarenal functions of NO, and much of the interpretation of data rests on assumptions made as to both the site of formation and the target of NO in each particular experiment. For this reason, a brief survey of the distribution of NOS isoforms in the kidney is given first. For more extensive information, the reader is referred to recent reviews (BACHMANN and MUNDEL 1994; BACHMANN 1997; KONE and BAYLIS 1997; STAR 1997).

B. Nitric Oxide Synthase Isoforms in the Kidney

Expression of all three main isoforms of nitric oxide synthase (NOS), i.e. endothelial NOS (eNOS, NOS III), neuronal NOS (nNOS, NOS I) and iNOS (NOS II), has been demonstrated in whole kidney extracts or dissected renal zones by Western blotting or enzyme-linked immunosorbent assay, by measuring isoform-specific mRNA in RNAse-protection assays, by Northern blotting or by the reverse transcriptase-polymerase chain reaction (RT-PCR). Histochemical localisation of NOS has been performed with the reduced nicotinamide adenine dinucleotide phosphate (NADPH)-diaphorase reaction (which, however, does not differentiate between isoforms) and by immunocytochemistry using isoform-specific antibodies, and of NOS mRNA by in situ hybridization with isoform-specific mRNA probes.

There are also reports on kidney-specific variants of NOS mRNA, one of which codes for a shortened protein (BACHMANN and OBERBÄUMER 1998). However, information on the expression, intrarenal distribution and possible cell-specific functions of these variants of NOS is not yet available.

In the rat kidney, iNOS (NOS II) is also represented by two isoforms, at least at the mRNA level. Because of their homology to macrophage or vascular smooth muscle (VSM) NOS II mRNA, these isoforms have been called Mac-NOS II and VSM-NOS II, respectively (KONE and BAYLIS 1997). As pointed out by KONE and BAYLIS, the functional and regulatory properties of the two isoforms of iNOS and their contribution to renal physiology and pathophysiology are as yet unknown. A recent study by NOIRI et al. (1996) indicates that the two isoforms may have opposing functions, i.e., protective and detrimental, in post-ischemic renal failure.

C. Distribution of NOS in the Kidney

I. Distribution of NOS in the Renal Vasculature

In the endothelium of interlobar, arcuate and interlobular arteries, eNOS (NOS III) expression has been demonstrated with the NADPH-diaphorase reaction and specific antibody staining and, at the mRNA level, by in situ hybridization or RT-PCR (TERADA et al. 1992; TOJO et al. 1994; UJIIE et al. 1994; BACHMANN et al. 1995). The glomerular afferent and efferent arterioles and the glomerular capillaries exhibit immunocytochemical staining, with the highest intensity located at the efferent arteriole and only weak staining of glomerular capillaries (TERADA et al. 1992; TOJO et al. 1994b; UJIIE et al. 1994; BACHMANN et al. 1995), while the mesangium is NOS III-negative (BACHMANN et al. 1995). Interestingly, in the podocyte lining of the glomerulus, immunoreactivity for the α_1-subunit of guanylate cyclase, the target of NO, was detected (BACHMANN 1997). The preferential staining of the efferent arteriole has been associated with a putative role of NOS in shear-stress-mediated relaxation (BACHMANN 1997). However, this view contrasts with functional data (see below).

In addition to NOS III, NOS I-specific immunoreactivity was also detected in the endothelium of the efferent arteriole (BACHMANN et al. 1995). The diaphorase reaction produces NOS-positive staining extending from the inner medulla to the point where the descending vasa recta branches from the efferent arteriole (BACHMANN et al. 1995). These immunocytochemical data were confirmed by RT-PCR in microdissected vasa recta bundles (TERADA et al. 1992). While the concentration of cyclic guanosine monophosphate (cGMP) was reported to be higher in the inner medulla compared with the outer medulla, diaphorase-positive staining was more pronounced in the outer medulla (BACHMANN et al. 1995).

The intrarenal distribution of iNOS (NOS II) is an area of conflicting results and opinions. By histochemical techniques, NOS II immunoreactivity could not be detected under basal conditions in any vascular segment, except for a weak signal in the afferent arteriole (AHN et al. 1994; TOJO et al. 1994). In contrast, "tonic" expression of a VSM-type iNOS mRNA was identified by quantitative PCR in microdissected arcuate and interlobular arteries (MOHAUPT et al. 1994). The signal was only moderately enhanced by lipopolysaccharide (LPS) stimulation. In isolated glomeruli and cultured mesangial cells, the Mac-type iNOS was the principal isoform found by PCR under basal conditions, while the VSM-type isoform of NOS II mRNA was inducible by cytokines (AHN et al. 1994; MOHAUPT et al. 1994; MÜHL et al. 1994; TOJO et al. 1994). Similar to the situation in pre-glomerular vessels, histochemical analysis failed to detect significant diaphorase staining or specific NOS II immunoreactivity in the mesangium of unstimulated kidneys in situ (AHN et al. 1994; MOHAUPT et al. 1994). The obvious discrepancy between immunocytochemical in situ data and PCR data from microdissected material with respect to the expression of NOS II under basal conditions, i.e. without an inducing agent, cannot be explained at present. In accordance with the concept of NOS II as an inducible, inflammation-related enzyme, NOS II expression can be identified by immunocytochemistry in acutely inflamed glomeruli, e.g. in immuncomplex glomerulonephritis, with a preferential localization to glomerular macrophages or polymorphonuclear leukocytes. NOS II mRNA was found in mesangial cells by in situ hybridization (JANSEN et al. 1994; GOTO et al. 1995). However, in endotoxin-treated rats, mesangial cell staining for NOS II mRNA by in situ hybridization and isoform-specific immunocytochemistry for NOS II protein has been demonstrated (AHN et al. 1994; BUTTERY et al. 1994). Furthermore, in cultured rat or human mesangial cells, NOS II can be induced by tumor necrosis factor α, interleukin 1 and interferon γ (IFNγ) (MARSDEN and BALLERMANN 1990; PFEILSCHIFTER and SCHWARZENBACH 1990; PFEILSCHIFTER et al. 1992; NICOLSON et al. 1993). This inducible NOS II has been identified as the Mac-type NOS II (PFEILSCHIFTER et al. 1992).

In summary, eNOS expression has convincingly been demonstrated to exist in major parts of the arterial and capillary system of the kidney, including arcuate and interlobular arteries, afferent and efferent arterioles and the medullary vasa recta. The renovascular distribution of iNOS (NOS II) in basal conditions, i.e. in the absence of inducing inflammatory reactions, remains to be clarified.

II. Distribution of NOS in Renal Tubules

In primary cultures of human proximal tubule cells, NOS I mRNA has been identified by Northern-blot analysis (HWANG et al. 1994) while, in LLC-PK1 kidney cells, which resemble proximal tubule cells, NOS III protein was found by Western blot (Ishii et al. 1991; Tracey et al. 1994). However, in studies using RT-PCR on microdissected proximal tubules of rat kidney (Terada et al. 1992), and in immunocytochemistry and in situ hybridization studies with isoform-specific antibodies or mRNA probes (Mundel et al. 1992; Bachmann et al. 1995), no evidence for the presence of NOS I or III mRNA or the respective proteins was obtained. Therefore, it is not possible at present to identify the source of NO, which is thought to influence ion transport in the proximal tubule (see below). Likewise, the presence of NOS II in the proximal tubule is still a matter of debate. Reports on the "tonic" expression of Mac-type NOS II mRNA in microdissected rat proximal tubule (Terada et al. 1992), and LPS- or IFNγ-inducible NOS II in the same preparation (Harris et al. 1994; Mayeux et al. 1995) contrast with negative findings by Thorup and Persson (1994), who could not detect NOS II mRNA in microdissected proximal tubule segments under basal or stimulated conditions by RT-PCR.

In the thick ascending limb (TAL) neither NOS I mRNA (Terada et al. 1992) nor NOS I protein (Bachmann et al. 1995) could be detected, except in a few scattered cells of the cortical TAL close to the macula densa (Bachmann et al. 1995). Interestingly, this pattern resembles findings on the tubular distribution of inducible cyclooxygenase (COX-2) mRNA (Bachmann 1997). iNOS mRNA has been found in the medullary and cortical TAL by RT-PCR and in situ hybridization and appears to be of the Mac-type NOS II (Terada et al. 1992; Mühl et al. 1994), whereas immunocytochemical analysis indicates the presence of smooth-muscle-type NOS II under basal conditions (Tojo et al. 1994b).

In contrast to other tubular segments, convincing evidence obtained by in situ hybridization, RT-PCR, the diaphorase reaction and immunocytochemical analysis is available for the presence of NOS I mRNA and NOS I protein in macula densa cells of all species studied (Mundel et al. 1992; Schmidt et al. 1992; Wilcox et al. 1992; Thorup and Persson 1994; Tojo et al. 1994b; Bachmann et al. 1995; Bachmann 1997). The NOS signal in macula densa cells is by far the strongest of all tubular segments and co-localizes with the signal for COX-2. Expression of both NOS I and COX-2 in macula densa cells is significantly increased during dietary salt reduction (Harris et al. 1994; Tojo et al. 1995). NOS II mRNA was not detected in macula densa cells by in situ hybridization (Ahn et al. 1994).

In the distal convoluted tubule, NOS I was not detected by immunocytochemistry, although a weak signal for NOS I mRNA has been reported (Ahn et al. 1994). As for the proximal tubule, the presence of inducible NOS II has also been reported for distal convoluted tubules, connecting tubules, cortical collecting ducts, intercalated cells and even the papillary surface epithelia, with little change in signal intensity upon LPS-stimulation (Tojo et al. 1994b; Ahn et al. 1994). The widespread distribution of iNOS in renal tubules requires confirmation before the possible functional significance can be assessed.

III. Distribution of NOS in Renal Nerves

In renal nerves, NOS I and diaphorase staining has been localized to the ganglionic plexus of the kidney hilus, renal nerves in the wall of the renal pelvis and somata near the interlobular artery (LIU and BARAJAS 1993) as well as to nerve fibers associated with other renal arteries (BACHMANN et al. 1995). A few nerve fibers approaching the JGA were also NOS I-positive; however, no direct association of NOS-containing neurons with the JGA could be detected (BACHMANN et al. 1995).

D. Physiological Roles of NO

I. Role of NO in the Regulation of Renal Blood Flow

1. Endogenous Mediators of NO Release

Among the mediators and conditions described to stimulate the generation and release of NO in the general circulation under physiological conditions, shear stress is most likely to play a major role as a mediator of NO release in the kidney vasculature. However, despite numerous studies suggesting a contribution of shear-stress-induced release of NO in the control of renal hemodynamics, this issue remains controversial. In fact, no direct evidence for this mechanism is available at present (see below).

Another signal for the release of NO in the general circulation is bradykinin (or other kinins). Bradykinin is both a circulating tissue hormone and an intrarenally generated paracrine mediator. Systemic or intrarenal infusion of bradykinin produces a significant decrease in renovascular resistance (RVR), which is mediated largely by the liberation of NO (LAHERA et al. 1990; KING and BRENNER 1991). However, the contribution of endogenous bradykinin to the maintenance of RBF seems to be small: the non-specific inhibitor of kallikrein, aprotinin, had no effect on RBF or glomerular filtration rate (GFR) in rats or pigs (KRAMER et al. 1979; MAIER et al. 1985) and induced only a small reduction in RVR in volume-expanded rats (SETO et al. 1983). Similarly, infusion of a non-selective or B_2-selective bradykinin-receptor antagonist failed to influence RBF, GFR or sodium excretion in conscious or anesthetized rats (ROMAN et al. 1988; MADEDDU et al. 1990) and reduced RBF only slightly in anesthetized dogs, without influencing RVR (BEIERWALTES et al. 1988). In salt-restricted conscious rats, bradykinin antagonists produced a small reduction in RBF but not in GFR (MADEDDU et al. 1992). These observations are in agreement with reports on low intrarenal levels of endogenous bradykinin, as measured by microdialysis, and the observation that bradykinin levels can be augmented by low sodium intake (SIRAGY 1993; SIRAGY et al. 1993, 1994). In conclusion, endogenous bradykinin appears to be a minor determinant of NO release under physiological conditions but may contribute to NO release in salt depletion or during treatment with angiotensin-converting enzyme (ACE) inhibitors. Likewise, the role of endogenously released acetylcholine (another mediator of NO release) in the control of renal hemodynamics is small and, if contributing at all, is probably mediated by the release of vasodilator prostaglandins rather than release of NO (SALOM et al. 1991; MAJID and NAVAR 1992; HAYASHI et al. 1994).

2. Inhibitors of NOS

In the normal kidney, tonically generated NO plays a major role in the control of RBF and GFR. It has been suggested that NO has a greater role in maintaining the low vascular resistance of the kidney than in other tissues since, for example, coronary, brain or femoral blood flow are less affected by NOS inhibitors than is RBF.

Systemic infusion of NOS inhibitors, such as N^G-momomethyl L-arginine (L-NMMA) or N^G-nitro L-arginine methyl ester (L-NAME), decreases RBF and increases RVR. This has been demonstrated in conscious and anesthetized dogs (LAHERA et al. 1991; ALBEROLA et al. 1992; BAUMANN et al. 1992; CHEVALIER et al. 1992; MAJID and NAVAR 1992; MAJID et al. 1993; PERSSON et al. 1993; MANNING and HU 1994) and rats (BREZIS et al. 1991; SHULTZ and TOLINS 1993; QIU et al. 1994). While part of the increase in RVR may be autoregulatory in response to the increase in arterial pressure, most of the effect is certainly intrarenal, since low-dose systemic infusion and intrarenal infusion of NOS-inhibitors decrease RBF in the absence of increases of arterial pressure (LAHERA et al. 1991; ALBEROLA et al. 1992; WOLTZ et al. 1997). Furthermore, infusion of NOS inhibitors into the isolated perfused kidney similarly reduces renal perfusate flow when the kidney is perfused at constant pressure (BREZIS et al. 1991; MÜNTER and HACKENTHAL 1991; RADERMACHER et al. 1991, 1992; HACKENTHAL et al. 1994). While basal RBF is usually reduced by NOS inhibitors by about 20–30%, their effect on GFR is smaller and less consistent because the filtration fraction rises (see below).

It is still unclear to what extent the renal vasoconstrictor response to NOS inhibition results from lack of NO vasodilation and to what extent it is the consequence of an augmented vasoconstrictor activity. Studies in conscious animals in which renal vasoconstrictor systems, such as renal sympathetic tone, the renin-angiotensin system and others operate at very low levels (BAUMANN et al. 1992; GRANGER et al. 1992; PERSSON et al. 1993; MANNING and HU 1994; QIU et al. 1994), animal studies in which the renal vasoconstrictor response to NOS inhibition persisted during combined α_1-adrenergic and angiotensin II blockade (RAIJ and BAYLIS 1995; KONE and BAYLIS 1997) and studies in isolated perfused kidneys, in which most of these systems are absent suggest that the role of activation of vasoconstrictor system in mediating the response to NOS inhibition is not prominent. This may be different in conditions where vasoconstrictor systems are already activated, such as anesthesia or hypovolemia (RAIJ and BAYLIS 1995; KONE and BAYLIS 1997; MADRID et al. 1997). Thus, under basal conditions, tonic endogenous generation of NO (possibly by eNOS) significantly contributes to the maintenance of the relatively low vasculature of the kidney by a direct effect on the renal vasculature. The renovascular sensitivity to NOS inhibitors increases in rats and healthy humans on a high-salt diet (SHULTZ and TOLINS 1993; DENG et al. 1994; BECH et al. 1998), probably as a consequence of increased medullary NO formation in the kidney (DENG et al. 1994). NOS-inhibitor-treated rats also develop higher arterial pressures (TOLINS and SHULTZ 1994). Renovascular sensitivity to NOS inhibitors also increases with age (RECKELHOFF and MANNING 1993; HILL et al. 1997), and it has been concluded that tonically generated NO has a more prominent role in maintaining renal perfusion in aging (HILL et al. 1997), which is consistent with the observation that supplementation with L-arginine protects rats against age-related reductions in renal function (RECKELHOFF et al. 1997).

II. Role of NO in Glomerular Circulation

Glomerular filtration, one of the fundamental functions of the kidney, is critically determined by the balance of vascular tone between the afferent and efferent arterioles. There is a continuing debate about the role of NO in the control of this balance, i.e. whether NO vasodilates predominantly the afferent or the efferent arteriole (NAVAR et al. 1996).

In rat micropuncture studies by ZATZ and DE NUCCI (1991), systemic blockade of NO synthesis increased efferent more than afferent arteriolar resistance and, as a consequence, glomerular pressure increased substantially. Others found no significant difference in afferent and efferent arteriolar resistance when NOS inhibitors were given systemically to dogs, rats or rabbits (LAHERA et al. 1991; PERRELLA et al. 1991; DE NICOLA et al. 1992; DENG and BAYLIS 1993; DENTON and ANDERSON 1994). In all reported studies, inhibition of NO synthesis decreased glomerular flow and hydraulic conductivity of filtration (K_f) (NAVAR et al. 1996). Since systemic application of NOS inhibitors invariably increased arterial pressure and may have influenced other inputs to the kidney, such as renal sympathetic tone, the preferential site of vasodilatation of endogenous NO in the glomerulus remained uncertain. This problem was addressed in a study by DENG and BAYLIS (1993). During systemic L-NAME infusion, afferent and efferent arteriolar resistances increased equally, glomerular pressure increased and glomerular flow decreased. In contrast, intrarenal infusion of L-NAME at a concentration which had no influence on arterial pressure induced a significant increase in afferent arteriolar resistance and no significant change in efferent arteriolar resistance, while glomerular pressure was not altered. In both conditions, i.e. systemic and intrarenal application of L-NAME, single-nephron glomerular filtration rate (SNGFR) and K_f decreased. The authors concluded that locally produced NO primarily affects afferent arteriolar resistance. They further speculated that the observed decrease in K_f by L-NAME, which had also been observed by others (ZATZ and DE NUCCI 1991; DENTON and ANDERSON 1994), reflects the withdrawal of the relaxing effect of NO on mesangial cells.

The role of NO in pre- and postglomerular vasodilation has also been studied in glomerular preparations (OHISHI et al. 1992; IMIG et al. 1993; ITO and REN 1993) and isolated perfused afferent and efferent arterioles (EDWARDS and TRIZNA 1993). In most of these studies, NOS inhibitors reduced afferent more than efferent arteriolar diameter (NAVAR et al. 1996; ITO et al. 1997). Furthermore, in the isolated perfused glomerular preparation, L-NAME augmented the sensitivity of the afferent but not the efferent arteriole to the vasoconstrictor effect of angiotensin II (ITO et al. 1997).

In conclusion, the majority of functional data suggest that NO plays a more important role in the modulation of vascular tone in the afferent than in the efferent arteriole (NAVAR et al. 1996; ITO et al. 1997; PETERSON et al. 1997). This view contrasts somewhat with a preferential location of eNOS and diaphorase staining in the efferent arteriole as compared with the afferent arteriole (see above). It should be noted, however, that the local concentration of NOS does not necessarily correlate with the magnitude of NO release. The functional significance of the action of NO in glomerular microcirculation is discussed in Sect. C.III.2.

III. Role of NO in Renal Autoregulation

The kidney has the ability to respond to changes in systemic arterial pressure and, hence, perfusion pressure in a way that keeps RBF and GFR constant. For example, when renal perfusion pressure in the conscious dog is gradually decreased, RBF remains constant until a perfusion pressure between 65 mmHg and 70 mmHg is reached, below which RBF falls dramatically. Similarly, GFR is autoregulated, although the lower autoregulatory limit is somewhat higher, about 80 mmHg (KIRCHHEIM et al. 1987) (Fig. 1). Autoregulation can also be demonstrated in anesthetized animals, but the regulatory breakpoint is usually higher, because anesthesia activates neuronal and humoral mechanisms, which reset autoregulation to higher pressures (PERSSON et al. 1990).

With the recognition of NO as a powerful vasodilator in arteries and arterioles and its release from adjacent endothelial cells, this paracrine agent was also considered as a mediator of autoregulation. However, it soon became apparent that a major stimulus for the synthesis and release of NO from endothelial cells is shear stress (or an increase in viscosity). Since the autoregulatory response of the preglomerular vasculature to an increase in perfusion pressure increases shear stress to the endothelial surface and, at the same time, calls for activation of a vasoconstrictor mechanism, NO was ruled out as a principal mediator. Rather, its role can be considered as a modulator that may prevent excessive renal vasoconstriction at elevated arterial pressures, perhaps with regional differences along the preglomerular vascular tree. However, the buffering capacity of endogenously released NO appears to be small, since several studies in conscious (BAUMANN et al. 1992; PERSSON et al. 1993) or anesthetized dogs (MAJID and NAVAR 1992; MAJID et al. 1992a,1993b) as well as in

The Role of Nitric Oxide in Kidney Function

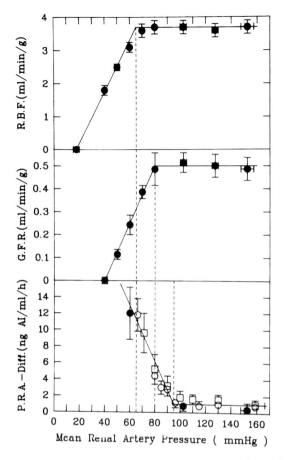

Fig. 1. Glomerular filtration rate (*G.F.R.*), renal blood flow (*R.B.F.*), and pressure-dependent renin release (*P.R.A.-diff.*) in the chronically instrumented conscious dog. Experimental details as described by KIRCHHEIM et al. (1987)

rats (ZATZ and DE NUCCI 1991; BEIERWALTES et al. 1992; IKENAGA et al. 1993) indicate that, while inhibition of NO synthesis significantly lowers the plateau of RBF and, to a smaller degree, the plateau of GFR in the autoregulatory range, the ability to autoregulate both RBF and GFR in response to changes in renal perfusion pressure is essentially preserved. In a recent study in conscious dogs, (JUST, 1997) it has been demonstrated that autoregulatory adaptations to spontaneous dynamic changes in blood pressure do not depend on NO synthesis.

Autoregulation of RBF and GFR is accomplished by two different mechanisms, the myogenic response and tubulo-glomerular feedback (TGF). The relative contribution of these two mechanisms is still debated. While some investigators claim TGF dominates the autoregulatory capacity (NAVAR et al.

1996), data from a recent study in conscious dogs suggest that the myogenic response alone seems to be sufficient to maintain autoregulation during changes in arterial pressure under physiologic resting conditions (JUST 1997). It is beyond the scope of this review to enter this debate. Since, however, both the myogenic response and TGF are modulated by NO and TGF is related to the macula densa mechanism of renin release (see below), these two mechanisms will be briefly discussed.

1. The Myogenic Response and NO

The myogenic response in the preglomerular vessels of the kidney is part of the autoregulation of RBF and GFR. In essence, an increase in the transluminal pressure gradient (by an increase in perfusion pressure) elicits a contractile (myogenic) response that decreases the diameter of the vascular lumen and, thus, helps to maintain downstream flow rate. The mechano-sensitive baroreceptor and its mechanism of transduction to the VSM cell have not yet been identified. Most of the myogenic autoregulatory response in the kidney seems to be located in the afferent arteriole, but upstream segments of the preglomerular vasculature also contribute to myogenic autoregulation. The role of NO in the myogenic response of renal vessels has been examined in isolated microperfused afferent arterioles of the rabbit, in which NOS inhibition augmented the vasoconstrictor response only in free-flow arterioles, suggesting that flow through the lumen stimulates the endothelium to release NO, which in turn attenuates the myogenic response (JUNCOS et al. 1995). Also, in the hydronephrotic rat kidney, NOS inhibition augmented the vascular response to an increase in perfusion pressure (STEINHAUSEN et al. 1989) and, in isolated rat afferent arterioles, NO donors attenuate the myogenic response (BOURIQUET and CASELLAS 1995). However, there is general agreement that the intrinsic mechanisms of the myogenic response of the preglomerular vasculature operate independently of the generation of NO and that the buffering capacity of NO synthesis is limited (NAVAR et al. 1996; ITO et al. 1997; KONE 1997; KONE and BAYLIS 1997).

2. NO and Tubuloglomerular Feedback

In each nephron of the mammalian kidney, the tubulus returns to the hilus of the parent glomerulus, forming the JGA as a unique arrangement of the afferent and efferent arteriole, interstitial cells (the extraglomerular mesangium) and the macula densa, a plaque of 20–30 specialized tubular cells in a tubular segment located between the TAL of the loop of Henle and the early distal convoluted tubule. This arrangement permits the kidney to control glomerular filtration by a TGF mechanism. The basic circuit of this TGF, which has been studied by sophisticated micropuncture techniques and in microdissected glomerular preparations, is well established. An increase in the tubular NaCl concentration at the exit from the TAL is sensed by the macula densa cells and translated into a signal, which increases the tone of the afferent arteriole.

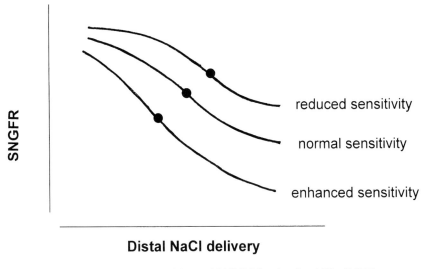

Fig. 2. Changes in tubulo-glomerular feedback sensitivity induced by nitric oxide (NO) and other agents. The response of single nephron glomerular filtration rate to changes in the delivery of sodium chloride to the early distal tubule in micropuncture experiments is attenuated by NO, prostaglandins (PGE_2, PGI_2), and atrial natriuretic peptide, and is enhanced by angiotensin II, thromboxane, and 20-hydroxyeicosatetraenoic acid. Adapted from NAVAR (1998)

This reduces glomerular capillary pressure, SNGFR and, consequently, the tubular load of NaCl. Thus, by oscillating around a given operating point, TGF acts as a minute-to-minute stabilizer of NaCl delivery to the distal tubule (Fig. 2). When the luminal NaCl concentration at the site of the macula densa deviates from the "normal" range for a prolonged time, e.g., by changes in dietary salt supply, this TGF short-term function is preserved by shifting the function curve and the operating point in the appropriate direction (TGF resetting). As the input signal for TGF, the rate of Na^+–K^+–$2Cl^-$ co-transport is accepted by most investigators (NAVAR et al. 1996). This transport system has been identified in macula densa cells both by electrophysiologic (NAVAR et al. 1996) and immunocytochemical techniques (OBERMÜLLER et al. 1996). It is still unknown, however, how macula densa cells communicate with the afferent arteriole. Over the past 30 years, several paracrine substances of potential tubular origin, such as prostaglandins, angiotensin II, dopamine or adenosine, have been proposed as "the" mediator of TGF.

All of these agents influence the TGF response, but none has yet emerged as "the" signal (NAVAR et al. 1996). These agents are, therefore, described as modulators of TGF and, as such, remain of interest in the analysis of TGF. In fact, the unique and strong expression of COX-2 in macula densa cells and its several-fold increase in salt-restricted animals (Harris et al. 1994) have renewed the interest in cyclooxygenase products as participants in the TGF response.

The unexpected finding of a strong expression of nNOS (NOS I) in macula densa cells (actually the strongest in the whole kidney) by SCHMIDT et al. (1992) and MUNDEL et al. (1992) has focussed interest on the role of NO in TGF. Several micropuncture studies in vivo have demonstrated that NOS inhibitors enhance the sensitivity of the TGF response in euvolemic and salt-loaded rats (WILCOX and WELCH 1996; VALLON and THOMSON 1995; KAWABATA et al. 1996; BRAAM and KOOMANS 1995; TOLINS and SHULTZ 1994). In studies by ITO and REN (1993) with the isolated perfused glomerulus–macula densa preparation, a change in macula densa perfusion from low to high NaCl induced the expected TGF response. When L-NAME was added to the high-salt tubular perfusion, the luminal diameter of the afferent arteriole decreased further, while L-NAME had no influence on afferent diameter in low-salt perfusion. This effect of L-NAME is thought to reflect selective inhibition of macula densa NOS I, since the vasodilator effect of acetylcholine infused into the afferent arteriole was not modified by tubular perfusion with L-NAME.

The principal finding in all these studies, that NOS inhibitors do not affect TGF responses in salt-depleted rats or during low-salt perfusion of the macula densa, is difficult to reconcile with the significant increase in NOS I gene expression in salt-depleted rats (TOJO et al. 1995). It has recently been proposed that this apparent functional inactivity of macula densa NOS during dietary salt restriction may result from the combined effects of a lower plasma and tubular concentration of L-arginine, a reduced transport capacity for L-arginine of the y^+ transport system in macula densa cells and an increased rate of formation of mono- or dimethylated arginines, which inhibit both L-arginine transport and NOS activity by competing with L-arginine (WILCOX and WELCH 1998). However, these effects do not explain the in vitro data of ITO and REN (1993).

The view of the role of macula densa NOS I is further obscured by the finding that TGF function in mice lacking expression of the NOS I gene (homozygous knock-out mice) is essentially intact whereas, in angiotensin II (AT_1)-receptor deficient or ACE-deficient knock-out mice, the TGF is completely lost (SCHNERMANN et al. 1998).

In summary, the available data demonstrate that NO acutely modulates TGF by attenuating the magnitude of the response to changes in distal salt delivery, thus facilitating salt excretion. The role of macula densa NOS I in the resetting of TGF sensitivity during chronic changes in salt balance remains to be clarified. It is conceivable that the observed adaptational changes in NOS I expression are more relevant for the long-term control of renin secretion than for TGF (Sect. E.V).

IV. Role of NO in the Control of Medullary Blood Flow and Pressure Natriuresis

In the past decade, considerable advance has been made in our understanding of the role of the renal medulla in blood pressure and volume homeostasis. As has been described previously, a rise in systemic arterial pressure is autoregulated by the preglomerular vasculature, resulting in unaltered RBF and GFR. Nevertheless, sodium and volume excretion increase, the response to an increase in perfusion pressure being more prominent in volume-expanded than in euvolemic animals (MAJID and NAVAR 1997). As an explanation of this phenomenon, called pressure natriuresis, it has been hypothesized that, as renal perfusion pressure rises, medullary capillary flow and, thus, renal interstitial pressure increase and, as a consequence, tubular sodium reabsorption falls (MAJID and NAVAR 1997; NAVAR 1998). This apparently non-adaptive mechanism is thought to be responsible for the long-term control of blood pressure and volume homeostasis, while malfunction of this system may be one of the main causes of hypertension (HALL et al. 1996) (see below).

Among the various paracrine agents thought to contribute to the functioning of pressure natriuresis and volume control by the renal medulla, NO appears to play a prominent role. In accordance with the rich supply of vasa recta with eNOS, several studies have shown that acute systemic or intrarenal infusion of NOS inhibitors in volume-expanded rats, dogs or primates preferentially decreases medullary blood flow, attenuates the diuretic and natriuretic response to volume expansion and abolishes pressure natriuresis (ATUCHA et al. 1994; LOCKHART et al. 1994; FENOY et al. 1995; PAREKH and ZOU 1996; MATTSON et al. 1997). In euvolemic rats, the response to NOS inhibition is less prominent, which is consistent with the observation that salt loading increases sensitivity of inhibition of NOS in the renal vasculature, probably by promoting NO synthesis (SHULTZ and TOLINS 1993; DENG et al. 1994; BECH et al. 1998).

The increasing awareness of NO as an important mediator or modulator of the renal adaptation to an increased salt load and or an elevated blood pressure has raised the question of whether a malfunctioning of renal NO synthesis may contribute to or even cause the pathogenesis of hypertension. Indeed, impaired NO synthesis, with the consequence of impaired salt excretion, appears to play a pathogenetic role in patients with salt-sensitive essential hypertension (MATTEI et al. 1990; RUILOPE et al. 1994; BAYLIS and VALLANCE 1996; HALL et al. 1996; HIGASHI et al. 1996; LAHERA et al. 1997; SCHNACKENBERG et al. 1997). Support for this concept also came from studies in rats with salt-sensitive genetic hypertension, in which normalization of the impaired pressure natriuresis, hypertension and renal tubular transport by supplementation with L-arginine was observed (BAYLIS and VALLANCE 1996; HALL et al. 1996; KONE 1997).

E. Tubular Functions of NO

In addition to the indirect natriuretic effect of NO mediated by the increase in medullary blood flow, NO may also contribute to sodium and volume regulation by direct tubular actions. A variety of in vitro studies (Table 1) suggest that NO influences tubular Na^+–K^+–adenosine triphosphatease (ATPase) activity (the driving force for several "secondary active" transport systems), H^+–ATPase activity, Na^+–H^+ exchange, Na^+–HCO_3^- transport, high- and low-conductance K^+ channels and alcohol dehydrogenase-stimulated water permeability. Some, but not all, of these studies suggest that NO may interfere with net Na^+ reabsorption. However, the diversity of methods and conditions used in these studies, the multitude of transport processes and ion channels affected by NO (some perhaps secondary to influences on a common target) and the heterogeneity and occasional disparity of results, do not at present allow us to delineate an integrating concept of how the diverse effects of NO may contribute to ion and fluid homeostasis under physiological or pathophysiological conditions. Therefore, these studies are listed descriptively in Table 1.

In some studies, the effect of NO or NO donors could be mimicked by lipid-soluble cGMP analogs, such as 8-bromo-cGMP, suggesting that these

Table 1. Tubular actions of nitric oxide (NO) in vitro

Preparation	Effects of NO[a]	References
Medullary slices	Inhibition of Na^+ transport	McKee et al. 1994
PCT, cell culture, iNOS stimulation by LPS	Autocrine inhibition of Na^+–K^+–ATPase	Guzman et al. 1995
PCT, isolated cells	Paracrine stimulation of Na^+–H^+ exchange	Amorena and Castro 1997
PCT, dissected	Inhibition of Na^+–H^+–exchange	Roczniak and Burns 1996
PCT, in situ perfused	Stimulation of Na^+–HCO_3^- transport	Buttery et al. 1994
mTAL, whole-cell patch clamp	Activation of apical 70-pS K^+ channel	Lu et al. 1998
CCD, isolated, perfused	Inhibition of Na^+ absorption	Stoos et al. 1995
CCD, isolated, perfused	Inhibition of H^+–ATPase	Tojo et al. 1994
CCD, isolated cells	Inhibition of Na^+ transport	Stoos and Garvin 1997
CCD, dissected, perfused	Inhibition of ADH-stimulated water permeability	Garcia et al. 1996
CCD, whole-cell patch clamp	Activation of basolateral 28 pS K^+-channel	Lu and Wang 1996; Lu et al. 1997

ADH, alcohol dehydrogenase; *ADP*, adenosine diphosphate; *ATP*, adenosine triphosphate; *CCD*, cortical collecting duct; *mTAL*, medullary thick ascending limb; *LPS*, lipopolysaccharide; *PCT*, proximal convoluted tubule.
[a] Deduced from effects of NO donors or NO synthase inhibitors.

effects are mediated by stimulation of the soluble guanylyl cyclase. In contrast, in the study by GUZMAN et al. (1995), inhibition of Na^+–K^+–ATPase activity was observed in LPS-and IFNγ-stimulated proximal tubule cells. In this case, the inhibitory effect of NO appears to depend on the induction of iNOS and the generation of peroxynitrite rather than on guanylyl cyclase stimulation, since it was attenuated by superoxide dismutase and was not mimicked by 8-bromo-cGMP.

F. NO, Renin Secretion and Renin Synthesis

Renin secretion from juxtaglomerular (JG) cells of the kidney is controlled by renal perfusion pressure, renal sympathetic nerve activity, the macula densa mechanism and a variety of humoral or paracrine agents. There is a remarkable correlation between the response of renin secretion and that of the preglomerular vasculature to changes in the setting of these control mechanisms: pressure-dependent renin release has physical and biochemical characteristics similar to those described for renal autoregulation, β-adrenergic renin stimulation corresponds to β-adrenergic renal vasodilatation, and macula-densa-mediated renin secretion is associated with TGF (KURTZ 1989; HACKENTHAL et al. 1990).

This relationship is not surprising when taking into account that renin-producing cells in the wall of the afferent arteriole are modified VSM cells. In fact, VSM cells can convert into JG cells and vice versa depending on the "demand" on renin storage and release, and a considerable portion of all renin-producing cells in the afferent arteriole are morphologically and functionally in a stage of transition to or from VSM cells (intermediate cells) (TAUGNER et al. 1984, 1988; TAUGNER and HACKENTHAL 1989; HACKENTHAL et al. 1990). Furthermore, there is a remarkable correlation between the vasoconstrictor properties and inhibition of renin secretion of many humoral signals, such as endothelins, neuropeptide Y or angiotensin II, while most vasodilators stimulate renin release.

This functional relationship between VSM and renin-producing cells is reflected in intracellular signaling pathways. It is well established that cyclic adenosine monophosphate (cAMP) mediates vasodilatation and stimulates renin secretion, while an increase in intracellular calcium mediates vasoconstriction and inhibits renin release. Although initially questioned, there is now agreement that this relationship also extends to NO, which relaxes the afferent arteriole and stimulates renin release. However, the role of its second messenger cGMP is less clear (see below).

In the following sections, the evidence for a stimulatory function of NO on renin secretion, its participation in the main controlling systems of renin secretion, the information available on the role of cGMP and, finally, the links between renin-gene expression and NOS-gene expression will be discussed.

I. NO as a Stimulator of Renin Secretion

Earlier studies performed with rat renal cortical slices exposed to acetylcholine, sodium nitroprusside (SNP) or NOS inhibitors suggested that NO may be an inhibitor of renin release (BEIERWALTES 1990, 1994; BEIERWALTES and CARRETERO 1992). However, the majority of studies in conscious and anesthetized animals suggest that NO is a stimulator of renin release (NAESS et al. 1993; PERSSON et al. 1993; JOHNSON and FREEMAN 1994; SCHRICKER et al. 1995; CHIU and REID 1996; KNOBLICH et al. 1996). Moreover, all studies performed in the isolated perfused rat kidney, a preparation free of systemic influences, have clearly demonstrated that NOS inhibitors, such as L-NAME or L-NMMA, inhibit renin release independent of changes in perfusion pressure (MÜNTER and HACKENTHAL 1991; GARDES et al. 1992, 1994; HACKENTHAL et al. 1994; KURTZ et al. 1998) and that stimulation of NO synthesis by acetylcholine or carbachol, as well as infusion of NO donors, such as SNP or the sydnonimine SIN-1, enhance renin release (ISHII et al. 1991; MÜNTER and HACKENTHAL 1991; HACKENTHAL et al. 1994). These effects were still observed when the possible interference of the macula densa mechanism was excluded by using the isolated, perfused, hydronephrotic rat kidney (HACKENTHAL et al. 1994) or the isolated kidney with blocked macula densa function (WAGNER et al. 1998).

A stimulatory effect of NO on renin release was also seen in isolated JG cells exposed to the NO donor SNP (SCHRICKER et al. 1993; SCHRICKER and KURTZ 1993; KRÄMER et al. 1994). In these and other studies with isolated cells (GREENBERG et al. 1995), an initial decrease of renin release by SNP has been observed, which was thought to represent an initial inhibitory effect of the elevated intracellular concentration of cGMP, while long-term stimulation by SNP was considered to represent a cGMP-independent effect of NO. However, this interpretation may have to be revised in view of newer findings demonstrating that the stimulatory effect of NO donors on renin secretion is attenuated or abolished by the specific inhibitor of guanylyl cyclase ^{1}H-[1,2,4]-oxadiazolo-[4,3-a]-quinoxalin-1-one (ODQ) in the isolated perfused rat kidney (KURTZ et al. 1998; Fig. 3). These observations indicate that NO exerts its stimulatory effect on renin secretion via generation of cGMP. However, this stimulation of renin secretion by cGMP may not be through the "conventional" pathways of activation of the cGMP-dependent protein kinase but rather by reducing the rate of hydrolysis of cAMP by the cGMP-inhibited cAMP phosphodiesterase (PDE III), thus causing stimulation by the elevated levels of cAMP (KURTZ et al. 1998). In conclusion, there is now general agreement that NO is a potent stimulator of renin release and that this effect is mediated by activation of the soluble guanylyl cyclase in renin-producing cells.

II. NO and Pressure Control of Renin Release

Several studies in conscious and anesthetized animals and in the isolated perfused kidney have demonstrated that graded reductions in perfusion pressure

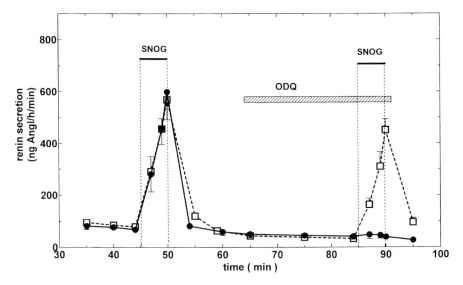

Fig. 3. Stimulation of renin secretion from the isolated perfused rat kidney by the nitric oxide donor S-nitrosoglutathion (*SNOG*, 1 μM), infused twice for 5 min, either alone (*open squares, broken line*) or in the presence of the guanylylcyclase inhibitor ^{1}H[1,2,4]-oxodiazolo-[4,3-a]-quinoxalin-1-one (*ODQ*, 1 μM) during the second infusion period (*full circles, full line*) (HACKENTHAL and MENZ, unpublished data)

enhance renin secretion (HACKENTHAL et al. 1990). This relationship has been described quantitatively in studies by FARHI et al. (1987), FINKE et al. (1983) and KIRCHHEIM et al. (1987) in conscious dogs. Variations in renal perfusion pressure above resting blood pressure had no influence on renin release. When renal perfusion pressure is reduced below a "threshold" pressure (about 95 mmHg), which is in the lower range of resting blood-pressure variations in the dog, descending autoregulatory vasodilatation reaches the distal part of the afferent arteriole, where renin-producing cells are located, and renin secretion increases dramatically (Fig. 1). This "autoregulatory" behavior of renin secretion is thought to function not only as an emergency system when blood pressure falls, but also as a short-term feedback system reacting to physiological fluctuations in resting blood pressure (EHMKE et al. 1987). The mechanism by which renin-producing cells sense changes in perfusion pressure is not known. There are data supporting either flow (SALOMONSSON et al. 1991) or pressure (NAFZ et al. 1997) as the major determinants. Still another position holds that it is the activation of macula-densa-controlled renin release that induces an increase in renin release when perfusion pressure is reduced (see below).

The role of NO in pressure-dependent renin release has been examined in the isolated perfused rat kidney (SCHOLZ and KURTZ 1993), in anesthetized dogs (NAESS et al. 1993) and in conscious dogs (PERSSON et al. 1993). In all three settings, renin release was little affected by inhibition of NOS activity

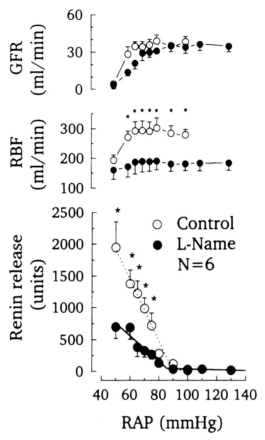

Fig. 4. Glomerular filtration rate (GFR), renal blood flow (RBF), and pressure-dependent renin release in chronically instrumented dogs before and after nitric oxide synthase blockade with N^G-nitro L-arginine methyl ester. *RAP*, renal arteial pressure. Reproduced from PERSSON et al. (1993)

when perfusion pressure was high. However, when renin secretion was stimulated by reduction of perfusion pressure below the threshold pressure for "autoregulation" or renin release, NOS inhibitors significantly reduced renin release, suggesting that NO augments the renin response to pressure reduction (Fig. 4). The source of NO and the releasing mechanisms are not known. Release of NO from endothelial cells by shear stress is unlikely to play a role, since static reductions of renal perfusion pressure reduce shear stress and, hence, NO release. A reduction in GFR and, consequently, activation of the macula densa mechanism of renin release via release of NO from macula densa cells could account for the stimulation of renin release at low perfusion pressures. However, autoregulation of GFR has a significantly lower threshold pressure than renin release, making it unlikely that the macula densa is the major determinant of this effect.

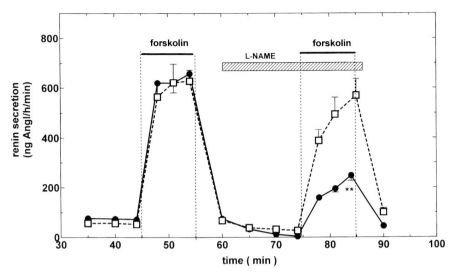

Fig. 5. Stimulation of renin secretion from the isolated perfused rat kidney by forskolin (0.25 μM), infused twice for 10 min either alone (*open squares, broken line*), or in the presence of N^G-nitro L-arginine methyl ester (10 μM) during the second infusion period (*full circles, full line*) (HACKENTHAL and MENZ, unpublished data)

In conclusion, the available data demonstrate that NO augments the stimulation of renin release induced by lowering perfusion pressure. The effect is proportional to the decrease in renal perfusion pressure. The source of NO mediating this effect and the mechanisms by which a decrease in pressure induces its release are still unknown.

III. NO, Renal Nerves and Renin Release

Mild activation of renal sympathetic nerves stimulates renin release via activation of β-adrenoceptors located on renin-producing cells. There is no evidence for a nitrergic innervation of the JGA. However, NOS inhibitors inhibit central and peripheral sympathetic activity, thus influencing renin release by reducing sympathetic activation of β-adrenoceptors and, indirectly, by reducing sympathetic vasoconstriction and modulating tubular functions (KUMAGAI et al. 1994; DIBONA and KOPP 1997).

There is no indication that the release of norepinephrine and its interaction with β-adrenoceptors is influenced by NO. However, stimulation of renin release by isoproterenol or forskolin, a receptor-independent stimulator of adenylylcyclase, in vivo (REID 1994) and in the isolated perfused rat kidney (KURTZ et al. 1998) is markedly attenuated by inhibitors of NOS or guanylyl cyclase (Figs. 5, 6). It has been proposed that this attenuation results from disinhibition of the cGMP-inhibitable isoform of cAMP PDE III and, consequently, from the accelerated breakdown of cAMP (CHIU and REID 1996;

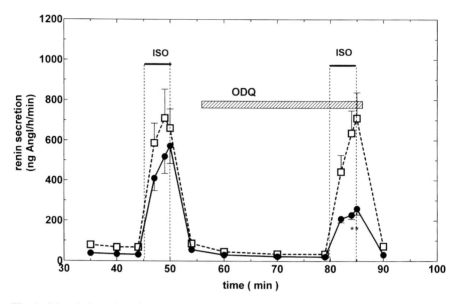

Fig. 6. Stimulation of renin secretion from the isolated perfused rat kidney by isoproterenol (1 nM), infused twice for 5 min either alone (*open squares, broken line*) or in the presence of ¹H-[1,2,4]-oxadiazolo-[4,3-a]-quinoxalin-1-one (1 µM) during the second infusion period (*full circles, full line*) see legend to Fig. 3 for details. *RAP*, renal arterial pressure; *RBF*, renal blood flow; *GFR*, glomerular filtration rate (HACKENTHAL amd MENZ, unpublished data)

KURTZ et al. 1998). This would imply that NO exerts a tonic inhibition on PDE III activity through the generation of cGMP.

IV. NO and Macula-Densa-Mediated Renin Secretion

It has long been known that chronic salt intake is inversely related to plasma renin activity. It was hypothesized that this inverse relationship somehow depends on the function of the macula densa and is associated with the TGF mechanism controlling GFR.

In vitro evidence for an acute inverse relationship between the tubular NaCl concentration at the site of the macula densa and renin release was first described by ITO and CARRETERO (1985), who observed that renin release from isolated glomeruli, to which the macula densa segment was attached, was stimulated by furosemide, presumably by inhibiting Na^+–K^+–$2Cl^-$ co-transport, while furosemide was ineffective in macula-densa-free glomeruli. Direct evidence for the existence of the macula densa mechanism of renin secretion was eventually obtained by SKOTT and BRIGGS (1987) who, using the dissected glomerulus–macula densa preparation, perfused the tubular segment with solutions containing different NaCl concentrations and observed inverse changes in renin release.

The tubular signal perceived by the macula densa cells is probably identical to that of the TGF response, i.e., the tubular chloride concentration, which determines the rate of Na^+–K^+–$2Cl^-$ co-transport located in the apical membrane of macula densa cells. However, the mechanisms by which macula densa cells communicate with renin-producing granulated cells of the afferent arteriole are still unclear. The recognition of a prominent and selective expression of NOS I in macula densa cells (MUNDEL et al. 1992; SCHMIDT et al. 1992; BACHMANN 1997) has prompted several studies examining the role of NO as the mediator of macula-densa-induced renin release. Thus, REID and CHOU (1995) have shown that infusion of L-NAME into conscious rabbits blunted the increase in plasma renin activity in response to infusion of furosemide. Similarly, L-NAME infusion in rats on a low-salt diet prevented an increase in plasma renin activity (SCHRICKER et al. 1995). The NOS inhibitor 7-nitroindazole, which is claimed to have some selectivity for nNOS, prevented the rise in plasma renin in response to furosemide or salt depletion when given for 5 days, but had no acute effect on plasma renin (BEIERWALTES 1995, 1997). Direct, conclusive evidence for the putative role of NO as a mediator of macula-densa-mediated renin release came from studies of (HE et al. 1995), in which tubular perfusion of the dissected macula densa–glomerulus preparation with a low-salt solution increased renin release. The addition of L-NAME to the low-salt perfusate attenuated renin release, while the lower renin secretion rate during high-salt perfusion was not affected by L-NAME.

These data would suggest that macula-densa-mediated renin secretion follows the simple scheme: decreased distal tubular load of NaCl → decreased Na^+–K^+–Cl^- co-transport → increased formation of NO by NOS I → diffusion of NO to granulated cells → activation of guanylyl cyclase → increased cGMP → increased renin secretion. However, several observations and arguments make one hesitate to accept this signaling sequence at the present time.

Although chronic dietary-salt restriction stimulates the expression of NOS I in macula densa cells, an acute effect of changes in tubular salt loading on the formation of NO has not yet been demonstrated. Actually, a decreased rate of Na^+–K^+–Cl^- co-transport lowers rather than enhances intracellular free calcium by increasing the activity of Na^+–Ca^{2+} exchange in macula densa cells (LAPOINTE et al. 1998) and would, therefore, attenuate rather than stimulate NOS I activity. Furthermore, in the experiments described by ITO and CARRETERO (1993), an effect of L-NAME on the TGF response was only observed during high-salt tubular perfusion, whereas L-NAME was without any effect during low-salt perfusion, suggesting that NO formation is only increased during high-salt perfusion.

In addition, in the experiments of HE et al. (1995), high concentrations of L-NAME inhibited the renin response to low-salt tubular perfusion by only 37% while, in the same preparation, low-salt-stimulated renin release was almost completely blocked by COX inhibitors (GREENBERG et al. 1993).

In summary, the available data clearly demonstrate that NO participates in the acute control of renin secretion by the macula densa. However, its precise role in this mechanism and its interaction with other paracrine or autocrine agents, in particular COX products, remain unclear. It may well be that the more important function of macula-densa-derived NO is the long-term adjustment of renin secretion by controlling renin gene expression (see following section).

V. NO, Prostaglandins and Renin Synthesis

Most conditions and agents known to modify renin secretion also induce parallel changes in renin synthesis when acting on renin-producing cells for a prolonged time. The obvious question of whether NO per se might induced renin synthesis and possibly modulate some of the systemic stimuli of renin synthesis has been addressed in several studies by examining the effects of NOS inhibitors in various well-known models of stimulated renin gene expression. It was observed that L-NAME, when given to rats for 2 days to block NOS activity, attenuated or completely abolished the increase in renin mRNA (measured by RNAse protection assay) in the clipped kidney of Goldblatt (two kidney, one clip) hypertensive rats (MACLEOD et al. 1987; SCHRICKER et al. 1994), in rats fed a low-salt diet (SCHRICKER and KURTZ 1996) and in rats treated with either furosemide (to block macula densa $Na^+-K^+-2Cl^-$ co-transport), ramipril (an ACE inhibitor; SCHRICKER et al.1995b) or losartan [an angiotensin II (AT_1)-receptor antagonist; THARAUX et al. 1997; SCHRICKER et al. 1995b]. The conclusion from these studies – that NO is an important mediator in the regulation of renin gene expression – was supported by the observation that the NO donor SNP induces an increase in renin mRNA in isolated mouse JG cells in primary culture (SCHRICKER et al. 1994; DELLA BRUNA et al. 1995).

As pointed out before, macula densa cells strongly express COX-2. Since NO has been demonstrated to induce COX-2 activity in a number of organs, including the kidney (SALVEMINI et al. 1993), and prostaglandins are powerful stimulators of renin secretion, the role of COX-2 products in the regulation of renin gene expression has been examined. Meclofenamate or indomethacin, nonspecific inhibitors of COX activity, attenuated the increase in renin mRNA in clipped kidneys to an extent similar to that caused by L-NAME (SCHRICKER et al. 1994) but only slightly reduced the rise in renin mRNA in ramipril- or losartan-treated rats (SCHRICKER et al. 1995b). In mice, the increase in renal renin content during chronic salt depletion was prevented by COX-2-specific inhibitors (HARDING et al. 1997).

There is not only a close association between NO and renin gene expression, but also between NOS I gene expression, COX-2 gene expression and renin gene expression. BOSSE et al. (1995) found both renin mRNA and NOS I mRNA and their respective proteins to be markedly increased in the clipped kidneys of two-kidney, one-clip Goldblatt-rats, in rats treated with furosemide and, to a lesser extent, in salt-depleted rats. Similar observations were reported by SCHRICKER et al. (1996). The effect of high- and low-salt feeding in rats was also examined by SINGH et al. (1996), who found parallel changes of cortical nNOS mRNA and renin mRNA, while neither iNOS nor eNOS mRNA in the renal medulla was influenced by the pretreatment. Taken together, these observations suggest that NOS activity and gene expression are regulated in parallel with renin gene expression and that both are inversely related to chronic changes in salt balance and distal tubule salt transport.

The two-kidney, one-clip model of renovascular hypertension was also used to compare the expression of COX-2 and renin (HARTNER et al. 1998). In the clipped kidney, COX-2 mRNA and protein increased significantly and in parallel with immunocytochemical staining intensity for renin. This aspect has recently been studied in more detail by YANG et al. (1998), who compared the regional expression of COX-1 and COX-2 in salt-restricted and salt-loaded rats. After 1 week of salt loading, the concentration of COX-2 mRNA was significantly increased in the inner medulla and decreased in the cortex while, in rats on a low-salt diet, COX-2 mRNA increased preferentially in macula densa cells and the cortical TAL. COX-1 expression was not altered by the change in salt balance, suggesting that prostaglandins in different kidney regions serve different functions.

In summary, there is a consistent and close association between the expression of the renin gene, the macula densa COX-2 gene and the macula densa nNOS gene. Since NOS inhibitors prevent a rise in renin mRNA in different experimental settings, it appears likely that renin gene expression in JG cells is controlled by nNOS activity and also that COX-2-derived prostaglandins are involved in controlling renin synthesis, perhaps by mediating the stimulatory effect of NO on renin gene expression.

It should be considered, however, that salt depletion and/or furosemide treatment not only increase the local concentration of renin mRNA and renin protein in JG cells but also promote the recruitment of smooth-muscle cells for transformation into renin-producing cells further upstream in the afferent arteriole, occasionally reaching the interlobular arteries. In this situation, a considerable portion of renin-producing cells, if not the major part, is very likely too far away from the macula densa to be influenced by any macula-densa-generated signal, in particular the short-lived NO molecule (TAUGNER and HACKENTHAL 1989; HACKENTHAL et al. 1990). Therefore, the interpretation of the previously described data as macula-related is probably an oversimplification. There is, at present, no information available on the reaction of eNOS in preglomerular vessels to changes in salt balance or other renin-stimulating maneuvers or of the participation of eNO in the control of renin synthesis.

The cellular mechanism by which NO stimulates renin gene expression are still unknown, as are the intracellular mechanisms controlling renin gene expression in general. With respect to NO, the speculative view that NO-induced cGMP formation may lower intracellular calcium, which then activates gene expression is compatible with many observations and gains some support from the recent report that the angiotensin-II-receptor antagonist

Table 2. Renal functions of nitric oxide (NO)

Function	Role of NO
Renovascular resistance	Reduction
Autoregulation of RBF and GFR	Attenuation of the response to pressure changes
TGF	Attenuation of efficiency, long-term resetting
Tubular sodium reabsorption	inhibition
Pressure natriuresis	Mediation through hemodynamic and tubular effects
Renal renin secretion	Stimulation
Pressure control of renin release	Enhanced sensitivity to pressure reduction
Macula-densa-controlled secretion of renin	Mediator of acute response to low salt
Macula-densa-controlled renin gene expression	Mediator of adaptation to salt depletion
Erythropoietin release	Mediator of hypoxia-stimulated release
Pathophysiological conditions (acute and chronic renal failure, diabetic nephropathy, acute and chronic glomerular inflammation)	Both beneficial and detrimental effects

GFR, glomerular filtration rate; *RBF*, renal blood flow; *TGF*, tubulo-glomerular feedback.

losartan induces parallel increases in renin mRNA and cGMP-dependent protein kinase 2 levels in JG cells (GAMBARYAN et al. 1996).

G. Concluding Remarks

The role of NO in physiological and pathophysiological functions in the kidney is an area of intensive research. Many open questions, as described in this review, are likely to find an answer in the near future. It can already be stated that NO is one of the most influential paracrine modulators and mediators of many renal functions, most of which converge in a functional network that protects the body from salt and volume overloading and their consequences. In fulfilling these functions, NO interacts with many other relevant systems in a very complex fashion which could not be discussed in detail in this review. The present state of information on the renal functions of NO is summarized in Table 2. It should be noted that, for some of the listed functions, final proof is still lacking.

Abbreviations

ACE	angiotensin-converting enzyme
ATPase	adenosine triphosphatase

cAMP	cyclic adenosine monophosphate
cGMP	cyclic guanosine monophosphate
COX	cyclo-oxygenase
eNOS	endothelial nitric oxide synthase (NOS III)
GFR	glomerular filtration rate
IFNγ	interferon γ
iNOS	inducible nitric oxide synthase (NOS II)
JGA	juxtaglomerular apparatus
L-NAME	N^G-nitro L-arginine methyl ester
LPS	lipopolysaccharide
NADPH	reduced nicotinamide adenine dinucleotide phosphate
nNOS	neuronal nitric oxide synthase (NOS I)
NO	nitric oxide
NOS	nitric oxide synthase
ODQ	1H-[1,2,4]-oxadiazolo-[4,3-a]-quinoxalin-1-one
PDE	phosphodiesterase
RBF	renal blood flow
RT-PCR	reverse transcriptase-polymerase chain reaction
RVR	renovascular resistance
SNGFR	single-nephron glomerular filtration rate
SNP	sodium nitroprusside
TAL	thick ascending limb of Henle's loop
TGF	tubulo-glomerular feedback
VSM	vascular smooth muscle

References

Ahn KY, Mohaupt MG, Madsen KM, Kone BC (1994) In situ hybridization localization of mRNA encoding inducible nitric oxide synthase in rat kidney. Am J Physiol 267:F748–F757

Alberola A, Pinilla JM, Quesada T, Romero JC, Salom MG, Salazar FJ (1992) Role of nitric oxide in mediating renal response to volume expansion. Hypertension Dallas 19:780–784

Amorena C, Castro AF (1997) Control of proximal tubule acidification by the endothelium of the peritubular capillaries. Am J Physiol 272:R691–R694

Atucha NM, Ramirez A, Quesada T, Garcia-Estan J (1994) Effects of nitric oxide inhibition on the renal papillary blood flow response to saline-induced volume expansion in the rat. Clin Sci 86:405–409

Bachmann S (1997) Distribution of NOSs in the kidney. In: Goligorski MS, Gross MS (eds) Nitric oxide and the kidney. Physiology and pathophysiology. Chapman and Hall, pp 133–157

Bachmann S, Mundel P (1994) Nitric oxide in the kidney: synthesis, localization, and function. Am J Kidney Dis 24:112–129

Bachmann S, Oberbäumer I (1998) Structural and molecular dissection of the juxtaglomerular apparatus: new aspects for the role of nitric oxide. Kidney Int 54 (Suppl. 67):S29–S33

Bachmann S, Bosse HM, Mundel P (1995) Topography of nitric oxide synthesis by localizing constitutive NO synthases in mammalian kidney. Am J Physiol 268:F885–F898

Baumann JE, Persson PB, Ehmke H, Nafz B, Kirchheim HR (1992) Role of endothelium-derived relaxing factor in renal autoregulation in conscious dogs. Am J Physiol 264:F208–F213

Baylis C, Vallance P (1996) Nitric oxide and blood pressure: effects of nitric oxide deficiency. Curr Opin Nephrol Hypertens 5:80–88

Bech JN, Nielsen CB, Ivarsen P, Jensen KT, Pedersen EB (1998) Dietary sodium affects systemic and renal hemodynamic response to NO inhibition in healthy humans. Am J Physiol 274:F914–F923

Beierwaltes WH (1990) Possible endothelial modulation of prostaglandin-stimulated renin release. Am J Physiol 258:F1363–F1371

Beierwaltes WH (1994) Nitric oxide participates in calcium-mediated regulation of renin release. Hypertension 23 (Suppl. I):I40–I-44

Beierwaltes WH (1995) Selective neuronal nitric oxide synthase inhibition blocks furosemide-stimulated renin secretion in vivo. Am J Physiol 269:F134–F139

Beierwaltes WH (1997) Macula densa stimulation of renin is reversed by selective inhibition of neuronal nitric oxide synthase. Am J Physiol 272:R1359–R1364

Beierwaltes WH, Carretero OA (1992) Nonprostanoid endothelium-derived factors inhibit renin release. Hypertension 19(Suppl. II):II68–II73

Beierwaltes WH, Carretero OA, Scicli AG (1988) Renal hemodynamics in response to a kinin analogue antagonist. Am J Physiol 255:F408–F414

Beierwaltes WH, Sigmon DH, Carretero OA (1992) Endothelium modulates renal blood flow but not autoregulation. Am J Physiol 262:F943–F949

Bosse HM, Böhm R, Resch S, Bachmann S (1995) Parallel regulation of constitutive NO synthase and renin at JGA of rat kidney under various stimuli. Am J Physiol 269:F793–F805

Bouriquet N, Casellas D (1995) Interaction between cGMP-dependent dilators and autoregulation in rat preglomerular vasculature. Am J Physiol 268:F338–F346

Braam B, Koomans HA (1995) Nitric oxide antagonizes the actions of angiotensin II to enhance tubuloglomerular feedback responsiveness. Kidney Int 48:1406–1411

Brezis M, Heyman SN, Dinour D, Epstein FH, Rosen S (1991) Role of nitric oxide in renal medullary oxygenation: studies in isolated and intact rat kidneys. J Clin Invest 88:390–395

Buttery LD, Evans TJ, Springall DR, Carpenter A, Cohen J, Polak JM (1994) Immunochemical localization of inducible nitric oxide synthase in endotoxin-treated rats. Lab Invest 71:755–764

Chevalier RL, Thornhill BA, Gomez RA (1992) EDRF modulates renal hemodynamics during unilateral ureteral obstruction in the rat. Kidney Int 42:400–406

Chiu T, Reid IA (1996) Role of cyclic GMP-inhibitable phosphodiesterase and nitric oxide in the beta adrenoceptor control of renin secretion. J Pharmacol Exp Ther 278:793–799

De Nicola L, Blantz RC, Gabbai FB (1992) Nitric oxide and angiotensin II. Glomerular and tubular interaction in the rat. J Clin Invest 89:1248–1256

Della Bruna R, Pinet F, Corvol P, Kurtz A (1995) Opposite regulation of renin gene expression by cyclic AMP and calcium in isolated mouse juxtaglomerular cells. Kidney Int 47:1266–1273

Deng X, Welch WJ, Wilcox CS (1994) Renal vasoconstriction during inhibition of NO synthase: effects of dietary salt. Kidney Int 46:639–646

Denton KM, Anderson WP (1994) Intrarenal haemodynamic and glomerular responses of inhibition of nitric oxide formation in rabbits. J Physiol London 475:159–167

Dibona GF, Kopp UC (1997) Neural control of renal function. Physiol Rev 77:75–197

Edwards RM, Trizna W (1993) Modulation of glomerular arteriolar tone by nitric oxide synthase inhibitors. J Am Soc Nephrol 4:1127–1132

Ehmke H, Persson P, Kirchheim H (1987) Pressure-dependent renin release: the kidney factor in long-term control of arterial blood pressure in conscious dogs. Clin Exp Hypert A9(Suppl. 1):181–195

Farhi ER, Cant JR, Pagnelli WC, Dzau VJ, Barger AC (1987) Stimulus-response curve of the renal baroreceptor: effect of converting enzyme inhibition and changes in salt intake. Circ Res 61:670–677

Fenoy FJ, Ferrer P, Carbonell L, Garcia-Salom M (1995) Role of nitric oxide on papillary blood flow and pressure natriuresis. Hypertension Dallas 25:408–414

Finke R, Gross R, Hackenthal E, Huber I, Kirchheim H (1983) Threshold pressure for the pressure-dependent renin release in the autoregulating kidney of conscious dogs. Pflügers Arch. 399:102–110

Gambaryan S, Häusler C, Markert T, Pöhler D, Jarchau T, Walter U, Haase W, Kurtz A, Lohmann SM (1996) Expression of type II cGMP-dependent protein kinase in rat kidney is regulated by dehydration and correlated with renin gene expression. J Clin Invest 98:662–670

Garcia NH, Pomposiello SI, Garvin JL (1996) Nitric oxide inhibits ADH-stimulated osmotic water permeability in cortical collecting ducts. Am J Physiol 270:F206–F210

Gardes J, Poux JM, Gonzalez MF, Alhenc-Gelas F, Ménard J (1992) Decreased renin release and constant kallikrein secretion after injection of L-NAME in isolated perfused rat kidney. Life Sci. 50:987–993

Gardes J, Gonzalez MF, Alhenc-Gelas F, Ménard J (1994) Influence of sodium diet on L-NAME effects on renin release and renal vasoconstriction. Am J Physiol 267:F798–F804

Goto S, Yamamoto T, Feng L, Yaoita E, Hirose S, Fujjinaka H, Kawasaki K, Hattori R, Yui Y, Wilson CB (1995) Expression and localization of inducible nitric oxide synthase in anti-Thy-1 glomerulonephritis. Am J Pathol 147:1133–1141

Granger JP, Alberola AM, Salazar FJ, Nakamura T (1992) Control of renal hemodynamics during intrarenal and systemic blockade of nitric oxide synthesis in conscious dogs. J Cardiovasc Pharmacol 20 (Suppl. 12):S160–S162

Greenberg SG, Lorenz JN, He XR, Schnermann JB, Briggs JP (1993) Effect of prostaglandin synthesis inhibition on macula densa-stimulated renin secretion. Am J Physiol 265:F578–F583

Greenberg SG, He X-R, Schnermann JB, Briggs JP (1995) Effect of nitric oxide on renin secretion. I. Studies in isolated juxtaglomerular granular cells. Am J Physiol 268:F948–F952

Guzman NJ, Fang MZ, Tang SS, Ingelfinger JR, Garg LC (1995) Autocrine inhibition of Na+/K(+)-ATPase by nitric oxide in mouse proximal tubule epithelial cells. J Clin Invest 95:2083–2088

Hackenthal E, Paul M, Ganten D, Taugner R (1990) Morphology, physiology, and molecular biology of renin secretion. Physiol Rev 70:1067–1116

Hackenthal E, Münter K, Fritsch S (1994) Role of nitric oxide in the control of renin release from the normal and hydronephrotic rat kidney. Endothelium 2:229–237

Hall JE, Guyton AC, Brands MW (1996) Pressure-volume regulation in hypertension. Kidney Int 55:535–541

Harding P, Sigmon DH, Alfie ME, Huang PL, Fishman MC, Beierwaltes WH, Carretero OA (1997) Cyclooxygenase-2 mediates increased renal renin content induced by low-sodium diet. Hypertension 29:297–302

Harris RC, McKanna JA, Akai Y, Jacobson HR, Dubois RN, Breyer MD (1994) Cyclooxygenase-2 is associated with the macula densa of rat kidney and increases with salt restriction. J Clin Invest 94:2504–2510

Hartner A, Goppelt-Struebe M, Hilgers KF (1998) Coordinate expression of cyclooxygenase-2 and renin in the rat kildney in renovascular hypertension. Hypertension 31:201–205

Hayashi K, Loutzenhiser R, Epstein M, Suzuki H, Saruta T (1994) Multiple factors contribute to acetylcholine-induced renal afferent arteriolar vasodilation during myogenic and norepinephrine- and KCl-induced vasoconstriction. Studies in the isolated perfused hydronephrotic kidney. Circ Res 75:821–828

He X-R, Greenberg SG, Briggs JP, Schnermann JB (1995) Effect of nitric oxide on renin secretion. II. Studies in the perfused juxtaglomerular apparatus. Am J Physiol 268:F953–F959

Higashi Y, Oshima T, Watanabe M, Matsuura H, Kajiyama G (1996) Renal response to L-arginine in salt-sensitive patients with essential hypertension. Hypertension 27:643–648

Hill C, Lateef AM, Engels K, Samsell L, Baylis C (1997) Basal and stimulated nitric oxide in control of kidney function in the aging rat. Am J Physiol 272:R1747–R1753

Hwang SM, Wilson PD, Laskin JD, Denhardt DT (1994) Age and development-related changes in osteopontin and nitric oxide synthase mRNA levels in human kidney proximal tubule epithelial cells: contrasting responses to hypoxia and reoxygenation. J Cell Physiol 160:61–68

Ikenaga H, Suzuki H, Ishi N, Ito H, Saruta T (1993) Role of NO on pressure-natriuresis in Wistar-Kyoto and spontaneously hypertensive rats. Kidney Int 43:205–211

Imig JD, Gebremedhin D, Harder DR, Roman RJ (1993) Modulation of vascular tone in renal microcirculation by erythrocytes: role of EDRF. Am J Physiol 264:H190–H195

Ishii K, Warner TD, Sheng H, Murad F (1991) Endothelin increases cyclic GMP levels in LLC-PK1 porcine kidney epithelial cells via formation of an endothelium-derived relaxing factor-like substance. J Pharmacol Exp Ther 259:1102–1108

Ito S, Carretero OA (1985) Role of the macula densa in renin release. Hypertension 7:49-I-54

Ito S, Carretero OA, Abe K (1997) Role of nitric oxide in the control of glomerular microcirculation. Clin Exp Pharmacol Physiol 24:578–581

Ito S, Ren Y (1993) Evidence for the role of nitric oxide in macula densa control of glomerular hemodynamics. J Clin Invest 92:1093–1098

Jansen A, Cook T, Taylor GM, Largen P, Riveros-Moreno V, Moncada S, Cattell V (1994) Induction of nitric oxide synthase in rat immune complex glomerulonephritis. Kidney Int 45:1215–1219

Johnson RA, Freeman RH (1994) Renin release in rats during blockade of nitric oxide synthesis. Am J Physiol 266:R1723–R1729

Juncos LA, Garvin J, Carretero OA, Ito S (1995) Flow modulates myogenic responses in isolated microperfused rabbit afferent arterioles via endothelium-derived nitric oxide. J Clin Invest 95:2741–2748

Just A (1997) Nitric oxide and renal autoregulation. Kidney Blood Press Res 20:201–204

Kawabata M, Han WH, Ise T, Kobayashi K, Takabatake T (1996) Role of endogenous endothelin anad nitric oxide in tubuloglomerular feedback. Kidney Int 55 (Suppl.):S135–S137

King AJ, Brenner BM (1991) Endothelium-derived vasoactive factors and the renal vasculature. Am J Physiol 260:R653–R662

Kirchheim HR, Ehmke H, Hackenthal E, Lowe W, Persson P (1987) Autoregulation of renal blood flow, glomerular filtration rate and renin release in conscious dogs. Pflügers Arch. 410:441–449

Knoblich PR, Freeman RH, Villareal D (1996) Pressure dependent renin release during chronic blockade of nitric oxide synthase. Hypertension 28:738–742

Kone BC (1997) Nitric oxide in renal health and disease. Am J Kidney Dis 30:311–333

Kone BC, Baylis C (1997) Biosynthesis and homeostatic roles of nitric oxide in the normal kidney. Am J Physiol 272:F561–F578

Krämer BK, Ritthaler T, Ackermann M, Holmer S, Schricker K, Riegger GAJ, Kurtz A (1994) Endothelium-mediated regulation of renin secretion. Kidney Int 46:1577–1579

Kramer HJ, Moch T, von Sicherer I, Dusing R (1979) Effects of aprotinin on renal function and urinary prostaglandin excretion in conscious rats after acute salt loading. Clin Sci 56:548–553

Kumagai K, Suzuki H, Ichikawa M, Jimbo M, Murakami M, Ryuzaki M, Saruta T (1994) Nitric oxide increases renal blood flow by interacting with the sympathetic nervous system. Hypertension 24:220–226

Kurtz A (1989) Cellular control of renin secretion. Rev Physiol Biochem Pharmacol 113:1–40

Kurtz A, Gotz KH, Hamann M, Wagner C (1998a) Stimulation of renin secretion by nitric oxide is mediated by phosphodiesterase 3. Proc Natl Acad Sci USA 95:4743–4747

Kurtz A, Götz K-H, Hamann M, Kieninger M, Wagner C (1998b) Stimulation of renin secretion by NO donors is related to the cAMP pathway. Am J Physiol 274:F709–F712

Lahera V, Salom MG, Fiksen-Olsen MJ, Raij L, Romero JC (1990) Effects of N^G-monomethyl-L-Arginine and L-arginine on acetylcholine renal response. Hypertension 15:659–663

Lahera V, Salom MG, Miranda-Guardiola F, Moncada S, Romero JC (1991) Effects of N^G-nitro-L-arginine methyl ester on renal function and blood pressure. Am J Physiol 261:F1033–F1037

Lahera V, Navarro-Cid J, Cachofeiro V, Garcia-Estan J, Ruilope LM (1997) Nitric oxide, the kidney, and hypertension. Am.J.Hypertens. 10:129–140

Lapointe JY, Laamarti A, Bell PD (1998) Ionic transport in macula densa cells. Kidney Int 54 (Suppl. 67):S58–S64

Liu L, Barajas L (1993) Nitric oxide synthase immunoreactive neurons in the rat kidney. Neurosci Lett 161:145–148

Lockhart JC, Larson TS, Knox FG (1994) Perfusion pressure and volume status determine the microvascular response of the rat kidney to N^G-monomethyl-arginine. Circ Res 75:829–835

Lu M, Wang W (1996) Protein kinase C stimulates the small conductance K^+ channel in the basolateral membrane of the CCD. Am J Physiol 271:F1045–F1051

Lu M, Giebisch G, Wang W (1997) Nitric oxide-induced hyperpolarization stimulates low-conductance Na^+ channel of rat CCD. Am J Physiol 272:F498–F504

Lu M, Wang X, Wang W (1998) Nitric oxide increases the activity of the apical 70-pS K^+ channel in TAL of rat kidney. Am J Physiol 274:F946–F950

Macleod KM, NG DDW, Harris KH, Diamond J (1987) Evidence that cGMP is the mediator of endothelium dependent inhibition of contractile responses of rat arteries to s-adrenoceptor stimulation. Molecular Pharmacol. 32:59–64

Madeddu P, Glorioso N, Soro A, Manunta P, Troffa C, Tonolo G, Melis MG, Pazzola A (1990) Effect of a kinin antagonist on renal function and haemodynamics during alterations in sodium balance in conscious normotensive rats. Clin Sci 78:165–168

Madeddu P, Anania V, Parpaglia PP, Demontis MP, Varoni MV, Pisanu G, Troffa C, Tonolo G, Glorioso N (1992) Effects of Hoe 140, a bradykinin B2-receptor antagonist, on renal function in conscious normotensive rats. Br J Pharmacol 106:380–386

Madrid MI, Garcia-Salom M, Tornel J, Gasparo de M, Fenoy FJ (1997) Interactions between nitric oxide and angiotensin II on renal cortical and papillary blood flow. Hypertension 30:1175–1182

Maier M, Starlinger M, Zhegu Z, Rana H, Binder BR (1985) Effect of the protease inhibitor aprotinin on renal hemodynamics in the pig. Hypertension Dallas 7:32–38

Majid DSA, Navar LG (1992) Suppression of blood flow autoregulation plateau during nitric oxide blockade in canine kidney. Am J Physiol 262:F40–F46

Majid DS, Navar LG (1997) Nitric oxide in the mediation of pressure natriuresis. Clin Exp Pharmacol Physiol 24:595–599

Majid DSA, Williams A, Navar LG (1993a) Renal responses to intra-arterial administration of nitric oxide donor in dogs. Hypertension Dallas 22:535–541

Majid DSA, Williams A, Navar LG (1993b) Inhibition of nitric oxide synthesis attenuates pressure-induced natriuretic responses in anesthetized dogs. Am J Physiol 264:F79–F87

Manning RD Jr, Hu L (1994) Nitric oxide regulates renal hemodynamics and urinary sodium excretion in dogs. Hypertension Dallas 23:619–625

Marsden PA, Ballermann BJ (1990) Tumor necrosis factor a activates soluble guanylate cyclase in bovine glomerular mesangial cells via an L-arginine-dependent mechanism. J Exp Med 172:1843–1852

Mattei P, Virdis A, Ghiadoni L, Taddei S, Salvetti A (1997) Endothelial function in hypertension. J Nephrol 10:192–197

Mattson DL, Lu S, Cowley AW Jr (1997) Role of nitric oxide in the control of the renal medullary circulation. Clin Exp Pharmacol Physiol 24:587–590

Mayeux PR, Garner HR, Gibson JD, Beanum VC (1995) Effect of lipopolysaccharide on nitric oxide synthase activity in rat proximal tubules. Biochem Pharmacol 49:115–118

McKee M, Scavone C, Nathanson JA (1994) Nitric oxide, cGMP, and hormone regulation of active sodium transport. Proc Natl Acad Sci USA 91:12056–12060

Mohaupt MG, Elzie JL, Ahn KY, Clapp WL, Wilcox CS, Kone BC (1994) Differential expression and induction of mRNAs encoding two inducible nitric oxide synthases in rat kidney. Kidney Int 46:653–665

Mühl H, Kunz D, Pfeilschifter J (1994) Expression of nitric oxide synthase in rat glomerular mesangial cells mediated by cyclic AMP. Br J Pharmacol 112:1–8

Mundel P, Bachmann S, Bader M, Fischer A, Kummer W, Mayer B, Kriz W (1992) Expression of nitric oxide synthase in kidney macula densa cells. Kidney Int 42:1017–1019

Münter K, Hackenthal E (1991) The participation of the endothelium in the control of renin release. J Hypertens 9 (Suppl. 6):S236–S237

Naess PA, Christensen G, Kirkeboen KA, Kiil F (1993) Effect on renin release of inhibiting renal nitric oxide synthesis in anesthetized dogs. Acta Physiol Scand 148:137–142

Nafz B, Berthold H, Ehmke H, Hackenthal E, Kirchheim HR, Persson PB (1997) Flöow versus pressure in the control of renin release in conscious dogs. Am J Physiol 273:F200–F205

Navar LG (1998) Integrating multiple paracrine regulators of renala microvascular dynamics. Am J Physiol 274:F433–F444

Navar LG, Inscho EW, Majid DSA, Imig JD, Harrison-Bernard LM, Mitchell KD (1996) Paracrine regulation of the renal microcirculation. Physiol Rev 76:425–536

Nicolson, Haites NE, McKay NG, Wilson HM, MacLeod AM, Benjamin N (1993) Induction of nitric oxide synthase in human mesangial cells. Biochem Biophys Res Commun 193:1269–1274

Noiri E, Peresleni T, Miller F, Goligorsky MS (1996) In vivo targeting of inducible NO synthase with oligodeoxynucleotides protects rat kidney against ischemia. J Clin Invest 97:2377–2383

Obermuller N, Kunchaparty S, Ellison DH, Bachmann S (1996) Expression of the Na-K-2Cl cotransporter by macula densa and thick ascending limb cells of rat and rabbit nephron. J Clin Invest 98:635–640

Ohishi K, Carmines PK, Inscho EW, Navar LG (1992) EDRF-angiotensin II interactions in rat juxtamedullary afferent and efferent arterioles. Am J Physiol 263:F900–F905

Parekh N, Zou AP (1996) Role of prostaglandins in renal medullary circulation: response to different vasoconstrictors. Am J Physiol 271:F653–F658

Perrella MA, Hildebrand FL Jr, Margulies KB, Burnett JC Jr (1991) Endothelium-derived relaxing factor in regulation of basal cardiopulmonary and renal function. Am J Physiol 261:R323–R328

Persson PB, Ehmke H, Nafz B, Kirchheim HR (1990) Resetting of renal autoregulation in conscious dogs: angiotensin II and α1-adrenoceptors. Pflügers Arch 417:42–47

Persson PB, Baumann JE, Ehmke H, Hackenthal E, Kirchheim HR, Nafz B (1993) Endothelium-derived NO stimulates pressure-dependent renin release in conscious dogs. Am J Physiol 264:F943–F947

Peterson TV, Carter AB, Miller RA (1997) Nitric oxide and renal effects of volume expansion in conscious monkeys. Am J Physiol 272:R1033–R1038

Pfeilschifter J, Schwarzenbach H (1990) Interleukin 1 and tumor necrosis factor stimulate cGMP formation in rat renal mesangial cells. FEBS Lett 273:185–187

Pfeilschifter J, Rob P, Mülsch A, Fandrey J, Vosbeck K, Busse R (1992) Interleukin 1b and tumour necrosis factor a induce a macrophage-type of nitric oxide synthase in rat renal mesangial cells. Eur J Biochem 203:251–255

Qiu C, Engels K, Baylis C (1994) Angiotensin II and a1-adrenergic tone in chronic nitric oxide blockade-induced hypertension. Am J Physiol 266:R1470–R1476

Radermacher J, Klanke B, Kastner S, Haake G, Schurek HJ, Stolte HF, Frolich JC (1991) Effect of arginine depletion on glomerular and tubular kidney function: studies in isolated perfused rat kidneys. Am J Physiol 261:F779–F786

Radermacher J, Klanke B, Schurek HJ, Stolte HF, Frolich JC (1992) Importance of NO/EDRF for glomerular and tubular function: studies in the isolated perfused rat kidney. Kidney Int 41:1549–1559

Raij L, Baylis C (1995) Glomerular actions of nitric oxide. Kidney Int 48:20–32

Reckelhoff JF, Manning RD Jr (1993) Role of endothelium-derived nitric oxide in control of renal microvasculature in aging male rats. Am J Physiol 265:R1126–R1131

Reckelhoff JF, Kellum JA Jr, Racusen LC, Hildebrandt DA (1997) Long-term dietary supplementation with L-arginine prevents age-related reduction in renal function. Am J Physiol 272:R1768–R1774

Reid IA (1994) Role of nitric oxide in the regulation of renin and vasopressin secretion. Front Neuroendocrinol 15:351–383

Reid IA, Chou L (1995) Effect of blockade of nitric oxide synthesis on the renin secretory response to frusemide in conscious rabbits. Clin Sci 88:657–663

Roczniak A, Burns KD (1996) Nitric oxide stimulates guanylate cyclase and regulates sodium transport in rabbit proximal tubule. Am J Physiol 270:F106–F115

Roman RJ, Kaldunski ML, Scicli AG, Carretero OA (1988) Influence of kinins and angiotensin II on the regulation of papillary blood flow. Am J Physiol 255:F690–F698

Ruilope LM, Lahera V, Rodicio JL, Romero JC (1994) Participation of nitric oxide in the regulation of renal function: possible role in the genesis of arterial hypertension. J Hypertens 12:625–631

Salom MG, Lahera V, Romero JC (1991) Role of prostaglandins and endothelium-derived relaxing factor on the renal response to acetylcholine. Am J Physiol 260:F145–F149

Salomonsson M, Skott O, Persson AEG (1991) Influence of intraluminal arterial pressure on renin release. Acta Physiol Scand 141:285–286

Salvemini D, Misko TP, Masferrer JL, Seibert K, Currie MG, Needleman P (1993) Nitric oxide activates cyclooxygenase enzymes. Proc Natl Acad Sci USA 90:7240–7244

Schmidt HH, Gagne GD, Nakane M, Pollock JS, Miller MF, Murad F (1992) Mapping of neural nitric oxide synthase in the rat suggests frequent co-localization with NADPH diaphorase but not with soluble guanylyl cyclase, and novel paraneural functions for nitrinergic signal transduction. J Histochem Cytochem 40:1439–1456

Schnackenberg C, Patel AR, Kirchner KA, Granger JP (1997) Nitric oxide, the kidney and hypertension. Clin Exp Pharmacol Physiol 24:600–606

Schnermann J, Traynor T, Yang T, Arend L, Huang YG, Smart A, Briggs JP (1998) Tubuloglomerular feedback: New concepts and developments. Kidney Int 54 (Suppl. 67):S40–S45

Scholz H, Kurtz A (1993) Involvement of endothelium-derived relaxing factor in the pressure control of renin secretion from isolated perfused kidney. J Clin Invest 91:1088–1094

Schricker K, Kurtz A (1993) Liberators of NO exert a dual effect on renin secretion from isolated mouse renal juxtaglomerular cells. Am J Physiol 265:F180–F186

Schricker K, Kurtz A (1996) Blockade of nitric oxide formation inhibits the stimulation of the renin system by a low salt intake. Pflügers Arch. 432:187–191

Schricker K, Ritthaler T, Krämer BK, Kurtz A (1993) Effect of endothelium-derived relaxing factor on renin secretion from isolated mouse renal juxtaglomerular cells. Acta Physiol Scand 149:347–354

Schricker K, Della Bruna R, Hamann M, Kurtz A (1994a) Endothelium derived relaxing factor is involved in the pressure control of renin gene expression in the kidney. Pflügers Arch. 428:261–268

Schricker K, Hamann M, Kaissling B, Kurtz A (1994b) Renal autocoids are involved in the stimulation of renin gene expression by low perfusion pressure. Kidney Int 46:1330–1336

Schricker K, Hamann M, Kurtz A (1995a) Nitric oxide and prostaglandins are involved in the macula densa control of the renin system. Am J Physiol 269:F825–F830

Schricker K, Hegyi I, Hamann M, Kaissling B, Kurtz A (1995b) Tonic stimulation of renin gene expression by nitric oxide is counteracted by tonic inhibition through angiotensin II. Proc Natl Acad Sci USA 92:8006–8010

Schricker K, Potzl B, Hamann M, Kurtz A (1996) Coordinate changes of renin and brain-type nitric-oxide-synthase (b-NOS) mRNA levels in rat kidneys. Pflügers Arch. 432:394–400

Seto S, Kher V, Scicli AG, Beierwaltes WH, Carretero OA (1983) The effect of aprotinin (a serine protease inhibitor) on renal function and renin release. Hypertension 5:893–899

Shultz PJ, Tolins JP (1993) Adaption to increased dietary salt intake in the rat: role of endogenous nitric oxide. J Clin Invest 91:642–650

Singh I, Grams M, Wang WH, Yang T, Killen P, Smart A, Schnermann J, Briggs JP (1996) Coordinate regulation of renal expression of nitric oxide synthase, renin, and angiotensinogen mRNA by dietary salt. Am J Physiol 270:F1027–F1037

Siragy HM (1993) Evidence that intrarenal bradykinin plays a role in regulation of renal function. Am J Physiol 265:E648–E654

Siragy HM, Jaffa AA, Margolius HS (1993) Stimulation of renal interstitial bradykinin by sodium depletion. Am.J.Hypertens. 6:863–866

Siragy HM, Ibrahim MM, Jaffa AA, Mayfield R, Margolius HS (1994) Rat renal interstitial bradykinin, prostaglandin E2, and cyclic guanosine 3',5'-monophosphate. Effects of altered sodium intake. Hypertension Dallas 23:1068–1070

Skott O, Briggs JP (1987) Direct demonstration of macula densa-mediated renin secretion. Science 237:1618–1620

Star RA (1997) Intrarenal localization of nitric oxide synthase isoforms and soluble guanylyl cyclase. Clin Exp Pharmacol Physiol 24:607–610

Steinhausen M, Blum M, Fleming JT, Holz FG, Parekh N, Wiegman DL (1989) Visualization of renal autoregulation in the split hydronephrotic kidney of rats. Kidney Int 35:1151–1160

Stoos BA, Garvin JL (1997) Actions of nitric oxide on renal epithelial transport. Clin Exp Pharmacol Physiol 24:591–594

Stoos BA, Garcia NH, Garvin JL (1995) Nitric oxide inhibits sodium reasorption in the isolated perfused cortical collecting duct. J Am Soc Nephrol 6:89–94

Taugner R, Hackenthal E (1989) The juxtaglomerular apparatus. Springer, Berlin Heidelberg New York

Taugner R, Bührle CP, Hackenthal E, Mannek E, Nobiling R (1984) Morphology of the juxtaglomerular apparatus and secretory mechanisms. Contr Nephrol 43: 76–101

Taugner R, Nobiling R, Metz R, Taugner F, Bührle C, Hackenthal E (1988) Hypothetical interpretation of the calcium paradox in renin secretion. Cell Tissue Res 252:687–690

Terada Y, Tomita K, Nonoguchi H, Marumo F (1992) Polymerase chain reaction localization of constitutive nitric oxide synthase and soluble guanylate cyclase messenger RNAs in microdissected rat nephron segments. J Clin Invest 90:659–665

Tharaux P-L, Dussaule J-C, Pauti M-D, Vassitch Y, Ardaillou R, Chatziantoniou C (1997) Activation of renin synthesis is dependent on intact nitric oxide production. Kidney Int 51:1780–1787

Thorup C, Persson AEG (1994) Inhibition of locally produced nitric oxide resets tubuloglomerular feedback mechanism. Am J Physiol 267:F606–F611

Tojo A, Gross SS, Zhang L, Tisher CC, Schmidt HHHW, Wilcox CS, Madsen KM (1994a) Immunocytochemical localization of distinct isoforms of nitric oxide synthase in the juxtaglomerular apparatus of normal rat kidney. J Am Soc Nephrol 4:1438–1447

Tojo A, Madsen KM, Wilcox CS (1995) Expression of immunoreactive nitric oxide synthase isoforms in rat kidney: Effects of dietary salt and losartan. Jpn Heart J 36:389–398

Tolins JP, Shultz PJ (1994) Endogenous nitric oxide synthesis determines sensitivity to the pressor effect of salt. Kidney Int 46:230–236

Tracey WR, Pollock JS, Murad F, Nakane M, Forstermann U (1994) Idenetification of an endothelial-like type III NO synthase in LLC–PK1 kidney epithelial cells. Am J Physiol 266:C22–C28

Ujiie K, Hogarth L, Danziger R, Star RA (1994) Localization and regulation of endothelial NO synthase mRNA expression in rat kidney. Am J Physiol 267:F296–F302

Vallon V, Thomson S (1995) Inhibition of local nitric oxide synthase increases homeostatic efficiency of tubuloglomerular feed back. Am J Physiol 269:F892–F899

Wagner C, Jensen BL, Krämer BK, Kurtz A (1998) Control of the renal renin system by local factors. Kidney Int 54 (Suppl. 67):S78–S83

Wilcox CS, Welch WJ (1996) TGF and nitric oxide: effects of salt intake and salt-sensitive hypertension. Kidney Int 49 (Suppl. 55):S9–S13

Wilcox CS, Welch WJ (1998) Macula densa nitric oxide synthase: expression, regulation, and function. Kidney Int 54 (Suppl.67):S53–S57

Wilcox CS, Welch WJ, Murad F, Gross SS, Taylor G, Levi R, Schmidt HHHW (1992) Nitric oxide synthase in macula densa regulates glomerular capillary pressure. Proc Natl Acad Sci USA 89:11993–11997

Woltz M, Schmetterer L, Ferber W, Artner E, Mensik C, Eishler HG, Krejcy K (1997) Effect of nitric oxide synthase inhibition on renal hemodynamics in man: reversal by L-arginine. Am J Physiol 272:F178–F182

Yang T, Singh I, Pham H, Sun D, Smart A, Schnermann JB, Briggs JP (1998) Regulation of cyclooxygenase expression in the kidney ba dietary salt intake. Am J Physiol 274:F481–F489

Zatz R, De Nucci G (1991) Effects of acute nitric oxide inhibition on rat glomerular microcirculation. Am J Physiol 261:F360–F363

Section IV
The Role of Pharmacological Action of NO in Human Disease

CHAPTER 15
Therapeutic Importance of Nitrovasodilators

G. KOJDA

A. Introduction

Nitrovasodilators are a group of drugs with heterogeneous chemical structures (Fig. 1). The first medical use of these drugs was reported more than 100 years ago. In 1879, WILLIAM MURREL described that the organic nitrate glyceryl trinitrate has beneficial effects in angina pectoris (MURREL 1879). Today, organic nitrates, such as glyceryl trinitrate, isosorbide-2,5-dinitrate, isosorbide-5-nitrate and, to a lesser extent, pentaerythritol tetranitrate, remain important drugs for the acute and preventive treatment of myocardial ischemia (ABRAMS 1995; PARKER and PARKER 1998). Likewise, glyceryl trinitrate is particularly useful in acute heart failure. Chemically, these drugs are esters of nitric acid and aliphatic alcohols. They are also explosives and are used as substitutes in special gasoline mixtures (AHLNER et al. 1991). In addition to organic nitrates, sydnonimines (such as molsidomine) and inorganic compounds (such as sodium nitroprusside) are also clinically used nitrovasodilators. Finally, a variety of different nitrovasodilators such as nitrosothiols or oxyhydrazines (Fig. 1) are now widely used in experimental medicine (IGNARRO et al. 1981; KEEFER et al. 1996). All nitrovasodilators act by the release of nitrogen monoxide (NO•). Nevertheless, they show profound differences in their pharmacokinetics, hemodynamic actions and, thus, their usefulness for either acute or preventive treatment of myocardial ischemia or acute heart failure. Nitrovasodilators reduce smooth-muscle tone. In the vasculature, this results in venodilation and relaxation of coronary arteries, coronary stenoses and coronary collateral vessels (BASSENGE and STUART 1986; AHLNER et al. 1991; ABRAMS 1995). These hemodynamic actions cause the well-established anti-ischemic activity. In addition, glyceryl trinitrate can also substantially reduce blood pressure, relax the smooth muscle of the gastrointestinal tract or the uterus and show antiplatelet activity. Organic nitrates elicit only a few side effects, of which transient headache and postural hypotension are the most prominent ones. An important shortcoming of these drugs is their loss of effectiveness after continuous treatment, an effect which is known as nitrate tolerance. Today, the most effective strategy to overcome this problem is a therapeutic regimen that includes a daily drug-free interval of at least 10h.

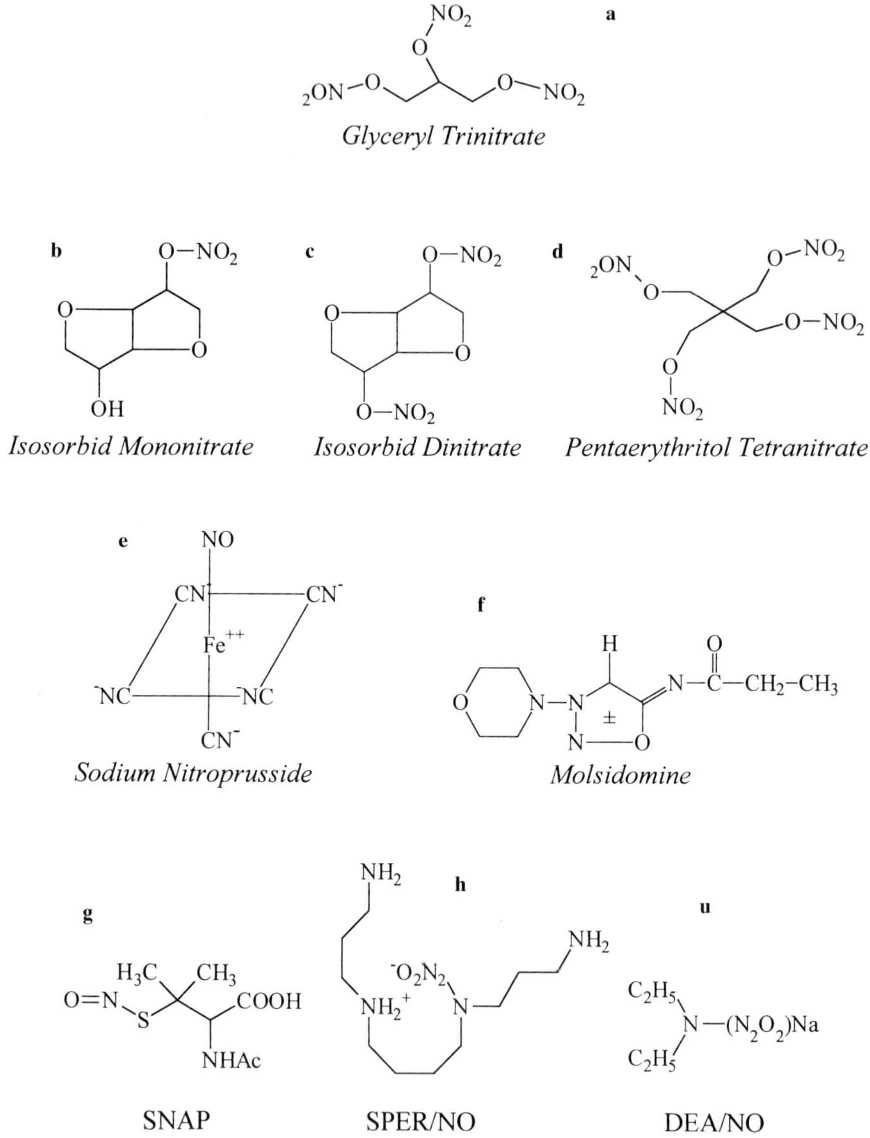

Fig. 1. Chemical structures of various nitrovasodilators. *SNAP*, *S*-nitroso-*N*-acetyl-D,L-penicillamine; *SPER/NO*, (*Z*)-1-(*N*-[3-aminopropyl]-N-[4-aminopropylammonio)butyl]-amino)diazen-1-ium-1,2-diolat; *DEA/NO*, sodium (*Z*)-1-(*N,N*-diethylamino)diazen-1-ium-1,2-diolat

B. Mechanisms of Action

The first experimental evidence for the mechanism of action of nitrovasodilators was evaluated only 20 years ago, when it was demonstrated that sodium nitroprusside increases the level of cyclic guanosine monophosphate (cGMP) in the smooth muscle (SCHULTZ et al. 1977). Subsequently, nitrovasodilators were shown to activate soluble guanylate cyclase, and a similar activation could be induced by NO• (MITTAL and MURAD 1982). The results of further investigations suggested a relationship between the formation of NO• generated by organic nitrates and their biological activity (IGNARRO et al. 1981). Based on these observations, it was proposed that, in smooth-muscle cells, nitrovasodilators produce nitrosothiols, which activate soluble guanylate cyclase. Indeed, formation of the denitrated metabolites of organic nitrates, e.g. 1,2- and 1,3-glyceryl dinitrate from glyceryl trinitrate, was shown to precede the increase of cGMP in the smooth muscle and its relaxation as well (BRIEN et al. 1986, 1988; KAWAMOTO et al. 1990). Direct measurement of NO• has demonstrated that organic nitrates release this radical after incubation with cysteine in vitro and induce an activation of isolated soluble guanylate cyclase (FEELISCH and NOACK 1987). Similar results were obtained with other nitrovasodilators such as sodium nitroprusside, various nitrosothiols or SIN-1, the active metabolite of molsidomine (NOACK and FEELISCH 1991). In contrast to organic nitrates, these nitrovasodilators generate NO• in the absence of free thiols. Thus, the mechanism of formation of NO• is one important difference between the various nitrovasodilators.

Generation of NO• from organic nitrates requires the reduction of a nitrate nitrogen from +5 to +2 and, thus, a flow of three electrons. In tissues, this reaction is most likely catalyzed by intracellular membrane bound enzymes (CHUNG and FUNG 1990; FEELISCH and KELM 1991; BENNETT et al. 1992; SALVEMINI et al. 1992). So far, the identity of these enzymes is unknown. Preliminary evidence indicates an involvement of cytochrome P450 enzymes (SERVENT et al. 1989; McDONALD and BENNETT 1990; SCHRÖDER and SCHRÖR 1990) and of glutathion-S-transferase (LAU and BENET 1990; HILL et al. 1992; NIGAM et al. 1993; KENKARE et al. 1994), but these findings are not consistent (SAKANASHI et al. 1991; KURZ et al. 1993; LIU et al. 1993). It is, therefore, most likely that a variety of enzymes or isoenzymes are capable of producing NO• by reducing a nitrate moiety of organic nitrates. Once generated, NO• activates soluble guanylate cyclase to produce cGMP. The second messenger activates cGMP-dependent protein kinases and initiates several effects, such as phosphorylation of myosin light chain, sequestration of intracellular calcium, reduction of calcium entry from the extracellular space, reduced release of intracellularly stored calcium and inhibition of formation of inositol-1,4,5-trisphosphate (PFITZER et al. 1984; COLLINS et al. 1986; TWORT and VAN BREEMEN 1988; LANG and LEWIS 1989). These changes result in vasorelaxation and reduce the ability of the smooth muscle to contract.

C. Hemodynamic Actions

Nitrovasodilators elicit a pattern of hemodynamic changes that is not matched by any other cardiovascular drug. Among organic nitrates, this pattern is almost comparable. The cornerstones of the hemodynamic changes causing clinical benefit are preferential venodilation and vessel-size-selective coronary vasodilation. Molsidomine has also been shown to preferentially reduce preload. In contrast, sodium nitroprusside is a balanced vasodilator causing profound and rapid reduction of blood pressure as well.

I. Preferential Venodilation

Organic nitrates induce a preferential reduction of preload that is caused by a dilatation of venous blood vessels. This vasodilation occurs at doses that have a minor effect on blood pressure (BASSENGE and STUART 1986). Interestingly, the preferential reduction of preload distinguishes organic nitrates from other nitrovasodilators, such as sodium nitroprusside, despite the common generation of NO$^•$ (ARMSTRONG et al. 1975). The mechanism underlying the preferential venodilation by organic nitrates is not yet clear. It has been shown that sodium nitroprusside undergoes a completely different bioactivation process in the vascular wall, which is considered to be important for the striking differences in its hemodynamic actions compared to organic nitrates (BATES et al. 1991; KOWALUK et al. 1992; MOHAZZAB et al. 1999). Indeed, production of NO$^•$ from glyceryl trinitrate in vivo is greater in venous compared to arterial vessels (MÜLSCH et al. 1995). According to a recent investigation, these observations might be explained by NO$^•$-induced inhibition of bioactivation of glyceryl trinitrate (Fig. 2) (KOJDA et al. 1998). In venous blood vessels, generation of endothelial NO$^•$ is considerably smaller than in arterial blood vessels (DE MEY and VANHOUTTE 1982; LÜSCHER et al. 1988), which is probably the result of a lesser extent of shear stress in veins (POHL et al. 1986).

Preferential venodilation induced by organic nitrates has several advantages for patients with coronary-artery disease. It initiates a redistribution of the circulating blood volume towards the splanchnic and mesenteric vascular beds. This results in a reduction of preload, which reduces left ventricular end-diastolic pressure and systolic ventricular-wall tension and may increase cardiac output (ABRAMS 1995). These changes improve the balance of cardiac oxygen demand and supply, which is of considerable importance for the ischemic heart. It should, however, be mentioned that, according to the Frank-Starling mechanism, reduction of normal preload might result in a reduction of cardiac output and, thus, aggravate left ventricular function. Preferential venodilation by organic nitrates also induces a reduction of the hepatic venous pressure gradient and of portal blood flow. These effects have been shown to be beneficial in patients with liver cirrhosis (FREEMAN et al. 1985; LEVACHER et al. 1995; MERKEL et al. 1995). Sodium nitroprusside is also an effective dilator

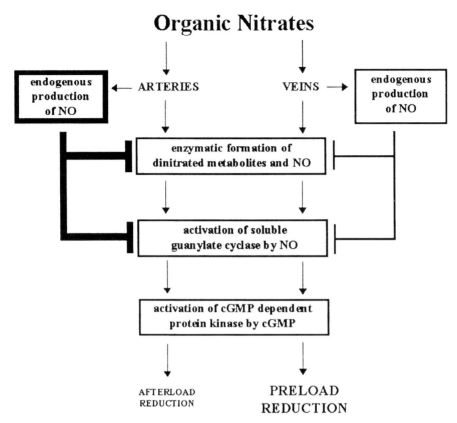

Fig. 2. Scheme of the suggested mechanism underlying the preferential venodilation elicited by organic nitrates (KOJDA et al. 1998; Sect. C.I.)

of capacitance vessels but strongly reduces blood pressure as well (RISOE et al. 1992).

II. Vessel-Size-Selective Coronary Vasodilation

Another unique feature of organic nitrates is the redistribution of coronary flow to ischemic regions of the cardiac muscle. The first observation of this favourable action was made almost 30 years ago by measuring the oxygen saturation in large and small coronary arteries in the dog in vivo and was explained by a vessel-size-selective coronary vasodilation (WINBURY et al. 1971). Therapeutic concentrations of organic nitrates dilate only large coronary arteries but have no effects on coronary resistance vessels (Fig. 3). In ischemic regions of the heart, resistance vessels are already maximally dilated by autoregulation. Pharmacologic dilatation of all arterial vessels within the coronary circulation, as achieved with dipyridamol, redistributes coronary

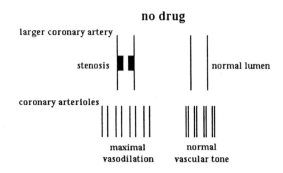
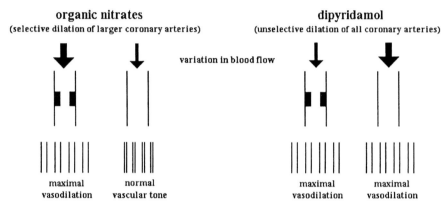

Fig. 3. Vessel-size-selective vasodilation of organic nitrates, resulting in redistribution of coronary flow to ischemic regions of the heart

blood flow away from the ischemic regions of myocardial muscle and elicits the well-known coronary steal phenomenon (Fig. 3). In striking contrast, dilatation of only large coronary arteries, as initiated by organic nitrates, facilitates coronary flow to the ischemic regions, where coronary resistance is already low. This hypothesis was confirmed by studies directly investigating the effects of organic nitrates on small versus large coronary arteries in organ chambers (SELLKE et al. 1990; HARRISON and BATES 1993). Thus, vessel-size-selective coronary vasodilation by organic nitrates improves oxygen supply to the ischemic regions of the heart.

III. Effects on Blood Pressure

Sodium nitroprusside almost equieffectively relaxes both venous and arterial blood vessels resulting in reduction of preload and afterload (RISOE et al. 1992). The rapid onset and short duration of this action allows titration of the blood pressure by adjusting the infused dose (AHLNER et al. 1991). Organic nitrates, such as glyceryl trinitrate, can also reduce blood pressure. At low concentrations that reduce preload almost maximally, this effect is usually less pronounced and a result of the redistribution of the circulating blood volume (BASSENGE and STUART 1986). In contrast, higher concentrations of glyceryl trinitrate can cause a substantial drop in blood pressure that is mediated by direct dilation of large arterioles and subsequent reduction of peripheral resistance. For two reasons, the resulting hypotension is a disadvantage for the ischemic heart. Reduction in blood pressure diminishes coronary perfusion pressure and blood flow and, thus, oxygen supply to the myocardium. Secondly, activation of the baroreceptor reflex increases the sympathetic tone in the heart and leads to enhanced inotropy and chronotropy. Both of these effects aggravate the oxygen balance of the heart by decreasing oxygen supply and increasing its demand. Thus, the relatively small effect of glyceryl trinitrate on blood pressure compared with its pronounced effect on preload is important and therapeutically valuable. Nevertheless, extensive clinical experience with glyceryl trinitrate has shown a great interindividual variability of the blood-pressure response (AHLNER et al. 1991), emphazising the necessity of titrating the drug according to its hypotensive action and other clinical parameters, such as anginal pain. This is especially important in emergency cases, such as acute myocardial infarction or acute decompensated heart failure (ABRAMS 1995).

IV. Other Effects on Hemodynamics

Organic nitrates increase aortic distensibility and improve the Windkessel function (WILLE et al. 1980). Consequently, the impedance to left ventricle ejection declines, resulting in decreased cardiac workload and oxygen demand While it is unclear how much this action of organic nitrates contributes to their therapeutic effectiveness, it is likely important. Finally, organic nitrates and other NO• donors have been shown to influence myocardial contractility. In vitro, low concentrations of NO• ($0.1–1\,\mu M$) facilitate – and high concentrations ($>30\,\mu M$) inhibit – myocardial contraction (KOJDA et al. 1996), while in vivo studies with nitrovasodilators in animals and man showed only positive inotropic actions (STRAUER and SCHERPE 1978; PRECKEL et al. 1997). In general, the effects of NO• on myocardial contractility are small and probably of minor importance.

V. Effects on Platelets

Therapeutic doses of isosorbide dinitrate have been shown to reduce platelet aggregation in patients with stable angina (DE CATERINA et al. 1984). Similar observations were made with sodium nitroprusside and glyceryl trinitrate in patients with unstable and stable angina (DIODATI et al. 1990, 1995). The inhibitory effects of nitrovasodilators on platelet function are most likely mediated by the same mechanism of action as vasorelaxation – namely, generation of NO• and activation of soluble guanylate cyclase (STAMLER and LOSCALZO 1991). Recent data suggest that NO• derived from nitrovasodilators can also induce activation of endothelial cyclooxygenase and subsequent release of prostacyclin, which might contribute to the antiplatelet effects in vivo (SALVEMINI et al. 1996). Although the importance of the antiplatelet activity of nitrates is still controversial, it is believed that inhibition of platelet activation and aggregation might have potential for preventing thrombus formation in unstable coronary syndromes (ABRAMS 1995).

D. Pharmacokinetics

Clinically used nitrovasodilators show important differences in their pharmacokinetics (Table 1). In general, there are short-acting drugs with a rapid onset of action, such as glyceryl trinitrate, sodium nitroprusside and isosorbide dinitrate, and long-acting drugs with a delayed onset of action, such as isosorbide mononitrate, pentaerythritol tetranitrate and molsidomine (AHLNER et al. 1991). Furthermore, sustained-release oral formulations and patches have been developed to prolong the duration of action of glyceryl trinitrate, isosorbide dinitrate and isosorbide mononitrate.

All nitrovasodilators are metabolized either in the liver or in target tissues, such as vascular smooth muscle (AHLNER et al. 1991; FUNG 1993; ABRAMS 1995). Rapid hepatic metabolism limits the bioavailibility of oral glyceryl trinitrate, pentaerythritol tetranitrate and isosorbide dinitrate. While the denitrated metabolites of glyceryl trinitrate are most likely not involved in its

Table 1. Doses, onsets of action and durations of action of nitrovasodilators in normal and slow-release formulations and as patches. Pharmacokinetic differences between nitrovasodilators

Nitrate	Usual dose (mg)	Onset of action (min)	Duration of action (h)
Glyceryl trinitrate			
Sublingual	0.3–0.8	2–5	0.3–0.5
Buccal (three times daily)	1–3	2–5	0.5–5
(only 8–12h daily)	0.4–0.8 mg/h	30–60	24 (maximal)
Isosorbide dinitrate			
Sublingual	2.5–10	5–20	0.75–2
Oral (2–3 times daily)	10–60	15–45	2–6
Oral, slow release (daily)	80–120	30–60	10–14
Isosorbide mononitrate			
Oral (twice daily)	20	30–60	3–6
Oral, slow release (daily)	60–120	15–45	10–14
Pentaerythritol tetranitrate			
Oral (2–3 times daily)	80	30–60	5–7
Molsidomine			
Oral (2–3 times daily)	4–8	30–60	3–5

activity, isosorbide mononitrate, as a hepatic metabolite of isosorbide dinitrate, and all denitrated metabolites of pentaerythritol tetranitrate are pharmacologically active (PARKER et al. 1975; WEBER et al. 1995). In contrast, oral isosorbide mononitrate is 100% bioavailable, and molsidomine needs to be hepatically bioactivated to linsidomine (SIN-1) (AHLNER et al. 1991).

E. Clinical Use

The parmacokinetics of nitrovasodilators largely determines their clinical use in cardiovascular disease. Nitrates, such as glyceryl trinitrate and isosorbide dinitrate, display a rapid onset and a short duration of action and are mainly used for acute treatment and short-term prevention of anginal attacks, while nitrates with delayed onset and a prolonged duration of action are used for long-term prevention of anginal attacks. Application of intravenous glyceryl trinitrate also represents an important pharmacologic intervention in acute myocardial infarction and acute decompensated heart failure. Interestingly, recent reports have shown beneficial effects of organic nitrates in diffuse esophageal spasm, functional dyspepsia, biliary colic, biliary endoscopy, liver cirrhosis, anal fissure, preterm labor, severe fetal distress and severe dysmenorrhoea (Table 2).

I. Effects in Stable Angina

Stable angina was the first disease treated with nitrovasodilators and is still the most important indication for these drugs. There are three different therapeutical strategies: (1) treatment of acute anginal attacks, (2) short-time prevention of anginal attacks in case of expected extraordinary physical or mental stain and (3) continuous treatment during the day to reduce and/or prevent the occurrence of anginal attacks. The goal of these approaches is to

Table 2. Clinical use of organic nitrates in non-cardiovascular disorders

Disorder	Nitrate	Reference
Diffuse esophagal spasm	Glyceryl trinitrate	ORLANDO and BOZYMSKI 1973
Functional dyspepsia	Glyceryl trinitrate	GILJA et al. 1997
Biliary colic	Glyceryl trinitrate	HASSEL 1993
Biliary endoscopy	Glyceryl trinitrate	LUMAN et al. 1997
Liver cirrhosis	Glyceryl trinitrate	LEVACHER et al. 1995
	Isosorbide dinitrate	FREEMAN et al. 1985
	Isosorbide mononitrate	ANGELICO et al. 1997; MERKEL et al. 1995
Anal fissure	Glyceryl trinitrate	BANERJEE 1997; LUND and SCHOLEFIELD 1997
Pre-term labor	Glyceryl trinitrate	LEES et al. 1994
Severe fetal distress	Glyceryl trinitrate	MERCIER et al. 1997
Severe dysmenorrhea	Glyceryl trinitrate	PITTROF et al. 1996

provide symptomatic relief. The overall effect of organic nitrates on survival in patients with stable angina is unknown (Sect. E.III.).

1. Treatment and Short-Term Prevention of Anginal Attacks

Short-acting organic nitrates showing a rapid onset of action, such as sublingual glyceryl trinitrate and the somewhat slower-acting sublingual isosorbide dinitrate (Table 1), are the drugs of choice for treatment of acute episodes of angina (AHLNER et al. 1991; ABRAMS 1995; PARKER and PARKER 1998). In most instances, these drugs promptly relieve pain within a few minutes. Sublingual formulations of organic nitrates are also effective in short-term prevention of anginal attacks if taken a few minutes before activities that are known to possibly cause an attack.

2. Long-Term Management of Chronic Stable Angina

Since many acute attacks in stable angina are not predictable, it is often necessary to provide effective long-term antianginal treatment. For this purpose, long-acting nitrates are indicated as first-line therapy (AHLNER et al. 1991; ABRAMS 1995; PARKER and PARKER 1998). To avoid the development of nitrate tolerance (Sect. F), nitrates are given as intermittent therapy by providing a daily nitrate-free interval of at least 10h. This is usually accomplished by continuous treatment during the day and withdrawal at night. Effective management of chronic stable angina throughout the day using an intermittent treatment regimen has been proven for glyceryl trinitrate patches (DEMOTS and GLASSER 1989), standard-formulation isosorbide mononitrate (PARKER and ISOSORBIDE-5-mononitrate Study Group 1993; THADANI et al. 1994) and sustained-formulation isosorbide mononitrate (CHRYSANT et al. 1993). Likewise, long-term intermittent treatment with standard- and sustained-formulation isosorbide dinitrate has shown antianginal activity without development of tolerance (PARKER et al. 1987; SILBER et al. 1987). A study with a small number of patients indicated that the beneficial effect of isosorbide dinitrate might decline during the day, which raises the necessity for further studies to prove its sustained action (PARKER and PARKER 1998). During intermittent treatment, withdrawal of glyceryl trinitrate patches at night has an adverse effect on exercise performance but is not associated with rebound myocardial ischemia (DEMOTS and GLASSER 1989; J.D. PARKER et al. 1995). In addition to their well-established effects in overt stable angina, organic nitrates are also capable of reducing the frequency of asymptomatic myocardial ischemia (silent angina) (SHELL et al. 1986). However, there is no indication for treatment of these ischemic episodes with nitrates (PURCELL et al. 1994).

II. Effects in Unstable Angina

Intravenous application of glyceryl trinitrate has been shown to reduce the frequency and duration of chest pain in patients with unstable angina

(JAFFRANI et al. 1993; GÖBEL et al. 1995; KARLBERG et al. 1998). Most likely, this beneficial effect is related to prevention of coronary constriction and antiplatelet effects (Sects. C.II., C.V.). During continuous monitoring of blood pressure and heart rate to avoid harmful reduction of blood pressure, the dosage of the drug is adjusted to the disappearance of pain. Since cessation of treatment might result in rebound ischemia (J.D. PARKER et al. 1995), sudden withdrawal or intermittent treatment is not recommended.

III. Effects in Acute Myocardial Infarction

In patients with acute myocardial infarction, infusion of glyceryl trinitrate reduces chest pain, pulmonary congestion and sustained hypertension. These effects are well established, and the drug is frequently used for this indication (AHLNER et al. 1991; ABRAMS 1995). Its efficiency relies on the favorable changes in hemodynamics and, probably, its antiplatelet effects (Sect. C).

Besides symptomatic relief, limitation of infarct size and the overall outcome after myocardial infarction is of major interest. A meta-analysis on this topic summarizing earlier trials with nitrates revealed a total reduction of mortality of as much as 35% (YUSUF et al. 1988). Most of the beneficial effect of nitrates has been attributed to the initial 24–48h of therapy. The favorable result could not be confirmed in two recent mega-trials, including 77,000 post-infarction patients (Gruppo Italiano per lo Studio della Sopravvivenza Nell'Infarto Miocardico 1994; ISIS-4 Collaborative Group 1995). In these studies, therapy with a glyceryl trinitrate patch or with oral sustained-formulation isosorbide mononitrate was initiated in the early post-infarction phase and continued for 5–6 weeks. Overall, only a trend (but not a significant improvement) of survival has been observed. In contrast, subgroup analysis showed a significant benefit of nitrates on survival of women and patients older than 70 years of age, and the combination of the glyceryl trinitrate patch with enalapril had a significantly greater effect than enalapril alone (Gruppo Italiano per lo Studio della Sopravvivenza Nell'Infarto Miocardico 1994). Although there are some limitations in both of these trials, such as permission of nitrate use in control groups, one can conclude that effects of nitrates on mortality after myocardial infarction are much lower than reported previously. One explanation for the discrepancy between earlier investigations and the mentioned mega-trials could be that the latter studies were done after establishment of the benefits of thrombolysis with streptokinase (ISIS-2 Second International Study of Infarct Survival Collaborative Group 1988).

IV. Effects in Heart Failure

Acute left ventricular failure is an indication for sodium nitroprusside and glyceryl trinitrate if blood pressure is normal or increased (AHLNER et al. 1991; COHN 1996). The goal of this intervention is to improve left ventricular function and cardiac output and to prevent or diminish the development of pulmonary edema. In the presence of myocardial ischemia or established coronary-artery disease, glyceryl trinitrate should be preferred because of its preferential venodilating activity.

Organic nitrates have also been shown to be effective in long-term treatment of congestive heart failure. Additional treatment of mild-to-moderate heart-failure patients already receiving digoxin and diuretics with the combination of isosorbide dinitrate and the arterial vasodilator hydralazine reduced symptoms and improved ejection fraction, exercise capacity and overall survival as compared with additional treatment with the α_1-adrenoceptor antagonist prazosin alone (COHN et al. 1986). Although a later trial showed that treatment with the angiotensin-converting-enzyme inhibitor enalapril is superior (COHN et al. 1991), the isosorbide dinitrate/hydralazine combination remains a valuable therapeutic approach for patients showing an insufficient response, poor tolerability or contraindications to angiotensin-converting-enzyme inhibitors (COHN 1996).

V. Effects in Gastrointestinal Disorders

There are several reports on clinically effective actions of glyceryl trinitrate mediated by relaxation of gastrointestinal smooth muscle. Topical application of a glyceryl trinitrate ointment relaxes the anal sphincter and lowers the resting pressure in the anus, which has been shown to be beneficial in the treatment of anal fissure (BANERJEE 1997; LUND and SCHOLEFIELD 1997). It was also demonstrated that local application of glyceryl trinitrate can relax the sphincter of Oddi by reducing tonic and phasic contractions, and this effect was suggested to be useful for diagnostic and therapeutic biliary endoscopy (LUMAN et al. 1997). Likewise, relaxation of the gall bladder and the bile duct by glyceryl trinitrate most likely underlie beneficial effects in biliary colic (HASSEL 1993). Another study in patients with functional dyspepsia outlined the possibility that sublingual glyceryl trinitrate improves accommodation of the proximal stomach to a meal and thereby reduces postprandial symptoms (GILJA et al. 1997). Glyceryl trinitrate was also shown to be helpful in diffuse esophageal spasm (ORLANDO and BOZYMSKI 1973). Finally, glyceryl trinitrate, isosorbide mononitrate and isosorbide dinitrate are known to induce a reduction of the hepatic venous pressure gradient and of portal blood flow, which is beneficial in patients with liver cirrhosis (FREEMAN et al. 1985; LEVACHER et al. 1995; MERKEL et al. 1995; ANGELICO et al. 1997).

VI. Effects on the Uterus

Some recent reports suggest that intravenous, sublingual and percutaneous glyceryl trinitrate has relaxant effects on the uterus. It was shown that treatment with this nitrovasodilator arrests pre-term labor and prolongs gestation (LEES et al. 1994). In 22 patients with uterine hyperactivity, small doses of glyceryl trinitrate were associated with resolution of severe fetal distress when standard interventions, such as oxygen administration and discontinuation of oxytocin infusion, failed to be effective (MERCIER et al. 1997). Likewise, glyceryl trinitrate can be effectively used to reduce umbilical–placental vascular resistance (GILES et al. 1992). Two other reports demonstrated relaxant effects of sublingual glyceryl trinitrate in one patient hyperstimulated by intravenous pitocin administration (BELL 1996) and in 10 patients undergoing external version of the fetus (REDICK et al. 1997). Finally, application of glyceryl trinitrate patches has been successfully used to control pain in women with severe dysmenorrhea (PITTROF et al. 1996).

F. Nitrate Tolerance

Continous administration of therapeutic doses of organic nitrates results in a loss of clinical efficiency in about 17h following onset of therapy. This is associated with disappearance of increased treadmill walking time and an increased frequency of anginal attacks (AHLNER et al. 1991). Nitrate tolerance is a multifactorial phenomenon to which several mechanisms seem to contribute. Factors most likely involved include increased plasma volume and neurohumoral counter-regulation leading to vascular supersensitivity to endogenous vasoconstrictors and increased vascular superoxide production (AHLNER et al. 1991; PARKER 1992; MANGIONE and GLASSER 1994; MÜNZEL and HARRISON 1997; PARKER and PARKER 1998).

Nitrate tolerance is associated with activation of the renin–angiotensin system and with increases in plasma levels of catecholamines and vasopressin (PARKER et al. 1991). According to these findings, it is conceivable that the vasodilator effects of organic nitrates are functionally antagonized in the setting of tolerance. In general, functional antagonism of vasorelaxation induced by organic nitrates might be mediated by at least three major mechanisms: (1) inhibition of vascular bioactivation of organic nitrates to NO$^•$, (2) counteracting vasoconstriction or (3) inactivation of NO$^•$ generated by these drugs. While diminished vascular bioactivation seems to be unlikely (LAURSEN et al. 1996), there is considerable experimental evidence to substantiate the involvement of the other two mechanisms, as extensively reviewed only recently (MÜNZEL and HARRISON 1997). Briefly, induction of nitrate tolerance in the rabbit in vivo was shown to increase both vascular sensitivity to vasoconstrictors and vascular superoxide production (MÜNZEL et al. 1995a,b). Interestingly, these investigators suggested that both effects might be the result of a common cause – that is, activation of protein kinase C mediated by angiotensin II (MÜNZEL and HARRISON 1997). Some important questions remain to be elucidated, e.g. (1) how can superoxide anions generated in endothelial cells inactivate NO$^•$ generated from organic nitrates in the vascular smooth muscle and (2) why does nitrate tolerance persist although the increase in circulating angiotensin II levels is only transient? The answers to these questions might also shed new light on our current understanding of the pharmacology of organic nitrates. Nevertheless, the extended neurohumoral counter-regulation hypothesis, including pronounced changes in vascular biology, is currently the most attractive and reasonable concept to explain nitrate tolerance.

In contrast, the most extensively discussed hypothesis, that a sulfhydryl-depletion of the vascular smooth muscle causes nitrate tolerance (NEEDLEMAN and JOHNSON 1973), is probably incorrect. It has been shown that restoration of thiol levels in the vascular wall does not reverse tolerance to glyceryl trinitrate (GRUETTER and LEMKE 1985). Similarly, it is possible to strongly reduce the thiol content of vascular smooth muscle without affecting the vasorelaxing activity of organic nitrates at all (KOJDA et al. 1993). In an attempt to investigate the sulfhydryl-depletion hypothesis in the rat in vivo, it was demonstrated that nitrate tolerance can occur in the absence of any changes in vascular sulfhydryl content (BOESGAARD et al. 1994). Thus, depletion of vascular free thiols might be associated with nitrate tolerance but is probably not the underlying cause.

G. Side Effects and Contraindications

Organic nitrates are remarkably safe drugs. Their most prominent side effects are headache, postural hypotension, nausea and flush. All of these are related to the vasodilatory activity. Headache usually disappears during the first week of therapy but reoccurs if therapy is interrupted for 1–2 days. Most likely, headache is caused by NO•-mediated cerebral vasodilation (OLESEN et al. 1994). However, the mechanism of cerebral vasodilation by nitrovasodilators seems to be different from that in non-cerebral vessels. It has been shown that activation of sensory fibers to release calcitonin gene-related peptide is an important intermediate step in cerebral vasodilation by organic nitrates (WEI et al. 1992). Another side effect of nitrate therapy is skin reactions caused by the patch. These include erythema and inflammation at the site of patch application.

Sodium nitroprusside contains 5 moles of cyanide anions per mole of NO•. Thus, formation of cyanide anions during degradation is not surprising and is known to limit the clinical use of this drug by the development of methemoglobinemia (HALL and GUEST 1992).

In view of their vasodilator activity, nitrovasodilators are contraindicated in severe hypotension, shock, obstructive cardiomyopathy and toxic pulmonary edema. Close cardiovascular monitoring is recommended if nitrates are administered in acute myocardial infarction with low filling pressure or in patients suffering from orthostatic dysregulation.

References

Abrams J (1995) The role of nitrates in coronary heart disease. Arch Intern Med 155:357–364

Ahlner J, Andersson RGG, Torfgård K, Axelsson KL (1991) Organic nitrate esters: Clinical use and mechanisms of actions. Pharmacol Rev 43:351–423

Angelico M, Carli L, Piat C, Gentile S, Capocaccia L (1997) Effects of isosorbide-5-mononitrate compared with propranolol on first bleeding and long-term survival in cirrhosis. Gastroenterology 113:1632–1639

Armstrong PW, Walker DC, Burton JR, Parker JO (1975) Vasodilator therapy in acute myocardial infarction. A comparison of sodium nitroprusside and nitroglycerin. Circulation 52:1118–1122

Banerjee AK (1997) Treating anal fissure – glyceryl trinitrate ointment may remove the need for surgery. Br Med J 314:1638–1639

Bassenge E, Stuart DJ (1986) Effects of nitrates in various vascular sections and regions. Z Kardiol 75 (Suppl. 3):1–7

Bates JN, Baker MT, Guerra R, Jr., Harrison DG (1991) Nitric oxide generation from nitroprusside by vascular tissue. Evidence that reduction of the nitroprusside anion and cyanide loss are required. Biochem Pharmacol 42 (Suppl.):S157–S165

Bell E (1996) Nitroglycerin and uterine relaxation. Anesthesiology 85:683

Bennett BM, McDonald BJ, St.James (1992) Hepatic cytochrome P-450-mediated activation of rat aortic guanylyl cyclase by glyceryl trinitrate. J Pharmacol Exp Ther 261:716–723

Boesgaard S, Aldershvile J, Poulsen HE, Loft S, Anderson ME, Meister A (1994) Nitrate tolerance in vivo is not associated with depletion of arterial or venous thiol levels. Circ Res 74:115–120

Brien JF, McLaughlin BE, Breedon TH, Bennett BM, Nakatsu K, Marks GS (1986) Biotransformation of GTN occurs concurrently with relaxation of rabbit aorta. J Pharmacol Exp Ther 237:608–614

Brien JF, McLaughlin BE, Kobus SM, Kawamoto JH, Nakatsu K, Marks GS (1988) Mechanism of glyceryl trinitrate induced vasodilation. I. Relationship between drug biotransformation, tissue cGMP elevation and relaxation of rabbit aorta. J Pharmacol Exp Ther 244:322–327

Chrysant SG, Glasser SP, Bittar N, Shahidi FE, Danisa K, Ibrahim R, Watts LE, Garutti RJ, Ferraresi R, Casareto R (1993) Efficacy and safety of extended-release isosorbide mononitrate for stable effort angina pectoris. Am J Cardiol 72:1249–1256

Chung S-J, Fung H-L (1990) Identification of the subcellular site for nitroglycerin metabolism to nitric oxide in bovine coronary smooth muscle cells. J Pharmacol Exp Ther 253:614–619

Cohn JN (1996) The management of chronic heart failure. N Engl J Med 335:490–498

Cohn JN, Archibald DG, Ziesche S, Fraciosa JA, Harston WE, Tristani FE, Dunkman WB, Jacobs W, Francis GS, Flohr KH, et al. (1986) Effect of vasodilator therapy on mortality in chronic congestive heart failure. N Engl J Med 314:1547–1552

Cohn JN, Johnson G, Ziesche S, Cobb F, Francis G, Tristani F, Smith R, Dunkman WB, Loeb H, Wong M, Bhat G, Goldman S, Fletcher RD, Doherty J, Hughes CV, Carson P, Cintron G, Shabetai R, Haakenson C (1991) A comparison of enalapril with hydralazine-isosorbide dinitrate in the treatment of chronic congestive heart failure. N Engl J Med 325:303–310

Collins P, Griffith TM, Hendersson AH, Lewis MJ (1986) Endothelium-derived relaxing factor alters calcium-fluxes in rabbit aorta: a cyclic guanosine monophosphate mediated effect. J Physiol (Lond) 381:427–437

De Caterina R, Giannessi D, Crea F, Chiercha S, Bernini W, Gazzetti P, L'Abbate A (1984) Inhibition of platelet function by injectable isosorbide dinitrate. Am J Cardiol 53:1683–1687

De Mey JG, Vanhoutte PM (1982) Heterogeneous behavior of the canine arterial and venous wall: importance of the endothelium. Circ Res 51:439–447

DeMots H, Glasser SP (1989) Intermittent transdermal nitroglycerin therapy in the treatment of chronic stable angina. J Am Coll Cardiol 13:786–795

Diodati J, Théroux P, Latour JG, Lacoste L, Lam JYT, Waters D (1990) Effects of nitroglycerin at therapeutic doses on platelet aggregation in unstable angina pectoris and acute myocardial infarction. Am J Cardiol 66:683–688

Diodati JG, Cannon RO, III, Hussain N, Quyyumi AA (1995) Inhibitory effect of nitroglycerin and sodium nitroprusside on platelet activation across the coronary circulation in stable angina pectoris. Am J Cardiol 75:443–448

Feelisch M, Kelm M (1991) Biotransformation of organic nitrates to nitric oxide by vascular smooth muscle and endothelial cells. Biochem Biophys Res Commun 180:286–293

Feelisch M, Noack E (1987) Correlation between nitric oxide formation during degradation of organic nitrates and activation of guanylate cyclase. Eur J Pharmacol 139:19–30

Freeman JG, Barton JR, Record CO (1985) Effect of isosorbide dinitrate, verapamil and labetalol on portal pressure in cirrhosis. BMJ 291:561–562

Fung H-L (1993) Clinical pharmacology of organic nitrates. Am J Cardiol 72:9C–15C

Giles W, O'Callaghan S, Boura A, Walters W (1992) Reduction in human fetal umbilical-placental vascular resistance by glyceryl trinitrate. Lancet 340:856

Gilja OH, Hausken T, Bang CJ, Berstad A (1997) Effect of glyceryl trinitrate on gastric accommodation and symptoms in functional dyspepsia. Dig Dis Sci 42:2124–2131

Göbel EJAM, Hautvast RWM, Van Gilst WH, Spanjaard JN, Hillege HL, DeJongste MJL, Molhoek GP, Lie KI (1995) Randomised, double-blind trial of intravenous diltiazem versus glyceryl trinitrate for unstable angina pectoris. Lancet 346:1653–1657

Gruetter CA, Lemke SL (1985) Dissociation of cystein and glutathione levels from nitroglycerin induced relaxation. Eur J Pharmacol 111:85–92

Gruppo Italiano per lo Studio della Sopravvivenza nell'Infarto Miocardico (1994) GISSI-3: effects of lisinopril and transdermal glyceryl trinitrate singly and together on 6-week mortality and ventricular function after acute myocardial infarction. Lancet 343:1115–1122

Hall VA, Guest JM (1992) Sodium nitroprusside induced cyanide intoxication and prevention with sodium thiosulfate prophylaxis. Am J Crit Care 1:19–25

Harrison DG, Bates JN (1993) The nitrovasodilators: New ideas about old drugs. Circulation 87:1461–1467

Hassel B (1993) Treatment of biliary colic with nitroglycerin. Lancet 342:1305

Hill KE, Hunt RW, Jr., Jones R, Hoover RL, Burk RF (1992) Metabolism of nitroglycerin by smooth muscle cells. Involvement of glutathione and glutathione S-transferase. Biochem Pharmacol 43:561–566

Ignarro LJ, Lipton H, Edwards JC, Barricos WH, Hyman AL, Kadowitz PJ, Gruetter CA (1981) Mechanism of vascular smooth muscle relaxation by organic nitrates, nitrites, nitroprusside and nitric oxide: evidence for the involvement of S-nitrosothiols as active intermediates. J Pharmacol Exp Ther 218:739–749

ISIS-2 Second International Study of Infarct Survival Collaborative Group (1988) Randomised trial of intravenous streptokinase, oral aspirin, both, or neither among 17,187 cases of suspected acute myocardial infarction: ISIS-2. Lancet 2(8607):349–360

ISIS-4 Collaborative Group (1995) ISIS-4: a randomised factorial trial assessing early oral captopril, oral mononitrate, and intravenous magnesium sulphate in 58 050 patients with suspected acute myocardial infarction. Lancet 345:669–685

Jaffrani NA, Ehrenpreis S, Laddu A, Somberg J (1993) Therapeutic approach to unstable angina: nitroglycerin, heparin, and combined therapy. Am Heart J 126:1239–1242

Karlberg KE, Saldeen T, Wallin R, Henriksson P, Nyquist O, Sylvén C (1998) Intravenous nitroglycerin reduces ischaemia in unstable angina pectoris: a double-blind placebo-controlled study. J Intern Med 243:25–31

Kawamoto JH, McLaughlin BE, Brien JF, Marks GS, Nakatsu K (1990) Biotransformation of glyceryl trinitrate and elevation of cGMP precede glyceryl trinitrate induced vasodilation. J Cardiovasc Pharmacol 15:714–719

Keefer LK, Nims RW, Davies KM, Wink DA (1996) "NONOates" (1-substituted diazen-1-ium-1,2-diolates) as nitric oxide donors: convenient nitric oxide dosage forms. Methods Enzymol 268:281–293

Kenkare SR, Han C, Benet LZ (1994) Correlation of the response to nitroglycerin in rabbit aorta with the activity of the μ-class glutathione S-transferase. Biochem Pharmacol 48:2231–2235

Kojda G, Meyer W, Noack E (1993) Influence of endothelium and nitrovasodilators on free thiols and disulfides in porcine coronary smooth muscle. Eur J Pharmacol 250:385–394

Kojda G, Kottenberg K, Nix P, Schlüter KD, Piper HM, Noack E (1996) Low increase in cGMP induced by organic nitrates and nitrovasodilators improves contractile response of rat ventricular myocytes. Circ Res 78:91–101

Kojda G, Patzner M, Hacker A, Noack E (1998) Nitric oxide inhibits vascular bioactivation of glyceryl trinitrate. A novel mechanism to explain preferential venodilation of organic nitrates. Mol Pharmacol 53:547–545

Kowaluk EA, Seth P, Fung H-L (1992) Metabolic activation of sodium nitroprusside to nitric oxide in vascular smooth muscle. J Pharmacol Exp Ther 262:916–922

Kurz MA, Boyer TD, Whalen R, Peterson TE, Harrison DG (1993) Nitroglycerin metabolism in vascular tissue: Role of glutathione S-transferases and relationship between NO^x and NO_2^- formation. Biochem J 292:545–550

Lang D, Lewis MJ (1989) Endothelium-derived relaxing factor inhibits the formation of inositol triphosphate by rabbit aorta. J Physiol (Lond) 411:45–52

Lau DT-W, Benet LZ (1990) Nitroglycerin metabolism in subcellular fractions of rabbit liver. Dose dependency of glyceryl dinitrate formation and possible involvement of multiple isoenzymes of glutathione S-transferases. Drug Metab Dispos 18:292–297

Laursen JB, Mülsch A, Boesgaard S, Mordvintcev P, Trautner S, Gruhn N, Nielsen-Kudsk JE, Busse R, Aldershvile J (1996) In vivo nitrate tolerance is not associated with reduced bioconversion of nitroglycerin to nitric oxide. Circulation 94:2241–2247

Lees C, Campbell S, Jauniaux E, Brown R, Ramsay B, Gibb D, Moncada S, Martin JF (1994) Arrest of preterm labour and prolongation of gestation with glyceryl trinitrate, a nitric oxide donor. Lancet 343:1325–1326

Levacher S, Letoumelin P, Pateron D, Blaise M, Lapandry C, Pourriat JL (1995) Early administration of terlipressin plus glyceryl trinitrate to control active upper gastrointestinal bleeding in cirrhotic patients. Lancet 346:865–868

Liu Z, Brien JF, Marks GS, McLaughlin BE, Nakatsu K (1993) Lack of evidence for the involvement of cytochrome P-450 or other hemoproteins in metabolic activation of glyceryl trinitrate in rabbit aorta. J Pharmacol Exp Ther 264:1432–1439

Luman W, Pryde A, Heading RC, Palmer KR (1997) Topical glyceryl trinitrate relaxes the sphincter of Oddi. Gut 40:541–543

Lund JN, Scholefield JH (1997) A randomised, prospective, double-blind, placebo-controlled trial of glyceryl trinitrate ointment in treatment of anal fissure. Lancet 349:11–14

Lüscher TF, Diedrich D, Siebenmann R, Lehmann K, Stülz P (1988) Difference between endothelium-dependent relaxation in arterial and in venous coronary bypass grafts. N Engl J Med 319:462–467

Mangione NJ, Glasser SP (1994) Phenomenon of nitrate tolerance. Am Heart J 128:137–146

McDonald BJ, Bennett BM (1990) Cytochrome P-450 mediated biotransformation of organic nitrates. Can J Physiol Pharmacol 68:1552–1557

Mercier FJ, Dounas M, Bouaziz H, Lhuissier C, Benhamou D (1997) Intravenous nitroglycerin to relieve intrapartum fetal distress related to uterine hyperactivity: a prospective observational study. Anesth Analg 84:1117–1120

Merkel C, Gatta A, Donada C, Enzo E, Marin R, Amodio P, Torboli P, Angeli P, Cavallarin G, Sebastianelli G, Susanna S, Mazzaro C, Beltrame P, Bartoli G, Borsato L, Caruso N, Cielo R, Dabroi L, DeVenuto G, Donadon V, Spandri P, Bellumat A, Brosolo P, Caregaro L (1995) Long-term effect of nadolol or nadolol plus isosorbide-5-mononitrate on renal function and ascites formation in patients with cirrhosis. Hepatology 22:808–813

Mittal CK, Murad F (1982) Guanylate cyclase: Regulation of cyclic GMP metabolism. In: Nathanson JA, Kebabian JW (eds) Cyclic nucleotides I. Springer, Berlin Heidelberg New York, pp 225–260

Mohazzab-H. KM, Kaminski PM, Agarwal R, Wolin MS (1999) Potential role of a membrane-bound NADH oxidoreductase in nitric oxide release and arterial relaxation to nitroprusside. Circ Res 84:220–228

Mülsch A, Mordvintcev P, Bassenge E, Jung F, Clement B, Busse R (1995) In vivo spin trapping of glyceryl trinitrate-derived nitric oxide in rabbit blood vessels and organs. Circulation 92:1876–1882

Münzel T, Harrison DG (1997) Evidence for a role of oxygen-derived free radicals and protein kinase C in nitrate tolerance. J Mol Med 75:891–900

Münzel T, Giaid A, Kurz S, Stewart DJ, Harrison DG (1995a) Evidence for a role of endothelin 1 and protein kinase C in nitroglycerin tolerance. Proc Natl Acad Sci USA 92:5244–5248

Münzel T, Sayegh H, Freeman BA, Tarpey MM, Harrison DG (1995b) Evidence for enhanced vascular superoxide anion production in nitrate tolerance. A novel mechanism underlying tolerance and cross-tolerance. J Clin Invest 95:187–194

Murrel W (1879) Nitro-glycerine as a remedy for angina pectoris. Lancet 1:80–81

Needleman P, Johnson EM (1973) Mechanism of tolerance development to organic nitrates. J Pharmacol Exp Ther 184:709–713

Nigam R, Whiting T, Bennett BM (1993) Effect of inhibitors of glutathione S-transferase on glyceryl trinitrate activity in isolated rat aorta. Can J Physiol Pharmacol 71:179–184

Noack E, Feelisch M (1991) Molecular mechanism of nitrovasodilator bioactivation. In: Drexler H, Zeiher AM, Bassenge E, et al. (eds) Endothelial mechanism of vasomotor control. Steinkopff Verlag, Darmstadt, pp 37–50

Olesen J, Thomsen LL, Iversen H (1994) Nitric oxide is a key molecule in migraine and other vascular headaches. Trends Pharmacol Sci 15:149–153

Orlando RC, Bozymski EM (1973) Clinical and manometric effects of nitroglycerin in diffuse esophageal spasm. N Engl J Med 289:23–25

Parker JC, DiCarlo FJ, Davidson IW (1975) Comparative vasodilator effects of nitroglycerin, pentaerythritol trinitrate and biometabolites, and other organic nitrates. Eur J Pharmacol 31:29–37

Parker JD, Parker JO (1998) Nitrate therapy for stable angina pectoris. N Engl J Med 338:520–531

Parker JD, Farrell B, Fenton T, Cohanim M, Parker JO (1991) Counter-regulatory responses to continuous and intermittent therapy with nitroglycerine. Circulation 84:2336–2345

Parker JD, Parker AB, Farrell B, Parker JO (1995) Intermittent transdermal nitroglycerin therapy: Decreased anginal threshold during the nitrate-free interval. Circulation 91:973–978

Parker JO (1992) Update on nitrate tolerance. Br J Clin Pharmacol 34 (Suppl. 1):11S–14S

Parker JO, Farrell B, Lahey KA, Moe G (1987) Effects of intervals between doses on the development of tolerance to isosorbide dinitrate. N Engl J Med 316:1440–1444

Parker JO, Isosorbide-5-mononitrate Study Group (1993) Eccentric dosing with isosorbide-5-mononitrate in angina pectoris. Am J Cardiol 72:871–876

Parker JO, Amies MH, Hawkinson RW, Heilman JM, Hougham AJ, Vollmer MC, Wilson RR, Minitran Efficacy Study Group (1995) Intermittent transdermal nitroglycerin therapy in angina pectoris: Clinically effective without tolerance or rebound. Circulation 91:1368–1374

Pfitzer G, Hofman F, Disalvo J, Ruegg JC (1984) cGMP and cAMP inhibit tension development in skinned coronary arteries. Pflügers Arch 401:277–280

Pittrof R, Lees C, Thompson C, Pickles A, Martin JF, Campbell S (1996) Crossover study of glyceryl trinitrate patches for controlling pain in women with severe dysmenorrhoea. BMJ 312:884,

Pohl U, Holtz J, Busse R, Bassenge E (1986) Crucial role of endothelium in the vascular response to increased flow in vivo. Hypertension 8:37–44

Preckel B, Kojda G, Schlack W, Ebel D, Kottenberg K, Noack E, Thämer V (1997) Inotropic effects of glyceryl trinitrate and spontaneous NO-donors in the dog heart. Circulation 96:2675–2682

Purcell H, Mulcahy D, Fox K (1994) Nitrates in silent ischemia. Cardiovasc Drugs Ther 8:727–734

Redick LF, Livingston E, Bell E (1997) Sublingual aerosol nitroglycerin for uterine relaxation in attempted external version. Am J Obstet Gynecol 176:496–497

Risoe C, Simonsen S, Rootwelt K, Sire S, Smiseth OA (1992) Nitroprusside and regional vascular capacitance in patients with severe heart failure. Circulation 85(3):997–1002

Sakanashi M, Matsuzaki T, Aniya Y (1991) Nitroglycerin relaxes coronary artery of the pig with no change in glutathione content or glutathione S-transferase activity. Br J Pharmacol 103:1905–1908

Salvemini D, Mollace V, Pistelli A, Anggard E, Vane J (1992) Metabolism of glyceryl trinitrate to nitric oxide by endothelial cells and smooth muscle cells and its induction by *Escherichia coli* lipopolysaccharide. Proc Natl Acad Sci USA 89:982–986

Salvemini D, Currie MG, Mollace V (1996) Nitric oxide-mediated cyclooxygenase activation – a key event in the antiplatelet effects of nitrovasodilators. J Clin Invest 97:2562–2568

Schröder H, Schrör K (1990) Inhibitors of cytochrome P-450 reduce cGMP stimulation by glyceryl trinitrate in LLC-PK1 kidney epithelial cells. Arch Pharmacol 342:616–618

Schultz KD, Schultz K, Schultz G (1977) Sodium nitroprusside and other smooth muscle relaxants increase cyclic GMP levels in rat ductus deferens. Nature 265:750–751

Sellke FW, Myers PR, Bates JN, Harrison DG (1990) Influence of vessel size on the sensitivity of porcine microvessels to nitroglycerin. Am J Physiol 258:H515-H520369

Servent D, Delaforge M, Ducrocq C, Mansuy D, Lenfant M (1989) Nitric oxide formation during microsomal hepatic denitration of glyceryl trinitrate: involvement of cytochrome P-450. Biochem Biophys Res Commun 163:1210–1216

Shell WE, Kivowitz CF, Rubins SB, See J (1986) Mechanisms and therapy of silent myocardial ischemia: the effect of transdermal nitroglycerin. Am Heart J 112:222–229

Silber S, Vogler AC, Krause KH, Vogel M, Theisen K (1987) Induction and circumvention of nitrate tolerance applying different dosage intervals. Am J Med 83:860–870

Stamler JS, Loscalzo J (1991) The antiplatelet effects of organic nitrates and related nitroso compounds in vitro and in vivo and their relevance to cardiovascular disorders. J Am Coll Cardiol 18:1529–1536

Strauer BE, Scherpe A (1978) Ventricular function and coronary hemodynamics after intravenous nitroglycerin in coronary artery disease. Am Heart J 95:210–219

Thadani U, Maranda CR, Amsterdam E, Spaccavento L, Friedman RG, Chernoff R, Zellner S, Gorwit J, Hinderaker PH (1994) Lack of pharmacologic tolerance and

rebound angina pectoris during twice-daily therapy with isosorbide-5-mononitrate. Ann Intern Med 120:353–359
Twort CHC, van Breemen C (1988) Cyclic guanosine monophosphate enhanced sequestration of calcium by sarcoplasmatic reticulum in vascular smooth muscle. Circ Res 62:961–964
Weber W, Michaelis K, Luckow V, Kuntze U, Stalleicken D (1995) Pharmacokinetics and bioavailability of pentaerythrityl tetranitrate and two of its metabolites. Arzneim Forsch 45:781–784
Wei EP, Moskowitz MA, Boccalini P, Kontos HA (1992) Calcitonin gene-related peptide mediates nitroglycerin and sodium nitroprusside-induced vasodilation in feline cerebral arterioles. Circ Res 70:1313–1319
Wille HH, Sauer G, Tebbe U, Neuhaus KL, Kreuzer H (1980) Nitroglycerin and afterload: effects of aortic compliance and capacity of the Windkessel. Eur Heart J 1:445–452
Winbury MM, Howe BB, Weiss HR (1971) Effect of nitroglycerin and dipyridamole on epicardial and endocardial oxygen tension-further evidence for redistribution of myocardial blood flow. J Pharmacol Exp Ther 176:184–199
Yusuf S, MAC Mahon S, Collins R, Peto R (1988) Effect of intravenous nitrates on mortality in acute myocardial infarction: an overview of the randomised trials. Lancet 1(8594):1088–1092

CHAPTER 16
Therapeutic Potential of NOS Inhibitors in Septic Shock

P. VALLANCE, D. REES, and S. MONCADA

A. Introduction

Septic shock is a syndrome characterised by vascular collapse presumed to be initiated by infection. It is the most common cause of death in intensive care units and is associated with an overall mortality rate of at least 50% (ASTIZ and RACKOW 1998). The cause of death is a combination of relentlessly progressive hypotension and multiple organ failure, and even survivors of the initial event have a relatively poor prognosis, with increased morbidity and mortality continuing for months after the acute syndrome has resolved (QUARTIN et al. 1997). In this chapter, we discuss the possible roles of nitric oxide (NO) in the pathogenesis of sepsis and its complications and consider the therapeutic potential of NO synthase (NOS) inhibitors.

B. Clinical Features of Sepsis

Although systemic infection is probably the underlying cause of most cases of septic shock, there are few data that directly support this assumption. Microbiological evidence of active infection is detected in under 30% of patients at the time at which the diagnosis of septic shock is made (ILKKA et al. 1998); bacterial lipopolysaccharide (endotoxin) is detected in the blood of only a minority of patients, and the concentration of cytokines in the plasma varies greatly among individual patients. Interleukins (IL) 1β, 2, 6, 8 and 10, γ interferon (IFNγ), tumour necrosis factor α (TNF), platelet-activating factor and other mediators have been implicated in the genesis of the systemic inflammatory syndrome and septic shock (VAN DISSEL et al. 1998). Many of the important cytokines may be generated locally within inflammatory cells or within the vessel wall and, therefore, their concentration in blood may correlate poorly with their concentration at relevant biological sites. Nonetheless, there is some evidence that most patients with sepsis have elevated circulating concentrations of TNFα and IL-6, and that the concentrations of these cytokines (or their persistence) is loosely predictive of outcome or progression to multiple organ dysfunction (MARTIN et al. 1997; RHODES et al. 1998; VAN DISSEL et al. 1998). Together, the systemic and local cytokine responses to infection are thought to underlie many of the pathological features of septic shock.

I. Cardiovascular Changes

Early in the course of sepsis, vasodilatation occurs, and there is a compensatory increase in cardiac output. Subsequently, blood pressure drops, and it seems as though the increase in cardiac output is insufficient to compensate for the greatly reduced peripheral resistance. The failure of cardiac output to increase adequately is initially due to intravascular fluid loss but, later, is probably due to a reversible defect in cardiac-muscle function, since patients appear to develop a dilated cardiomyopathy, which returns to normal in survivors. Vasodilatation is seen in veins and resistance vessels and leads to pooling of blood in capacitance vessels and a substantial drop in systemic peripheral resistance. However, not all blood vessels dilate; indeed, some seem to constrict; in many patients, there is a rise in pulmonary vascular resistance due to constriction of pulmonary small arteries and arterioles. Although the significance of the cardiovascular changes is not fully understood, they probably contribute to maldistribution of blood flow and impaired delivery of oxygen and nutrients to vital organs or to vulnerable areas within organs. The degree of hypotension is an independent predictor of outcome in patients with sepsis (RHODES et al. 1998), but whether the vasodilatation serves some useful physiological function early in the course of sepsis is a matter of speculation.

II. Tissue Oxygenation

Oxygen delivery to tissues is determined by the state of oxygenation of the blood and the blood flow to the tissues. Early in the course of sepsis, overall oxygen delivery seems to be normal or even enhanced, due to the increased flow secondary to vasodilatation and increased cardiac output. Later, or in the most severe cases, as cardiac output drops, oxygen delivery ultimately falls. However, even during the hyperdynamic phase, tissue oxygenation seems to be impaired (ANNING et al. 1998). One possible reason is that the enhanced oxygen delivery assessed at a whole-body or organ level hides significant defects in blood flow and oxygenation at a local tissue level as blood is shunted away from vital tissues. An alternative explanation is that, although oxygen delivery is adequate, oxygen extraction and/or utilisation is in some way impaired (FINK and PAYEN 1996).

III. Tissue and Organ Damage

Any of the vital organs may be damaged during sepsis, but damage to the liver, lung, gut and kidney are usually particularly evident. In some studies, it has been estimated that significant foci of infection will be found in more than 90% of patients developing failure of two or more organ systems and that the development of organ dysfunction post-operatively is a reliable sign of occult infection. However, many studies indicate that, even in the absence of microbiologically proven infection, the sepsis syndrome is associated with a high rate of multiple organ failure. Both apoptosis and necrosis have been implicated in the organ dysfunction and, whilst some of the damage is reversible, irreversible damage can also occur.

C. NO in Experimental Models of Shock

Endotoxin from gram-negative bacteria and gram-positive wall fragments (including lipoteichoic acid) or pro-inflammatory cytokines (including IL-1β, IFNγ and TNFα) induce NO generation in a wide variety of cells and tissues (VALLANCE and MONCADA 1993). Injection of endotoxin into rats leads to expression of messenger-RNA-encoding inducible NOS (iNOS) in lung, liver, kidney, blood vessels, heart, gut and brain, with subsequent expression of iNOS protein and generation of large amounts of NO, as measured by production of nitrite and/or nitrate (NOx) (VALLANCE and MONCADA 1993; REES et al. 1998; Fig. 1). This sequence of events is thought to contribute to the pathogenesis of many of the features of septic shock.

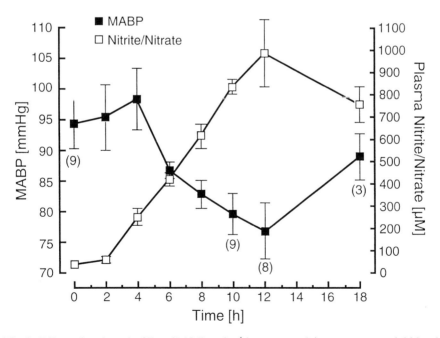

Fig. 1. Effect of endotoxin (*E. coli*, 12.5 mg kg^{-1} intravenously) on mean arterial blood pressure (*MABP*) and on concentrations of plasma nitrite and nitrate in the conscious mouse. Each *point* is the mean of 3–9 animals, where (*n*) represents numbers of survivors; *vertical lines* show the standard error of the mean

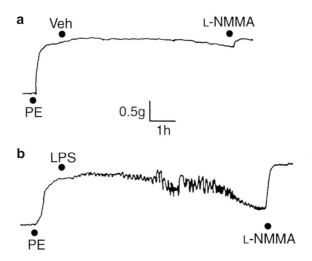

Fig. 2a,b. Induction of nitric oxide synthase in isolated rat thoracic aorta mounted in organ baths and preconstricted with phenylephrine (*PE*). **a** Rings incubated with vehicle (*Veh*) control for 8h followed by addition of N^G-monomethyl-L-arginine (*L-NMMA*; 3×10^{-4}M). **b** Rings incubated with lipopolysaccharide (*LPS*, $5\mu g\,ml^{-1}$) for 8h followed by addition of L-NMMA (3×10^{-4}M)

I. Cardiovascular Changes

Early experiments were undertaken using rings of rat aorta suspended in organ baths (REES et al. 1990; BOGLE and VALLANCE 1996; Fig. 2). Incubation with bacterial endotoxin induced a slowly developing vascular relaxation and hyporesponsiveness to vasoconstrictors. This change in tone was associated with de novo expression of iNOS and reversed by NOS inhibitors, such as N^G-monomethyl-L-arginine (L-NMMA; Chap. 7). Later, experiments were undertaken using different vessels – conduit arteries, resistance vessels and veins – and in models in vivo. These studies confirmed that the vasodilatation, hypotension and vascular hyporeactivity induced by endotoxin or certain pro-inflammatory cytokines are mediated by NO, occur in many different vessel types and are reversed by inhibitors of NOS. More recently, studies in mice genetically modified to lack the iNOS gene (iNOS knockout mice) (REES et al. 1998; Fig. 3), together with those using selective inhibitors of iNOS (REES 1999), have confirmed that NO generated from iNOS mediates most of the vasodilatation induced by endotoxin or cytokines in rodents. Although iNOS is expressed in the pulmonary vessels, the vasodilator effects of the NO generated seem to be outweighed by the induction of constrictor tone, which may be mediated by endothelin (YAMAMOTO et al. 1997).

II. Tissue Oxygenation

One of the ways in which NO may be transformed from a physiological mediator to a pathophysiological entity may be through the actions it has on

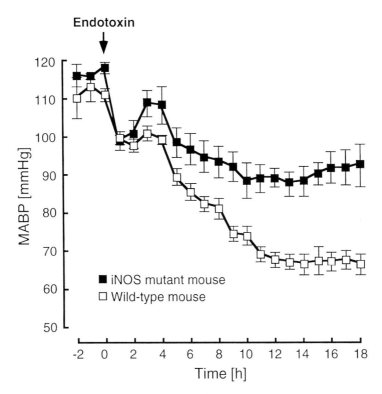

Fig. 3. Effect of endotoxin (*E. coli*, 3 mg kg^{-1} intravenously) on mean arterial blood pressure (*MABP*) in conscious wild-type mice ($n = 5$) and mice lacking inducible nitric oxide synthase ($n = 4$). Each *point* is the mean of 4–5 animals; *vertical lines* show the standard error of the mean (REES et al. 1998)

mitochondrial function. At low physiological concentrations, NO inhibits cytochrome c oxidase (complex IV) in a reversible manner that is competitive with oxygen (CLEETER et al. 1994). Prolonged exposure to NO, however, results in a gradual and persistent inhibition of complex I that appears to result from *S*-nitrosation of critical thiols in the enzyme complex (CLEMENTI et al. 1998). At high concentrations of NO, peroxynitrite can be formed from the interaction between NO and superoxide anion. Peroxynitrite can contribute to the cytotoxic actions of NO by inhibiting complexes I–III irreversibly (LIZASOAIN et al. 1996). It is likely that such a sequence of events is responsible for the defective extraction of oxygen by tissues and organs during septic shock (FINK and PAYEN 1996), a defect that can be reproduced in cells in vitro (ODDIS and FINKEL 1995).

III. Tissue and Organ Damage

Increased generation of NO has the potential to cause tissue and organ damage through a number of distinct mechanisms (Fig. 4). Cellular damage

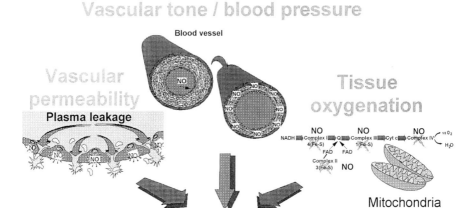

Fig. 4. Schematic representation of the potential mechanisms leading to multiple organ failure and death following overproduction of nitric oxide

might occur through effects on mitochondrial enzymes (see above), interaction of NO with iron–sulphur clusters in enzymes (HIBBS et al. 1990), direct damage to DNA (GREEN et al. 1982), or nitration of tyrosine residues on critical proteins (BECKMAN and KOPPENOL 1996). NO may either induce or inhibit apoptosis.

D. NO in Clinical Sepsis

Nitrate is a stable breakdown product of NO, and the concentration of nitrite and/or nitrate (NOx) in the blood has been used to estimate NO generation. Elevated plasma nitrate levels are seen in patients with fever (GREEN et al. 1982), gastroenteritis (DYKHUIZEN et al. 1996) or sepsis (OCHOA et al. 1991; EVANS et al. 1993) and following immunotherapy with IL-2 (HIBBS et al. 1992). In the latter situation, studies with ^{15}N-arginine have confirmed that the nitrate derives from arginine and, therefore, from NO. In some studies of patients with sepsis, the circulating concentrations of NOx correlate with other markers of systemic inflammation, disease severity or outcome (ARNALICH et al. 1996; DOUGHTY et al. 1996, 1998; ENDO et al. 1996; GROENEVELD et al. 1996; RIXEN et al. 1997). However, in other studies, NOx concentrations seem to be independent of disease severity, and the increase could be fully accounted for by changes in renal function leading to impaired excretion of nitrate (OCHOA et al. 1991; ZWAVELING et al. 1996; DUKE et al. 1997). In most studies, the concentration of NOx in plasma seems to be in the order of 30–40 μM in control subjects and in the order of 60–100 μM in patients with severe septic shock.

This maximal two- to threefold increase in NOx levels seen in septic shock in humans is much smaller than the 10-fold or greater increase seen in rodent models of endotoxaemia (VALLANCE and CHARLES 1998). The ability of certain microbes to convert nitrite and nitrate to other products may explain the lower plasma concentrations of NOx in adult human septic shock and in animal studies with live organisms compared with studies using endotoxin or heat-inactivated organisms (BRAUN and ZUMFT 1991). It is also possible that timing of sample collection is critical, since NOx concentrations may be increased during the initial phase of sepsis and decrease thereafter as anti-inflammatory cytokines are generated (GROENEVELD et al. 1997). Interestingly, in septic neonates, very high levels of NOx have been reported (SHI et al. 1993), and these are similar to the levels seen in rodent models.

I. iNOS Induction in Humans

Liver biopsies of patients with sepsis have confirmed the presence of mRNA encoding iNOS in this organ (VALLANCE and CHARLES 1998), but there are no reports of induction of iNOS in other tissues of septic patients. Polymorphonuclear cells from patients with sepsis have been found to have either decreased (CARRERAS et al. 1994) or increased (GOODE et al. 1995) NO generation, but the enzymatic source of the NO was not characterised. However, iNOS mRNA and protein have been detected in children in areas in which malaria is endemic (ANSTEY et al. 1996) and in inflammatory cells in the urine of patients with urinary-tract infection (WHEELER et al. 1997). Furthermore, mononuclear cells of patients with hepatitis C infection express virtually no mRNA for iNOS but, once the patients are treated with IFNα, NOS activity, iNOS protein and mRNA for iNOS increase, consistent with induction of functionally active iNOS in these cells (SHARARA et al. 1997). Thus, it seems as though iNOS induction does occur in certain cells and tissues during certain types of infection.

II. Cardiovascular Changes

Several clinical studies have confirmed that NOS inhibitors cause substantial vasoconstriction and restore blood pressure in patients with septic shock (Fig. 5). L-NMMA produces a dose-dependent increase in systemic vascular resistance and arterial blood pressure, and this is accompanied by a fall in cardiac output. It is assumed that the fall in cardiac output is a reflex response, secondary to the increased resistance. Although NO may have negative inotropic effects and induction of iNOS in the heart is thought to be a mechanism of cardiac dysfunction in models of endotoxaemia (VALLANCE and MONCADA 1993), there is no evidence, from clinical studies, that inhibition of NOS improves cardiac performance. In addition to causing constriction of systemic vessels, inhibition of NOS causes pulmonary vasoconstriction, leading to an increase in pulmonary vascular resistance. As in the systemic circulation,

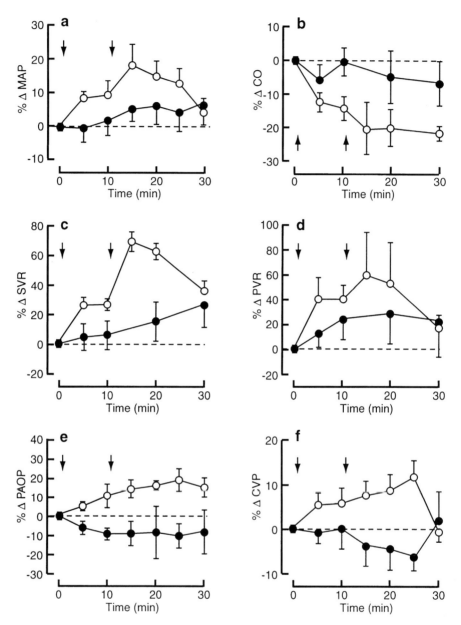

Fig. 5a–f. Haemodynamic effects of bolus administration of N^G-monomethyl-L-arginine (L-NMMA). Effects of two injections of L-NMMA (0.3 mg kg^{-1} and 1.0 mg kg^{-1}; *open circles*) or placebo (physiological saline; *closed circles*) on mean arterial pressure (**a**), cardiac output (**b**), systemic vascular resistance (**c**), pulmonary vascular resistance (**d**), pulmonary artery occlusion pressure (**e**) and central venous pressure (**f**). The time of each injection is marked by an *arrow* (PETROS et al. 1994)

the fall in cardiac output partially offsets the increase in resistance, so the overall effect on pulmonary arterial pressure is relatively small. The constrictor response to L-NMMA is not confined to the arterial circulation, and venoconstriction is also seen. Treatment with L-NMMA causes a dose-dependent increase in central venous pressure which appears to be caused by venoconstriction (PETROS et al. 1994). Since venoconstriction to L-NMMA is not seen in healthy individuals, in whom veins do not appear to generate NO basally (VALLANCE et al. 1989), the effect of L-NMMA is consistent with enhanced generation of NO in the venous circulation during sepsis. Overall, inhibition of NOS generation in septic shock causes a potentially favourable change in haemodynamics and allows reduction in concomitant pressor therapy (PETROS et al. 1994). However, the increase in pulmonary vascular resistance and the fall in cardiac output might be disadvantageous in certain situations.

The inhibitors used in the treatment of septic shock are not isoform-selective. Since L-NMMA will inhibit constitutive endothelial NOS (eNOS), neuronal NOS and iNOS, it has not been possible to conclude with certainty that the haemodynamic effects seen were due to inhibition of iNOS. Studies of human endothelial cells in vitro suggest that cytokines can induce NO generation from endothelial cells through activation of constitutive eNOS. Certain pro-inflammatory cytokines induce expression of the enzyme guanosine triphosphate cyclohydrolase 1, the first step in the formation of tetrahydrobiopterin (BH_4), which is an essential co-factor for all three isoforms of NOS (BHAGAT et al. 1999). It appears as though the increased generation of BH_4 is sufficient to stimulate NO generation from eNOS. Recently it has been shown that a similar mechanism occurs in human veins exposed to IL-β in vivo, and the NO generated is sufficient to cause hyporesponsiveness to vasoconstrictors and blunt the effects of the sympathetic nervous system (BHAGAT et al. 1999). Thus, it is possible that some of the excess NO generated in human sepsis originates from eNOS rather than from iNOS.

III. Tissue Oxygenation

There have been no detailed studies of tissue oxygenation following administration of NOS inhibitors to patients with septic shock. However, following administration of L-NMMA, oxygen delivery falls due to the increased vascular resistance and fall in tissue blood flow; nevertheless, oxygen consumption appears to remain constant (PETROS et al. 1994). These data are consistent with an increased efficiency of oxygen extraction following administration of L-NMMA, but further studies are required to explore this directly.

IV. Tissue and Organ Damage

The evidence that NO contributes to cellular damage in patients with septic shock is limited. In the lungs of patients with sepsis-induced lung injury, nitrotyrosine residues are found in the interstitium, epithelium, inflammatory cells and vascular endothelium (Kooy et al. 1995). Nitrotyrosine staining is considered a marker of peroxynitrite formation, and these data are consistent with a role for NO in lung damage in sepsis. The clinical trials undertaken so far do not indicate any significant benefit of NOS inhibition in terms of tissue or organ damage.

E. Outcome Studies

Initial studies of L-NMMA in septic shock in humans were too small to draw any conclusions about survival. However, it was demonstrated that, whilst L-NMMA produced significant haemodynamic effects, it did not appear to alter platelet function and had no major effects on renal or hepatic function. These early findings were subsequently reproduced in a large phase-II study in 312 patients. In this double-blind placebo-controlled study in which patients were randomly allocated to L-NMMA or placebo in addition to conventional therapy, there were no apparent adverse effects on several indices of organ function (Grover et al. 1995). Overall, the outcome in this study did not differ significantly between patients receiving L-NMMA and the control group. A more recent, phase-III study has been undertaken in a larger number of patients. This phase-III study was stopped after a trend to excess mortality was observed in the L-NMMA-treatment group. Although the full details of the study have yet to be published, there appeared to be a trend towards benefit at doses of 10mg/kg/h or below. Above this dose, there was a trend towards excess mortality.

F. Conclusions

NO is generated in large amounts in animal models of sepsis and in patients with septic shock. Studies in animals and in vitro indicate that the increased production of NO contributes to vascular collapse and has the potential to affect tissue oxygen utilisation and induce tissue damage. In animal models (including in primates) (Schlag et al. 1997), even late treatment with NOS inhibitors may be beneficial to survival if the drugs are given in low doses and in concert with other appropriate measures. Selective inhibitors of iNOS seem to be easier to use, may produce a greater effect on survival in animal models and have the theoretical advantage of targeting the "pathological" NO whilst leaving the "physiological" NO unaffected. Studies in patients are complicated by the heterogeneity of the patient population and the diverse causes of the septic shock syndrome. However, it seems clear that the enhanced generation of NO in sepsis contributes to cardiovascular instability and is potentially detrimental. NOS inhibitors that are not isoform selective, such as L-NMMA, restore cardiovascular stability but, at least when used in large doses, may increase mortality. The mechanisms underlying these effects are not clear but might include impaired tissue perfusion due to excessive vasoconstriction, excessive suppression of cardiac contractility or inhibition of "physiological" NO, which might be exerting cytoprotective effects. Selective inhibitors of iNOS may prove more effective in septic shock, but it is possible that these drugs will also have detrimental effects if used in doses that fully suppress NO generation from iNOS. Furthermore, there is the possibility that some of the "pathological" NO may originate from eNOS rather than iNOS, and it may be that other strategies will need to be adopted to decrease damaging formation of NO in human sepsis.

References

Anning PB, Sair M, Evans TW (1998) Microvascular regulation of tissue oxygenation in sepsis. In: Vincent JL (ed) 1998 Yearbook of intensive care and emergency medicine. Springer, Berlin Heidelberg New York, pp 153–160

Anstey NM, Weinberg JB, Hassanali MY, Mwaikambo ED, Manyenga D, Misukonis MA, Arnelle DR, Hollis D, McDonald MI, Granger DL (1996) Nitric oxide in Tanzanian children with malaria: inverse relationship between malaria severity and nitric oxide production/nitric oxide synthase type 2 expression. J Exp Med 184:557–567

Arnalich F, Hernanz A, Jimenez M, Lopez J, Tato E, Vazquez JJ, Montiel C (1996) Relationship between circulating levels of calcitonin gene-related peptide, nitric oxide metabolites and hemodynamic changes in human septic shock. Regul Pept 65: 115–121

Astiz ME, Rackow EC (1998) Septic shock. Lancet 351:1501–1505

Beckman JS, Koppenol WH (1996) Nitric oxide, superoxide, and peroxynitrite: the good, the bad, and ugly. Am J Physiol 271:C1424–1437

Bhagat K, Hingorani AD, Charles IG, Vallance P (1999) Cytokine induced venodilatation in humans in vivo: eNOS masquerading as iNOS. Cardiovascular Research 41:754–764

Bogle RG, Vallance P (1996) Functional effects of econazole on inducible nitric oxide synthase: production of a calmodulin-dependent enzyme. Br J Pharmacol 117: 1053–1058

Braun C, Zumft WG (1991) Marker exchange of the structural genes for nitric oxide reductase blocks the denitrification pathway of *Pseudomonas stutzeri* at nitric oxide. J Biol Chem 266:22785–22788

Carreras MC, Catz SD, Pargament GA, Del Bosco CG, Poderoso JJ (1994) Decreased production of nitric oxide by human neutrophils during septic multiple organ dysfunction syndrome. Comparison with endotoxin and cytokine effects on normal cells. Inflammation 18:151–161

Cleeter MW, Cooper JM, Darley Usmar VM, Moncada S, Schapira AH (1994) Reversible inhibition of cytochrome c oxidase, the terminal enzyme of the mitochondrial respiratory chain, by nitric oxide. Implications for neurodegenerative diseases. FEBS Lett 345:50–54

Clementi E, Brown GC, Feelisch M, Moncada S (1998) Persistent inhibition of cell respiration by nitric oxide: crucial role of S-nitrosylation of mitochondrial complex I and protective action of glutathione. Proc Natl Acad Sci USA 95:7631–7636

Doughty LA, Kaplan SS, Carcillo JA (1996) Inflammatory cytokine and nitric oxide responses in pediatric sepsis and organ failure. Crit Care Med 24:1137–1143

Doughty LA, Carcillo JA, Kaplan S, Janosky J (1998) Plasma nitrite and nitrate concentrations and multiple organ failure in pediatric sepsis. Crit Care Med 26: 157–162

Duke T, South M, Stewart A (1997) Activation of the L-arginine nitric oxide pathway in severe sepsis. Arch Dis Child 76:203–209

Dykhuizen RS, Masson J, McKnight G, Mowat AN, Smith CC, Smith LM, Benjamin N (1996) Plasma nitrate concentration in infective gastroenteritis and inflammatory bowel disease. Gut 39:393–395

Endo S, Inada K, Nakae H, Arakawa N, Takakuwa T, Yamada Y, Shimamura T, Suzuki T, Taniguchi S, Yoshida M (1996) Nitrite/nitrate oxide (NOx) and cytokine levels in patients with septic shock. Res Commun Mol Pathol Pharmacol 91: 347–356

Evans T, Carpenter A, Kinderman H, Cohen J (1993) Evidence of increased nitric oxide production in patients with the sepsis syndrome. Circ Shock 41:77–81

Fink MP, Payen D (1996) The role of nitric oxide in sepsis and ARDS: synopsis of a roundtable conference held in Brussels on 18–20 March 1995. Intensive Care Med 22:158–165

Goode HF, Howdle PD, Walker BE, Webster NR (1995) Nitric oxide synthase activity is increased in patients with sepsis syndrome. Clin Sci Colch 88:131–133

Green LC, Wagner DA, Glogowski J, Skipper PL, Wishnok JS, Tannenbaum SR (1982) Analysis of nitrate, nitrite, and [15N]nitrate in biological fluids. Anal Biochem 126:131–138

Groeneveld PH, Kwappenberg KM, Langermans JA, Nibbering PH, Curtis L (1996) Nitric oxide (NO) production correlates with renal insufficiency and multiple organ dysfunction syndrome in severe sepsis. Intensive Care Med 22:1197–1202

Groeneveld PH, Kwappenberg KM, Langermans JA, Nibbering PH, Curtis L (1997) Relation between pro- and anti-inflammatory cytokines and the production of nitric oxide (NO) in severe sepsis. Cytokine 9:138–142

Grover R, Zaccardelli D, Colice G, Guntupalli K, Watson D, Vincent J (1995) The cardiovascular effects of 546C88 in human septic shock. Intensive Care Med 21: S21

Hibbs JB Jr, Taintor RR, Vavrin Z, et al (1990) Synthesis of nitric oxide from a terminal guanidino nitrogen atom of L-arginine: a molecular mechanism regulating cellular proliferation that targets intracellular iron. In: Moncada S, Higgs EA (eds) Nitric oxide from L-arginine: a bioregulatory system. Elsevier, Amsterdam, pp 189–223

Hibbs JB Jr, Westenfelder C, Taintor R, Vavrin Z, Kablitz C, Baranowski RL, Ward JH, Menlove RL, McMurry MP, Kushner JP, et al (1992) Evidence for cytokine-inducible nitric oxide synthesis from L-arginine in patients receiving interleukin-2 therapy . J Clin Invest 89:867–877

Ilkka L, Parviainen I, Takala J (1998) Do we really know the epidemiology and clinical course of sepsis? In: Vincent JL (ed) 1998 Yearbook of intensive care and emergency medicine. Springer, Berlin Heidelberg New York, pp 229–237

Kooy NW, Royall JA, Ye YZ, Kelly DR, Beckman JS (1995) Evidence for in vivo peroxynitrite production in human acute lung injury. Am J Respir Crit Care Med 151:1250–1254

Lizasoain I, Moro MA, Knowles RG, Darley Usmar V, Moncada S (1996) Nitric oxide and peroxynitrite exert distinct effects on mitochondrial respiration which are differentially blocked by glutathione or glucose. Biochem J 314:877–880

Martin C, Boisson C, Haccoun M, Thomachot L, Mege JL (1997) Patterns of cytokine evolution (tumor necrosis factor-alpha and interleukin-6) after septic shock, hemorrhagic shock, and severe trauma. Crit Care Med 25:1813–1819

Ochoa JB, Udekwu AO, Billiar TR, Curran RD, Cerra FB, Simmons RL, Peitzman AB (1991) Nitrogen oxide levels in patients after trauma and during sepsis. Ann Surg 214:621–626

Oddis CV, Finkel MS (1995) Cytokine-stimulated nitric oxide production inhibits mitochondrial activity in cardiac myocytes. Biochem Biophys Res Commun 213: 1002–1009

Petros A, Lamb G, Leone A, Moncada S, Bennett D, Vallance P (1994) Effects of a nitric oxide synthase inhibitor in humans with septic shock. Cardiovasc Res 28: 34–39

Quartin AA, Schein RM, Kett DH, Peduzzi PN (1997) Magnitude and duration of the effect of sepsis on survival. Department of Veterans Affairs Systemic Sepsis Cooperative Studies Group. JAMA 277:1058–1063

Rees DD (1999) Inhibition of the overproduction of nitric oxide in septic shock using N^G-methyl-L-arginine. In: Schlag G, Redl H (eds) Shock, sepsis and organ failure. Springer, Berlin Heidelberg New York, pp 1–21

Rees DD, Cellek S, Palmer RM, Moncada S (1990) Dexamethasone prevents the induction by endotoxin of a nitric oxide synthase and the associated effects on vascular tone: an insight into endotoxin shock. Biochem Biophys Res Commun 173:541–547

Rees DD, Monkhouse JE, Cambridge D, Moncada S (1998) Nitric oxide and the haemodynamic profile of endotoxin shock in the conscious mouse. Br J Pharmacol 124:540–546

Rhodes A, Newman PJ, Bennett ED (1998) Prognostic markers in sepsis. In: Vincent JL (ed) 1998 Yearbook of intensive care and emergency medicine. Springer, Berlin Heidelberg New York, pp 238–246

Rixen D, Siegel JH, Espina N, Bertolini M (1997) Plasma nitric oxide in post-trauma critical illness: a function of "sepsis" and the physiologic state severity classification quantifying the probability of death. Shock 7:17–28

Schlag G, Redl H, Gasser H, et al (1997) Treatment with the NO synthase inhibitor, 546C88 following *E. coli* infusion is beneficial in a baboon model of septic shock. In: Schlag G, Redl H (eds) Shock, sepsis and organ failure. Springer, Berlin Heidelberg New York, pp 23–30

Sharara AI, Perkins DJ, Misukonis MA, Chan SU, Dominitz JA, Weinberg JB (1997) Interferon (IFN)-alpha activation of human blood mononuclear cells in vitro and in vivo for nitric oxide synthase (NOS) type 2 mRNA and protein expression: possible relationship of induced NOS2 to the anti-hepatitis C effects of IFN-alpha in vivo. J Exp Med 186:1495–1502

Shi Y, Li HQ, Shen CK, Wang JH, Qin SW, Liu R, Pan J (1993) Plasma nitric oxide levels in newborn infants with sepsis. J Pediatr 123:435–438

Vallance P, Charles I (1998) Nitric oxide as an antimicrobial agent: does NO always mean NO? Gut 42:313–314

Vallance P, Moncada S (1993) Role of endogenous nitric oxide in septic shock. New Horiz 1:77–86

Vallance P, Collier J, Moncada S (1989) Nitric oxide synthesised from L-arginine mediates endothelium dependent dilatation in human veins in vivo. Cardiovasc Res 23:1053–1057

Van Dissel JT, van Langevelde P, Westendorp RG, Kwappenberg K, Frolich M (1998) Anti-inflammatory cytokine profile and mortality in febrile patients. Lancet 351:950–953

Wheeler MA, Smith SD, Garcia Cardena G, Nathan CF, Weiss RM, Sessa WC (1997) Bacterial infection induces nitric oxide synthase in human neutrophils. J Clin Invest 99:110–116

Yamamoto S, Burman HP, O'Donnell CP, Cahill PA, Robotham JL (1997) Endothelin causes portal and pulmonary hypertension in porcine endotoxemic shock. Am J Physiol 272:H1239–49

Zwaveling JH, Maring JK, Moshage H, Van Ginkel RJ, Hoekstra HJ, Schraffordt Koops H, Donse IF, Girbes AR (1996) Role of nitric oxide in recombinant tumor necrosis factor-alpha-induced circulatory shock: a study in patients treated for cancer with isolated limb perfusion. Crit Care Med 24:1806–1810

CHAPTER 17
Inhalation Therapy with Nitric Oxide Gas

D. Keh, H. Gerlach, and K. Falke

A. Introduction

Different diseases, such as acute lung injury (ALI) or its aggravated form, acute respiratory-distress syndrome (ARDS), persistent pulmonary hypertension of the newborn (PPHN), and chronic pulmonary hypertension, are characterized by an increase of pulmonary vascular resistance (PVR) and concomitant impairment of gas exchange. Aiming to dilate pulmonary vessels, systemic application of vasodilators, such as nitroprusside, prostacyclin, or nitroglycerin, was used but was found to be less beneficial or even harmful in patients with respiratory failure. Systemic infusion of vasodilators induced both pulmonary and systemic vasodilation, promoting arterial hypotension; dilation of pulmonary vessels was accompanied by deterioration of arterial oxygenation due to an increase in venous admixture by unselective perfusion of less ventilated areas (Radermacher et al. 1989). When, in 1987, nitric oxide (NO) was identified as the biological equivalent of endothelium-derived relaxing factor (Ignarro et al. 1987; Palmer et al. 1987), investigators soon focused on NO for use as a more selective pulmonary vasodilator. Released to ventilated alveoli, lipophilic NO should, like endothelium-derived NO, rapidly diffuse into adjacent smooth-muscle cells, stimulating cyclic guanosine 3′,5′-monophosphate (cGMP) and relaxing vascular smooth muscles near ventilated areas. Unlike systemic vasodilators, NO was also presumed to be rapidly scavenged by hemoglobin (Hb), thus limiting the effects on pulmonary circulation. Since the beginning of this decade, tremendous progress has been made in introducing NO-inhalation therapy as a therapeutic option for the treatment of pulmonary hypertension and acute respiratory failure, and commercially available NO-delivery systems and monitoring devices now allow its use outside of specialized centers. In the following sections, the effects of low doses of inhaled NO on hemodynamics and pulmonary gas exchange under different conditions are described, concentrating on the most widespread use in ARDS and PPHN. Special attention is given to potential adverse effects of NO inhalation, delivery, and monitoring issues, and formation of toxic higher nitrogen oxides.

B. Therapy with NO Gas

I. NO Inhalation in ARDS Patients

1. Introduction

ARDS, first described by ASHBAUGH in 1967, is characterized by a generalized inflammation of lung parenchyma, triggered by exogenous (aspiration) or endogenous (sepsis) insults. Major characteristics are non-cardiogenic pulmonary edema, pulmonary hypertension, reduction of lung compliance, and systemic hypoxemia due to pulmonary ventilation/perfusion mismatch and increased right-to-left shunt. In addition to treatment of the underlying causes of the disease, common therapeutic strategies focus on minimizing iatrogenic lung damage due to mechanical ventilation, i.e., high inspiratory oxygen concentrations (F_IO_2) and barotrauma/volutrauma are avoided by using pressure-controlled ventilation with positive end expiratory pressure (PEEP), permissive hypercapnia, prone position, fluid restriction, and veno-venous extracorporeal membrane oxygenation (ECMO). Treatment of pulmonary hypertension is critical, since increased PVR aggravates both lung edema and right ventricular afterload.

In 1988, the use of NO as a pulmonary vasodilator was first described by HIGENBOTTAM (HIGENBOTTAM et al. 1988; PEPKE-ZABA et al. 1991). They reported that, in patients with primary pulmonary hypertension, inhalation of 40 ppm NO reduced PVR without systemic effects, whereas infusion of prostacyclin reduced both pulmonary and systemic vascular resistance (SVR). These observations led to the concept that NO acts as a selective pulmonary vasodilator, since induction of cGMP formation and smooth-muscle relaxation should be locally restricted by rapid scavenging of NO in the circulation by Hb. This hypothesis was confirmed with several animal models. In awake lambs, effects of inhaled NO (5–80 ppm) on hypoxic pulmonary vasoconstriction (HPV) and thromboxane-induced acute pulmonary hypertension, were investigated (FRATACCI et al. 1991; FROSTELL et al. 1991). When animals were exposed to 40–80 ppm NO, pulmonary hypertension was reversed within 3 min and returned to baseline within 6 min after cessation of NO. SVR remained unchanged during NO inhalation but decreased during systemic application of the NO-donor nitroprusside, which similarly prevented pulmonary hypertension. Furthermore, 5–80 ppm NO dose-dependently induced pulmonary vasodilation, which was maintained without tachyphylaxis whereas, in controls without pulmonary hypertension, NO inhalation produced no hemodynamic changes. Using the multiple inert-gas-elemination technique (MIGET) in mechanically ventilated hypoxic sheep, it could be demonstrated that improvement of oxygenation during NO inhalation was due to reversal of HPV and, importantly, redistribution of blood flow to better ventilated alveoli (PISON et al. 1993). Further experiments confirmed selective reversal of increased PVR in different animal models of acute pulmonary hypertension and lung injury (BERGER et al. 1993; TÖNZ et al. 1993; WEITZBERG et al. 1993). Furthermore, in awake spontaneously breathing humans, inhalation of 40ppm NO in normal air for 6min produced no changes in pulmonary or systemic hemodynamics. On inhalation of hypoxic air (12% O_2), pulmonary-artery pressure (PAP) increased from 14.7 ± 0.8mmHg to 19.8 ± 0.9mmHg, and arterial oxygen tension (PaO_2) decreased from 106 ± 4mmHg to 47 ± 2mmHg. Inhalation of 40ppm NO during hypoxia completely reversed HPV and improved oxygenation without changing SVR or mean arterial pressure (MAP) (FROSTELL et al. 1993a). Taken together, there was increasing evidence that inhaled NO could also be used to treat pulmonary hypertension in patients with ARDS.

2. Acute Effects of NO Inhalation in Patients with ARDS

In 1991, FALKE et al. first reported the application of inhaled NO to treat pulmonary hypertension in a patient with severe ARDS. During the inhalation of NO (18ppm and 36ppm), hemodynamic and gas exchange parameters were registered online, and the acute effects of NO were compared with those of prostacyclin infusion (Fig. 1). Both prostacylin and NO induced reduction of pulmonary resistance to a similar extent but, unexpectedly and in contrast to prostacyclin, a rise of arterial oxygenation and decrease of pulmonary right-to-left shunt occurred during NO inhalation. In a subsequent prospective study by ROSSAINT et al. (1993), these first observations were confirmed in a larger series of patients with severe ARDS (Fig. 2). Inhalation of 18ppm NO significantly reduced PVR by 71 ± 17 dyne $\times$ s $\times$ cm^{-5} ($P = 0.008$) and decreased right-to-left shunt from $36 \pm 5\%$ to $31 \pm 5\%$ ($P = 0.028$), while the ratio of PaO_2 to the inspired oxygen fraction (PaO_2/F_IO_2) increased from 152 ± 15mmHg to 199 ± 23mmHg ($P = 0.008$) without effects on cardiac output (CO) or SVR. During prostacyclin infusion, PVR also decreased by 102 ± 30 dyne$\times$s $\times$cm^{-5} ($P = 0.011$) but, in contrast to the case with NO inhalation, SVR decreased by 152 ± 34 dyne $\times$ s $\times$ cm^{-5} ($P = 0.002$), PaO_2/F_IO_2 decreased by 26 ± 7mmHg ($P = 0.005$), right-to-left shunt increased by $8 \pm 2\%$ ($P = 0.011$), and CO increased by 1.3 ± 0.4l/min. In this study, no differences were found between the effects of 18ppm and 36ppm NO. Using the inert-gas analysis, it was demonstrated that the beneficial effects on oxygenation were caused by redistribution of pulmonary blood flow away from non-ventilated regions of the lungs toward ventilated areas, thereby improving the matching of ventilation and perfusion (Fig. 3). These beneficial effects of NO were maintained during long-term observation without tachyphylaxis (Fig. 4); the longest treatment period in our patients was 53days. Further studies confirmed the hypothesis that modulation (by inhaled NO) of ventilation–perfusion mismatch in ARDS is selectively caused in vascular smooth muscles of ventilated regions by increased cGMP synthesis (ROVIRA et al. 1994).

In a sheep model of lavage-induced ALI, inhalation of 60ppm NO decreased pulmonary hypertension and venous admixture and increased PaO_2 without causing systemic effects. When endogenous NO synthesis was blocked with N^G-nitro-L-arginine methyl ester (L-NAME), PVR and SVR increased.

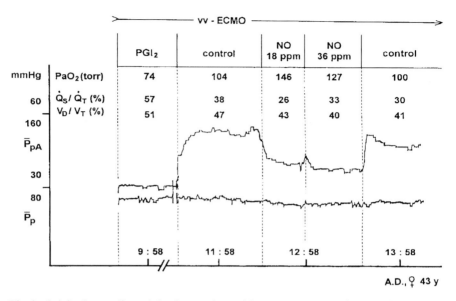

Fig. 1. Original recording of the first patient with severe acute respiratory distress syndrome due to aspiration pneumonia receiving nitric oxide (NO) inhalation. On the y-axis, mean systemic and pulmonary arterial pressure (PAP) are indicated (0–60 mmHg for PAP, 0–160 mmHg for systemic pressure). On the x-axis, the time course of the pressures is indicated (*upper curve*: PAP; *lower curve*: systemic pressure), subdivided for the following protocol: first, prostacyclin (4 ng/kg/min) was infused systemically, then the infusion was stopped (control). This was then followed by the inhalation of 18 ppm NO and 36 ppm NO. Finally, inhalation was stopped for another control. The monitored data for oxygenation (arterial oxygen tension, PaO_2), intrapulmonary right-to-left shunt (Q_s/Q_t), and dead-space ventilation (V_D/V_T) after reaching a steady state for each treatment are inserted as a time-dependent table. As demonstrated, both systemic prostacyclin and inhaled NO are able to reduce PAP, whereas inhaled NO improves the systemic oxygenation. In contrast, systemic prostacyclin decreases PaO_2 (compared with the control data) by increasing the intrapulmonary right-to-left shunt

NO inhalation in L-NAME-treated animals produced pulmonary vasodilation and decreased right-to-left shunt to the same degree as NO inhalation alone, and transpulmonary cGMP formation increased to the same extent as without L-NAME infusion. Improvement of oxygenation and reduction of pulmonary hypertension in adults and children with severe ARDS were consistently confirmed by numerous investigators (BENZING and GEIGER 1994; BIGATELLO et al. 1994; PUYBASSET et al. 1994a; DAY et al. 1996a; KRAFFT et al. 1996a). Some studies demonstrated enhanced NO effect by the additional systemic administration of vasoconstricting drugs (almitrine) potentiating intrapulmonary blood shifting from non-ventilated to ventilated and NO-exposed alveoli (WYSOCKI et al. 1994; LU et al. 1995). Other results indicated that similar effects on oxygenation or reduction of acute PVR could be achieved by the use of

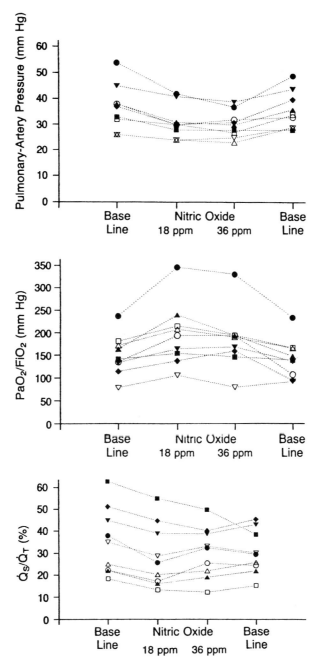

Fig. 2. Mean pulmonary artery pressure (*PAP*), arterial oxygenation efficiency (PaO_2/F_IO_2), and intrapulmonary shunting (Q_s/Q_t) in nine patients with acute respiratory distress syndrome during inhalation of nitric oxide. *Solid symbols* represent patients treated with extracorporeal membrane oxygenation (ROSSAINT et.al. 1993)

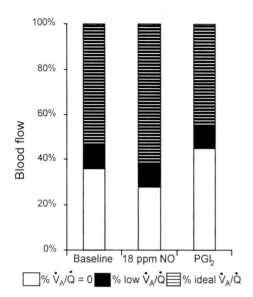

Fig. 3. Blood flow to regions of different ventilation/perfusion (V_A/Q) during nitric oxide inhalation and prostacyclin infusion. $V_A/Q = 0$ is the intrapulmonary right-to-left shunt (Q_s/Q_t) in regions with perfusion but without any ventilation. In regions with low V_A/Q ($0 \leq V_A/Q < 0.1$), ventilation is minimal (venous admixture, $n = 6$) (ROSSAINT et al. 1993)

aerosolized prostacyclin (PAPPERT et al. 1995; WALMRATH et al. 1996) or the NO-releasing compound SIN-1 (SCHUTTE et al. 1997).

3. NO Inhalation and Non-Cardiogenic Pulmonary Edema

Transvascular fluid filtration in the lung depends mainly on hydrostatic pulmonary capillary pressure (PCP), which increases with intrapulmonary vasoconstriction, promoting transvascular fluid filtration and lung edema. BENZING and GEIGER (1994) investigated the effects of NO inhalation (40 ppm) on PCP in patients with ALI. NO significantly reduced PCP from 25 ± 6 mmHg to 22 ± 5 mmHg ($P < 0.0001$), mainly due to a vasodilating effect on pulmonary venous vasculature. In a follow-up study, the effects of 40 ppm NO on transvascular albumin flux was measured by a double-radioisotope method using ^{99m}Tc-labeled albumin and ^{51}Cr-labeled erythrocytes (BENZING et al. 1995). In this study, NO reduced pulmonary transvascular albumin flux, indicating that, in patients with ALI, short-term NO inhalation may have beneficial effects on edema formation and resolution. However, long-term effects of NO inhalation on pulmonary edema and dose dependency have yet to be established.

4. Dose–Response Relationship of NO Inhalation

Studies on dose-response relationships showed that the most effective NO concentration is much lower for improvement of oxygenation than for reduction of PVR. GERLACH et al. (1993b) investigated the time course and dose response of NO inhalation in 12 patients with severe ARDS. While NO was delivered intermittently at different concentrations (0.01–100 ppm), PAP

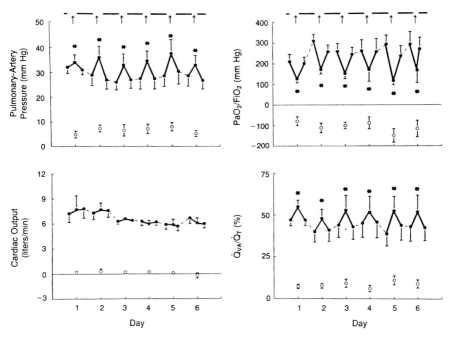

Fig. 4. Hemodynamic function and gas exchange before, during, and after brief interruptions (*arrows*) of nitric oxide inhalation (*bars*) during the first 6 days of treatment in seven patients with acute respiratory distress syndrome. Values are means ± SE (*solid symbols*); also shown (*open symbols*) are the means ± SE of the individual differences between the values for the effect of treatment and the means of the values determined before and after interruption of nitric oxide therapy. The standard errors for the treatment effects were small, indicating that the effects of withdrawal of nitric oxide were clear and precisely estimated. Each *asterisk* denotes a significant difference from the means of the values determined before and after interruption of nitric oxide therapy (ROSSAINT et al. 1993)

was reduced with increasing NO concentrations which was significant at 1 ppm and 50% maximal at 2–3 ppm (ED_{50}). In contrast, PaO_2 showed a biphasic response, with a dose-dependent increase (significant at 100 ppb) that peaked at 10 ppm and switched towards baseline values at 100 ppm NO (Fig. 5). The estimated ED_{50} for improvement of oxygenation was only 100 ppb, i.e., more than ten times lower than the ED_{50} found for the reduction of PAP. In two patients, the increase of PaO_2 at 0.01 ppm, which was in the range of the ambient NO concentration, was already more than 40% of the maximum (1 ppm), and exposure to 100 ppm NO even led to a deterioration in arterial oxygenation, with PaO_2 values below those of the respective baselines. The time course for different NO concentrations showed a delayed response of PAP similar to that of PaO_2, giving further evidence that improvement of oxygenation was due to redistribution of blood flow to ventilated regions rather than enhancement of pulmonary perfusion.

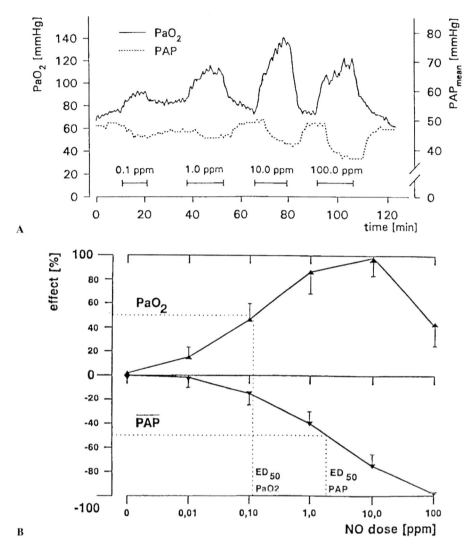

Fig. 5. Time course and dose response of nitric oxide (NO) inhalation for oxygenation and pulmonary artery pressure (PAP). **A** Representative example for a continuous registration of arterial oxygen tension (PaO_2; *solid line, left y-axis*) and mean PAP (*dotted line, right y-axis*) during inhalation of NO. Inhalation periods with inspiratory doses of NO are indicated above the *x-axis*, which represents the time of measurement. **B** Dose response for PaO_2 (*upper part*) and PAP (*lower part*) to inspiratory NO doses (*x-axis*). Values are means ± SD ($n = 12$), expressed as percentage (*y-axis*) compared with the initial value (=0% effect) and the highest registered alteration (=100% effect). The estimated ED_{50} of NO for improvement of arterial oxygenation (*dotted line, upper part*) and for reduction of PAP (*dotted line, lower part*) are indicated on the *x-axis*. Note the different ED_{50} for PaO_2 and PAP and the diminished effect of 100 ppm NO compared with 1 ppm and 10 ppm NO on arterial oxygenation (GERLACH et al. 1993)

Although the physiologic responses to different concentrations of NO are speculative, the phenomena may be explained by hypothetical models. At low concentrations, the very lipophylic NO diffuses only into capillaries in the very near distance to ventilated alveoli, promoting a strictly selective vasodilation, balancing ventilation/perfusion mismatch, and increasing systemic oxygenation. At higher NO doses, the range of action also covers small vessels in non-ventilated shunt areas, promoting perfusion in these regions and relativizing the beneficial effect on systemic oxygenation. However, both low- and high-dose NO will reduce the total PVR ("diffusion theory"). Alternatively (or additionally), the range of action of inhaled NO may be expanded by formation of e.g. low-molecular-weight nitrosothiols or by binding to albumin ("transport theory"). At low concentrations, NO acts predominantly on vascular smooth-muscle cells near ventilated alveoli; however, some of the NO that diffuses into the vascular space will be immediately captured by Hb, ensuring selective vasodilation; perhaps small amounts will also be bound to NO carriers. At high NO doses, enhanced formation of NO carriers increases the lifetime of NO, which may indirectly dilate "shunt" vessels after their reunion with "ventilated" vessels, increasing blood flow in shunt areas due to afterload reduction of both types of vessels and leading to a deterioration of oxygenation. Since nitrosothiols may easily transverse cell membranes, diffusion to non-ventilated regions may contribute. A subsequent study during long-term NO inhalation in three patients with ARDS showed that NO levels as low as 60–230 ppb was sufficient to reach a 30% improvement of PaO_2/F_iO_2 without tachyphylaxis and without significant changes in PAP (GERLACH et al. 1993a). Thus, improvement of oxygenation with such low doses of NO does not necessarily require reduction of pulmonary resistance, although a reduction of PAP was observed even at low concentrations of NO when PAP was high (>45 mmHg). Most importantly, our results and others demonstrate inter- and intra-individual variability (LUNDIN et al. 1996). For example, in one study, 0.1–2 ppm NO induced a dose-dependent increase of PaO_2 and a decrease of PAP, but 5 ppm did not further reduce PAP and, in two patients, PAP was reduced to 91% and 74% by only 0.1 ppm (PUYBASSET et al. 1994a). Others reported a dose-dependent PAP reduction with 5–40 ppm NO but not a clear dose-response effect on intrapulmonary shunt and PaO_2/F_iO_2; however, lower doses of NO were not tested (BIGATELLO et al. 1994). Thus, individual sensitivity to NO must be considered before starting NO therapy.

Depending on the primary goal, improvement of oxygenation or reduction of PVR, different concentrations of NO may be advantageous, i.e. if oxygenation is favored, increasing NO above an "optimal" value bears the risk of worse oxygenation. Since mechanical ventilation disrupts the patient from inhaling NO and since significant improvement of oxygenation in the majority of patients with ARDS can be achieved with concentrations of NO that are nearly in the range of autoinhaled doses, low-dose NO inhalation may be considered as a physiological replacement therapy. Furthermore, low-dose NO inhalation attenuates toxic effects and may reduce the risk of creating NO dependency. When NO must be delivered without delay, inhalation could be initiated with 10ppm; however, the concentration with the maximum effect must be titrated and used for further therapy as soon as possible. For reduction of pulmonary hypertension, higher doses (20–40ppm) might be necessary. The NO response should be controlled daily and, if it disappears, NO administration should be discontinued until a response can be shown again. In neonates, maximum effects on oxygenation may require higher doses of NO (20ppm) (DEMIRAKCA et al. 1996).

5. Effects of NO Inhalation on Right Heart Function

Pulmonary hypertension in ARDS is due to the combined effects of mechanical occlusion (thrombosis in the pulmonary microcirculation) and functional vasoconstriction (HPV, release of vasoconstricting mediators). As a result, right-ventricular afterload increases, which may lead to reduction of the right-ventricular ejection fraction (RVEF), increase in right-ventricular volume and atrial pressure and, finally, in right heart insufficiency and failure. In the study by ROSSAINT et al. (1995b), effects of inhaled NO (18ppm and 36ppm) and infusion of prostacyclin on right-ventricular function parameters assessed by the thermodilution technique were compared. With 18ppm NO, PAP decreased from 33 ± 2 mmHg to 28 ± 1 mmHg ($P = 0.008$), RVEF increased from $28 \pm 2\%$ to $32 \pm 2\%$ ($P = 0.005$), right-ventricular end-diastolic-volume index (RVEDVI) decreased from 114 ± 6 ml/m^2 to 103 ± 8 ml/m^2 ($P = 0.005$), and right-ventricular end-systolic-volume index (RVESVI) decreased from 82 ± 4 ml/m^2 to 70 ± 5 ml/m^2 ($P = 0.009$) without changes in cardiac index (CI) or MAP. Inhalation of 36ppm NO revealed no further improvement. During prostacyclin infusion, PAP decreased from 34 ± 2 mmHg to 30 ± 2 mmHg ($P = 0.02$) and RVEF increased from $29 \pm 2\%$ to $32 \pm 2\%$ ($P = 0.02$) but, in contrast to inhaled NO, MAP decreased from 80 ± 4 mmHg to 70 ± 5 mmHg ($P = 0.03$) and CI increased from 4.0 ± 0.5 l/min $\times$ m^2 to 4.5 ± 0.5 l/min $\times$ m^2 ($P = 0.02$) without changes of RVESVI or RVEDVI. A similar study investigated the correlation of improvement of oxygenation and reduction of PVR during inhalation of 5ppm NO in 13 patients with ARDS (FIEROBE et al. 1995). PAP decreased from 36 ± 5 mmHg to 31 ± 6 mmHg ($P < 0.01$) and PVR decreased from 211 ± 43 dyne $\times$ s $\times$ cm^{-5} to 180 ± 59 dyne $\times$ s $\times$ cm^{-5} ($P < 0.05$), a reduction which was associated with an increase of RVEF from $32 \pm 5\%$ to $36 \pm 6\%$ ($P < 0.05$) and a trend towards decreased right-ventricular volumes. Stroke volumes and heart rate were unchanged. There was no correlation between improvement of oxygenation and decrease in PVR.

These studies indicate that, at least in ARDS patients without severe right heart failure, improvement of RVEF during NO inhalation is not necessarily associated with an increase in CI but rather reflects a normal response of the right ventricle in the presence of different loading conditions, i.e. the ventricle is working on a different part of the pressure–volume curve. In contrast, prostacyclin infusion increased CI, probably due to reduction of left-ventricular afterload. However, in ARDS patients with severe right heart failure, NO-induced right-sided afterload reduction may lead to marked improvement of CI (BENZING et al. 1997). Finally, NO inhalation may ameliorate pulmonary vasoconstriction and increased right-to-left shunt occurring during permissive hypercapnia, leading to reduction of right-ventricular workload and improvement of transpulmonary vascular efficiency (PUYBASSET et al. 1994b; CHEIFETZ et al. 1996).

6. NO Non-Responders

For reasons that are not known, some patients with ARDS may not respond to inhaled NO (MIRA et al. 1994). In a series of 30 ARDS patients, first-time delivery of 10–20ppm NO increased PaO_2/F_IO_2 by over 10% in only 83% of patients, reduced venous admixture by over 10% in 87%, reduced PAP by at least 3mmHg in only 63%, and there was no effect on arterial oxygenation in five patients (ROSSAINT et al. 1995a). The analysis of different variables which might influence NO response revealed no correlation between PaO_2/F_IO_2 and PVR, PAP decrease after NO inhalation, venous admixture before NO inhalation, PAP levels before NO inhalation, or the amount of PAP reduction by NO. Others reported that, during NO inhalation, reduction of both PVR and venous admixture correlated with respective degrees of impairment without NO (BIGATELLO et al. 1994), that the NO effect depends on PEEP-induced alveolar recruitment (PUYBASSET et al. 1995) or the degree of HPV in non-ventilated areas (KOIZUMI et al. 1994), and that venous admixture improved with low CO but deteriorated with high CO (BENZING et al. 1996). In patients with ARDS and septic shock, inhaled NO was reported to be more (MOURGEON et al. 1997) or less effective (PUYBASSET et al. 1994b; MANKTELOW et al. 1997). In a rat model of endotoxin-induced lung injury, diminished vasodilatory response to inhaled NO could be partly attributed to increased cGMP breakdown due to enhanced cGMP-phosphodiesterase activity (HOLZMANN et al. 1996). However, in a population of septic-shock patients, the response to inhaled NO might differ. In 35 patients with ARDS and septic shock, only 40% were NO responders. Patients who had marked improvement of oxygenation and hemodynamics during NO inhalation where characterized by an increased RVEF, which was accompanied by higher CI, oxygen delivery, and oxygen-extraction ratio whereas, in NO non-responders, RVEF and correlated parameters remained depressed, reflecting impaired cardiac reserve (KRAFFT et al. 1996a). The reason for NO insufficiency of some patients remains unclear and, so far, there are no definite parameters of predictive value. Importantly, non-responders may become responders in the follow-up, strongly indicating that NO response should be tested repetitively.

7. NO Dependency

It was observed that, especially after long-term NO inhalation, some patients had to be weaned from NO therapy over several days to avoid rebound phenomena. It is reasonable to suppose that rebound is caused by an endogenous lack of NO due to feedback inhibition of NO synthases (NOS) by exogenously supplied NO. This is supported by in vitro experiments, in which NO donors inhibited inducible NOS (iNOS) and constitutive NOS in cell cultures (Assreuy et al. 1993), gaseous NO attenuated iNOS activity in rat aortic ring exposed to endotoxin (Kiff et al. 1994), and chronic NO inhalation decreased endothelium NO-dependent and NO-independent pulmonary vasodilation in rats (Roos et al. 1996). In isolated rat lungs, 48-h inhalation of 30ppm NO reversibly enhanced pulmonary vasoreactivity to angiotensin II and thromboxan analogue U-46619 immediately after cessation of NO inhalation, indicating a decrease in endogenous NO release (Combes et al. 1997).

Therefore, a sudden absence of NO in ARDS patients may induce marked vasoconstriction in ventilated areas and deterioration of oxygenation due to blood shifting to non-ventilated shunt regions. Fortunately, in most patients, sudden rebound phenomena will be followed again by a decrease of both PVR and oxygenation; however, some patients may remain NO dependent, and rebound effects may lead to severe hypoxemia and right-ventricular failure (Chiche et al. 1995; Miller et al. 1995; Cueto et al. 1997). These observations, combined with other studies showing that the absolute level of pulmonary arterial pressure correlates with the severity of pulmonary vascular injury in ARDS (Villar et al. 1989) and that pulmonary hypertension may be associated with impaired NO production in patients with chronic obstructive pulmonary diseases (Dinh-Xuan et al. 1991) and ARDS (Brett and Evans 1998), support the hypothesis that even low doses of NO might reduce PAP in most severe cases of ALI.

8. Recent Studies of NO Inhalation in ARDS

Since the first description by Ashbaugh et al. (1967), extensive clinical and research efforts have been made to improve outcome; however, analysis of outcome studies had a sobering effect and indicated no improvement of mortality, which has remained at approximately 50% during recent decades (Krafft et al. 1996b). Retrospective matched paired analysis of our own data revealed that there was no difference in survival rate between patients who received NO therapy and those who did not (both 69%), and no differences were found for the duration of ventilation or intensive-care unit (ICU) stay (Rossaint et al. 1995a). Recently, the first prospective randomized, blinded, and placebo-controlled phase-II multicenter trial on 177 ARDS patients was published; this trial was designed to evaluate safety, physiologic response, and outcome parameters of different doses of inhaled NO (1.25–80ppm) (Dellinger et al. 1998). To minimize the influence of differing equipment among the participating centers, the same type of ventilator and NO-delivery system was used, and NO treatment was performed according to a strict protocol. During the first 4h after initiation of treatment, ventilator settings were not changed, and improvement of oxygenation reflected the acute effects of

NO inhalation whereas, in the follow-up, changes of ventilator settings were used to calculate the intensity of ventilation, using the oxygenation index. Sixty percent of patients responded to NO with an increase of PaO_2 of at least 20%, which caused a reduction of F_IO_2. However, this was only significant on the first day of treatment, and the reduction of F_IO_2 was only modest (0.71 ± 0.14 vs 0.69 ± 0.13). In the NO group, the oxygenation index remained lower during the first 4 days, and mean PAP remained lower (~ 2 mmHg) for 2 days. During the 4-h observation period, 24% of the placebo group also had an increase in PaO_2 of at least 20%. There were no differences between the pooled inhaled groups and placebo groups with respect to mortality rate or the number of days alive after meeting oxygenation criteria for extubation. A post hoc analysis revealed that, in the group receiving 5 ppm NO, the percentage of patients alive and off mechanical ventilation at day 28 was higher (62% vs 44%) than in the placebo group. However, when 56 patients who (mainly due to death) had their gas treatment discontinued before meeting oxygenation threshold criteria were included in the post hoc analysis, mortality in the NO group was even higher (38% vs 30%) (MATTHAY et al. 1998).

Another randomized trial investigated acute effects of inhaled NO on oxygenation parameters in 40 patients with ARDS during the first three days of treatment (MICHAEL et al. 1998). Similar to the study above, NO therapy, as compared with the conventional therapy, increased PaO_2/F_IO_2 only during the first day; beyond 24 h, the two groups had an equivalent improvement in PaO_2/F_IO_2 and, during the 72 h following randomization, reduction of F_IO_2 (≥ 0.15) was not different with or without NO. Another 30 ARDS patients were enrolled in a randomized, controlled pilot study (TRONCY et al. 1998). Improvement of arterial oxygenation was significant only during the first 24 h of NO inhalation, and there was no significant difference between the NO group and controls with regard to 30-day mortality or days on mechanical ventilation. Furthermore, preliminary results of a European multicenter study in 260 patients with early ALI [of whom 172 (66%) were NO responders ($PaO_2 > 25\%$) and all of whom were randomly allocated to receive NO inhalation or conventional therapy] showed no improvement in 30-day mortality in the NO group (45% in the NO group vs 38% in controls) (LUNDIN et al. 1997). However, it must be noted that recent prospective outcome studies enrolled only patients with early ARDS who would probably survive and improve without NO therapy. Focusing on patients who do not respond to standard therapy might elucidate effects on outcome more clearly. Additionally, NO inhalation is targeted toward lung pathophysiology, but most patients do not die of lung disease but of multiple organ failure; hence, inhaled NO can not reduce mortality.

In conclusion, when focusing on outcome, NO inhalation seems to reveal no advantage. However, when focusing on the primary targets of NO therapy, i.e. selective pulmonary vasodilation, reduction of pulmonary resistance, and improvement of oxygenation, there was a sustained effect – even more pronounced with low doses of NO – in the majority of patients; this result might suggest gentler ventilation, at least during the first day(s) of treatment. In our department, a prospective, placebo-controlled trial of NO inhalation in 20 severe ARDS patients was performed. In the NO group, 10ppm NO was delivered in addition to the conventional therapy. The results (as yet unpublished) of this trial revealed that: (a) a peak effect for oxygenation was achieved with 10ppm NO; (b) the initial reduction of F_IO_2 was only significant on the first day; (c) outcome, ICU stay, or ventilation days were not different for either group; and (d) NO effects on PAP reduction diminished after 72h treatment, i.e., PAP values during prolonged NO inhalation were no longer significantly different from the baseline PAP in the control group. Interestingly, halting administration of NO after 72h resulted in higher PAP values compared with the control group, indicating rebound phenomena and sensitization by inhaled NO. Thus, beneficial long-term effects of NO inhalation in patients with ARDS has not been established. However, none of the patients in the NO group required ECMO therapy, whereas four patients in the control group did. This might indicate that NO inhalation could be considered as a useful bridging therapy to bypass more invasive and expensive strategies. Interestingly, similar results were found for NO inhalation in newborns (see below). Furthermore, first studies on ARDS survivors who received NO therapy indicate that NO inhalation has so far had no long-term harmful effects on lung function or other parameters (LUHR et al. 1998). Discussion about the usefulness of NO inhalation in patients with ARDS has deranged clinical studies (MATTHAY et al. 1998; ZAPOL 1998).

II. NO Inhalation in PPHN

Immediately after birth, an oxygen-associated decrease of pulmonary vascular tone and increase of pulmonary blood flow occurs; these are modulated by endogenous vasoconstrictors (endothelin, thromboxan) and vasodilators (prostaglandin I_2, NO) (ABMAN et al. 1990). In PPHN, which is mostly due to hypoxia, increased PVR results in right-to-left shunting of blood across the patent ductus arteriosus and foramen ovale, further increasing hypoxia (KINSELLA et al. 1992). Specific therapy is aimed at PVR; however, the use of systemic vasodilator drugs is limited due to profound arterial hypotension, which increases right-to-left-shunt and hypoxemia. Thus, ECMO is often indicated when conventional therapy fails to maintain oxygenation. It was postulated that the pulmonary vasoconstriction might result from inadequate endogenous NO production by the vascular endothelium; therefore, NO inhalation could replace endogenous NO release and should reveal beneficial effects. Studies in different newborn animals with pulmonary hypertension due to hypoxia, acidosis, prenatal ductus arteriosus legation, and thromboxane or bacteria infusion demonstrated that NO inhalation (6–100ppm), depending on the respective model, increased lung cGMP concentration, reduced PVR, PAP, and right-to-left shunt, and improved pulmonary blood flow and arterial oxygenation without producing systemic or toxic effects (BERGER et al. 1993;

ROBERTS et al. 1993a; ZAYEK et al. 1993; DEMARCO et al. 1994). First clinical results were reported by KINSELLA et al. (1992) and ROBERTS et al. (1992), who demonstrated improvement in preductal oxygen saturation in hypoxic neonates with severe PPHN inhaling 10–20ppm and 80ppm NO, respectively. In the study by Kinsella, clinical improvement was sustained at 6ppm NO for 24h in 6 of 9 infants. Subsequent studies by the same group on nine neonates evidenced that prolonged treatment of PPHN with low doses of NO (6ppm) might reduce the need for ECMO therapy (JELLINEK et al. 1997). Another study investigated the dose-response effects of 5–80ppm NO on oxygenation in 23 hypoxic [oxygen index (OI) > 20; OI = MAP × F_1O_2 × $100/PaO_2$] near-term neonates with and without PPHN, who remained candidates for ECMO after surfactant-replacement therapy. They demonstrated a significant improvement of oxygenation in 13 infants, which was not significantly different for low and high doses of inhaled NO; however, in neonates with isolated PPHN, the response was more effective (FINER et al. 1994).

Several trials on neonates with PPHN either alone or in combination with hypoxic respiratory failure gave further evidence of beneficial effects of inhaled NO. A prospective placebo-controlled, randomized multicenter trial was conducted to determine whether inhaled NO (20ppm followed by a 80ppm trial in non-responders) could reduce mortality or the need for ECMO in 235 neonates of at least 34weeks gestation with hypoxic respiratory failure (OI ≥ 25). NO treatment resulted in a significant improvement in PaO_2 (58 ± 85mmHg vs 10 ± 52mmHg for controls, mean ± SD, $P < 0.001$) and OI (a decrease of 14 ± 21 in the NO group vs an increase of 1 ± 21 in the control group, $P < 0.001$). The incidence of the primary outcome (death in under 120 days or initiation of ECMO) was significantly lower in the NO group (46% vs 64%; $P = 0.005$; relative risk = 0.71, 95%; CI = 0.56, 0.90), as was the initiation of ECMO (39% vs 54%, $P = 0.014$); however, NO inhalation had no apparent effect on final mortality (14% vs 17% for controls). Infants with the lowest OI appeared more likely to have complete response to the initial NO administration and to survive without ECMO. In 53 of 112 of infants who did not respond completely to 20ppm NO (PaO_2 increase ≤20mmHg), only 15% had improved responses to 80ppm NO, suggesting that limited numbers of infants will benefit from higher doses of NO (Neonatal Inhaled Nitric Oxide Study Group 1997a). Another large prospective multicenter crossover study investigated the efficacy of either inhaled NO in combination with conventional ventilation, high-frequency oscillatory ventilation (HFOV) without NO, or a combination of both in 205 infants with PPHN of either idiopathic origin or due to respiratory distress, meconium aspiration, or pulmonary hypoplasia. Treatment failure ($PaO_2 < 60$mmHg) in either group resulted in crossover to the alternative treatment, and treatment failure after crossover led to combination treatment of HFOV plus NO. There was no significant difference in response to NO and HFOV alone in infants who recovered after successful initial treatment or after failure and crossover. Sixty percent of all infants responded to either therapy alone or to the combination of both. Of 125 infants in whom both treatments failed, 32% responded to a combination of both, indicating that a combined therapy might improve efficacy, especially in respiratory failure and after meconium aspiration. No significant differences were found for mortality and initiation of ECMO, both of which varied markedly depending on the participating center (KINSELLA et al. 1997).

Additionally, in a prospective placebo-controlled randomized multicenter study on 58 full-term neonates with PPHN and severe hypoxia, prolonged inhalation of 20–80 ppm NO doubled oxygenation in 53% of infants treated with NO in contrast to improved oxygenation in only 7% of infants in the control group. This was accompanied by a significantly reduced need for ECMO therapy (40% in NO group vs 71% in the control group, $P = 0.02$); however, mortality was not different in either group (ROBERTS et al. 1997). In contrast, a reduction of the incidence of ECMO was not observed in a study on 17 infants suffering from hypoxic respiratory failure (BAREFIELD et al. 1996). It was suggested that the acute response to inhaled NO might reflect the severity of the lung disease and might be a predictive marker for the use of ECMO. In a prospective trial on PPHN, newborns with OIs between 25 and 40 were randomized to receive conventional therapy with or without 20 ppm NO or received NO when OI was over 40. Infants with focal disease or with normal lung fields in the chest X-ray had the greatest degree of oxygenation, whereas NO failed to improve oxygenation in lung hypoplasia and severe diffuse lung disease. Additionally, oxygenation improved in 87% of patients in whom exposure to NO induced an acute drop of OI from above 40 to below 40, whereas 90% of infants without that response to NO needed ECMO therapy or died (DAY et al. 1996b). However, NO inhalation seemed to be less effective in neonates with congenital diaphragmatic hernia, although individual cases indicated transient improvement of oxygenation and beneficial effects in the immediate postoperative course (FROSTELL et al. 1993b). To investigate whether inhaled NO would reduce the occurrence of death and/or the initiation of ECMO, 53 infants with congenital diaphragmatic hernia and hypoxemic respiratory failure were enrolled in parallel with the Neonatal Inhaled Nitric Oxide Study Group study in a randomized placebo-controlled trial (see above). Although short-term improvements in oxygenation in some infants revealed transient beneficial effects, neither the increase of PaO_2 nor the decrease of OI or mortality was significantly different in either group, and the incidence of ECMO therapy was even greater in infants who received NO (54% vs 80%) (Neonatal Inhaled Nitric Oxide Study Group 1997b).

III. NO Inhalation in Other Diseases

Case reports and observational studies indicate potential beneficial effects of NO inhalation on pulmonary hypertension and oxygenation under various clinical conditions, such as chronic obstructive lung disease (ADATIA et al. 1993; ADNOT et al. 1993), asthma (HÖGMAN et al. 1993b; KACMAREK et al. 1996; PFEFFER et al. 1996), lung transplantation (ADATIA et al. 1994), congenital heart diseases (ROBERTS et al. 1993b; ATZ et al. 1996), cardiac surgery (RICH et al. 1993; BENDER et al. 1997; BICHEL et al. 1997), and high-altitude pulmonary edema (SCHERRER et al. 1996). In contrast to beneficial effects on right-sided cardiac failure or dysfunction associated with pulmonary hypertension, the use of inhaled NO in left-sided heart failure or biventricular failure may be associated with an increase of left-ventricular preload, paradoxical increase of systolic PAP, and further deterioration of left-ventricular performance (LOH et al. 1994; SEMIGRAN et al. 1994; CUJEC et al. 1997). The use of inhaled NO therapy was defined as a relative contraindication for New York Heart Association grade-III and -IV patients with left-ventricular failure (CUTHBERTSON et al. 1997).

IV. NO Autoinhalation

First studies in humans and animals determining NO in exhaled air indicated that the origin of endogenously synthesized and exhaled NO was the respiratory tract and that little exhaled NO derived from alveoli; rather, it came predominantly from terminal and respiratory bronchioles (GUSTAFSSON et al. 1991; PERSSON et al. 1993). GERLACH et al. (1994) measured exhaled NO in volunteers and found maximum values of NO in the nasopharynx but found prelaryngeal and tracheal concentrations that were lower during expiration than during inspiration, indicating that authentic NO was produced predominantly in the nasopharynx and was autoinhaled; of this, 50–70% was resorbed by the lower respiratory tract. Nasopharyngeal NO synthesis was lower in smokers than in non-smokers, probably due to inhibition of endogenous NO synthesis. Studies on intubated patients demonstrated disruption of natural NO autoinhalation from the upper respiratory tract (Fig. 6); during long-term ventilation, nasopharyngeal NO synthesis was depressed, but it increased when bacteria were present in the nose. It was postulated that NO autoinhalation contributed to matching of the ventilation/perfusion ratio in the lung and that low-dose NO inhalation in ARDS patients might be considered as NO replacement therapy, since the inspiratory NO concentrations in volunteers were similar to those that were effective in ARDS patients (approximately 0.1 ppm).

The hypothesis that the nasopharynx was the preliminary source of exhaled NO was supported by the detection of constitutively expressed NOS resembling the inducible isoform in paranasal sinus epithelial cells and the fact that high concentrations of NO in the nasopharynx contributed to over 90% of the exhaled NO concentration in patients and healthy individuals (LUNDBERG et al. 1995; SCHEDIN et al. 1995). Furthermore, replacement therapy with physiologically low doses of NO (10–100 ppb) in intubated patients with and without ARDS improved oxygenation (KELLY et al. 1996). LUNDBERG et al. (1996) investigated the effects of nasally derived NO on pulmonary function. In healthy volunteers during mouth breathing, transcutaneous oxygen tension was 10% lower than during nose breathing. In intubated patients, nasally suc-

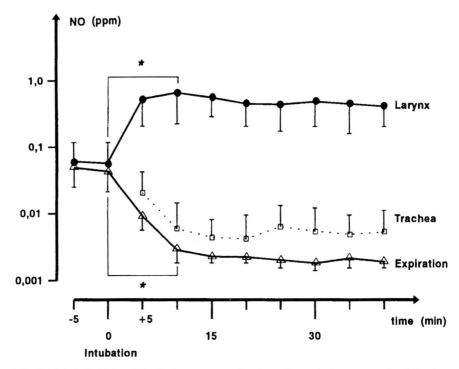

Fig. 6. Exhaled nitric oxide (NO concentration in patients during orotracheal intubation. NO concentrations from the expiratory limb of the ventilator (expiration), from the trachea (after intubation), and from the lower oropharynx above the larynx during inspiration. Mean (95% CI, $n = 10$, $P < 0.05$). Note parts per million for NO concentration. After intubation, trachea and expiratory-limb NO concentrations decreased, whereas the pharyngeal concentration increased, probably due to accumulation. The inspiratory NO concentration in the trachea was slightly higher than that during expiration or from the inspiratory limb of the ventilator. Thus, NO may also be produced in the lower airways (GERLACH et al. 1994)

tioned NO delivered to the inspiratory limb of the ventilator increased PaO_2 to 18% and reduced PVR to 11%. Taken together, these observations may implicate a new role for paranasal sinuses, i.e. production of NO to modulate ventilation/perfusion in the lower respiratory tract. Indeed, ALI and PPHN might be aggravated by impaired endogenous pulmonary NO synthesis, and intubation of these patients further impedes endogenous inhaled NO. Thus, low-dose NO inhalation could be viewed as a physiologic replacement therapy. Interestingly, in animals, upper-respiratory-tract NO production was found to be species specific and dependent on the presence of paranasal sinuses (LEWANDOWSKI et al. 1996, 1998; SCHEDIN et al. 1997). The physiological relevance of paranasal NO production might be further supported by measurement of nasal NO production in newborns. Nasal peak concentrations up to 4.6 ppm NO immediately after birth increased further to 30% during the first day, indicating that autoinhalation of upper-airway NO is involved in the adaptation of the respiratory system to postnatal life, at least in humans (SCHEDIN et al. 1996).

C. NO Metabolism, Toxicology, and Adverse Effects
I. NO Uptake and Clearance

Inhalation of NO gas is not restricted to medical conditions in the ICU. NO is an air pollutant which is produced during the combustion of fossil fuels and may be inhaled in considerable amounts with smog and cigarette smoke (which contain 400–1000ppm NO) without acute toxic effects. Much lower concentrations of NO are physiologically autoinhaled after synthesis in the nasopharynx (Sect. B.IV). However, continuous delivery of 1–100ppm NO to critically ill patients might bear the risk of toxic side effects of NO itself and, even more important, nitrogen oxides. When inhaled, the hydrophobic NO molecule rapidly crosses cell membranes within the alveolar space and diffuses into pulmonary capillaries as rapidly as oxygen to act as a local messenger. As a free radical with a single unpaired electron, NO is short lived, being oxidized, reduced, or complexed with other biomolecules depending on the micro-environmental conditions (GASTON et al. 1994). In particular, the increased affinity for oxygen, superoxide, and Hb promotes rapid oxidation of NO to nitrogen dioxide and peroxynitrite as well as methemoglobin (metHb) formation, respectively. In healthy subjects, 75–95% of exogenously delivered NO reaches ventilated and perfused alveoli and is absorbed (BORLAND and HIGENBOTTAM 1989; WESTFELT et al. 1995; YOUNG et al. 1996) whereas, in patients with ARDS, only 35% of the total amount of NO, 70% of NO delivered to the alveolar space, and 95% delivered to the perfused alveolar space was retained (WESTFELT et al. 1997). The major sink for NO in vivo is the irreversible reaction with oxyhemoglobin to form metHb and nitrate (BECKMAN 1996). Conversion of inhaled NO to nitrate is the predominant metabolic pathway, covering more than 70% of its inactivation, the remaining NO being excreted as urea via unknown metabolism (WESTFELT et al. 1995). In healthy volunteers, inhalation of 25ppm NO for 1h increased serum nitrate to 46% (WENNMALM et al. 1993), inhalation of 100ppm NO for 3h increased serum nitrate to approximately 300% (YOUNG et al. 1996) and, in a patient with ARDS receiving 90ppm NO for 2days, a 13-fold increase of serum nitrate was reported (VALVINI and YOUNG 1995). Most nitrates are excreted by the kidneys with a clearance of 20ml/min in healthy subjects; in patients with renal failure, clearance of nitrogen oxides might be impaired (WENNMALM et al. 1993).

II. NO and Nitrogen Dioxide

In the presence of oxygen, NO is oxidized to nitrogen dioxide. Inhalation of gas mixtures containing very high concentrations of NO and NO_2 can be rapidly lethal due to acute pulmonary edema, acid pneumonitis, and profound methemoglobinemia (CLUTTON-BROCK 1967; GREENBAUM et al. 1967). Short-term inhalation of high concentrations of NO_2 is much more dangerous than long-term exposure to low concentrations of NO_2 (LEHNERT et al. 1994). Exposure of rats to at least 50ppm NO_2 for only 5min induced significant lung injury, with edema formation and accumulation of leukocytes in the alveolar space (STAVERT and LEHNERT 1990). In healthy humans exposed to 2.3ppm NO_2 for 5h, a delayed decrease of glutathione peroxidase activity and alveolar permeability without obvious signs of mucous-membrane irritation or deterioration of lung function was observed (RASMUSSEN et al. 1992). Inhalation of 2ppm NO_2 in rats led to terminal bronchial epithelial hypertrophy and alveolar cell hyperplasia; NO_2 levels in excess of 1.5ppm increased granulocyte and macrophage recruitment in the lungs of hamsters, 0.4–0.5ppm NO_2 narrowed surfactant hysteresis (GASTON et al. 1994), and 0.3ppm NO_2 was reported to potentiate bronchospasm in asthmatics (BAUER et al. 1986). Compared with NO_2, NO itself seems to be less toxic. Rats exposed to 1500ppm NO for 15min showed no evidence of ALI (STAVERT and LEHNERT 1990), inhalation of 80ppm NO for 1h or 3h did not increase extravascular lung water (FROSTELL et al. 1991), and electron microscopy of lung specimens of rats exposed to 43ppm NO and 3.6ppm NO_2 for 6days gave no evidence of acute NO toxicity (HUGOD 1979). However, harmful effects of NO-induced inhibition of DNA synthesis and deamination of DNA and the inhibition of DNA repair proteins and cytochrome P450 must be considered during long-term NO inhalation, especially in neonates (WINK et al. 1993; KEEFER and WINK 1996; LAVAL et al. 1997).

With regard to NO inhalation therapy, it is critical to note that the rate of NO_2 formation depends on the concentrations of NO and O_2 and the contact time and temperature (formation is faster at lower temperature) but is independent of humidity (MIYAMOTO et al. 1994). With long intrapulmonary residence times for NO (small minute ventilation, large lung volume), there is a potential for NO_2 formation within the lungs. At high concentrations of NO (80ppm) and O_2 (90%), 2ppm NO_2 is generated within seconds, whereas it takes 20min for this amount to form in 90% O_2 when the NO level is 10ppm or less (NISHIMURA et al. 1995). Recommendations for occupational safety and health standards have set the upper limit for NO_2 inhalation at 5ppm. Formulas were provided to calculate the time required to generate 5ppm NO from different concentrations of NO and oxygen, assuming that the change in NO concentration is completely due to oxidation to NO_2: $1/(NO)_t - 1/(NO)_i = 2k(O_2)_t$, where $(NO)_i$ is the initial NO concentration, $(NO)_t$ is that at time t, and k (rate constant) is $1.93 \times 10^{-38} cm^6 mol^{-2} s^{-1}$ at 300°K (FOUBERT et al. 1992; Table 1). Others found a 2.5 times faster rate constant when NO_2 generation was measured using chemiluminescence (BOUCHET et al. 1993). Recent recommendations for NO inhalation therapy have set the upper limit for NO_2 at under 3ppm (CUTHBERTSON et al. 1997; Table 2). However, these recommendations are based on data for environmental exposure of individuals at work (exhaust from diesel engines). Effects of continuous exposure to even much

Table 1. Formation of nitrogen dioxide (NO_2). Time (min) required to yield 5 ppm NO_2 with diffrerent mixtures of nitric oxide (NO) and oxygen (O_2) (FOUBERT et al. 1992)

O_2 (%)	20 ppm NO	40 ppm NO	80 ppm NO	120 ppm NO
20	60.08	12.86	3.00	1.32
30	40.05	8.56	2.00	0.88
40	30.03	6.43	1.75	0.66
50	24.03	5.15	1.20	0.52
60	20.01	4.28	1.00	0.44
70	17.16	3.66	0.85	0.37
80	15.01	3.20	0.75	0.33
90	13.35	2.85	0.66	0.29
100	12.01	2.56	0.60	0.26

lower concentrations of NO_2 in patients with lung diseases might be different. Thus, the concentration of NO should be kept as low as possible, F_IO_2 should not be higher than clinically indicated, and the contact time between NO and oxygen as short as possible. The NO and NO_2 concentrations in inhaled gas must be monitored.

III. NO, Superoxide, and Peroxynitrite

The complex reactivity of peroxynitrite was extensively reviewed by Beckman (BECKMAN 1996). NO reacts irreversibly in a nearly diffusion-controlled manner with superoxide (O_2^-) to form peroxynitrite ($ONOO^-$). Peroxynitrite is a powerful, relatively long-lived, and toxic oxidant which plays a key role in tissue damage during cerebral and myocardial ischemia, inflammation, sepsis, ALI, reperfusion injury, and many other conditions.

Since granulocytes, alveolar macrophages, endothelial and epithelial type-II cells, and many others produce both NO and superoxide when activated, peroxinitrite formation is thought to be crucially involved in inflammatory cell-mediated tissue injury in ALI or sepsis (ISCHIROPOULOS et al. 1992; ROYALL et al. 1995). This concept is supported by many in vitro and in vivo observations. In vitro, peroxynitrite inhibits oxygen uptake and amiloride-sensitive sodium transport channels, which regulate alveolar fluid reabsorption in alveolar type-II cells (HU et al. 1994). Peroxynitrite also induces lipid peroxidation (BECKMAN 1996), damages surfactant lipids, nitrates surfactant protein A tyrosine residues (HADDAD et al. 1994a; ZHU et al. 1996), inhibits adenosine triphosphate and surfactant synthesis in alveolar type-II cells (HADDAD et al. 1996), and inactivates neutrophil protease inhibitor α_1-antiprotease (MORENO and PRYOR 1992). In vivo, rats exposed to endotoxin expressed iNOS in lung alveolar and interstitial macrophages within hours, resulting in increased peroxynitrite and nitrotyrosine formation in the lung (WIZEMANN et al. 1994) whereas, in dogs, selective inhibition of endotoxin-induced iNOS expression

Table 2. UK guidelines for the use of inhaled nitric oxide (NO) in adult intensive-care units (CUTHBERTSON et.al. 1997)

Indications	
1	Severe ARDS, optimally ventilated, PaO_2 90 mmHg on F_IO_2 1.0
2	Right-sided cardiac failure, significant RSCF; MPAP > 24 mmHg, TPG > 15, PVR > 400 dyne × s × cm^{-5}; must support systemic circulation (inotropes etc.); beware adverse effects on the left ventricle
Dose	
1	Maximum dose 40 ppm
2	Dose titration: 20–10–5–0 ppm for 30 min
3	A 20% rise in PaO_2 on F_IO_2 1.0 required
4	Use minimum effective dose
5	RSCF: 20–40 ppm
Delivery	
1	Continuous injection or synchronised inspiratory injection devices suitable with injection near to ventilator
2	Medical NO/N_2 gas mixture
3	Stainless-steel pressure regulators, connectors and flow-meter needle valves
4	Calibrated flow meter
5	Position of humidifier unimportant
Monitoring	
1	Continuous inspiratory NO and NO_2 at Y piece
2	Electrochemical monitoring adequate
3	Monitors correctly calibrated
4	Methaemoglobin levels: time 0, 1 and, 6 h then daily
5	Expiratory monitoring not necessary
Exposure	
1	Maximum inhaled NO < 40 ppm
2	Maximum inhaled NO_2 < 3 ppm
3	Maximum environmental NO < 25 ppm for 8-h TWA
4	Maximum environmental NO_2 < 3 ppm for 8-h TWA
5	Minimum effective dose for shortest periods advised (safety data up to 28 days available)
Scavenging	
1	Not required in a well-ventilated unit
2	Environmental monitoring required in units with less than 10–12 air changes per hour and scavenging if exposure limits exceeded
3	Scavenging techniques: filtration, active or passive scavenging
Contraindications	
Absolute	Methaemoglobinemia
Relative	Bleeding diathesis, intracranial hemorrhage, severe LVF

ARDS, acute respiratory distress syndrome; F_IO_2, fractional inspired oxygen; *LVF*, left ventricular failure; *MPAP*, mean pulmonary artery pressure; PaO_2, partial pressure of oxygen in arterial blood; *PVR*, pulmonary vascular resistance; *RSCF*, right-sided cardiac failure; *TPG*, transpulmonary gradient; *TWA*, time-weight average.

with aminoguanidine or *S*-methylisothiourea attenuated nitrotyrosine formation and ALI (NUMATA et al. 1998). Furthermore, in lung sections of patients with ALI, increased nitrotyrosine formation was found throughout the lung, including interstitium, alveolar epithelium, and inflammatory cells; additionally, nitrotyrosine residues were found in vascular endothelium and subendothelial tissues in sepsis-induced lung injury (HADDAD et al. 1994b; KOOY et al. 1995). Hence, in view of the generation of even more reactive species, the role of NO as a scavenger of superoxide or other oxidants to protect from tissue injury must be questioned (BECKMAN and KOPPENOL 1996).

However, in vitro, NO has been shown to attenuate cytotoxic effects of reactive oxygen species in cell cultures (WINK et al. 1994), and delivery of 50 ppm NO to isolated rat lungs reduced neutrophil-mediated, oxygen-radical-dependent endothelial capillary leak, indicating anti-inflammatory properties of inhaled NO (GUIDOT et al. 1995). Additionally, excess NO was found to inhibit superoxide-induced lipid peroxidation, but impairment of surfactant function due to persistent generation of peroxynitrite remained unchanged, indicating that pro-oxidant and anti-oxidant effects of NO depend predominantly on the micro-environmental conditions and the balance between reactive oxygen species and NO (RUBBO et al. 1994; ZHU et al. 1996). This might be further complicated when additive effects of inhaled NO and hyperoxia must be considered; unfortunately, most patients who receive inhaled NO also need a high F_IO_2. In vitro, synergistic non-apoptotic cytotoxic effects of NO and hyperoxia (95% O_2) were demonstrated in cultured human lung alveolar epithelial and microvascular endothelial cells (NARULA et al. 1998). In addition, significant interactions between NO and inspired oxygen concentration were reported in newborn piglets, which were ventilated for 48h with either 100ppm NO and 21% O_2 or 100ppm NO and 90% O_2 (ROBBINS et al. 1995). These interactions affected surfactant function, pulmonary inflammation, and neutrophil chemotactive activity in brocho-alveolar fluid. Another set of experiments then revealed protective effects of systemic application of superoxide dismutase on pulmonary inflammation, lipid peroxidation, and neutrophil chemotactic activity but not on surfactant function (ROBBINS et al. 1997). However, for improvement of oxygenation in patients with ARDS, lower concentrations of inhaled NO are generally used. In newborn lambs ventilated with NO for 6h, surfactant function was unchanged when NO concentration was as low as 20ppm, but was markedly impaired with concentrations of 80ppm and 200ppm (MATALON et al. 1996).

Taken together, the discussion about beneficial or harmful effects of inhaled or endogenously produced NO on inflammatory tissue damage remains controversial, i.e. NO might be protective or harmful, depending on the balance between NO and reactive oxygen species or other target molecules. In ALI, there may be no uniform inflammatory response in all regions at the same time, and delivery of NO to the whole lung may have different effects in different parts at different times. However, in view of the generation of highly toxic compounds, it is evident that both the NO and oxygen concentrations must be held as low as possible to achieve the desired therapeutic effects.

IV. NO and S-Nitrosothiols

Nitric oxide reacts with thiols to form S-nitrosothiols (RSNO). Sulfhydryl-containing compounds such as glutathione, cysteine, and albumin are biologically important targets for RSNO formation, as they stabilize NO in a bioactive form, serve as NO carriers, protect NO from inactivation, and mitigate the toxicity arising from its reaction with oxygen and superoxide (GASTON et al. 1994). In plasma, NO as S-nitroso-albumin was detected (STAMLER et al. 1992a), which may regulate vascular tone and platelet function in vivo (KEANEY et al. 1993). In the lung, RSNOs derived from NO sources within the lung are involved in airway and vascular smooth-muscle relaxation, bacteriostasis, platelet inhibition, tumor cell lysis, and modulation of enzyme activity (GASTON et al. 1994). S-nitrosoglutathione (GSNO), the predominant nitrosothiol in the airways, was found in broncho-alveolar lavage fluid and upper-airway aspirates of normal subjects, suggesting a physiological role of RSNOs in the regulation of airway resistance and pulmonary function. RSNOs also form upon inhalation of NO gas, indicating RSNO formation in the extracellular airway environment (GASTON et al. 1993). In vitro, injury of hamster lung fibroblasts exposed to nitroxyl was markedly enhanced in the presence of oxygen and attenuated in the presence of glutathione, indicating that glutathione is critical for cellular protection (WINK et al. 1998). However, RSNO formation in the lung will depend on several factors, such as pH, thiol concentration, and the redox state of the airway. Recent data imply that oxidation or nitrosation of thiols is regulated by the balance between superoxide and NO (WINK et al. 1997) and that superoxide promotes decomposition of GSNO to glutathione disulfide, nitrite, and nitrate (JOURDQHEUIL et al. 1998). This might be of clinical relevance during inhalation with NO and high inspired oxygen concentration.

1. Methemoglobin

The major route for NO inactivation in vivo is the fast and irreversible reaction with oxyhemoglobin to form metHb and nitrate: $Hb[Fe_{II}]O_2 + NO \rightarrow Hb[Fe_{II}]OONO \rightarrow Hb[Fe_{III}] + NO_3^-$ (BECKMAN 1996; MOTTERLINI et al. 1996). Approximately 3% of Hb undergoes daily spontaneous auto-oxidation, and many drugs, nitrates, nitrites, and inhaled NO or NO_2 may increase metHb formation. Under normal physiological conditions, several reducing systems maintain the level of metHb below 1% (MANSOURI 1985). The major route for metHb reduction in red blood cells is the reduced nicotinamide adenine dinucleotide (NADH)–cytochrome b_5–metHb reductase system; alternative routes, such as the reduced nicotinamide adenine dinucleotide phosphate-dependent metHb reductase, ascorbic acid, and reduced glutathione contribute to elimination, but they are much less effective and can not fully compensate NADH-reductase deficiency. Methemoglobinemia may develop due to: (a) hereditary decreased reducing capacity where, in heterozygotes, NADH-reductase deficiency is commonly masked until exposure to oxidants; (b) increased oxidative stress, where the rate of Hb oxidation exceeds the rate of reduction; and (c) Hb abnormalities. MetHb itself is not toxic but produces a functional anemia, which decreases oxygen delivery to the tissues; however, in patients with congenital reductase deficiency, metHb levels up to 20% are well tolerated. Several studies indicate that NO inhalation in adults with commonly used dosages of NO (<40ppm) may not raise metHb to levels of concern. For example, metHb values in volunteers during inhalation 40ppm NO for 10min increased from 0.61% to 0.77% (FROSTELL et al. 1993a), from 0.87% to 2.65% during inhalation of 100ppm NO for 3h (YOUNG et al. 1996), and from 0.74% to 0.98% in 30 patients with ARDS inhaling 10–15ppm NO for 3weeks (ROSSAINT et al. 1995a).

To evaluate the kinetics of metHb formation and elimination, volunteers inhaled 32, 64, and 128ppm NO for 3h or 512ppm NO until metHb exceeded 5%, respectively (YOUNG et al. 1994b). The pharmacokinetic analysis revealed both the metHb formation (where levels increased to 1.04%, 1.75%, and 3.75% NO) and metHb elimination (where the peak value of 6.93% declined continuously below 2% within 3h after inhalation of 512ppm NO) to be of first order. Since metHb elimination followed first order kinetics of levels of up to 20%, it was assumed that maximum metHb levels with clinically used dosages of NO (<100ppm) would be reached within 3–5h after the start of inhalation. However, this might be different in neonates who have an immature metHb reductase system and whose fetal Hb is sensitive to oxidation. An accidental exposure to high concentrations of inhaled NO due to a faulty delivering system was reported in a newborn, leading to a metHb level of 14% which declined to 2% within 4h after cessation of inhaled NO. In a study on neonates with PPHN, inhalation of 80ppm NO for 30min had no effect on metHb concentration (ROBERTS et al. 1992), and metHb remained below 1.5% during inhalation of 10–20ppm NO for 24h (KINSELLA et al. 1992; ROBERTS et al. 1992). Another study on term infants with hypoxic respiratory failure who received 20–80ppm NO for 1h reported a rise of metHb to 7% (BAREFIELD et al. 1996). In a study on 58 term newborns with PPHN, prolonged NO inhalation with decreasing concentrations from 80ppm to 10ppm over a period of 8days was accompanied by metHb formation, which increased continuously to approximately 4–5%. Maximum values during the first two days remained almost below 10%; however, in one infant, a transient increase to 18.2% occurred (ROBERTS et al. 1997). Finally, in 121 neonates with hypoxic respiratory failure treated with NO concentrations up to 80ppm for different times, mean metHb was 2.4 ± 1.8%; values between 5% and 10% were observed in 11 infants (Neonatal Inhaled Nitric Oxide Study Group 1997a).

During NO inhalation, routine monitoring of metHb formation with a co-oximeter is recommended, especially in newborns and when high doses of inhaled NO are delivered. MetHb formation might be increased in patients with reduced metHb clearance capacity, in critically ill patients with increased levels of Hb[Fe$_{II}$] where NO affinity is high, or during extracorporeal circulation (DOTSCH et al. 1997). However, as demonstrated by several studies, metHb formation is rarely a limiting factor for NO inhalation. Finally, methemoglobinemia may not be as suitable as a marker for red cell turnover or long-term toxicity of NO, since animals exposed to inhaled NO for a long time had normal metHb levels but increased spleen weights, total bilirubin and Heinz bodies and reduced haptoglobin levels (YOUNG et al. 1994b).

D. NO Administration
I. The NO/Nitrogen Gas Mixture

For a patient ventilated with a minute ventilation of 10 l/min, 1–100 ppm would correspond to a delivery of 0.01–1.0 ml/min of pure NO (YOUNG and DYAR 1996). To avoid such low flow rates, dilute mixtures of NO in an inert carrier gas (nitrogen) are used. The flow rate must be in a manageable range, i.e. the mixture must not be so dilute as to avoid unacceptable reduction of F_IO_2 and fast exhaustion of the cylinder, and must not be too high, since low flow rates are more difficult to control. An applicable concentration for custom-designed delivery systems may be 800–1000 ppm NO. The required flow of NO mixture (V_{mix}) can be calculated using the following formula (YOUNG and DYAR 1996):

$$\text{Final } F_IO_2 = \frac{\text{Initial } F_IO_2 \times V}{V + V_{mix}} \quad V_{mix} = \frac{V}{(F_{mix}NO/F_INO) - 1} \quad (1)$$

where V is the flow rate of the gas to which the NO mixture is added (gas flow around the circuit on a pediatric ventilator, minute ventilation on an adult ventilator), $F_{mix}NO$ is the concentration of NO in the NO/N$_2$ mixture, and F_INO is the desired final inspired NO concentration.

The subsequent reduction in F_IO_2 due to admixture of NO/N$_2$ can be calculated as:

$$\text{Final } F_IO_2 = \frac{\text{Initial } F_IO_2 \times V}{V + V_{mix}} \quad (2)$$

The exact concentration of NO in the NO/N$_2$ mixture and contamination by nitrogen dioxide is given in a certificate provided by the company. In accordance with safety recommendations of inhaled nitrogen dioxide in a range of 1–5 ppm, the nitrogen dioxide concentration in the mixture must not exceed 1–5%. NO is a corrosive gas, especially at high partial pressures and in the presence of water, which is kept to very low levels in cylinders (<3 ppm water vapor) to avoid cylinder and regulator damage. Before each use, regulators and tubings must be flushed to exhaust nitrogen dioxide, which may form in the presence of oxygen. All NO used on patients should be of medical and not industrial grade.

II. Delivery of NO

Since the discovery of inhaled NO as a drug, several approaches have been designed to deliver NO during different modes of ventilatory support; however, not all meet the requirements necessary to administer NO precisely and without the risk of increased formation of nitrogen dioxide. The amount of nitrogen dioxide formation depends on contact time and concentrations of NO and oxygen (Table 1); hence, delivery of NO to the inspired gas at the most distal point in the ventilator circuit minimizes the time available for oxidation.

This strategy can be realized in *pediatric ventilators* with a continuous gas flow and high gas velocity, where any gas added to the inspiratory limp will be inspired within seconds and NO_2 formation will be minimal (LAGUENIE et al. 1993). The continuous flow of the NO mixture can be metered with precision rotameters and will lead to the delivery of constant NO concentrations to the infant, independent of the minute ventilation. Changes of the desired inspired NO concentration must be regulated by the NO flow, which can be calculated with the formula above.

In *adult ventilators*, the gas flow through the circuit depends on tidal volume, flow characteristics (decelerating flow), respiratory rate, and mode of ventilation (synchronized intermittent mandatory ventilation, pressure support ventilation). Changes of minute ventilation and ventilator settings may alter the volume of the gas NO is diluted into and, thus, the final NO concentration in the delivered tidal volume and the contact time between NO and oxygen (MOORS et al. 1994).

To overcome these problems, several custom-designed delivery systems have been described (YOUNG and DYAR 1996; IMANAKA et al. 1997; KACMAREK 1998). In a *pre-mixing system*, NO is mixed with N_2 or air, and the NO mixture source is connected, together with an air/oxygen blender, to the low- or high-pressure inlet port of the ventilator (BLOMQVIST et al. 1993; GERLACH et al. 1993b; STENQVIST et al. 1993; TIBBALLS et al. 1993; BIGATELLO et al. 1994; CHANNICK et al. 1994; MOORS et al. 1994; WESSEL et al. 1994; MCINTYRE et al. 1995; NISHIMURA et al. 1995; PUTENSEN et al. 1995). If a Siemens Servo 900C ventilator is used, the gas supply to the high-pressure input can be disconnected, and the low-pressure input can be connected to the NO mixture source and to an air/oxygen blender. It is important that the breathing gas and the NO mixture are blended before the low-pressure input; if the NO mixture alone is connected to the low-pressure input and the breathing gas is left connected to the high pressure input, there is a risk that the reservoir could be filled with high concentrations of NO. To adjust the flow of breathing gas and NO precisely, custom-designed mass-flow-controlled NO admixture to ensure constant NO concentrations with changing flow of breathing gas was described (STENQVIST et al. 1993). After pre-mixing, the ventilator draws gas from the reservoir below containing-constant concentrations of NO diluted in the breathing gas; thus, delivery of NO to the lung will be independent of minute ventilation, mode of ventilation, or flow waveforms. Of concern with this method is the increased formation of NO_2 at high F_IO_2 (NISHIMURA et al. 1995). The rate of NO oxidation conversion was determined in a test lung model with Siemens Servo 900C and Puritan-Bennett 7200 ventilators using a NO/N_2 or NO/air source connected to the high-pressure port of the ventilator, and NO_2 was measured via chemiluminescence 20 cm from the Y piece. NO_2 concentration increased with increased F_IO_2 and NO concentration, decreased minute ventilation, blending with air, and increased lung volumes. NO_2 was below 1 ppm in the Puritan-Bennet 7200 but was greater with the Siemens Servo 900C at similar settings (NO blended with nitrogen, 5 l min ventilation, $FiO_2 = 0.87$, NO $\geq$ 70 ppm) due to the large internal reservoir. When NO was diluted with air, clinically important NO_2 values were measured with both ventilators at high NO and F_IO_2, but less than 2 ppm NO_2 was observed with the Puritan-Benett 7200 when NO was under 20 ppm.

To reduce the inspired concentration of NO_2, incorporation of soda-lime canisters in the inspiratory limb of the ventilator has been described (BLOMQVIST et al. 1993; STENQVIST et al. 1993). The efficacy of nitrogen dioxide removal by soda lime varies with the type of preparation and the potassium hydroxide content (MOORS et al. 1994); however, some questioned the use of soda lime, since NO was removed as well, and absorption depends on the type of indicator and not on soda lime itself (PICKETT et al. 1994). Others reported removal of 60–75% NO_2 and 7% NO by an ethyl violet indicator/soda lime absorber (ISHIBE et al. 1995) or used NO_2-selective scavengers (GILLY et al. 1996).

Another method is the use of a *continuous flow of NO into the inspiratory limb* of the ventilator (STENQVIST et al. 1993; TIBBALLS et al. 1993; WATKINS et al. 1993; YOUNG et al. 1994a; LU et al. 1995; SAMAMA et al. 1995; DAY et al. 1996a; LOWSON et al. 1996). The advantage of this technique is that it is easy to use, applicable to every type of ventilator and, due to the shorter contact time between NO and oxygen, oxidation of NO is less important. As with pediatric ventilators, the mean inspired NO concentration can be calculated from the minute ventilation and NO flow rate. However, there are several clinically important aspects. In contrast to continuous-flow pediatric ventilators, where continuous admixture of NO does not accumulate in the tubing, continuous injection of NO into phasic-flow ventilators results in NO accumulation during expiration in addition to marked fluctuation of inspiratory NO concentration, which is enhanced by both decelerating flow and increased minute ventilation (SYDOW et al. 1997a). Thus, the inspiratory peak NO concentration may markedly exceed the calculated dose (Fig. 7) and, with prolonged expiration time, accumulation of oxygen-free gas may considerably decrease the F_IO_2 (TIBBALLS et al. 1993; IMANAKA et al. 1997). The fluctuations of inspired NO

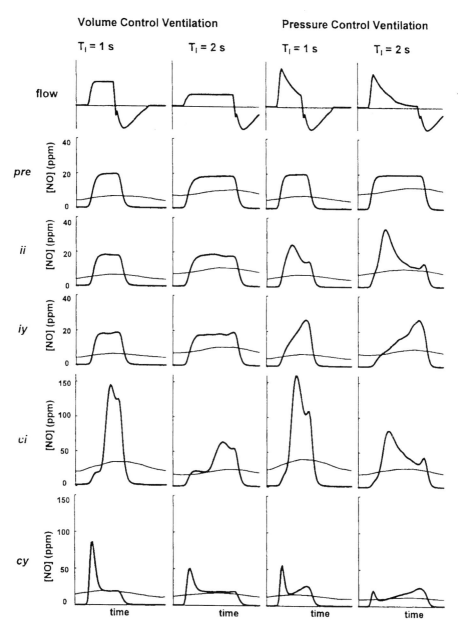

Fig. 7. Nitric oxide sampled at simulated mid-trachea during volume-controlled ventilation (*VCV*) and pressure-controlled ventilation (*PCV*) (respiratory rate rate = 15 breaths/min, tidal volume = 500 ml, target [NO] = 20 ppm). *Thick lines* represent [NO] measured using the fast-response analyzer, and *thin lines* represent [NO] measured using the slow-response analyzer. *ci*, continuous injection into inspiratory limb; *cy*, continuous injection into Y piece; *ii*, inspiratory-phase injection into inspiratory limb; *iy*, inspiratory-phase injection into Y piece; *pre*, premixing systems (IMANAKA et al. 1997)

concentration can only be detected with fast-response chemiluminescence analyzers and not with slow-response chemiluminescence or electrochemical analyzers. Additionally, any change in minute ventilation requires adjustment of the NO flow to ensure constant NO concentrations. Therefore, this method is only applicable during controlled ventilation with constant minute volumes in patients without spontaneous breathing (SYDOW et al. 1997a). Finally, the inspired oxygen concentration must be monitored by an additional analyzer, since the respirator display will not reflect the true value.

If a continuous flow of NO is injected into the Y piece, NO accumulation during expiration can be avoided. However, with this technique, measurement of NO is difficult due to contamination with exhaled gas. Calculation of the inspired NO concentration depends on minute ventilation, NO flow, inspiratory-to-expiratory ratio (I:E ratio), and the geometry of the Y piece and connectors (YOUNG and DYAR 1996); thus, peak inspiratory concentrations may markedly exceed those calculated (Fig. 7). If a continuous flow of NO is delivered directly into the trachea, there might be accumulation of oxygen-free gas during prolonged expiration.

An *inspiratory timed-injection technique*, which overcomes some problems of continuous delivery systems, was described by ROSSAINT et al. (1993) and PUYBASSET et al. (1994b). Using a cylinder containing 800ppm NO in N_2, the flow of the NO source was adjusted with a flow rotameter and delivered to a nebulizer attached to a Siemens Servo 900C ventilator. Triggered by the ventilator, the pneumatic-valve nebulizer delivers NO only during inspiration into the inspiratory limb of the ventilator. Changing the minute ventilation requires readjustment of the NO flow, since the nebulizer flow is constant. With this technique, constant and precise NO concentrations can be achieved during controlled ventilation with square-wave flow patterns; however, with decelerating flows (pressure-controlled ventilation), the peak inspiratory NO concentration exceeds the estimated value when measured with fast-response chemiluminescence. Other groups also reported considerable fluctuations of NO concentration with this technique during constant-flow ventilation in a test lung model where the desired NO concentration was about 30ppm and that measured ranged from 11ppm to 42ppm (SYDOW et al. 1998).

When flow-controlled injection systems are used, a constant inspired NO concentration can be maintained despite changing minute and tidal volumes, peak inspiratory flow rates, ventilator rates, and different flow profiles (YOUNG 1994; SYDOW et al. 1997b). There are several *company-designed delivery systems* now available, which inject NO into the inspiratory limb of the ventilator in proportion to the inspiratory flow (SYDOW et al. 1998). The I-Novent (Ohmeda, USA) and the Pulmonox mini (Messer Griesheim, Austria) can be used with each type of ventilator. The NOdomo (Dräger, Germany) requires flow-based data from the ventilator and, therefore, can only be used with Dräger ventilators (Evita, Babylog). Due to the signal-processing time and external valve-response time, NO delivery may be delayed at high inspiratory flow, resulting in inspiratory NO concentration fluctuation; however, the mean inspiratory NO concentrations are stable and accurate, regardless of the flow levels and patterns. In the Servo 300 ventilator (Siemens-Elema, Sweden) the NO-delivery system is incorporated, using an additional gas-inlet port for the NO source, which is blended with air/oxygen within the respirator. In contrast to the devices above, the F_IO_2 is accurately displayed by the ventilator, i.e. additional oxygen analyzers are not mandatory. A comparison of NO_2 formation in the Servo 300 and NOdomo during ventilation with different tidal volumes and NO concentrations revealed that, if NO is below 10ppm, NO_2 formation remains below the toxic range even for small tidal volumes. For NO concentrations of at least 10ppm, NO_2 formation is higher in the Servo 300 due to the longer contact after mixing within the respirator. Furthermore, measurement of NO_2 in the mixing chamber behind the expiratory valve leads to much higher NO_2 levels compared with the inspired concentrations, which may be misinterpreted (KUHLEN et al. 1997). All delivery systems are equipped with an alarm and monitor NO and NO_2 with electrochemical analyzers.

III. Monitoring NO Inhalation

The two approaches that are currently used to measure NO and NO_2 clinically and experimentally are chemiluminescence and electrochemical analyzers. Each has distinct advantages and disadvantages for use in clinical settings (KACMAREK 1998).

1. Chemiluminescence

The detection of NO by chemiluminescence is based on the observation that ozone interacts with NO to generate light; the luminescence, amplified by a sensitive photomultiplier tube, is directly proportional to the concentration of NO. NO_2 is measured after conversion to NO in a thermal chamber containing molybdenum or chromium trioxide. The reaction formulas are:

$$NO + O_3 \rightarrow NO_2^* + O_2 \qquad (3)$$
$$NO_2^* \rightarrow NO_2 + h\nu \qquad (4)$$
$$3NO_2 + Mo + 325°C \rightarrow 3NO + MoO_3 \qquad (5)$$

where * indicates a highly excited state and $h\nu$ indicates electromagnetic radiation.

The sum of NO and converted NO_2 is referred to as NO_x (NO_x = NO + NO_2). Most analyzers measure NO and NO_x simultaneously and display NO_x, NO, and NO_2. Chemiluminescent analyzers are very accurate and precise, but they are large, expensive, and cumbersome to use. While measurement of NO as an atmospheric pollutant or during continuous injection of NO in a continuous-flow pediatric ventilator can be performed with every chemiluminescence analyzer, during phasic-flow mechanical ventilation, rapid changes of the NO concentrations and gas flow may complicate accurate measurement of NO due to the long transport delay and response time of the analyzer. Transport delay is defined as the time delay between receipt of the initial signal by the analyzer and the initial display of NO. Response time is the time necessary to display a specific percentage of the maximum NO concentration (a 95% response time of 4s means that it takes 4s to display 95% of the maximal NO concentration). Thus, analyzers can only precisely measure the NO concentration within a single tidal volume, when the response time is shorter than its duration. Examples of fast-response analyzers with low sample flow rates (100–300 ml/min), short response times (≤ 0.2s), and high sensitivity (<1 ppb NO) which allow breath-by-breath and exhaled gas measurements are the 280 NOA (Sievers Instruments Inc., Boulder, Colorado) and CLD 77AM (ECO Physics, Switzerland). Analyzers with slower response times (>3s) and higher sample flow rates include the CLD 700 AL (ECO Physics) and the CLA 510S (Horiba Industries, Kyoto, Japan). The devices tend to underestimate peak NO concentrations, and overestimate minimal NO concentrations. As sidestream analyzers, the sample flow is removed from the breathing circuit, which might be a problem with fixed tidal volumes and high sample flow rates.

2. Electrochemical Analyzers

In electrochemical cells, a current proportional to the oxidation of NO to nitric acid is generated. The electrochemical cell consists of sensing, counter, and reference electrodes, which provide a bias voltage to keep the sensing electrodes at the correct operating voltage. To detect NO, the following reactions occur at the sensor and counter electrodes, respectively (KACMAREK 1998):

$$NO + 2H_2O \rightarrow HNO_3 + 3H^+ + 3e^- \tag{6}$$

$$O_2 + 4H^+ + 4e^- \rightarrow 2H_2O \tag{7}$$

The overall reaction for the NO cell is:

$$4NO + 2H_2O + 3O_2 \rightarrow 4HNO_3^- \tag{8}$$

At the NO_2 sensing electrode, NO_2 is reduced to NO:

$$NO_2 + 2H^+ + 2e^- \rightarrow NO + H_2O \tag{9}$$

At the counter electrode, water oxidizes:

$$2H_2O \rightarrow 4H^+ + 4e^- + O_2 \tag{10}$$

The overall reaction for the NO_2 cell is:

$$2NO_2 \rightarrow 2NO + O_2 \tag{11}$$

The analyzers are sensitive to water vapor, which may shorten the half life of the electrodes and increases the likelihood of inaccurate measurements. The accuracy of measurements is less than that of the chemiluminescence analyzers

Inhalation Therapy with Nitric Oxide Gas

at low NO concentrations (1 ppm); thus, measurements at the ppb level cannot be performed. Due to the long response time (30s), the devices are not suitable for single-breath or exhaled NO measurements. However, they are small, portable, less expensive, and accurate with specific NO-delivery systems.

Abbrevations

ALI	acute lung injury
ARDS	acute respiratory-distress syndrome
cGMP	cyclic guanosine 3',5'-monophosphate
CI	cardiac index
CO	cardiac output
ECMO	extracorporeal membrane oxygenation
F_IO_2	inspired oxygen fraction
GSNO	S-nitrosoglutathione
Hb	hemoglobin
Hb[Fe_{II}]	deoxyhemoglobin
Hb[Fe_{II}]NO	nitrosyl hemoglobin
HFOV	high-frequency oscillatory ventilation
HPV	hypoxic pulmonary vasoconstriction
ICU	intensive-care unit
IL	interleukin
iNOS	inducible nitric oxide synthase
L-NAME	N^G-nitro-L-arginine methyl ester
MAP	mean arterial pressure
metHb	methemoglobin
MIGET	multiple inert-gas-elimination technique
NADH	reduced nicotinamide adenine dinucleotide
NO	nitric oxide
NOS	nitric oxide synthase
OI	oxygen index
PaO_2	arterial oxygen tension
PAP	pulmonary-artery pressure
PEEP	positive end expiratory pressure
PMN	polymorphonuclear neutrophil
PPHN	persistent pulmonary hypertension of the newborn
PVR	pulmonary vascular resistance
RSNO	S-nitrosothiol
RVEDVI	right-ventricular end-diastolic-volume index
RVEF	right-ventricular ejection fraction
RVESVI	right-ventricular end-systolic-volume index
SVR	systemic vascular resistance

References

Abman SH, Chatfield BA, Hall SL, McMurtry IF (1990) Role of endothelium-derived relaxing factor during transition of pulmonary circulation at birth. Am J Physiol 259:1921–1927

Adams DH, Shaw S (1994) Leucocyte-endothelial interactions and regulation of leucocyte migration. Lancet 343:831–836

Adatia I, Thompson J, Landzberg M, Wessel DL (1993) Inhaled nitric oxide in chronic obstructive lung disease. Lancet 341:307–308

Adatia I, Lillehei C, Arnold JH, Thompson JE, Palazzo R, Fackler JC, Wessel DL (1994) Inhaled nitric oxide in the treatment of postoperative graft dysfunction after lung transplantation. Ann Thorac Surg 57:1311–1318

Adnot S, Kouyoumdjian C, Defouilloy C, Andrivet P, Sediame S, Herigault R, Fratacci MD (1993) Hemodynamic and gas exchange responses to infusion of acetylcholine and inhalation of nitric oxide in patients with chronic obstructive lung disease and pulmonary hypertension. Am Rev Respir Dis 148:310–316

Adrie C, Bloch KD, Moreno PR, Hurford WE, Guerrero JL, Holt R, Zapol WM, Gold HK, Semigran MJ (1996) Inhaled nitric oxide increases coronary artery patency after thrombolysis. Circulation 94:1919–1926

Ashbaugh DG, Bigelow DB, Petty TL, Levine BE (1967) Acute respiratory distress in adults. Lancet 2:319–323

Assreuy J, Cunha FQ, Liew FY, Moncada S (1993) Feedback inhibition of nitric oxide synthase activity by nitric oxide. Br J Pharmacol 108:833–837

Atz AM, Adatia I, Jonas RA, Wessel DL (1996) Inhaled nitric oxide in children with pulmonary hypertension and congenital mitral stenosis. Am J Cardiol 77:316–319

Barefield ES, Karle VA, Phillips JB III, Carlo WA (1996) Inhaled nitric oxide in term infants with hypoxemic respiratory failure. J Pediatr 129:279–286

Bauer MA, Utell MJ, Morrow PE, Speers DM, Gibb FR (1986) Inhalation of 0.30 ppm nitrogen dioxide potentiates exercise-induced bronchospasm in asthmatics. Am Rev Respir Dis 134:1203–1208

Beckman JS (1996) The physiological and pathophysiological chemistry of nitric oxide. In: Nitric oxide. Principles and actions. 1–82. Edited by Lancaster, J. San Diego, Academic Press

Beckman JS, Koppenol WH (1996) Nitric oxide, superoxide, and peroxynitrite: the good, the bad, and ugly. Am J Physiol 271:1424–1437

Bender KA, Alexander JA, Enos JM, Skimming JW (1997) Effects of inhaled nitric oxide in patients with hypoxemia and pulmonary hypertension after cardiac surgery. Am J Crit Care 6:127–131

Benzing A, Geiger K (1994) Inhaled nitric oxide lowers pulmonary capillary pressure and changes longitudinal distribution of pulmonary vascular resistance in patients with acute lung injury. Acta Anaesthesiol Scand 38:640–645

Benzing A, Brautigam P, Geiger K, Loop T, Beyer U, Moser E (1995) Inhaled nitric oxide reduces pulmonary transvascular albumin flux in patients with acute lung injury. Anesthesiology 83:1153–1161

Benzing A, Loop T, Mols G, Geiger K (1996) Effect of inhaled nitric oxide on venous admixture depends on cardiac output in patients with acute lung injury and acute respiratory distress syndrome. Acta Anaesthesiol Scand 40:466–474

Benzing A, Mols G, Beyer U, Geiger K (1997) Large increase in cardiac output in a patient with ARDS and acute right heart failure during inhalation of nitric oxide. Acta Anaesthesiol Scand 41:643–646

Berger JI, Gibson RL, Redding GJ, Standaert TA, Clarke WR, Truog WE (1993) Effect of inhaled nitric oxide during group B streptococcal sepsis in piglets. Am Rev Respir Dis 147:1080–1086

Bichel T, Spahr-Schopfer I, Berner M, Jaeggi E, Velkovski Y, Friedli B, Kalangos A, Faidutti B, Rouge JC (1997) Successful weaning from cardiopulmonary bypass after cardiac surgery using inhaled nitric oxide. Paediatr Anaesth 7:335–339

Bigatello LM, Hurford WE, Kacmarek RM, Roberts JD Jr, Zapol WM (1994) Prolonged inhalation of low concentrations of nitric oxide in patients with severe adult respiratory distress syndrome. Effects on pulmonary hemodynamics and oxygenation. Anesthesiology 80:761–770

Blomqvist H, Wickerts CJ, Andreen M, Ullberg U, Ortqvist A, Frostell C (1993) Enhanced pneumonia resolution by inhalation of nitric oxide? Acta Anaesthesiol Scand 37:110–114

Borland CD, Higenbottam TW (1989) A simultaneous single breath measurement of pulmonary diffusing capacity with nitric oxide and carbon monoxide. Eur Respir J 2:56–63

Bouchet M, Renaudin MH, Raveau C, Mercier JC, Dehan M, Zupan V (1993) Safety requirement for use of inhaled nitric oxide in neonates. Lancet 341:968–969

Brett SJ, Evans TW (1998) Measurement of endogenous nitric oxide in the lungs of patients with the acute respiratory distress syndrome. Am J Respir Crit Care Med 157:993–997

Channick RN, Newhart JW, Johnson FW, Moser KM (1994) Inhaled nitric oxide reverses hypoxic pulmonary vasoconstriction in dogs. A practical nitric oxide delivery and monitoring system. Chest 105:1842–1847

Cheifetz IM, Craig DM, Kern FH, Black DR, Hillman ND, Greeley WJ, Ungerleider RM, Smith PK, Meliones JN (1996) Nitric oxide improves transpulmonary vascular mechanics but does not change intrinsic right ventricular contractility in an acute respiratory distress syndrome model with permissive hypercapnia. Crit Care Med 24:1554–1561

Chiche JD, Canivet JL, Damas P, Joris J, Lamy M (1995) Inhaled nitric oxide for hemodynamic support after postpneumonectomy ARDS. Intensive Care Med 21:675–678

Clutton-Brock J (1967) Two cases of poisoning by contamination of nitrous oxide with higher oxides of nitrogen during anaesthesia. Br J Anaesth 39:388–392

Combes X, Mazmanian M, Gourlain H, Herve P (1997) Effect of 48 hours of nitric oxide inhalation on pulmonary vasoreactivity in rats. Am J Respir Crit Care Med 156:473–477

Cueto E, Lopez-Herce J, Sanchez A, Carrillo A (1997) Life-threatening effects of discontinuing inhaled nitric oxide in children. Acta Paediatr 86:1337–1339

Cujec B, Hurst T, McCuaig R, Antecol D, Mayers I, Johnson D (1997) Inhaled nitric oxide reduction in systolic pulmonary artery pressure is less in patients with decreased left ventricular ejection fraction. Can J Cardiol 13:816–824

Cuthbertson BH, Dellinger P, Dyar OJ, Evans TE, Higenbottam T, Latimer R, Payen D, Stott SA, Webster NR, Young JD (1997) UK guidelines for the use of inhaled nitric oxide therapy in adult ICUs. American-European Consensus Conference on ALI/ARDS. Intensive Care Med 23:1212–1218

Day RW, Guarin M, Lynch JM, Vernon DD, Dean JM (1996a) Inhaled nitric oxide in children with severe lung disease: results of acute and prolonged therapy with two concentrations. Crit Care Med 24:215–221

Day RW, Lynch JM, White KS, Ward RM (1996b) Acute response to inhaled nitric oxide in newborns with respiratory failure and pulmonary hypertension. Pediatrics 98:698–705

De Marco V, Skimming J, Ellis TM, Cassin S (1994) Nitric oxide inhalation. Effects on the ovine neonatal pulmonary and systemic circulations. Chest 105:91–92

Dellinger RP, Zimmerman JL, Taylor RW, Straube RC, Hauser DL, Criner GJ, Davis K Jr, Hyers TM, Papadakos P (1998) Effects of inhaled nitric oxide in patients with acute respiratory distress syndrome: results of a randomized phase II trial. Inhaled Nitric Oxide in ARDS Study Group. Crit Care Med 26:15–23

Demirakca S, Dotsch J, Knothe C, Magsaam J, Reiter HL, Bauer J, Kuehl PG (1996) Inhaled nitric oxide in neonatal and pediatric acute respiratory distress syndrome: dose response, prolonged inhalation, and weaning. Crit Care Med 24:1913–1919

Dinh-Xuan AT, Higenbottam TW, Clelland CA, Pepke Zaba J, Cremona G, Butt AY, Large SR, Wells FC, Wallwork J (1991) Impairment of endothelium-dependent pulmonary-artery relaxation in chronic obstructive lung disease. N Engl J Med 324:1539–1547

Dotsch J, Demirakca S, Hamm R, Knothe C, Bauer J, Kuhl PG, Rascher W (1997) Extracorporeal circulation increases nitric oxide-induced methemoglobinemia in vivo and in vitro. Crit Care Med 25:1153–1158

Falke K, Rossaint R, Pison U, Slama K, Lopez F, Santak B, Zapol WM (1991) Inhaled nitric oxide selectively reduces pulmonary hypertension in severe ARDS and improves gas exchange as well as right heart ejection fraction: a case report. Am Rev Respir Dis 143:Suppl A248

Fierobe L, Brunet F, Dhainaut JF, Monchi M, Belghith M, Mira JP, Dall'ava Santucci J, Dinh-Xuan AT (1995) Effect of inhaled nitric oxide on right ventricular function in adult respiratory distress syndrome. Am J Respir Crit Care Med 151:1414–1419

Finer NN (1997) Inhaled nitric oxide in neonates. Arch Dis Child Fetal Neonatal Ed 77:81–84

Finer NN, Etches PC, Kamstra B, Tierney AJ, Peliowski A, Ryan CA (1994) Inhaled nitric oxide in infants referred for extracorporeal membrane oxygenation: dose response. J Pediatr 124:302–308

Foubert L, Fleming B, Latimer R, Jonas M, Oduro A, Borland C, Higenbottam T (1992) Safety guidelines for use of nitric oxide. Lancet 339:1615–1616

Fratacci MD, Frostell CG, Chen TY, Wain JC Jr, Robinson DR, Zapol WM (1991) Inhaled nitric oxide. A selective pulmonary vasodilator of heparin-protamine vasoconstriction in sheep. Anesthesiology 75:990–999

Frostell C, Fratacci MD, Wain JC, Jones R, Zapol WM (1991) Inhaled nitric oxide. A selective pulmonary vasodilator reversing hypoxic pulmonary vasoconstriction. Circulation 83:2038–2047

Frostell CG, Blomqvist H, Hedenstierna G, Lundberg J, Zapol WM (1993a) Inhaled nitric oxide selectively reverses human hypoxic pulmonary vasoconstriction without causing systemic vasodilation. Anesthesiology 78:427–435

Frostell CG, Lonnqvist PA, Sonesson SE, Gustafsson LE, Lohr G, Noack G (1993b) Near fatal pulmonary hypertension after surgical repair of congenital diaphragmatic hernia. Successful use of inhaled nitric oxide. Anaesthesia 48:679–683

Gaston B, Reilly J, Drazen JM, Fackler J, Ramdev P, Arnelle D, Mullins ME, Sugarbaker DJ, Chee C, Singel DJ, et al. (1993) Endogenous nitrogen oxides and bronchodilator S-nitrosothiols in human airways. Proc Natl Acad Sci USA 90:10957–10961

Gaston B, Drazen JM, Loscalzo J, Stamler JS (1994) The biology of nitrogen oxides in the airways. Am J Respi Crit Care Med 149:538–551

Gerlach H, Pappert D, Lewandowski K, Rossaint R, Falke KJ (1993a) Long-term inhalation with evaluated low doses of nitric oxide for selective improvement of oxygenation in patients with adult respiratory distress syndrome. Intensive Care Med 19:443–449

Gerlach H, Rossaint R, Pappert D, Falke KJ (1993b) Time-course and dose-response of nitric oxide inhalation for systemic oxygenation and pulmonary hypertension in patients with adult respiratory distress syndrome. Eur J Clin Invest 23:499–502

Gerlach H, Rossaint R, Pappert D, Knorr M, Falke KJ (1994) Autoinhalation of nitric oxide after endogenous synthesis in nasopharynx. Lancet 343:518–519

Gilly H, Krebs C, Nowotny T (1996) Reduction of NO2 by a NO2-selective scavenger. Acta Anaesthesiol Scand Suppl 109:83–84

Greenbaum R, Bay J, Hargreaves MD, Kain ML, Kelman GR, Nunn JF, Prys-Roberts C, Siebold K (1967) Effects of higher oxides of nitrogen on the anaesthetized dog. Br J Anaesth 39:393–404

Guidot DM, Repine MJ, Hybertson BM, Repine JE (1995) Inhaled nitric oxide prevents neutrophil-mediated, oxygen radical-dependent leak in isolated rat lungs. Am J Physiol 269:2–5

Gustafsson LE, Leone AM, Persson MG, Wiklund NP, Moncada S (1991) Endogenous nitric oxide is present in the exhaled air of rabbits, guinea pigs and humans. Biochem Biophys Res Commun 181:852–857

Haddad IY, Crow JP, Hu P, Ye Y, Beckman J, Matalon S (1994a) Concurrent generation of nitric oxide and superoxide damages surfactant protein A. Am J Physiol 267:242–249

Haddad IY, Pataki G, Hu P, Galliani C, Beckman JS, Matalon, S (1994b) Quantitation of nitrotyrosine levels in lung sections of patients and animals with acute lung injury. J Clin Invest 94:2407–2413

Haddad IY, Zhu S, Crow J, Barefield E, Gadilhe T, Matalon S (1996) Inhibition of alveolar type II cell ATP and surfactant synthesis by nitric oxide. Am J Physiol 270:898–906

Higenbottam T, Pepke-Zaba J, Scott J, Woolman P, Coutts C, Wallwork J (1988) Inhaled "endothelium derived relaxing factor" (EDRF) in primary pulmonary hypertension. Am Rev Respir Dis 137:107

Högman M, Frostell CG, Hedenstrom H, Hedenstierna G (1993b) Inhalation of nitric oxide modulates adult human bronchial tone. Am Rev Respir Dis 148:1474–1478

Holzmann A, Bloch KD, Sanchez LS, Filippov G, Zapol WM (1996) Hyporesponsiveness to inhaled nitric oxide in isolated, perfused lungs from endotoxin-challenged rats. Am J Physiol 271:981–986

Hu P, Ischiropoulos HH, Beckman JS, Matalon S (1994) Peroxynitrite inhibition of oxygen consumption and sodium transport in alveolar type II cells. Am J Physiol 266:628–634

Hugod C (1979) Effect of exposure to 43 ppm nitric oxide and 3.6 ppm nitrogen dioxide on rabbit lung. A light and electron microscopic study. Int Arch Occup Environ Health 42:159–167

Ignarro LJ, Buga GM, Wood KS, Byrns RE, Chaudhuri G (1987) Endothelium-derived relaxing factor produced and released from artery and vein is nitric oxide. Proc Natl Acad Sci USA 84:9265–9269

Imanaka H, Hess D, Kirmse M, Bigatello LM, Kacmarek RM, Steudel W, Hurford WE (1997) Inaccuracies of nitric oxide delivery systems during adult mechanical ventilation. Anesthesiology 86:676–688

Ischiropoulos H, Zhu L, Beckman JS (1992) Peroxynitrite formation from macrophage-derived nitric oxide. Arch Biochem Biophys 298:446–451

Ishibe T, Sato T, Hayashi T, Kato N, Hata T (1995) Absorption of nitrogen dioxide and nitric oxide by soda lime. Br J Anaesth 75:330–333

Jellinek H, Krafft P, Hiesmayr M, Steltzer H (1997) Measurement of right ventricular performance during apnea in patients with acute lung injury. J Trauma 42:1062–1067

Jourdqheuil D, Mai CT, Laroux FS, Wink DA, Grisham MB (1998) The reaction of S-nitrosoglutathione with superoxide. Biochem Biophys Res Commun 244:525–530

Kacmarek RM (1998) The administration and measurement of nitric oxide during mechanical ventilation. In: Yearbook of intensive care and emergency medicine, 545–560. Edited by Vincent, JL Berlin, Springer

Kacmarek RM, Ripple R, Cockrill BA, Bloch KJ, Zapol WM, Johnson DC (1996) Inhaled nitric oxide. A bronchodilator in mild asthmatics with methacholine-induced bronchospasm. Am J Respir Crit Care Med 153:128–135

Keaney JF Jr, Simon DI, Stamler JS, Jaraki O, Scharfstein J, Vita JA, Loscalzo J (1993) NO forms an adduct with serum albumin that has endothelium-derived relaxing factor-like properties. J Clin Invest 91:1582–1589

Keefer LK, Wink DA (1996) DNA damage and nitric oxide. Adv Exp Med Biol 387:177–185

Kelly KP, Busch T, Gerlach H, Rossaint R (1996) Arterial oxygenation during replacement of a physiological concentration of nitric oxide in intubated patients. Intensive Care Med S137

Kiff RJ, Moss DW, Moncada S (1994) Effect of nitric oxide gas on the generation of nitric oxide by isolated blood vessels: implications for inhalation therapy. Br J Pharmacol 113:496-498

Kinsella JP, Neish SR, Shaffer E, Abman SH (1992) Low-dose inhalation nitric oxide in persistent pulmonary hypertension of the newborn. Lancet 340:819-820

Kinsella JP, Truog WE, Walsh WF, Goldberg RN, Bancalari E, Mayock DE, Redding GJ, De Lemos RA, Sardesai S, McCurnin DC, Moreland SG, Cutter GR, Abman SH (1997) Randomized, multicenter trial of inhaled nitric oxide and high-frequency oscillatory ventilation in severe, persistent pulmonary hypertension of the newborn. J Pediatr 131:55-62

Koizumi T, Gupta R, Banerjee M, Newman JH (1994) Changes in pulmonary vascular tone during exercise. Effects of nitric oxide (NO) synthase inhibition, L-arginine infusion, and NO inhalation. J Clin Invest 94:2275-2282

Kooy NW, Royall JA, Ye YZ, Kelly DR, Beckman JS (1995) Evidence for in vivo peroxynitrite production in human acute lung injury. Am J Respir Crit Care Med 151:1250-1254

Krafft P, Fridrich P, Fitzgerald RD, Koc D, Steltzer H (1996a) Effectiveness of nitric oxide inhalation in septic ARDS. Chest 109:486-493

Krafft P, Fridrich P, Pernerstorfer T, Fitzgerald RD, Koc D, Schneider B, Hammerle AF, Steltzer H (1996b) The acute respiratory distress syndrome: definitions, severity and clinical outcome. An analysis of 101 clinical investigations. Intensive Care Med 22:519-529

Kuhlen R, Busch T, Volckers U, Gerlach H, Falke K, Rossaint R (1997) Nitrogen dioxide (NO2) production for different doses of inhaled nitric oxide (NO) during mechanical ventilation with different tidal volumes using two prototypes for the administration of NO. Crit Care 1 [Suppl. 1], 27-28

Laguenie G, Berg A, Saint-Maurice JP, Dinh-Xuan AT (1993) Measurement of nitrogen dioxide formation from nitric oxide by chemiluminescence in ventilated children. Lancet 341:969

Laval F, Wink DA, Laval J (1997) A discussion of mechanisms of NO genotoxicity: implication of inhibition of DNA repair proteins. Rev Physiol Biochem Pharmacol 131:175-191

Lehnert BE, Archuleta DC, Ellis T, Session WS, Lehnert NM, Gurley LR, Stavert DM (1994) Lung injury following exposure of rats to relatively high mass concentrations of nitrogen dioxide. Toxicology 89:239-277

Lewandowski K, Busch T, Lewandowski M, Keske U, Gerlach H, Falke KJ (1996) Evidence of nitric oxide in the exhaled gas of Asian elephants (*Elephas maximus*). Respir Physiol 106:91-98

Lewandowski K, Busch T, Lohbrunner H, Rensing S, Keske U, Gerlach H, Falke KJ (1998) Low nitric oxide concentrations in exhaled gas and nasal airways of mammals without paranasal sinuses. J Appl Physiol 82:405-410

Loh E, Stamler JS, Hare JM, Loscalzo J, Colucci WS (1994) Cardiovascular effects of inhaled nitric oxide in patients with left ventricular dysfunction. Circulation 90: 2780-2785

Lowson SM, Rich GF, McArdle PA, Jaidev J, Morris GN (1996) The response to varying concentrations of inhaled nitric oxide in patients with acute respiratory distress syndrome. Anesth Analg 82:574-581

Lu Q, Mourgeon E, Law Koune JD, Roche S, Vezinet C, Abdennour L, Vicaut E, Puybasset L, Diaby M, Coriat P, et al. (1995) Dose-response curves of inhaled nitric oxide with and without intravenous almitrine in nitric oxide-responding patients with acute respiratory distress syndrome. Anesthesiology 83:929-943

Luhr O, Aardal S, Nathorst-Westfelt U, Berggren L, Johansson LA, Wahlin L, Frostell C (1998) Pulmonary function in adult survivors of severe acute lung injury treated with inhaled nitric oxide. Acta Anaesthesiol Scand 42:391–398

Lundberg JO, Farkas Szallasi T, Weitzberg E, Rinder J, Lidholm J, Anggaard A, Hokfelt T, Lundberg JM, Alving K (1995) High nitric oxide production in human paranasal sinuses. Nat Med 1:370–373

Lundberg JO, Settergren G, Gelinder S, Lundberg JM, Alving K, Weitzberg E (1996) Inhalation of nasally derived nitric oxide modulates pulmonary function in humans. Acta Physiol Scand 158:343–347

Lundin S, Westfelt UN, Stenqvist O, Blomqvist H, Lindh A, Berggren L, Arvidsson S, Rudberg U, Frostell CG (1996) Response to nitric oxide inhalation in early acute lung injury. Intensive Care Med 22:728–734

Lundin S, Mang H, Smithies M, Stenqvist O, Frostell C (1997) Inhalation of nitric oxide in acute lung injury: preliminary results of a European multicenter study. Intensive Care Med 23:S2

Manktelow C, Bigatello LM, Hess D, Hurford WE (1997) Physiologic determinants of the response to inhaled nitric oxide in patients with acute respiratory distress syndrome. Anesthesiology 87:297–307

Mansouri A (1985) Methemoglobinemia. Am J Med Sci 289:200–209

Matalon S, De Marco V, Haddad IY, Myles C, Skimming JW, Schurch S, Cheng S, Cassin S (1996) Inhaled nitric oxide injures the pulmonary surfactant system of lambs in vivo. Am J Physiol 270:L273–80

Matthay MA, Pittet JF, Jayr C (1998) Just say NO to inhaled nitric oxide for the acute respiratory distress syndrome. Crit Care Med 26:1–2

McIntyre RC Jr, Moore FA, Moore EE, Piedalue F, Haenel JS, Fullerton DA (1995) Inhaled nitric oxide variably improves oxygenation and pulmonary hypertension in patients with acute respiratory distress syndrome. J Trauma 39:418–425

Michael JR, Barton RG, Saffle JR, Mone M, Markewitz BA, Hillier K, Elstad MR, Campbell EJ, Troyer BE, Whatley RE, Liou TG, Samuelson WM, Carveth HJ, Hinson DM, Morris SE, Davis BL, Day RW (1998) Inhaled nitric oxide versus conventional therapy: effect on oxygenation in ARDS. Am J Respir Crit Care Med 157:1372–1380

Miller OI, Tang SF, Keech A, Celermajer DS (1995) Rebound pulmonary hypertension on withdrawal from inhaled nitric oxide. Lancet 346:51–52

Mira JP, Monchi M, Brunet F, Fierobe L, Dhainaut JF, Dinh-Xuan AT (1994) Lack of efficacy of inhaled nitric oxide in ARDS. Intensive Care Med 20:532

Miyamoto K, Aida A, Nishimura M, Nakano T, Kawakami Y, Ohmori Y, Ando S, Ichida T (1994) Effects of humidity and temperature on nitrogen dioxide formation from nitric oxide. Lancet 343:1099–1100

Moors AH, Pickett JA, Mahmood N, Latimer RD, Oduro A (1994) Nitric oxide administration. Anaesth Intensive Care 22:310–312

Moreno JJ, Pryor WA (1992) Inactivation of alpha 1-proteinase inhibitor by peroxynitrite. Chem Res Toxicol 5:425–431

Motterlini R, Vandegriff KD, Winslow RM (1996) Hemoglobin-nitric oxide interaction and its implications. Transfus Med Rev 10:77–84

Mourgeon E, Puybasset L, Law-Koune JD, Lu Q, Abdennour L, Gallart L, Malassine P, Rao GU, Cluzel P, Bennani A, Coriat P, Rouby JJ (1997) Inhaled nitric oxide in acute respiratory distress syndrome with and without septic shock: a dose-response study. Crit Care 1:25–29

Narula P, Xu J, Kazzaz JA, Robbins CG, Davis JM, Horowitz S (1998) Synergistic cytotoxicity from nitric oxide and hyperoxia in cultured lung cells. Am J Physiol 274:411–416

Neonatal Inhaled Nitric Oxide Study Group (1997a) Inhaled nitric oxide in full-term and nearly full-term infants with hypoxic respiratory failure. The Neonatal Inhaled Nitric Oxide Study Group. N Engl J Med 336:597–604

Neonatal Inhaled Nitric Oxide Study Group (1997b) Inhaled nitric oxide and hypoxic respiratory failure in infants with congenital diaphragmatic hernia. The Neonatal Inhaled Nitric Oxide Study Group (NINOS). Pediatrics 99:838–845

Nishimura M, Hess D, Kacmarek RM, Ritz R, Hurford WE (1995) Nitrogen dioxide production during mechanical ventilation with nitric oxide in adults. Effects of ventilator internal volume, air versus nitrogen dilution, minute ventilation, and inspired oxygen fraction. Anesthesiology 82:1246–1254

Numata M, Suzuki S, Miyazawa N, Miyashita A, Nagashima Y, Inoue S, Kaneko T, Okubo T (1998) Inhibition of inducible nitric oxide synthase prevents LPS-induced acute lung injury in dogs. J Immunol 160:3031–3037

Palmer RM, Ferrige AG, Moncada S (1987) Nitric oxide release accounts for the biological activity of endothelium-derived relaxing factor. Nature 327:524–526

Pappert D, Busch T, Gerlach H, Lewandowski K, Radermacher P, Rossaint R (1995) Aerosolized prostacyclin versus inhaled nitric oxide in children with severe acute respiratory distress syndrome. Anesthesiology 82:1507–1511

Pepke-Zaba J, Higenbottam TW, Dinh-Xuan AT, Stone D, Wallwork J (1991) Inhaled nitric oxide as a cause of selective pulmonary vasodilatation in pulmonary hypertension. Lancet 338:1173–1174

Persson MG, Wiklund NP, Gustafsson LE (1993) Endogenous nitric oxide in single exhalations and the change during exercise. Am Rev Respir Dis 148:1210–1214

Pfeffer KD, Ellison G, Robertson D, Day RW (1996) The effect of inhaled nitric oxide in pediatric asthma. Am J Respir Crit Care Med 153:747–751

Pickett JA, Moors AH, Latimer RD, Mahmood N, Ghosh S, Oduro A (1994) The role of soda lime during administration of inhaled nitric oxide. Br J Anaesth 72:683–685

Pison U, Lopez FA, Heidelmeyer CF, Rossaint R, Falke KJ (1993) Inhaled nitric oxide reverses hypoxic pulmonary vasoconstriction without impairing gas exchange. J Appl Physiol 74:1287–1292

Putensen C, Rasanen J, Thomson MS, Braman RS (1995) Method of delivering constant nitric oxide concentrations during full and partial ventilatory support. J Clin Monit 11:23–31

Puybasset L, Rouby JJ, Mourgeon E, Stewart TE, Cluzel P, Arthaud M, Poete P, Bodin L, Korinek AM, Viars P (1994a) Inhaled nitric oxide in acute respiratory failure: dose-response curves. Intensive Care Med 20:319–327

Puybasset L, Stewart T, Rouby J, Cluzel P, Mourgeon E, Belin M, Arthaud M, Landault C, Viars P (1994b) Inhaled nitric oxide reverses the increase in pulmonary vascular resistance induced by permissive hypercapnia in patients with acute respiratory distress syndrome. Anesthesiology 80:1254–1267

Puybasset L, Rouby JJ, Mourgeon E, Cluzel P, Souhil Z, Law Koune JD, Stewart T, Devilliers C, Lu Q, Roche S, et al. (1995) Factors influencing cardiopulmonary effects of inhaled nitric oxide in acute respiratory failure. Am J Respir Crit Care Med 152:318–328

Radermacher P, Santak B, Becker H, Falke KJ (1989) Prostaglandin E1 and nitroglycerin reduce pulmonary capillary pressure but worsen ventilation-perfusion distributions in patients with adult respiratory distress syndrome. Anesthesiology 70:601–606

Rasmussen TR, Kjaergaard SK, Tarp U, Pedersen OF (1992) Delayed effects of NO2 exposure on alveolar permeability and glutathione peroxidase in healthy humans. Am Rev Respir Dis 146:654–659

Rich GF, Murphy GD Jr, Roos CM, Johns RA (1993) Inhaled nitric oxide. Selective pulmonary vasodilation in cardiac surgical patients. Anesthesiology 78:1028–1035

Robbins CG, Davis JM, Merritt TA, Amirkhanian JD, Sahgal N, Morin FC III, Horowitz S (1995) Combined effects of nitric oxide and hyperoxia on surfactant function and pulmonary inflammation. Am J Physiol 269:545–550

Robbins CG, Horowitz S, Merritt TA, Kheiter A, Tierney J, Narula P, Davis JM (1997) Recombinant human superoxide dismutase reduces lung injury caused by inhaled nitric oxide and hyperoxia. Am J Physiol 272:903–907

Roberts JD, Polaner DM, Lang P, Zapol WM (1992) Inhaled nitric oxide in persistent pulmonary hypertension of the newborn. Lancet 340:818–819

Roberts JD Jr, Chen TY, Kawai N, Wain J, Dupuy P, Shimouchi A, Bloch K, Polaner D, Zapol WM (1993a) Inhaled nitric oxide reverses pulmonary vasoconstriction in the hypoxic and acidotic newborn lamb. Circ Res 72:246–254

Roberts JD Jr, Lang P, Bigatello LM, Vlahakes GJ, Zapol WM (1993b) Inhaled nitric oxide in congenital heart disease. Circulation 87:447–453

Roberts JD Jr, Fineman JR, Morin FC III, Shaul PW, Rimar S, Schreiber MD, Polin RA, Zwass MS, Zayek MM, Gross I, Heymann MA, Zapol WM (1997) Inhaled nitric oxide and persistent pulmonary hypertension of the newborn. The Inhaled Nitric Oxide Study Group. N Engl J Med 336:605–610

Roos CM, Frank DU, Xue C, Johns RA, Rich GF (1996) Chronic inhaled nitric oxide: effects on pulmonary vascular endothelial function and pathology in rats. J Appl Physiol 80:252–260

Rossaint R, Falke KJ, Lopez F, Slama K, Pison U, Zapol WM (1993) Inhaled nitric oxide for the adult respiratory distress syndrome. N Engl J Med 328:399–405

Rossaint R, Gerlach H, Schmidt Ruhnke H, Pappert D, Lewandowski K, Steudel W, Falke K (1995a) Efficacy of inhaled nitric oxide in patients with severe ARDS. Chest 107:1107–1115

Rossaint R, Slama K, Steudel W, Gerlach H, Pappert D, Veit S, Falke K (1995b) Effects of inhaled nitric oxide on right ventricular function in severe acute respiratory distress syndrome. Intensive Care Med 21:197–203

Rovira I, Chen TY, Winkler M, Kawai N, Bloch KD, Zapol WM (1994) Effects of inhaled nitric oxide on pulmonary hemodynamics and gas exchange in an ovine model of ARDS. J Appl Physiol 76:345–355

Royall JA, Kooy NW, Beckman JS (1995) Nitric oxide-related oxidants in acute lung injury. New Horiz 3:113–122

Rubbo H, Radi R, Trujillo M, Telleri R, Kalyanaraman B, Barnes S, Kirk M, Freeman BA (1994) Nitric oxide regulation of superoxide and peroxynitrite-dependent lipid peroxidation. Formation of novel nitrogen-containing oxidized lipid derivatives. J Biol Chem 269:26066–26075

Samama CM, Diaby M, Fellahi JL, Mdhafar A, Eyraud D, Arock M, Guillosson JJ, Coriat P, Rouby JJ (1995) Inhibition of platelet aggregation by inhaled nitric oxide in patients with acute respiratory distress syndrome. Anesthesiology 83:56–65

Schedin U, Frostell C, Persson MG, Jakobsson J, Andersson G, Gustafsson LE (1995) Contribution from upper and lower airways to exhaled endogenous nitric oxide in humans. Acta Anaesthesiol Scand 39:327–332

Schedin U, Norman M, Gustafsson LE, Herin P, Frostell C (1996) Endogenous nitric oxide in the upper airways of healthy newborn infants. Pediatr Res 40:148–151

Schedin U, Roken BO, Nyman G, Frostell C, Gustafsson LE (1997) Endogenous nitric oxide in the airways of different animal species. Acta Anaesthesiol Scand 41:1133–1141

Scherrer U, Vollenweider L, Delabays A, Savcic M, Eichenberger U, Kleger GR, Fikrle A, Ballmer PE, Nicod P, Bartsch P (1996) Inhaled nitric oxide for high-altitude pulmonary edema. N Engl J Med 334:624–629

Schutte H, Grimminger F, Otterbein J, Spriestersbach R, Mayer K, Walmrath D, Seeger W (1997) Efficiency of aerosolized nitric oxide donor drugs to achieve sustained pulmonary vasodilation. J Pharmacol Exp Ther 282:985–994

Semigran MJ, Cockrill BA, Kacmarek R, Thompson BT, Zapol WM, Dec GW, Fifer MA (1994) Hemodynamic effects of inhaled nitric oxide in heart failure. J Am Coll Cardiol 24:982–988

Shah NS, Nakayama DK, Jacob TD, Nishio I, Imai T, Billiar TR, Exler R, Yousem SA, Motoyama EK, Peitzman AB (1997) Efficacy of inhaled nitric oxide in oleic acid-induced acute lung injury. Crit Care Med 25:153–158

Stamler JS, Jaraki O, Osborne J, Simon DI, Keaney J, Vita J, Singel D, Valeri CR, Loscalzo J (1992a) Nitric oxide circulates in mammalian plasma primarily as an S-nitroso adduct of serum albumin. Proc Natl Acad Sci USA 89:7674–7677

Stavert DM, Lehnert BE (1990) Nitrogen oxide and nitrogen dioxide as inducers of acute pulmonary injury when inhaled at relatively high concentrations. Inhal Toxicol 2:53–67

Stenqvist O, Kjelltoft B, Lundin S (1993) Evaluation of a new system for ventilatory administration of nitric oxide. Acta Anaesthesiol Scand 37:687–691

Sydow M, Bristow F, Zinserling J, Allen SJ (1997a) Variation of nitric oxide concentration during inspiration. Crit Care Med 25:365–371

Sydow M, Bristow F, Zinserling J, Allen SJ (1997b) Flow-proportional administration of nitric oxide with a new delivery system: inspiratory nitric oxide concentration fluctuation during different flow conditions. Chest 112:496–504

Sydow M, Zinserling J, Allen SJ (1998) Inspiratory NO Concentration Fluctuation during inhalational NO administration. In: Yearbook of intensive care and emergency medicine, 561–568. Edited by Vincent, JL. Berlin, Springer

Tibballs J, Hochmann M, Carter B, Osborne A (1993) An appraisal of techniques for administration of gaseous nitric oxide. Anaesth Intensive Care 21:844–847

Tönz M, von Segesser LK, Turina M (1993) Selective pulmonary vasodilatation with inhaled nitric oxide. J Thorac Cardiovasc Surg 105:760–762

Troncy E, Collet JP, Shapiro S, Guimond JG, Blair L, Ducruet T, Francoeur M, Charbonneau M, Blaise G (1998) Inhaled nitric oxide in acute respiratory distress syndrome: a pilot randomized controlled study. Am J Respir Crit Care Med 157:1483–1488

Valvini EM, Young JD (1995) Serum nitrogen oxides during nitric oxide inhalation. Br J Anaesth 74:338–339

Villar J, Blazquez MA, Lubillo S, Quintana J, Manzano JL (1989) Pulmonary hypertension in acute respiratory failure. Crit Care Med 17:523–526

Walmrath D, Schneider T, Schermuly R, Olschewski H, Grimminger F, Seeger W (1996) Direct comparison of inhaled nitric oxide and aerosolized prostacyclin in acute respiratory distress syndrome. Am J Respir Crit Care Med 153:991–996

Watkins DN, Jenkins IR, Rankin JM, Clarke GM (1993) Inhaled nitric oxide in severe acute respiratory failure–its use in intensive care and description of a delivery system. Anaesth Intensive Care 21:861–866

Weitzberg E, Rudehill A, Lundberg JM (1993) Nitric oxide inhalation attenuates pulmonary hypertension and improves gas exchange in endotoxin shock. Eur J Pharmacol 233:85–94

Wennmalm A, Benthin G, Edlund A, Jungersten L, Kieler-Jensen N, Lundin S, Westfelt UN, Petersson AS, Waagstein F (1993) Metabolism and excretion of nitric oxide in humans. An experimental and clinical study. Circ Res 73:1121–1127

Wessel DL, Adatia I, Thompson JE, Hickey PR (1994) Delivery and monitoring of inhaled nitric oxide in patients with pulmonary hypertension. Crit Care Med 22:930–938

Westfelt UN, Benthin G, Lundin S, Stenqvist O, Wennmalm A (1995) Conversion of inhaled nitric oxide to nitrate in man. Br J Pharmacol 114:1621–1624

Westfelt UN, Lundin S, Stenqvist O (1997) Uptake of inhaled nitric oxide in acute lung injury. Acta Anaesthesiol Scand 41:818–823

Wink DA, Osawa Y, Darbyshire JF, Jones CR, Eshenaur SC, Nims RW (1993) Inhibition of cytochromes P450 by nitric oxide and a nitric oxide-releasing agent. Arch Biochem Biophys 300:115–123

Wink DA, Hanbauer I, Laval F, Cook JA, Krishna MC, Mitchell JB (1994) Nitric oxide protects against the cytotoxic effects of reactive oxygen species. Ann N Y Acad Sci 738:265–278

Wink DA, Cook JA, Kim SY, Vodovotz Y, Pacelli R, Krishna MC, Russo A, Mitchell JB, Jourdqheuil D, Miles AM, Grisham MB (1997) Superoxide modulates the oxidation and nitrosation of thiols by nitric oxide-derived reactive intermediates. Chemical aspects involved in the balance between oxidative and nitrosative stress. J Biol Chem 272:11147–11151

Wink DA, Feelisch M, Fukuto J, Chistodoulou D, Jourdqheuil D, Grisham MB, Vodovotz Y, Cook JA, Krishna M, DeGraff WG, Kim S, Gamson J, Mitchell JB (1998) The cytotoxicity of nitroxyl: possible implications for the pathophysiological role of NO. Arch Biochem Biophys 351:66–74

Wizemann TM, Gardner CR, Laskin JD, Quinones S, Durham SK, Goller NL, Ohnishi ST, Laskin DL (1994) Production of nitric oxide and peroxynitrite in the lung during acute endotoxemia. J Leukoc Biol 56:759–768

Wysocki M, Delclaux C, Roupie E, Langeron O, Liu N, Herman B, Lemaire F, Brochard L (1994) Additive effect on gas exchange of inhaled nitric oxide and intravenous almitrine bismesylate in the adult respiratory distress syndrome. Intensive Care Med 20:254–259

Young JD (1994) A universal nitric oxide delivery system. Br J Anaesth 73:700–702

Young JD, Dyar OJ (1996) Delivery and monitoring of inhaled nitric oxide. Intensive Care Med 22:77–86

Young JD, Brampton WJ, Knighton JD, Finfer SR (1994a) Inhaled nitric oxide in acute respiratory failure in adults. Br J Anaesth 73:499–502

Young JD, Dyar O, Xiong L, Howell S (1994b) Methaemoglobin production in normal adults inhaling low concentrations of nitric oxide. Intensive Care Med 20:581–584

Young JD, Sear JW, Valvini EM (1996) Kinetics of methaemoglobin and serum nitrogen oxide production during inhalation of nitric oxide in volunteers. Br J Anaesth 76:652–656

Zapol WM (1998) Nitric oxide inhalation in acute respiratory distress syndrome: It works, bur can we prove it? Crit Care Med 26:2–3

Zayek M, Cleveland D, Morin FC (1993) Treatment of persistent pulmonary hypertension in the newborn lamb by inhaled nitric oxide. J Pediatr 122:743–750

Zhu S, Haddad IY, Matalon S (1996) Nitration of surfactant protein A (SP-A) tyrosine residues results in decreased mannose binding ability. Arch Biochem Biophys 333:282–290

CHAPTER 18
The Function of Nitric Oxide in the Immune System

C. BOGDAN

A. Introduction

Nitric oxide (NO) research in immunology started in 1985 when Stuehr and Marletta discovered that mouse macrophages produce nitrite and nitrate in a L-arginine-dependent manner after stimulation by lipopolysaccharide (LPS) and/or cytokines in vitro (STUEHR and MARLETTA 1985). Although the biochemistry of the underlying pathway was not known at that time, the critical role of inflammatory stimuli for its induction soon led immunologists, cancer researchers, and infectious-diseases specialists to search for functions of the newly discovered macrophage metabolites. In 1987, the *Schistosoma mansoni* schistosomula were the first parasites shown to be destroyed by mammalian macrophages only in the presence of L-arginine (MALKIN et al. 1987). The same was found to be true for the killing of some tumor cells (HIBBS et al. 1987). With the stepwise elucidation of the NO-synthesis pathway and the cloning of the cytokine-inducible macrophage NO synthase (iNOS, macNOS or NOS-II) in 1992 (LOWENSTEIN et al. 1992; XIE et al. 1992), the interest in NO as an effector molecule of the immune system rapidly grew. By the end of 1999, NOS-II-derived NO was fully established as an important modulator in the immune system and had become a subject in virtually every area of immunology research. Furthermore, evidence for a functional role of the constitutive isoforms of NOS (ecNOS/NOS-III; ncNOS/NOS1) in the mammalian immune system is now also emerging. Finally, NOSs have been purified and/or cloned from non-mammalian organisms, such as bacteria (CHEN and ROSAZZA 1995), protozoa (GHIGO et al. 1995; PAVETO et al. 1995), slime molds (WERNER-FELMAYER et al. 1994), invertebrates (*Drosophila*, *Anopheles* mosquitoes, bugs, silkworms) (CHOI et al. 1995; REGULSKI and TULLY 1995; KUZIN et al. 1996; YUDA et al. 1996; LUCKHART et al. 1998), and plants (DURNER et al. 1998; HAUSLADEN and STAMLER 1998). At least some of these "primordial" NOSs seem to also fulfill important defense tasks (DELLEDONNE et al. 1998; DURNER et al. 1998; LUCKHART et al. 1998).

In this chapter, we present an overview of the current knowledge of the function of NO in the immune system. For detailed aspects of the regulation of NO biosynthesis and the role of NO in apoptosis, inflammatory diseases, diabetes mellitus, atherosclerosis, and septic shock, the reader is referred to the respective sections in this volume.

B. Type-2 NOS (NOS-II, iNOS) and the Immune System

I. Cell Types

The prototypic cell type expressing NOS-II in the immune system is the macrophage. Production of NO by macrophages has been documented for many mammalian species, including mouse, rat, cattle, goat, sheep, dog, and humans (ADLER et al. 1996; JUNGI et al. 1997; MACMICKING et al. 1997a; PANARO et al. 1998). Not surprisingly, for many years, expression of NOS-II was mainly viewed as a potent antimicrobial defense mechanism. With the discovery of additional (mainly regulatory and autotoxic) functions of NO, it was realized that many other types of cells in the immune system also produce NO (Table 1). In some cases (for mouse T cells, epidermal Langerhans' cells, and human macrophages, microglia and neutrophils), conflicting results on the expression of NOS-II have been published, so definitive statements on the existence and role of NOS-II-derived NO in these types of cells are not yet possible (Table 1).

II. Induction and Regulation

1. Overview

A characteristic feature of NOS-II is its regulation by cytokines. Several levels of regulation of NOS-II expression have been described (NATHAN and XIE 1994). These include the increase or decrease of NOS-II gene transcription, NOS-II mRNA stability, NOS-II mRNA translation and NOS-II protein degradation. Prototypic activating or inhibitory cytokines that were found to act on these levels are interferon-γ (IFN-γ), transforming growth factor-β (TGF-β), interleukin (IL)-4, and IL-13 (VODOVOTZ et al. 1993; XIE et al. 1993; MELILLO et al. 1994; SAURA et al. 1996; BOGDAN et al. 1997; MACMICKING et al. 1997a; WRIGHT et al. 1997). Immunosuppressive drugs, such as dexamethasone, were also found to affect NOS-II via transcriptional and post-transcriptional mechanisms (KUNZ et al. 1996). Furthermore, the enzymatic activity of NOS-II can be modulated by substrate and/or cofactor availability. Arginase, for instance, degrades arginine to ornithine and urea. This enzyme, which is upregulated by cytokines that usually suppress NOS2 activity, competes with NOS-II for the substrate and thereby indirectly impedes NO production (MUNDER et al. 1998a). Likewise, de novo synthesis of tetrahydrobiopterin, an essential cofactor of NOS-II involved in NOS-II dimerization (TZENG et al. 1995), requires guanosine triphosphate cyclohydrolase I, which is also a target of cytokine action (WERNER et al. 1998). Finally, there is evidence for a post-translational inactivation of NOS-II that involves neither substrate/cofactor depletion nor degradation of NOS-II protein (VODOVOTZ et al. 1994).

As another mechanism of NOS-II regulation, the end product (NO) was reported to increase or decrease the expression of NOS-II mRNA, to limit the intracellular assembly of NOS-II dimers by preventing heme insertion

Table 1. Examples of cell types expressing nitric oxide (NO) synthase type 2 during immune reactions

Cell type	(Putative) functions of NO	Key references
Macrophages	Inhibition of T-cell activation and proliferation; apoptosis of macrophages or (autoreactive) T cells; antimicrobial activity; tissue damage	Stuehr and Marletta 1987; Hoffman et al. 1990; Albina et al. 1991, 1993; Kröncke et al. 1991; Mills 1991; Bogdan 1997; Zettl et al. 1997; Bingisser et al. 1998; Tarrant et al. 1999
Microglia	Antimicrobial activity; toxicity towards neurons or oligodendrocytes (during ischemic/infectious/demyelinating processes)	Betz-Corradin et al. 1993b; Chao et al. 1993; Merrill et al. 1993; Zheng et al. 1993; Wong et al. 1996; Kim and Täuber 1996; Lu et al. 1996; Ding et al. 1997; Hooper et al. 1997
Kupffer cells	Hepatocellular dysfunction (inhibition of protein synthesis) during sepsis	Billiar et al. 1989
Neutrophilic granulocytes	Antimicrobial activity	McCall et al. 1991; Padgett and Pruett 1995; Evans et al. 1996; Wheeler et al. 1997; Eisenich et al. 1998
Eosinophils	Eosinophil locomotion; antimicrobial activity; inhibition of apoptosis	Del Pozo et al. 1997; Zanardo et al. 1997; Hebestreit et al. 1998; Oliveira et al. 1998
Mast cells	Participation in inflammatory responses during allergic reactions	Bidri et al. 1997; Lin and Befus 1997
Lymphoid/myeloid dendritic cells	Inhibition of T-cell proliferation; apoptosis of dendritic cells; antimicrobial activity	Lu et al. 1996; Stenger et al. 1996
Epidermal dendritic cells (Langerhans' cells)	Skin swelling in contact hypersensitivity?	Blank et al. 1996; Qureshi et al. 1996; Ross et al. 1998
Natural killer cells (mouse, rat, human)	Cytotoxic activity (?); inhibition of lytic potential (?); regulation of IL-12 responsiveness and IFNγ production	Cifone et al. 1994, 1999; Diefenbach et al. 1998; Salvucci et al. 1998; Diefenbach et al. 1999
Mouse type-1 T-helper lymphocytes	Inhibition of IL-2/IFN-γ production by Th1 cells?	Taylor-Robinson et al. 1994; Thüring et al. 1995; Niedbala et al. 1999
Human T cell lines	Suppression of apoptosis	Mannick et al. 1997, 1999
Human B cell lines	Control of latent viruses (EBV); suppression of apoptosis	Mannick et al. 1994, 1997, 1999

Table 1. Continued

Cell type	(Putative) functions of NO	Key references
Keratinocytes	Regulation of cell growth (anti-proliferative effect); pro-inflammatory effect in skin lesions (psoriasis)	Heck et al. 1992; Bruch-Gerharz et al. 1996; Sirsjo et al. 1996
Fibroblasts	antimicrobial activity?	Werner-Felmayer et al. 1990; Turco et al. 1998
Endothelial cells	Regulation of adhesion of immune cells; regulation of inflammation; antimicrobial activity	Gross et al. 1991; Suschek et al. 1993; Oswald et al. 1994
Bronchial epithelial cells	Antimicrobial activity; airway inflammation; damage of ciliated cells	Asano et al. 1994; Guo et al. 1995; Flak and Goldman, 1999
Vascular smooth-muscle cells	Vasodilatation during septic shock	Fleming et al. 1991; Imai et al. 1994; Spink et al. 1995
Mesangial cells	Pro-inflammatory effect in glomerulonephritis; regulation of cell proliferation and matrix synthesis	Pfeilschifter et al. 1992; Saura et al. 1995; Beck and Sterzel 1996
Hepatocytes	Antimicrobial activity; inhibition of hepatovascular thrombosis; hypotension during septic shock	Harbrecht et al. 1992; Hortelano et al. 1992; Nüssler et al. 1993; Geller et al. 1995
Chondrocytes	Anticatabolic effect on human cartilage; regulation of PGE_2 and gelatinase synthesis	Stadler et al. 1991; Charles et al. 1993; Häuselmann et al. 1998; Shalom-Barak et al. 1998
Schwann cells	Anti-proliferative effect on T-cell/APC co-cultures	Gold et al. 1996
Astrocytes (astroglia)	Neuronal damage during ischemia/demyelinating/infectious processes; antimicrobial activity	Galea et al. 1992; Endoh et al. 1994; Lee et al. 1994; Peterson et al. 1995; Liu et al. 1996
Neurons	Neuronal damage during inflammatory/ischemic processes	Minc-Golomb et al. 1996; Wong et al. 1996

APC, antigen-presenting cells; *EBV*, Epstein-Barr virus; *IFN*, interferon; *IL*, interleukin; *PGE*, prostaglandin E; *Th1*, T-helper type-1 cells.

and decreasing heme availability, and to inactivate argininosuccinate synthetase which regenerates L-arginine from L-citrulline (ASSREUY et al. 1993; GRISCAVAGE et al. 1993; SHEFFLER et al. 1995; ALBAKRI and STUEHR 1996; XIE et al. 1999). These findings suggest that NO can act as an endogenous feedback regulator of NOS-II. Finally, the biological activity of NO itself is controlled by a variety of factors, e.g., by compounds that stabilize, transport, store, or detoxify/scavenge NO or react with the NO molecule via the formation of other (more active or stable) nitrogen intermediates (BECKMAN et al. 1990; STAMLER et al. 1992; COONEY et al. 1993; DENICOLA et al. 1993; CHRISTEN et al. 1997; MNAIMNEH et al. 1997; GOW and STAMLER 1998).

2. Transcriptional Regulation

In resident, non-inflammatory macrophages, NOS-II is transcriptionally induced by IFN-γ. NOS-II is not a primary response gene. Its transcription clearly requires de novo synthesis of protein (NATHAN and XIE 1994). Induction of NOS-II by IFN-γ is initiated by IFN-γ-receptor-mediated activation of Jak1 tyrosine kinase, which causes tyrosine-phosphorylation of the 91kDa cytosolic protein signal transducer and activator of transcription (STAT) 1α. Tyrosine-phosphorylated STAT1α is translocated into the nucleus, where it leads to transcription of IFN regulatory factor (IRF)-1. IRF-1 then binds to the promoter of NOS-II, thereby activating NOS-II gene transcription (MARTIN et al. 1994; GAO et al. 1997).

The importance of this pathway for the expression of NOS-II and the critical role of the different signaling molecules has been most convincingly shown by studies with gene-deleted mice. Macrophages from mice lacking the IFN-γ receptor (HUANG et al. 1993), Jak1 (RODIG et al. 1998), STAT1α (DURBIN et al. 1996; MERAZ et al. 1996), or IRF-1 (KAMIJO et al. 1994) failed to produce NO in response to IFN-γ. In contrast, IFN-γ-stimulated macrophages deficient for IRF-2, the IFN consensus-sequence-binding protein, or the nuclear factor NF-IL-6 expressed normal amounts of NOS-II mRNA and produced NO (TANAKA et al. 1995; SALKOWSKI et al. 1996; FEHR et al. 1997; GIESE et al. 1997). Furthermore, in accordance with previous analyses on macrophage-modulating cytokines (DING et al. 1988), IFN-γ turned out to be the single most important inducer of NOS-II (DALTON et al. 1993). However, despite a severely impaired production of NO, IFN-$\gamma^{-/-}$ mice were able to express NOS-II. In fact, during the innate response to the protozoan parasite *Leishmania major*, NOS-II protein was found at similar levels in lesions from wild-type and IFN-$\gamma^{-/-}$ mice. In this situation, type-I IFN (IFN-α/β) (in conjunction with the parasites) turned out to be the primary stimulus for the early (day-1) expression of NOS-II at the cutaneous site of infection (DIEFENBACH et al. 1998). An IFN-γ-independent expression of NOS-II mRNA has also been observed in the liver of *L. donovani*-infected mice treated with IL-12 (TAYLOR and MURRAY 1997), in the spleens of IFN-γ-receptor$^{-/-}$ mice infected with *Listeria monocytogenes* (DAI et al. 1997), and in the spinal cords of mice with experimental allergic encephalomyelitis (EAE) (SEGAL et al. 1998). These findings are in accordance with reports on IRF-1-independent expression of NOS-II mRNA and/or protein in articular chondrocytes [stimulated with IL-1β (or IFN-γ) ± LPS] (SHIRAISHI et al. 1997) and in the spinal cords of mice with EAE (TADA et al. 1997).

The IFN-γ/Jak1/Stat1α/IRF-1 signaling cascade is a critical (but not the only) pathway for NOS-II induction. Based on the structure of the mouse NOS-II promoter, a number of other transcription factors have been implicated in the transcriptional regulation of NOS-II, including nuclear factor (NF) κB and the activation protein AP-1 (NATHAN and XIE 1994; TAYLOR et al. 1998; MARKS-KONCZALIK et al. 1998). In inflammatory peritoneal macrophages, which already constitutively expressed IRF-1, IFN-γ-stimulated expression of NOS-II was paralleled by the activation and nuclear translocation of NF-κB, indicating that, under certain conditions, activation of NF-κB by IFN-γ is a prerequisite for IFN-γ-induced NO production (LOPEZ-COLLAZO et al. 1998). Certain microbial stimuli such as LPS, which leads to the induction of NOS-II in the absence of exogenous cytokine factors, were thought to act via nuclear translocation of NF-κB (LOWENSTEIN et al. 1993; XIE et al. 1993, 1994). However, other studies demonstrated that stimulation of primary macrophages or macrophage cell lines with LPS caused autocrine and/or paracrine production of type-I IFNs (IFN-α/β), a phenomenon which was closely paralleled by tyrosine-phosphorylation of STAT1α (RICHES and UNDERWOOD 1991; FUJIHARA et al. 1994b; GAO et al. 1998). Neutralization of IFN-α/β abolished the activation of STAT1α and reduced the production of NO in LPS-stimulated macrophages, indicating a pivotal role of endogenous IFN-α/β and STAT1α in the LPS-induced expression of NOS-II (GAO et al. 1998). Thus, although LPS alone can activate the NOS-II promoter in cell lines transfected with NOS-II promoter constructs containing NF-κB-binding elements (XIE et al. 1993), it is unlikely that the induction of NOS-II by LPS in primary macrophages results solely from the activation of NF-κB. The contribution of endogenously released cytokines to the expression of NOS-II in response to LPS has also been documented in vivo (SALKOWSKI et al. 1997).

In addition to NF-κB and STAT1, several other signaling molecules participate in the cytokine-triggered pathways that regulate NOS-II. Examples include: (a) various isoforms of *protein kinase C*, which are activated in response to LPS (± IFN-γ) and contribute to the expression of NOS-II in macrophages (SEVERN et al. 1992; BOSCA and LAZO 1994; FUJIHARA et al. 1994a); (b) the *protein phosphatases 1 and 2A*, whose activity is a prerequisite for the induction of NOS-II activity by IFN-γ plus LPS in macrophages but not in astrocytes (DONG et al. 1995; PAHAN et al. 1998); (c) the *mitogen-activated protein kinase* (MAPK) pathways (p38 MAPK, extracellular-signal-regulated kinase, Jun N-terminal kinase/stress-activated protein kinase), which are required for the induction of NOS-II by IL-1 in islet cells, chondrocytes, cardiac myocytes, and cardiac microvascular endothelial cells (BADGER et al. 1998; GUAN et al. 1997; LARSEN et al. 1998; SINGH et al. 1996) or by IFN-γ and

LPS or tumor necrosis factor (TNF)-α in mouse macrophages (AJIZIAN et al. 1999; CHAN et al. 1999; CHEN et al. 1999); (d) *phosphatidylinositol-3 kinase*, which mediates the inhibitory effect of macrophage-stimulating protein (MSP) or IL-13 on the expression of NOS-II (WRIGHT et al. 1997; CHEN et al. 1998b) and functions as a negative-feedback regulator of NOS-II during stimulation of macrophages, glial cells or astrocytes with LPS (DIAZ-GUERRA et al. 1999a; PAHAN et al. 1999); (e) protein tyrosine phosphatases, which negatively affect the expression of NOS2 in response to IFN-γ (OLIVIER et al. 1998; DIAZ-GUERRA et al. 1999b); and (f) *cyclic adenosine monophosphate*, which was reported to enhance or inhibit the expression of NOS-II, depending on the cell type (BULUT et al. 1992; KOIDE et al. 1993; ALONSO et al. 1995). For all these molecules, the molecular mode of action and the exact position of the molecule within the signaling cascade that are necessary for the expression or inhibition of NOS-II remain to be determined.

An interesting extension to the known immunological (cytokine) regulators of NOS-II has recently emerged with the discovery of non-proteinaceous activators of NOS-II transcription. Chelation (depletion) of non-ferritin-bound iron (Fe^{2+}) or a decrease of oxygen tension (hypoxia or anoxia) were both shown to activate the NOS-II gene transcription via a hypoxia-responsive enhancer element in the NOS-II promoter (WEISS et al. 1994; ALBINA et al. 1995; MELILLO et al. 1997). This suggests the existence of unexpected mechanisms of NOS-II regulation during chronic infections, malignancies, and other inflammatory processes, all of which are frequently accompanied by hypoferremia, increased incorporation of iron into the storage protein ferritin, and tissue hypoxia.

3. Positive and Negative Regulation of NOS-II by Cytokines, Ligand–Receptor Interactions, and Microbial Products

The expression of NOS-II during immune reactions is regulated by soluble host factors (cytokines), cell–cell contact and cross-linking of surface receptors, as well as by products of microbial pathogens. In addition, the expression of NOS-II is also positively or negatively affected by a number of drugs (glucocorticoids, taxol, aspirin, macrolides, and tetracyclines), which are frequently used for the treatment of autoimmune, infectious, or malignant diseases in humans (MANTHEY et al. 1994; AMIN et al. 1995, 1996; D'AGOSTINO et al. 1998; DOHERTY et al. 1998; TAMAOKI et al. 1999). The extent to which the induction or suppression of NOS-II contributes to the therapeutic value of these compounds remains to be established.

a) Cytokines

Amongst the activating cytokines known to regulate NOS-II, IFN-γ is certainly the best-studied factor. Important types of cells that can function as inducers of NOS-II via their production of IFN-γ include $CD4^+$ type-1 T-helper lymphocytes (Th1) (TAO and STOUT 1993; TAUB and COX 1995; MUNDER et al. 1998a), natural killer (NK) cells (DIEFENBACH et al. 1998), γ/δ T lymphocytes (JONES-CARSON et al. 1995; RAJAN et al. 1998), cytotoxic (CD8$^+$) T cells (STEFANI et al. 1994), and macrophages (PUDDU et al. 1997; MUNDER et al. 1998b).

Other than IFN-γ, very few cytokines have been found to induce NOS-II on their own. Examples in vitro include: (a) IL-1β, which stimulated the expression of NOS-II in human chondrocytes or rat islet β-cells (CORBETT and McDANIEL 1995; SHIRAISHI et al. 1997; LARSEN et al. 1998); (b) IL-6, which was described to cause NOS-II gene induction in a myeloid leukemia cell line during its differentiation into macrophages and in a human enterocyte cell-line (SAWADA et al. 1997; OUADRHIRI et al. 1999); (c) IFN-α2b, which led to an increase in NOS-II activity in human peripheral blood monocytes (SHARARA et al. 1997); (d) IL-17, which induced expression of NOS-II and NO production in normal human chondrocytes or in chondrocytes from osteoarthritic human cartilage (ATTUR et al. 1997; SHALOM-BARAK et al. 1998); (e) macrophage-colony-stimulating factor, which augmented β-amyloid-induced production of NO by an immortalized mouse microgial cell line (MURPHY et al., 1998); and (f) stem cell factor, which caused mucosal mast cells to differentiate into connective tissue mast cells expressing NOS-II protein (BIDRI et al. 1997). In vivo, both IL-1β and IL-2 led to the accumulation of nitrite/nitrate when applied to tumor-bearing mice or to humans with cancer (HIBBS et al. 1992; YIM et al. 1995; OGILVIE et al. 1996). Similar observations were made in patients with hepatitis C, where treatment with IFN-α2b was associated with a significant elevation of monocyte NOS-II expression (SHARARA et al. 1997). In none of these cases, however, has it been disproven that the stimulatory effect is, in fact, due to secondary endogenous cytokines.

The list of activating factors synergizing with IFN-γ (or LPS) for the up-regulation of NOS-II is long (BOGDAN and NATHAN 1993; BOGDAN et al. 1994; MACMICKING et al. 1997a). It includes IL-2, IFN-α/β (DING et al. 1988; GAO et al. 1998), exogenously or endogenously produced TNF-α, IL-7, IL-8 (BRUCH-GERHARZ et al. 1996; OLIVEIRA et al. 1998), macrophage migration inhibitory factor (BERNHAGEN et al. 1994; JÜTTNER et al. 1998) and, under certain stimulation conditions, the macrophage-deactivating cytokine IL-10 (BETZ-CORRADIN et al. 1993a; APPELBERG 1995; SUNYER et al. 1996; JACOBS et al. 1998).

The number of cytokines that have been found to antagonize the action of IFN-γ is similarly extensive and continues to grow. The most potent amongst them is TGF-β, the only factor to date that was shown to down-regulate NOS-II in an already fully activated macrophage (VODOVOTZ et al. 1993). Other factors that were reported to suppress the induction of NO release from macrophages (or other cell types) are IL-4, IL-6, IL-10, IL-11, IL-13, monocyte chemotactic protein-1, MSP, bactericidal/permeability-increasing protein, α-melanocyte-stimulating hormone, osteopontin, calcitonin-gene-related peptide, taurine chloramine, (partial) agonists of class-I or -II purinoreceptors (extracellular adenosine or adenine nucleotides, respectively), and ligands of the peroxisome proliferator-activated receptor-γ (15-deoxy-$\Delta^{12,14}$-prostaglandin J$_2$) (BOGDAN et al. 1994, 1997; MELILLO et al. 1994; DENLINGER et al. 1996; HASKO et al. 1996; KWON et al., 1999; TAKEDA et al. 1996; TREPICCHIO et al. 1996, 1997; MACMICKING et al. 1997a; MARSHALL et al. 1997; OHMORI and HAMILTON 1997; PETROVA et al., 1999; VOULDOUKIS et al. 1997; WRIGHT et al. 1997; LOPEZ-CALLAZO et al., 1998; XAUS et al., 1999; COLVILLE-NASH et al. 1998; RICOTE et al. 1998; TAYLOR et al. 1998). These mediators act, in part, by antagonizing the activities of the transcription factors AP-1, STAT1, and NF-kB (via induction of STAT6), and by post-transcriptional and/or translational mechanisms.

b) Cross-Linking of Cell-Surface Receptors

Whereas the regulation of NOS-II by soluble mediators (cytokines) has been studied by immunologists in great detail, much less attention has been paid to the induction of NOS-II via cross-linking of non-cytokine cell-surface receptors. Considering the fact that (in addition to locally secreted humoral factors) cell–cell contacts are critical for the activation of immune cells, simultaneous induction of NOS-II via cognate interactions might have important influences on the shaping of the ensuing immune response. Table 2 summarizes the receptors, the ligation of which was shown to lead to the induction of NOS-II in macrophages. Fc$_\varepsilon$-receptor-mediated NO production was also observed with cell types other than macrophages, such as keratinocytes, eosinophils, and mast cells (BIDRI et al. 1997). In human monocytes, direct cell–cell contact with certain tumor cell lines also induced expression of NOS-II, but the respective receptor–ligand interactions have not yet been defined (SIEDLAR et al., 1999).

For several of the receptors listed in Table 2, expression of NOS-II has only been demonstrated by cross-linking antibodies, but not yet with the respective natural ligands. Furthermore, it is largely unclear whether transcription of the NOS-II gene is a direct consequence of the activation of the receptor or indirectly mediated by endogenously produced secondary cytokines. At least for the type-III complement receptor (CR3), expression of NOS-II after receptor ligation is due to autocrine TNF-α (GOODRUM et al. 1995).

c) Microbial Products

Although host factors are generally thought to be the principal stimuli for NO production, a large body of studies shows that microbial pathogens are also potent inducers of NOS-II. Due to the antimicrobial function of NO, induction of NOS-II, at first glance, appears to be disadvantageous to infectious pathogens. However, enhanced production of NO can also promote their survival, e.g., via suppression of T-cell proliferation, via induction of macrophage or T-cell apoptosis, or via host-damaging modulation of the cytokine response. The list of microbial agents inducing NOS-II either alone or in synergy with

Table 2. Receptor molecules on macrophages which after ligation transmit activating signals for the induction of nitric oxide (NO) synthase 2 (NOS-II)

Macrophage receptor (cell type)	NOS-II expression via	Possible implications	Reference
MHC class-II (mouse bone-marrow macrophages)	Anti-MHC class II antibodies; superantigens (?)	T-cell-mediated activation of macrophages	Hauschildt et al. 1993
CD40 (mouse splenic macrophages)	CD40L (gp39) on T cells	T-cell-mediated activation of macrophages	Tian et al. 1995; Stout et al. 1996
LFA-1 (mouse splenic macrophages)	ICAM-family adhesion molecules?	T-cell-mediated activation of macrophages	Tian et al. 1995
CD8 (α or β chain) (rat alveolar macrophages)	Anti-CD8 antibodies	Killing of intracellular parasites	Hirji et al. 1998
Fcγ-receptor (mouse J774.16 cells, rat peritoneal macrophages)	IgG immune complexes	Indirect antimicrobial activity of IgG	Mozaffarian et al. 1995; Bayon et al. 1997
TNF receptor p55 (CD120a) and p75 (CD120b) (mouse bone-marrow macrophages)	Anti-TNF-R p55- or anti-TNF-R p75-antibody in the presence of IFN-γ	costimulatory effect of TNF-α on the induction of NOS (primarily via TNF-R p55)	Riches et al., 1998
Fcϵ-receptor IIb (CD23) (human monocytes or rat peritoneal macrophages)	Anti-CD23 antibody or IgE immune complexes	Indirect antimicrobial activity of IgE	Alonso et al. 1995; Vouldoukis et al. 1995
CD69 (human monocytes)[a]	Anti-CD69 antibody	NO-dependent monocyte cytotoxicity	De Maria et al. 1995
CD53 (rat peritoneal macrophages)[b]	Anti-CD53 antibody (MRC OX-44)	Antimicrobial/cytotoxic activity?	Bosca and Lazo 1994
Type-III complement receptor (CR3, CD11b/CD18; mouse macrophages)	Anti-CR3 antibodies; LPS, group-B streptococci; *L. major* (?)	Induction of NOS-II by phagocytosis of microbial pathogens	Betz-Corradin and Mauël 1991; Goodrum et al. 1995; Matsuno et al. 1998
TNF receptor p55 (CD120a) and p75 (CD120b); mouse bone-marrow macrophages	Anti-TNF-R p55- or anti-T NF-R p75-antibody in the presence of IFN-γ	Co-stimulatory effect of TNF-α on the induction of NOS (primarily via TNF-R p55)	Riches et al. 1998

ICAM, intracellular adhesion molecule; *IFN*, interferon; *Ig*, immunoglobulin; *LFA*, leukocyte-function antigen; *LPS*, lipopolysaccharide; *MHC*, major histocompatibility complex; *TNF*, tumor necrosis factor.
[a] CD69 is a type-II integral membrane protein and member of the natural-killer-cell gene complex expressed on a wide variety of hematopoietic cells.
[b] CD53 is a panleukocyte antigen present on all mature cells of the immune system. It is a member of the transmembrane-4 superfamily of membrane proteins, which is implicated in the control of cell proliferation.

cytokines in various types of cells (macrophages, endothelial and epithelial cells, hepatocytes, and glial cells) includes:

1. Viruses and viral products (human immunodeficiency virus 1, gp41, gp120)
2. Bacteria (*Borrelia burgdorferi*, *Chlamydia trachomatis*, *Streptococcus pyogenes*) and bacterial products (LPS, **19 kDa *M. tuberculosis* lipoprotein**, ***Borrelia burgdorferi* OspA**, tracheal cytotoxin of *Bordetella pertussis*, lipoteichoic acid of gram-positive bacteria, peptidoglycan of *Staphylococcus aureus*, **Pneumolysin**, bacterial DNA)
3. Protozoa (*Leishmania major*) and protozoan products (*Trypanosoma cruzi* glycoconjugates, *T. cruzi* Tc52 protein)

and has been reviewed elsewhere (BOGDAN 1997; CAMARGO et al. 1997; FERNANDEZ-GOMEZ et al. 1998; KENGATHARAN et al. 1998; BRAUN et al., 1999; BRIGHTBILL et al., 1999; GAO et al., 1999). Interestingly, several of the microbes and microbial products (LPS, *L. major*, glycoinositol phospholipids, *T. cruzi*, *Candida albicans*, *Cryptococcus neoformans*), some of which were shown to induce NOS-II were also found to suppress NOS-II, depending on the precise stimulation conditions (BOGDAN 1997; KAWAKAMI et al. 1997; BOGDAN and RÖLLINGHOFF 1999; CHINEN et al. 1999). This indicates the complexity of NO function during infections, which will be discussed in more detail below.

III. Functions

1. Overview

Soon after the description of the cytokine-inducible L-arginine/NO pathway, it was recognized that NOS-II-derived NO, in addition to its antimicrobial, anti-tumor, and tissue-destructive functions, also has the potential to act as an immunoregulatory molecule. To date, four different categories of NO functioning in the immune system have been described (Table 3); these will be discussed below.

2. Antimicrobial Functions

a) Results from Host-Cell-Free Experiments and Studies in Rodents

An antimicrobial role of NO (and its derivatives) has been observed for all groups of infectious pathogens (viruses, bacteria, protozoa, helminths, fungi), with a broad spectrum of host and effector cells (monocytes and macrophages, microglia, neutrophils, eosinophils, fibroblasts, hepatocytes, endothelial cells, epithelial cells, and astroglia; Table 1), in almost every organ (skin, lymph node, spleen, lung, heart, brain, gastrointestinal system, peritoneum, and urogenital tract), during both acute and chronic infections, and in various mammalian species, most prominently in mice and rats (BOGDAN 1997; FANG 1997). Commonly chosen experimental strategies to document an involvement of NOS-II-derived NO in the control of infectious agents include: (a) the use of

Table 3. Functions of nitric oxide (NO) synthase 2-derived NO in the immune system

Category	Effector functions	Regulatory function
Host-protective functions	Antimicrobial activity	Modulation of cytokine production
	Anti-tumor activity	Feedback control of excessive lymphocyte responses
	Apoptosis of autoreactive T cells	
Host-damaging functions	Immunopathology (tissue destruction during autoimmune reactions)	Immunosuppression (inhibition of lymphocyte proliferation or cytokine production) Immunopathology (anti-apoptotic effect on inflammatory cells; immune deviation)

cellular NO sources (activated macrophages) or NO-donating drugs for in vitro killing assays of extracellular microbes; (b) the analysis of intracellular microbial growth or killing in cytokine-activated infected host cells lacking the NOS-II gene or treated with inhibitors of NOS-II activity; (c) the comparison of resistance to infectious agents in wild-type cells vs host cells transfected with the NOS-II gene; (d) the in situ analysis of NOS-II expression vs viral/microbial burden in infected tissues; (e) the evaluation of the course of infection in vivo in mice lacking the NOS-II gene or treated with NOS-II inhibitors; and (f) the effect of the application of NO donors on the course of infection in vivo (BOGDAN 1997). The extensive use of these approaches has certainly helped to establish NO as an important killing mechanism. The use of NOS-II$^{-/-}$ mice instead of non-selective NOS inhibitors with unknown side effects (N^{ω}-monomethyl-L-arginine, N^{ω}-nitro-L-arginine methyl ester, aminoguanidine; BOGDAN 1998) has provided especially firm genetic evidence for or against antimicrobial roles of NOS-II (Table 4; Sect. C.III.4.). However, there are several unresolved questions and findings that make us realize that NO is not nature's universal killing molecule.

1. Many putative microbial targets (free thiols, proteins with heme groups or iron–sulfur clusters, SH– and NH$_2$– groups of proteins and enzymes, membrane lipids, DNA) of NO and NO-derived reactive species (NO$_2$, peroxynitrite ONOO$^-$, S-nitrosothiols) have been identified to date, providing sufficient explanation for the broad-spectrum antimicrobial activity of reactive nitrogen intermediates. However, in almost all cases (see below), it remains to be determined to what extent mutation (deamination) of DNA, inhibition of DNA repair and synthesis, alteration of protein functions by S-nitrosylation, adenosine diphosphate (ADP)–ribosylation or tyrosine nitration, or inactivation of enzymes by disruption of Fe–S clusters accounts for NO-mediated microbial damage, growth arrest, or killing (FANG 1997).

Table 4. The effect of the deletion of the nitric oxide synthase 2 (NOS-II) gene on the course of viral, bacterial or parasite infections in mice

Infectious agent	Mouse disease model	Effect of NOS-II deletion on the severity of the disease	Reference
Herpes simplex virus type 1	Virus reactivation in the dorsal root ganglia	↑	MacLean et al. 1998
Coxsackie virus B3	Myocarditis	↑	Zaragoza et al. 1998
	Pankreatitis	↑	Zaragoza et al. 1999
Influenza virus A	Pneumonitis	↓	Karupiah et al. 1998a
Ectromelia virus	Mouse pox (liver, spleen, lung)	↑	Karupiah et al. 1998b
Chlamydia trachomatis (MoPn strain)	Vaginal infection	±	Perry et al. 1998; Ramsey et al. 1998
Chlamydia trachomatis (MoPn strain)	Mouse pneumonitis, infection of the spleen	↑	Igietseme et al. 1998
Chlamydia pneumoniae	Pneumonitis	↑	Rottenberg et al. 1999
Listeria monocytogenes	Listeriosis (liver, spleen)	↑	MacMicking et al. 1995
Mycobacterium tuberculosis	Tuberculosis (lung, liver, spleen)	↑	MacMicking et al. 1997b
Mycobacterium avium	Systemic infection (lung, spleen, liver)	± or ↓	Doherty and Sher, 1997; Gomes et al. 1999; Ehlers et al. 1999
Mycoplasma pulmonis	Pneumonia	↑	Hickman-Davis et al. 1999
Salmonella typhimurium	Systemic infection	± or ↑	Shiloh et al. 1999
Shigella flexneri	Bronchopulmonary infection	±	Way and Goldberg, 1999
Staphylococcus aureus	Septic arthritis	↑	McInnes et al. 1998
Leishmania major	Cutaneous leishmaniosis	↑(non-healing) ↑(healing)[1]	Diefenbach et al. 1998; Wei et al. 1995; Huang et al. 1998b; Niedbala et al. 1999
Leishmania donovani	Visceral leishmaniosis	↑(non-healing)	Murray and Nathan, 1999
Plasmodium berghei XAT	Blood stage malaria	±	Yoneto et al. 1999
Plasmodium berghei ANKA	murine cerebral malaria	±	Favre et al. 1999
Trypanosoma cruzi (Tulahuen)	experimental Chagas' disease (spleen, liver, heart)	↑	Hölscher et al. 1998
Trypanosoma brucei	experimental sleeping sickness	↓ or ±	Hertz and Mansfield 1999
Toxoplasma gondii	cerebral toxoplasmosis	↑	Scharton-Kersten 1997
	inflammation of small intestine and liver	↓	Khan et al. 1997; Liesenfeld et al. 1999
Schistosoma mansoni	challenge of immunized NOS2$^{+/+}$ vs. NOS2$^{-/-}$ mice with live cercariae	(↓)	Coulson et al. 1998; James et al. 1998

[1] Healing was only observed in a NOS2 mutant mouse strain (Wei et al., 1995) in which the NOS2 gene was not deleted and residual NOS2 bioactivity was expressed (Niedbala et al., 1999)

2. Although, in a host-cell-free system, NO donors are undoubtedly able to kill the vast majority of prokaryotic and eukaryotic microbes provided the concentration of NO is high enough, there is little experimental evidence to date that this kind of *direct* microbicidal effect also takes place within NOS-II-positive activated host cells or during infectious diseases in vivo. Recently, however, evidence was presented showing that, in HeLa epithelial cells infected with coxsackie virus, exogenous NO (derived from activated macrophages in the vicinity) *S*-nitrosylates the cysteine residue in the active site of the viral protease 3C, thereby inhibiting the protease activity and viral life cycle (SAURA et al., 1999). The reported *indirect* (immunomodulatory) effects of NO [inhibition of parasite spreading by up-regulation of IFN-γ(DIEFENBACH et al. 1998)] alone are unlikely to cause efficient control of microbial pathogens in vivo.
3. Susceptibility of an infectious agent to NO in vitro does not imply that NO is always indispensable for the control of the microorganism in vivo. For example, the protozoa *Leishmania major* and *Toxoplasma gondii* are prototypes of parasites that are killed or inhibited by NOS-II-positive host cells in vitro. Furthermore, functional suppression or deletion of NOS-II exacerbated the course of both infections in certain strains of mice (WEI et al. 1995; SCHARTON-KERSTEN et al. 1997; DIEFENBACH et al. 1998). On the other hand, *T. gondii* parasites were recently found to remain in a clinically latent state in BALB/c mice even in the absence of NOS2 activity (SCHLÜTER et al., 1999).
4. In various NO-susceptible bacterial species (*Salmonella typhimurium*, *Mycobacterium tuberculosis*), genes and mechanisms have been discovered which can confer phenotypic NO resistance (EHRT et al. 1997; FANG 1997; CHEN et al. 1998a; RUAN et al., 1999). The ability to (partially) evade nitrosative stress might account for the virulence of certain bacterial isolates (*M. tuberculosis*) and could also contribute to the long-term persistence of these pathogens.

b) NO as an Antimicrobial Molecule in Humans

Despite a large number of studies, it is still not clear to what extent NOS-II-derived NO fulfills antimicrobial functions in humans. This uncertainty originates from a number of circumstances. First, the majority of researchers has obviously been unable to efficiently stimulate monocytes or human peripheral blood mononuclear cells for high-output production of NO, at least when applying stimulation conditions adapted from a murine system, such as IFN-γ plus LPS. Second, even in cases where the authors succeeded in the induction of NOS-II, NO was not regularly shown to mediate enhanced killing. Third, cytokines, which in the mouse system failed to induce (or even inhibited) NOS-II, paradoxically were reported to activate human monocytes/macrophages. Fourth, expression of NOS-II was observed in infected human tissues and in inflammatory cells isolated from human body fluids but, for obvious reasons, final proof of its functional relevance could not be provided (ANSTEY et al. 1996; NICHOLSON et al. 1996; WHEELER et al. 1997; KHANOLKAR-YOUNG et al. 1998). However, a recent immunohistological study of the expression of NOS-II in the skin of patients with leprosy provided indirect evidence for an antimicrobial function of NO. Paucibacillary tuberculoid lesions showed strong staining for NOS-II, whereas multibacillary lepromatous lesions were NOS-II-negative (KHANOLKAR-YOUNG et al. 1998). Likewise, in Tanzanian children with *Plasmodium falciparum* infection, the expression of NOS-II in peripheral blood monocytes and the plasma levels of NO_x were highest in subclinical (asymptomatic) disease and lowest or absent in patients with cerebral malaria (ANSTEY et al. 1996).

It is possible that NO-dependent control of intracellular microorganisms, such as *Mycobacterium*, *Toxoplasma*, *Leishmania* spp. and *Trypanosoma cruzi*, primarily takes place in human host cells other than blood monocytes. These include differentiated macrophages (derived from blood monocytes by prolonged in vitro culture) (BONECINI-ALMEIDA et al. 1998; VILLALTA et al. 1998), certain populations of tissue macrophages (alveolar macrophages) (NOZAKI et al. 1997), astrocytes (LEE et al. 1994; PETERSON et al. 1995), hepatocytes (MELLOUK et al. 1994), and perhaps epithelial cells (GUO et al. 1995; SALZMAN et al. 1998) and neutrophils (EVANS et al. 1996; WHEELER et al. 1997). Furthermore, although some groups (reviewed in: ALBINA 1995) have reported induction of NOS-II and NO-dependent antimicrobial activity in human monocytes/macrophages by IFN-γ plus LPS, alternative (cytokine) stimuli might be more relevant (Table 5). IFN-γ repeatedly failed to activate human blood monocytes/macrophages or human microglia for the expression of NOS-II and/or NO-dependent antimicrobial killing activity (MURRAY and TEITELBAUM 1992; HATZIGEORGIOU et al. 1993; SCHNEEMANN et al. 1993; CHAO et al. 1994; WARWICK-DAVIES et al. 1994; LEE et al. 1995), whereas cytokines, such as type-I IFN, IL-4, or chemokines, were able to do so (VOULDOUKIS et al. 1995; SHARARA et al. 1997; VILLALTA et al. 1998).

c) Interaction Between NO and Other Antimicrobial Effector Pathways

In addition to the production of NO by NOS-II, other antimicrobial effector mechanisms are known to operate inside phagocytes. These include: the generation of reactive oxygen intermediates (O_2^-, H_2O_2) by the multicomponent enzyme reduced nicotinamide adenine dinucleotide phosphate (NADPH) oxidase, the formation of HOCl and other chlorinated oxidants by the myeloperoxidase (MPO)/H_2O_2/Cl^- system, and the depletion of tryptophan by induction of indoleamine-2,3-dioxygenase (IDO). NO has been demonstrated to interact with each of these pathways.

In mammalian neutrophils, NO was shown to inhibit the activity of NADPH oxidase, leading to a reduced production of O_2^- (CLANCY et al. 1992; FUJII et al. 1997). Conversely, O_2^- was shown to scavenge NO, thereby reducing NO-dependent antimicrobial activity (ASSREUY et al. 1994). NO, however, is also able to increase the release of O_2^- and H_2O_2 by neutrophils (ANDONEGUI et al. 1999) and to agonistically react with O_2^-, leading to the generation of

Table 5. Cytokine-induced nitric oxide (NO)-dependent control of intracellular microorganisms in human phagocytes. NO was implicated in the killing process, based on the increased number of intracellular microorganisms found after treatment of the phagocytes with L-arginine analogues

Microorganism	Phagocyte population	In vitro stimulus for the induction of killing	Reference
Mycobacterium bovis BCG	Human alveolar macrophages (from patients with pulmonary fibrosis)	None (preactivation by cytokines in vivo?)	Nozaki et al. 1997
Mycobacterium tuberculosis (H37Rv)	Human macrophage-like cell line (derived from HL-60 cells)	1,25-Dihydroxy vitamin D_3 (stimulation for 6 days)	Rockett et al. 1998
Leishmania spp.	Adherent human blood monocytes	IL-4 + anti-CD23	Vouldoukis et al. 1995, 1997
Trypanosoma cruzi	Blood monocyte-derived macrophages (4 weeks of differentiation)	MIP-1α, MIP-1β or RANTES	Villalta et al. 1998
Toxoplasma gondii	Human astrocytes	IFN-γ plus IL-1β	Peterson et al. 1995

BCG, bacille Calmette-Guerin; *HL*, human leukemia; *IFN*, interferon; *IL*, interleukin; *MIP*, macrophage inflammatory protein; *RANTES*, regulated on activation, normal T cell expressed and secreted.

peroxynitrite (ONOO⁻) (Beckman et al. 1990), a potent oxidant which might directly damage microbes via nitration of tyrosine residues critical for their survival (Denicola et al. 1993; Evans et al. 1996; Vazquez-Torres et al. 1996). In addition, some microbes that partially resist the toxic activity of NO or H_2O_2 are efficiently killed when exposed to both oxidants simultaneously (Pacelli et al. 1995).

Recently, it was demonstrated that activated human polymorphonuclear neutrophils (PMN) can utilize the MPO pathway for conversion of nitrite (NO_2^-) into the oxidants nitryl chloride (NO_2Cl) and nitrogen dioxide (NO_2), which can cause tyrosine nitration and chlorination of target molecules (Eiserich et al. 1998). NOS-II-positive cells, such as activated macrophages, might function as exogenous sources of NO_2^- and thereby promote the antimicrobial activity of neighboring neutrophils. However, as NOS-II protein was shown to co-localize with MPO in the primary granules of neutrophils (Evans et al. 1996), the NO_2^--dependent formation of NO_2Cl and NO_2 in the phago(lyso)some of neutrophils is likely to also occur in the absence of macrophages. In any case, the conversion of the NOS-II and MPO pathways might represent a novel mechanism of the antimicrobial machinery of PMNs.

The IDO-catalyzed degradation of L-tryptophan to kynurenine can be efficiently induced by IFN-γ in human phagocytes, but only poorly in mouse cells. NOS-II, in contrast, is strongly expressed in response to IFN-γ in mouse but not in human macrophages (DÄUBENER and HADDING 1997). Recent studies demonstrated that NO is a potent inhibitor of IDO activity in macrophages because it binds to the heme complex of IDO (THOMAS et al. 1994; ALBERATI-GIANI et al. 1997). Thus, a lack of endogenous NOS-II might explain the prominent expression of IDO by IFN-γ-activated human cells. Conversely, IDO activity was partially restored in mouse macrophages after inhibition of NO production (ALBERATI-GIANI et al. 1997). To date, the biological implications of this relationship between L-arginine and L-tryptophan metabolism remain largely unknown. However, as picolinic acid, a catabolite of L-tryptophan, acts as a co-stimulus for the induction of NOS-II in mouse macrophages (MELILLO et al. 1993), one could speculate that NO functions as a feedback regulator of IDO activity. Furthermore, high-output production of NO might impair the IDO-dependent control of microbial pathogens within the same or a neighboring cell. Indeed, in a *T. gondii*-infected epithelial cell line, expression of NOS-II after stimulation with IL-1β antagonized the IDO-dependent toxoplasmostasis induced by IFN-γ (DÄUBENER et al. 1999).

3. Anti-Tumor Function

The observation that macrophage cytotoxicity towards tumor cells requires L-arginine was one of the key findings that led to the discovery of the NOS-II pathway (HIBBS et al. 1987). A series of studies demonstrated that NO (produced by macrophages or applied as authentic gas) causes cytostasis of various tumor cell lines (HIBBS et al. 1988; MARLETTA et al. 1988; STUEHR and NATHAN 1989; SVEINBJORNSSON et al. 1996). Subsequently, various mechanisms were identified by which NO or NO-derived reactive nitrogen intermediates might cause death (apoptosis) of tumor cells. These include: the inhibition of mitochondrial iron–sulfur enzymes of the respiratory chain, inhibition of the citric-acid-cycle enzyme *cis*-aconitase (DRAPIER and HIBBS 1988), the inactivation of ribonucleotide reductase (LEPOIVRE et al. 1990; KWON et al. 1991), the consumption of cellular adenosine triphosphate by triggering the nuclear DNA-repair enzyme poly(ADP–ribose) polymerase (SZABÓ et al. 1994), the activation of IL-1β-converting enzyme (ICE)-like proteases (caspases) (BROCKHAUS and BRUNE 1998; CHLICHIA et al. 1998) and the accumulation of wild-type p53, a central regulator of cell growth and apoptosis (FORRESTER et al. 1996). In keeping with these in vitro analyses, high level production of NO by NOS-II-transfected tumor cell lines or activation of endogenous NOS-II in tumor cells led to apoptosis, suppression of tumorigenicity and metastasis, and regression of established hepatic metastases in vivo (XIE et al. 1995, 1996). A central role of NOS-II in the antitumor immune response of CD4[+] T cells was also observed in a B16 melanoma mouse model (HUNG et al., 1998).

Despite these encouraging anti-tumor effects of high levels of NO, the function of NO/NOS-II in tumor immunology is far more complex.

Several studies have demonstrated resistance to NO in certain tumor cells and the existence of NO-independent mechanisms of macrophage tumor cytotoxicity (MATEO et al. 1996; LAVNIKOVA et al. 1997). Furthermore, constitutive expression of NOS-II has been detected in human breast, brain, head, neck, and colon cancers with a high frequency of p53 mutations, where NO appears to promote tumor growth, vascularization, and invasiveness (JENKINS et al. 1995; THOMSEN et al. 1997). The pro-tumor function of NO might result from its stimulatory effect on the expression of vascular endothelial growth factor (VEGF) by tumor cells (CHIN et al. 1997) and from a strong selective pressure for NOS-II-positive tumor cells expressing mutant p53. Whereas wild-type p53 induces growth arrest of tumor cells, serves as a negative regulator of NOS-II, and inhibits tumor angiogenesis, NOS-II-positive tumor cells with mutant p53 showed accelerated growth, enhanced expression of VEGF, and increased neovascularization in vivo (AMBS et al. 1998).

4. Autotoxic Functions

Recognition of the ability of the radical NO to induce cellular dysfunctions and apoptotic or necrotic cell death originally sparked a large number of detailed analyses on the role of NO in autoimmune and degenerative diseases. These studies mostly provided evidence for a disease-promoting effect of NO in diseases, such as diabetes, autoimmune arthritis or myocarditis, EAE, host-vs-graft reaction (graft rejection) in mice, and multiple sclerosis, acquired immunodeficiency syndrome dementia, and Alzheimer's and Parkinson's diseases in humans (WORRALL et al. 1995; BOGDAN 1998; KOLB and KOLB-BACHOFEN 1998; KRÖNCKE et al. 1998). However, the classic view that, in the immune system, NOS-II-derived NO resembles a double-edged sword that protects against infectious agents but damages tissue during autoimmune processes, requires careful revision.

In various experimental autoimmune models (EAE, experimental allergic uveitis, autoimmune arthritis, autoimmune interstitial nephritis) deletion or functional inhibition of NOS-II failed to ameliorate or even exacerbated the course of disease (GILKESON et al. 1997; BOGDAN 1998). As discussed below, a number of anti-inflammatory properties of NO are likely to explain the more severe clinical phenotype seen in the absence of NOS-II.

In several experimental infections, expression of NOS-II did not confer protection of the host but was associated with more severe pathology (cytopathic effects and tissue damage) and overt clinical disease. Herpes simplex virus type 1, cytomegalovirus, or influenza virus caused consolidating pneumonitis and death in wild-type (NOS-II$^{+/+}$) mice, whereas NOS-II$^{-/-}$ hosts (or wild-type mice treated with an inhibitor of NOS-II) survived with little histopathologic evidence of pneumonitis irrespective whether the virus was cleared from the lungs or not (AKAIKE et al. 1996; ADLER et al. 1997; TANAKA et al. 1997; KARUPIAH et al. 1998a). A positive correlation between disease progression and the production of NO was also observed during viral infections of the central nervous systems. In mice infected with Borna disease virus or lymphocytic choriomeningitis virus, the expression of NOS-II paralleled the inflammatory damage in the brain (KOPROWSKI et al. 1993; ZHENG et al. 1993; CAMPBELL et al. 1994). In mice chronically infected with *Mycobacterium avium*, the infiltration of monocytic cells and the formation of granulomata in the spleen were associated with the presence of NOS-II activity; in NOS-II$^{-/-}$ mice splenomegaly and immunopathology was much less prominent (DOHERTY and SHER 1997). In two other studies, in contrast, NOS-II-deficiency led to an increased number, cellularity and size of the granulomatous lesions, to elevated levels of IFN-γ in the serum, and to a reduced bacterial burden in the spleens and lungs (GOMES et al. 1999; EHLERS et al. 1999). A counterprotective effect of NOS-II is likely to occur in pertussis, because the tracheal cytotoxin of *Bordetella pertussis* was shown to damage ciliated epithelial cells via induction of NOS-II in neighbouring non-ciliated, secretory epithelial cells (HEISS et al., 1994; FLAK et al. 1999). In the gastrointestinal tract, oral infection of C57BL/6 mice with *T. gondii* led to extensive tissue necrosis, which was largely prevented by genetic deletion or functional inactivation of NOS-II (KHAN et al. 1997; LIESENFELD et al., 1999). In *Trypanosoma brucei* infection of mice, NOS-II deficiency was associated with an increased proliferation of T cells, enhanced production of IFN-γ (HERTZ and MANSFIELD 1999; MILLAR et al. 1999) and, in one study, with a more rapid clearance of the parasites from the blood stream (MILLAR et al., 1999). Thus, despite well-established antimicrobial and host-protective effects, NOS-II-derived NO clearly can also mediate disease in various organs and in response to viral, bacterial or protozoan pathogens.

Another possible untoward effect of NO during chronic infections relates to the capacity of NO to deaminate DNA (WINK et al. 1991) and to induce functional modifications of p53, the protein product of a major cellular tumor-suppressor gene (CALMELS et al. 1997). NO production in macrophages or exposure of macrophages to exogenous sources of NO results in increased rates of gene mutations (ZHUANG et al. 1998). NO-induced mutagenesis might be even further enhanced when, during chronic infections, viral or microbial gene products inactivate p53, which otherwise acts as an inhibitor of NOS-II gene transcription (FORRESTER et al. 1996). The hepatitis-B virus transactivator protein X, for example, is not only an inducer of NOS-II, but also inactivates p53. Both processes might account for the prominent expression of NOS-II in the hepatocytes of patients with chronic hepatitis B and might contribute to the development of secondary hepatocellular carcinoma (MAJANO et al. 1998).

5. Regulatory Functions

The history of NO as an immunoregulatory molecule started with the discovery of its strong anti-proliferative effect on T lymphocytes (ALBINA et al. 1991; HOFFMAN et al. 1990; MILLS 1991). Since then, many other cells of the immune system (B lymphocytes, NK cells, macrophages, eosinophils; Table 1) have been described as potential targets of NO. Furthermore, NO was also shown to exert various immunostimulatory effects.

a) Regulation of Proliferation, Apoptosis and Survival, and Cytotoxic Activity of Lymphocytes

It has been known for more than 30 years that macrophages not only stimulate T- and B-lymphocytes but can also inhibit immune responses (ALLISON 1978). One important mechanism by which these "suppressor macrophages" impair the proliferation of mitogen- or antigen-activated T-lymphocytes is the production of NO by NOS-II (ALBINA et al. 1991; HOFFMAN et al. 1990; MILLS 1991). There is evidence that T lymphocytes themselves regulate their inhibition by reactive nitrogen intermediates. Activated, but not resting T lymphocytes express the ectoenzyme γ-glutamyl transpeptidase. This enzyme catalyzes the decomposition of S-nitrosoglutathione leading to the release of NO that rapidly inhibits DNA synthesis in the T cells (HENSON et al., 1999). This NO-mediated inhibition of T-cell proliferation and activity can have multiple consequences.

1. In various bacterial, protozoan, fungal or helminthic infectious disease models, NOS-II-derived NO was shown to cause immunosuppression, which might lead to increased microbial burdens or to strongly reduced antibody responses to heterologous antigens (GREGORY et al. 1993; SCHLEIFER and MANSFIELD 1993; CANDOLFI et al. 1994; ABRAHAMSOHN and COFFMAN 1995; AHVAZI et al. 1995; BOCCA et al. 1998; MABBOTT et al. 1998; SCHWACHA et al. 1998; WU-HSIEH et al. 1998).
2. Similar observations were also made in several tumor mouse models, where intratumoral macrophages were found to (hyper-)express NO and to suppress alloantigen-driven T-cell proliferation (ALLEVA et al. 1994; LEJEUNE et al. 1994; YOUNG et al. 1996), which might promote tumor growth. In graft-vs-host reactions, NO was shown to contribute to the immunosuppressive state (HOFFMAN et al. 1993).
3. The deletion of T cells or the inhibition of their proliferation or function (cytokine production) can also be a host-protective, counter-regulatory (feedback) effect of NO during infections, bacterial superantigen-induced shock, and autoimmune diseases, where it serves to abate or stop inflammatory processes (FLORQUIN et al. 1994; WEI et al. 1995; BOGDAN 1998; TARRANT et al. 1999). NO might also be involved in the elimination of autoreactive T cells in the thymus (negative selection) (TAI et al. 1997; DOWNING et al. 1998; VIRAG et al. 1998).

Proposed molecular mechanisms by which NO achieves control of T-lymphocytes include: the apoptosis of antigen-presenting cells (macrophages, dendritic cells) (ALBINA et al. 1993; LU et al. 1996), the modulation of IL-12 production by antigen-presenting cells (Sects. B.III.5.b, B.III.5.c), the down-regulation of major histocompatibility complex class-II antigens or intercellular adhesion molecules (ICAMs) operating between macrophages and T cells via expression and stabilization of IκB (SICHER et al. 1995; PENG et al. 1998), the disruption of the Jak3 kinase/Stat5 (and perhaps also of the Jak2) T-cell activation pathway (BINGISSER et al. 1998; DUHÉ et al. 1998), the suppression of IL-2 gene expression by abrogation of the nuclear binding of the zinc finger transcription factor Sp1 (BERENDJI et al. 1999), and the induction of T-cell apoptosis (BRITO et al., 1999; FEHSEL et al. 1995; TAI et al. 1997; ZETTL et al. 1997; VIRAG et al. 1998; TARRANT et al. 1999).

Other types of lymphocytes are also subject to regulation by NOS-II-derived NO. In B lymphocytes, NOS-II-derived NO was shown to inhibit apoptosis and to prevent the reactivation of Epstein-Barr virus (MANNICK et al. 1994; GENARO et al. 1995). The anti-apoptotic effect of NO in lymphocytes and other leukocytes involves overexpression of death domain proteins (Fas-associated death domain protein), and inhibition of the TNF-α-or Apo-1/Fas-triggered apoptosis pathways by S-nitrosylation of members of the proteolytic cascade (caspase-8, caspase-1, and caspase-3) (LI et al. 1997; MANNICK et al. 1997; MOHR et al. 1997; DIMMELER et al. 1997, 1998; HEBESTREIT et al. 1998). In human B and T cell lines expressing NOS-II, induction of apoptotic cell death by cross-linking of Fas was paralleled by denitrosylation of caspase-3 (MANNICK et al. 1999). Exogenous NO (derived from chemical NO donors) was shown to inhibit the cytotoxic activity of NK cells and lymphokine-activated killer cells (ITO et al. 1996; SALVUCCI et al. 1998).

b) Modulation of Cytokine Responses

For a number of cell types (macrophages, T cells, endothelial cells, fibroblasts), the production of certain cytokines was reported to be dependent on the presence of endogenous NO (Table 6). The conclusions are mainly based on the use of chemical NO donors and of non-selective inhibitors of all NOS isoforms, some of which are also known to exert effects unrelated to the inhibition of NO production. Therefore, we must be cautious not to overestimate the role of NOS-II-derived NO until the published results are confirmed by the currently available transgenic systems. Nevertheless, there is currently no doubt that NO has the capacity to regulate components of a broad spectrum of signaling pathways, including those that are critical for the production of cytokines. Examples are ion channels, G proteins (the proto-oncogene $p21^{ras}$), protein tyrosine kinases, Janus kinases (Jak1, Jak2, Jak3, Tyk2), mitogen-activated protein kinases, caspase-1 (ICE), redox-sensitive kinases, and transcription factors (NF-κB, AP-1, Sp1, vitamin D-receptor, retinoid X-receptor) (LANDER 1997; DUHÉ et al. 1998; HIERHOLZER et al. 1998; KIM et al. 1998; KLATT et al., 1999; UMANSKY et al. 1998; BERENDJI et al. 1999; DIEFENBACH et al. 1999; KRÖNCKE et al., 2000). However, for most types of cells, the pathway that is regulated by NO and thereby mediates the altered cytokine phenotype remains to be determined. Interestingly, in several instances, both activating (co-stimulatory) and inactivating effects of NO have been described, depending on the concentration of NO, the source of NO, or the exact stimulation conditions (Table 6). Mechanisms that are likely to account for the inhibitory or stimulatory action of NO on transcription factors in cells include the

Table 6. Regulatory effects of nitric oxide (NO) synthase 2 (NOS-II)-derived NO on the expression of cytokines

Cytokine	Cell type	Effect of NO[a]	Reference
IL-1α	Human colonic epithelial cells (HT29); mouse peritoneal macrophages	↑	Marcinkiewicz et al. 1995; Vallette et al. 1996
IL-1β	Mouse peritoneal macrophages	↓[b]	Kim et al. 1998
	RAW 264.7 cells	↑ or ↓	Hill et al. 1996; Kim et al. 1998
	Human alveolar macrophages	↓	Persoons et al. 1996
	HL-60 cells (monocytic differentiation)	↑	Magrinat et al. 1992
IL-3	Spleen cells	↓	Marcinkiewicz and Chain 1993
IL-4	Th2 lymphocytes	(↑)	Chang et al. 1997
IL-6	Lung, liver (in mice); human keratinocytes and monocytes	↓	Becherel et al. 1994; Mossalayi et al. 1994; Hierholzer et al. 1998
	Rat Kupffer cells; mouse J774 macrophage line	↑	Stadler et al. 1993; Deakin et al. 1995
IL-8	Keratinocyte cell lines; human melanoma cell lines; human whole blood; human neutrophils	↓	Andrew et al. 1995; Remick and Villarete 1996; Andrew et al. 1999; Corriveau et al. 1998
IL-10	Human macrophages	↓	Dugas et al. 1996
	Mouse inflammatory lymph-node cells	↑	Ianaro et al. 1994
IL-12	Mouse macrophages (J774, RAW264.7, peritoneal exudate)	↓ or ±[b] or ↑	Diefenbach et al. 1998; Huang et al. 1998a; Rothe et al. 1996; Mullins et al. 1999
IL-18	Mouse peritoneal macrophages	↓*	Kim et al. 1998
IFN-γ	Human PBMNC; mouse lymph-node cells	↑	Ianaro et al. 1994; Sriskandan et al. 1996
	Th1 lymphocytes	↓ or ↑	Taylor-Robinson et al. 1994; Niedbala et al. 1999
	Natural killer cells	↑	Diefenbach et al. 1998, 1999; Cifone et al. 1999; Salvucci et al. 1999
TGF-β	Mouse peritoneal macrophages	↑	Diefenbach et al. 1998
Granulocyte CSF	Lung, liver (in vivo)	↓	Hierholzer et al. 1998
Macrophage CSF	Vascular endothelial cells	↓	Peng et al. 1995b
TNF-α<??2>	HL-60 cells (monocytic); human PBMNC; mouse macrophage line J774; mouse peritoneal macrophages	↑	Magrinat et al. 1992; Eigler et al. 1993; Lander et al. 1993; Deakin et al. 1995; Dugas et al. 1996; Huang et al. 1998a
	Human neutrophils	↓	Van Dervort et al. 1994
	Mouse macrophage line RAW264.7	↓	Eigler et al. 1995
Macrophage inflammatory protein-1α	Human monocytes	↑	Mühl and Dinarello 1997
Monocyte chemoattractant protein-1	Human endothelial cells	(↓)	Zeiher et al. 1995

↑, increase of cytokine production; ↓, decrease of cytokine production; *CSF*, colony-stimulating factor; *HL*, human leukemia; *IFN*, interferon; *IL*, interleukin; *PBMNC*, peripheral blood mononuclear cells; *Th2*, T-helper cell type 2; *TGF*, transforming growth factor; *TNF*, tumor necrosis factor.

S-nitrosylation of NF-κB (dela TORRE et al., 1999), the S-glutathionylation of c-Jun (KLATT et al., 1999), or the disruption of zinc finger domains (KRÖNCKE et al. 2000).

α. NO and IL-12

IL-12 is a heterodimeric 70-kDa cytokine (consisting of p35 and a p40 subunit) which is of utmost importance for the immune response towards intracellular (viral, bacterial, and protozoan) microbial pathogens. In particular, antibody-neutralization and gene-deletion studies have demonstrated that, in the absence of IL-12, the activation of NK and Th1 cells, and hence the production of IFN-γ, are severely impaired (BIRON and GAZZINELLI 1995; TRINCHIERI and GEROSA 1996).

Several interactions between NO and IL-12 have been described. IL-12 induces NOS-II in an IFN-γ-dependent or -independent manner during infections, autoimmune diseases, or alloimmunization. Expression of NOS-II in response to IL-12 can lead to antimicrobial activity, tissue destruction, immunosuppression, or immunoprotection by feedback inhibition or apoptosis of inflammatory cells (SCHWACHA and EISENSTEIN 1997; TAYLOR and MURRAY 1997; ZHANG et al. 1997; KOBLISH et al. 1998; SEGAL et al. 1998; TARRANT et al. 1999). However, there is now compelling evidence that NO also regulates the expression of IL-12, although the published results are discrepant and, at present, hard to reconcile. In a mouse macrophage cell line (IC-21), NOS-II-derived NO was required for the IFN-γ-induced up-regulation of IL-12 p40 mRNA. Furthermore, exogenous NO-donors increased the level of IL-12 p40 mRNA but did not induce IL-12 p35 mRNA (ROTHE et al. 1996). In accordance with these observations, NOS-II$^{-/-}$ mice exhibited a markedly reduced baseline expression of IL-12 p40 mRNA in the footpad or popliteal lymph node when compared with NOS-II$^{+/+}$ wild-type mice, whereas the level of IL-12 p35 mRNA was comparable in both strains of mice. Inflammatory peritoneal macrophages from NOS-II$^{+/+}$ and NOS-II$^{-/-}$ mice, in contrast, yielded comparable amounts of IL-12 p70 protein when stimulated with IFN-γ plus LPS. Likewise, the levels of IL-12 p40 mRNA in the footpad and draining lymph node of *L. major*-infected NOS-II$^{-/-}$ mice were indistinguishable from those of NOS-II$^{+/+}$ mice from day 7 of infection onwards, indicating that NOS-II deficiency does not lead to a persistent lack of IL-12 in vivo (DIEFENBACH et al. 1998). Quite different conclusions were reached in a recent study by Huang et al., who found an elevated production of IL-12 p40 or p70 in vitro by IFN-γ/LPS-stimulated peritoneal macrophages derived from NOS2 mutant mice as well as in vivo when the NOS2 mutant mice had been infected with *L. major* for 15 days (HUANG et al., 1998a). It is, however, important to note that this strain of mice does not have a true deletion of the NOS2 gene but expresses residual NOS2 bioactivity (NIEDBALA et al., 1999). Finally, Mullins et al. reported that, in the presence of the NOS-inhibitor *N*-monomethyl-L-arginine, the anticancer agent paclitaxel (Taxol) failed to upregulate the production of IL-12 p70 by peritoneal exudate macrophages from normal or tumor-bearing mice (MULLINS et al. 1999). Thus, NOS-II-derived NO enhances, inhibits, or does not alter the production of IL-12, depending on parameters that remain to be elucidated.

A novel aspect of interaction between NOS-II-derived NO and IL-12 emerged recently during the analysis of the innate immune response to *L. major*. Total lymph-node cells or partially purified NK cells from day-1 infected NOS-II$^{-/-}$ mice completely failed to produce IFN-γ in response to IL-12. Similar results were obtained with highly purified splenic NK cells or lymphokine-activated killer cells, which also did not respond to IL-12 in the absence of NOS-II activity (DIEFENBACH et al. 1999). NOS-II activity turned out to be an indispensable cofactor for the activation of Tyk2 kinase and, thus, for IL-12 signalling in NK cells; the activation of T cells by IL-12, however, did not require NOS-II-derived NO (DIEFENBACH et al. 1999). In rat NK cells activated by IL-2, cytotoxic activity and production of IFN-γ were at least partially dependent on the expression of NOS-II (CIFONE et al. 1999).

c) Leukocyte Chemotaxis and Adhesion

Endothelial adhesion, transmigration and locomotion of leukocytes into the periphery are central to all inflammatory processes. The overwhelming experimental evidence points to a primarily anti-adhesive effect of NO on leukocytes (monocytes, neutrophils), which is likely to be due to a reduced expression of adhesion molecules (vascular cell adhesion molecule-1, E-selectin, ICAM-1) and a diminished production of pro-inflammatory and chemotactic cytokines by vascular endothelial and smooth-muscle cells (KUBES et al. 1991; DE CATERINA et al. 1995; PENG et al. 1995a; KHAN et al. 1996; SHIN et al. 1996; PENG et al. 1998). Accordingly, neutrophil infiltration was enhanced in NOS-II$^{-/-}$ mice with zymosan-induced peritonitis compared with wild-type mice (AJUEBOR et al. 1998).

d) Immune (T-Helper Cell) Deviation

It has been proposed that NOS-II-derived NO induced by IFN-γ down-regulates Th1 responses (characterized by a preponderance of IFN-γ production), and thereby indirectly promotes the expansion of Th2 cells (producing IL-4 and IL-10), e.g. in allergic asthma (BARNES and LIEW 1995; LIEW 1995). The original hypothesis has been mainly based on an apparently "selective" immunoregulatory effect of NO observed with a small panel of T helper cell clones: exogenous NO was shown to suppress the IFN-γ production by Th1 cells, whereas the production of IL-4 by Th2 cells was reported to be unaffected (TAYLOR-ROBINSON et al. 1994). The hypothesis has been more recently supported by the observation that, in NOS-II$^{-/-}$ mice, a number of infections led to an enhanced Th1 response compared with NOS-II$^{+/+}$ mice (GOMES et al., 1999; HERTZ and MANSFIELD, 1999; JAMES et al., 1998; MACLEAN et al., 1998; MCINNES et al., 1998; MILLAR et al., 1999; WEI et al., 1995). Furthermore, one study showed that sodium nitroprusside slightly enhanced the production of IL-4 by two established Th2 clones and markedly enhanced the production of IL-4 by EL-4 thymoma cells (the latter was accompanied by a reduced expression of IL-2) (CHANG et al. 1997). Finally, as discussed above, there is evidence that, in the absence of NOS-II-derived NO, macrophages express more mature IL-12 protein (p70 heterodimer) (HUANG et al. 1998a) and perhaps less IL-12 p40 homodimers (a known IL-12 receptor antagonist) (ROTHE et al. 1996), which might further promote the development of Th1 cells. There are a number of caveats that should be considered before joining the concept that expression of NOS-II results in a Th2 bias of the immune response (KOLB and KOLB-BACHOFEN 1998). First, continuous expression of NOS-II during a Th1 response towards intracellular pathogens is not known to be followed by a subsequent Th2 conversion. Second, in several mouse models, inhibition or genetic deletion of NOS-II was shown to cause an unaltered or even diminished Th1 response in vivo (HOGABOAM et al. 1997; DIEFENBACH et al. 1998; HÖLSCHER et al. 1998; IGIETSEME et al. 1998). Third, with macrophages derived from a different NOS-II$^{-/-}$ strain, the production of IL-12 p70 was not found to be regulated by NO (DIEFENBACH et al. 1998). Fourth, no strong molecular evidence has been presented explaining why NO should be inhibitory for Th1 but stimulatory for Th2 cells. In fact, small amounts of NO were demonstrated to promote rather than to inhibit the development of Th1 cells, whereas the differentiation of Th2 cells or the activity of fully committed Th1 cells remained unaffected by NO (NIEDBALA et al., 1999). Another study demonstrated that small amounts of exogenous NO equally well inhibited the proliferation of mouse Th1 and Th2 clones without a differential effect on cytokine production (VAN DER VEEN et al., 1999). Fifth, exogenous and endogenous NO was shown to inhibit the IgE mediated degranulation of mast cells and basophils, which might be important for the control of Th2-dependent allergic reactions (EASTMOND et al. 1997; IIKURA et al. 1998).

C. Other NOS Isoforms and Perspective

After 10 years of analysis of NO function in the immune system, which has been dominated by studies on NOS-II, it is now time to consider the other two isoforms of NOS (NOS1 or ncNOS, NOS-III or ecNOS). Both types of NOS are not constitutively expressed, though this is implied by their names (c = constitutive or Ca^{2+}/calmodulin-dependent). They are not only regulated by calcium fluxes but also by cytokines, (neuroimmuno-) hormones, and various types of stress (FÖRSTERMANN et al. 1998). Furthermore, their expression is not restricted to neurons or endothelial cells, though this is also suggested by the commonly used acronyms. For example, Ca^{2+}-dependent NOS activity was found in rat thymocytes (CRUZ et al. 1998), and NOS-III was present in human monocytes/macrophages and in B and T lymphocytes (REILING et al. 1994, 1996). Finally, they can also assume typical immunological functions previously assigned to NOS-II, such as the induction of apoptotic cell death (WILLIAMS et al. 1998) and the control of viruses (BARNA et al. 1996; KOMATSU et al. 1996). Similarly, a recently described Ca^{2+}-dependent NOS isoform of plants was shown to exert antibacterial effects against *Pseudomonas syringae*; these effects were inhibitable by cNOS inhibitors (DELLEDONNE et al. 1998).

Will there be a the future for NO in clinical immunology? Clearly, we have learned a lot about the functions of NO in the immune system but, even more than in the case of cytokines, we are puzzled by its pleiotropic effects. Most problematic is the fact that virtually every cell and every immunological parameter can be modulated by NO, often in two opposing ways. To precisely define the parameters that make NO pro-inflammatory (immunostimulatory, immunopathological, anti-apoptotic) or anti-inflammatory (immunosuppressive, pro-apoptotic), host-protective or host-damaging during infections, autoimmune diseases and malignancies will be one of the future targets of NO-directed therapeutic approaches. Other streams of research will be the molecular analysis of NO-regulated intracellular signaling cascades, the further elucidation of the antimicrobial role of NOS-II in humans, the characterization of putative NOS (promoter) allotypes mediating resistance of patients to certain pathogens (BURGNER et al. 1998; KUN et al. 1998), and the development of organ- and cell-specific NO donors and isoform-selective NOS inhibitors.

Acknowledgements. I wish to thank Martin Röllinghoff for critical comments on this chapter and for his continuous support. I apologize to those investigators whose primary papers could only be cited indirectly by reference to reviews owing to space limitations. The preparation of this review and of some of the studies reviewed was supported by The Deutsche Forschungsgemeinschaft (SFB263, project A5).

References

Abrahamsohn IA, Coffman RL (1995) Cytokine and nitric oxide regulation of the immunosuppression in *Trypanosoma cruzi* infection. J Immunol 155:3955–3963

Adler H, Adler B, Peveri P, Werner ER, Wachter H, Peterhans E, Jungi TW (1996) Differential regulation of inducible nitric oxide synthase production in bovine and caprine macrophages. J Infect Dis 173:971–978

Adler H, Beland JL, Del-Pan NC, Kobzik L, Brewer JP, Martin JR, Rimm IJ (1997) Suppression of herpes simplex virus type 1 (HSV-1)-induced pneumonia in mice by inhibition of inducible nitric oxide synthase (iNOS, NOS2). J Exp Med 185:1533–1540

Ahvazi BC, Jacobs P, Stevenson MM (1995) Role of macrophage-derived nitric oxide in suppression of lymphocyte proliferation during blood-stage malaria. J Leukoc Biol 58:23–31

Ajizian SJ, English BK, Meals EA (1999) Specific inhibitors of p38 and extracellular signal-related kinase mitogen-activated protein kinase pathways block inducible nitric oxide synthase and tumor necrosis factor accumulation in murine macrophages stimulated with lipopolysaccharide and interferon-γ. J Infect Dis 179:939–944

Ajuebor MN, Virag L, Flower RJ, Perretti M, Szabo C (1998) Role of inducible nitric oxide synthase in the regulation of neutrophil migration in zymosan-induced inflammation. Immunology 95:625–630

Akaike T, Noguchi Y, Ijiri S, Setoguchi K, Suga M, Zheng YM, Dietzschold B, Maeda H (1996) Pathogenesis of influenza virus-induced pneumonia: involvement of both nitric oxide and oxygen radicals. Proc Natl Acad Sci USA 93:2448–2453

Albakri QA, Stuehr DJ (1996) Intracellular assembly of inducible NO synthase is limited by NO-mediated changes in heme insertion and availability. J Biol Chem 271:5414–5421

Alberati-Giani D, Malherbe P, Riccardi-Castagnoli P, Köhler C, Denis-Donini S, Cesura AM (1997) Differential regulation of indoleamine 2,3-dioxygenase expression by nitric oxide and inflammatory mediators in IFN-γ activated murine macrophages and microglial cells. J Immunol 159:419–426

Albina JE (1995) On the expression of nitric oxide synthase by human macrophages. Why no NO? J Leukoc Biol 58:643–649

Albina JE, Abate JA, Henry WL Jr (1991) Nitric oxide production is required for murine resident peritoneal macrophages to suppress mitogen-stimulated T cell proliferation. Role of IFN-γ in the induction of the nitric oxide-synthesizing pathway. J Immunol 147:144–148

Albina JE, Cui S, Mateo RB, Reichner JS (1993) Nitric oxide-mediated apoptosis in murine peritoneal macrophages. J Immunol 150:5080–5085

Albina JE, Henry WL, Mastrofrancesco B, Martin B-A, Reichner JS (1995) Macrophage activation by culture in an anoxic environment. J Immunol 155:4391–4396

Alleva DG, Burger CJ, Elgert KD (1994) Tumor-induced regulation of suppressor macrophage nitric oxide and TNF-α production. Role of tumor-derived IL-10, TGF-β and prostaglandin E2. J Immunol 153:1674–1686

Allison AC (1978) Mechanisms by which activated macrophages inhibit lymphocyte responses. Immunological Rev 40:3–27

Alonso A, Carvalho J, Alonso-Torre SR, Nunez L, Bosca L, Crespo MS (1995) Nitric oxide synthesis in rat peritoneal macrophages is induced by IgE/DNP complexes and cyclic AMP analogues. J Immunol 154:6475–6483

Ambs S, Merriam WG, Ogunfusika MO, Bennett WP, Ishibe N, Hussain P, Tzeng EE, Geller DA, Billiar TR, Harris CC (1998) p53 and vascular endothelial growth factor regulate tumor growth of NOS2-expressing human carcinoma cells. Nat Medicine 4:1371–1376

Amin AR, Vyas P, Attur M, Leszczynska-Piziak J, Patel IR, Weissmann G, Abramson SB (1995) The mode of action of aspirin-like drugs: effect on inducible nitric oxide synthase. Proc Natl Acad Sci USA 92:7926–7930

Amin AR, Attur MG, Thakker GD, Patel PD, Vyas PR, Patel RN, Patel IR, Abramson SB (1996) A novel mechanism of action of tetracyclines: effects on nitric oxide synthases. Proc Natl Acad Sci USA 93:14014–14019

Andonegui G, Trevani AS, Gamberale R, Carreras MC, Poderoso JJ, Giordano M, Geffner JR (1999) Effect of nitric oxide donors on oxygen-dependent cytotoxic responses by neutrophils. J Immunol, in press

Andrew PJ, Harant H, Lindley IJD (1999) Upregulation of interleukin-1β-stimulated interleukin-8 in human keratinocytes by nitric oxide. Biochem Pharmacol 57:1423–1429

Andrew PJ, Harant H, Lindley IJ (1995) Nitric oxide regulates IL-8 expression in melanoma cells at the transcriptional level. Biochem Biophys Res Commun 214:949–956

Anstey NM, Weinberg JB, Hassanali MY, Mwaikambo ED, Manyenga D, Misukonis MA, Arnelle DR, Hollis D, McDonald MI, Granger DL (1996) Nitric oxide in Tanzanian children with malaria: inverse relationship between malaria severity and nitric oxide production/nitric oxide synthase type 2 expression. J Exp Med 184:557–567

Appelberg R (1995) Opposing effects of interleukin-10 on mouse macrophage functions. Scand J Immunol 41:539–544

Asano K, Chee CBE, Gaston B, Lilly CM, Gerard C, Drazen JM, Stamler JS (1994) Constitutive and inducible nitric oxide synthase gene expression, regulation, and activity in human lung epithelial cells. Proc Natl Acad Sci USA 91:10089–10093

Assreuy J, Cunha FQ, Liew FY, Moncada S (1993) Feedback inhibition of nitric oxide synthase by nitric oxide. Br J Pharmacol 108:833–837

Assreuy J, Cunha FQ, Epperlein M, Noronha-Dutra A, O'Donnell CA, Liew FY, Moncada S (1994) Production of nitric oxide and superoxide by activated macrophages and killing of *Leishmania major*. Eur J Immunol 24:672–676

Attur MG, Patel RN, Abramson SB, Amin SR (1997) Interleukin-17 up-regulation of nitric oxide production in human osteoarthritis cartilage. Arthritis Rheum 40:1050–1053

Badger AM, Cook MN, Lark MW, Newman-Tarr TM, Swift BA, Nelson AH, Barone FC, Kumar S (1998) SB203580 inhibits p38 mitogen-activated protein kinase, nitric oxide production, and inducible nitric oxide synthase in bovine cartilage-derived chondrocytes. J Immunol 161:467–473

Barna M, Komatsu T, Reiss CS (1996) Activation of type III nitric oxide synthase in astrocytes following a neurotropic viral infection. Virology 15:332–343

Barnes PJ, Liew FY (1995) Nitric oxide and asthmatic inflammation. Immunol Today 3:128–130

Bayon Y, Alonso A, Crespo MS (1997) Stimulation of Fcγ receptors in rat peritoneal macrophages induces the expression of nitric oxide synthase and chemokines by mechanisms showing different sensitivities to antioxidants and nitric oxide donors. J Immunol 159:887–894

Becherel P-A, Mossalayi MD, Quaaz F, Le Goff L, Dugas B, Paul-Eugene N, Frances C, Chosidow O, Kilchherr E, Guillosson J-J, Debre P, Arock M (1994) Involvement of cyclic AMP and nitric oxide in IgE-dependent activation of FcεRII/CD23$^+$ normal human keratinocytes. J Clin Invest 93:2275–2279

Beck K-F, Sterzel RB (1996) Cloning and sequencing of the proximal promoter of the rat iNOS gene: activation of NFκB is not sufficient for transcription of the iNOS gene in rat mesangial cells. FEBS Letters 394:263–267

Beckman JS, Beckman TW, Chen J, Marshall PA, Freeman BA (1990) Apparent hydroxyl radical production by peroxynitrite: implications for endothelial injury from nitric oxide and superoxide. Proc Natl Acad Sci USA 87:1620–1624

Berendji D, Kolb-Bachofen V, Zipfel PF, Skerka C, Carlberg C, Kröncke K-D (1999) Zinc finger transcription factor as molecular target for nitric oxide-mediated immunosuppression: inhibition of IL-2 gene expression in lymphocytes. Mol Med 5:721–730

Bernhagen J, Mitchell RA, Calandra T, Voelter W, Cerami A, Bucala R (1994) Purification, bioactivity and secondary structure analysis of mouse and human macrophage migration inhibitory factor (MIF). Biochemistry 33:14144–14155

Betz-Corradin S, Mauël J (1991) Phagocytosis of *Leishmania* enhances macrophage activation by IFN-γ and lipopolysaccharide. J Immunol 146:279–285

Betz-Corradin S, Fasel N, Buchmüller-Rouiller Y, Ransijn A, Smith J, Mauël J (1993a) Induction of macrophage nitric oxide production by interferon-γ and tumor necrosis factor-α is enhanced by interleukin-10. Eur J Immunol 23:2045–2048

Betz-Corradin S, Mauel J, Donini SD, Quattrochi E, Ricciardi-Castagnoli P (1993b) Inducible nitric oxide synthase activity of cloned murine microglial cells. Glia 7:255–262

Bidri M, Ktorza S, Vouldoukis I, Le Goff L, Debre P, Guillosson J-J, Arock M (1997) Nitric oxide pathway is induced by FcεRI and up-regulated by stem cell factor in mouse mast cells. Eur J Immunol 27:2907–2913

Billiar TR, Curran RD, Stuehr DJ, West MA, Bentz BG, Simmons RL (1989) An L-arginine-dependent mechanism mediates Kupffer cell inhibition of hepatocyte protein synthesis in vitro. J Exp Med 169:1467–1472

Bingisser RM, Tilbrook PA, Holt PG, Kees UR (1998) Macrophage-derived nitric oxide regulates T cell activation via reversible disruption of the Jak3/Stat5 signaling pathway. J Immunol 160:5729–5734

Biron CA, Gazzinelli RT (1995) Effects of IL-12 in immune responses to microbial infections: a key mediator in regulating disease outcome. Curr Opin Immunol 7:485–496

Blank C, Bogdan C, Bauer C, Erb K, Moll H (1996) Murine epidermal Langerhans cells do not express inducible nitric oxide synthase. Eur J Immunol 26:792–796

Bocca AL, Hayashi EE, Pinheiro AG, Furlanetto AB, Campanelli AP, Cunha FQ, Figueiredo F (1998) Treatment of *Paracoccidioides brasiliensis*-infected mice with a nitric oxide inhibitor prevents the failure of cell-mediated immune response. J Immunol 161:3056–3063

Bogdan C (1997) Of microbes, macrophages and NO. Behring Inst Res Commun 99:58–72

Bogdan C (1998) The multiplex function of nitric oxide in (auto)immunity. J Exp Med 9:1361–1365

Bogdan C, Nathan C (1993) Modulation of macrophage function by transforming growth factor-β, interleukin 4 and interleukin 10. Ann New York Acad Sci 685:713–739

Bogdan C, Röllinghoff M (1999) How do protozoan parasites survive inside macrophages? Parasitol Today 15:22–28

Bogdan C, Vodovotz Y, Xie Q-W, Nathan C, Röllinghoff M (1994) Regulation of inducible nitric oxide synthase in macrophages by cytokines and microbial products. In: Masihi N (eds) Immunotherapy of Infections. Marcel Dekker, New York, pp 37–54

Bogdan C, Thüring H, Dlaska M, Röllinghoff M, Weiss G (1997) Mechanism of suppression of macrophage nitric oxide release by IL-13. J Immunol 159:4506–4513

Bonecini-Almeida MG, Chitale S, Boutsikakis I, Geng J, Doo H, He S, Ho JL (1998) Induction of in vitro human macrophage anti-Mycobacterium tuberculosis activity: requirement for IFN-γ and primed lymphocytes. J Immunol 160:4490–4499

Bosca L, Lazo PA (1994) Induction of nitric oxide release by MRC OX-44 (anti-CD53) through a protein kinase C-dependent pathway in rat macrophages. J Exp Med 179:1119–1126

Braun JS, Novak R, Gao G, Murray PJ, Shenep JL (1999) Pneumolysin, a protein toxin of Streptococcus pneumoniae, induces nitric oxide production from macrophages. Infect Immun 67:3750-3756

Brightbill HD, Libraty DH, Krutzik SR, Yang R-B, Belisle JT, Bleharski, Maitland M, Norgard MV, Plevy SE, Smale ST, Brennan PJ, Bloom BR, Godowski PJ, Modlin RL (1999) Host defense mechanisms triggered by microbial lipoproteins through toll-like receptors. Science 285:732–734

Brito C, Naviliat M, Tiscornia AC, Vuillier F, Gualco G, Dighiero G, Radi R, Cayota AM (1999) Peroxynitrite inhibits T lymphocyte activation and proliferation by promoting impairment of tyrosine phosphorylation and peroxinitrite-driven apoptotic death. J Immunol 162:3356–3366

Brockhaus F, Brune B (1998) U937 apoptotic cell death by nitric oxide: bcl-2 down-regulation and caspase activation. Exp Cell Res 238:33–41

Bruch-Gerharz D, Fehsel K, Suschek C, Michel G, Ruzicka T, Kolb-Bachofen V (1996) A pro-inflammatory activity of interleukin 8 in human skin: expression of the inducible nitric oxide synthase in psoriatic lesions and cultured keratinocytes. J Exp Med 184:2007–2012

Bulut V, Severn A, Liew FY (1992) Nitric oxide production by murine macrophages is inhibited by prolonged elevation of cyclic AMP. Biochem Biophys Res Commun 195:1134–1138

Burgner D, Xu W, Rockett K, Gravenor M, Charles IG, Hill AV, Kwiatkowski D (1998) Inducible nitric oxide synthase polymorphism and fatal cerebral malaria. Lancet 352:1193–1194

Calmels S, Hainaut P, Ohshima H (1997) Nitric oxide induces conformational and functional modifications of wild-type p53 tumor suppressor protein. Cancer Res 57:3365–3369

Camargo MM, Andrade AC, Almeida IC, Travassos LR, Gazzinelli RT (1997) Glycoconjugates isolated from *Trypanosoma cruzi* but not from *Leishmania* species membranes trigger nitric oxide synthesis as well as microbicidal activity in IFN-γ-primed macrophages. J Immunol 159:6131–6139

Campbell IL, Samimi A, Chiang C-S (1994) Expression of the inducible nitric oxide synthase: correlation with neuropathology and clinical features in mice with lymphocytic choriomeningitis. J Immunol 153:3622–3629

Candolfi E, Hunter CA, Remington JS (1994) Mitogen- and antigen-specific proliferation of T cells in murine toxoplasmosis is inhibited by reactive nitrogen intermediates. Infect Immun 62:1995–2001

Chan ED, Winston BW, Uh S-T, Wynes MW, Rose DM, Riches DWH (1999) Evaluation of the role of mitogen-activated protein kinases in the expression of inducible nitric oxide synthase by IFN-α and TNF-α in mouse macrophages. J Immunol 162:415-422

Chang R-H, Lin Feng M-H, Liu W-H, Lai M-Z (1997) Nitric oxide increased interleukin-4 expression in T lymphocytes. Immunology 90:364–369

Chao CC, Anderson WR, Hu S, Gekker G, Martella A, Peterson PK (1993) Activated microglia inhibit multiplication of *Toxoplasma gondii* via a nitric oxide mechanism. Clin Immunol Immunopath 67:178–183

Chao CC, Gekker G, Hu S, Peterson PK (1994) Human microglial cell defense against *Toxoplasma gondii*. The role of cytokines. J Immunol 152:1246–1252

Charles IG, Palmer RMJ, Hickery MS, Bayliss MT, Chubb AP, Hall VS, Moss DW, Moncada S (1993) Cloning, characterization, and expression of a cDNA encoding an inducible nitric oxide synthase from the human chondrocyte. Proc Natl Acad Sci (USA) 90:11419–11423

Chen B-C, Chen Y-H, Lin W-W (1999) Involvement of p38 mitogen-activated protein kinase in lipopolysaccharide-induced iNOS and COX-2 expression in J774 macrophages. Immunology 97:124–129

Chen L, Xie Q-W, Nathan C (1998a) Alkyl hydroperoxide reductase subunit C (AhpC) protects bacterial and human cells against reactive nitrogen intermediates. Mol Cell 1:795–805

Chen Y, Rosazza JP (1995) Purification and characterization of nitric oxide synthase (NOSNoc) from a *Nocardia* species. J Bacteriol 177:5122–5128

Chen Y-Q, Fisher JH, Wang M-H (1998b) Activation of the RON receptor tyrosine kinase inhibits inducible nitric oxide synthase (iNOS) expression by murine peritoneal exudate macrophages: phosphatidylinositol-3 kinase is required for RON-mediated inhibition of iNOS expression. J Immunol 161:4950–4959

Chin K, Kurashima Y, Ogura T, Tajiri H, Yoshida S, Esumi H (1997) Induction of vascular endothelial growth factor by nitric oxide in human glioblastoma and hepatocellular carcinoma cells. Oncogene 15:437–442

Chinen T, Qureshi MH, Koguchi Y, Kawakami K (1999) Candida albicans suppresses nitric oxide (NO) production by interferon (IFN)-γ and lipopolysacchride (LPS)-stimulated murine peritoneal macrophages. Clin Exp Immunol 115:491–497

Chlichlia K, Peter ME, Rocha M, Scaffidi C, Bucur M, Krammer PH, Schirrmacher V, Umansky V (1998) Caspase activation is required for nitric oxide-mediated, CD95 (APO-1/Fas)-dependent and independent apoptosis in human neoplastic lymphoid cells. Blood 91:4311–4320

Choi SK, Choi HK, Kadono-Okuda K, Taniai K, Kato Y, Yamamoto M, Chowdhury S, Xu J, Miyanoshita A, Debnath NC (1995) Occurrence of novel types of nitric oxide synthase in the silkworm, *Bombyx mori*. Biochem Biophys Res Commun 207:452–459

Christen S, Woodall AA, Shigenaga MK, Sothwell-Keely PT, Duncan MW, Ames BN (1997) γ-Tocopherol traps mutagenic electrophiles such as NO_x and complements α-tocopherol: physiological implications. Proc Natl Acad Sci USA 94:3217–3222

Cifone MG, D'Alo S, Parroni R, Millimaggi D, Biordi L, Martinotti S, Santoni A (1999) Interleukin-2 activated rat natural killer cells express inducible nitric oxide synthase that contributes to cytotoxic function and interferon-γ production. Blood 93:3876-3884

Cifone MG, Festuccia C, Cironi L, Cavallo G, Chessa MA, Pensa V, Tubaro E, Santoni A (1994) Induction of the nitric oxide-synthesizing pathway in fresh and interleukin-2 cultured rat natural killer cells. Cell Immunol 157:181–194

Clancy RM, Leszczynska-Piziak J, Abramson SB (1992) Nitric oxide, an endothelial cell relaxation factor, inhibits neutrophil superoxide anion production via a direct action on the NADPH oxidase. J Clin Invest 90:1116–1121

Colville-Nash PR, Qureshi SS, Willis D, Willoughby DA (1998) Inhibition of inducible nitric oxide synthase by peroxisome proliferator-activated receptor agonists: correlation with induction of heme oxygenase 1. J Immunol 161:978–984

Cooney RV, Franke AA, Harwood PJ, Hatch-Pigott V, Custer LJ, Mordan LJ (1993) γ-Tocopherol detoxification of nitrogen dioxide: superiority to α-tocopherol. Proc Natl Acad Sci USA 90:1771–1775

Corbett JA, McDaniel ML (1995) Intraislet release of interleukin 1 inhibits β cell function by inducing β cell expression of inducible nitric oxide synthase. J Exp Med 181:559–568

Corriveau CC, Madara PJ, Van Dervort AL, Tropea MM, Wesley RA, Danner RL (1998) Effects of nitric oxide on chemotaxis and endotoxin-induced interleukin-8 production in human neutrophils. J Infect Dis 177:116–126

Coulson PS, Smythies LE, Betts C, Mabbott NA, Sternberg JM, Wei X-G, Liew FY (1998) Nitric oxide produced in the lungs of mice immunized with the radiation-attenuated schistosome vaccine is not the major agent casuing challenge parasite elimination. Immunology 93:55–63

Cruz MT, Carmo A, Carvalho AP, Lopes MC (1998) Calcium-dependent nitric oxide synthase activity in rat thymocytes. Biochem Biophys Res Commun 248:98–103

D'Agostino P, La Rosa M, Barbera C, Arcoleo F, Di Bella G, Milano S, Cillari E (1998) Doxycycline reduces mortality to lethal endotoxemia by reducing nitric oxide synthesis via an interleukin 10-independent mechanism. J Infect Dis 177:489–492

Dai WJ, Gottstein B (1999) Nitric oxide-mediated immunosuppression following murine *Echinococcus multilocularis* infection. Immunology 97:107–116

Dai WJ, Bartens W, Köhler G, Hufnagel M, Kopf M, Brombacher F (1997) Impaired macrophage listericidal and cytokine activities are responsible for the rapid death of *Listeria monocytogenes*-infected IFN-γ receptor-deficient mice. J Immunol 158:5297–5304

Dalton DK, Pitts-Meek S, Keshav S, Figari IS, Bradley A, Stewart TA (1993) Multiple defects of immune cell function in mice with disrupted interferon-γ genes. Science 259:1739–1742

Däubener W, Hadding U (1997) Cellular immune reactions directed against *Toxoplasma gondii* with special emphasis on the central nervous system. Med Microbiol Immunol 185:195–206

Däubener W, Mackenzie CR, Posdziech V, Hadding U (1999) Inducible anti-parasitic effector mechanisms in human uroepithelial cells: tryptophan degradation vs NO production. Med Microbiol Immunol 187:143–147

De Caterina R, Libby P, Peng H-B, Thannickal VJ, Rajavashisth TB, Gimbrone MA, Shin WS, Liao JK (1995) Nitric oxide decreases cytokine-induced endothelial activation. Nitric oxide selectively reduces endothelial expression of adhesion molecules and pro-inflammatory cytokines. J Clin Invest 96:60–68

De Maria R, Cifone MG, Trotta R, Rippo MR, Festuccia C, Santoni A, Testi R (1995) Triggering of human monocyte activation through CD69, a member of the natural killer cell gene complex family of signal transducing receptors. J Exp Med 180:1999–2004

Deakin AM, Payne AN, Whittle BJR, Moncada S (1995) The modulation of IL-6 and TNF-α release by nitric oxide following stimulation of J774 cells with LPS and IFN-γ. Cytokine 7:408–416

Del Pozo V, de Arruda-Chaves E, De Andres B, Cardaba B, Lopez-Farre A, Gallardo S, Cortegano I, Vidarte L, Jurado A, Sastre J, Palomino P, Lahoz C (1997) Eosinophils transcribe and translate messenger RNA for inducible nitric oxide synthase. J Immunol 158:859–864

dela Torre A, Schroeder RA, Punzalan C, Kuo PC (1999) Endotoxin-mediated S-nitrosylation of p50 alters NF-κB-dependent gene transcription in ANA-1 murine macrophages. J Immunol 162:4101–4108

Delledonne M, Xia Y, Dixon RA, Lamb C (1998) Nitric oxide functions as a signal in plant disease resistance. Nature 394:585–588

Denicola A, Rubbo H, Rodriguez D, Radi R (1993) Peroxinitrite-mediated cytotoxicity to *Trypanosoma cruzi*. Arch Biochem Biophys 304:279–286

Denlinger LC, Fisette PL, Garis KA, Kwon G, Vazquez-Torres A, Simon AD, Nguyen B, Proctor RA, Bertics PJ, Corbett JA (1996) Regulation of inducible nitric oxide synthase expression by macrophage purinoreceptors and calcium. J Biol Chem 271:337–342

Diaz-Guerra MJM, Castrillo A, Martin-Sanz P, Bosca L (1999a) Negative regulation by phosphatidyl-inositol 3-kinase of inducible nitric oxide synthase expression by macrophages. J Immunol 162:6184–6190

Diaz-Guerra MJM, Castrillo A, Martin-Sanz P, Bosca L (1999b) Negative regulation by protein tyrosine phosphatase of IFN-γ-dependent expression of inducible nitric oxide synthase. J Immunol 162:6776-6783

Diefenbach A, Schindler H, Donhauser N, Lorenz E, Laskay T, MacMicking J, Röllinghoff M, Gresser I, Bogdan C (1998) Type 1 interferon (IFN-α/β) and type 2 nitric oxide synthase regulate the innate immune response to a protozoan parasite. Immunity 8:77–87

Diefenbach A, Schindler H, Röllinghoff M, Yokoyama W, Bogdan C (1999) Requirement for type 2 NO-synthase for IL-12 responsiveness in innate immunity. Science 284:951–955

Dimmeler S, Haendeler J, Nehls M, Zeiher AM (1997) Suppression of apoptosis by nitric oxide via inhibition of interleukin-1β-converting enzyme (ICE)-like and cysteine protein (CPP)-32-like proteases. J Exp Med 185:601–607

Dimmeler S, Haendeler J, Sause A, Zeiher AM (1998) Nitric oxide inhibits APO-1/Fas-mediated cell death. Cell Growth Differ 9:415–422

Ding AH, Nathan CF, Stuehr DJ (1988) Release of reactive nitrogen intermediates and reactive oxygen intermediates from mouse peritoneal macrophages. Comparison of activating cytokines and evidence for independent production. J Immunol 141:2407–2412

Ding M, St. Pierre BA, Parkinson JF, Medberry P, Wong JL, Rogers NE, Ignarro LJ, Merrill JE (1997) Inducible nitric oxide synthase and nitric oxide production in human fetal astrocytes and microglia. A kinetic analysis. J Biol Chem 272:11327–11335

Doherty TM, Sher A (1997) Defects in cell-mediated immunity affect chronic, but not innate resistance of mice to *Mycobacterium avium* infection. J Immunol 158:4822–4831

Doherty TM, Sher A, Vogel SN (1998) Paclitaxel (taxol)-induced killing of *Leishmania major* in murine macrophages. Infect Immun 66:4553–4556

Dong Z, Yang X, Xie K, Juang S-H, Llansa N, Fidler IJ (1995) Activation of inducible nitric oxide synthase gene in murine macrophages requires protein phosphatases 1 and 2A. L Leukoc Biol 58:725–732

Downing JEG, Virag L, Jones IW (1998) NADPH diaphorase-positive dendritic profiles in rat thymus are discrete from autofluorescent cells, immunoreactive for inducible nitric oxide synthase, and show strain-specific abundance differences. Immunology 95:148–155

Drapier J-C, Hibbs JB (1988) Differentiation of murine macrophages to express non-specific cytotoxicity for tumor cells results in L-arginine-dependent inhibition of mitochondrial iron-sulfur enzymes in the macrophage effector cells. J Immunol 140:2829–2838

Dugas N, Vouldoukis I, Becherel P, Arock M, Debre P, Tardieu M, Mossalayi DM, Delfraissy JF, Kolb JP, Dugas B (1996) Triggering of CD23b antigen by anti-CD23 monoclonal antibodies induces interleukin-10 production by human macrophages. Eur J Immunol 26:1394–1398

Duhé RJ, Evans GA, Erwin RA, Kirken RA, Cox GW, Farrar WL (1998) Nitric oxide and thiol redox regulation of Janus kinase activity. Proc Natl Acad Sci USA 95:126–131

Durbin JE, Hackenmüller R, Simon MC, Levy DE (1996) Targeted disruption of the mouse Stat1 gene results in compromised innate immunity to viral disease. Cell 84:443–450

Durner J, Wendehenne D, Klessig DF (1998) Defense gene induction in tobacco by nitric oxide, cyclic AMP, and cyclic ADP–ribose. Proc Natl Acad Sci USA 95:10328–10333

Eastmond NC, Banks EMS, Coleman JW (1997) Nitric oxide inhibits IgE-mediated degranulation of mast cells and is the principal intermediate in IFN-γ-induced suppression of exocytosis. J Immunol 159:1444–1450

Ehlers S, Kutsch S, Benini J, Cooper A, Hahn C, Gerdes J, Orme I, Martin C, Rietschel ET (1999) NOS2-derived nitric oxide regulates the size, quantity and quality of granuloma formation in *Mycobacterium avium*-infected mice without affecting bacterial loads. Immunol 98:313–323

Ehrt S, Shiloh MU, Ruan J, Choi M, Gunzburg S, Nathan C, Xie Q-W, Riley LW (1997) A novel antioxidant gene from *Mycobacterium tuberculosis*. J Exp Med 186:1885–1896

Eigler A, Sinha B, Endres S (1993) Nitric oxide-releasing agents enhance cytokine-induced tumor necrosis factor synthesis in human mononuclear cells. Biochem Biophys Res Commun 196:494–501

Eigler A, Moeller J, Endres S (1995) Exogenous and endogenous nitric oxide attenuates tumor necrosis factor synthesis in the murine macrophage cell line RAW 264.7. J Immunol 154:4048–4054

Eiserich JP, Hristova M, Cross CE, Jones AD, Freeman BA, Halliwell B, Van der Vliet A (1998) Formation of nitric oxide-derived inflammatory oxidants by myeloperoxidase in neutrophils. Nature 391:393–397

Endoh M, Maiese K, Wagner J (1994) Expression of the inducible form of nitric oxide synthase by reactive astrocytes after transient global ischemia. Brain Res 651:92–100

Evans TJ, Buttery LDK, Carpenter A, Springall DR, Polak JM, Cohen J (1996) Cytokine-treated human neutrophils contain inducible nitric oxide synthase that produces nitration of ingested bacteria. Proc Natl Acad Sci USA 93:9553–9558

Fang FC (1997) Mechanisms of nitric oxide-related antimicrobial activity. J Clin Invest 99:2818–2825

Favre N, Ryffel B, Rudin W (1999) The development of murine cerebral malaria does not require nitric oxide production. Parasitology 118:135–138

Fehr T, Schoedon G, Odermatt B, Holtschke T, Schneemann M, Bachmann ME, Mak TW, Horak I, Zinkernagel RM (1997) Crucial role of interferon consensus sequence binding protein, but neither of interferon regulatory factor 1 nor of nitric oxide synthesis for protection against murine listeriosis. J Exp Med 185:921–931

Fehsel K, Kröncke K-D, Meyer KL, Huber H, Wahn V, Kolb-Bachofen V (1995) Nitric oxide induces apoptosis in mouse thymocytes. J Immunol 155:2858–2865

Fernandez-Gomez R, Esteban S, Gomez-Corvera R, Zoulika K, Quaissi A (1998) *Trypanosoma cruzi*: Tc52 released protein-induced increased expression of nitric oxide synthase and nitric oxide production by macrophages. J Immunol 160:3471–3479

Flak TA, Goldman WE (1999) Signalling and cellular specificity of airway nitric oxide production in pertussis. Cell Microbiol 1:51–60

Fleming I, Gray GA, Schott C, Stoclet JC (1991) Inducible but not constitutive production of nitric oxide by vascular smooth muscle cells. Eur J Pharmacol 200:375–376

Florquin S, Amraoui Z, Dubois C, Decuyper J, Goldman M (1994) The protective role of endogenously synthesized nitric oxide in staphylococcal enterotoxin B-induced shock in mice. J Exp Med 180:1153–1158

Forrester K, Ambs S, Lupold SE, Kapust RB, Spillare EA, Weinberg WC, Felley-Bosco E, Wang XW, Geller DA, Tzeng E, Billiar TR, Harris CC (1996) Nitric oxide-induced p53 accumulation and regulation of inducible nitric oxide synthase expression by wild-type p53. Proc Natl Acad Sci USA 93:2442–2447

Förstermann U, Boissel JP, Kleinert H (1998) Expressional control of the "constitutive" isoforms of nitric oxide synthase (NOSI and NOSIII). Faseb J 12:773–790

Fujihara M, Connolly N, Ito N, Suzuki T (1994a) Properties of protein kinase C isoforms (βII, ε, and ζ) in a macrophage cell line (J774) and their roles in LPS-induced nitric oxide production. J Immunol 152:1898–1906

Fujihara M, Ito N, Pace JL, Watanabe Y, Russell SW, Suzuki T (1994b) Role of endogenous IFN-β in lipopolysaccharide-triggered activation of the inducible nitric oxide synthase gene in a mouse macrophage cell line, J774. J Biol Chem 269:12773–12778

Fujii H, Ichimori K, Hoshiai K, Nakazawa H (1997) Nitric oxide inactivates NADPH oxidase in pig neutrophils by inhibiting its assembling process. J Biol Chem 272:32773–32778

Galea E, Feinstein DL, Reis DJ (1992) Induction of calcium-independent nitric oxide synthase activity in primary rat glial cultures. Proc Natl Acad Sci USA 89:10945–10949

Gao J, Morrison DC, Parmely TJ, Russell SW, Murphy WJ (1997) An interferon-γ activated site (GAS) is necessary for full expression of the mouse iNOS gene in response to interferon-γ and lipopolysaccharide. J Biol Chem 272:1226–1230

Gao JJ, Filla MB, Fultz MJ, Vogel SN, Russell SW, Murphy WJ (1998) Autocrine/paracrine IFN-α/β mediates the lipopolysaccharide-induced activation of transcription factor Stat1α in mouse macrophages: pivotal role of Stat1α in induction of the inducible nitric oxide synthase gene. J Immunol 161:4803–4810

Gao JJ, Zuvanich EG, Xue Q, Horn DL, Silverstein R, Morrison DC (1999) Cutting Edge: Bacterial DNA and LPS act in synergy in inducing nitric oxide production in RAW264.7 macrophages. J Immunol 163:4095–4099

Geller DA, de Vera ME, Russell DA, Shapiro RA, Nussler AK, Simmons RL, Billiar TR (1995) A central role for IL-1β in the in vitro and in vivo regulation of hepatic inducible nitric oxide synthase: IL-1β induces hepatic nitric oxide synthesis. J Immunol 155:4890–4898

Genaro AM, Hortelano S, Alvarez A, Martinez C, Bosca L (1995) Splenic B lymphocyte programmed cell death is prevented by nitric oxide release through mechanisms involving sustained Bcl-2 levels. J Clin Invest 95:1884–1890

Ghigo D, Todde R, Ginsburg H, Costamagna C, Gautret P, Bussolino F, Ulliers D, Giribaldi G, Deharo E, Gabrielli G, Pescarmona G, Bosia A (1995) Erythrocyte stages of *Plasmodium falciparum* exhibit a high Nitric oxide synthase (NOS) activity and release an NOS-inducing soluble factor. J Exp Med 182:677–688

Giese NA, Gabriele L, Doherty TM, Klinman DM, Tadesse-Heath L, Contursi C, Epstein SL, Morse HC (1997) Interferon (IFN) consensus sequence-binding protein, a transcription factor of the IFN regulatory factor family, regulates immune responses in vivo through control of interleukin 12 expression. J Exp Med 186:1535–1546

Gilkeson GS, Mudgett JS, Seldin MF, Ruiz P, Alexander AA, Misukonis MA, Pisetsky DS, Weinberg JB (1997) Clinical and serologic manifestations of autoimmune disease in MRL-lpr/lpr mice lacking nitric oxide synthase type 2. J Exp Med 186:365–373

Gold R, Zielasek J, Kiefer R, Toyka KV, Hartung H-P (1996) Secretion of nitrite by Schwann cells and its effect on T-cell activation in vitro. Cell Immunol 168:69–77

Gomes MS, Florido M, Pais TF, Appelberg R (1999) Improved clearance of *Mycobacterium avium* upon disruption of the inducible nitric oxide synthase gene. J Immunol 162:6734–6739

Goodrum KJ, Dierksheide J, Yoder BJ (1995) Tumor necrosis factor α acts as an autocrine second signal with γ interferon to induce nitric oxide in group B streptococcus-treated macrophages. Infect Immun 63:3715–3717

Gow AJ, Stamler JS (1998) Reactions between nitric oxide and haemoglobin under physiological conditions. Nature 391:169–173

Gregory SH, Wing EJ, Hoffman RA, Simmons RL (1993) Reactive nitrogen intermediates suppress the primary immunologic response to *Listeria*. J Immunol 150:2901–2909

Griscavage JM, Rogers NE, Sherman MP, Ignarro LJ (1993) Inducible nitric oxide synthase from a rat alveolar macrophage cell line is inhibited by nitric oxide. J Immunol 151:6329–6337

Gross SS, Jaffe EA, Levi R, Kilbourn RG (1991) Cytokine-activated endothelial cells express an isotype of nitric oxide synthase which is tetrahydrobiopterin-dependent, calmodulin-independent and inhibited by arginine analogs with a rank order of potency characteristic of activated macrophages. Biochem Biophys Res Commun 178:823–829

Guan Z, Baier LD, Morrison AR (1997) p38 Mitogen-activated protein kinase down-regulates nitric oxide and up-regulates prostaglandin E2 biosynthesis stimulated by interleukin-1β. J Biol Chem 272:8083–8089

Guo FH, de Raeve HR, Rice TW, Stuehr DJ, Thunnissen FBJM, Erzurum SC (1995) Continuous nitric oxide synthesis by inducible nitric oxide synthase in normal human airway epithelium in vivo. Proc Natl Acad Sci USA 92:7809–7813

Harbrecht BG, Billiar TR, Stadler J, Demetris AJ, Ochoa J, Curran RD, Simmons RL (1992) Inhibition of nitric oxide synthesis during endotoxemia promotes intrahepatic thrombosis and an oxygen radical-mediated hepatic injury. J Leukoc Biol 52:390–394

Hasko G, Szabo C, Nemeth ZH, Kvetan V, McCarthy Pastores S, Vizi ES (1996) Adenosine receptor agonists differentially regulate IL-10, TNF-α, and nitric oxide production in RAW264.7 macrophages and in endotoxemic mice. J Immunol 157:4634–4640

Hatzigeorgiou DE, He S, Sobel J, Grabstein KH, Haffner A, Ho JL (1993) IL-6 downmodulates the cytokine-enhanced antileishmanial activity in human macrophages. J Immunol 151:3682–3692

Hauschildt S, Bessler WG, Scheipers P (1993) Engagement of major histocompatibility complex class II molecules leads to nitrite production in bone marrow-derived macrophages. Eur J Immunol 23:2988–2992

Häuselmann HJ, Stefanovic-Racic M, Michel BA, Evans CH (1998) Differences in nitric oxide production by superficial and deep human articular chondrocytes: implications for proteoglycan turnover in inflammatory joint diseases. J Immunol 160:1444–1448

Hausladen A, Stamler JS (1998) Nitric oxide in plant immunity. Proc Natl Acad Sci USA 95:10345–10347

Hebestreit H, Dibbert B, Balatti I, Braun D, Schapowal A, Blaser K, Simon H-U (1998) Disruption of Fas receptor signaling by nitric oxide in eosinophils. J Exp Med 187:415–425

Heck DE, Laskin DL, Gardner CR, Laskin JD (1992) Epidermal Growth Factor suppresses nitric oxide and hydrogen peroxide formation by keratinocytes. J Biol Chem 267:21277–21280

Heiss LN, Lancaster JR, Corbett JA, Goldman WE (1994) Epithelial autotoxicity of nitric oxide: role in the respiratory cytopathology of pertussis. Proc Natl Acad Sci USA 91:267–270

Henson SE, Nichols TC, Holers VM, Karp DR (1999) The ectoenzyme g-glutamyl transpeptidase regulates antiproliferative effects of S-nitrosoglutathione on human T and B lymphocytes. J Immunol 163:1845–1852

Hertz CJ, Mansfield JM (1999) IFN-γ-dependent nitric oxide production is not linked to resistance in experimental african trypanosomiasis. Cell Immunol 192:24–32

Hibbs JB, Taintor RR, Vavrin Z (1987) Macrophage cytotoxicity: role of L-arginine deiminase and imino nitrogen oxidation to nitrite. Science 235:473–476

Hibbs JB, Taintor RR, Vavrin Z, Rachlin EM (1988) Nitric oxide: a cytotoxic activated macrophage effector molecule. Biochem Biophys Res Commun 157:87–94

Hibbs JB Jr, Westenfelder C, Taintor R, Vavrin Z, Kablitz C, Baranowski RL, Ward JH, Menlove RL, McMurry MP, Kushner JP, Samlowski WE (1992) Evidence for cytokine-inducible nitric oxide synthesis from L-arginine in patients receiving interleukin-2 therapy. J Clin Invest 89:867–877

Hierholzer C, Harbrecht B, Menezes J, Kane J, MacMicking J, Nathan CF, Peitzman A, Billiar TR, Tweardy DJ (1998) Essential role of induced nitric oxide in the initiation of the inflammatory response following hemorrhagic shock. J Exp Med 187:917–928

Hill JR, Corbett JA, Kwon G, Marshall CA, McDaniel ML (1996) Nitric oxide regulates IL-1 bioactivity released from murine macrophages. J Biol Chem 271:22672–22678

Hirji N, Lin T-J, Bissonnette E, Belosevic M, Befus AD (1998) Mechanisms of macrophage stimulation through CD8:macrophage CD8α and CD8β induce nitric oxide production and associated killing of the parasite *Leishmania major*. J Immunol 160:6004–6011

Hoffman RA, Langrehr JM, Billiar TR, Curran RD, Simmons RL (1990) Alloantigen-induced activation of rat splenocytes is regulated by the oxidative metabolism of L-arginine. J Immunol 145:2220–2226

Hoffman RA, Langrehr JM, Wren SM, Dull KE, Ildstad ST, McCarthy SA, Simmons RL (1993) Characterization of the immunosuppressive effects of nitric oxide in graft vs host disease. J Immunol 151:1508–1518

Hogaboam CM, Chensue SW, Steinhauser ML, Huffnagle GB, Lukacs NW, Strieter RM, Kunkel SL (1997) Alteration of the cytokine phenotype in an experimental lung granuloma model by inhibiting nitric oxide. J Immunol 159:5585–5593

Hölscher C, Köhler G, Möller U, Mossmann H, Schaub GA, Brombacher F (1998) Defective nitric oxide effector functions lead to extreme susceptibility of *Trypanosoma cruzi*-infected mice deficient in γ interferon receptor or inducible nitric oxide synthase. Infect Immun 66:1208–1215

Hooper DC, Bagasra O, Marini JC, Zborek A, Ohnishi ST, Kean R, Champion JM, Sarker AB, Bobroski L, Farber JL, Akaike T, Maeda H, Koprowski H (1997) Prevention of experimental allergic encephalomyelitis by targeting nitric oxide and peroxinitrite: implications for the treatment of multiple sclerosis. Proc Natl Acad Sci USA 94:2528–2533

Hortelano S, Genaro AM, Bosca L (1992) Phorbol esters induce nitric oxide synthase activity in rat hepatocytes. Antagonism with the induction elicited by lipopolysaccharide. J Biol Chem 267:24937–24940

Huang F-P, Niedbala W, Wei X-Q, Xu D, Feng G-J, Robinson JH, Lam C, Liew FY (1998a) Nitric oxide regulates Th1 cell development through the inhibition of IL-12 synthesis by macrophages. Eur J Immunol 28:4062–4070

Huang F-P, Xu D, Esfandiari E-O, Sands W, Wei X-Q, Liew FY (1998b) Mice defective in Fas are highly susceptible to *Leishmania major* infection despite elevated IL-12 synthesis, strong Th1 responses, and enhanced nitric oxide production. J Immunol 160:4143–4147

Huang S, Hendriks W, Althage A, Hemmi S, Bluethmann H, Kamijo R, Vilcek J, Zinkernagel RM, Aguet M (1993) Immune response in mice that lack the interferon-γ receptor. Science 259:1742–1744

Hung K, Hayashi R, Lafond-Walker A, Lowenstein C, Pardoll D, Levitsky H (1998) The central role of CD4+ T cells in the antitumor immune response. J Exp Med 188:2357–2368

Ianaro A, O'Donnell CA, Di Rosa M, Liew FY (1994) A nitric oxide synthase inhibitor reduces inflammation, down-regulates inflammatory cytokines and enhances IL-10 production in carrageenin-induced oedema in mice. Immunology 82:370–375

Igietseme JU, Perry LL, Ananaba GA, Uriri IM, Ojior O, Kumar SN, Caldwell HD (1998) Chlamydial infection in inducible nitric oxide synthase knockout mice. Infect Immun 66:1282–1286

Iikura M, Takaishi T, Hirai K, Yamada H, Iida M, Koshino T, Morita Y (1998) Exogenous nitric oxide regulates the degranulation of human basophils and rat peritoneal mast cells. Int Arch Allergy Immunol 115:129–136

Imai T, Hirata Y, Kanno K, Marumo F (1994) Induction of nitric oxide synthase by cyclic AMP in rat vascular smooth muscle cells. J Clin Invest 93:543–549

Ito M, Watanabe M, Kamiya H, Sakurai M (1996) Inhibition of natural killer cell activity against cytomegalovirus-infected fibroblasts by nitric oxide-releasing agents. Cell Immunol 174:13–18

Jacobs F, Chaussabel D, Truyens C, Leclerq V, Carlier Y, Goldman M, Vray B (1998) IL-10 up-regulates NO synthesis by LPS-activated macrophages: improved control of *Trypanosoma cruzi* infection. Clin Exp Immunol 113:59–64

James SL, Cheever AW, Caspar P, Wynn TA (1998) Inducible nitric oxide synthase-deficient mice develop enhanced type 1 cytokine-associated cellular and humoral immune responses after vaccination with attenuated *Schistosoma mansoni* cercariae but display partially reduced resistance. Infect Immun 66:3510–3518

Jenkins DC, Charles IG, Thomsen LL, Moss DW, Holmes LS, Baylis SA, Rhodes P, Westmore K, Emson PC, Moncada S (1995) Roles of nitric oxide in tumor growth. Proc Natl Acad Sci USA 92:4392–4396

Jones-Carson J, Vazquez-Torres A, Van der Heyde HC, Warner T, Wagner RD, Balish E (1995) $\gamma\delta$-T-cell-induced nitric oxide production enhances resistance to mucosal candidiasis. Nat Med 1:552–557

Jungi TW, Pfister H, Sager H, Fatzer R, Vandevelde M, Zurbriggen A (1997) Comparison of inducible nitric oxide synthase expression in the brains of *Listeria monocytogenes*-infected cattle, sheep, and goats and in macrophages stimulated in vitro. Infect Immun 65:5279–5288

Jüttner S, Bernhagen J, Metz CN, Röllinghoff M, Bucala R, Gessner A (1998) Migration inhibitory factor induces killing of *Leishmania major* by macrophages: dependence on reactive nitrogen intermediates and endogenous TNF-α. J Immunol 161:2383–2390

Kamijo R, Harada H, Matsuyama T, Bosland M, Gerecitano J, Shapiro D, Le J, Koh SI, Kimura T, Green SJ, Mak TW, Taniguchi T, Vilcek J (1994) Requirement for transcription factor IRF-1 in NO synthase induction in macrophages. Science 263:1612–1615

Karupiah G, Chen J-H, Mahalingam S, Nathan CF, MacMicking JD (1998a) Rapid interferon γ-dependent clearance of influenza A virus and protection from consolidating pneumonitis in nitric oxide 2-deficient mice. J Exp Med 188:1541–1546

Karupiah G, Chen JH, Nathan CF, Mahalingam S, MacMicking JD (1998b) Identification of nitric oxide synthase 2 as an innate resistance locus against ectromelia virus infection. J Virol 72:7703–7706

Kawakami K, Zhang T, Qureshi MH, Saito A (1997) Cryptococcus neoformans inhibits nitric oxide production by murine peritoneal macrophages stimulated with interferon-γ and lipopolysaccharide. Cell Immunol 180:47–54

Kengatharan KM, de Kimpe S, Robson C, Foster SJ, Thiermann C (1998) Mechanism of gram-positive shock: identification of peptidoglycan and lipoteichoic acid moieties essential in the induction of nitric oxide synthase, shock, and multiple organ failure. J Exp Med 188:305–315

Khan BV, Harrison DG, Olbrych MT, Alexander RW, Medford RM (1996) Nitric oxide regulates vascular cell adhesion molecule 1 gene expression and redox-sensitive transcriptional events in human vascular endothelial cells. Proc Natl Acad Sci USA 93:9114–9119

Khan IA, Schwartzman JD, Matsuura T, Kasper LH (1997) A dichotomous role for nitric oxide during acute *Toxoplasma gondii* infection in mice. Proc Natl Acad Sci USA 94:13955–13960

Khanolkar-Young S, Snowdon D, Lockwood DNJ (1998) Immunocytochemical localization of inducible nitric oxide synthase and transforming growth factor-β (TGF-β) in leprosy lesions. Clin Exp Immunol 113:438–442

Kim Y-M, Talanian RV, Li J, Billiar TR (1998) Nitric oxide prevents IL-1β and IFN-γ-inducing factor (IL-18) release from macrophages by inhibiting caspase-1 (IL-1β-converting enzyme). J Immunol 161:4122–4128

Kim YS, Täuber MG (1996) Neurotoxicity of glia activated by gram-positive bacterial products depends on nitric oxide production. Infect Immun 64:3148–3153

Klatt P, Molina EP, Lamas S (1999) Nitric oxide inhibits c-Jun DNA binding by specifically targeted S-glutathionylation. J Biol Chem 274:15857–15864

Koblish HK, Hunter CA, Wysocka M, Trinchieri G, Lee WMF (1998) Immune suppression by recombinant IL-12 involves IFN-γ induction of NOS2 (iNOS) activity: inhibitors of NO generation reveal the extent or rIL-12 vaccine adjuvant effect. J Exp Med 188:1603–1610

Koide M, Kawahara Y, Nakayama I, Tsuda T, Yokoyama M (1993) Cyclic AMP-elevating agents induce an inducible type of nitric oxide synthase in cultured vascular smooth muscle cells. Synergism with the induction elicited by inflammatory cytokines. J Biol Chem 268:24959–24966

Kolb H, Kolb-Bachofen V (1998) Nitric oxide in autoimmune disease: cytotoxic or regulatory mediator. Immunol Today 19:556–561

Komatsu T, Bi Z, Reiss CS (1996) IFN-γ induced type 1 nitric oxide synthase activity inhibits viral replication in neurons. J Neuroimmunol 68:101–108

Koprowski H, Zheng YM, Heber-Katz E, Fraser N, Rorke L, Fu ZF, Hanlon C, Dietzschold B (1993) *In vivo* expression of inducible nitric oxide synthase in experimentally induced neurologic diseases. Proc Natl Acad Sci USA 90:3024–3027

Kröncke KD, Kolb-Bachofen V, Berschick B, Burkart V, Kolb H (1991) Activated macrophages kill pancreatic syngeneic islet cells via arginine-dependent nitric oxide generation. Biochem Biophys Res Commun 175:752–758

Kröncke KD, Fehsel K, Kolb-Bachofen V (1998) Inducible nitric oxide synthase in human diseases. Clin Exp Immunol 113:147–156

Kröncke K-D, Carlberg C (2000) Inactivation of zinc finger transcription factors provides a mechanism for a gene regulatory role of nitric oxide. FASEB J 14:166–173

Kubes P, Suzuki M, Granger DN (1991) Nitric oxide: an endogenous modulator of leukocyte adhesion. Proc Natl Acad Sci USA 88:4651–4655

Kun JFJ, Mordmüller B, Lell B, Lehman LG, Luckner D, Kremsner PG (1998) Polymorphism in promoter region of inducible nitric oxide synthase gene and protection against malaria. Lancet 351:265–266

Kunz D, Walker G, Eberhardt W, Pfeilschifter J (1996) Molecular mechanisms of dexamethasone inhibition of nitric oxide synthase expression in interleukin 1b-stimulated mesangial cells: evidence for the involvement of transcriptional and post-transcriptional regulation. Proc Natl Acad Sci USA 93:255–259

Kuzin B, Roberts I, Peunova N, Enikolopov G (1996) Nitric oxide regulates cell proliferation during *Drosophila* development. Cell 87:639–649

Kwon G, Xu G, Marshall CA, McDaniel ML (1999) Tumor necrosis factor-a-induced pancreatic β-cell insulin resistance is mediated by nitric oxide and prevented by 15-deoxy- $\Delta^{12,14}$-prostaglandin J_2 and aminoguanidine. J Biol Chem 274:18702–18708

Kwon NS, Stuehr DJ, Nathan CF (1991) Inhibition of tumor cell ribonucleotide reductase by macrophage-derived nitric oxide. J Exp Med 174:761–767

Lander HM (1997) An essential role for free radicals and derived species in signal transduction. FASEB J 11:118–124

Lander HM, Sehajpal P, Levine DM, Novogrodsky A (1993) Activation of human peripheral blood mononuclear cells by nitric oxide-generating compounds. J Immunol 150:1509–1516

Larsen CM, Wadt KAW, Juhl LF, Andersen HU, Karlsen AE, Su MS-S, Seedorf K, Shapiro L, Dinarello CA, Mandrup-Poulsen T (1998) Interleukin-1β-induced rat pancreatic islet nitric oxide synthesis requires both the p38 and extracellular signal-regulated kinase 1/2 mitogen activated protein kinases. J Biol Chem 273:15294–15300

Lavnikova N, Burdelya L, Lakhotia A, Patel N, Prokhorova S, Laskin DL (1997) Macrophage and interleukin-1 induced nitric oxide production and cytostasis in hamster tumor cells varying in malignant potential. J Leukoc Biol 61:452–458

Lee SC, Dickson DW, Brosnan CF, Casadevall A (1994) Human astrocytes inhibit *Cryptococcus neoformans* growth by a nitric oxide-mediated mechanism. J Exp Med 180:365–369

Lee SC, Kress Y, Dickson DW, Casadevall A (1995) Human microglia mediate anti-*Cryptococcus neoformans* activity in the presence of specific antibody. J Neuroimmunol 62:43–52

Lejeune P, Lagadec P, Onier N, Pinard D, Ohshima H, Jeannin J-F (1994) Nitric oxide involvement in tumor-induced immunosuppression. J Immunol 152:5077–5083

Lepoivre M, Chenais B, Yapo A, Lemaire G, Thelander L, J.-P. T (1990) Alterations of ribonucleotide reductase activity following induction of the nitrite-generating pathway in adenocarcinoma cells. J Biol Chem 265:14143–14149

Li J, Billiar TR, Talanian RV, Kim YM (1997) Nitric oxide reversibly inhibits seven members of the caspase family via S-nitrosylation. Biochem Biophys Res Commun 240:419–424

Liesenfeld O, Kang H, Park D, Nguyen TA, Parkhe CV, Watanabe H, Abo T, Sher A, Remington JS, Suzuki Y (1999) TNF-α, nitric oxide, and IFN-γ are all critical for the development of necrosis in the small intestine and early mortality in genetically susceptible mice infected perorally with Toxoplasma gondii. Parasite Immunol 21:365–376

Liew FY (1995) Regulation of lymphocyte functions by nitric oxide. Curr Opin Immunol 7:396–399

Lin T-J, Befus D (1997) Differential regulation of mast cell function by IL-10 and stem cell factor. J Immunol 159:4015–4023

Liu J, Zhao M-L, Brosnan CF, Lee SC (1996) Expression of type II nitric oxide synthase in primary human astrocytes and microglia. Role of IL-1β and IL-1 receptor antagonist. J Immunol 157:3569–3576

Lopez-Collazo E, Hortelano S, Rojas A, Bosca L (1998) Triggering of peritoneal macrophages with IFN-α/β attenuates the expression of inducible nitric oxide through a decrease in NF-κB activation. J Immunol 160:2889–2895

Lowenstein CJ, Glatt CS, Bredt DS, Snyder SH (1992) Cloned and expressed macrophage nitric oxide synthase contrasts with the brain synthase. Proc Natl Acad Sci USA 89:6711–6715

Lowenstein CJ, Alley EW, Raval P, Snowman AM, Snyder SH, Russell SW, Murphy WJ (1993) Macrophage nitric oxide synthase gene: two upstream regions mediate induction by interferon γ and lipopolysaccharide. Proc Natl Acad Sci USA 90:9730–9734

Lu L, Bonham CA, Chambers FG, Watkins SC, Hoffman RA, Simmons RL, Thomson AW (1996) Induction of nitric oxide synthase in mouse dendritic cells by IFN-γ, endotoxin, and interaction with allogeneic T cells. Nitric oxide production is associated with dendritic cell apoptosis. J Immunol 157:3577–3586

Luckhart S, Vodovotz Y, Cui L, Rosenberg R (1998) The mosquito *Anopheles stephensi* limits malaria parasite development with inducible nitric oxide synthesis. Proc Natl Acad Sci USA 95:5700–5705

Mabbott NA, Coulson PS, Smythies LE, Wilson RA, Sternberg JM (1998) African trypanosome infections in mice that lack the interferon-γ receptor gene: nitric oxide-dependent and -independent suppression of T-cell proliferative responses and the development of anemia. Immunol 94:476–480

MacLean A, Wei X-Q, Huang FP, Al-Alem UA, Chan WL, Liew FY (1998) Mice lacking inducible nitric oxide synthase are more susceptible to herpes simplex virus infection despite enhanced Th1 cell responses. J Gen Virol 79:825–830

MacMicking JD, Nathan C, Hom G, Chartrain N, Fletcher DS, Trumbauer M, Stevens K, Xie Q-W, Sokol K, Hutchinson N, Chen H, Mudgett JS (1995) Altered responses to bacterial infection and endotoxic shock in mice lacking inducible nitric oxide synthase. Cell 81:641–650

MacMicking JD, Xie Q-W, Nathan C (1997a) Nitric oxide and macrophage function. Ann Rev Immunol 15:323–350

MacMicking JD, North RJ, LaCourse R, Mudgett JS, Shah SK, Nathan CF (1997b) Identification of nitric oxide synthase as a protective locus against tuberculosis. Proc Natl Acad Sci USA 94:5243–5248

Magrinat G, Mason SN, Shami PJ, Weinberg JB (1992) Nitric oxide modulation of human leukemia cell differentiation and gene expression. Blood 1992:1880–1884

Majano PL, Garcia-Monzon C, Lopez-Cabrera M, Lara-Pezzi E, Fernandez-Ruiz E, Garcia-Iglesias C, Borque MJ, Moreno-Otero R (1998) Inducible nitric oxide synthase expression in chronic viral hepatitis. Evidence for a virus-induced gene up-regulation. J Clin Invest 101:1343–1352

Malkin R, Flescher E, J. L, Keisari Y (1987) On the interactions between macrophages and the developmental stages of *Schistosoma mansoni*: the cytotoxic mechanism involved in macrophage-mediated killing of schistosomula in vitro. Immunobiology 176:63–72

Mannick JB, Asano K, Izumi K, Kieff E, Stamler JS (1994) Nitric oxide produced by human B lymphocytes inhibits apoptosis and Epstein-Barr virus reactivation. Cell 79:1137–1146

Mannick JB, Miao XQ, Stamler JS (1997) Nitric oxide inhibits Fas-induced apoptosis. J Biol Chem 272:24125–24128

Mannick JB, Hausladen A, Liu L, Hess DT, Zeng M, Miao QX, Kane LS, Gow AJ, Stamler JS (1999) Fas-induced caspase denitrosylation. Science 284:651–654

Manthey CL, Perera P-Y, Salkowski CA, Vogel SN (1994) Taxol provides a second signal for murine macrophage tumoricidal activity. J Immunol 152:825–831

Marcinkiewicz J, Chain BM (1993) Differential regulation of cytokine production by nitric oxide. 1993 Immunology 80:146–150

Marcinkiewicz J, Grabowska A, Chain B (1995) Nitric oxide up-regulates the release of inflammatory mediators by mouse macrophages. Eur J Immunol 25:947–951

Marks-Konczalik J, Chu SC, Moss J (1998) Cytokine-mediated transcriptional induction of the human inducible nitric oxide synthase gene requires both activator protein 1 and nuclear factor κB-binding sites. J. Biol. Chem. 273:22201–22208

Marletta MA, Yoon PS, Iyengar R, Leaf CD, Wishnok JS (1988) Macrophage oxidation of L-arginine to nitrite and nitrate: nitric oxide is an intermediate. Biochemistry 27:8706–8711

Marshall BG, Chambers MA, Wangoo A, Shaw RJ, Young DB (1997) Production of tumor necrosis factor and nitric oxide by macrophages infected with live and dead mycobacteria and their suppression by an interleukin-10-secreting recombinant. Infect Immun 65:1931–1935

Martin E, Nathan C, Xie Q-W (1994) Role of interferon regulatory factor-1 in induction of nitric oxide synthase. J Exp Med 180:977–984

Mateo RB, Reichner JS, Albina JS (1996) NO is not sufficient to explain maximal cytotoxicity of tumoricidal macrophages against an NO-sensitive cell line. J Leukoc Biol 60:245–252

Matsuno R, Aramaki Y, Arima H, Adachi Y, Ohno N, Yadomae T, Tsuchiya S (1998) Contribution of CR3 to nitric oxide production from macrophages stimulated with high-dose LPS. Biochem Biophys Res Commun 244:115–119

McCall TB, Palmer RMJ, Moncada S (1991) Induction of nitric oxide synthase in rat peritoneal neutrophils and its inhibition by dexamethasone. Eur J Immunol 21: 2523–2527

McInnes IB, Leung B, Wei X-Q, Gemmell CC, Liew FY (1998) Septic arthritis following *Staphylococcus aureus* infection in mice lacking inducible nitric oxide synthase. J Immunol 160:308–315

Melillo G, Cox GW, Radzioch D, Varesio L (1993) Picolinic acid, a catabolite of L-tryptophan, is a co-stimulus for the induction of reactive nitrogen intermediate production in murine macrophages. J Immunol 150:4031–4040

Melillo G, Cox GW, Biragyn A, Sheffler LA, Varesio L (1994) Regulation of nitric oxide synthase mRNA expression by interferon-γ and picolinic acid. J Biol Chem 269:8128–8133

Melillo G, Taylor LS, Brooks A, Musso T, Cox GW, Varesio L (1997) Functional requirement of the hypoxia-responsive element in the activation of the inducible nitric oxide synthase promoter by the iron chelator desferrioxamine. J Biol Chem 272:12236–12243

Mellouk S, Hoffman SL, Liu Z-Z, de la Vega P, Billiar TR, Nüssler AK (1994) Nitric oxide-mediated antiplasmodial activity in human and murine hepatocytes induced by γ interferon and the parasite itself: enhancement by exogenous tetrahydrobiopterin. Infect Immun 62:4043–4046

Meraz MA, White JM, Sheehan KCF, Bach EA, Rodig SJ, Dighe AS, Kaplan DH, Riley JK, Greenlund AC, Campbell D, Carver-Moore K, DuBois RN, Clark R, Aguet M, Schreiber RD (1996) Targeted disruption of the Stat1 gene in mice reveals unexpected physiologic specificity in the Jak-STAT signaling pathway. 84:431–442

Merrill JE, Ognarro LJ, Sherman MP, Melinek J, Lane TE (1993) Microglial cell cytotoxicity of oligodendrocytes is mediated through nitric oxide. J Immunol 151:2132–2141

Millar AE, Sternberg J, McSharry C, Wei X-Q, Liew FY, Turner MR (1999) T cell responses during Trypanosoma brucei infections in mice deficient in inducible nitric oxide synthase. Infect Immun 67:3334–3338

Mills CD (1991) Molecular basis of "suppressor" macrophages. J Immunol 146:2719–2723

Minc-Golomb D, Yadid G, Tsarfaty I, Resau JH, Schwartz JP (1996) In vivo expression of inducible nitric oxide synthase in cerebellar neurons. J Neurochem 66:1504–1509

Mnaimneh S, Geffard M, Veyret B, Vincendeau P (1997) Albumin nitrosylated by activated macrophages possesses antiparasitic effects neutralized by anti-NO-acetylated-cysteine antibodies. J Immunol 158:308–314

Mohr S, Zech B, Lapetina EG, Brüne B (1997) Inhibition of caspase-3 by S-nitrosation and oxidation caused by nitric oxide. Biochem Biophys Res Commun 238:387–391

Mossalayi MD, Pail-Eugéne N, Ouaaz F, Arock M, Kolb JP, Kilcherr E, Debré P, Dugas B (1994) Involvement of FcεRII/CD23 and L-arginine-dependent pathway in IgE-mediated stimulation of human monocyte functions. Int Immunol 6:931–934

Mozaffarian N, Berman JW, Casadevalli A (1995) Immune complexes increase nitric oxide production by interferon-γ-stimulated murine macrophage-like J774.16 cells. J Leukocyt Biol 57:657–662

Mühl H, Dinarello C (1997) Macrophage inflammatory protein-1a production in lipopolysaccharide-stimulated human adherent blood mononuclear cells is inhibited by the NO synthase inhibitor N^G-monomethyl-L-arginine. J Immunol 159:5063–5069

Mullins DW, Burger CJ, Elgert KD (1999) Paclitaxel enhances macrophage IL-12 production in tumor-bearing hosts through nitric oxide. J Immunol 162:6811–6818

Munder M, Eichmann K, Modolell M (1998a) Alternative metabolic states in murine macrophages reflected by the nitric oxide synthase/arginase balance: competitive regulation by CD4+ T cells correlates with Th1/Th2 phenotype. J Immunol 160:5347–5354

Munder M, Mallo M, Eichmann K, Modolell M (1998b) Murine macrophages secrete interferon-g upon combined stimulation with IL-12 and IL-18:a novel pathway of autocrine macrophage activation. J Exp Med 187:2103–2108

Murphy GM, Yang L, Cordell B (1998) Macrophage colony-stimulating factor augments b-amyloid-induced interleukin-1, interleukin-6, and nitric oxide production by microglial cells. J Biol Chem 273:20967–20971

Murray HW, Nathan CF (1999) Macrophage microbicidal mechanisms in vivo: reactive nitrogen vs. oxygen intermediates in the killing of intracellular visceral Leishmania donovani. J Exp Med 189:741–746

Murray HW, Teitelbaum RF (1992) L-arginine-dependent reactive nitrogen intermediates and the antimicrobial effect of activated human mononuclear phagocytes. J Infect Dis 165:513–517

Nathan C, Xie Q-W (1994) Regulation of biosynthesis of nitric oxide. J Biol Chem 269:13725–13728

Nicholson S, Da Gloria Bonecini-Almeida M, Lapa e Silva JR, Nathan C, Xie Q-W, Mumford R, Weidner JR, Calaycay J, Geng J, Boechat N, Linhares C, Rom W, Ho JL (1996) Inducible nitric oxide synthase in pulmonary alveolar macrophages from patients with tuberculosis. J Exp Med 183:2293–2302

Niedbala W, Wei X-Q, Piedrafita D, Xu D, Liew FY (1999) Effects of nitric oxide on the induction and differentiation of Th1 cells. Eur J Immunol 29:2498–2505

Nozaki Y, Hasegawa Y, Ichiyama S, Nakashima I, Shimokata K (1997) Mechanism of nitric oxide-dependent killing of *Mycobacterium bovis* BCG in human alveolar macrophages. Infect Immun 65:3644–3647

Nüssler AK, Rénia L, Pasquetto V, Miltgen F, Matile H, Mazier D (1993) In vivo induction of the nitric oxide pathway in hepatocytes after injection with irradiated malaria sporozoites, malaria blood parasites or adjuvants. Eur J Immunol 23:882–887

Ogilvie AC, Hack E, Wagstaff J, Van Mierlo GJ, Eerenberg AJM, Thomsen LL, Hoekman K, Rankin EM (1996) IL-1β does not cause neutrophil degranulation but does lead to IL-6, IL-8 and nitrite/nitrate release when used in patients in cancer. J Immunol 156:389–394

Ohmori Y, Hamilton TA (1997) IL-4-induced STAT6 suppresses IFN-γ-stimulated STAT1-dependent transcription in mouse macrophages. J Immunol 159:5474–5482

Oliveira SHP, Fonseca SG, Romao PRT, Figueiredo F, Ferreira SH, Cunha FQ (1998) Microbicidal activity of eosinophils is associated with activation of the arginine-NO pathway. Parasite Immunol 20:405–412

Olivier M, Romero-Gallo B-J, Matte C, Blanchette J, Posner BI, Tremblay MJ, Faure R (1998) Modulation of interferon-γ-induced macrophage activation by phosphotyrosine phosphatases inhibition. Effect on murine leishmaniasis progression. J Biol Chem 273:13944–13949

Oswald IP, Eltoum I, Wynn TA, Schwartz B, Caspar P, Paulin D, Sher A, James SL (1994) Endothelial cells are activated by cytokine treatment to kill an intravascular parasite, *Schistosoma mansoni*, through the production of nitric oxide. Proc Natl Acad Sci USA 91:999–1003

Ouadrhiri Y, Sibille Y, Tulkens PM (1999) Modulation of intracellular growth of *Listeria monocytogenes* in human enterocyte Caco-2 cells by interferon-g and interleukin-6: role of nitric oxide and cooperation with antibiotics. J Infect Dis 180:1195–1204

Pacelli R, Wink DA, Cook JA, Krishna MC, W. D, Friedman N, Tsokos M, Samuni A, Mitchell JB (1995) Nitric oxide potentiates hydrogen peroxide-induced killing of *Escherichia coli*. J Exp Med 182:1469–1479

Padgett EL, Pruett SB (1995) Rat, mouse and human neutrophils stimulated by a variety of activating agents produce much less nitrite than rodent macrophages. Immunology 84:135–141

Pahan K, Sheikh FG, Namboodiri AM, Singh I (1998) Inhibitors of protein phosphatase 1 and 2A differentially regulate the expression of inducible nitric oxide synthase in rat astrocytes and macrophages. J Biol Chem 273:12219–12226

Pahan K, Raymond JR, Singh I (1999) Inhibition of phosphatidylinositol 3-kinase induces niric oxide synthase in lipopolysaccharide- or cytokine-stimulated C_6 glial cells. J Biol Chem 274:7528–7536

Panaro MA, Fasanella A, Lisi S, Mitolo V, Andriola A, Brandonisio O (1998) Evaluation of nitric oxide production by *Leishmania infantum*-infected dog macrophages. Immunopharmacol Immunotoxicol 20:147–158

Paveto C, Pereira C, Espinosa J, Montagna AE, Farber M, Esteva M, Flawia MM, Torres HN (1995) The nitric oxide transduction pathway in *Trypanosoma cruzi*. J Biol Chem 270:16576–16579

Peng H-B, Libby P, Liao JK (1995a) Induction and stabilization of IκBα by nitric oxide mediates inhibition of NF-κB. J Biol Chem 270:14214–14219

Peng H-P, Rajavashisth TB, Libby P, Liao JK (1995b) Nitric oxide inhibits macrophage-colony-stimulating factor gene transcription in vascular endothelial cells. J Biol Chem 270:17050–17055

Peng H-B, Spiecker M, Liao JK (1998) Inducible nitric oxide: an autoregulatory feedback inhibitor of vascular inflammation. J Immunol 161:1970–1976

Perry LL, Feilzer K, Caldwell HD (1998) Neither interleukin-6 nor inducible nitric oxide synthase is required for clearance of *Chlamydia trachomatis* from the mujrine genital-tract epithelium. Infect Immun 66:1265–1269

Persoons JH, Schornagel K, Tilders FF, De Vente J, Berkenbosch F, Kraal G (1996) Alveolar macrophages autoregulate IL-1 and IL-6 production by endogenous nitric oxide. Am J Resp Cell Mol Biol 14:272–278

Peterson PK, Gekker G, Hu S, Chao CC (1995) Human astrocytes inhibit intracellular multiplication of *Toxoplasma gondii* by a nitric oxide-mediated mechanism. J Infect Dis 171:516

Petrova TV, Akama KT, van Eldik LJ (1999) Cyclopentenone prostaglandins suppress activation of microglia: down-regulation of inducible nitric oxide synthase by 15-deoxy-$\Delta^{12,14}$-prostaglandin J_2. Proc Natl Acad Sci USA 96:4668–4673

Pfeilschifter J, Rob P, Mülsch A, Fandrey J, Vosbeck K, Busse R (1992) Interleukin-1β and tumor necrosis factor-α induce a macrophage type of nitric oxide synthase in rat renal mesangial cells. Eur J Biochem 203:251–255

Puddu P, Fantuzzi L, Borghi P, Varano B, Rainaldi G, Guillemard E, Malorni W, Nicaise P, Wolf SF, Belardelli F, Gessani S (1997) IL-12 induces IFN-γ expression and secretion in mouse peritoneal macrophages. Infect Immun 159:3490–3497

Qureshi AA, Hosoi J, Xu S, Takashima A, Granstein RD, Lerner EA (1996) Langerhans' cells express inducible nitric oxide synthase and produce nitric oxide. J Invest Dermatol 107:815–821

Rajan AJ, Klein JDS, Brosnan CF (1998) The effect of $\gamma\delta$T cell depletion on cytokine gene expression in experimental allergic encephalomyelitis. J Immunol 160:5955–5962

Ramsey KH, Miranpuri GS, Poulson CE, Marthakis NB, Braune LM, Byrne GI (1998) Inducible nitric oxide synthase does not affect resolution of murine chlamydial genital tract infections or eradication of chlamydiae in primary murine cell culture. Infect Immun 66:835–838

Regulski M, Tully T (1995) Molecular and biochemical characterization of dNOS: a *Drosophila* Ca^{2+}/calmodulin-dependent nitric oxide synthase. Proc Natl Acad Sci USA 92:9072–9076

Reiling N, Ulmer AJ, Duchrow M, Ernst M, Flad H-D, Hauschildt S (1994) Nitric oxide synthase: mRNA expression of different isoforms in human monocytes/macrophages. Eur J Immunol 24:1941–1944

Reiling N, Kröncke R, Ulmer AJ, Gerdes J, Flad H-D, Hauschildt S (1996) Nitric oxide synthase: expression of the endothelial, Ca^{2+}/calmodulin-dependent isoform in human B and T lymphocytes. Eur J Immunol 26:511–516

Remick DG, Villarete L (1996) Regulation of cytokine gene expression by reactive oxygen and reactive nitrogen intermediates. J Leukoc Biol 59:471–475

Riches DW, Underwood GA (1991) Expression of interferon-β during the triggering phase of macrophage cytocidal activation. Evidence for an autocrine/paracrine role in the regulation of this state. J Biol Chem 266:24785–24792

Riches DWH, Chan ED, Zahradka EA, Winston BW, Remigio LK, Lake FR (1998) Cooperative signaling by tumor necrosis factor receptors CD120a (p55) and CD120b (p75) in the expression of nitric oxide and inducible nitric oxide synthase by mouse macrophages. J Biol Chem 273:22800–22806

Ricote M, Li AC, Willson TM, Kelly CJ, Glass CK (1998) The peroxisome proliferator-activated receptor γ is a negative regulator of macrophage activation. Nature 391:79–82

Rockett KA, Brookes R, Udalova I, Vidal V, Hill AVS (1998) 1,25-dihydroxyvitamin D3 induces nitric oxide synthase and suppresses growth of *Mycobacterium tuberculosis* in a human macrophage-like cell line. Infect Immun 66:5314–5321

Rodig SJ, Meraz MA, White JM, Lampe PA, Riley JK, Arthur CD, King KL, Sheehan KCF, Yin L, Pennica D, Johnson EM, Schreiber RD (1998) Disruption of the Jak1 gene demonstrates obligatory and nonredundant roles of the Jaks in cytokine-induced biologic responses. Cell 93:373–383

Ross R, Gillitzer C, Kleinz R, Schwing J, Kleinert H, Förstermann U, Reske-Kunz A (1998) Involvement of NO in contact hypersensitivity. Int Immunol 10:61–69

Rothe H, Hartmann B, Geerlings P, Kolb H (1996) Interleukin-12 gene expression of macrophages is regulated by nitric oxide. Biochem Biophys Res Commun 224:159–163

Rottenberg ME, Rothfuchs ACG, Gigliotti D, Svanholm C, Bandholtz L, Wigzell H (1999) Role of innate and adaptive immunity in the outcome of primary infection with *Chlamydia pneumoniae*, as analyzed in genetically modified mice. J Immunol 162:2829–2836

Ruan J, John GS, Ehrt S, Riley L, Nathan C (1999) *noxR3*, a novel gene from *Mycobacterium tuberculosis*, protects *Salmonella typhimurium* from nitrosative and oxidative stress. Infect Immun 67:3276–3283

Salkowski CA, Barber SA, Detore GR, Vogel SN (1996) Differential dysregulation of nitric oxide production in macrophages with targeted disruptions in IFN regulatory factor-1 and -2 genes. J Immunol 156:3107–3110

Salkowski CA, Detore G, McNally R, Van Rooijen N, Vogel SN (1997) Regulation of inducible nitric oxide synthase messenger RNA expression and nitric oxide production by lipopolysaccharide in vivo: the roles of macrophages, endogenous IFN-γ, and TNF-receptor-1-mediated signaling. J Immunol 158:905–912

Salvucci O, Kolb JP, Dugas B, Dugas N, Chouaib S (1998) The induction of nitric oxide by interleukin-12 and tumor necrosis factor-α in human natural killer cells: relationship with the regulation of lytic activity. Blood 92:2093–2102

Salzman AL, Eaves-Pyles T, Linn SC, Denenberg AG, Szabo C (1998) Bacterial induction of inducible nitric oxide synthase in cultured human intestinal epithelial cells. Gastroenterology 114:93–102

Saura M, Lopez S, Puyol MR, Puyol DR, Lamas S (1995) Regulation of inducible nitric oxide synthase expression in rat mesangial cells and isolated glomeruli. Kidney Int 47:500–509

Saura M, Martinez-Dalmau R, Minty A, Pérez-Sala D, Lamas S (1996) Interleukin-13 inhibits inducible nitric oxide synthase expression in human mesangial cells. Biochem J 313:641–646

Saura M, Zaragoza C, McMillan A, Quick RA, Hohenadl C, Lowenstein JM, Lowenstein CJ (1999) An antiviral mechanism of nitric oxide: inhibition of a viral protease. Immunity 10:21–28

Sawada T, Falk LA, Rao P, Murphy WJ, Pluznik DV (1997) IL-6 induction of protein-DNA complexes via a novel regulatory region of the inducible nitric oxide synthase gene promoter. Role of octamer-binding proteins. J Immunol 158:5267–5276

Scharton-Kersten TM, Yap G, Magram J, Sher A (1997) Inducible nitric oxide is essential for host control of persistent but not acute infection with the intracellular pathogen *Toxoplasma gondii*. J Exp Med 185:1261–1273

Schleifer KW, Mansfield JM (1993) Suppressor macrophages in african trypanosomiasis inhibit T cell proliferative responses by nitric oxide and prostaglandins. J Immunol 151:5492–5503

Schlüter D, Deckert-Schlüter M, Lorenz E, Meyer T, Röllinghoff M, Bogdan C (1999) Inhibition of inducible nitric oxide synthase exacerbates chronic cerebral toxoplasmosis in *Toxoplasma gondii*-susceptible C57BL/6 mice but does not reactivate the latent disease in *T. gondii*-resistant BALB/c mice. J Immunol 162:3512–3518

Schneemann M, Schoedon G, Hofer S, Blau N, Guerrero L, Schaffner A (1993) Nitric oxide synthase is not a constituent of the antimicrobial armature of human mononuclear phagocytes. J Infect Dis 167:1358–1363

Schwacha MG, Eisenstein TK (1997) Interleukin-12 is critical for induction of nitric oxide-mediated immunosuppression following vaccination of mice with attenuated *Salmonella typhimurium*. Infect Immun 65:4897–4903

Schwacha MG, Meissler JJ, Eisenstein TK (1998) *Salmonella typhimurium* infection in mice induces nitric oxide-mediated immunosuppression through a natural killer cell-dependent pathway. Infect Immun 66:5862–5866

Segal BM, Dwyer BK, Shevach EM (1998) An interleukin (IL)-10/IL-12 immunoregulatory circuit controls susceptibility to autoimmune disease. J Exp Med 187:537–546

Severn A, Wakelam MJO, Liew FY (1992) The role of protein kinase C in the induction of nitric oxide synthesis by murine macrophages. Biochem Biophys Res Commun 188:997–1002

Shalom-Barak T, Quach J, Lotz M (1998) Interleukin-17-induced gene expression in articular chondrocytes is associated with activation of mitogen-activated protein kinases and NF-κB. J Biol Chem 273:27467–27473

Sharara AI, Perkins DJ, Misukonis MA, Chan SU, Dominitz JA, Weinberg BJ (1997) Interferon-α activation of human blood mononuclear cells in vitro and in vivo for nitric oxide synthase (NOS) type 2 mRNA and protein expression: possible relationship of induced NOS2 to the anti-hepatitis C effects of IFN-α in vivo. J Exp Med 186:1495–1502

Sheffler LA, Wink DA, Melillo G, Cox GW (1995) Exogenous nitric oxide regulates IFN-γ plus lipopolysaccharide-induced nitric oxide synthase expression in mouse expression. J Immunol 155:886–894

Shiloh MU, MacMicking JD, Nicholson S, Brause JE, Potter S, Marino M, Fang F, Dinauer M, and Nathan C (1999) Phenotype of mice and macrophages deficient in both phagocyte oxidase and inducible nitric oxide synthase. Immunity 10:29–38

Shin WS, Hong Y-H, Peng H-B, de Caterina R, Libby P, Liao JK (1996) Nitric oxide attenuates vascular smooth muscle cell activation by interferon-γ. The role of constitutive NF-κB activity. J Biol Chem 271:11317–11324

Shiraishi A, Dudler J, Lotz M (1997) The role of IFN regulatory factor-1 in synovitis and nitric oxide production. J Immunol 159:3549–3554

Sicher SC, Chung GW, Vazquez MA, Lu CY (1995) Augmentation or inhibition of IFN-γ-induced MHC class II expression by lipopolysaccharides. The roles of TNF-α and nitric oxide and the importance of the sequence of signaling. J Immunol 155:5826–5834

Siedlar M, Mytar B, Krzeszowiak A, Baran J, Hyszko M, Ruggiero I, Wieckiewicz J, Stachura J, Zembala M (1999) Demonstration of iNOS mRNA and iNOS in human monocytes stimulated with cancer cells in vitro. J Leukoc Biol 65:597–604

Singh K, Balligand JL, Fischer TA, Smith TW, Kelly RA (1996) Regulation of cytokine-inducible nitric oxide synthase in cardiac myocytes and microvascular endothelial cells. Role of extracellular signal-regulated kinases 1 and 2 (ERK1/ERK2) and Stat1α. J Biol Chem 271:1111–1117

Sirsjo A, Karlsson M, Gidlöf A, Rollman O, Törmä H (1996) Increased expression of inducible nitric oxide synthase in psoriatic skin and cytokine-stimulated cultured keratinocytes. Brit J Dermatol 134:643–648

Spink J, Cohen J, Evans TJ (1995) The cytokine-responsive vascular smooth-muscle cell enhancer of inducible nitric oxide synthase: activation by nuclear factor κB. J Biol Chem 270:29541–29546

Sriskandan S, Evans TJ, Cohen J (1996) Bacterial superantigen-induced human lymphocyte responses are nitric-oxide dependent and mediated by IL-12 and IFN-γ. J Immunol 156:2430–2435

Stadler J, Stefanovic-Racic M, Billiar TR, Curran RD, McIntyre LA, Georgescu HI, Simmons RL, Evans CH (1991) Articular chondrocytes synthesize nitric oxide in response to cytokines and lipopolysaccharide. J Immunol 147:3915–3920

Stadler J, Harbrecht BG, Di Silvio M, Curran RD, Jordan ML, Simmons RL, Billiar TR (1993) Endogenous nitric oxide inhibits the synthesis of cyclooxygenase products and interleukin 6 by rat Kupffer cells. J Leukocyte Biol 53:165–172

Stamler JS, Jaraki O, Osborne J, Simon DI, Keaney J, Vita J, Singel D, Valeri CR, Loscalzo J (1992) Nitric oxide circulates in mammalian plasma primarily as an S-nitroso adduct of serum albumin. Proc Natl Acad Sci USA 89:7674–7677

Stefani MMA, Müller I, Louis J (1994) *Leishmania-major*-specific CD8+ T cells are inducers and targets of nitric oxide produced by parasitized macrophages. Eur J Immunol 24:746–752

Stenger S, Donhauser N, Thüring H, Röllinghoff M, Bogdan C (1996) Reactivation of latent leishmaniasis by inhibition of inducible nitric oxide synthase. J Exp Med 183:1501–1514

Stout RD, Suttles J, Xu J, Grewal I, Flavell RA (1996) Impaired T-cell-mediated macrophage activation in CD40 ligand-deficient mice. J Immunol 156:8–11

Stuehr D, Marletta MA (1985) Mammalian nitrite biosynthesis: mouse macrophages produce nitrite and nitrate in response to *Escherichia coli* lipopolysaccharide. Proc Natl Acad Sci USA 82:7738–7742

Stuehr DJ, Marletta MA (1987) Induction of nitrite/nitrate synthesis in murine macrophages by BCG infection, lymphokines or interferon-γ. J Immunol 139:518–525

Stuehr DJ, Nathan CF (1989) Nitric oxide: a macrophage product responsible for cytostasis and respiratory inhibition in tumor cell growth. J Exp Med 169:1543–1555

Sunyer T, Rothe L, Jiang X, Osdoby P, Collin-Osdoby P (1996) Pro-inflammatory agents, IL-8 and IL-10, up-regulate inducible nitric oxide synthase expression and nitric oxide production in avian osteoclast-like cells. J Cell Biochem 60:469–483

Suschek C, Rothe H, Fehsel K, Enczmann J, Kolb-Bachofen V (1993) Induction of a macrophage-like nitric oxide synthase in cultured rat aortic endothelial cells. IL-1β-mediated induction regulated by tumor necrosis factor-α and IFN-γ. J Immunol 151:3283–3291

Sveinbjornsson B, Olsen R, Seternes OM, Seljelid R (1996) Macrophage cytotoxicity against murine Meth A sarcoma involves nitric-oxide-mediated apoptosis. Biochem Biophysic Res Commun 223:643–649

Szabó C, Southan GJ, Thiemermann C (1994) Beneficial effects and improved survival in rodent models of septic shock with S-methylisothiourea sulfate, a potent and selective inhibitor of inducible nitric oxide synthase. Proc Natl Acad Sci USA 91:12472–12476

Tada Y, Ho A, Matsuyama T, Mak TW (1997) Reduced incidence and severity of antigen-induced autoimmune diseases in mice lacking interferon-regulatory-factor-1. J Exp Med 185:231–238

Tai X-G, Toyo-oka K, Yamamoto N, Yashiro Y, Mu J, Hamaoka T, Fujiwara H (1997) Expression of an inducible type of nitric oxide (NO) synthase in the thymus and involvement of NO in deletion of TCR-stimulated double-positive thymocytes. J Immunol 158:4696–4703

Takeda K, Kamanaka M, Tanaka T, Kishimoto T, Akira S (1996) Impaired IL-13-mediated functions of macrophages in Stat6-deficient mice. J Immunol 157:3220–3222

Tamaoki J, Kondo M, Kohri K, Aoshiba K, Tagaya E, Nagai A (1999) Macrolide antibiotics protect against immune complex-induced lung injury in rats: role of nitric oxide from alveolar macrophages. J Immunol 163:2909–2915

Tanaka K, Nakazawa H, Okada K, Umezawa K, Fukuyama N, Koga Y (1997) Nitric oxide mediates murine cytomegalovirus-associated pneumonitis in lungs that are free of the virus. J Clin Invest 100:1822–1830

Tanaka T, Akira S, Yoshida K, Umemoto M, Yoneda Y, Shirafuji N, Fujiwara H, Suematsu S, Yoshida N, Kishimoto T (1995) Targeted disruption of the NF-IL6 gene discloses its essential role in bacteria killing and tumor cytotoxicity by macrophages. Cell 80:353–361

Tao X, Stout RD (1993) T cell-mediated cognate signaling of nitric oxide production by macrophages. Requirements for macrophage activation by plasma membranes from T cells. Eur J Immunol 23:2916–2921

Tarrant TK, Silver PB, Wahlsten JL, Rizzo LV, Chan C-C, Wiggert B, Caspi RR (1999) Interleukin-12 protects from a Th1-mediated autoimmune disease, experimental autoimmune uveitis, through a mechanism involving IFN-γ, NO and apoptosis. J Exp Med 189:219–230

Taub DD, Cox GW (1995) Murine Th1 and Th2 cell clones differentially regulate macrophage nitric oxide production. J Leukoc Biol 58:80–89

Taylor AP, Murray HW (1997) Intracellular antimicrobial activity in the absence of interferon-γ: effect of interleukin-12 in experimental visceral leishmaniasis in interferon-γ gene-disrupted mice. J Exp Med 185:1231–1239

Taylor AW, Yee DG, Streilein JW (1998) Suppression of nitric oxide generated by inflammatory macrophages by calcitonin gene-related peptide in aequous humor. Invest Ophthalmol Vis Sci 39:1372–1378

Taylor BS, de Vera ME, Ganster RW, Wang Q, Shapiro RA, Morris SM, Billiar TR, Geller DA (1998) Multiple NF-kB enhancer elements regulate cytokine induction of the human inducible nitric oxide synthase gene. J Biol Chem 273:15148–15156

Taylor-Robinson AW, Liew FY, Severn A, Xu D, McScorley SJ, Garside P, Padron J, Phillips RS (1994) Regulation of the immune response by nitric oxide differentially produced by T helper type 1 and T helper type 2 cells. Eur J Immunol 24:980–984

Thomas SR, Mohr D, Stocker R (1994) Nitric oxide inhibits indoleamine 2,3-dioxygenase activity in interferon-γ-primed mononuclear phagocytes. J Biol Chem 269:14457–14464

Thomsen LL, Scott JM, Topley P, Knowles RG, Keerie AJ, Frend AJ (1997) Selective inhibition of inducible nitric oxide synthase inhibits tumor growth in vivo: studies with 1400W, a novel inhibitor. Cancer Res 57:3300–3304

Thüring H, Stenger S, Gmehling D, Röllinghoff M, Bogdan C (1995) Lack of inducible nitric oxide synthase activity in T cell-clones and T lymphocytes from naive and *Leishmania major*-infected mice. Eur J Immunol 25:3229–3234

Tian L, Noelle RJ, Lawrence DA (1995) Activated T cells enhance nitric oxide production by murine splenic macrophages through gp39 and LFA-1. Eur J Immunol 25:306–309

Trepicchio WL, Bozza M, Pedneault G, Dorner AJ (1996) Recombinant human IL-11 attenuates the inflammatory response through down-regulation of pro-inflammatory cytokine release and nitric oxide production. J Immunol 157:3627–3634

Trepicchio WL, Wang L, Bozza M, Dorner AJ (1997) IL-11 regulates macrophage effector function through the inhibition of nuclear factor-κ. J Immunol 159:5661–5670

Trinchieri G, Gerosa F (1996) Immunoregulation by interleukin-12. J Leukoc Biol 59:505–511

Turco J, Liu H, Gottlieb SF, Winkler HH (1998) Nitric oxide-mediated inhibition of the ability of Rickettsia prowazekii to infect mouse fibroblasts and mouse macrophage-like cells. Infect Immun 66:558–566

Tzeng E, Billiar TR, Robbins PD, Loftus M, Stuehr DJ (1995) Expression of human inducible nitric oxide synthase in a tetrahydrobiopterin (H_4B)-deficient cell line: H_4B promotes assembly of enzyme subunits into an active dimer. Proc Natl Acad Sci USA 92:11771–11775

Umansky V, Hehner SP, Dumont A, Hofmann TG, Schirrmacher V, Dröge W, Schmitz ML (1998) Co-stimulatory effect of nitric oxide on endothelial NF-κB implies a physiological self-amplifying mechanism. Eur J Immunol 28:2276–2282

Vallette G, Jarry A, Branka J-E, Laboisse CL (1996) A redox-based mechanism for induction of IL-1 production by nitric oxide in a human colonic epithelial cell line (HT29-Cl.16E). Biochem J 313:35–38

van der Veen RC, Dietlin TA, Pen L, Gray JD (1999) Nitric oxide inhibits the proliferation of T-helper 1 and 2 lymphocytes without reduction in cytokine secretion. Cell Immunol 193:194–201

Van Dervort AL, Yan L, Madara PJ, Cobb JP, Wesley RA, Corriveau CC, Tropea MM, Danner RL (1994) Nitric oxide regulates endotoxin-induced TNF-α production by human neutrophils. J Immunol 152:4102–4109

Vazquez-Torres A, Jones-Carson J, Balish E (1996) Peroxinitrite contributes to the candidacidal activity of nitric oxide-producing macrophages. Infect Immun 64:3127–3133

Villalta F, Zhang Y, Bibb KE, Kappes JC, Lima MF (1998) The cysteine-cysteine family of chemokines RANTES, MIP-1α, and MIP-1β induce trypanocidal activity in human macrophages via nitric oxide. Infect Immun 66:4690–4695

Virag L, Scott GS, Cuzzocrea S, Marmer D, Salzman AL, Szabo C (1998) Peroxynitrite-induced thymocyte apoptosis: the role of caspases and poly (ADP–ribose) synthetase (PARS) activation. Immunol 94:345–355

Vodovotz Y, Kwon NS, Pospischil M, Manning J, Paik J, Nathan C (1994) Inactivation of nitric oxide synthase following prolonged incubation of mouse macrophages with IFN-γ and bacterial lipopolysaccharide. J Immunol 152:4110–4118

Vodovotz Y, Bogdan C, Paik J, Xie Q-W, Nathan C (1993) Mechanisms of suppression of macrophage nitric oxide release by transforming growth factor-β. J Exp Med 178:605–613

Vouldoukis I, Riveros-Moreno V, Dugas B, Quaaz F, Bécherel P, Debré P, Moncada S, Mossalayi MD (1995) The killing of *Leishmania major* by human macrophages is mediated by nitric oxide induced after ligation of the FcεRII/CD23 surface antigen. Proc Natl Acad Sci USA 92:7804–7808

Vouldoukis I, Bécherel P-A, Riveros-Moreno V, Arock M, da Silva O, Debré P, Mazier D, Mossalayi MD (1997) Interleukin-10 and interleukin-4 inhibit intracellular killing of *Leishmania infantum* and *Leishmania major* by human macrophages by decreasing nitric oxide generation. Eur J Immunol 27:860–865

Warwick-Davies J, Dhillon J, O'Brien L, Andrew PW, Lowrie DB (1994) Apparent killing of Mycobacterium tuberculosis by cytokine-activated human monocytes can be an artefact of a cytotoxic effect on the monocytes. Clin Exp Immunol 96:214–217

Way SS, Goldberg MB (1999) Clearance of *Shigella flexneri* infection occurs through a nitric oxide-independent mechanism. Infect Immun 66:3012–3016

Wei X-Q, Charles IG, Smith A, Ure J, Feng G-J, Huang F-P, Xu D, Müller W, Moncada S, Liew FY (1995) Altered immune responses in mice lacking inducible nitric oxide synthase. Nature 375:408–411

Weiss G, Werner-Felmayer G, Werner ER, Grünewald K, Wachter H, Hentze MW (1994) Iron regulates nitric oxide synthase activity by controlling nuclear transcription. J Exp Med 180:969–976

Werner ER, Werner-Felmayer G, Mayer B (1998) Minireview: tetrahydro-biopterin, cytokines and nitric oxide synthesis. Proc Soc Exp Biol Med 219:171–182

Werner-Felmayer G, Werner ER, Fuchs D, Hausen A, Reibnegger G, Wachter H (1990) Tetrahydrobiopterin-dependent formation of nitrite and nitrate in murine fibroblasts. J Exp Med 172:1599–1607

Werner-Felmayer G, Golderer G, Werner ER, Gröbner P, Wachter H (1994) Pteridine biosynthesis and nitric oxide synthase in *Physarum polycephalum*. Biochem J 304:105–111

Wheeler MA, Smith SD, Garcia-Cardena G, Nathan CF, Weiss RM, Sessa WC (1997) Bacterial infection induces nitric oxide synthase in human neutrophils. J Clin Invest 99:110–116

Williams MS, Noguchi S, Henkart PS, Osawa Y (1998) Nitric oxide synthase plays a signalling role in TCR-triggered apoptotic death. J Immunol 161:6526–6531

Wink DA, Kasprzak KS, Maragos CM, Elespuru RK, Misra M, Dunams TM, Cebula TA, Koch WH, Andrews AW, Allen JS, Keefer LK (1991) DNA-deaminating ability and genotoxicity of nitric oxide and its progenitors. Science 254:1001–1003

Wong ML, Rettori V, al-Shekhlee A, Bongiorno PB, Canteros G, McCann SM, Gold PW, Licinio J (1996) Inducible nitric oxide synthase gene expression in the brain during systemic inflammation. Nat Med 2:581–584

Worrall NK, Lazenby WD, Misko TP, Lin T-S, Rodi CP, Manning PT, Tilton RG, Williamson JR, Ferguson TB (1995) Modulation of in vivo alloreactivity by inhibition of inducible nitric oxide synthase. J Exp Med 181:63–70

Wright K, Ward SG, Kolios G, Westwick J (1997) Activation of phosphatidylinositol 3-kinase by interleukin-13. An inhibitory signal for inducible nitric oxide synthase expression in the epithelial cell line HT-29. J Biol Chem 272:12626–12633

Wu-Hsieh BA, Chen W, Lee H-J (1998) Nitric oxide synthase expression in macrophages of *Histoplasma capsulatum*-infected mice is associated with splenocyte apoptosis and unresponsiveness. Infect Immun 66:5520–5526

Xaus J, Mirabet M, Lloberas J, Soler C, Lluis C, Franco R, Celada A (1999) IFN-γ up-regulates the A_{2B} adenosine receptor expression in macrophages: a mechanism of macrophage deactivation. J Immunol 162:3607–3614

Xie K, Huang S, Dong Z, Juang S-H, Gutman M, Xie Q-W, Nathan C, Fidler IJ (1995) Transfection with the inducible nitric oxide synthase gene suppresses tumorigenicity and abrogates metastasis by K-1735 murine melanoma cells. J Exp Med 181:1333–1343

Xie K, Dong Z, Fidler IJ (1996) Activation of nitric oxide gene for inhibition of cancer metastasis. J Leukoc Biol 797:797–803

Xie Q-W, Cho HJ, Calaycay J, Mumford RA, Swiderek KM, Lee TD, Ding A, Troso T, Nathan C (1992) Cloning and characterization of inducible nitric oxide synthase from mouse macrophages. Science 256:225–228

Xie Q-W, Whisnant R, Nathan C (1993) Promoter of the mouse gene encoding calcium-independent nitric oxide synthase confers inducibility by interferon-γ and bacterial lipopolysaccharide. J Exp Med 177:1779–1784

Xie Q-W, Kasshiwabara Y, Nathan C (1994) Role of transcription factor NF-κB/Rel in induction of nitric oxide synthase. J Biol Chem 269:4705–4708

Xie L, Tume N, Ho G, Smith J, Lu M, Gross SS (1999) Evolution of a specific cysteine residue in mammalian argininosuccinate synthetases that confers reversible inactivation by S-nitrosation. Acta Physiologica Scandinavica 167:4 (abstract O-5, supplementum 645).

Yim C-Y, McGregor JR, Kwon O-D, Bastian NR, Rees M, Mori M, Hibbs JB, Samlowski WE (1995) Nitric oxide synthesis contributes to IL-2-induced antitumor responses against intraperitoneal Meth-A tumor. J Immunol 155:4382–4390

Yoneto T, Yoshimoto T, Wang CR, Takahama Y, Tsuji M, Waki S, Nariuchi H (1999) Gamma interferon production is critical for protective immunity to infection with blood-stage Plasmodium berghei XAT but neither NO production nor NK cell activation is critical. Infect Immun. 67:2349–2356

Young MRI, Wright MA, Matthews JP, Malik I, Prechel M (1996) Suppression of T cell proliferation by tumor-induced granulocyte–macrophage progenitor cells producing transforming growth factor-β and nitric oxide. J Immunol 156:1916–1922

Yuda M, Hirai M, Miura K, Matsumura H, Ando K, Chinzei Y (1996) cDNA cloning, expression and characterization of nitric oxide synthase from the salivary glands of the blood-sucking insect *Rhodnius prolixus*. Eur J Biochem 242:807–812

Zanardo RCO, Costa E, Ferreira HHA, Antunes E, Martins AR, Murad F, de Nucci G (1997) Pharmacological and immunohistochemical evidence for a functional nitric oxide synthase system in rat peritoneal macrophages. Proc Natl Acad Sci USA 94:14111–14114

Zaragoza C, Ocampo C, Saura M, Leppo M, Wei X-Q, Quick R, Moncada S, Liew FY, Lowenstein CJ (1998) The role of inducible nitric oxide synthase in the host response to *Coxsackievirus myocarditis*. Proc Natl Acad Sci USA 95:2469–2474

Zaragoza C, Ocampo CJ, Saura M, Bao C, Leppo M, Lafond-Walker A, Thiemann DR, Hruban R, Lowenstein CJ (1999) Inducible nitric oxide synthase protection against Coxsackievirus pancreatitis. J Immunol 163:5497–5504

Zeiher AM, Fisslthaler B, Schray-Utz B, Busse R (1995) Nitric oxide modulates the expression of MCP-1 in cultured human endothelial cells. Circ Res 76:980–986

Zettl UK, Mix E, Zielasek J, Stangel M, Hartung HP, Gold R (1997) Apoptosis of myelin-reactive T cells induced by reactive oxygen and nitrogen intermediates in vitro. Cell Imunol 178:1–8

Zhang T, Kawakami K, Qureshi MH, Okamura H, Kurimoto M, Saito A (1997) IL-12 and IL-18 synergistically induce the fungicidal activity of murine peritoneal exudate cells against *Cryptococcus neoformans* through production of γ interferon by natural killer cells. Infect Immun 65:3594–3599

Zheng YM, Schäfer MK-H, Weihe E, Sheng H, Corisdeo S, Fu ZF, Koprowski H, Dietzschold B (1993) Severity of neurological signs and degree of inflammatory lesions in the brains of rats with Borna disease correlate with the induction of nitric oxide synthase. J Virol 67:5786–5791

Zhuang JC, Lin C, Lin D, Wogan GN (1998) Mutagenesis associated with nitric oxide production in macrophages. Proc Natl Acad Sci USA 95:8286–8291

CHAPTER 19
Nitric Oxide: A True Inflammatory Mediator

R. Zamora and T.R. Billiar

A. Introduction

Nitric oxide (NO) is a colorless gas at room temperature and is one of the simplest molecules known. For more than 300 years, it was considered mainly as an environmental pollutant and product of bacterial metabolism. Important discoveries in the 1980s showed that NO can be synthesized by mammalian cells, and it has since been implicated in a wide variety of regulatory mechanisms ranging from vasodilatation and blood-pressure control to neurotransmission (Moncada and Higgs 1993). NO is also involved in non-specific immunity and participates in the complex mechanism of tissue injury, acting as a major mediator of inflammatory processes and apoptosis. This work reviews the data showing that NO plays an important and complex role in inflammatory and autoimmune diseases.

I. Biosynthesis of NO

It was already known in 1981 that nitrate excretion in germ-free and conventional rats exceeded its intake (Green et al. 1981) and, in other studies, a marked increase in urinary nitrate excretion was observed in humans with diarrhea and fever (Wagner et al. 1983). Nitrate formation was then believed to be a result of microbial metabolism, but these observations suggested that mammalian cells also form nitrogen oxides and that a correlation between immunostimulation and nitrate synthesis may exist (Moncada et al. 1991). In 1985, production of nitrite- and nitrate-generating compounds by mammalian cells in vitro was first demonstrated in the mouse macrophage (Stuehr and Marletta 1985). Furthermore, cytotoxicity of activated macrophages against tumor target cells was shown to be dependent on the presence of L-arginine (Hibbs et al. 1987; Iyengar et al. 1987). After the demonstration of the synthesis of NO from L-arginine by endothelial cells (Palmer et al. 1987), it was shown that NO was indeed the precursor of nitrogen oxides in activated macrophages (Hibbs et al. 1988; Marletta et al. 1988; Stuehr and Nathan 1989).

The production of NO by mammalian cells constitutes a metabolic pathway known as the L-arginine:NO pathway, and the enzyme responsible for the process has been named NO synthase (NOS). It has become apparent, however, that the NO synthase is not just one enzyme but a family of isoenzymes (MACMICKING et al. 1997). Three genes encoding NO synthases have been identified and cloned in mammals; two of these synthases are generally believed to be cytosolic-calcium-dependent and constitutively expressed (cNOS), e.g., in vascular endothelium (eNOS or type III) and nervous tissue (nNOS or type I). The third type is calcium-independent and can be induced in many cells, such as macrophages (iNOS or type II) following activation by bacterial endotoxin and/or inflammatory cytokines (MACMICKING et al. 1997). Although the three isoforms vary considerably in cellular location, kinetics, regulation and functional roles, all of them catalyze the production of NO by the same biochemical pathway. This process is an overall five-electron oxidation of nitrogen. The process consists of two sequential mono-oxygenase reactions: oxidation at a guanidino nitrogen of one molecule of L-arginine to produce N^{ω}-hydroxy-L-arginine as an intermediate and further oxidation of the latter to yield one molecule each of NO and L-citrulline (MACMICKING et al. 1997).

B. NO and Inflammation

Inflammation can be shortly defined as a physiological defense response of the body to any kind of injurious stimulus. Although there is no clear dividing line between acute and chronic inflammation, the former refers to a response that is abrupt in onset and of short duration. Thus, acute inflammation may become chronic (in the temporal sense) if the injurious agent is persistent. However, chronic inflammation is characterized by a proliferation of fibroblasts and small blood vessels and an influx of chronic inflammatory cells, namely granulocytes (neutrophils, eosinophils, and basophils), lymphocytes, plasma cells, and macrophages. A comprehensive work providing an up-to-date look at the basics of inflammatory processes has recently been published (TROWBRIDGE and EMLING 1997).

In recent years, a role for NO as a major mediator of inflammation has been clearly demonstrated, and the number of related publications is increasing vertiginously. Although considerable evidence implicates NO in the pathophysiology of inflammatory processes, there are contradictory reports in the literature concerning its role as an anti-inflammatory or pro-inflammatory agent. The inconsistencies reported are probably due the multiple cellular actions of this molecule, the level and site of NO production and the redox milieu into which it is released. NO activates soluble guanylyl cyclase, leading to synthesis of cyclic guanosine monophosphate (cGMP), which constitutes a common pathway in many processes, including vascular smooth-muscle cell relaxation, inhibition of platelet activity, inhibition of neutrophil chemotaxis, and signal transduction in the central and peripheral nervous systems. Moreover, under physiological conditions, the reaction of NO with reactive oxygen intermediates, e.g., superoxide, results in more potent species like peroxyni-

Table 1. Regulatory and anti-inflammatory actions of nitric oxide (NO)

Tissue organ	Physiological action of NO
Vascular endothelium	Maintains vasodilator tone; inhibits smooth muscle cell migration and proliferation; inhibition of blood cell–vessel wall interactions; adhesion to endothelium
Blood cells	Inhibition of platelet adhesion and aggregation; inhibition of microvascular thrombosis; prevents aggregation and adhesion of white cells; mediates cytostatic and cytotoxic activity of macrophages for antimicrobial and antitumour defense; inhibition of mast cell degranulation
Heart	Maintains coronary perfusion; regulates cardiac contractility
Lung	Maintains ventilation/perfusion ratio; regulates bronchociliar motility and mucus secretion
Pancreas	Modulates endocrine secretion
Kidney	Regulates glomerular perfusion and renin secretion
Intestinal system	Modulates peristalsis and exocrine secretion; contributes to protection of mucosa
Central and peripheral nervous system	Modulates and mediates cell-to-cell communication; regulates visual transduction, neuro-endocrine secretion, cerebral blood flow, synaptic plasticity, feeding behavior and olfaction; mediates penile erection

trite, which mediate nitrosylation/nitrosation of a number of molecular targets. These chemical reactions are directly involved in the activation or inhibition of key enzymes in different metabolic processes (like mitochondrial respiration), the inhibition of DNA synthesis, and the cytotoxic activity of macrophages as a part of the non-specific immune response (MONCADA and HIGGS 1993; NÜSSLER and BILLIAR 1993; EVANS 1995; LYONS 1995; MACMICKING et al. 1997; NATHAN 1997; CLANCY et al. 1998). Many of the regulatory and physiological functions of NO can be considered as protective or "anti-inflammatory" (Table 1).

Furthermore, it is important to consider that the release of NO is not limited to a certain type of cell or tissue. Expression of iNOS and formation of NO have been shown in many cell types important in inflammation, including leukocytes, endothelial cells, and sensory nerves. The role of NO will thus vary considerably in different inflammatory events, depending on the cell/tissue in which it is produced and the complex interactions with the different components of the inflammatory process.

I. The Chemical Mediators of the Vascular Response

Inflammation is controlled by the presence of a group of chemical mediators, each with a specific role at some definite stage of the inflammatory reaction. These mediators may be exogenous (arising from bacteria or chemical irritants) or endogenous in origin. The most important endogenous mediators identified include the vasoactive amines histamine and serotonin, the kinin system, the fibrinolytic system, the complement system, the arachidonic-acid metabolites (prostaglandins and leukotrienes), platelet-activating factor (PAF), neuropeptides, reactive oxygen species, and inflammatory cytokines (TROWBRIDGE and EMLING 1997). We will discuss only those components of inflammation directly related to the actions of NO.

II. NO and the Vascular Response to Injury

Injury to an organ or tissue results in progressive changes in the damaged area. As results of vascular alterations in the area, three main signs of vascular response appear: redness, heat, and swelling.

The redness and heat result from an increase in blood flow, which is the result of local vasodilatation first involving arterioles and then capillaries and venules. NO production by cNOS in endothelial cells activates soluble guanylyl cyclase, leading to the synthesis of cGMP, which in turn leads to relaxation of vascular smooth-muscle cells. This pathway has been extensively investigated and constitutes a common process in both human and many animal tissues (MONCADA 1997). Swelling is the result of alterations in vascular permeability. The endothelial cells become leaky, leading to exudation of fluid, plasma proteins, and white blood cells (inflammatory edema).

To assess the contribution of mediators involved in vascular changes associated with acute inflammation, carrageenan-induced edema has been a useful experimental tool and standard model for screening for anti-inflammatory drugs (CUZZOCREA et al. 1998a). The development of carrageenan-induced edema in the rat hindpaw is a biphasic event in which the early phase is related to the production of histamine, leukotrienes, PAF, bradykinin, and possibly cyclooxygenase (COX) products, while the delayed phase has been linked to local neutrophil infiltration and activation. The contribution to edema of NO, superoxide, and peroxynitrite has also been demonstrated in this model (SALVEMINI et al. 1996b). Both the non-selective NOS inhibitors N-nitro-L-arginine-methyl-ester (L-NAME) and N^G-monomethyl-L-arginine (L-NMMA) (at the early phase), and the selective iNOS inhibitors L-N-(1-iminoethyl) lysine (L-NIL) and mercaptoethylguanidine (at the late phase) had a potent inhibitory effect strongly suggesting a pro-inflammatory effect for both constitutive and induced NO production (SALVEMINI et al. 1996c; CUZZOCREA et al. 1998a). However, the location and identity of the NOS isoforms responsible for NO synthesis at the site of inflammation still needs elucidation. In this context, a recent study using the selective nNOS inhibitor 7-nitroindazole suggested that NO synthethised by a nNOS isoform located in sensory nerves plays an important part in the early-phase response to carrageenan in this model of inflammation and that NO synthesized by an iNOS isoform located in inflammatory leukocytes contributes to the late-phase response (HANDY and MOORE 1998). In acute inflammation, neutrophils are the first leukocytes to emerge from the vessels in significant numbers. While the carrageenan-induced paw edema is neutrophil-dependent and both the NOS and COX pathways appear to operate together to amplify the inflammatory response (SALVEMINI et al. 1996c), in a model of dermal inflammation, no evidence was found for the involvement of either COX products or neutrophils in mediating the iNOS inflammatory component (RIDGER et al. 1997). As part of the zymosan-induced inflammatory response in the rat skin, NO contributed to edema formation by increasing blood flow, but the sources of iNOS appeared to be something other than neutrophils. It was suggested that other cell types like dermal fibroblasts and keratinocytes are also known to express iNOS and could be important sources of NO in the skin (WANG et al. 1996a).

Inducible NO is produced at sites of vascular inflammation by resident and nonresident vascular-wall cells, but its exact role is not known. One of the central mechanisms of the inflammatory response is the leukocyte recruitment from the main stream of blood to the endothelial layer. This process initiates with the adhesion of leukocytes to the endothelium, an event regulated by a series of adhesive mechanisms. Activated endothelium expresses surface adhesion molecules such as vascular cell adhesion molecule 1 (VCAM-1), intercellular adhesion molecule 1 (ICAM-1), and E-selectin, which interact with peripheral blood leukocytes, facilitating their attachment to the endothelial cell surface. The inhibitory effects of endogenous NO in endothelial–leukocyte interactions were previously shown (KUBES et al. 1991). Recently, a novel physiologic mechanism has been identified by which macrophage-derived NO can inhibit endothelial VCAM-1 expression and modulate the activation of resident and non-resident vascular-wall cells in an autocrine and paracrine manner (SPIECKER et al. 1997; PENG et al. 1998). The use of "knockout" technology has been very important for understanding the role of inducible NO in regulating the leukocyte–endothelium interactions. While administration of lipopolysaccharide (LPS) to wild-type mice increased sequestration of neutrophils in the lung and their adhesion to the endothelium, these responses were markedly exaggerated in mice lacking iNOS (HICKEY et al. 1997). Although this suggests a beneficial effect for iNOS expression, in other experiments, iNOS had no impact on the mobilization of leukocytes into the peritoneal cavity induced by a number of inflammatory irritants like thioglycollate broth and LPS plus interferon-γ (IFN-γ) (MACMICKING et al. 1995).

Formation of peroxynitrite from NO during nitrosative stress causes DNA single-strand breakage, which stimulates the activation of the nuclear enzyme poly(adenosine diphosphate-ribose) synthetase (PARS). Rapid activation of PARS depletes the intracellular concentration of its substrate, nicotinamide adenine dinucleotide (NAD^+), slowing the rate of glycolysis, electron transport, and subsequently adenosine triphosphate formation. This process can result in acute cell dysfunction and cell death (SZABO 1998; SZABO and DAWSON 1998). PARS also regulates the expression of a number of genes, including the genes for ICAM-1, collagenase, and the iNOS which appears to modulate the course of inflammation. Inhibition of PARS has been shown to protect against zymosan- or endotoxin-induced multiple-organ failure, arthritis, allergic encephalomyelitis, and diabetic islet cell destruction (SZABO 1998). Inhibition of PARS reduced neutrophil recruitment and reduced the extent of edema in zymosan- and carrageenan-triggered models of local inflammation. Furthermore, PARS knockout mice were more resistant against inflammation and organ injury than wild-type animals. Part of the anti-inflammatory effects of PARS inhibition was attributed to a reduced neutrophil recruitment, which may be related to maintained endothelial integrity (SZABO et al. 1997).

III. NO in Acute Inflammatory Responses

Sepsis and septic shock are caused by bacterial infection and represent an acute systemic inflammatory response. Septic shock is characterized by systemic hypotension, vascular smooth-muscle hyporeactivity to adrenergic mimics, and myocardial depression (GLAUSER et al. 1991; PARRILLO 1993). Cellular activation by cell-wall components of gram-negative or gram-positive bacteria results in the production of a variety of inflammatory mediators that are essential for the development of septic shock and its complications (GROENEVELD et al. 1997). The crucial role of NO in the pathogenesis of septic shock and the effects of inducible NOS-mediated NO release have been widely reported (KILBOURN et al. 1990; PETROS et al. 1991; LIU et al. 1993; MACMICKING et al. 1995; LIU et al. 1997). For a recent review on the involvement and dual effects of NO in septic shock, see WOLKOW (1998).

Infusion of LPS or tumor necrosis factor (TNF) into experimental animals is used to mimic septic shock and systemic NO production (NÜSSLER and BILLIAR 1993). Several studies showing an antihypotensive and protective effect for NOS inhibitors in rodent models of septic shock (SZABO et al. 1994; WU et al. 1995; PERRELLA et al. 1996) suggested that the inhibition of NO production could be useful in the treatment of this condition. In contrast, in a murine model of chronic hepatic inflammation (administration of *Corynebacterium parvum* followed by LPS), our group has consistently shown the deleterious effect of NOS inhibition, which causes liver damage, intravascular thrombosis, and oxygen-radical-mediated hepatic injury (BILLIAR et al. 1990; HARBRECHT et al. 1992, 1994; OU et al. 1997). The use of NOS inhibitors has also led to controversial results. In a murine model of endotoxemia, the nonselective NOS inhibitor L-NAME enhanced liver damage and tended to accelerate the time of death, while the iNOS-selective inhibitor L-canavanine significantly reduced mortality and had no deleterious effects in terms of organ damage (LIAUDET et al. 1998). L-NAME also aggravated liver damage in a rat model of endotoxemia, while the iNOS-selective inhibitor S-methylisothiourea did not increase LPS-induced damage (VOS et al. 1997). Recently, mice lacking the iNOS gene were conferred some protection against LPS-induced mortality (MACMICKING et al. 1995; WEI et al. 1995). Another study, however, showed that mice lacking iNOS were not resistant to LPS-induced death (LAUBACH et al. 1995). Nonspecific iNOS inhibition has also been reported to be detrimental rather than beneficial (BILLIAR et al. 1990; NÜSSLER and BILLIAR 1993; FUKATSU et al. 1995; PARK et al. 1996). Recent results showing that iNOS-deficient mice have enhanced leukocyte–endothelium interactions in endotoxemia raised the possibility that induction of iNOS is a homeostatic regulator for leukocyte recruitment (HICKEY et al. 1997).

While iNOS expression can protect the liver in acute hepatic inflammation, it may account for hepatic necrosis in ischemia/reperfusion and hemorrhagic shock (LI and BILLIAR 1999). In a murine model of hemorrhagic shock, it was found that expression of iNOS and NO production caused an increase in polymorphonuclear neutrophil influx, activation of the transcriptional factor nuclear factor κB (NF-κB) and upregulation of interleukin-6 (IL-6) and granulocyte colony-stimulating factor mRNA levels, resulting in a marked lung and liver injury (HIERHOLZER et al. 1998). Thus, in addition to the selectivity for the different NOS isoforms, other factors, (such as the cellular redox status and the production of reactive oxygen species and pro-inflammatory cytokines) are altered in certain conditions, and the type of insult will be a determinant in the possible therapeutic use of NOS inhibitors in the future. The dual effects of NO both as a hepatoprotective and as a hepatotoxic agent (LI and BILLIAR 1999) and its role as a bifunctional regulator of apoptosis (KIM et al. 1999) have been recently reviewed.

A critical role for the transcription factor NF-κB has been demonstrated in the transcriptional regulation of the murine and human iNOS gene induced by LPS and cytokines in cultured cells (XIE et al. 1994; SPINK et al. 1995; WONG et al. 1996; TAYLOR et al. 1998). More recently, it has also been demonstrated that LPS activates NF-κB in vivo, which in turn induces transcription of the iNOS gene and expression of the iNOS protein in a rat model of septic shock. The authors suggested that targeting of NF-κB may be a more effective strategy in the treatment of septic shock, because inhibition of NF-κB activation selectively prevented the increase in iNOS activity and iNOS-mediated NO production (LIU et al. 1997).

Recently, COBB et al. used iNOS-gene knockout mice to examine the effect of inducible NO production in a clinically relevant model of polymicrobial abdominal sepsis treated with antibiotics. The survival study showed that iNOS gene deficiency increased the mortality of sepsis in mice, suggesting a beneficial role for iNOS-gene function in septic mice (COBB et al. 1999).

IV. NO and Inflammatory Cytokines

Cytokines are small-molecular-weight proteins comprising regulatory factors of the immune system, hematopoiesis, tissue repair, and inflammation. The biological activities of cytokines have been investigated in vitro and in vivo mainly using purified or recombinant proteins and neutralizing antibodies, but the use of genetic models gives more definite answers to the implication of cytokines in experimental diseases. It has been shown, for example, that IL-2 and IL-10-mutant mice developed a severe inflammatory bowel disease resembling ulcerative colitis. Moreover, disruption of the transforming growth factor-β1 (TGF-β1) gene in mice resulted in a severe wasting syndrome with multifocal inflammation and early death (RYFFEL 1997). It has been reported in different cell types in vitro that inflammatory cytokines can have both enhancing and suppressing effects on the expression of iNOS and NO production. The transcriptional regulation of the expression of the human iNOS by cytokines has been recently demonstrated (DE VERA et al. 1996; LINN et al. 1997; TAYLOR et al. 1998).

The activated macrophage is one of the most important effector cells in the inflammatory response. In addition to NO, macrophages secrete pro-inflammatory cytokines, including TNF-α and IL-1β, and immunomodulatory cytokines, such as IL-2, IL-10, TGF-β1, and IL-6 (NATHAN and XIE 1994; TREPICCHIO et al. 1996). IL-4 (BOGDAN et al. 1994), IL-6 (BARTON and JACKSON 1993), IL-10 (HOWARD et al. 1993), and TGF-β1 (VODOVOTZ et al. 1993) have been reported to suppress the induction of NO from macrophages or to downregulate the expression of iNOS in activated macrophages, but the list continues to grow. In a murine model of endotoxemia, human recombinant IL-11 attenuated the inflammatory response through downregulation of proinflammatory cytokine release and NO production (TREPICCHIO et al. 1996). Furthermore, IL-13 was recently found to suppress macrophage NO production in both mouse peritoneal macrophages and the J774 macrophage cell line. Regulation of iNOS occurred at both the mRNA and translational levels, depending on the macrophage population (BOGDAN et al. 1997).

Since the release of cytokines constitute a major event in inflammatory and immune responses, their opposing effects on the production of NO may partially explain why pro-inflammatory cytokines induce their detrimental effects while anti-inflammatory cytokines may have beneficial effects on inflammation. A recent study showed that the levels of pro-inflammatory (TNF-α, IL-6, IL-8) and anti-inflammatory cytokines (IL-10, TNFsrI, TNFsrII) relate to serum nitrate levels in patients with severe sepsis. In the acute phase of sepsis, an excessive production of pro-inflammatory cytokines was related to an excessive production of NO while, during the secondary-phase, the production of NO was reduced and the anti-inflammatory cytokines were predominantly present (GROENEVELD et al. 1997). Although administration of exogenous anti-inflammatory cytokines, such as IL-10, to septic patients will possibly lead to diminished NO production because of the dual role of NO, the real efficacy of this treatment remains to be established.

TGF-β1 negatively regulates iNOS expression both in vitro and in vivo (VODOVOTZ et al. 1996a), but endogenous and exogenous TGF-β1 can act differently to suppress NO production (VODOVOTZ et al. 1996b). Exogenous TGF-β1 has been reported to reduce the expression of iNOS, improve hemodynamic parameters, and decrease mortality of endotoxemic rats (PERRELLA et al. 1996; PENDER et al. 1996). In contrast, a recent study showed increased mortality in endotoxemic albumin/TGF-β1 transgenic mice despite, or because of, greatly reduced systemic NO production (VODOVOTZ et al. 1998). Thus, the fact that a single cytokine may display opposite effects in different experimental models has to be considered when evaluating a possible therapeutic use of recombinant cytokines.

NO is also a potent inhibitor of cysteine proteases, such as IL-1β-converting enzyme, and therefore it may have an important regulatory role in the process of cytokine maturation. In this respect, a recent study with activated RAW 264.7 mouse macrophages showed that inhibition of caspase-1 activity by NO suppressed IL-1β and INF-γ-inducing factor (IL-18) processing (KIM et al. 1998). Furthermore, stimulated peritoneal macrophages from wild-type mice released more IL-1β if exposed to the NOS inhibitor L-NMMA, whereas macrophages from iNOS knockout mice did not (KIM et al. 1998). This indicates that regulation of pro-inflammatory-cytokine release by iNOS may contribute to the pathogenesis of certain inflammatory processes.

V. NO and Arachidonic Acid Metabolites

Expression of inducible NO synthase after stimulation by bacterial endotoxin and other cytokines is accompanied by the release of other mediators, such as prostaglandin E_2 (PGE_2) and prostacyclin, via the COX pathway (NATHAN 1987; SALVEMINI et al. 1993). This synergistic production has been subject of several studies (DI ROSA et al. 1996; SALVEMINI et al. 1996a), and a crucial role for the link between the NOS and COX pathways in certain pathological conditions, such as nephrosis, sepsis, and rheumatoid arthritis, has been suggested (SALVEMINI et al. 1993). Most studies have focused on the role of NO in the expression and/or activity of COX (SALVEMINI et al. 1993; SWIERKOSZ et al. 1995), and NO has been reported to increase prostaglandin production via activation of both constitutive and inducible forms of COX in a number of cell types (CORBETT et al. 1993a; INOUE et al. 1993; SALVEMINI et al. 1993; ZAMORA et al. 1998). Moreover, a recent study showed the existence of both NO-dependent and -independent pathways of prostaglandin synthesis after cytokine stimulation of rat osteoblasts in vitro (HUGHES et al. 1999). However, NOS inhibitors increased PGE_2 synthesis in Kupffer cells (STADLER et al. 1993) and chondrocytes (STADLER et al. 1991) and, more recently, two unrelated NO donors, GEA 3175 and S-nitroso-N-acetylpenicillamine (SNAP), were shown to inhibit prostacyclin production in human umbilical endothelial cells (KOSONEN et al. 1998b).

The effect of eicosanoids on NO synthesis by the activation of iNOS has also been studied (MAROTTA et al. 1992; BULUT et al. 1993; MILANO et al. 1995). After stimulation of macrophages with bacterial endotoxin plus IFN-γ, induction of iNOS and NO production is also accompanied by the release of prostaglandins via the COX pathway (NATHAN 1987; SALVEMINI et al. 1993). Like many other laboratories, we have shown that incubation with LPS plus IFN-γ led to a dose-dependent production of NO_2^- in murine J774 macrophage-like cells, an effect prevented by the inducible NOS inhibitor L-NMMA. Addition of the COX inhibitor indomethacin had no significant effect on the NO_2^- production (ZAMORA et al. 1998). This indicates that the products of the COX pathway do not play a major role in the regulation of NOS activity and confirms previous studies showing that the endogenous release of prostanoids from the RAW 264.7 and J774.2 murine macrophages is insufficient to affect the activity of NOS (SALVEMINI et al. 1993; SWIERKOSZ et al. 1995). However, the effects of prostaglandins on NOS activity are still controversial. Low concentrations of indomethacin have been reported to significantly reduce NO formation (MILANO et al. 1995) and the amount of iNOS protein (PANG and HOULT 1996) in LPS-stimulated J774 macrophages. Also, a significant reduction in the NO_2^- formation could only be found when indomethacin was used at very high concentrations in LPS plus IFN-γ stimulated J774 macrophages (ZAMORA et al. 1998). Similarly, several anti-inflammatory drugs, like aspirin and sodium salicylate, have been shown to inhibit induced NO production by immunostimulated RAW 264.7 cells at the high end of therapeutic concentrations. Moreover, this effect was not simply the result of inhibition of prostaglandin synthesis, because exogenous PGE_2 failed to overcome the effects of both drugs (BROUET and OHSHIMA 1995; KEPKA-LENHART et al. 1996). It is known that PGE_2 is a regulator of macrophage functions and displays a functional dualism in immunoinflammatory conditions (BONTA and BEN-EFRAIM 1993). In a recent study, inhibition of endogenous PGE_2 synthesis with indomethacin or ibuprofen had no effect on NO synthesis (HARBRECHT et al. 1997). Thus, the inhibitory effects of high concentrations of COX inhibitors like indomethacin must be interpreted with caution. Interestingly, exogenous but not endogenous PGE_2 decreased the levels of iNOS mRNA and iNOS protein in LPS-stimulated RAW 264.7 cells. This inhibition of macrophage iNOS expression was shown to be dependent on the time and concentration of prostaglandin exposure (HARBRECHT et al. 1997).

C. NO in Immunity and Chronically Inflammatory Diseases

I. NO and the Immune Response

Release of NO has been reported in inflammatory responses initiated by microbial products or autoimmune reactions. Although the role of NO in non-specific immunity has been well established in animal models (MACMICKING et al. 1997), it still awaits definitive confirmation in humans.

The effects of NO on specific immunity, however, need extensive investigation. In the generation of an inflammatory response, the defensive machinery of the immune system is mainly based in the activity of effector cells, such as T lymphocytes, macrophages, and neutrophils. The activation of these cells results in the production of immune modulators, including cytokines, chemokines and reactive oxygen and nitrogen species that form a complex regulatory network and determine the intensity and duration of inflammation.

Based on the cytokine secretion pattern of CD4⁺-helper T lymphocytes, two main subsets of T-helper cells have been defined (ABBAS et al. 1996): T-helper type I (Th1) and T-helper type II (Th2). The former are mainly implicated in cell-mediated immune reactions, macrophage activation, and the production of opsonizing antibodies; Th1 cells secrete IL-2, IFN-γ, and TNF-β. Th2 cells, on the other hand, secrete IL-4, IL-5, IL-6, IL-10, and IL-13 and are key players in humoral immunity and activate mast cells and eosinophils (ABBAS et al. 1996; MOSMANN and SAD 1996).

There is a body of literature showing an invariable correlation between the presence of pro-inflammatory Th1-type reactivities and the local cytokine expression or cytokine response profiles in relevant diseases positive for iNOS (KRÖNCKE et al. 1998). Studies of the effects of NO on lymphocyte functions provide insights into the mechanism by which NO could contribute to the pathogenesis of autoimmune disease. The elevated expression of iNOS in affected tissues suggests that iNOS is involved in the pathogenesis of certain immune diseases, but there are a number of controversial reports in the literature that reflect the complexity of the problem. While NO generation from L-arginine was shown to be required for DNA synthesis in human peripheral blood lymphocytes (EFRON et al. 1991), another study showed that human lymphocytes do not produce amounts of NO sufficient to affect lymphocyte mitogenesis (SHOKER et al. 1997). Furthermore, in the same study, the inhibitory effects of two NO donors (sodium nitroprusside and nitroglycerin) on lymphocyte function were shown to be nonspecific and unrelated to NO production. Recently, allogeneic (mixed leukocyte cultures), mitogenic, and superantigenic stimulation of bovine-blood mononuclear cells induced NO production at a low level and without having any effect on cellular activation and proliferation (SCHUBERTH et al. 1998).

However, most of the existing data suggests that NO suppresses rather than enhances lymphocyte activation and proliferation. Antigen-stimulated mouse Th1 cells produced high levels of NO, resulting in a concomitant reduction of IL-2 secretion and lymphocyte proliferation. This was reversed by addition of recombinant IL-2 (TAYLOR-ROBINSON 1997). Moreover, there is evidence that NO exerts different effects on discrete sub-populations of T cells, for example, by inhibiting secretion of IL-2 by murine Th1 cells and increasing secretion of IL-4 by Th2 cells (CHANG et al. 1997). This preferential effect, however, was not observed in activated human T cells and human T-cell clones in vitro, where the Th1- and Th2-associated cytokine production was equally impaired by the NO donors SIN-1 and SNAP (BAUER et al. 1997). The question of whether NO differentially affects T-lymphocyte function merits special attention, because the outcome of numerous diseases appears to depend critically on the Th1/Th2 balance in accompanying immune responses (MOSMANN and SAD 1996; MUNDER et al. 1998).

Both cGMP-independent and -dependent pathways have been described to explain the antiproliferative effects of NO. In human lymphocytes activated by lectin mitogen, concanavalin (ConA), two oxotriazole derivatives (GEA

3162 and 3175) and the nitrosothiol SNAP caused inhibition of cell proliferation and enhanced cGMP production. While the NO-donor-induced cGMP production was inhibited by a guanylyl cyclase inhibitor, the antiproliferative action remained unaltered (KOSONEN et al. 1998a). In contrast, other study showed that T cells activated in the presence of alveolar macrophages were unable to proliferate despite the expression of IL-2 receptor and secretion of IL-2. The NO-mediated T-cell suppression was reversible by the guanylyl cyclase inhibitors methylene blue and LY-83583, and was reproduced by a cell-permeable analogue of cGMP. In addition, this effect could be reproduced by the addition of SNAP and inhibited by the NOS inhibitor L-NAME (BINGISSER et al. 1998).

The involvement of NO production by iNOS in important autoimmune diseases, such as immunologically induced diabetes, inflammatory arthritis, and graft-vs-host disease, appears to be unquestionable. However, because of differences in the experimental animals and disease-induction methods, it is unclear whether NO is beneficial or detrimental. In several studies, administration of selective iNOS inhibitors to rodents with autoimmune diseases led to conflicting results. Experimental allergic encephalomyelitis (EAE), for example, is a well-studied animal model of organ-specific autoimmunity that mimics human multiple sclerosis. Treatment with aminoguanidine ameliorated EAE in both mice and rats (CROSS et al. 1994; ZHAO et al. 1996; BRENNER et al. 1997), but it led to aggravation and prolongation of disease in myelin- and T-cell-mediated EAE (ZIELASEK et al. 1995). Similarly, L-NIL administration caused a marked worsening in disease expression in myelin basic protein (MBP)-immunized Lewis rats, but ameliorated the severity of disease following adoptive transfer of MBP-reactive T cells into L-NIL-treated recipients. It is believed that iNOS inhibition is detrimental to the host during priming of pathogenic T-cell responses in the periphery but largely protective at the site of disease (GOLD et al. 1997). Also in Lewis rats, aminoguanidine was shown to ameliorate experimental autoimmune myocarditis (SHIN et al. 1998). Detrimental effects for iNOS inhibition have been also reported in a model of autoimmune interstitial nephritis (GABBAI et al. 1997) and experimental autoimmune uveoretinitis (EAU) (HOEY et al. 1997). More recently, it was reported that IL-12 protected mice from Th1-mediated EAU through a mechanism involving IFN-γ-induced NO production and *bcl-2*-regulated apoptotic deletion of the antigen-specific T cells (TARRANT et al. 1999).

To address the limitations of the administration of NOS inhibitors, iNOS knockout mice and antisense nucleotides have also been used to identify the functional roles of iNOS in different pathologies. In a model of EAE, mice deficient in or with reduced iNOS activity showed more disease and less remission than wild-type mice (FENYK-MELODY et al. 1998). In contrast, intraventricular administration to SJL/S mice of antisense oligodeoxynucleotides complementary to iNOS significantly reduced the clinical score of EAE and blocked the iNOS mRNA, protein synthesis, and iNOS activity within the central nervous system (DING et al. 1998). More recently, inhibition of allergic airway inflammation was observed in mice lacking iNOS (XIONG et al. 1999).

The consequences of iNOS disruption have also been studied in MRL-lpr/lpr mice. These mice produce an excess of NO and develop a systemic autoimmune disease associated with a number of inflammatory manifestations like glomerulonephritis, arthritis, and vasculitis. iNOS-disrupted and wild-type mice displayed equivalent degrees of nephritis and arthritis, but the former showed markedly reduced vasculitis, suggesting heterogeneity in mechanisms of inflammation in MRL-lpr/lpr mice (GILKESON et al. 1997). Similar results showing that, in iNOS knockout mice, glomerulonephritis did not differ from that in mice with an intact iNOS gene, suggested that iNOS does not play an essential role in this autoimmune disease in the mouse (CATTELL et al. 1998).

Acute rejection is an immunoinflammatory process characterized by an intense, inflammatory cell infiltrate and progressive destruction of the grafted organ. Although it has become apparent that NO contributes to allograft rejection, graft versus host disease (GVHD), and tissue damage in alloimmune responses (LANGREHR et al. 1993; BILLIAR 1995), the complex regulatory and effector mechanisms underlying the rejection process are not completely understood. Production of NO has been shown to partially account for the destruction of both lymphoid and erythroid host tissue and the reduced lymphoproliferative responses associated with the acute phase of GVHD in mice (HOFFMAN et al. 1996). Also, in mice with acute GVHD, suppression of B-cell proliferation in response to LPS was reported to be mediated through induction of the NOS pathway by TNF-α (FALZARANO et al. 1996). Administration of NOS inhibitors in models of GVHD has variously been reported to prolong graft survival in a mouse model of GVHD (HOFFMAN et al. 1997) or to have no effect at all in mice receiving allogeneic heterotopic heart transplants (BASTIAN et al. 1994). In a model of heterotopic cardiac transplantation in the rat, L-NMMA produced only a small increase in graft survival (WINLAW et al. 1995). Generation of NO has also been observed in acute rejection of rat hepatic allografts (GOTO et al. 1997) and of pancreas allografts in hyperglycemic rats, where electron-spin resonance measurement of NO was suggested as a useful marker for the diagnosis of acute rejection in pancreas transplantation (TANAKA et al. 1995).

In humans, expression of iNOS has been localized in lung-transplant recipients with obliterative bronchiolitis (MCDERMOTT et al. 1997) and in coronary arteries of transplanted human hearts with accelerated graft arteriosclerosis, but the role of NO in the pathogenesis remains unknown (LAFOND-WALKER et al. 1997a). In transplanted rat aortic allografts, the inhibition of NO production significantly increased the intimal thickening, suggesting that NO suppresses the development of allograft arteriosclerosis (SHEARS et al. 1997). In the same study, transduction with iNOS using an adenoviral vector completely suppressed the development of allograft arteriosclerosis, indicating that iNOS may be important for the suppression of transplant vasculopathy in chronic rejection associated with cardiac transplantation (SHEARS et al. 1997). It was recently reported that cardiac myocyte apoptosis is closely associated with expression of iNOS in macrophages and myocytes and with nitration of myocyte proteins by peroxynitrite during human cardiac allograft rejection (SZABOLCS et al. 1998). Moreover, it was suggested that NO plays a role in modulating the localized bone resorption that accompanies the aseptic loosening of prosthetic joints (WATKINS et al. 1997). The identification of iNOS will certainly open novel possibilities for the use of selective inhibitors and pharmacological intervention in the treatment of these alloimmune conditions. Other relevant immunologic effects attributed to NO are: the inhibition of lymphokine-activated killer cell induction by inducing apoptosis of cytolytic lymphocyte precursors (SAMLOWSKI et al. 1998); inhibition of major histocompatibility complex class-II expression on mouse peritoneal macrophages and antigen presentation by lung dendritic cells; tumor-induced immunosuppression; and the reduced immunological response resulting from administration of morphine (EVANS 1995).

The demonstration of the involvement of NO in a number of autoimmune and inflammatory diseases does not necessarily imply that NO itself is the effector molecule. Whether these pathologies are directly mediated by NO or NO-related forms, such as peroxynitrite, needs further investigation. Inflammatory mediators that enhance the cellular production of NO also increase cellular superoxide production from different cellular sources, including NOS (POU et al. 1992). Nitration of tyrosine to give nitrotyrosine is used as a footprint of peroxynitrite activity (WINK et al. 1996; BECKMAN and KOPPENOL 1996; MAYER and HEMMENS 1997; HALLIWELL 1997 for recent reviews). The putative role of peroxynitrite as an inflammatory mediator has been suggested after detection of nitrotyrosine in animal models of endotoxemia (WIZEMANN et al. 1994; SZABO et al. 1995; KAMISAKI et al. 1997), lung injury (NUMATA et al. 1998; CUZZOCREA et al. 1999), ileitis (MILLER et al. 1995; SADOWSKA-KROWICKA et al. 1998), experimental artoimmune encephalomyelitis (VAN DER VEEN et al. 1997; CROSS et al. 1997), myocardial ischemia-reperfusion injury (LIU et al. 1997; LIU et al. 1998), myocardial dysfunction (OYAMA et al. 1998), glomerulonephritis (HEERINGA et al. 1998), human atherosclerotic plaques (BECKMANN et al. 1994; LUOMA et al. 1998), adult respiratory distress syndrome (KOOY et al. 1995), the airways of asthmatic patients (SALEH et al. 1998), multiple sclerosis (BAGASRA et al. 1995; CROSS et al. 1998), and human sepsis and myocarditis (KOOY et al. 1997).

II. NO and Chronic Inflammatory Processes

Chronic inflammation is characterized by a proliferation of fibroblasts and small blood vessels and an influx of chronic inflammatory cells (lymphocytes, plasma cells, macrophages). In certain immunologic conditions, chronic inflammation is primary and not preceded by an acute inflammatory response. It also differs from acute inflammation in that it is orchestrated almost entirely by cells of the immune system (TROWBRIDGE and EMLING 1997).

The involvement of NO in chronic localized inflammatory diseases has been clearly demonstrated in a number of experimental animal models. NO stimulates TNF-α production by synoviocytes, and its catabolic effects on chondrocyte function promote the degradation of articular cartilage implicated in certain rheumatic diseases (CLANCY et al. 1998). A recent study in human chondrocytes has identified IL-18 as a cytokine that regulates chondrocyte responses and contributes to cartilage destruction through stimulation of the expression of several genes, including iNOS, inducible COX, IL-6, and stromelysin (OLEE et al. 1999). The beneficial effects of NOS inhibition have been shown in murine systemic lupus erythematosus (SLE) (OATES et al. 1997; CLANCY et al. 1998), suppression of rat adjuvant arthritis by *N*-iminoethyl-L-ornithine (SANTOS et al. 1997), attenuation of streptococcal cell-wall-induced arthritis in rats by L-NMMA (MCCARTNEY-FRANCIS et al. 1993), and the suppression by diphenylene diodonium chloride of potassium peroxocromate-induced arthritis in mice (MIESEL et al. 1996). Furthermore, in a model of osteoarthritis in dogs, inhibition of NOS reduced the progression of cartilage lesions and the production of metalloproteinases and IL-1 (CLANCY et al. 1998). Recently, iNOS-deficient mice were used to investigate the role of NO and IL-1 in joint inflammation and cartilage destruction in a non-immunologic model of inflamamation, the zymosan-induced gonarthritis (VAN DE LOO et al. 1998). In this study, IL-1 and NO played only a minor role in edema and neutrophil influx but a major role in cartilage destruction. Moreover, the results obtained from anti-IL-1 treatment of wild-type mice were comparable to those found in iNOS knockout mice, suggesting that most IL-1 related effects in arthritis are mediated by NO (VAN DE LOO et al. 1998). In a recent study, intravenous inoculation with *Staphylococcus aureus* induced significantly increased clinical severity of septic arthritis, with attendant septicemia in iNOS-deficient mice compared with similarly infected heterozygous or wild-type mice. This was associated with enhanced production of IFN-γ and TNF-α in vivo and in vitro, indicating that Th1 polarization of the cells mediated immune response (MCINNES et al. 1998).

The NO-mediated destruction of both rat and mouse islets of Langerhans and its effects on insulin secretion provide strong evidence for the involvement of NO in human diabetes (KRÖNCKE et al. 1998). However, it is not known whether the inhibition of human iNOS will reduce the destruction of over 90% of the pancreatic islets, as occurs in type-1 diabetes. NO also contributes to mucosal damage in inflammatory bowel disease (WHITTLE 1997; GUSLANDI 1998), and the beneficial effects of NOS inhibitors in reducing intestinal inflammation have been shown in various models of colitis (MOURELLE et al. 1996; KISS et al. 1997). The presence of iNOS in atherosclerotic plaques also suggests a role for NO in atherosclerosis (BULT 1996), but its exact role is still unknown. A recent study showed a vasculoprotective role of iNOS against inflammatory cytokine-induced proliferative vasospatic change of the coronary artery in a pig model of coronary arteriosclerosis (FUKUMOTO et al. 1997). Other experimental animal models where NO has been implicated include:

radiation pneumonitis and fibrosis (in rats), which are mediated by alveolar macrophages and epithelial cell-derived NO and can be inhibited by NOS inhibitors (NOZAKI et al. 1997); foreign-body-induced granulomatous lung inflammation (TSUJI et al. 1995); pathogenesis of cyclophosphamide-induced hemorrhagic cystitis (SOUZA-FIHO et al. 1997); and endotoxin-induced ocular inflammation (WANG et al. 1996b).

Cerebrospinal-fluid concentrations of the stable NO metabolite nitrite are elevated in animal models of bacterial meningitis (BM) and in human patients with the disease (BUSTER et al. 1995). In a model of BM, production of nitrite in the cerebrospinal fluid of rats correlated with elevated blood–brain barrier permeability, suggesting that NO contributes to the pathophysiology of BM (BUSTER et al. 1995). However, NO produced by iNOS may be beneficial as well. Recently, inhibition of iNOS primarily localized to the cerebral vasculature and inflammatory cells in the subarachnoid and ventricular space increased cortical hypoperfusion and ischemic neuronal injury in an infant-rat model of meningitis caused by group-B streptococci (LEIB et al. 1998).

All the findings showing that inhibition of NO production attenuates some inflammatory conditions provide compelling evidence that both NO and the enzyme(s) that produce it should be considered as potential therapeutic targets in human disease. However, the use of NOS inhibitors has also been shown to be ineffective (RIBBONS et al. 1997) or detrimental. In rats with adjuvant arthritis, for example, administration of the NOS inhibitor L-NIL was without effect (FLETCHER et al. 1998), and L-NAME worsened acute edematous and necrotizing pancreatitis, whereas NO donors reduced pancreatic injury (WERNER et al. 1997). There are also evidences that iNOS is beneficial rather than detrimental in resolving intestinal inflammation (McCAFFERTY et al. 1997). Moreover, the protective role of NO in the host response to infections with *S. aureus* strongly cautions against the clinical use of selective NOS-inhibitor therapy in diseases such as septic arthritis (McINNES et al. 1998).

One of the primary functions of the inflammatory response is to heal wounded tissue. Healing commences soon after injury, while acute inflammation in still in full swing. Aseptic wounding induces iNOS. Our group showed a delay in closure of excisional wounds in iNOS-deficient mice as compared to wild-type mice. This defect in healing of excisional wounds can be quantitatively corrected by a single topical administration of an adenoviral vector containing iNOS complementary DNA (YAMASAKI et al. 1998). Although more tightly regulated than the rodent iNOS gene, expression of human iNOS has been found in chronic inflammatory diseases of the airways, the vessels, the bowels, the kidney, the heart, the skin, and the apices of teeth (KRÖNCKE et al. 1998), suggesting that NO plays an important role in the pathogenesis of inflammation in humans (Table 2).

III. Induced NO in Antimicrobial Defense Mechanisms

Macrophages constitute a primary line of defense. In rodents, the cytotoxic and anti-microbial effector functions of inflammatory macrophages are largely

Table 2. Nitric oxide in human autoimmune and chronic inflammatory diseases

Disease	Reference
Rheumatic diseases	
Systemic lupus erythematosus	Belmont et al. 1997; Wigand et al. 1997
Vasculitis	Kausalya and Nath 1998
Rheumatoid arthritis	McInnes et al. 1996; Grabowski et al. 1997; Wigand et al. 1997
Osteoarthritis	Amin et al. 1995; McInnes et al. 1996; Grabowski et al. 1997
Inflammatory airway disease	
Asthma	Hamid et al. 1993; Barnes 1996; Saleh et al. 1998
Respiratory tract infections	McInnes et al. 1998
Idiopathic pulmonary fibrosis	Saleh et al. 1997
Bronchiectasis	Tracey et al. 1994
Gastrointestinal system	
Inflammatory bowel disease	
Ulcerative colitis	Mourelle et al. 1995; Godkin et al. 1996; Singer et al. 1996; Gupta et al. 1998; Herulf et al. 1998
Crohn's disease	Singer et al. 1996; Gupta et al. 1998; Herulf et al. 1998
Diverticulitis	Singer et al. 1996
Necrotizing enterocolitis	Ford et al. 1997
Celiac disease	Ter Steege et al. 1997
Helicobacter pylori-associated chronic non-atrophic gastritis	Mannick et al. 1996
Kidney	
Glomerulonephritis	Kashem et al. 1996
Lupus nephritis	Wang et al. 1997
Pancreas	
Diabetes	Corbett et al. 1993b; Vara et al. 1995
Pancreatitis	
Liver	
Chronic hepatitis	Cuzzocrea et al. 1998b
Bladder	
Infectious and noninfectious cystitis	Lundberg et al. 1996
Central and peripheral nervous system	
Parkinson's disease	Hirsch et al. 1998
Multiple sclerosis	Johnson et al. 1995; De Groot et al. 1997; Hooper et al. 1997
Severe AIDS dementia	Adamson et al. 1996
Vasculitic neuropathies	Satoi et al. 1998
Skin	
Psoriasis	Bruch-Gerharz et al. 1996
Cutaneous lupus erythematosus	Kuhn et al. 1998
Systemic sclerosis	Yamamoto et al. 1998
Dermatitis	Rowe et al. 1997
Atherosclerosis	Buttery et al. 1996; Lafond-Walker et al. 1997b; Ross 1999
Periapical periodontitis	Takeichi et al. 1998
Sjögren's syndrome	Konttinen et al. 1997

AIDS, acquired immune deficiency syndrome.

based on their ability to produce NO (JAMES 1995; FANG 1997; MACMICKING et al. 1997; NATHAN 1997). After the demonstration of the requirement of L-arginine for macrophage inhibition of *Cryptococcus neoformans* growth, the antimicrobial activity of the NO produced by the L-arginine:NO pathway has been confirmed as a nonspecific defense mechanism against a number of pathogens, including *Schistosoma mansoni, Trypanosoma brucei, Trypanosoma cruzi, Toxoplasma gondii, Mycobacterium avium, Leishmania major, Plasmodium yoelii, Plasmodium berghei, Plasmodium vinckei,* and *Plasmodium falciparum* (NÜSSLER and BILLIAR 1993). It is very likely that the list will continue growing. In the last several years, the use of knockout mice has served to gain more information on the exact contribution of iNOS to host protection. Depending on the pathogen, the role of iNOS will vary from beneficial (*Toxoplasma gondii, Listeria monocytogenes*) to detrimental (*Mycobacterium avium*). In other infections, such as Chagasic trypanosomiasis and *Plasmodium chabaudi*, however, iNOS deficiency has not been established (NATHAN 1997 for a recent review).

There are infections in which iNOS is critical for host survival. For example, failure to induce iNOS can explain the sensitivity of knockout mice to infection with *Mycobacterium tuberculosis* (MACMICKING et al. 1997a). Similarly, deletion or functional inactivation of iNOS abolished the IFN-γ and natural-killer cell response and caused parasite spreading in mice infected with *Leishmania major*, suggesting that iNOS is a critical regulator of the innate response to this protozoan parasite (MACMICKING et al. 1997b).

NO-mediated killing of *Trypanosoma cruzi, Mycobacterium avium,* and *Leishmania major* by activated human macrophages has also been found, as has growth inhibition of *Cryptococcus neoformans* by human astrocytes. Moreover, human iNOS expression has been shown in a number of viral and bacterial infections, including malaria, tuberculosis, and human immunodeficiency virus (KRÖNCKE et al. 1998).

D. Conclusions

It is now clear that NO cannot be rigidly catalogued as anti-inflammatory or pro-inflammatory molecule, but it can be considered a true inflammatory mediator. Expression of iNOS and NO production appears to mediate a number of inflammatory and infectious diseases by acting both as a direct effector and as a regulator of other effector pathways. The dichotomous role of NO in inflammation, often referred to as the NO paradox, is mainly based on the conflicting data of the effects of NOS inhibitors in different animal models. In addition, the use of iNOS-knockout animals for exploring the role of NO has reinforced this view and clearly demonstrates that caution has to be taken when extrapolating experimental results to possible therapeutic benefits.

Only the identification of the roles of NO and the cells that produce it and understanding of the mechanisms that regulate its cellular production in inflammation will help to develop therapeutic applications for both the use of NO-donor/NOS gene transfer and the specific NOS inhibition in acute and localized chronic inflammatory diseases. It is not surprising that targeting a versatile molecule like NO will have an impact on human diseases, but the strategy will necessarily have to consider important factors, such as the rate and duration of its release, the presence of reactive substrates, and the NO-induced cell-specific response.

References

Abbas AK, Murphy KM, Sher A (1996) Functional diversity of helper T lymphocytes. Nature 383:787–793

Adamson DC, Wildemann B, Sasaki M, Glass JD, McArthur JC, Christov VI, Dawson TM, Dawson VL (1996) Immunologic NO synthase: elevation in severe AIDS dementia and induction by HIV-1 gp41. Science 274:1917–1921

Amin A, Di Cesare PE, Vyas P, Attur M, Tzeng E, Billiar TR, Stuchin SA, Abramson SB (1995) The expression and regulation of nitric oxide synthase in human osteoarthritis-affected chondrocytes: evidence for up-regulated neuronal nitric oxide synthase. J Exp Med 182:2097–2102

Bagasra O, Michaels FH, Zheng YM, Bobroski LE, Spitsin SV, Fu ZF, Tawadros R, Koprowski H (1995) Activation of the inducible form of nitric oxide synthase in the brains of patients with multiple sclerosis. Proc Natl Acad Sci USA 92:12041–12045

Barnes PJ (1996) Pathophysiology of asthma. Br J Clin Pharmacol 42:3–10

Barton BE, Jackson JV (1993) Protective role of interleukin 6 in the lipopolysaccharide–galactosamine septic shock model. Infect Immun 61:1496–1499

Bastian NR, Xu S, Shao XL, Shelby J, Granger DL, Hibbs JBJ (1994) N^{ω}-monomethyl-L-arginine inhibits nitric oxide production in murine cardiac allografts but does not affect graft rejection. Biochim Biophys Acta 1226:225–231

Bauer H, Jung T, Tsikas D, Stichtenoth DO, Frolich JC, Neumann C (1997) Nitric oxide inhibits the secretion of T-helper 1- and T-helper 2-associated cytokines in activated human T cells. Immunology 90:205–211

Beckman JS, Koppenol WH (1996) Nitric oxide, superoxide, and peroxynitrite: the good, the bad, and ugly. Am J Physiol 271:C1424–C1437

Beckmann JS, Ye YZ, Anderson PG, Chen J, Accavitti MA, Tarpey MM, White CR (1994) Extensive nitration of protein tyrosines in human atherosclerosis detected by immunohistochemistry. Biol Chem Hoppe Seyler 375:81–88

Belmont HM, Levartovsky D, Goel A, Amin A, Giorno R, Rediske J, Skovron ML, Abramson SB (1997) Increased nitric oxide production accompanied by the up-regulation of inducible nitric oxide synthase in vascular endothelium from patients with systemic lupus erythematosus. Arthritis Rheum 40:1810–1816

Billiar TR (1995) Nitric oxide. Novel biology with clinical relevance. Ann Surg 221: 339–349

Billiar TR, Curran RD, Harbrecht BG, Stuehr DJ, Demetris AJ, Simmons RL (1990) Modulation of nitrogen oxide synthesis in vivo: N^{G}-monomethyl-L-arginine inhibits endotoxin-induced nitrate/nitrate biosynthesis while promoting hepatic damage. J Leukoc Biol 48:565–569

Bingisser RM, Tilbrook PA, Holt PG, Kees UR (1998) Macrophage-derived nitric oxide regulates T cell activation via reversible disruption of the Jak3/STAT5 signaling pathway. J Immunol 160:5729–5734

Bogdan C, Vodovotz Y, Paik J, Xie QW, Nathan C (1994) Mechanism of suppression of nitric oxide synthase expression by interleukin-4 in primary mouse macrophages. J Leukoc Biol 55:227–233

Bogdan C, Thuring H, Dlaska M, Rollinghoff M, Weiss G (1997) Mechanism of suppression of macrophage nitric oxide release by IL-13: influence of the macrophage population. J Immunol 159:4506–4513

Bonta IL, Ben-Efraim S (1993) Involvement of inflammatory mediators in macrophage antitumor activity. J Leukoc Biol 54:613–626

Brenner T, Brocke S, Szafer F, Sobel RA, Parkinson JF, Perez DH, Steinman L (1997) Inhibition of nitric oxide synthase for treatment of experimental autoimmune encephalomyelitis. J Immunol 158:2940–2946

Brouet I, Ohshima H (1995) Curcumin, an anti-tumour promoter and anti-inflammatory agent, inhibits induction of nitric oxide synthase in activated macrophages. Biochem.Biophys Res Commun 206:533–540

Bruch-Gerharz D, Fehsel K, Suschek C, Michel G, Ruzicka T, Kolb-Bachofen V (1996) A proinflammatory activity of interleukin 8 in human skin: expression of the inducible nitric oxide synthase in psoriatic lesions and cultured keratinocytes. J Exp Med 184:2007–2012

Bult H (1996) Nitric oxide and atherosclerosis: possible implications for therapy. Mol Med Today 2:510–518

Bulut V, Severn A, Liew FY (1993) Nitric oxide production by murine macrophages is inhibited by prolonged elevation of cyclic AMP. Biochem Biophys Res Commun 195:1134–1138

Buster BL, Weintrob AC, Townsend GC, Scheld WM (1995) Potential role of nitric oxide in the pathophysiology of experimental bacterial meningitis in rats. Infect Immun 63:3835–3839

Buttery LD, Springall DR, Chester AH, Evans TJ, Standfield EN, Parums DV, Yacoub MH, Polak JM (1996) Inducible nitric oxide synthase is present within human atherosclerotic lesions and promotes the formation and activity of peroxynitrite. Lab Invest 75:77–85

Cattell V, Cook HT, Ebrahim H, Waddington SN, Wei XQ, Assmann KJ, Liew FY (1998) Anti-GBM glomerulonephritis in mice lacking nitric oxide synthase type 2. Kidney Int 53:932–936

Chang RH, Feng MH, Liu WH, Lai MZ (1997) Nitric oxide increased interleukin-4 expression in T lymphocytes. Immunology 90:364–369

Clancy RM, Amin AR, Abramson SB (1998) The role of nitric oxide in inflammation and immunity. Arthritis Rheum 41:1141–1151

Cobb JP, Buchman TG, Chang K, Qiu Y, Laubach VE, Karl IE, Hotchkiss RS. iNOS gene deficiency increases the mortality of sepsis in mice. Sixtieth Annual Meeting of The Society of University Surgeons, 77:2–11–1999. New Orleans

Corbett JA, Kwon G, Turk J, McDaniel ML (1993a) IL-1 beta induces the coexpression of both nitric oxide synthase and cyclooxygenase by islets of Langerhans: activation of cyclooxygenase by nitric oxide. Biochemistry 32:13767–13770

Corbett JA, Sweetland MA, Wang JL, Lancaster JRJ, McDaniel ML (1993b) Nitric oxide mediates cytokine-induced inhibition of insulin secretion by human islets of Langerhans. Proc Natl Acad Sci USA 90:1731–1735

Cross AH, Misko TP, Lin RF, Hickey WF, Trotter JL, Tilton RG (1994) Aminoguanidine, an inhibitor of inducible nitric oxide synthase, ameliorates experimental autoimmune encephalomyelitis in SJL mice. J Clin Invest 93:2684–2690

Cross AH, Manning PT, Stern MK, Misko TP (1997) Evidence for the production of peroxynitrite in inflammatory CNS demyelination. J Neuroimmunol 80:121–130

Cross AH, Manning PT, Keeling RM, Schmidt RE, Misko TP (1998) Peroxynitrite formation within the central nervous system in active multiple sclerosis. J Neuroimmunol 88:45–56

Cuzzocrea S, Zingarelli B, Hake P, Salzman AL, Szabo C (1998a) Antiinflammatory effects of mercaptoethylguanidine, a combined inhibitor of nitric oxide synthase and peroxynitrite scavenger, in carrageenan-induced models of inflammation. Free Radic Biol Med 24:450–459

Cuzzocrea S, Zingarelli B, Villari D, Caputi AP, Longo G (1998b) Evidence for in vivo peroxynitrite production in human chronic hepatitis. Life Sci 63:L25–L30

Cuzzocrea S, Zingarelli B, Costantino G, Caputi AP (1999) Beneficial effects of Mn(III)tetrakis (4-benzoic acid) porphyrin (MnTBAP), a superoxide dismutase mimetic, in carrageenan-induced pleurisy. Free Radic Biol Med 26:25–33

De Groot CJ, Ruuls SR, Theeuwes JW, Dijkstra CD, Van der Valk P (1997) Immunocytochemical characterization of the expression of inducible and constitutive isoforms of nitric oxide synthase in demyelinating multiple sclerosis lesions. J Neuropathol Exp Neurol 56:10–20

De Vera ME, Shapiro RA, Nüssler AK, Mudgett JS, Simmons RL, Morris SMJ, Billiar TR, Geller DA (1996) Transcriptional regulation of human inducible nitric oxide synthase (NOS2) gene by cytokines: initial analysis of the human NOS2 promoter. Proc Natl Acad Sci USA 93:1054–1059

Di Rosa M, Ialenti A, Ianaro A, Sautebin L (1996) Interaction between nitric oxide and cyclooxygenase pathways. Prostagl Leukotr Essential Fatty Acids 54:229–238

Ding M, Zhang M, Wong JL, Rogers NE, Ignarro LJ, Voskuhl RR (1998) Antisense knockdown of inducible nitric oxide synthase inhibits induction of experimental autoimmune encephalomyelitis in SJL/J mice. J Immunol 160:2560–2564

Efron DT, Kirk SJ, Regan MC, Wasserkrug HL, Barbul A (1991) Nitric oxide generation from L-arginine is required for optimal human peripheral blood lymphocyte DNA synthesis. Surgery 110:327–334

Evans CH (1995) Nitric oxide: what role does it play in inflammation and tissue destruction? Agents Actions Suppl 47:107–16, 107–116

Falzarano G, Krenger W, Snyder KM, Delmonte JJ, Karandikar M, Ferrara JL (1996) Suppression of B-cell proliferation to lipopolysaccharide is mediated through induction of the nitric oxide pathway by tumor necrosis factor- alpha in mice with acute graft-versus-host disease Blood 87:2853–2860

Fang FC (1997) Perspectives series: host/pathogen interactions. Mechanisms of nitric oxide-related antimicrobial activity. J Clin Invest 99:2818–2825

Fenyk-Melody JE, Garrison AE, Brunnert SR, Weidner JR, Shen F, Shelton BA, Mudgett JS (1998) Experimental autoimmune encephalomyelitis is exacerbated in mice lacking the NOS2 gene. J Immunol 160:2940–2946

Fletcher DS, Widmer WR, Luell S, Christen A, Orevillo C, Shah S, Visco D (1998) Therapeutic administration of a selective inhibitor of nitric oxide synthase does not ameliorate the chronic inflammation and tissue damage associated with adjuvant-induced arthritis in rats. J Pharmacol Exp Ther 284:714–721

Ford H, Watkins S, Reblock K, Rowe M (1997) The role of inflammatory cytokines and nitric oxide in the pathogenesis of necrotizing enterocolitis. J Pediatr Surg 32:275–282

Fukatsu K, Saito H, Fukushima R, Inoue T, Lin MT, Inaba T, Muto T (1995) Detrimental effects of a nitric oxide synthase inhibitor (N^{ω}-nitro-L-arginine-methyl-ester) in a murine sepsis model. Arch Surg 130:410–414

Fukumoto Y, Shimokawa H, Kozai T, Kadokami T, Kuwata K, Yonemitsu Y, Kuga T, Egashira K, Sueishi K, Takeshita A (1997) Vasculoprotective role of inducible nitric oxide synthase at inflammatory coronary lesions induced by chronic treatment with interleukin-1beta in pigs in vivo. Circulation 96:3104–3111

Gabbai FB, Boggiano C, Peter T, Khang S, Archer C, Gold DP, Kelly CJ (1997) Inhibition of inducible nitric oxide synthase intensifies injury and functional deterioration in autoimmune interstitial nephritis. J Immunol 159:6266–6275

Gilkeson GS, Mudgett JS, Seldin MF, Ruiz P, Alexander AA, Misukonis MA, Pisetsky DS, Weinberg JB (1997) Clinical and serologic manifestations of autoimmune disease in MRL-lpr/lpr mice lacking nitric oxide synthase type 2. J Exp Med 186:365–373

Glauser MP, Zanetti G, Baumgartner JD, Cohen J (1991) Septic shock: pathogenesis. Lancet 338:732–736

Godkin AJ, De Belder AJ, Villa L, Wong A, Beesley JE, Kane SP, Martin JF (1996) Expression of nitric oxide synthase in ulcerative colitis. Eur J Clin Invest 26: 867–872

Gold DP, Schroder K, Powell HC, Kelly CJ (1997) Nitric oxide and the immunomodulation of experimental allergic encephalomyelitis. Eur J Immunol 27:2863–2869

Goto M, Yamaguchi Y, Ichiguchi O, Miyanari N, Akizuki E, Matsumura F, Matsuda T, Mori K, Ogawa M (1997) Phenotype and localization of macrophages expressing inducible nitric oxide synthase in rat hepatic allograft rejection. Transplantation 64:303–310

Grabowski PS, Wright PK, Van't H, Helfrich MH, Ohshima H, Ralston SH (1997) Immunolocalization of inducible nitric oxide synthase in synovium and cartilage in rheumatoid arthritis and osteoarthritis. Br J Rheumatol 36:651–655

Green LC, Tannenbaum SR, Goldman P (1981) Nitrate synthesis in the germ-free and conventional rat Science 212:56–58

Groeneveld PH, Kwappenberg KM, Langermans JA, Nibbering PH, Curtis L (1997) Relation between pro- and anti-inflammatory cytokines and the production of nitric oxide (NO) in severe sepsis. Cytokine 9:138–142

Gupta SK, Fitzgerald JF, Chong SK, Croffie JM, Garcia JG (1998) Expression of inducible nitric oxide synthase (iNOS) mRNA in inflamed esophageal and colonic mucosa in a pediatric population. Am J Gastroenterol 93:795–798

Guslandi M (1998) Nitric oxide and inflammatory bowel diseases. Eur J Clin Invest 28: 904–907

Halliwell B (1997) What nitrates tyrosine? Is nitrotyrosine specific as a biomarker of peroxynitrite formation in vivo? FEBS Lett 411:157–160

Hamid Q, Springall DR, Riveros-Moreno V, Chanez P, Howarth P, Redington A, Bousquet J, Godard P, Holgate S, Polak JM (1993) Induction of nitric oxide synthase in asthma. Lancet 342:1510–1513

Handy RL, Moore PK (1998) A comparison of the effects of L-NAME, 7-NI and L-NIL on carrageenan-induced hindpaw oedema and NOS activity. Br J Pharmacol 123:1119–1126

Harbrecht BG, Billiar TR, Stadler J, Demetris AJ, Ochoa J, Curran RD, Simmons RL (1992) Inhibition of nitric oxide synthesis during endotoxemia promotes intrahepatic thrombosis and an oxygen radical-mediated hepatic injury. J Leukoc Biol 52:390–394

Harbrecht BG, Stadler J, Demetris AJ, Simmons RL, Billiar TR (1994) Nitric oxide and prostaglandins interact to prevent hepatic damage during murine endotoxemia. Am J Physiol 266:G1004–G1010

Harbrecht BG, Kim YM, Wirant EA, Simmons RL, Billiar TR (1997) Timing of prostaglandin exposure is critical for the inhibition of LPS-or IFN-γ-induced macrophage NO synthesis by PGE_2. J Leukoc Biol 61:712–720

Heeringa P, van Goor H, Moshage H, Klok PA, Huitema MG, de Jager A, Schep AJ Kallenberg CG (1998) Expression of iNOS, eNOS, and peroxynitrite-modified proteins in experimental anti-myeloperoxidase associated crescentic glomerulonephritis. Kidney Int 53:382–393

Herulf M, Ljung T, Hellstrom PM, Weitzberg E, Lundberg JO (1998) Increased luminal nitric oxide in inflammatory bowel disease as shown with a novel minimally invasive method. Scand J Gastroenterol 33:164–169

Hibbs JB, Taintor RR, Vavrin Z (1987) Macrophage cytotoxicity: Role for L-arginine deiminase and imino nitrogen oxidation to nitrite. Science 235:473–476

Hibbs JBJr, Taintor RR, Vavrin Z, Rachlin EM (1988) Nitric Oxide: a cytotoxic activated macrophage effector molecule. Biochem Biophys Res Commun 157:87–94

Hickey MJ, Sharkey KA, Sihota EG, Reinhardt PH, MacMicking JD, Nathan C, Kubes P (1997) Inducible nitric oxide synthase-deficient mice have enhanced leukocyte-endothelium interactions in endotoxemia. FASEB J 11:955–964

Hierholzer C, Harbrecht B, Menezes JM, Kane J, MacMicking J, Nathan CF, Peitzman AB, Billiar TR, Tweardy DJ (1998) Essential role of induced nitric oxide in the initiation of the inflammatory response after hemorrhagic shock. J Exp Med 187: 917–928

Hirsch EC, Hunot S, Damier P, Faucheux B (1998) Glial cells and inflammation in Parkinson's disease: a role in neurodegeneration? Ann Neurol 44:S115-S120

Hoey S, Grabowski PS, Ralston SH, Forrester JV, Liversidge J (1997) Nitric oxide accelerates the onset and increases the severity of experimental autoimmune uveoretinitis through an IFN-gamma-dependent mechanism. J Immunol 159:5132–5142

Hoffman RA, Langrehr JM, Berry LM, White DA, Schattenfroh NC, McCarthy SA, Simmons RL (1996) Bystander injury of host lymphoid tissue during murine graft-verus-host disease is mediated by nitric oxide. Transplantation 61:610–618

Hoffman RA, Nüssler NC, Gleixner SL, Zhang G, Ford HR, Langrehr JM, Demetris AJ, Simmons RL (1997) Attenuation of lethal graft-versus-host disease by inhibition of nitric oxide synthase. Transplantation 63:94–100

Hooper DC, Bagasra O, Marini JC, Zborek A, Ohnishi ST, Kean R, Champion JM, Sarker AB, Bobroski L, Farber JL, Akaike T, Maeda H, Koprowski H (1997) Prevention of experimental allergic encephalomyelitis by targeting nitric oxide and peroxynitrite: implications for the treatment of multiple sclerosis. Proc Natl Acad Sci USA 94:2528–2533

Howard M, Muchamuel T, Andrade S, Menon S (1993) Interleukin 10 protects mice from lethal endotoxemia. J Exp Med 177:1205–1208

Hughes FJ, Buttery LK, Hukkanen MJ, O'Donnell A, Maclouf J, Polak JM (1999) Cytokine-induced prostaglandin E2 synthesis and cyclooxygenase-2 activity are regulated both by a nitric oxide-dependent and – independent mechanism in rat osteoblasts in vitro. J Biol Chem 274:1776–1782

Inoue T, Fukuo K, Morimoto S, Koh E, Ogihara T (1993) Nitric oxide mediates interleukin-1-induced prostaglandin E2 production by vascular smooth muscle cells. Biochem Biophys Res Commun 194:420–424

Iyengar R, Stuehr DJ, Marletta MA (1987) Macrophage synthesis of nitrite, nitrate, and N-nitrosamines: Precursors and role of the respiratory burst. Proc Natl Acad Sci USA 84:6369–6373

James SL (1995) Role of nitric oxide in parasitic infections. Microbiol Rev 59:533–547

Johnson AW, Land JM, Thompson EJ, Bolanos JP, Clark JB, Heales SJ (1995) Evidence for increased nitric oxide production in multiple sclerosis. J Neurol Neurosurg Psychiatry 58:107

Kamisaki Y, Wada K, Ataka M, Yamada Y, Nakamoto K, Ashida K, Kishimoto Y (1997) Lipopolysaccharide-induced increase in plasma nitrotyrosine concentrations in rats. Biochim Biophys Acta 1362:24–28

Kashem A, Endoh M, Yano N, Yamauchi F, Nomoto Y, Sakai H (1996) Expression of inducible-NOS in human glomerulonephritis: the possible source is infiltrating monocytes/macrophages. Kidney Int 50:392–399

Kausalya S, Nath J (1998) Interactive role of nitric oxide and superoxide anion in neutrophil-mediated endothelial cell injury. J Leukoc Biol 64:185–191

Kepka-Lenhart D, Chen L-C, Morris SM (1996) Novel actions of aspirin and sodium salicylate: discordant effects on nitric oxide synthesis and induction of nitric oxide synthase mRNA in a murine macrophage cell line. J Leukoc Biol 59:840–846

Kilbourn RG, Gross SS, Jubran A, Adams J, Griffith OW, Levi R, Lodato RF (1990) N^G-methyl-L-arginine inhibits tumor necrosis factor-induced hypotension: implications for the involvement of nitric oxide. Proc Natl Acad Sci USA 87:3629–3632

Kim YM, Talanian RV, Li J, Billiar TR (1998) Nitric oxide prevents IL-1beta and IFN-gamma-inducing factor (IL-18) release from macrophages by inhibiting caspase-1 (IL-1beta-converting enzyme) J Immunol 161, 4122–4128

Kim YM, Bombeck CA, Billiar TR (1999) Nitric Oxide as a Bifunctional Regulator of Apoptosis. Circ Res (In press)

Kiss J, Lamarque D, Delchier JC, Whittle BJ (1997) Time-dependent actions of nitric oxide synthase inhibition on colonic inflammation induced by trinitrobenzene sulphonic acid in rats. Eur J Pharmacol 336:219–224

Konttinen YT, Platts LA, Tuominen S, Eklund KK, Santavirta N, Tornwall J, Sorsa T, Hukkanen M, Polak JM (1997) Role of nitric oxide in Sjogren's syndrome. Arthritis Rheum. 40:875–883

Kooy NW, Royall JA, Ye YZ, Kelly DR, Beckman JS (1995) Evidence for in vivo peroxynitrite production in human acute lung injury. Am J Respir Crit Care Med 151:1250–1254

Kooy NW, Lewis SJ, Royall JA, Ye YZ, Kelly DR, Beckman JS (1997) Extensive tyrosine nitration in human myocardial inflammation: evidence for the presence of peroxynitrite. Crit Care Med 25:812–819

Kosonen O, Kankaanranta H, Lahde M, Vuorinen P, Ylitalo P, Moilanen E (1998a) Nitric oxide-releasing oxatriazole derivatives inhibit human lymphocyte proliferation by a cyclic GMP-independent mechanism. J Pharmacol Exp Ther 286: 215–220

Kosonen O, Kankaanranta H, Malo-Ranta U, Ristimaki A, Moilanen E (1998b) Inhibition by nitric oxide-releasing compounds of prostacyclin production in human endothelial cells. Br J Pharmacol 125:247–254

Kröncke KD, Fehsel K, Kolb-Bachofen V (1998) Inducible nitric oxide synthase in human diseases. Clin Exp Immunol 113:147–156

Kubes P, Suzuki M, Granger DN (1991) Nitric oxide: an endogenous modulator of leukocyte adhesion. Proc Natl Acad Sci USA 88:4651–4655

Kuhn A, Fehsel K, Lehmann P, Krutmann J, Ruzicka T, Kolb-Bachofen V (1998) Aberrant timing in epidermal expression of inducible nitric oxide synthase after UV irradiation in cutaneous lupus erythematosus. J Invest Dermatol 111:149–153

Lafond-Walker A, Chen CL, Augustine S, Wu TC, Hruban RH, Lowenstein CJ (1997a) Inducible nitric oxide synthase expression in coronary arteries of transplanted human hearts with accelerated graft arteriosclerosis. Am J Pathol 151:919–925

Lafond-Walker A, Chen CL, Augustine S, Wu TC, Hruban RH, Lowenstein CJ (1997b) Inducible nitric oxide synthase expression in coronary arteries of transplanted human hearts with accelerated graft arteriosclerosis. Am J Pathol 151:919–925

Langrehr JM, Hoffman RA, Lancaster JRJ, Simmons RL (1993) Nitric oxide–a new endogenous immunomodulator. Transplantation 55:1205–1212

Laubach VE, Shesely EG, Smithies O, Sherman PA (1995) Mice lacking inducible nitric oxide synthase are not resistant to lipopolysaccharide-induced death. Proc Natl Acad Sci USA 92:10688–10692

Leib SL, Kim YS, Black SM, Tureen JH, Tauber MG (1998) Inducible nitric oxide synthase and the effect of aminoguanidine in experimental neonatal meningitis. J Infect Dis 177:692–700

Li J, Billiar TR (1999) Determinants of Nitric Oxide Protection and Toxicity in Liver. Am J Physiol (In Press)

Liaudet L, Rosselet A, Schaller MD, Markert M, Perret C, Feihl F (1998) Nonselective versus selective inhibition of inducible nitric oxide synthase in experimental endotoxic shock. J Infect Dis 177:127–132

Linn SC, Morelli PJ, Edry I, Cottongim SE, Szabo C, Salzman AL (1997) Transcriptional regulation of human inducible nitric oxide synthase gene in an intestinal epithelial cell line. Am J Physiol 272:G1499–G1508

Liu S, Adcock IM, Old RW, Barnes PJ, Evans TW (1993) Lipopolysaccharide treatment in vivo induces widespread tissue expression of inducible nitric oxide synthase mRNA. Biochem Biophys Res Commun 196:1208–1213

Liu SF, Ye X, Malik AB (1997a) In vivo inhibition of nuclear factor-kappa B activation prevents inducible nitric oxide synthase expression and systemic hypotension in a rat model of septic shock. J Immunol 159:3976–3983

Liu P, Hock CE, Nagele R, Wong PY (1997b) Formation of nitric oxide, superoxide, and peroxynitrite in myocardial ischemia-reperfusion injury in rats. Am J Physiol 272:H2327–H2336

Liu P, Yin K, Nagele R, Wong PY (1998) Inhibition of nitric oxide synthase attenuates peroxynitrite generation, but augments neutrophil accumulation in hepatic ischemia-reperfusion in rats. J Pharmacol Exp Ther 284:1139–1146

Lundberg JO, Ehren I, Jansson O, Adolfsson J, Lundberg JM, Weitzberg E, Alving K, Wiklund NP (1996) Elevated nitric oxide in the urinary bladder in infectious and noninfectious cystitis. Urology 48:700–702

Luoma JS, Stralin P, Marklund SL, Hiltunen TP, Sarkioja T, Yla-Herttuala S (1998) Expression of extracellular SOD and iNOS in macrophages and smooth muscle cells in human and rabbit atherosclerotic lesions: colocalization with epitopes characteristic of oxidized LDL and peroxynitrite-modified proteins. Arterioscler Thromb Vasc Biol 18:157–167

Lyons CR (1995) The role of nitric oxide in inflammation. Adv Immunol 60:323–71, 323–371

MacMicking JD, Nathan C, Hom G, Chartrain N, Fletcher DS, Trumbauer M, Stevens K, Xie QW, Sokol K, Hutchinson N (1995) Altered responses to bacterial infection and endotoxic shock in mice lacking inducible nitric oxide synthase. Cell 81:641–650

MacMicking JD, North RJ, LaCourse R, Mudgett JS, Shah SK, Nathan CF (1997a) Identification of nitric oxide synthase as a protective locus against tuberculosis. Proc Natl Acad Sci USA 94:5243–5248

MacMicking,J, Xie Qw, Nathan C (1997c) Nitric oxide and macrophage function. Annu Rev Immunol 15:323–350

Mannick EE, Bravo LE, Zarama G, Realpe JL, Zhang XJ, Ruiz B, Fontham ET, Mera R, Miller MJ, Correa P (1996) Inducible nitric oxide synthase, nitrotyrosine, and apoptosis in Helicobacter pylori gastritis: effect of antibiotics and antioxidants. Cancer Res 56:3238–3243

Marletta MA, Yoon PS, Iyengar R, Leaf CD, Wishnok JS (1988) Macrophage oxidation of L-arginine to nitrite and nitrate: nitric oxide is an intermediate. Biochemistry 27:8706–8711

Marotta P, Sautebin L, Di Rosa M (1992) Modulation of the induction of nitric oxide synthase by eicosanoids in the murine macrophage cell line J774. Br J Pharmacol 107:640–641

Mayer B, Hemmens B (1997) Biosynthesis and action of nitric oxide in mammalian cells. Trends Biochem Sci 22:477–481

McCafferty DM, Mudgett JS, Swain MG, Kubes P (1997) Inducible nitric oxide synthase plays a critical role in resolving intestinal inflammation. Gastroenterology 112:1022–1027

McCartney-Francis N, Allen JB, Mizel DE, Albina JE, Xie Qw, Nathan CF, Wahl SM (1993) Suppression of arthritis by an inhibitor of nitric oxide synthase. J Exp Med 178:749–754

McDermott CD, Gavita SM, Shennib H, Giaid A (1997) Immunohistochemical localization of nitric oxide synthase and the oxidant peroxynitrite in lung transplant recipients with obliterative bronchiolitis. Transplantation 64:270–274

McInnes IB, Leung BP, Field M, Wei XQ, Huang FP, Sturrock RD, Kinninmonth A, Weidner J, Mumford R, Liew FY (1996) Production of nitric oxide in the synovial membrane of rheumatoid and osteoarthritis patients. J Exp Med 184:1519–1524

McInnes IB, Leung B, Wei XQ, Gemmell CC, Liew FY (1998) Septic arthritis following *Staphylococcus aureus* infection in mice lacking inducible nitric oxide synthase. J Immunol 160:308–315

Miesel R, Kurpisz M, Kroger H (1996) Suppression of inflammatory arthritis by simultaneous inhibition of nitric oxide synthase and NADPH oxidase. Free Radic Biol Med 20:75–81

Milano S, Arcoleo F, Dieli M, D'Agostino R, D'Agostino P, De Nucci G, Cillari E (1995) Prostaglandin E_2 regulates inducible nitric oxide synthase in the murine macrophage cell line J774. Prostaglandins 49:105–115

Miller MJ, Thompson JH, Zhang XJ, Sadowska-Krowicka H, Kakkis JL, Munshi UK, Sandoval M, Rossi JL, Eloby-Childress S, Beckman JS (1995) Role of inducible nitric oxide synthase expression and peroxynitrite formation in guinea pig ileitis. Gastroenterology 109:1475–1483

Moncada S (1997) Nitric oxide in the vasculature: physiology and pathophysiology. Ann NY Acad Sci 811:60–67

Moncada S, Higgs EA (1993) The L-arginine-nitric oxide pathway. N Engl J Med 329: 2002–2012

Moncada S, Palmer RMJ, Higgs EA (1991) Nitric oxide: physiology, pathophysiology, and pharmacology. Pharmacol Rev 43:109–142

Mosmann TR, Sad S (1996) The expanding universe of T-cell subsets: Th1, Th2 and more. Immunol Today 17:138–146

Mourelle M, Casellas F, Guarner F, Salas A, Riveros-Moreno V, Moncada S, Malagelada JR (1995) Induction of nitric oxide synthase in colonic smooth muscle from patients with toxic megacolon. Gastroenterology 109:1497–1502

Mourelle M, Vilaseca J, Guarner F, Salas A, Malagelada JR (1996) Toxic dilatation of colon in a rat model of colitis is linked to an inducible form of nitric oxide synthase. Am J Physiol 270:G425–G430

Munder M, Eichmann K, Modolell M (1998) Alternative metabolic states in murine macrophages reflected by the nitric oxide synthase/arginase balance: competitive regulation by CD4+ T cells correlates with Th1/Th2 phenotype. J Immunol 160: 5347–5354

Nathan C (1987) Secretory products of macrophages. J Clin Invest 79:319–326

Nathan C (1997) Inducible nitric oxide synthase: what difference does it make? J Clin Invest 100:2417–2423

Nathan C, Xie QW (1994) Nitric oxide synthases: roles, tolls, and controls. Cell 78: 915–918

Nozaki Y, Hasegawa Y, Takeuchi A, Fan ZH, Isobe KI, Nakashima I, Shimokata K (1997) Nitric oxide as an inflammatory mediator of radiation pneumonitis in rats. Am J Physiol 272:L651–L658

Numata M, Suzuki S, Miyazawa N, Miyashita A, Nagashima Y, Inoue S, Kaneko T, Okubo T (1998) Inhibition of inducible nitric oxide synthase prevents LPS-induced acute lung injury in dogs. J Immunol 160:3031–3037

Nüssler AK, Billiar TR (1993) Inflammation, immunoregulation, and inducible nitric oxide synthase. J Leukoc Biol 54:171–178

Oates JC, Ruiz P, Alexander A, Pippen AM, Gilkeson GS (1997) Effect of late modulation of nitric oxide production on murine lupus. Clin Immunol Immunopathol 83:86–92

Olee T, Hashimoto S, Quach J, Lotz M (1999) IL-18 is produced by articular chondrocytes and induces proinflammatory and catabolic responses. J Immunol 162:1096–1100

Ou J, Carlos TM, Watkins SC, Saavedra JE, Keefer LK, Kim YM, Harbrecht BG, Billiar TR (1997) Differential effects of nonselective nitric oxide synthase (NOS) and selective inducible NOS inhibition on hepatic necrosis, apoptosis, ICAM-1 expression, and neutrophil accumulation during endotoxemia. Nitric Oxide 1:404–416

Oyama J, Shimokawa H, Momii H, Cheng X, Fukuyama N, Arai Y, Egashira K, Nakazawa H, Takeshita A (1998) Role of nitric oxide and peroxynitrite in the cytokine-induced sustained myocardial dysfunction in dogs in vivo. J Clin Invest 101:2207–2214

Palmer RMJ, Ferrige AG, Moncada S (1987) Nitric oxide release accounts for the biological activity of endothelium-derived relaxing factor. Nature 327:524–526

Pang L, Hoult JRS (1996) Induction of cyclooxygenase and nitric oxide synthase in endotoxin-activated J774 macrophages is differentially regulated by indomethacin: Enhanced cyclooxygenase-2 protein expression but reduction of inducible nitric oxde synthase. Eur J Pharmacol 317:151–155

Park JH, Chang SH, Lee KM, Shin SH (1996) Protective effect of nitric oxide in an endotoxin-induced septic shock. Am J Surg 171:340–345

Parrillo JE (1993) Pathogenetic mechanisms of septic shock. N Engl J Med 328:1471–1477

Pender BS, Chen H, Ashton S, Wise WC, Zingarelli B, Cusumano V, Cook JA (1996) Transforming growth factor beta 1 alters rat peritoneal macrophage mediator production and improves survival during endotoxic shock. Eur Cytokine Netw 7:137–142

Peng HB, Spiecker M, Liao JK (1998) Inducible nitric oxide: an autoregulatory feedback inhibitor of vascular inflammation. J Immunol 161:1970–1976

Perrella MA, Hsieh CM, Lee WS, Shieh S, Tsai JC, Patterson C, Lowenstein CJ, Long NC, Haber E, Shore S, Lee ME (1996) Arrest of endotoxin-induced hypotension by transforming growth factor beta1. Proc Natl Acad Sci USA 93:2054–2059

Petros A, Bennett D, Vallance, P (1991) Effect of nitric oxide synthase inhibitors on hypotension in patients with septic shock. Lancet 338:1557–1558

Pou S, Pou WS, Bredt DS, Snyder SH, Rosen GM (1992) Generation of superoxide by purified brain nitric oxide synthase. J Biol Chem 267:24173–24176

Ribbons KA, Currie MG, Connor JR, Manning PT, Allen PC, Didier P, Ratterree MS, Clark DA, Miller MJ (1997) The effect of inhibitors of inducible nitric oxide synthase on chronic colitis in the rhesus monkey. J Pharmacol Exp Ther 280:1008–1015

Ridger VC, Pettipher ER, Bryant CE, Brain SD (1997) Effect of the inducible nitric oxide synthase inhibitors aminoguanidine and L-N6-(1-iminoethyl)lysine on zymosan-induced plasma extravasation in rat skin. J Immunol 159:383–390

Ross R (1999) Atherosclerosis–an inflammatory disease. N Engl J Med 340:115–126

Rowe A, Farrell AM, Bunker CB (1997) Constitutive endothelial and inducible nitric oxide synthase in inflammatory dermatoses. Br J Dermatol 136:18–23

Ryffel B (1997) Impact of knockout mice in toxicology. Crit Rev Toxicol 27:135–154

Sadowska-Krowicka H, Mannick EE, Oliver PD, Sandoval M, Zhang XJ, Eloby-Childess S, Clark DA, Miller MJ (1998) Genistein and gut inflammation: role of nitric oxide. Proc Soc Exp Biol Med 217:351–357

Saleh D, Barnes PJ, Giaid A (1997) Increased production of the potent oxidant peroxynitrite in the lungs of patients with idiopathic pulmonary fibrosis. Am J Respir Crit Care Med 155:1763–1769

Saleh D, Ernst P, Lim S, Barnes PJ, Giaid A (1998) Increased formation of the potent oxidant peroxynitrite in the airways of asthmatic patients is associated with induction of nitric oxide synthase: effect of inhaled glucocorticoid. FASEB J 12:929–937

Salvemini D, Misko TP, Masferrer JL, Seibert K, Currie MG, Needleman P (1993) Nitric oxide activates cyclooxygenase enzymes. Proc Natl Acad Sci USA 90:7240–7244

Salvemini D, Seibert K, Marino MH (1996a) PG release, as a consequence of NO-driven COX activation, contributes to the proinflammatory effects of NO. Drugs News Perspect. 4:204–219

Salvemini D, Wang ZQ, Bourdon DM, Stern MK, Currie MG, Manning PT (1996b) Evidence of peroxynitrite involvement in the carrageenan-induced rat paw edema. Eur J Pharmacol 303:217–220

Salvemini D, Wang ZQ, Wyatt PS, Bourdon DM, Marino MH, Manning PT, Currie MG (1996c) Nitric oxide: a key mediator in the early and late phase of carrageenan-induced rat paw inflammation. Br J Pharmacol 118:829–838

Samlowski WE, Yim CY, McGregor JR (1998) Nitric oxide exposure inhibits induction of lymphokine-activated killer cells by inducing precursor apoptosis. Nitric Oxide 2:45–56

Santos LL, Morand EF, Yang Y, Hutchinson P, Holdsworth SR (1997) Suppression of adjuvant arthritis and synovial macrophage inducible nitric oxide by N-iminoethyl-L-ornithine, a nitric oxide synthase inhibitor. Inflammation 21:299–311

Satoi H, Oka N, Kawasaki T, Miyamoto K, Akiguchi I, Kimura J (1998) Mechanisms of tissue injury in vasculitic neuropathies. Neurology 50:492–496

Schuberth HJ, Hendricks A, Leibold W (1998) There is no regulatory role for induced nitric oxide in the regulation of the in vitro proliferative response of bovine mononuclear cells to mitogens, alloantigens or superantigens. Immunobiology 198: 439–450

Shears LL, Kawaharada N, Tzeng E, Billiar TR, Watkins SC, Kovesdi I, Lizonova A, Pham SM (1997) Inducible nitric oxide synthase suppresses the development of allograft arteriosclerosis. J Clin Invest 100:2035–2042

Shin T, Tanuma N, Kim S, Jin J, Moon C, Kim K, Kohyama K, Matsumoto Y, Hyun B (1998) An inhibitor of inducible nitric oxide synthase ameliorates experimental autoimmune myocarditis in Lewis rats. J Neuroimmunol 92:133–138

Shoker AS, Yang H, Murabit MA, Jamil H, al-Ghoul A, Okasha K (1997) Analysis of the in vitro effect of exogenous nitric oxide on human lymphocytes. Mol Cell Biochem 171:75–83

Singer II, Kawka DW, Scott S, Weidner JR, Mumford RA, Riehl TE, Stenson WF (1996) Expression of inducible nitric oxide synthase and nitrotyrosine in colonic epithelium in inflammatory bowel disease. Gastroenterology 111:871–885

Souza-Fiho MV, Lima MV, Pompeu MM, Ballejo G, Cunha FQ, Ribeiro RD (1997) Involvement of nitric oxide in the pathogenesis of cyclophosphamide- induced hemorrhagic cystitis. Am J Pathol 150:247–256

Spiecker M, Peng HB, Liao JK (1997) Inhibition of endothelial vascular cell adhesion molecule-1 expression by nitric oxide involves the induction and nuclear translocation of IκBα. J Biol Chem 272:30969–30974

Spink J, Cohen J, Evans TJ (1995) The cytokine responsive vascular smooth muscle cell enhancer of inducible nitric oxide synthase. Activation by nuclear factor-κB. J Biol Chem 270:29541–29547

Stadler J, Stefanovic-Racic M, Billiar TR, Curran RD, McIntyre LA, Georgescu HI, Simmons RL, Evans CH (1991) Articular chondrocytes synthesize nitric oxide in response to cytokines and lipopolysaccharide. J Immunol 147:3915–3920

Stadler J, Harbrecht BG, Di Silvio M, Curran RD, Jordan ML, Simmons RL, Billiar TR (1993) Endogenous nitric oxide inhibits the synthesis of cyclooxygenase products and interleukin-6 by rat Kupffer cells. J Leukoc Biol 53:165–172

Stuehr DJ, Marletta MA (1985) Mammalian nitrate biosynthesis: mouse macrophages produce nitrite and nitrate in response to *Escherichia coli* lipopolysaccharide. Proc Natl Acad Sci USA 82:7738–7742

Stuehr DJ, Nathan CF (1989) Nitric oxide, a macrophage product responsible for cytostasis and respiratory inhibition in tumor target cells. J Exp Med 169: 1543–1545

Swierkosz TA, Mitchell JA, Warner TD, Botting RM, Vane JR (1995) Co-induction of nitric oxide synthase and cyclo-oxygenase: interactions between nitric oxide and prostanoids. Br J Pharmacol 114:1335–1342

Szabo C (1998) Role of poly(ADP-ribose)synthetase in inflammation. Eur J Pharmacol 350:1–19

Szabo C, Dawson VL (1998) Role of poly(ADP-ribose) synthetase in inflammation and ischaemia reperfusion. Trends Pharmacol Sci 19:287–298

Szabo C, Southan GJ, Thiemermann C (1994) Beneficial effects and improved survival in rodent models of septic shock with S-methylisothiourea sulfate, a potent and selective inhibitor of inducible nitric oxide synthase. Proc Natl Acad Sci USA 91:12472–12476

Szabo C, Salzman AL, Ischiropoulos H (1995) Endotoxin triggers the expression of an inducible isoform of nitric oxide synthase and the formation of peroxynitrite in the rat aorta in vivo. FEBS Lett 363:235–238

Szabo C, Lim LH, Cuzzocrea S, Getting SJ, Zingarelli B, Flower RJ, Salzman AL, Perretti M (1997) Inhibition of poly (ADP-ribose) synthetase attenuates neutrophil recruitment and exerts antiinflammatory effects. J Exp Med 186:1041–1049

Szabolcs MJ, Ravalli S, Minanov O, Sciacca RR, Michler RE, Cannon PJ (1998) Apoptosis and increased expression of inducible nitric oxide synthase in human allograft rejection. Transplantation 65:804–812

Takeichi O, Saito I, Hayashi M, Tsurumachi T, Saito T (1998) Production of human-inducible nitric oxide synthase in radicular cysts. J Endod 24:157–160

Tanaka S, Kamiike W, Ito T, Uchikoshi F, Matsuda H, Nozawa M, Kumura E, Shiga T, Kosaka H (1995) Generation of nitric oxide as a rejection marker in rat pancreas transplantation. Transplantation 60:713–717

Tarrant TK, Silver PB, Wahlsten JL, Rizzo LV, Chan CC, Wiggert B, Caspi RR (1999) Interleukin 12 protects from a T-helper type-1-mediated autoimmune disease, experimental autoimmune uveitis, through a mechanism involving interferon γ, nitric oxide, and apoptosis. J Exp Med 189:219–230

Taylor BS, de Vera ME, Ganster RW, Wang Q, Shapiro RA, Morris SMJ, Billiar TR, Geller DA (1998) Multiple NF-κB enhancer elements regulate cytokine induction of the human inducible nitric oxide synthase gene. J Biol Chem 273:15148–15156

Taylor-Robinson AW (1997) Counter-regulation of T-helper-1 cell proliferation by nitric oxide and interleukin-2. Biochem Biophys Res Commun 233:14–19

Ter Steege J, Buurman W, Arends JW, Forget P (1997) Presence of inducible nitric oxide synthase, nitrotyrosine, CD68, and CD14 in the small intestine in celiac disease. Lab Invest 77:29–36

Tracey WR, Xue C, Klinghofer V, Barlow J, Pollock JS, Forstermann U, Johns RA (1994) Immunochemical detection of inducible NO synthase in human lung. Am J Physiol 266:L722-L727

Trepicchio WL, Bozza M, Pedneault G, Dorner AJ (1996) Recombinant human IL-11 attenuates the inflammatory response through down-regulation of proinflammatory cytokine release and nitric oxide production. J Immunol. 157:3627–3634

Trowbridge HO, Emling RC (1997) Inflammation: a review of the process. Quintessence, Chicago

Tsuji M, Dimov VB, Yoshida T (1995) In vivo expression of monokine and inducible nitric oxide synthase in experimentally induced pulmonary granulomatous inflammation. Evidence for sequential production of interleukin-1, inducible nitric oxide synthase, and tumor necrosis factor. Am J Pathol 147:1001–1015

Van de Loo FA, Arntz OJ, van Enckevort FH, van Lent PL, van den Berg WB (1998) Reduced cartilage proteoglycan loss during zymosan induced gonarthritis in NOS2-deficient mice and in anti-interleukin-1-treated wild-type mice with unabated joint inflammation. Arthritis Rheum 41:634–646

Van der Veen RC, Hinton DR, Incardonna F, Hofman FM (1997) Extensive peroxynitrite activity during progressive stages of central nervous system inflammation. J Neuroimmunol 77:1–7

Vara E, Arias-Diaz J, Garcia C, Hernandez J, Garcia-Carreras C, Cuadrado A, Balibrea JL (1995) Production of TNF alpha, IL-1, IL-6 and nitric oxide by isolated human islets. Transplant Proc 27:3367–3371

Vodovotz Y, Bogdan C, Paik J, Xie QW, Nathan C (1993) Mechanisms of suppression of macrophage nitric oxide release by transforming growth factor beta. J Exp Med 178:605–613

Vodovotz Y, Geiser AG, Chesler L, Letterio JJ, Campbell A, Lucia MS, Sporn MB, Roberts AB (1996a) Spontaneously increased production of nitric oxide and aberrant expression of the inducible nitric oxide synthase in vivo in the transforming growth factor beta 1 null mouse. J Exp Med 183:2337–2342

Vodovotz Y, Letterio JJ, Geiser AG, Chesler L, Roberts AB, Sparrow J (1996b) Control of nitric oxide production by endogenous TGF-beta1 and systemic nitric oxide in retinal pigment epithelial cells and peritoneal macrophages. J Leukoc Biol 60:261–270

Vodovotz Y, Kopp JB, Takeguchi H, Shrivastav S, Coffin D, Lucia MS, Mitchell JB, Webber R, Letterio J, Wink D, Roberts AB (1998) Increased mortality, blunted production of nitric oxide, and increased production of TNF-α in endotoxemic TGF-β1 transgenic mice. J Leukoc Biol 63:31–39

Vos TA, Gouw AS, Klok PA, Havinga R, van Goor H, Huitema S, Roelofsen H, Kuipers F, Jansen PL, Moshage H (1997) Differential effects of nitric oxide synthase inhibitors on endotoxin-induced liver damage in rats. Gastroenterology 113:1323–1333

Wagner DA, Young VR, Tannenbaum SR (1983) Mammalian nitrate biosynthesis: Incorporation of $^{15}NH_3$ into nitrate is enhanced by endotoxin treatment. Proc Natl Acad Sci USA 80:4518–4521

Wang JS, Tseng HH, Shih DF, Jou HS, Ger LP (1997) Expression of inducible nitric oxide synthase and apoptosis in human lupus nephritis. Nephron 77:404–411

Wang R, Ghahary A, Shen YJ, Scott PG, Tredget EE (1996a) Human dermal fibroblasts produce nitric oxide and express both constitutive and inducible nitric oxide synthase isoforms. J Invest Dermatol 106:419–427

Wang ZY, Alm P, Hakanson R (1996b) The contribution of nitric oxide to endotoxin-induced ocular inflammation: interaction with sensory nerve fibres. Br J Pharmacol 118:1537–1543

Watkins SC, Macaulay W, Turner D, Kang R, Rubash HE, Evans CH (1997) Identification of inducible nitric oxide synthase in human macrophages surrounding loosened hip prostheses. Am J Pathol 150:1199–1206

Wei XQ, Charles IG, Smith A, Ure J, Feng GJ, Huang FP, Xu D, Muller W, Moncada S, Liew FY (1995) Altered immune responses in mice lacking inducible nitric oxide synthase. Nature 375:408–411

Werner J, Rivera J, Fernandez-del CC, Lewandrowski K, Adrie C, Rattner DW, Warshaw AL (1997) Differing roles of nitric oxide in the pathogenesis of acute edematous versus necrotizing pancreatitis. Surgery 121:23–30

Whittle BJ (1997) Nitric oxide–a mediator of inflammation or mucosal defence. Eur J Gastroenterol Hepatol 9:1026–1032

Wigand R, Meyer J, Busse R, Hecker M (1997) Increased serum N^G-hydroxy-L-arginine in patients with rheumatoid arthritis and systemic lupus erythematosus as an index of an increased nitric oxide synthase activity. Ann Rheum Dis 56:330–332

Wink DA, Hanbauer I, Grisham MB, Laval F, Nims RW, Laval J, Cook J, Pacelli R, Liebmann J, Krishna M, Ford PC, Mitchell JB (1996) Chemical biology of nitric oxide: regulation and protective and toxic mechanisms. Curr Top Cell Regul 34:159–87, 159–187

Winlaw DS, Schyvens CG, Smythe GA, Du ZY, Rainer SP, Lord RS, Spratt PM, Macdonald PS (1995) Selective inhibition of nitric oxide production during cardiac allograft rejection causes a small increase in graft survival. Transplantation 60:77–82

Wizemann TM, Gardner CR, Laskin JD, Quinones S, Durham SK, Goller NL, Ohnishi ST, Laskin DL (1994) Production of nitric oxide and peroxynitrite in the lung during acute endotoxemia. J Leukoc Biol 56:759–768

Wolkow PP (1998) Involvement and dual effects of nitric oxide in septic shock. Inflamm Res 47:152–166

Wong HR, Finder JD, Wasserloos K, Lowenstein CJ, Geller DA, Billiar TR, Pitt BR, Davies P (1996) Transcriptional regulation of iNOS by IL-1 beta in cultured rat pulmonary artery smooth muscle cells. Am J Physiol 271:L166-L171

Wu CC, Chen SJ, Szabo C, Thiemermann C, Vane JR (1995) Aminoguanidine attenuates the delayed circulatory failure and improves survival in rodent models of endotoxic shock. Br J Pharmacol 114:1666–1672

Xie QW, Kashiwabara Y, Nathan C (1994) Role of transcription factor NF-κB/Rel in induction of nitric oxide synthase. J Biol Chem 269:4705–4708

Xiong Y, Karupiah G, Hogan SP, Foster PS, Ramsay AJ (1999) Inhibition of allergic airway inflammation in mice lacking nitric oxide synthase 2. J Immunol 162:445–452

Yamamoto T, Katayama I, Nishioka K (1998) Nitric oxide production and inducible nitric oxide synthase expression in systemic sclerosis. J Rheumatol 25:314–317

Yamasaki K, Edington HD, McClosky C, Tzeng E, Lizonova A, Kovesdi I, Steed DL, Billiar TR (1998) Reversal of impaired wound repair in iNOS-deficient mice by topical adenoviral-mediated iNOS gene transfer. J Clin Invest 101:967–971

Zamora R, Bult H, Herman AG (1998) The role of prostaglandin E2 and nitric oxide in cell death in J774 murine macrophages. Eur J Pharmacol 349:307–315

Zhao W, Tilton RG, Corbett JA, McDaniel ML, Misko TP, Williamson JR, Cross AH, Hickey WF (1996) Experimental allergic encephalomyelitis in the rat is inhibited by aminoguanidine, an inhibitor of nitric oxide synthase. J Neuroimmunol 64:123–133

Zielasek J, Jung S, Gold R, Liew FY, Toyka KV, Hartung HP (1995) Administration of nitric oxide synthase inhibitors in experimental autoimmune neuritis and experimental autoimmune encephalomyelitis. J Neuroimmunol 58:81–88

CHAPTER 20
Nitric Oxide in the Immunopathogenesis of Type 1 Diabetes

V. Burkart and H. Kolb

A. Introduction

During the past decade, research on the pathogenesis of human type 1 diabetes considerably increased our insight into the mechanisms involved in the destruction of insulin-producing β cells of pancreatic islets of Langerhans. Cell-mediated immune reactivity against islet constituents is assumed to play a major role in the development of this disease (Eisenbarth 1986). In an attempt to identify cellular effector mechanisms involved in β-cell destruction, Kröncke et al. (1991) demonstrated that nitric oxide (NO) released from activated macrophages is able to exert cytotoxic activity against islet cells. Since then, increasing research activities focused on the source, the target structures and the metabolic or cytotoxic effects of NO in the process of β-cell destruction. From the growing number of observations, a picture emerges which identifies NO as a major regulatory and cytotoxic mediator during islet inflammation. This qualifies the mechanisms involved in NO-induced β-cell death as important targets for intervention strategies aiming at the prevention of type 1 diabetes (Corbett and McDaniel 1992; Kolb and Kolb-Bachofen 1992; Burkart et al. 1994).

This review focuses on the present knowledge about the role of NO as a pathogenetic factor involved in the destruction of pancreatic β cells during the development of type 1 diabetes. It will deal with the identification of the cellular sources of β-cell-damaging NO, its potential target structures in the β cell, the possible pathways of NO-induced β-cell death and, finally, with β-cell protective strategies which have been developed based on these findings.

B. Type 1 Diabetes
I. Clinical Characteristics

Human type 1 diabetes, previously termed insulin-dependent diabetes mellitus, is characterized by an absolute deficiency of insulin secretion due to a loss of the insulin-producing β cells from the pancreatic islets of Langerhans (Atkinson and MacLaren 1994; Expert Committee on the Diagnosis and Classification of Diabetes Mellitus 1997). Type 1 diabetes commonly occurs in children and young adults, but it can occur at any age. Generally, after a sudden manifestation of hyperglycemia, the affected individuals become strictly insulin-dependent for the rest of their life (The Expert Committee on the Diagnosis and Classification of DIABETES MELLITUS 1997). Familiar accumulation and the association with certain human leukocyte antigen haplotypes (CANTOR et al. 1995) indicate that predisposing genes determine disease susceptibility. Substantial geographic variations of diabetes incidence strongly suggest that environmental factors (KARVONEN et al. 1993), such as dietary components (SCOTT et al. 1994) or exposure to certain toxins or infectious agents, can act as disease-inducing or -precipitating events (VERGE et al. 1994). Diabetic patients often suffer from late complications, including microvascular and neurological dysfunctions, and reduced quality of life and life expectancy (Expert Committee on the Diagnosis and Classification of DIABETES MELLITUS 1997).

II. Studies on the Immunopathogenesis of Type 1 Diabetes

Intensive research to elucidate the mechanisms involved in the pathogenesis of human type 1 diabetes revealed the presence of islet-specific antibodies in the sera of individuals in the so-called prediabetic phase, long before the appearance of any clinical signs of diabetes (EISENBARTH 1986). The antibodies are directed against autologous β-cell constituents, including insulin (ATKINSON et al. 1986), glutamic acid decarboxylase (KAUFMANN et al. 1992), and tyrosine-phosphatase-like protein (LAN et al. 1996). The detection of these autoantibodies gave rise to the assumption that the pathogenesis of type 1 diabetes involves autoaggressive reactions of the immune system. Today, sensitive detection systems allow the use of autoantibodies as reliable markers for the identification of persons at risk to develop diabetes. However, islet-directed autoantibodies may also be observed in individuals who never develop diabetes. It was not possible to transfer the disease (with human immunoglobulin) to animals, whereas the disease reappeared in several recipients of bone-marrow cells from diabetic donors (LAMPETER et al. 1998). Therefore, it is believed that islet autoantibodies do not play a major role in the destructive processes directed against pancreatic β cells during the pathogenesis of type 1 diabetes.

Further detailed analysis to elucidate the pathogenesis of type 1 diabetes became feasible due to the availability of animal models of the human disease. The BioBreeding (BB) rat (MORDES et al. 1987) and the non-obese diabetic (NOD) mouse (KOLB 1987) spontaneously develop diabetes with many basic features resembling type 1 diabetes. A disease state similar to human diabetes can be induced in susceptible mouse strains by the administration of multiple low doses of the β-cell-selective toxin streptozotocin (SZ) (KOLB 1987). In analogy to the findings in human prediabetic individuals, the islet-directed autoantibodies detected in these animal models do not seem to play a critical role in β-cell destruction. However, the use of animal models revealed that, in addition to humoral autoreactivity, cell-mediated immune reactions that considerably contribute to the destruction of islet cells evolve.

III. Cellular Immune Reactions Against Islet Cells

A few reports from pancreas biopsy specimens of newly diagnosed type 1 diabetic patients showed mononuclear cells, predominantly CD8$^+$ T-lymphocytes and macrophages, invading the islets (ITOH et al. 1993). This mononuclear cell infiltration into the islets, termed insulitis, is regarded as a characteristic morphological feature of recent-onset type 1 diabetes (GEPTS 1965; FOULIS et al. 1986) and strongly indicates the involvement of cell-mediated inflammatory reactions against the endocrine tissue.

Prospective studies in animal models of type 1 diabetes allowed researchers to define distinct stages of islet inflammation. The appearance of islet-infiltrating macrophages (LEE et al. 1988; HANENBERG et al. 1989) and the shift in the dominance of the T-helper (Th) lymphocyte subtypes from Th2 to Th1 (RABINOVITCH 1998), seem to be decisive events in the progression of the inflammatory process. In the early phase of "peri-insulitis" the islet-infiltrating immune cells are located in the periphery of the islet. NOD mice with this state of "benign" insulitis can live over a long period of time without showing any clinical signs of islet dysfunction. In NOD mice and BB rats, the predominance of the Th2-associated cytokines interleukin 4 (IL-4) and IL-10 and a relatively low expression of the Th1-specific cytokine interferon γ (IFNγ) are characteristic for this stage. The shift from the Th2 to the Th1 phase in insulitis is characterized by the upregulation of the cytokines IFNγ and tumor necrosis factor α (TNFα) and indicates the beginning of the autoaggressive phase of insulitis (ZIPRIS et al. 1996; KOLB 1997; RABINOVITCH 1998).

All animal models investigated revealed the presence of activated macrophages as the first islet-infiltrating immune cells in the prediabetic phase before diabetes manifestation (LEE et al. 1988; HANENBERG et al. 1989). At this stage of islet infiltration, the earliest signs of β-cell destruction can be detected by electron microscopy in close vicinity to the macrophages (KOLB-BACHOFEN et al. 1988). Interestingly, macrophage-directed interventions can induce almost complete protection from diabetes development in animal models (OSCHILEWSKI et al. 1985). Macrophage infiltration into the islet is followed by immigration of T-lymphocytes and, finally, B-lymphocytes and natural-killer cells (HANENBERG et al. 1989).

The complex processes involved in islet inflammation, which end in the complete destruction of all insulin-producing β cells, require the controlled release of a variety of regulatory and cytotoxic mediators. Meanwhile, numerous studies have identified NO as a major mediator (KOLB and KOLB-BACHOFEN 1992) which obviously plays a predominant role in the autoaggressive phase during the inflammatory processes leading to β-cell death.

C. NO as a Major Pathogenetic Factor in Immune-Mediated Diabetes

Numerous studies suggested NO as a major cytotoxic effector molecule responsible for β-cell death in model systems of human type 1 diabetes (KOLB and KOLB-BACHOFEN 1992; McDANIEL et al. 1996). Therefore, recent research efforts focused on the identification of the sources of β-cell-damaging NO, its primary target structures, and the pathways of NO-induced β-cell death. At present, most observations are based on in vitro and in vivo studies performed in rodents. Recently, a few studies with human material indicating a lesser toxicity of NO in humans versus rodent β cells became available.

I. Cellular Sources of β-Cell-Damaging NO

Detailed investigations on the potential cellular sources of NO in the early phase of islet destruction revealed that, under conditions of islet inflammation, inducible nitric oxide synthase (iNOS) can be expressed by infiltrating macrophages, endothelial cells lining the islet capillary system, and the β cell itself.

1. Macrophages

As already mentioned (Sect. B.III), activated macrophages play an important role in the initiation of islet destruction, and lysed β cells have been detected in their neighborhood. Since activated macrophages are known to express iNOS, several studies were performed to show that macrophages are able to lyse islet cells via the release of NO. Macrophages were found to exert a strong cytotoxic activity towards syngeneic islet cells in co-cultivation experiments (APPELS et al. 1989). This islet-cell-directed cytotoxicity was mediated via arginine-dependent NO generation (KRÖNCKE et al. 1991). Separation of the macrophages from the islet cells over a short distance by a membrane (BURKART and KOLB 1996) or by co-encapsulation of islets and erythrocytes in an alginate matrix (WIEGAND et al. 1993) protected the islet cells from lysis. These findings pointed to the short-lived mediator NO as a major islet-cell-toxic macrophage product.

Studies in animals with spontaneous diabetes development strongly indicate that the formation of NO from islet-infiltrating macrophages also plays an important role in islet destruction during the pathogenesis of diabetes in vivo. Immunocytochemical analysis of pancreata from diabetes-prone BB rats revealed strong iNOS protein expression in islet-infiltrating macrophages (KLEEMANN et al. 1993; KOLB-BACHOFEN 1996). Analysis of islets from diabetes-prone NOD mice localised iNOS expression in macrophages (ROTHE et al. 1994) and, additionally, in β cells (RABINOVITCH et al. 1996; REDDY et al. 1997). Interestingly, macrophages from diabetes-prone BB rats, which also exhibit T-lymphopenia, show an elevated production of NO (LAU et al. 1998).

2. Endothelial Cells

Pancreatic islets are densely capillarized, and each insulin-producing cell is in close proximity to at least one endothelial capillary cell. Upon stimulation with IL-1β and TNFα, cultivated endothelial cells were found to release high amounts of NO generated by a macrophage-like NO synthase (SUSCHEK et al. 1993). Therefore, the potential of endothelial cells to exert cytotoxic activity against β cells was investigated by co-cultivating endocrine islet cells and endothelial cells isolated from islet capillaries. Exposure to inflammatory cytokines induced the release of NO (SUSCHEK et al. 1994) and lysis of syngeneic islet β cells, which could be completely blocked by the iNOS inhibitor N^G-monomethyl-L-arginine (L-NMMA) (STEINER et al. 1997). These findings demonstrate that islet capillary endothelial cells may act as cytotoxic effector cells, which are able to lyse islet cells via the release of NO after stimulation with inflammatory cytokines.

3. β Cells

Interestingly, several reports demonstrated the formation of NO in the β cell itself, which represents the primary target of the autoimmune reaction in the inflamed islet. A cytokine-inducible form of NO synthase encoded by the same transcript as iNOS in other cell types was cloned and expressed from rat islets and insulinoma cells (KARLSEN et al. 1995). Exposure to inflammatory cytokines was found to induce an accumulation of iNOS mRNA in insulinoma cells and rodent β cells (CORBETT et al. 1992b, 1994; EIZIRIK et al. 1992). The cytokine IL-1β, which may be released from activated macrophages within the islet (ARNUSH et al. 1998), appears to be the most potent inducer of NO formation, but its stimulatory effect can be enhanced in combination with TNFα and IFNγ (CETKOVIC-CVRLJE and EIZIRIK 1994). This potentiating effect can be explained by the results of a detailed analysis of the iNOS promoter in cells of the rat insulinoma RINm5F and in primary rat β cells, which revealed the presence of IL-1β-, TNFα-, and IFNγ-responsive elements located in a proximal and a distal region of the promoter (DARVILLE and EIZIRIK 1998). IL-1β alone is able to induce cyclic guanosine monophosphate formation (CORBETT et al. 1992a) and to inhibit glucose-stimulated insulin secretion (CORBETT et al. 1991, 1993b). In this context, it may be noted that low concentrations of inflammatory cytokines appear to stimulate insulin secretion from β cells (CORBETT et al. 1993b; EIZIRIK et al. 1993). Interestingly, transgenic mice that overexpress iNOS selectively in the pancreatic β cells develop hypoinsulinemia and diabetes. The pancreatic islets of these animals show a marked reduction of the β-cell mass but no signs of immune cell infiltration (TAKAMURA et al. 1998).

In summary, the presently available data from in vitro and in vivo studies suggest that, during islet inflammation, NO is released from several cellular sources. The release of high amounts of NO from islet-invading macrophages and from islet endothelial cells is able to exert immediate, strong toxicity on the β cells from the outside, while cytokine-induced NO formation within the β cell has a detrimental effect on the NO-producing cell itself.

II. Primary Target Structures of NO in the β Cell

High amounts of NO, as are released during islet inflammation, are able to impair a multitude of structures critical for the maintenance of vital functions and the survival of the β cell. NO is able to inhibit SH-dependent enzymes and proteins containing heme groups or Fe–S clusters. Furthermore, NO was found to disrupt zinc-finger domains of transcription factors and to mediate DNA-strand breaks (KRÖNCKE et al. 1997). Based on these findings, the search for NO-sensitive target structures in the β cell focused on mitochondrial enzyme systems and on the DNA as vital cell components.

1. Mitochondria

Recent research revealed that mitochondrial DNA, enzymes involved in glucose catabolism, and the components of the mitochondrial respiratory chain are highly sensitive to NO-induced damage (DELANEY and EIZIRIK 1996; KRÖNCKE et al. 1997). In isolated pancreatic islets, IL-1β-induced NO production caused metabolic dysfunctions by inhibiting the activity of the citric-acid-cycle enzyme aconitase, which contains an Fe–S cluster in its catalytic site (WELSH et al. 1991; WELSH and SANDLER 1992). In the respiratory chain, cytochrome oxidase (complex IV) seems to possess an increased sensitivity to NO damage due to its heme group, which represents a target for NO attacks (KRÖNCKE et al. 1997). More recent investigations on NO-induced mitochondrial damage showed that the mitochondrial DNA of the β cell is also a sensitive target for NO generated exogenously by a chemical NO donor or intracellularly by IL-1β exposure (WILSON et al. 1997). Interestingly, this study even described an increased vulnerability of mitochondrial DNA compared with nuclear DNA.

2. Nuclear DNA

The sensitivity of the nuclear DNA of β cells was studied in cultivated rat islet cells or insulinoma cells. NO generated exogenously by activated macrophages or by a chemical NO donor induced DNA-strand breaks in rat islet cells within 1h of exposure (FEHSEL et al. 1993). In islets, significant DNA damage could be observed already 15min after the induction of endogenous NO formation by IL-1β (DELANEY et al. 1993). The DNA-damaging effects observed in the presence of macrophages or IL-1β could be prevented by the iNOS inhibitor L-NMMA. Combination of the cytokines IL-1β, TNFα, and IFNγ also induced rapid DNA fragmentation in isolated islets and insulinoma cells (RABINOVITCH et al. 1994; DUNGER et al. 1996b). DNA damage was found to precede cell lysis (RABINOVITCH et al. 1994). Interestingly, in non-islet cell lines, the cytokines induced cell death, but DNA fragmentation was not significant or occurred only after cell lysis (RABINOVITCH et al. 1994).

The in vitro studies to identify primary NO targets demonstrated that NO is able to cause considerable damage to vital structures of the β cell. Interestingly, the studies further showed that the initial damage precedes islet cell death for several hours.

III. Pathways of NO-Induced β-Cell Death

The observation that primary damage to mitochondrial enzymes or DNA does not result in the immediate complete breakdown of the cell functions but precedes cell death for several hours strongly indicates that initial NO-mediated damage triggers further processes that finally lead to the death of the β cells. Recent research revealed experimental evidence for several pathways possibly involved in β-cell death triggered by primary NO damage.

1. Mitochondrial Damage

NO, which may be released during pancreatic islet inflammation, can induce severe mitochondrial dysfunction in the β cell; this may cause or accelerate cell death. NO-mediated damage of the Fe–S cluster in the citric-acid-cycle enzyme aconitase seems to be responsible for the reduced glucose oxidation rate and decreased adenosine triphosphate (ATP) production observed in IL-1β-exposed islets (SANDLER et al. 1991). As a consequence, these islets exhibit a defective insulin release upon glucose stimulation. In the presence of iNOS inhibitors, the suppressive effects of IL-1β were reversed (SOUTHERN et al. 1990; WELSH et al. 1991). Furthermore, since NO is able to inhibit heme-containing enzymes, it could be expected that the radical interacts with the heme-binding site of cytochrome oxidase (complex IV) of the mitochondrial respiratory chain, thereby impairing electron transfer and ATP generation (KRÖNCKE et al. 1997). In fact, exposure of isolated rat islet cells to exogenously generated NO resulted in a strong decrease of the respiratory activity (BURKART et al. 1996). In addition to its enzyme-inhibiting effects, NO is able to damage mitochondrial DNA. Islet cells exposed to NO, generated exogenously by an NO-releasing compound or endogenously by IL-1β treatment, showed considerable damage to the mitochondrial DNA. IL-1β-induced DNA fragmentation could be reduced by incubation of the cells in the presence of aminoguanidine (AG) (WILSON et al. 1997). Taken together, these findings show that NO is able to damage various target structures in the mitochondria; this results in an impairment of basic mitochondrial functions followed by a critical reduction of the cellular energy supply.

2. Apoptotic Pathway

As mentioned above (Sect. C.II.2), an early event in NO-mediated islet cell death is the induction of nuclear DNA damage. Severe DNA fragmentation was detected in isolated rat islet cells (KANETO et al. 1995) and in cells of the pancreatic β-cell line HIT (LOWETH et al. 1997) after exposure to NO, which was generated either exogenously by chemical NO donors or endogenously by IL-1β. In cells of the rat insulinoma line RINm5F (ANKARKRONA et al. 1994) and in rat islets (DUNGER et al. 1996a) IL-1β stimulated NO production, which correlated with the appearance of DNA fragments. The iNOS inhibitor N^G-nitro-L-arginine-methyl ester (L-NAME) inhibited IL-β-induced NO formation and DNA fragmentation. Since the DNA fragments detected in these studies resemble the cleavage patterns found in apoptotic cells, the observations gave rise to the assumption that NO induces apoptosis in islet cells. However, it is rather difficult to clearly discriminate apoptosis from other cell-death pathways (COLUMBANO 1995) using the single parameter of DNA fragmentation. Unless further criteria, such as morphological or biochemical alterations of nuclear structures, are included in the investigations, a reliable decision on the type of NO-induced islet cell death will hardly be possible.

Further approaches to investigate a possible role of NO-induced apoptosis in islet cells in vivo were performed in animal models and in patients with type 1 diabetes. At present, the most conclusive indications for the involvement of NO in the induction of apoptotic processes in vivo come from studies in islets of diabetic patients. In healthy individuals, pancreatic β cells were not found to express the cell-surface receptor Fas (Apo-1, CD95), which is involved in the transduction of a signal leading to apoptosis. In contrast, β cells from patients with newly diagnosed type 1 diabetes express Fas and show a combination of apoptosis-associated features, such as typical alterations of the nuclear morphology and a specific DNA-fragmentation pattern (STASSI et al. 1997). More detailed in vitro analysis implied a role for NO in the induction of human β-cell apoptosis, since functional Fas expression and apoptosis could be induced in isolated islet cells by exposure to IL-1β and to chemical NO donors. Furthermore, IL-1β-induced Fas expression was suppressed by the iNOS inhibitor L-NMMA in vitro (STASSI et al. 1997).

3. Poly(Adenosine Diphosphate–Ribose)Polymerase-Dependent Pathway

Several years ago, a hypothesis of radical-induced β-cell death was proposed, which described the excessive activation of the enzyme poly(adenosine diphosphate–ribose)polymerase (PARP) as a critical step in a chain of events leading to β-cell death (YAMAMOTO et al. 1981). According to this model, the exposure of β cells to reactive radicals results in the induction of widespread damage of nuclear DNA. The appearance of DNA-strand breaks immediately triggers the activation of the abundant nuclear enzyme PARP as an initial step of the DNA-repair sequence. Activated PARP synthesizes large polymers of adenosine diphosphate (ADP)–ribose from its substrate, nicotinamide adenine dinucleotide ion (NAD^+). The rapid depletion of intracellular NAD^+ pools finally leads to cell death.

Further in vitro analysis demonstrated that NO represents an effective trigger of the PARP-dependent pathway of β-cell death in rodent islet cells. Primary NO-induced DNA damage resulted in a fast activation of PARP, as assessed by the rapid appearance of (ADP–ribose) polymers in the nuclei of isolated rodent islet cells (RADONS et al. 1994). As a consequence, a strong decline in intracellular NAD$^+$ levels could be detected. NAD$^+$-depleted cells finally died, as judged by the loss of their membrane integrity, a clear sign of irreversible cell damage (RADONS et al. 1994). Recent studies with islet cells isolated from mice lacking the PARP gene confirmed these findings. Compared with islet cells with normal PARP expression, the PARP-deficient islet cells showed an increased resistance against lysis induced by low doses of chemically generated NO (HELLER et al. 1995). However, at higher doses of NO, both cell types were lysed to the same high degree, indicating the activation of an alternative, PARP-independent pathway of cell death under these conditions.

The PARP-dependent pathway of β-cell death may also be involved in the development of spontaneous diabetes in the NOD mouse (YAMADA et al. 1982) and the induction of diabetes in animals treated with the chemicals alloxan or SZ. These substances are known to release β-cell-toxic radicals (UCHIGATA et al. 1982; TURK et al. 1993; KRÖNCKE et al. 1995), which may trigger PARP activation by inducing primary DNA damage. The presently available data from the in vitro and in vivo studies clearly indicate an important role for PARP in the death of rodent β cells. In humans, however, experimental evidence for the PARP dependent pathway of β-cell death is still lacking.

D. Open Issues

The studies outlined in Sect. C clearly show that NO-mediated primary damage to β-cell structures acts as a highly effective trigger of cell-death pathways, which may be important in the development of autoimmune diabetes. However, in the research on the role of NO in β-cell-death pathways, several issues remain to be clarified; e.g., since mitochondria have recently been identified to play an important role in the induction of the apoptotic pathway (KROEMER et al. 1997), the possible contribution of NO-mediated damage to β-cell mitochondria remains to be elucidated. Furthermore, the characterization of the PARP-independent pathway of β-cell death, which dominates after exposure to high doses of NO, may also yield important information about alternative pathways of radical-induced β-cell death. A further issue with potential implications for the understanding of the pathogenesis of human type 1 diabetes may arise from the observation of considerable species differences in the NO sensitivity of β cells and in the signals required for the induction of β-cell iNOS (EIZIRIK et al. 1994a, 1994b). A growing number of studies show that rodent islets or islet cells exhibit a higher sensitivity to NO-mediated toxicity when compared with human islets. In rat and mouse islets, chemically generated NO reduced the glucose oxidation rate, the glucose-stimulated insulin release, and viability; under the same conditions, the functions and viability of human islets remained unaffected (EIZIRIK et al. 1994a).

However, after prolonged exposure to elevated NO concentrations, human pancreatic islets also show signs of nuclear DNA fragmentation and cell death (EIZIRIK et al. 1996a). The comparison of human and rodent β cells further revealed differences in the signal requirement for the induction of iNOS (EIZIRIK et al. 1996b). Whereas IL-1β alone generally suffices to induce transcription of iNOS mRNA in rat islets, iNOS mRNA expression in human islets obviously requires a combination of the cytokines IL-1β, IFNγ, and TNFα (EIZIRIK et al. 1992). These findings show that, despite considerable progress in the last years, a number of partially unsolved and controversial issues remain to be clarified in order to complete our understanding of NO-induced pathways of β-cell death (EIZIRIK and PAVLOVIC 1997).

E. Strategies to Protect Islets Cells from NO-Induced Damage

The increasing knowledge about the effector mechanisms and the pathways leading to β-cell death allowed the development of strategies to protect the insulin-producing cells from autoimmune-mediated damage, with the aim of preventing the development of diabetes. In recent years, basically two approaches were performed to protect islet cells from NO-induced damage: the suppression of NO formation and the reduction or prevention of the damaging effects of NO in the target cell.

I. Suppression of NO Formation

In vitro studies designed to suppress the formation of β-cell-toxic NO revealed a strong protection of pharmacological iNOS inhibitors on islet cells exposed to NO-generating effector cells or to cytokines inducing NO in the β cell itself. Isolated rat islet cells were completely protected from the lysis induced by syngeneic activated macrophages when the co-incubation was performed in the presence of the iNOS inhibitor L-NMMA (BURKART and KOLB 1993). In co-cultivation studies of islet cells and cytokine-stimulated endothelial cells isolated from rat islet capillaries, the addition of L-NMMA resulted in an improved survival of islet cells (STEINER et al. 1997). Furthermore, the cytotoxic activity of IL-1β against cultured whole pancreatic islets of the rat, which is mediated via the formation of NO, could be inhibited by L-NMMA (BERGMANN et al. 1992). Inhibition of endogenous NO formation was also found to protect β cells from functional disturbances and structural damage. In cultivated β cells, the IL-1β-induced inhibition of glucose-stimulated insulin release could be prevented by L-NMMA (SOUTHERN et al. 1990; CORBETT et al. 1991, 1992b).

Consequently, several attempts were performed to prevent the development of diabetes in animals by administration of iNOS inhibitors (HOLSTAD et al. 1997). In a model of low-dose SZ-induced diabetes of the mouse, the administration of L-NMMA, L-NAME, or AG showed no or only moderate protective effects, including a slight reduction of islet inflammation and a delay or a partial suppression of the development of hyperglycemia (KOLB et al. 1991; LUKIC et al. 1991; HOLSTAD and SANDLER 1993; PAPACCIO et al. 1995; KARABATAS et al. 1996). The administration of AG to diabetes-prone NOD mice had no effect on diabetes incidence (BOWMAN et al. 1996) and caused merely a delay but not a suppression of autoimmune diabetes transferred by spleen cells from acutely diabetic NOD mice (CORBETT et al. 1993a). Furthermore, introduction of a disrupted iNOS gene into the NOD mouse strain had no effect on the development of diabetes in these animals (Mudgett, personal communication). In spontaneously diabetic BB rats, the administration of L-NMMA, L-NAME or AG resulted in a reduction of diabetes incidence and a delay of disease onset (LINDSAY et al. 1995; WU 1995). The inhibition of NO production by application of L-NMMA was also found to protect transplanted rat and mouse islets from early dysfunction. This finding indicates that early dysfunction of transplanted rodent islets is also mediated via the release of NO (STEVENS et al. 1996). Inhibition of NO formation by AG had a similar protective effect and prolonged islet xenograft survival in rats (BEHBOO et al. 1997).

II. Improvement of β-Cell Defense Mechanisms

Pancreatic β cells are known to be highly sensitive to the damaging effects of reactive radicals; this may in part be explained by their reduced capacity to scavenge these compounds. Therefore, several protective strategies were developed with the aim of protecting β cells from NO-induced damage by improving their defense mechanisms.

In an attempt to improve the weak radical-scavenger potential found in β cells (MALAISSE et al. 1982), isolated rat islet cells were incubated in the presence of vitamin E and then exposed to chemically generated NO. The antioxidant vitamin partially protected the cells from lysis. It was found to interfere at early steps of NO-induced damage by preventing the occurrence of DNA-strand breaks (BURKART et al. 1995). In another approach aiming at the improvement of the low expression of antioxidant enzymes (LENZEN et al. 1996), it was shown that stable overexpression of manganese superoxide dismutase in insulinoma cell lines reduced NO formation and cytotoxicity during exposure to IL-1β (HOHMEIER et al. 1998).

A highly efficient protective response of many cell types to various stress conditions is the increased expression of heat-shock proteins (HSPs). Among the HSPs, the upregulation of the inducible form HSP70 is most prominent in β cells of rodents, whereas human β cells spontaneously express high levels of the protein under normal conditions, which may explain their increased resistance to the damaging effects of inflammatory mediators (EIZIRIK et al. 1994a; EIZIRIK 1996). The potential role of HSPs in β-cell protection against NO was investigated in isolated islets and in cells of a genetically modified rat β-cell line. In both experimental systems, increased expression of HSP70 conferred strong protection against NO induced toxicity (BELLMANN et al. 1995, 1996). Furthermore, fusion of islet cells with HSP70-containing liposomes transiently protected the cells from the depressive effect of IL-1β on glucose-induced insulin secretion (MARGULIS et al. 1991). Interestingly, when compared with islets of normal rats, islets from diabetes-prone BB rats have a reduced capacity to increase HSP70 expression under heat stress and are more vulnerable to NO-induced injury (BELLMANN et al. 1997). These observations show that the increase of HSP70 may be a promising approach to protection of β cells against NO-induced damage.

III. Inhibition of the PARP-Dependent Pathway

To protect islet cells from NO-mediated lysis, macrophage–islet cell co-cultures were performed in the presence of the PARP inhibitor nicotinamide (NA). The substance was found to reduce islet cell lysis in a dose-dependent manner and, at higher concentrations, provided complete protection from macrophage-mediated killing (KOLB et al. 1990), obviously via preservation of the intracellular NAD^+ pool. Indeed, in isolated islet cells exposed to chemically generated NO, the PARP inhibitors NA and 3-aminobenzamide completely prevented the formation of (ADP–ribose) polymers and the decline of intracellular NAD^+ (RADONS et al. 1994). Finally, PARP inhibition protected islet cells from NO-induced death without preventing primary DNA damage (RADONS et al. 1994).

Based on the protective effect of PARP inhibition in vitro, NA was also used in animal models of human type 1 diabetes. A diabetes-protective effect was observed in NOD mice (YAMADA et al. 1982) and mice with diabetes induced by multiple injections of low doses of SZ (ROSSINI et al. 1977). Studies in the BB rat showed no clear protective effects of NA (HERMITTE et al. 1989; KOLB et al. 1989; SARRI et al. 1989).

IV. Regulation of Th1/Th2 Balance in Islet Inflammation

As mentioned above (Sect. B.III), the initiation of the destructive phase of insulitis is characterized by a shift from the Th2-type peri-insulitis to the Th1-type intra-insulitis. The shift in the Th subset is generally known to be driven by an increased expression of Th1 cytokines, such as IFNγ and TNF α. The expression of iNOS is closely associated with Th1-type immune responses but, at the same time, is part of a feedback-inhibition circuit limiting Th1 responses (KOLB and KOLB-BACHOFEN 1998). Studies in the model of cyclophosphamide-accelerated diabetes of the NOD mouse revealed an early up-regulation of iNOS in the pancreas shortly before diabetes manifestation (ROTHE et al. 1994). Increased pancreatic iNOS expression correlated with up-regulation of the Th1-specific cytokines IFNγ, TNF α and IL-12 (ROTHE et al. 1996a).

However, NO was reported to inhibit Th1 cell reactivity in mice (TAYLOR-ROBINSON et al. 1994) and to induce expression of the IL-12p40 but not the IL-12p35 gene (ROTHE et al. 1996b). The homodimer of the p40 subunit of IL-12 [IL-12(p40)$_2$] is known to counteract the Th1-promoting activity of the IL-12 heterodimer (MATTNER et al. 1993). Daily injections of IL-12(p40)$_2$ starting in the prediabetic phase were found to suppress destructive insulitis and significantly reduced diabetes incidence in NOD mice (ROTHE et al. 1997). Therefore, it was concluded that NO from iNOS may dampen Th1 responses during islet inflammation.

Further studies of the pathogenetic role of NO in type 1 diabetes will have to consider the immunoregulatory dimension of NO action. A complex picture is about to emerge in which damaging or immunoregulatory roles of NO will prevail, depending on the disease state and genetic background.

F. Concluding Remarks

Numerous studies on the pathogenesis of human type 1 diabetes have revealed a complex role for NO as a β-cell-toxic and immunmodulatory mediator during the process of islet inflammation. The identification of the cellular sources of NO and the increasing knowledge of potential target structures and pathways of β-cell death allow the development of procedures designed to protect β cells from NO-induced damage with the aim of preventing the development of type 1 diabetes.

Abbreviations

AG	aminoguanidine
ATP	adenosine triphosphate
BB rat	BioBreeding rat
HSP	heat-shock protein
IFNγ	interferon γ
IL	interleukin
iNOS	inducible nitric oxide synthase
L-NAME	N^G-nitro-L-arginine-methyl ester
L-NMNA	N-monomethyl-L-arginine
NA	nicotinamide
NAD$^+$	nicotinamide adenine dinucleotide ion
NAME	N^G-nitro-L-arginine-methyl ester
NO	nitric oxide
NOD mouse	non-obese diabetic mouse
PARP	poly(adenosine diphosphate–ribose)polymerase
SZ	streptozotocin
Th	T-helper lymphocyte
TNF α	tumor necrosis factor α

References

Ankarcrona M, Dypbukt JM, Brune B, Nicotera P (1994) Interleukin-1β-induced nitric oxide production activates apoptosis in pancreatic RINm5F cells. Exp Cell Res 213:172–177

Appels B, Burkart V, Kantwerk-Funke G, Funda J, Kolb-Bachofen V, Kolb H (1989) Spontaneous cytotoxicity of macrophages against pancreatic islet cells. J Immunol 142:3803–3808

Arnush M, Scarim AL, Heitmeier MR, Kelly CB, Corbett JA (1998) Potential role of resident islet macrophage activation in the initiation of autoimmune diabetes. J Immunol 160:2684–2691

Atkinson MA, MacLaren NK (1994) The pathogenesis of insulin dependent diabetes. N Engl J Med 331:1428–1436

Atkinson MA, MacLaren NK, Riley WJ, Winter WE, Fisk DD, Spillar RP (1986) Are insulin autoantibodies markers for insulin-dependent diabetes mellitus? Diabetes 35:894–898

Behboo R, Ricordi C, Lumachi F, Tedeschi U, Urso E, Cillo U, Bonariol L, Favia G, D'Amico DF (1997) Aminoguanidine inhibits the generation of nitric oxide and prolongs islet xenograft survival in rats. Transplant Proc 29:2152–2153

Bellmann K, Wenz A, Radons J, Burkart V, Kleemann R, Kolb H (1995) Heat shock induces resistance in rat pancreatic islet cells against nitric oxide, oxygen radicals and streptozotocin toxicity in vitro. J Clin Invest 95:2840–2845

Bellmann K, Jäättelä M, Wissing D, Burkart V, Kolb H (1996) Heat shock protein 70 overexpression confers resistance against nitric oxide. FEBS Lett 391:185–186

Bellmann K, Liu H, Radons J, Burkart V, Kolb H (1997) Low stress response enhances vulnerability of islet cells in diabetes-prone BB rats. Diabetes 46:232–236

Bergmann L, Kröncke K-D, Kolb H, Kolb-Bachofen V (1992) Cytotoxic action of IL-1β against pancreatic islets is mediated via nitric oxide formation and is inhibited by N^G-monomethyl-L-arginine. FEBS Lett 299:103–106

Bowman MA, Simell OG, Peck AB, Cornelius J, Luchetta R, Look Z, MacLaren NK, Atkinson MA (1996) Pharmacokinetics of aminoguanidine administration and effects on the diabetes frequency in nonobese diabetic mice. J Pharmacol Exp Ther 279:790–794

Burkart V, Kolb H (1993) Protection of islet cells from inflammatory cell death in vitro. Clin Exp Immunol 93:273–278

Burkart V, Kolb H (1996) Macrophages in islet destruction in autoimmune diabetes mellitus. Immunobiol 195:601–613

Burkart V, Kröncke K-D, Kolb-Bachofen V, Kolb H (1994) Nitric oxide as an inflammatory mediator in insulin-dependent diabetes mellitus. A new therapeutic target? Clin Immunother 2:233–239

Burkart V, Gross-Eick A, Bellmann K, Radons J, Kolb H (1995) Suppression of nitric oxide toxicity in islet cells by α-tocopherol. FEBS Lett 364:259–263

Burkart V, Brenner H-H, Hartmann B, Kolb H (1996) Metabolic activation of islet cells improves resistance against oxygen radicals or streptozotocin, but not nitric oxide. J Clin Endocrinol Metab 81:3966–3971

Cantor AB, Krischer JP, Cuthbertson DD, Schatz DA, Riley WJ, Malone J, Schwartz S, Quattrin T, MacLaren NK (1995) Age and family relationship accentuate the risk of IDDM in relatives of patients with insulin dependent diabetes. J Clin Endocrinol Metabol 80:3739–3743

Cetkovic-Cvrlje M, Eizirik DL (1994) TNF-α and IFN-γ potentiate the deleterious effects of IL-1β on mouse pancreatic islets mainly via generation of nitric oxide. Cytokine 6:399–406

Columbano A (1995) Cell death: current difficulties in discriminating apoptosis from necrosis in the context of pathological processes in vivo. J Cell Biochem 58:2239–2244

Corbett JA, McDaniel ML (1992) Does nitric oxide mediate autoimmune destruction of β-cells? Possible therapeutic interventions in IDDM. Diabetes 41:897–903

Corbett JA, Lancaster JR, Sweetland MA, McDaniel ML (1991) Interleukin-1β-induced formation of EPR-detectable iron-nitrosyl-complexes in islets of Langerhans. J Biol Chem 266:21351–21354

Corbett JA, Wang JL, Hughes JH, Wolf BA, Sweetland MA, Lancaster JR, McDaniel ML (1992a) Nitric oxide and cyclic GMP formation induced by interleukin 1β in islets of Langerhans. Evidence for an effector role of nitric oxide in islet dysfunction. Biochem J 287:229–235

Corbett JA, Wang JL, Sweetland MA, Lancaster JR, McDaniel ML (1992b) Interleukin 1β induces the formation of nitric oxide by β-cells purified from rodent islets of Langerhans. J Clin Invest 90:2384–2391

Corbett JA, Mikhael A, Shimizu J, Frederick K, Misko TP, McDaniel ML, Kanagawa O, Unanue ER (1993a) Nitric oxide production in islets from nonobese diabetic mice: aminoguanidine-sensitive and -resistant stages in the immunological diabetic process. Proc Natl Acad Sci USA 90:8992–8995

Corbett JA, Sweetland MA, Wang JL, Lancaster JRj, McDaniel ML (1993b) Nitric oxide mediates cytokine-induced inhibition of insulin secretion by human islets of Langerhans. Proc Natl Acad Sci USA 90:1731–1735

Corbett JA, Kwon G, Misko TP, Rodi CP, McDaniel ML (1994) Tyrosine kinase involvement in IL-1b-induced expression of iNOS by β-cells purified from islet of Langerhans. Am J Physiol 267:C48-C54

Darville MI, Eizirik DL (1998) Regulation by cytokines of the inducible nitric oxide synthase promoter in insulin-producing cells. Diabetologia 41:1101–1108

Delaney CA, Eizirik DL (1996) Intracellular targets for nitric oxide toxicity to pancreatic β cells. Braz J Med Biol Res 29:569–579

Delaney CA, Green MHL, Green IC (1993) Endogenous nitric oxide induced by interleukin-1β in rat islets of Langerhans and HIT-T15 cells causes significant DNA damage as measured by the "comet" assay. FEBS Lett 333:291–295

Dunger A, Augstein P, Schmidt S, Fischer U (1996a) Identification of interleukin 1-induced apoptosis in rat islets using in situ specific labelling of fragmented DNA. J Autoimmunity 9:309–313

Dunger A, Cunningham JM, Delaney CA, Lowe JE, Green MH, Bone AJ, Green IC (1996b) Tumor necrosis factor-alpha and interferon-gamma inhibit insulin secretion and cause DNA damage in unweaned-rat islets. Extent of nitric oxide involvement. Diabetes 45:183–189

Eisenbarth GS (1986) Type I diabetes mellitus: a chronic autoimmune disease. New Engl J Med 314:1360–1368

Eizirik DL (1996) β cell defense and repair mechanisms in human pancreatic islets. Horm Metab Res 28:302–305

Eizirik DL, Pavlovic D (1997) Is there a role for nitric oxide in β-cell dysfunction and damage in IDDM? Diab Metabol Rev 13:293–307

Eizirik DL, Cagliero E, Bjorklund A, Welsh M (1992) Interleukin-1β induces the expression of an isoform of nitric oxide synthase in insulin-producing cells, which is similar to that observed in activated macrophages. FEBS Lett 308:249–252

Eizirik DL, Welsh N, Hellerström C (1993) Predominance of stimulatory effects of interleukin-1β on isolated human pancreatic islets. J Clin Endocrinol Metab 76:399–403

Eizirik DL, Pipeleers DG, Zhidong L, Welsh N, Hellerström C, Andersson A (1994a) Major species differences between humans and rodents in the susceptibility to pancreatic β-cell injury. Proc Natl Acad Sci USA 91:9253–9256

Eizirik DL, Sandler S, Welsh N, Cetkovic-Cvrlje M, Nieman A, Geller DA, Pipeleers DG, Bendtzen K, Hellerström C (1994b) Cytokines suppress human islet function irrespective of their effects on nitric oxide generation. J Clin Invest 93:1968–1974

Eizirik DL, Delaney CA, Green MH, Cunningham JM, Thorpe JR, Pipeleers DG, Hellerström C, Green IC (1996a) Nitric oxide donors decrease the function and survival of human pancreatic islets. Mol Cell Endocrinol 118:71–83

Eizirik DL, Flodström M, Karlsen AE, Welsh N (1996b) The harmony of the spheres: inducible nitric oxide synthase and related genes in pancreatic β cells. Diabetologia 39:875–890

Expert Committee on the Diagnosis and Classification of Diabetes Mellitus (1997) Report of the expert committee on the diagnosis and classification of diabetes mellitus. Diab Care 20:1183–1197

Fehsel K, Jalowy A, Sun Q, Burkart V, Hartmann B, Kolb H (1993) Islet cell DNA is a target of inflammatory attack by nitric oxide. Diabetes 42:496–500

Foulis AK, Liddle CN, Farquharson MA, Richmond JA, Weir RS (1986) The histopathology of the pancreas in type-1 (insulin-dependent) diabetes mellitus: a 25-yr review of deaths in patients under 20 years of age in the United Kingdom. Diabetologia 29:267–274

Gepts W (1965) Pathological anatomy of the pancreas in juvenile diabetes. Diabetes 14:619–633

Hanenberg H, Kolb-Bachofen V, Kantwerk-Funke G, Kolb H (1989) Macrophage infiltration precedes and is a prerequisite for lymphocytic insulitis in pancreatic islets of prediabetic BB rats. Diabetologia 32:126–134

Heller B, Wang Z-Q, Wagner EF, Radons J, Bürkle A, Fehsel K, Burkart V, Kolb H (1995) Inactivation of the poly(ADP–ribose)polymerase gene affects oxygen radical and nitric oxide toxicity in islet cells. J Biol Chem 270:11176–11180

Hermitte L, Vialettes B, Atlef N, Payan MJ, Doll N, Scheinmann A, Vague P (1989) High dose nicotinamide fails to prevent diabetes in BB rats. Autoimmunity 5:79–86

Hohmeier H-E, Thigpen A, Tran VV, Davis R, Newgard CB (1998) Stable expression of manganese superoxide dismutase (MnSOD) in insulinoma cells prevents IL-1β-induced cytotoxicity and reduces nitric oxide production. J Clin Invest 101:1811–1820

Holstad M, Sandler S (1993) Aminoguanidine, an inhibitor of nitric oxide formation, fails to protect against insulitis and hyperglycemia induced by multiple low dose streptozotocin injections in mice. Autoimmunity 15:311–314

Holstad M, Jansson L, Sandler S (1997) Inhibition of nitric oxide formation by aminoguanidine: an attempt to prevent insulin-dependent diabetes mellitus. Gen Pharmacol 29:697–700

Itoh N, Hanafusa T, Miyazaki A, Miyagawa J, Yamagata K, Yamamoto K, Waguri M, Imagawa A, Tamura S, Inada M, Kawata S, Tarui S, Kono N, Matsuzawa Y (1993) Mononuclear cell infiltration and its relation to the expression of major histocompatibility complex antigens and adhesion molecules in pancreas biopsy specimens from newly diagnosed insulin dependent diabetes mellitus patients. J Clin Invest 92:2313–2322

Kaneto H, Fujii J, Seo HG, Suzuki K, Matsuoka T, Nakamura M, Tatsumi H, Yamasaki Y, Kamada T, Taniguchi N (1995) Apoptotic cell death triggered by nitric oxide in pancreatic β cells. Diabetes 44:733–738

Karabatas LM, Fabiano-de-Bruno L, Pastorale CF, Cullen C, Basabe JC (1996) Inhibition of nitric oxide generation: normalization of in vitro insulin secretion in mice with multiple low-dose streptozotocin diabetes and in mice injected with mononuclear splenocytes from diabetic syngeneic donors. Metabolism 45:940–946

Karlsen AE, Andersen HU, Vissing H, Larsen PM, Fey SJ, Cuartero BG, Madsen O, Petersen JS, Mortensen SB, Mandrup-Poulsen T, Boel E, Nerup J (1995) Cloning and expression of cytokine-inducible nitric oxide synthase cDNA from rat islets of Langerhans. Diabetes 44:753–758

Karvonen M, Tuomilehto J, Libman I, LaPorte R (1993) A review of the recent epidemiological data on the worldwide incidence of type-1 (insulin-dependent) diabetes mellitus. Diabetologia 36:883–892

Kaufmann D, Erlander M, Clare-Salzler M, Atkinson M, MacLaren NK, Tobin A (1992) Autoimmunity to two forms of glutamate decarboxylase in insulin-dependent diabetes mellitus. J Clin Invest 89:283–292

Kleemann R, Rothe H, Kolb-Bachofen V, Xie Q, Nathan C, Martin S, Kolb H (1993) Transcription and translation of inducible nitric oxide synthase in the pancreas of prediabetic BB rats. FEBS Lett 328:9–12

Kolb H (1987) Mouse models of insulin-dependent diabetes: low dose streptozotocin-induced diabetes and non-obese diabetic (NOD) mice. Diabetes Metab Rev 3:751–778

Kolb H (1997) Benign versus destructive insulitis. Diabetes Metab Rev 13:139–146

Kolb H, Kolb-Bachofen V (1992) Nitric oxide: a pathogenetic factor in autoimmunity. Immunol Today 13:157–160

Kolb H, Kolb-Bachofen V (1998) Nitric oxide in autoimmune disease: cytotoxic or regulatory mediator? Immunol Today 19:556–561

Kolb H, Schmidt M, Kiesel U. (1989) Immunomodulatory drugs in type-1 diabetes. In: Eisenbarth GS (ed) Immunotherapy of type-1 diabetes and selected autoimmune diseases. CRC, Boca Raton, FL, p 111

Kolb H, Burkart V, Appels B, Hanenberg H, Kantwerk-Funke G, Kiesel U, Funda J, Schraermeyer U, Kolb-Bachofen V (1990) Essential contribution of macrophages to islet cell destruction in vivo and in vitro. J Autoimmunity 3(Suppl.):117–120

Kolb H, Kiesel U, Kröncke K-D, Kolb-Bachofen V (1991) Suppression of low dose streptozotocin induced diabetes in mice by administration of a nitric oxide synthase inhibitor. Life Sci 25:PL213–PL217

Kolb-Bachofen V (1996) Intraislet expression of inducible nitric oxide synthase and islet cell death. Biochem Soc Trans 24:233–234

Kolb-Bachofen V, Epstein S, Kiesel U, Kolb H (1988) Low dose streptozotocin-induced diabetes in mice. Electron microscopy reveals single-cell insulitis before diabetes onset. Diabetes 37:21–27

Kroemer G, Zamzami N, Susin SA (1997) Mitochondrial control of apoptosis. Immunol Today 18:44–51

Kröncke K-D, Kolb-Bachofen V, Berschick B, Burkart V, Kolb H (1991) Activated macrophages kill pancreatic syngeneic islet cells via arginine-dependent nitric oxide generation. Biochem Biophys Res Commun 175:752–758

Kröncke K-D, Fehsel K, Sommer A, Rodriguez M-L, Kolb-Bachofen V (1995) Nitric oxide generation during cellular metabolization of the diabetogenic N-methyl-N-nitroso-urea streptozotocin contributes to islet cell DNA damage. Biol Chem Hoppe-Seyler 376:179–185

Kröncke K-D, Fehsel K, Kolb-Bachofen V (1997) Nitric oxide: cytotoxicity versus cytoprotection: how, why, when and where? Nitric Oxide: Biology and Chemistry 1:107–120

Lampeter EF, McCann SR, Kolb H (1998) Transfer of diabetes type 1 by bone-marrow transplantation. Lancet 351:568–569

Lan MS, Wasserfall C, MacLaren NK, Notkins AL (1996) IA-2, a transmembrane protein of the protein tyrosine phosphatase family, is a major autoantigen in insulin-dependent diabetes mellitus. Proc Natl Acad Sci USA 93:6367–6370

Lau A, Ramanathan S, Poussier P (1998) Excessive production of nitric oxide by macrophages from DP-BB rats is secondary to the T-lymphopenic state of these animals. Diabetes 47:197–205

Lee K-U, Amano K, Yoon J-W (1988) Evidence for initial involvement of macrophage in development of insulitis in NOD mice. Diabetes 37:989–991

Lenzen S, Drinkgern J, Tiedge M (1996) Low antioxidant enzyme gene expression in pancreatic islets compared with various other mouse tissues. Free Rad Biol Med 20:463–466

Lindsay RM, Smith W, Rossiter SP, McIntyre MA, Williams BC, Baird JD (1995) N^{ω}-nitro-L-arginine methyl ester reduces the incidence of IDDM in BB/E rats. Diabetes 44:365–368

Loweth AC, Williams GT, Scarpello JH, Morgan NG (1997) Evidence for the involvement of cGMP and protein kinase G in nitric oxide-induced apoptosis in the pancreatic B-cell line, HIT-T15. FEBS Lett 400:285–288

Lukic ML, Stosic-Grujicic S, Ostojic N, Chan WL, Liew FY (1991) Inhibition of nitric oxide generation affects the induction of diabetes by streptozotocin in mice. Biochem Biophys Res Commun 178:913–920

Malaisse WJ, Malaisse-Lagae F, Sener A, Pipeleers DG (1982) Determinants of the selective toxicity of alloxan to the pancreatic β cell. Proc Natl Acad Sci USA 79:927–930

Margulis BA, Sandler S, Eizirik DL, Welsh N, Welsh M (1991) Liposomal delivery of purified heat shock protein 70 into rat pancreatic islets as protection against interleukin 1β-induced impaired β cell function. Diabetes 40:1418–1422

Mattner F, Fischer S, Guckes S, Jin S, Kaulen H, Rüde E, Germann T (1993) The interleukin-12 p40 subunit specifically inhibits effects of the interleukin-12 heterodimer. Eur J Immunol 23:2203–2208

McDaniel ML, Kwon G, Hill JR, Marshall CA, Corbett JA (1996) Cytokines and nitric oxide in islet inflammation and diabetes. Proc Soc Exp Biol Med 211:24–32

Mordes JP, Desemone J, Rossini AA (1987) The BB rat. Diabetes Metab Rev 3:725–750

Oschilewski U, Kiesel U, Kolb H (1985) Adminstration of silica prevents diabetes in BB rats. Diabetes 34:197–199

Papaccio G, Esposito V, Latronico MVG, Pisanti FA (1995) Administration of a nitric oxide synthase inhibitor does not suppress low-dose streptozotocin diabetes in mice. Biochem Biophys Res Commun 178:913–919

Rabinovitch A (1998) An update on cytokines in the pathogenesis of insulin-dependent diabetes mellitus. Diabetes Metab Rev 14:129–151

Rabinovitch A, Suarez-Pinzon WL, Shi Y, Morgan AR, Bleackley RC (1994) DNA fragmentation is an early event in cytokine-induced islet β cell destruction. Diabetologia 37:733–738

Rabinovitch A, Suarez-Pinzon WL, Sorensen O, Bleackley RC (1996) Inducible nitric oxide synthase (iNOS) in pancreatic islets of nonobese diabetic mice: identification of iNOS-expressing cells and relationships to cytokines expressed in the islets. Endocrinology 137:2093–2099

Radons J, Heller B, Bürkle A, Hartmann B, Rodriguez M-L, Kröncke K-D, Burkart V, Kolb H (1994) Nitric oxide toxicity in islet cells involves poly(ADP–ribose)polymerase activation and concomitant NAD$^+$ depletion. Biochem Biophys Res Commun 199:1270–1277

Reddy S, Kaill S, Poole CA, Ross J (1997) Inducible nitric oxide synthase in pancreatic islets of the non-obese diabetic mouse: a light and confocal microscopical study of its ontogeny, co-localization and up-regulation following cytokine administration. Histochem J 29:53–64

Rossini AA, Like AA, Chick WL, Appel MC, Cahill GF (1977) Studies of streptozotocin insulitis and diabetes. Proc Natl Acad Sci USA 74:2485–2489

Rothe H, Faust A, Schade U, Kleemann R, Bosse G, Hibino T, Martin S, Kolb H (1994) Acceleration of diabetes development in NOD mice by cyclophosphamide is associated with a shift from IL-4 to IFN-γ production and with enhanced expression of inducible NO-synthase in pancreatic lesions. Diabetologia 37:1154–1158

Rothe H, Burkart V, Faust A, Kolb H (1996a) Interleukin-12 gene expression is associated with rapid diabetes development in NOD mice. Diabetologia 39:119–122

Rothe H, Hartmann B, Geerlings P, Kolb H (1996b) Interleukin-12 gene expression of macrophages is regulated by nitric oxide. Biochem Biophys Res Commun. 224:159–163

Rothe H, O'Hara RM, Martin S, Kolb H (1997) Suppression of cyclophosphamide induced diabetes development and pancreatic Th1 reactivity in NOD mice treated with the interleukin (IL)-12 antagonist IL-12(p40)$_2$. Diabetologia 40:641–646

Sandler S, Eizirik DL, Svensson C, Strandell E, Welsh M, Welsh N (1991) Biochemical and molecular actions of interleukin-1 on pancreatic β-cells. Autoimmunity 10:241–253

Sarri Y, Mendola J, Ferrer J, Gomis R (1989) Preventive effects of nicotinamide administration on spontaneous diabetes of BB rats. Med Sci Res 17:987–988

Scott F, Cui J, Rowsell P (1994) Food and the development of autoimmune disease. Trends in Food Sci and Technol 5:111–116

Southern C, Schulster D, Green IG (1990) Inhibition of insulin secretion by interleukin 1β and tumour necrosis factor-α via an L-arginine-dependent nitric oxide generating mechanism. FEBS Lett 276:42–44

Stassi G, DeMaria RD, Trucco G, Rudert W, Testi R, Galluzzo A, Giordano C, Trucco M (1997) Nitric oxide primes pancreatic β cells for Fas-mediated destruction in insulin-dependent diabetes mellitus. J Exp Med 186:1193–1200

Steiner L, Kröncke K-D, Fehsel K, Kolb-Bachofen V (1997) Endothelial cells as cytotoxic effector cells: cytokine-activated rat islet endothelial cells lyse syngeneic islet cells via nitric oxide. Diabetologia 40:150–155

Stevens RB, Ansite JD, Mills CD, Lokeh A, Rossini TJ, Saxena M, Brown RR, Sutherland DER (1996) Nitric oxide mediates early dysfunction of rat and mouse islets after transplantation. Transplantation 61:1740–1749

Suschek C, Rothe H, Fehsel K, Enczmann J, Kolb-Bachofen V (1993) Induction of a macrophage-like nitric oxide synthase in cultured rat aortic endothelial cells. J Immunol 151:3283–3291

Suschek C, Fehsel K, Kröncke K-D, Sommer A, Kolb-Bachofen V (1994) Primary cultures of rat islet capillary endothelial cells. Constitutive and cytokine-inducible macrophagelike nitric oxide synthases are expressed and activities are regulated by glucose concentration. Am J Pathol 145:685–695

Takamura T, Kato I, Kimura N, Nakazawa T, Yonekura H, Takasawa S, Okamoto H (1998) Transgenic mice overexpressing type 2 nitric oxide synthase in pancreatic β cells develop insulin-dependent diabetes without insulitis. J Biol Chem 273:2493–2496

Taylor-Robinson AW, Liew FY, Severn A, Xu D, McSorley SJ, Garside P, Padron J, Phillips RS (1994) Regulation of the immune response by nitric oxide differentially produce by T helper type-1 and T helper type 2 cells. Eur J Immunol 24:980–984

Turk J, Corbett JA, Ramanadham S, Bohrer A, McDaniel ML (1993) Biochemical evidence for nitric oxide formation from streptozotocin in isolated pancreatic islets. Biochem Biophys Res Commun 197:1458–1464

Uchigata Y, Yamamoto H, Kawamura A, Okamoto H (1982) Protection by superoxide dismutase, catalase, and poly(ADP–ribose)synthetase inhibitors against alloxan- and streptozotocin-induced islet DNA strand breaks and against the inhibition of proinsulin synthesis. J Biol Chem 257:6084–6088

Verge CF, Howard NJ, Irwig L, Simpson JM, Mackerras D, Silink M (1994) Environmental factors in childhood IDDM. Diab Care 17:1381–1389

Welsh N, Sandler S (1992) Interleukin-1β induces nitric oxide production and inhibits the activity of aconitase without decreasing glucose oxidation rates in isolated mouse pancreatic islets. Biochem Biophys Res Commun 182:333–340

Welsh N, Eizirik DL, Bendtzen K, Sandler S (1991) Interleukin-1β-induced nitric oxide production in isolated rat pancreatic islets requires gene transcription and may lead to inhibition of the Krebs cycle enzyme aconitase. Endocrinology 129:3167–3173

Wiegand F, Kröncke K-D, Kolb-Bachofen V (1993) Macrophage generated nitric oxide as cytotoxic factor in destruction of alginate-encapsulated islets. Protection by arginine analogs and/or co-encapsulated erythrocytes. Transplantation 56:1206–1212

Wilson GL, Patton NJ, LeDoux SP (1997) Mitochondrial DNA is a sensitive target for damage by nitric oxide. Diabetes 48:1291–1295

Wu G (1995) Nitric oxide synthesis and the effect of aminoguanidine and N^G-monomethyl-L-arginine on the onset of diabetes in the spontaneously diabetic BB-rat. Diabetes 44:360–365

Yamada K, Nonaka K, Hanafusa T, Miyazaki A, Toyoshima H, Tarui S (1982) Preventive and therapeutic aspects of large dose nicotinamide injections on diabetes associated with insulitis: an observation in non-obese diabetic (NOD) mice. Diabetes 31:749–753

Yamamoto H, Uchigata Y, Okamoto H (1981) Streptozotocin and alloxan induce DNA strand breaks and poly(ADP–ribose)synthetase in pancreatic islets. Nature 294:284–286

Zipris D, Greiner DL, Malkani S, Whalen B, Mordes JP, Rossini AA (1996) Cytokine gene expression in islets and thyroids of BB rats. IFN-γ and IL-12p40 mRNA increase with age in both diabetic and insulin-treated nondiabetic BB rats. J Immunol 156:1315–1321

CHAPTER 21
The Role of Nitric Oxide in Cardiac Ischaemia–Reperfusion

P.A. MacCarthy and A.M. Shah

A. Introduction

Myocardial ischaemia, resulting mainly from atherosclerotic coronary artery disease, is the most common pathology in clinical cardiology and is a major cause of mortality and morbidity in the developed world. Iatrogenic myocardial ischaemia is also commonly encountered clinically, especially in the context of cardiac surgery and interventional cardiology procedures, such as percutaneous balloon angioplasty. There have been great advances in the clinical management of myocardial ischaemia over the last 2–3 decades. In particular, the benefits of early reperfusion following coronary occlusion have become well established. Such reperfusion is usually achieved in the clinical setting with thrombolytic drugs (or with balloon angioplasty, in recent years) and translates into a substantial improvement in patient survival (Gissi 1986; Isis 1988). However, it has also been realised that reperfusion of ischaemic myocardium may itself cause detrimental effects in the tissue to which blood flow is being restored (Braunwald and Kloner 1985).

Over the last 15 years, there have been major advances in our understanding of the physiological and pathophysiological roles of the ubiquitous messenger molecule nitric oxide (NO) in the cardiovascular system. Apart from its now well-established role in regulating vascular smooth-muscle tone and blood flow, it plays an important part in the inhibition of platelet and white blood cell aggregation and adhesion to the endothelium and may be involved in the pathophysiology of vascular smooth-muscle proliferation and atherosclerosis (Harrison 1997; Mayer and Hemmens 1997; Michel and Feron 1997). NO is now also known to have several direct actions on the myocardium, independent of its effects on vascular function and blood flow (Kelly et al. 1996; Shah 1996). An important aspect of the biochemistry of NO is its free-radical nature and its potential for interaction with reactive oxygen species (ROS), such as the superoxide anion (O_2^-) (Beckman and Koppenol 1996). Indeed, interactions between NO and ROS under conditions of "oxidative stress" are believed to be pathophysiologically important in many cardiovascular diseases (Darley-Usmar and Halliwell 1996). It is perhaps not surprising, therefore, that NO has a role in the pathophysiology of myocardial ischaemia–reperfusion. However, the experimental evidence regarding the function of NO in myocardial ischaemia–reperfusion is quite conflicting in nature and remains controversial. In this review, we attempt to summarise the literature to date, offer some explanations for the conflicting results, and provide some rational hypotheses concerning the precise pathophysiology of NO in cardiac ischaemia–reperfusion syndromes.

B. Consequences of Myocardial Ischaemia–Reperfusion

Myocardial ischaemia generally refers to an abnormally diminished coronary blood supply relative to the oxygen requirements of the myocardium. A major component of ischaemia is the reduction in oxygen supply (hypoxia), but other disturbances include a reduced delivery of substrates, accumulation of tissue metabolites and reduction in vascular turgor. The severity and duration of myocardial ischaemia are important determinants of the degree of functional abnormality and tissue damage. Thus, brief myocardial ischaemia (<20 min or so) results in contractile dysfunction during ischaemia but is usually followed (ultimately) by full recovery of function, whereas more prolonged ischaemia results in some permanent damage due to myocardial cell death or infarction (REIMER and JENNINGS 1979). Myocardial necrosis occurring during infarction proceeds in a "wavefront" from subendocardium to subepicardium in a matter of hours. The occurrence of arrhythmia is an important additional aspect of the functional abnormalities accompanying ischaemia–reperfusion (MANNING and HEARSE 1984).

Reperfusion, especially early after the onset of ischaemia, is beneficial in so much as it restores blood supply. However, it is now recognised that the process of reperfusion itself can result in deleterious effects. That is, impaired myocardial function following reperfusion is often the result of both ischaemic and post-ischaemic (reperfusion) injury. Abnormalities that may be induced by cardiac reperfusion include: (1) coronary endothelial dysfunction (TSAO et al. 1990); (2) adherence of neutrophils to the endothelium and transendothelial migration such that these blood cells can release their mediators (ROS, proteases, elastases and collagenases; INAUEN et al. 1990) and cause cell damage (LEFER and LEFER 1996); (3) impaired myocardial blood flow, the so called "no-reflow" phenomenon (KLONER et al. 1974); (4) ischaemic cell swelling, accelerated necrosis and reperfusion-induced haemorrhage, leading to infarct extension; (5) prolonged post-ischaemic contractile depression in the absence of irreversible myocyte damage, a phenomenon termed "myocardial stunning" (BRAUNWALD and KLONER 1982); and (6) arrhythmogenesis. The mechanisms involved in these processes are complex, but it is generally accepted that important aspects include: (1) the generation of a burst of ROS, in particular O_2^-, immediately upon reperfusion (ZWEIER 1988; BOLLI et al. 1989; TSAO and LEFER 1990; WANG and ZWEIER 1996), which causes tissue damage via lipid peroxidation, protein oxidation and DNA damage; (2) the induction of cellular calcium overload (THOMPSON and HESS 1986); and (3) re-energisation-induced myocyte hypercontracture (PIPER et al. 1998).

C. Interactions Between NO and ROS

A critical factor that may determine the influence of NO on myocardial function in the context of ischaemia–reperfusion is its interaction with ROS generated during reperfusion (Fig. 1). Potential sources of these ROS include endothelial enzymes (xanthine oxidase, reduced nicotinamide adenine dinucleotide/reduced nicotinamide adenine dinucleotide phosphate (NADPH) oxidases, cyclooxygenase, NO synthases), myocyte mitochondria, arachidonic acid/catecholamine metabolism and activated neutrophils (DARLEY-USMAR and HALLIWELL 1996). NO reacts with equimolar amounts of O_2^- in a diffusion-limited, essentially irreversible manner to form peroxynitrite (ONOO$^-$) (BECKMAN and KOPPENOL 1996). Since tissues have a large capacity to remove O_2^- (mainly via superoxide dismutase, SOD), and since NO competes with SOD for O_2^-, a critical issue becomes whether the concentration of NO approaches that of SOD; i.e., an increase both in NO and O_2^- production will generally be necessary to form ONOO$^-$. In addition to its direct interaction with O_2^-, however, NO can also reduce the generation of O_2^- by neutrophil NADPH oxidase, which could be an important feedback mechanism (CLANCY et al. 1992).

The removal of O_2^- by reaction with NO may be considered beneficial (WINK et al. 1993) in that O_2^- is thought to be responsible for much of the endothelial dysfunction (TSAO and LEFER 1990) and myocyte damage (BOLLI et al. 1989; FERRARI et al. 1998) caused by reperfusion. In addition, low concentrations of ONOO$^-$ may be beneficial via effects such as vasodilatation and an increase in coronary flow and by inhibiting endothelial cell–neutrophil interaction (LEFER et al. 1997; NOSSULI et al. 1997). However, the protonation of ONOO$^-$ to form peroxynitrous acid, especially at low pH, results in the generation of a strong oxidant (BECKMAN et al. 1990). Both the direct reaction of peroxynitrous acid with biological molecules as well as the generation of other species with characteristics similar to the hydroxyl radical may result in toxic effects secondary to protein nitration and oxidation. Other reactive nitrogen species formed from chemical reactions between neutrophil-derived ROS, NO and nitrite may also be involved in these processes (EISERICH et al. 1998). These species can induce negative inotropic actions, Ca^{2+} overload, nuclear and mitochondrial toxicity, and apoptosis.

D. Potential Ways in Which NO and ONOO$^-$ May Influence Myocardial Ischaemia–Reperfusion

Before reviewing the experimental studies that have investigated the role of NO in myocardial ischaemia–reperfusion, it may be useful to consider potential ways in which NO and ONOO$^-$ can influence cardiac function (Fig. 2).

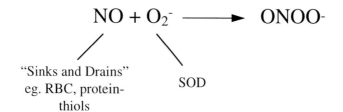

BENEFICIAL EFFECTS	DETRIMENTAL EFFECTS
LESS AVAILABLE O_2^-	**LESS AVAILABLE NO**
• Less cell damage	• Endothelial dysfunction
• ↓ PMN-endothelial interaction	• ↓ coronary flow
• Improved post-ischaemic contractile function	• ↑ endothelial-PMN interaction
	• ↑ endothelial-platelet interaction
	• Contractile dysfunction
LOW LEVEL ONOO-	**HIGH LEVEL ONOO-/RNS**
• Vasodilator effect	• Protein nitration/oxidation and cellular damage
• ↓ platelet aggregation	• Ca^{++} overload
• ↓ PMN-endothelial interaction	• Negative inotropic effects
• Preserved endothelial function	• Inhibition of mitochondrial respiration
• Improved myocardial contractile function	• Apoptosis
• ↓ infarct size	

Fig. 1. Interaction between nitric oxide (NO) and reactive oxygen species: good or bad? O_2^-, superoxide anion; *ONOO-*, peroxynitrite; *PMN*, polymorphonuclear leukocyte; *RBC*, red blood cells; *SOD*, superoxide dismutase

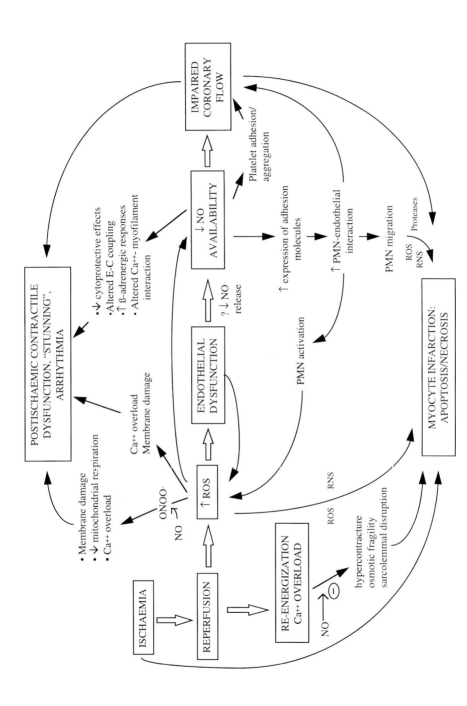

Fig. 2. Potential effects of nitric oxide (NO) or the pathophysiology of myocardial ischaemia–reperfusion. *E–C*, excitation–contraction; *ONOO⁻*, peroxynitrite; *PMN*, polymorphonuclear leukocyte; *RNS*, reactive nitrogen species; *ROS*, reactive oxygen species

I. Changes in Coronary Blood Flow and Vessel–Blood Cell Interactions

The status of myocardial oxygen supply–demand balance is clearly a critical determinant of contractile function and the extent of myocardial infarction and injury. NO-dependent changes in native coronary and collateral vessel blood flow and in transmural flow distribution will, therefore, have an important indirect influence on myocardial function. In addition to its effects on vascular tone, the actions of NO on interactions between blood cells and the vessel wall have a major impact on coronary blood flow. Deficiency of NO promotes platelet adhesion and aggregation, which significantly impairs blood flow (GOLINO et al. 1992). A reduction in NO also promotes the expression of adhesion molecules, such as P-selectin and intracellular adhesion molecule 1 (ICAM-1), on endothelial cells, resulting in increased adhesion of leukocytes (MA et al. 1993; LEFER 1995; LEFER and LEFER 1996). The "no-reflow" phenomenon seen at reperfusion following ischaemia is thought to be due in part to diminished release of NO, resulting in the plugging of the microvascular bed by leukocytes (ENGLER et al. 1983). As mentioned previously, low levels of ONOO⁻ are reported to have effects similar to NO with respect to endothelial–leukocyte interaction (LEFER et al. 1997). In addition to the effects of enhanced vessel wall–blood cell interaction on coronary flow, the mediators released by aggregating platelets and activated leukocytes may be directly damaging to the myocardium.

II. Direct Effects of NO and ONOO⁻ on Myocardium

Several potentially beneficial effects of "physiological" levels of NO on myocardial function are now recognised. NO enhances myocardial relaxation and increases diastolic distensibility in a range of different preparations, including isolated rat cardiac myocytes (SHAH et al. 1994), isolated ferret (SMITH et al. 1991), cat (MOHAN et al. 1996) and human (FLESCH et al. 1997) papillary muscles, isolated ejecting guinea-pig hearts (GROCOTT-MASON et al. 1994a, 1994b), and in human subjects in vivo (PAULUS et al. 1994, 1995). These effects are thought to result from a cyclic guanosine monophosphate (cGMP)-induced reduction in the relative myofilament response to calcium (SHAH et al. 1994). NO was also recently shown to facilitate the Frank-Starling response in the isolated heart (PRENDERGAST et al. 1997), an effect which could conceivably occur on a beat-to-beat basis (PINSKY et al. 1997). Both these actions would be expected to improve cardiac filling. Studies from the Boston group (KELLY et al. 1996; HARE et al. 1998) and others (BARTUNEK et al. 1997) have shown that NO can modulate β-adrenergic responsiveness in animals as well as humans, especially those with diseased hearts. The subcellular mechanism of this effect is believed to involve cGMP-mediated modulation of cyclic adenosine monophosphate phosphodiesterase activity (MERY et al. 1997). A reduction in β-adrenergic response may be beneficial in the context of acute ischaemia–reperfusion by limiting oxygen consumption and infarct size. NO is also reported to slightly reduce myocardial oxygen consumption independent of changes in contractile function, an effect which would improve efficiency (ZHANG et al. 1997). NO-mediated reduction in myofilament response to Ca^{++} (SHAH et al. 1994) and in Ca^{++} influx (KELLY et al. 1996) could reduce reperfusion-induced hypercontracture (SCHLUTER et al. 1994).

However, higher levels of NO may have significant negative inotropic effects, both via cGMP-dependent and -independent actions. The potential deleterious effects of $ONOO^-$-mediated oxidation and nitration have been mentioned earlier and include membrane damage, Ca^{2+} overload, an irreversible reduction in mitochondrial respiration (XIE et al. 1998), and the induction of apoptosis (SZABOLCS et al. 1996).

E. Experimental Studies

I. Post-Ischaemic Endothelial Dysfunction

Reperfusion-induced endothelial dysfunction was first demonstrated in 1982 (KU 1982) and has subsequently been demonstrated in many different species and experimental preparations (VAN BENTHUYSEN et al. 1987; MEHTA et al. 1989a, 1989b; TSAO et al. 1990), usually in terms of loss of endothelium-dependent vasodilatation. This has been shown to be due, at least in part, to diminished NO bioactivity (LEFER et al. 1991; ENGELMAN et al. 1995). TSAO et al. (1990) documented the time course of endothelial dysfunction by studying the impaired response to acetylcholine of coronary artery rings obtained from cats subjected to 90min of coronary artery occlusion followed by varying periods of reperfusion. In these studies, endothelial dysfunction was demonstrable as early as 2.5min after the onset of reperfusion but, interestingly, no dysfunction was apparent prior to reperfusion. PEARSON et al. (1992) also showed, in a canine model of global cardiac ischaemia, that endothelium-dependent relaxations to aggregating platelets, serotonin and adenosine diphosphate (ADP) were reduced in reperfused coronary arteries but not in ischaemic non-reperfused vessels.

The reasons for the diminished NO bioactivity are not entirely clear. GIRALDEZ et al. (1997) reported that in vitro biochemical endothelial-type nitric oxide synthase (eNOS) activity was substantially reduced after more than 60min global ischaemia of isolated rat hearts and that reperfusion resulted in only partial restoration of activity. This was attributed, in part, to loss of eNOS protein during ischaemia. However, other groups have reported an increase in eNOS activity during ischaemia, e.g., within 5min of ischaemia of isolated rabbit hearts (DEPRE et al. 1997). NODE et al. (1995) found a substantial increase in coronary nitrate and nitrite (NO_x) during myocardial ischaemia in dogs in vivo in proportion to the decrease in coronary blood flow. This increase was blocked by concomitant administration of a NOS inhibitor and restored by administration of L-arginine. Likewise, endothelial cell NOS activity has been reported to be increased in brain vessels subjected to anoxic injury (KUMAR et al. 1996). NO production by isolated cardiomyocytes was also reported to be increased by hypoxia (KITAKAZE et al. 1995). However, WILDHIRT et al. (1995) reported that there was no change in eNOS activity in rabbit myocardium in vivo 48h after myocardial infarction. However, these authors found that inducible NOS (iNOS) activity was significantly increased, and this was attributed to infiltrating macrophages. A likely explanation for the reduced biological actions of NO, regardless of the level of NO production by endothelial cells, is that its effective concentrations are reduced due to degradation by ROS released during reperfusion. It has also been suggested that ROS may selectively impair the activation of NOS via receptor-G protein-mediated signal transduction (SECCOMBE and SCHAFF 1995). These authors found that coronary arteries exposed to oxygen radicals in vitro failed to relax in response to the endothelium-dependent, receptor-dependent agonist ADP or to an activator of pertussis-toxin-sensitive G proteins or sodium fluoride but responded normally to the receptor-independent calcium ionophore A23187.

A series of elegant studies by LEFER and colleagues has shown that an important consequence of endothelial dysfunction and reduction in NO bioactivity is an increase in leukocyte–endothelial cell interaction (MA et al. 1993) due, at least in part, to an upregulation of the cellular-adhesion molecules P-selectin and ICAM-1 (LEFER 1995; LEFER and LEFER 1996). This is followed by neutrophil accumulation in the ischaemic tissue approximately 3–4.5h into reperfusion as well as an increase in myeloperoxidase activity in the ischaemic zone. Pharmacological manipulation of the NO pathway or of ROS production is capable of improving or preventing reperfusion-induced endothelial dysfunction and its consequences. SIEGFRIED et al. (1992a, 1992b) showed that NO donors preserved endothelium-dependent vasodilatation in isolated coronary arterial segments after ischaemia–reperfusion in cats. NO donors also inhibited endothelial expression of ICAM-1 (LEFER et al. 1993b), downregulated P-selectin in ischaemia–reperfusion (GAUTHIER et al. 1994) and reduced neutrophil accumulation and tissue myeloperoxidase activity. Endothelial dysfunction is also preventable by the anti-oxidant SOD (MEHTA et al. 1989b; LAWSON et al. 1990). Interestingly, this was also the case in the isolated, buffer-perfused rat heart (WANG and ZWEIER 1996), indicating that neutrophil-derived ROS production is not essential for the induction of endothelial dysfunction.

II. Myocardial Function

1. NO as a Beneficial Agent

a) Post-Ischaemic Contractile Function

A large number of studies have reported beneficial effects of NO on post-ischaemic contractile function both in buffer-perfused hearts (in vitro) and in blood/neutrophil-perfused preparations (in vitro and in vivo; Table 1).

Table 1. Examples of studies reporting beneficial effects of exogenous and endogenous nitric oxide (NO) in ischaemia–reperfusion

Effect	Species	Drug used	Experimental preparation	Potential mechanism	Reference
Preservation of endothelial function (all studies in vivo)	Cat	NO donors: SPM 5185, C87-3754, SIN-1	90 min RI, 4.5 h RP	O_2^- scavenging, anti-PMN	Siegfried et al. 1992a, 1992b
	Cat	L-arginine	90 min RI, 4.5 h RP	Improved EF, anti-PMN	Weyrich et al. 1992
	Dog	NO donor (CAS-1609)	90 min RI, 4.5 h RP	Improved EF and CF, anti-PMN	Pabla et al. 1995
	Dog	NO donor (SPM 5185)	30 min GI, 1 h CPA, 1 h RP	Anti-PMN	Lefer et al. 1993c
	Dog	L-arginine	60 min RI, 4.5 h RP	Improved EF, anti-PMN	Nakanishi et al. 1992
	Dog	L-arginine	90 min RI, 1 h CPA	Anti-PMN, improved EF	Sato et al. 1995
Improved post-ischaemic contractile function					
Buffer-perfused hearts	Rat	Bradykinin/ramiprilat	15 min RI, 30 min RP IH	Increased CF	Linz et al. 1992
	Rat	L-NAME	60 min GI, RP IH	Altered CF	Pabla and Curtis 1996
	Rat	L-NNA	35 min GI, 40 min RP IH	Independent of CF	Beresewicz et al. 1995
	Rat	L-arginine	4 h CPA, RP IH	Improved coronary reflow	Amrani et al. 1995
	Rat	L-arginine	60 min RI, RP IH	cGMP-dependent mechanism	Engelman et al. 1996
	Rat	NO donor (GEA-3162)	30 min GI, 40 min RP IH	Improved transmural flow distribution, direct myocardial effect	Szekeres et al. 1997
	Rat	NO donor (SNP)	5 min H, RO IH	Direct myocardial effect	Draper and Shah 1997

Table 1. Continued

Effect	Species	Drug used	Experimental preparation	Potential mechanism	Reference
	Rat	NO donor (SNP)	25 min GI, 25 min RP IH	Direct cGMP-dependent myocardial effect	Du Toit et al. 1998
	Guinea pig	Ramiprilat, bradykinin, SNP	15 min GI, 35 min RP IH	cGMP-independent effect of NO, O_2^- scavenging	Massoudy et al. 1995
Blood/PMN-perfused hearts	Dog	NO donor (SPM-5185)	30 min GI, 1 h CPA, 1 h RP in vivo	O_2^- scavenging, anti-PMN	Lefer et al. 1993c
	Dog	NO donor (CAS-1609)	90 min RI, 4.5 h RP in vivo	Improved EF, improved CF, anti-PMN	Pabla et al. 1995
	Dog	L-NNA	10 min RI, RP in vivo	Direct myocardial effect of NO independent of CF	Hasebe et al. 1993
	Rat	NO donor (CAS 754), L-arginine	20 min GI, 45 min RP IH perfused with PMNs	O_2^- scavenging, anti-PMN	Pabla et al. 1996
	Rat	$ONOO^-$	20 min GI, 45 min RP IH perfused with PMNs	Anti-PMN effect of $ONOO^-$, increased CF	Lefer et al. 1997
Decreased myocardial infarct size	Cat	Authentic NO	1.5 h RI, 4.5 h RP in vivo	Anti-platelet, anti-PMN, improved EF	Johnson et al. 1990, 1991

	Cat	NO donors (SIN-1, C88-3754)	1.5 h RI, 4.5 h RP in vivo	Improved EF, O_2^- scavenging	Siegfried et al. 1992a
	Cat	L-arginine	1.5 h RI, 4.5 h RP in vivo	O_2^- scavenging, anti-PMN, improved EF	Weyrich et al. 1992
	Dog	L-arginine	1 h RI, 4.5 h RP in vivo	Anti-PMN, improved EF	Nakanishi et al. 1992
	Rabbit	L-NNA	30 min RI, 120 min RP in vivo	O_2^- production by PMNs, PMN adherence	Williams et al. 1995
	Rabbit	L-NAME	30 min RI, 48 h RP in vivo	Improved CF, anti-PMN	Hoshida et al. 1995
	Cat	ONOO⁻	1.5 h RI, 4.5 h RP in vivo	Anti-PMN, improved EF	Nossuli et al. 1997
	Others	NO donors		Cytoprotective effect of endogenous NO	Siegfried et al. 1992b; Lefer et al. 1993a; Pabla et al. 1995
Decreased reperfusion-induced arrhythmias	Rat	L-NAME	5, 35, 60 min RI, 10 min RP IH		Pabla and Curtis 1995
	Rat	Bradykinin/ramiprilat	15 min RI, 30 min RP	Improved CF	Linz et al. 1992

CF, coronary flow; *CPA*, cardioplegic arrest; *EF*, endothelial function; *GI*, global ischaemia; *H*, hypoxia; *IH*, isolated heart; L-NAME, N^G-nitro-L-arginine methyl ester; L-NMMA, N^G-monomethyl-L-arginine; L-NNA, N^G-nitro-L-arginine; O_2^-, superoxide anion; ONOO⁻, peroxynitrite; *PMN*, polymorphonuclear leukocyte; *RI*, regional ischaemia; *RO*, reoxygenation; *RP*, reperfusion; *SNP*, sodium nitroprusside.

α. Buffer-Perfused Preparations

The stimulated release of endogenous NO, either by bradykinin or the angiotensin-converting enzyme (ACE) inhibitor ramiprilat, augmented post-ischaemic left ventricle (LV) developed pressure and high-energy phosphates in isolated rat hearts subjected to 15 min regional ischaemia, effects which were abolished by a NOS inhibitor and were attributed to an NO-induced increase in coronary flow (LINZ et al. 1992). PABLA and CURTIS (1996) reported that the NOS inhibitor N^G-nitro-L-arginine methyl ester (L-NAME) hastened the onset of ischaemic contracture in isolated rat hearts subjected to 60 min global ischaemia and caused a delay in the initial post-ischaemic recovery of contractile function. These effects were closely correlated to the NOS-inhibitor-induced decrease in coronary flow. In the isolated working rat heart, AMRANI et al. (1995) found that the NO precursor L-arginine improved low coronary reflow and post-ischaemic contractile function after a period of global ischaemia.

In a similar preparation, BERESEWICZ et al. (1995) found that inhibition of NOS impaired post-ischaemic cardiac output after 35 min global ischaemia. Interestingly, this effect appeared to be independent of changes in coronary flow. SZEKERES et al. (1997) found an improvement in post-ischaemic contractile function of isolated rat hearts perfused at constant flow and treated with the NO donor GEA-3162. This was speculated to be due either to a beneficial transmural redistribution of coronary flow or to a direct myocardial effect. A direct beneficial effect of NO on myocardium was also suggested by DRAPER and SHAH (1997), who found that an NO donor improved post-hypoxic contractile function of isolated rat hearts subjected to brief hypoxia at constant coronary flow. The same group reported that a cGMP analogue, 8-bromo-cGMP (a potential downstream messenger for NO), fully prevented impaired post-hypoxic relaxation of isolated rat cardiac myocytes subjected to 10 min anoxia (SHAH et al. 1995), confirming a direct beneficial effect of cGMP in cardiac myocytes during hypoxia-reoxygenation. This effect of cGMP appeared to involve a reduction in myofilament response to calcium, which was otherwise augmented at reperfusion. In a study in isolated rat hearts subjected to global ischaemia–reperfusion, Du TOIT et al. (1998) found that the beneficial effect of NO donors on ischaemic contracture and post-ischemic LV function was related to the rise in tissue cGMP content.

An alternative mechanism for a beneficial effect of NO was suggested by MASSOUDY et al. (1995), who studied 15 min global ischaemia in isolated working guinea-pig hearts. They reported that an ACE inhibitor (ramiprilat), bradykinin and an NO donor – but not 8-bromo-cGMP – improved post-ischaemic external heart function and decreased glutathione release into the coronary effluent, a measure of oxidative stress. These authors suggested that NO was beneficial, acting as an ROS scavenger during reperfusion.

The *timing* of NO supplementation may be important in terms of the efficacy of its beneficial effect. ENGELMAN et al. (1996) reported that, in isolated working rat hearts subjected to 60 min regional ischaemia and cardioplegic arrest, post-ischaemic contractile function was improved by L-arginine treatment prior to ischaemia but was worsened by L-arginine administered at reperfusion. DRAPER and SHAH (1997) also found that the NO donor sodium nitroprusside improved post-hypoxic contractile function of isolated rat hearts when administered prior to and during hypoxia but was detrimental when given at reoxygenation alone.

β. Blood/Neutrophil-Perfused Preparations

In a blood-perfused experimental preparation, the effective level of NO and its interaction with ROS, leukocytes and other blood cells is likely to be different. Enhancement of the NO pathway has been shown to ameliorate post-ischaemic contractile dysfunction in several studies. LEFER et al. (1993c) studied the effects of an NO donor (SPM-5185) on recovery of contractile function in a canine model of global myocardial ischaemia and cardioplegic arrest and found that this agent markedly improved post-ischaemic function. PABLA and colleagues (1995) demonstrated that the NO donor CAS-1609 significantly improved post-ischaemic (at 4.5h reperfusion) coronary flow and contractile function, reduced myocardial necrosis and neutrophil infiltration in the ischaemic zone, and preserved coronary artery endothelial function in a canine model of regional myocardial ischaemia. These authors went on to demonstrate the importance of neutrophils in such effects by studying isolated rat hearts subjected to 20min global ischaemia and 45min reperfusion (PABLA et al. 1996). Human neutrophils were infused over the first 5min of reperfusion and caused an impairment of coronary flow and LV function at the end of the reperfusion period. Both the NO donor CAS-754 and L-arginine improved the post-ischaemic coronary flow and contractile state, whereas the NOS inhibitor L-NAME exacerbated contractile dysfunction. Cardioprotective effects of CAS-754 and L-arginine were not apparent in the absence of neutrophils in this study. In a similar study in isolated rat hearts with neutrophil infusion at reperfusion, the addition of nanomolar concentrations of $ONOO^-$ almost completely prevented the reduction in coronary flow and contractile dysfunction and markedly attenuated neutrophil accumulation (LEFER et al. 1997). It should be remembered, however, that neutrophil adhesion and accumulation during myocardial ischaemia in vivo is a relatively late event (occurring 3–4.5h into reperfusion) (LEFER and LEFER 1996)

It has also been reported that endogenous NO may reduce myocardial stunning in vivo. In a model of canine regional ischaemia–reperfusion (10min coronary occlusion), post-ischaemic recovery of regional myocardial function (ultrasonic wall thickening) was significantly delayed by a NOS inhibitor, an effect thought to be independent of coronary flow (HASEBE et al. 1993).

b) Myocardial Infarction

Similar to the effects on contractile function, interventions that augment the NO pathway have been found to also reduce myocardial infarct size. In an in vivo model of regional myocardial ischaemia in cats, infusion of authentic NO (JOHNSON et al. 1991), a NO donor (SIEGFRIED et al. 1992a) or intravenous L-arginine (WEYRICH et al. 1992) decreased myocardial necrosis and neutrophil accumulation. Similar effects were observed in a canine model of left anterior descending coronary artery occlusion with L-arginine (NAKANISHI et al. 1992) or NO donors (LEFER et al. 1993a; PABLA et al. 1995).

Concurrent with these findings are observations that *inhibition* of endogenous NO production worsens ischaemic injury. In the regional ischaemic-reperfused rabbit heart in vivo, the NOS inhibitor N^G-nitro-L-arginine (L-NNA) significantly increased infarct size whether administered prior to ischaemia or only at reperfusion (WILLIAMS et al. 1995). This was confirmed in a similar model in the same species with a different NOS inhibitor, L-NAME, which significantly increased infarct size (HOSHIDA et al. 1995) – an effect attenuated by concurrent treatment with L-arginine.

A protective effect of $ONOO^-$ in terms of reduction of infarct size has also been reported by Lefer and colleagues in studies in cats subjected to 90 min regional ischaemia (NOSSULI et al. 1997). These effects were associated with a reduction in expression of P-selectin and neutrophil adherence and accumulation, and an improvement in endothelium-dependent relaxation.

A number of possible mechanisms could account for the above beneficial effects on infarct size: (1) inhibition of neutrophil–endothelial interaction and, therefore, inhibition of neutrophil accumulation in the ischaemic zone and of neutrophil-mediated damage (LEFER et al. 1991); (2) scavenging of ROS released by endothelial cells and neutrophils (WINK et al. 1993); (3) a reduction in myocardial oxygen demand and, thus, an improvement in oxygen supply–demand balance (ZHANG et al. 1997); and (4) possible direct cytoprotective effects of NO (SCHLUTER et al. 1994, 1996). An effect of NO on coronary reflow at reperfusion seems to be less important with respect to the myocardial infarction studies (NAKANISHI et al. 1992) than some of the studies looking at post-ischemic contractile function. With respect to direct cytoprotective effects of NO on myocytes, a number of potential mechanisms have recently been suggested by Schluter and colleagues (SCHLUTER et al. 1994, 1996). Using an isolated cardiac myocyte model of 135 min substrate-free anoxia, these authors found that an NO donor infused during the last 15 min of anoxia and the first 15 min of reoxygenation significantly inhibited reoxygenation-induced hypercontracture (SCHLUTER et al. 1994). In a separate study, they showed that NO donors also reduced osmotic fragility of isolated cardiac myocytes at reoxygenation, independent of cGMP (SCHLUTER et al. 1996). The latter effect was reproduced by an anti-lipid peroxidant and might, therefore, have involved an ROS-scavenging effect of NO.

c) Reperfusion-Induced Arrhythmia

Another manifestation of the protective effect of NO is a reduction in the incidence of ischaemia–reperfusion-induced arrhythmia. Reperfusion induces cardiac arrhythmias but, interestingly, ventricular-fibrillation susceptibility declines markedly if the duration of ischaemia is extended beyond 30min, implying that there may be a protective endogenous substance produced by the heart (CURTIS and PABLA 1997). In isolated rat hearts subjected to regional ischaemia followed by 10min reperfusion, the NOS inhibitor L-NAME increased the incidence of reperfusion-induced ventricular fibrillation, an effect which was abolished by co-perfusion of L-arginine (PABLA and CURTIS 1995). Stimulation of endogenous NO with bradykinin or ramiprilat in isolated, regionally ischaemic rat hearts also reduced the incidence and duration of ventricular fibrillation (LINZ et al. 1992). However, such effects of NO may be specific to reperfusion, because neither endogenous nor exogenous NO reduced ischaemia-induced arrhythmias in the absence of reperfusion in rats in vivo (SUN and WAINWRIGHT 1997).

2. NO as a Deleterious Agent

Contrary to the above evidence, some investigators have suggested that endogenous NO is detrimental to the ischaemic–reperfused heart and that inhibition of NOS leads to a reduction in ischaemia–reperfusion injury (Table 2). The first study to suggest this possibility was reported by MATHIES et al. (1992), who studied hypoxia–reoxygenation on cardiopulmonary bypass in piglets. These authors found that post-hypoxic contractile function was markedly improved either by a NOS inhibitor or by a ROS scavenger, mercaptopropionylglycine. WANG and ZWEIER (1996) studied LV contractile function as well as the production of NO and O_2^- (both by electron paramagnetic resonance) and $ONOO^-$ (by luminol chemiluminescence) in isolated rat hearts subjected to 30min global ischaemia. They found that the release of NO, O_2^- and $ONOO^-$ was increased within the first 2min of reperfusion. Furthermore, either a NOS inhibitor or SOD greatly enhanced post-ischaemic contractile function. It was postulated, therefore, that NO was deleterious via production of $ONOO^-$, a hypothesis supported by the demonstration of protein nitration (nitrotyrosine) on immunohistochemistry. Likewise, YASMIN et al. (1997) demonstrated $ONOO^-$ formation (by dityrosine fluorescence) in isolated buffer-perfused rat hearts upon reperfusion and found that post-ischaemic LV function was improved either by a NOS inhibitor or by the NO donor S-nitroso-N-acetylpenicillamine, which was suggested to act by reducing $ONOO^-$ levels. The same group also reported a protective effect of NOS inhibitors in the isolated working rabbit heart subjected to global ischaemia (SCHULZ and WAMBOLT 1995). LIU et al. (1997) have provided evidence for myocardial protein nitration (using anti-nitrotyrosine antibodies) in an in vivo model of regional myocardial ischaemia–reperfusion in the rat, although the effects of this on LV function or infarct size were not investigated. XIE et al. (1998) studied the effects of $ONOO^-$ in isolated rat papillary muscle subjected to hypoxia–reoxygenation and reported that it appeared to irreversibly reduce both mitochondrial respiration and contractile function, in contrast to NO

Table 2. Examples of studies reporting detrimental effects of endogenous nitric oxide in ischaemia–reperfusion

Effect	Species	Drug used	Experimental preparation	Potential mechanism	Reference
Post-ischaemic contractile dysfunction	Rat	L-NNA	37.5 min GI, 20 min RP IH	Preconditioning-type effect	Naseem et al. 1995
	Rat	L-NNA	15 min H, RO PapM	Formation of ONOO⁻	Xie et al. 1998
	Rat	L-NAME	30 min GI, 15 min RP IH	Formation of ONOO⁻	Wang and Zweier 1996
	Rat	L-NMMA	20 min GI, RP IH	Formation of ONOO⁻	Yasmin et al. 1997
	Pig	L-NAME	120 min WBH, 30 min RO (CPB)	Formation of ONOO⁻	Matheis et al. 1992
	Pig	L-NAME	120 min WBH, 30 min RO	Formation of ONOO⁻	Morita et al. 1995
	Rabbit	L-NMMA, L-NAME	35 min GI, 30 min RP IH	Formation of ONOO⁻	Schulz and Wambolt 1995
Increased myocardial infarct size	Rabbit	L-NAME	30–50 min RI, 2–3 h RP in vivo	L-NAME-induced ischaemic preconditioning	Patel et al. 1993
	Rabbit	L-NAME	45 min RI, 180 min RP IH	Adenosine-dependent cytoprotective mechanism	Woolfson et al. 1995

CPB, cardiopulmonary bypass; *GI*, global ischaemia; *IH*, isolated heart; L-NAME, N^G-nitro-L-arginine methyl ester; L-NMMA, N^G-monomethyl-L-arginine; L-NNA, N^G-nitro-L-arginine; ONOO⁻, peroxynitrite; *PapM*, papillary muscle preparation; *RI*, regional ischaemia; *RP*, reperfusion; *WBH*, whole-body hypoxia.

donors, which had modest, reversible effects on these parameters (XIE et al. 1996).

Alternative mechanisms for a beneficial effect of NOS inhibitors during ischaemia–reperfusion have also been suggested. Pre-treatment with L-NAME significantly reduced infarct size following regional coronary occlusion in rabbits in vivo (PATEL et al. 1993), an effect which appeared to be dependent on L-NAME-induced release of adenosine (WOOLFSON et al. 1995). Thus, it is possible that NOS inhibitors could be beneficial by inducing pharmacological preconditioning. NASEEM et al. (1995) found a beneficial effect of L-NNA on post-ischaemic function and arrhythmias following 37.5 min global ischaemia in the isolated rat heart. In this study, L-NNA treatment was associated with a significant reduction in pre-ischaemic coronary flow and LV developed pressure. Thus, L-NNA could have induced a preconditioning-like phenomenon. Recently, DEPRE et al. (1995) reported that non-vasoactive concentrations of NOS inhibitors were beneficial during ischaemia by stimulating glycolysis. DU TOIT et al. (1998) also reported a beneficial effect of low (but not high) doses of L-NNA in model of global ischaemia–reperfusion in the isolated rat heart, an effect which was suggested to be independent of NOS or cGMP.

F. Reasons for Conflicting Results

In considering the experimental studies that have investigated the role of NO during ischaemia–reperfusion, a number of factors are important to bear in mind: (1) the experimental preparation employed, in particular whether it is buffer-perfused in vitro or blood-containing (in vitro or in vivo); (2) the nature, duration and severity of the ischaemia; (3) consideration of effects resulting during ischaemia separately from effects occurring as a result of reperfusion; (4) distinction between effects observed early upon reperfusion (when ROS-mediated actions may be particularly prominent) from effects observed later; and (5) the end-point(s) studied, i.e. contractile function, coronary flow, infarct size or arrhythmia. It is clear from the studies discussed above that NO can exert several beneficial effects both during ischaemia and reperfusion and that this may be the dominant role of NO in vivo. However, under certain conditions, NO also has the potential to be deleterious. The latter is most likely immediately following reperfusion if levels of $ONOO^-$ high enough to result in significant protein nitration and oxidation are generated. Whether or not $ONOO^-$ formation is a mechanism for significant damage during myocardial reperfusion in vivo remains to be established. A significant limitation to our current understanding of this area is the absence of reliable data regarding the levels of $ONOO^-$ achieved during myocardial reperfusion in vivo. Although a number of in vitro studies have provided evidence to suggest a deleterious effect of $ONOO^-$ (WANG and ZWEIER 1996; YASMIN et al. 1997; XIE et al. 1998), it is important to remember that in vitro studies vastly overestimate the cytotoxic potential and chemical reactivity of NO and $ONOO^-$ in vivo (BECKMAN and KOPPENOL 1996). This is because (1) many of the sinks and drains that remove NO in vivo are absent in vitro, (2) the level of ROS generation is considerably higher in physiological buffers than in vivo and (3) antioxidant defences in vitro are usually lower (BECKMAN and KOPPENOL 1996). In addition, it has been suggested that the reactivity of ONOO$^-$ may be altered by interaction with bicarbonate, which is present in millimolar concentrations in blood plasma (DENICOLA et al. 1996). In vivo studies that have employed extremes of hypoxia–reoxygenation (MATHIES et al. 1992) may likewise have exaggerated ONOO$^-$ formation. Nevertheless, the potential for deleterious effects of ONOO$^-$ is important to bear in mind and may be particularly relevant in situations where antioxidant capacity is reduced and/or NO production is excessive (with expression of iNOS).

G. NO and Ischaemic Preconditioning

Ischaemic cardiac preconditioning is the phenomenon whereby one or more brief episodes of ischaemia/reperfusion result in an increased tolerance of the myocardium to subsequent, more prolonged ischaemia (YELLON et al. 1998). In addition to the reduction of myocardial infarct size initially reported (MURRY et al. 1986), improved post-ischaemic contractile function (BOLLI 1996) and suppression of reperfusion-induced arrhythmias (SHIKI and HEARSE 1987) have also been reported. More recently, it has become apparent that ischaemic preconditioning can induce both early (manifest within minutes to a few hours) and late (apparent 12–24h after the preconditioning ischaemia) myocardial "protection" (YELLON et al. 1998). Despite intensive efforts by many laboratories, the underlying basis for these phenomena remains incompletely understood. Factors thought to be important in the early phase include the release of adenosine, activation of protein kinase C, the opening of adenosine triphosphate-sensitive potassium channels, and the release of endothelial factors, while the late phase probably involves the synthesis of new proteins (YELLON et al. 1998).

A number of studies have investigated the possible role of NO in preconditioning. In an in vivo canine model of regional myocardial ischaemia, Vegh and colleagues reported that the beneficial effects of preconditioning on ischaemic arrhythmias was attenuated either by L-NAME (VEGH et al. 1992a) or by an inhibitor of guanylyl cyclase, methylene blue (VEGH et al. 1992b). They also demonstrated that L-NAME eradicated the anti-arrhythmic effect of intracoronary bradykinin during myocardial ischaemia in the same canine model, suggesting that bradykinin may be a "primary mediator" of ischaemic preconditioning, acting via release of NO and elevation of cGMP (VEGH et al. 1993). It was notable, however, that the effect of bradykinin on reperfusion arrhythmias was not NO-dependent. In a model of regional ischaemic preconditioning in anaesthetised rats, Lu et al. (1995) found that neither L-NAME nor N^G-monomethyl-L-arginine significantly affected the protective effect of preconditioning on reperfusion-induced arrhythmia. Likewise, in the isolated rat heart perfused at constant flow, WESELCOUCH et al. (1995) found that L-NAME did not influence the beneficial effect of preconditioning on LV contractile function after 30 min global ischaemia.

Recent studies by BOLLI et al. (1997) have implicated NO as a trigger of delayed ischaemic preconditioning. These workers reported that L-NNA blocked the development of late preconditioning against myocardial stunning in conscious rabbits subjected to repeated brief coronary occlusion/reperfusion cycles on three consecutive days. The mechanism for this effect remains to be elucidated, but an interaction with ROS has been suggested to be involved (BOLLI et al. 1997).

H. Summary and Conclusions

NO is involved in the physiological control of several aspects of cardiac function, including the regulation of coronary flow, interaction between blood cells and the vessel wall, myocardial contractile function and, probably, oxygen consumption. During myocardial ischaemia–reperfusion, both endogenously produced NO and exogenously augmented NO may have beneficial effects on infarct size, post-ischaemic contractile function and arrhythmia. The mechanisms responsible for these effects include augmentation of coronary flow, anti-neutrophil and anti-platelet effects, the scavenging of ROS at reperfusion and direct effects on cardiac myocytes. Recent studies suggest that NO may also be involved in some aspects of the phenomena of ischaemic preconditioning. However, NO has the potential to exert deleterious effects via the generation of $ONOO^-$ from its interaction with ROS at reperfusion. The significance of this as a mechanism for reperfusion damage in vivo remains to be ascertained; important factors that will influence $ONOO^-$-induced myocardial dysfunction include antioxidant capacity in vivo, and the amount of NO available for reaction with O_2^- after taking into account its interactions with other biological molecules, such as haemoglobin and thiols. Appropriate pharmacological manipulation of the NO pathway and its second messengers could have therapeutic potential in the context of myocardial ischaemia–reperfusion injury.

Acknowledgements. The research in the authors' laboratory is supported by the UK Medical Research Council (MRC) and the British Heart Foundation. PAM is the recipient of an MRC Clinical Training Fellowship and AMS of an MRC Clinical Senior Fellowship.

Abbreviations

ACE　　　angiotensin-converting enzyme
ADP　　　adenosine diphosphate

cGMP	cyclic guanosine monophosphate
eNOS	endothelial-type nitric oxide synthase
ICAM-1	intracellular adhesion molecule 1
iNOS	inducible nitric oxide synthase
L-NAME	N^G-nitro-L-arginine methyl ester
L-NNA	N^G-nitro-L-arginine
LV	left ventricle
NADPH	reduced nicotinamide adenine dinucleotide phosphate
NO	nitric oxide
O_2^-	superoxide anion
ONOO$^-$	peroxynitrite
ROS	reactive oxygen species
SOD	superoxide dismutase

References

Amrani M, Chester AH, Jayakumar J, Schyns CJ, Yacoub MH (1995) L-arginine reverses low coronary reflow and enhances post-ischaemic recovery of cardiac mechanical function. Cardiovasc Res 30:200–204

Bartunek J, Shah AM, Vanderheyden M, Paulus WJ (1997) Dobutamine enhances cardiodepressant effects of receptor-mediated coronary endothelial stimulation. Circulation 95:90–96

Beckman JS, Koppenol WH (1996) Nitric oxide, superoxide, and peroxynitrite: the good, the bad, and the ugly. Am J Physiol 271:C1424-C1437

Beckman JS, Beckman TW, Chen J, Marshall PA, Freeman BA (1990) Apparent hydroxyl radical production by peroxynitrite: implications for endothelial injury from nitric oxide and superoxide. Proc Natl Acad Sci USA 87:1620–1624

Beresewicz A, Karwatowska-Prokopczuk E, Lewartowski B, Cedro-Ceremuzynska K (1995) A protective role of nitric oxide in isolated ischaemic/reperfused rat heart. Cardiovasc Res 30:1001–1008

Bolli R (1996) The early and late phases of preconditioning against myocardial stunning and the essential role of oxyradicals in the late phase: an overview. Basic Res Cardiol 91:57–63

Bolli R, Jeroudi MO, Patel BS, Aruoma OI, Halliwell B, Lai EK, McCay PB (1989) Marked reduction of free radical generation and contractile dysfunction by antioxidant therapy begun at the time of reperfusion. Circ Res 65:607–622

Bolli R, Bhatti ZA, Tang X-L, Qui Y, Zhang Q, Guo Y, Jadoon AK (1997) Evidence that late preconditioning against myocardial stunning in conscious rabbits is triggered by the generation of nitric oxide. Circ Res 81:42–52

Braunwald E, Kloner RA (1982) The stunned myocardium: prolonged, post-ischaemic ventricular dysfunction. Circulation 66:1146–1149

Braunwald E, Kloner RA (1985) Myocardial reperfusion: a double-edged sword? J Clin Invest 76:1713–1719

Clancy RM, Leszczynska-Piziak J, Abramson SB (1992) Nitric oxide, an endothelial cell relaxation factor, inhibits neutrophil superoxide anion production via a direct action on the NADPH oxidase. J Clin Invest 90:1116–1121

Curtis MJ, Pabla R (1997) Nitric oxide supplementation or synthesis block – which is the better approach to treatment of heart disease? Trends Pharmacol Sci 18:239–244

Darley-Usmar V, Halliwell B (1996) Reactive nitrogen species, reactive oxygen species, transition metal ions, and the vascular system. Pharmaceut Res 13:649–662

Denicola A, Freeman BA, Trujillo M, Radi R (1996) Peroxynitrite reaction with carbon dioxide/bicarbonate kinetics and influence on peroxynitrite-mediated oxidations. Arch Biochem Biophys 333:49–58

Depré C, Vanoverschelde J-L, Goudemant J-F, Mottet I, Hue L (1995) Protection against ischaemic injury by nonvasoactive concentrations of nitric oxide synthase inhibitors in the perfused rabbit heart. Circulation 92:1911–1918

Depré C, Fiérain L, Hue L (1997) Activation of nitric oxide synthase by ischaemia in the perfused heart. Cardiovasc Res 33:82–87

Draper NJ, Shah AM (1997) Beneficial effects of a nitric oxide donor on recovery of contractile function following brief hypoxia in isolated rat heart. J Moll Cell Cardiol 29: 1195–1205

Du Toit EF, McCarthy J, Miyashiro J, Opie LH, Brunner F (1998) Effect of nitrovasodilators and inhibitors of nitric oxide synthase on ischaemic and reperfusion function of rat isolated hearts. Br J Pharmacol 123:1159–1167

Eiserich JP, Hristova M, Cross CE, Jones AD, Freeman BA, Halliwell B, Van der Vliet A (1998) Formation of nitric oxide-derived inflammatory oxidants by myeloperoxidase in neutrophils. Nature 391:393–397

Engelman DT, Watanabe M, Engelman RM, Rousou JA, Flack JE, Deaton DW, Das DK (1995) Constitutive nitric oxide release is impaired after ischaemia and reperfusion. J Thorac Cardiovasc Surg 110:1047–1053

Engelman DT, Watanabe M, Maulik N, Engelman RM, Rousou JA, Flack JE, Deaton DW, Das DK (1996) Critical timing of nitric oxide supplementation in cardioplegic arrest and reperfusion. Circulation 94 (suppl II): II-407-II-411

Engler RL, Schmid-Schonbein GW, Pravelec RS (1983) Leukocyte capillary plugging in myocardial ischaemia and reperfusion in the dog. Am J Pathol 111:98–111

Ferrari R, Agnoletti L, Comini L, Gaia G, Bachetti T, Cargnoni A, Ceconi C, Curello S, Visioli O (1998) Oxidative stress during myocardial ischaemia and heart failure. Eur Heart J 19 (suppl B): B2-B11

Flesch M, Kilter H, Cremers B, Lenz O, Sudkamp M, Kuhn-Regnier F, Bohm M (1997) Acute effects of nitric oxide and cyclic GMP on human myocardial contractility. J Pharmacol Exp Ther 281:1340–1349

Gauthier TW, Davenpeck KL, Lefer AM (1994) Nitric oxide attenuates leukocyte-endothelial interaction via P-selectin in splanchnic ischaemia–reperfusion. Am J Physiol 267:H562-H568

Giraldez RR, Panda A, Xia Y, Sanders SP, Zweier JL (1997) Decreased nitric oxide synthase activity causes impaired endothelium-dependent relaxation in the post-ischaemic heart. J Biol Chem 272 (34): 21420–21426

Golino P, Ambrosio G, Pascucci I, Ragni M, Russolillo E, Chiariello M (1992) Experimental carotid stenosis and endothelial injury in the rabbit: an in vivo model to study intravascular platelet aggregation. Thromb Haemost 67 (3): 302–305

Grocott-Mason R, Anning P, Evans HG, Lewis MJ, Shah AM (1994a) Modulation of left ventricular relaxation in the isolated ejecting heart by endogenous nitric oxide. Am J Physiol 267:H1804-H1813

Grocott-Mason RM, Fort S, Lewis MJ, Shah AM (1994b) Myocardial relaxant effect of exogenous nitric oxide in isolated ejecting hearts. Am J Physiol 266:H1699–H1705

Gruppo Italiano Per lo Studio della Streptochinasi nell'Infarto miocardico (GISSI) (1986) Effectiveness of intravenous thrombolytic treatment in acute myocardial infarction. Lancet 1 (8478): 397–402

Hare JM, Givertz MM, Creager MA, Colucci WS (1998) Increased sensitivity to nitric oxide synthase inhibition in patients with heart failure. Potentiation of β-adrenergic inotropic responsiveness. Circulation 97:161–166

Harrison DG (1997) Cellular and molecular mechanisms of endothelial cell dysfunction. J Clin Invest 100:2153–2157

Hasebe N, Shen Y-T, Vatner SF (1993) Inhibition of endothelium-derived relaxing factor enhances myocardial stunning in conscious dogs. Circulation 88:2862–2871

Hoshida S, Yamashita N, Igarashi J, Nishida M, Hori M, Kamada T, Kuzuya T, Tada M (1995) Nitric oxide synthase protects the heart against ischaemia–reperfusion injury in rabbits. J Pharmacol Exp Ther 274:413–418

Inauen ME, Granger DN, Meininger CJ, Schelling ME, Granger HJ, Kvietys PR (1990) Anoxia-reoxygenation-induced, neutrophil-mediated endothelial cell injury: role of elastase. Am J Physiol 28:H925–931

ISIS-2 (Second International Study of Infarct Survival) Collaborative Group (1988) Randomized trial of intravenous streptokinase, oral aspirin, both, or neither among 17187 cases of acute myocardial infarction: ISIS-2. Lancet 2 (8607): 349–360

Johnson G, Tsao PS, Mulloy D, Lefer AM (1990) Cardioprotective effects of acidified sodium nitrite in myocardial ischaemia with reperfusion. J Pharmacol Exp Ther 252 (1): 35–41

Johnson G, Tsao PC, Lefer AM (1991) Cardioprotective effects of authentic nitric oxide in myocardial ischaemia with reperfusion. Crit Care Med 19: 244–252

Kelly RA, Balligand JL, Smith TW (1996) Nitric oxide and cardiac function. Circ Res: 79: 363–380

Kitakaze M, Node K, Komamura K, Minamino T, Inoue M, Hori M, Kamada T (1995) Evidence for nitric oxide generation in the cardiomyocytes: its augmentation by hypoxia. J Mol Cell Cardiol 27:2149–2154

Kloner RA, Ganote CE, Jennings RB (1974) The "no-reflow" phenomenon after temporary coronary occlusion in the dog. J Clin Invest 54:1496–1508

Ku DD (1982) Coronary vascular reactivity after acute myocardial ischaemia. Science 218:576–578

Kumar M, Liu G-J, Floyd RA, Grammas P (1996) Anoxic injury of endothelial cells increases production of nitric oxide and hydroxyl radicals. Biochem Biophys Res Comm 219: 497–501

Lawson DL, Mehta JL, Nichols WW (1990) Coronary reperfusion in dogs inhibits endothelium-dependent relaxation: role of superoxide radicals. Free Radic Biol Med 8:373–380

Lefer DJ (1995) Myocardial protective actions of nitric oxide donors after myocardial ischaemia and reperfusion. New Horizons 3 (1): 105–112

Lefer AM, Lefer DJ (1996) The role of nitric oxide and cell adhesion molecules on the microcirculation in ischaemia–reperfusion. Cardiovasc Res 32:743–751

Lefer AM, Tsao PS, Lefer DJ, Ma X-L (1991) Role of endothelial dysfunction in the pathogenesis of reperfusion injury after myocardial ischaemia. FASEB J 5:2029–2034

Lefer DJ, Nakanishi K, Johnson WE, Vinten-Johansen J (1993a) Antineutrophil and myocardial protection actions of a novel nitric oxide donor after acute myocardial ischaemia and reperfusion in dogs. Circulation 88:2337–2350

Lefer DJ, Klunk DA, Lutty GA (1993b) Nitric oxide (NO) donors reduce basal ICAM-1 expression on human aortic endothelial cells (HAECs). Circulation 88:I-565

Lefer DJ, Nakanishi K, Vinten-Johansen J (1993c) Endothelial and myocardial cell protection by a cysteine-containing nitric oxide donor after myocardial ischaemia and reperfusion. J Cardiovasc Pharmacol 22 (suppl 7): S34–S43

Lefer DJ, Scalia R, Campbell B, Nossuli T, Hayward R, Salamon M, Grayson J, Lefer AM (1997) Peroxynitrite inhibits leukocyte-endothelial cell interactions and protects against ischaemia–reperfusion injury in rats. J Clin Invest 99: 684–691

Linz W, Weimer G, Scholkens BA (1992) ACE-inhibition induces NO-formation in cultured bovine endothelial cells and protects isolated ischaemic rat hearts. J Mol Cell Cardiol 24:909–919

Liu P, Hock CE, Nagele R, Wong PY-K (1997) Formation of nitric oxide, superoxide and peroxynitrite in myocardial ischaemia–reperfusion injury in rats. Am J Physiol 272:H2327–H2336

Lu HR, Remeysen P, de Clerck F (1995) Does the anti-arrhythmic effect of ischaemic preconditioning in rats involve the L-arginine nitric oxide pathway. J Cardiovasc Pharmacol: 25:524–530

Ma X-L, Weyrich AS, Lefer DJ, Lefer AN (1993) Diminished basal nitric oxide release after myocardial neutrophil adherence to coronary endothelium. Circ Res 72:403–412

Manning AS, Hearse DJ (1984) Reperfusion-induced arrhythmias: mechanisms and prevention. J Moll Cell Cardiol 16:497–518

Massoudy P, Becker BF, Gerlach E (1995) Nitric oxide accounts for post-ischaemic cardioprotection resulting from angiotensin-converting enzyme inhibition: indirect evidence for a radical scavenger effect in isolated guinea pig heart. J Cardiovasc Pharmacol 25:440–447

Matheis G, Sherman MP, Buckberg GD, Haybron DM, Young HH, Ignarro LJ (1992) Role of L-arginine-nitric oxide pathway in myocardial reoxygenation injury. Am J Physiol 262:H616–H620

Mayer B, Hemmens B (1997) Biosynthesis and action of nitric oxide in mammalian cells. Trends Biochem Sci 22 (12): 477–481

Mehta JL, Lawson DL, Nichols WW (1989a) Attenuated coronary relaxation after reperfusion: effects of superoxide dismutase and TxA_2 inhibitor U63557. Am J Physiol 257:H1240–H1246

Mehta JL, Nichols WW, Donnelly WH, Lawson DL, Thompson L, Riet Mt, Saldeen TGP (1989b) Protection by superoxide dismutase from myocardial dysfunction and attenuation of vasodilator reserve after coronary occlusion and reperfusion in the dog. Circ Res 65:1283–1295

Méry PF, Abi-Gerges N, Vandecasteele G, Jurevicius J, Eschenhagen T, Fischmeister R (1997) Muscarinic regulation of the L-type calcium current in isolated cardiac myocytes. Life Sci 60:1113–1120

Michel T, Feron O (1997) Nitric oxide synthases: which, where, how and why? J Clin Invest 100:2146–2152

Mohan P, Brutsaert DL, Paulus WJ, Sys SU (1996) Myocardial contractile response to nitric oxide and cGMP. Circulation 93:1223–1229

Morita K, Sherman NSP, Buckberg GD, Ihnken K, Matheis G, Young HH, Ignarro LJ (1995) Studies of hypoxaemic/reperfusion injury: without aortic clamping. Role of the L-arginine-nitric oxide pathway: the nitric oxide paradox. J Thorac Cardiovasc Surg 110:1200–1211

Murry CE, Jennings RB, Reimer KA (1986) Preconditioning with ischaemia: a delay of lethal cell injury in ischaemic myocardium. Circulation 74:1124–1136

Nakanishi K, Vinten-Johansen J, Lefer DJ, Zhao Z, Fowler WC, McGee S, Johnston WF (1992) Intracoronary L-arginine during reperfusion improves endothelial function and reduces infarct size. Am J Physiol 263:H1650–H1658

Naseem SA, Kontos MC, Rao PS, Jesse RL, Hess ML, Kukreja RC (1995) Sustained inhibition of nitric oxide by N^G-nitro-L-arginine improves myocardial function following ischaemia/reperfusion in isolated perfused rat heart. J Moll Cell Cardiol 27:419–426

Node K, Kitakaze M, Kosaka H, Komamura K, Minamino T, Tada M, Inoue M, Hori M, Kamada T (1995) Plasma nitric oxide end-products are increased in the ischaemic canine heart. Biochem Biophys Res Comm 211 (2): 370–374

Nossuli TO, Hayward R, Scalia R, Lefer AM (1997) Peroxynitrite reduces myocardial infarct size and preserves coronary endothelium after ischaemia and reperfusion in cats. Circulation 96:2317–2324

Pabla R, Curtis MJ (1995) Effects of NO modulation on cardiac arrhythmias in the rat isolated heart. Circ Res 77:984–992

Pabla R, Curtis NJ (1996) Effect of endogenous nitric oxide on cardiac systolic and diastolic function during ischaemia and reperfusion in the rat isolated perfused heart. J Mol Cell Cardiol 28:2111–2121

Pabla R, Buda AJ, Flynn DM, Salzberg DB, Lefer DJ (1995) Intracoronary nitric oxide improves post-ischaemic coronary blood flow and myocardial contractile function. Am J Physiol 269: H1113–H1121

Pabla R, Buda AJ, Flynn DM, Blessé SA, Shin AM, Curtis MJ, Lefer DJ (1996) Nitric oxide attenuates neutrophil-mediated myocardial contractile dysfunction after ischaemia and reperfusion. Circ Res 78:65–72

Patel VC, Yellon DM, Singh KJ, Neild GH, Woolfson RG (1993) Inhibition of nitric oxide limits infarct size in the in situ rabbit heart. Biochem Biophys Res Comm 194:234–238

Paulus WJ, Vantrimpont PJ, Shah AM (1994) Acute effects of nitric oxide on left ventricular relaxation and diastolic distensibility in man. Circulation 89: 2070–2078

Paulus WJ, Vantrimpont PJ, Shah AM (1995) Paracrine coronary endothelial control of left ventricular function in humans. Circulation 92:2119–2126

Pearson PJ, Lin PJ, Schaff HV (1992) Global myocardial ischaemia and reperfusion impair endothelium-dependent relaxations to aggregating platelets in the canine coronary artery. J Thorac Cardiovasc Surg 103:1147–1154

Pinsky DJ, Patton S, Mesaros S, Brovkovych V, Kubaszewski E, Grunfeld S, Malinski T (1997) Mechanical transduction of nitric oxide synthesis in the beating heart. Circ Res 81:372–379

Piper HM, Garcia-Dorado D, Ovize M (1998) A fresh look at reperfusion injury. Cardiovasc Res 38:291–300

Prendergast BD, Sagach VF, Shah AM (1997) Basal release of nitric oxide augments the Frank-Starling response in the isolated heart. Circulation 96:1320–1329

Reimer KA, Jennings RB (1979) The "wavefront phenomenon" of myocardial ischaemic cell death. II. Transmural progression of necrosis within the framework of ischaemic bed size (myocardium at risk) and collateral flow. Lab Invest 40:633–644

Sato H, Zhao ZQ, McGee D, Williams MW, Hammon JW, Vinten-Johansen J (1995) Supplemental L-arginine during cardioplegic arrest and reperfusion avoids regional post-ischaemic injury. J Thorac Cardiovasc Surg 110:302–314

Schluter KD, Weber M, Schraven E, Piper HM (1994) NO donor SIN-1 protects against reoxygenation-induced cardiomyocyte injury by a dual action. Am J Physiol 267:H1461–H1466

Schluter KD, Jakob G, Ruiz-Meana M, Garcia-Dorado D, Piper HM (1996) Protection of reoxygenated cardiomyocytes against osmotic fragility by nitric oxide donors. Am J Physiol 271:H428–H434

Schulz R, Wambolt R (1995) Inhibition of nitric oxide synthesis protects the isolated working rabbit heart from ischaemia–reperfusion injury. Cardiovasc Res 30:432–439

Seccombe JF, Schaff HV (1995) Coronary artery endothelial function after myocardial ischaemia and reperfusion. Ann Thorac Surg 60:778–788

Shah AM (1996) Paracrine modulation of heart cell function by endothelial cells. Cardiovasc Res 31:847–867

Shah AM, Spurgeon H, Sollott SJ, Talo A, Lakatta EG (1994) 8-Bromo cyclic GMP reduces the myofilament response to calcium in intact cardiac myocytes. Circ Res 74:970–978

Shah AM, Silverman HS, Griffiths EJ, Spurgeon HA, Lakatta EG (1995) cGMP prevents delayed relaxation at reoxygenation after brief hypoxia in isolated cardiac myocytes. Am J Physiol 268:H2396–H2404

Shiki K, Hearse DJ (1987) Preconditioning of ischaemic myocardium: reperfusion-induced arrhythmias. Am J Physiol 253:H1470–H1476

Siegfried MR, Erhardt J, Rider T, Ma X-L, Lefer AM (1992a) Cardioprotection and attenuation of endothelial dysfunction by organic nitric oxide donors in myocardial ischaemia–reperfusion. J Pharmacol Exp Ther 260:668–675

Siegfried MR, Carey C, Ma X-L, Lefer AM (1992b) Beneficial effects of SPM 5185, a cysteine-containing NO donor in myocardial ischaemia–reperfusion. Am J Physiol 263:H771–H777

Smith JA, Shah AM, Lewis MJ, (1991) Factors released from endocardium of the ferret and pig modulate myocardial contraction. J Physiol Lond 439: 1–14

Sun W, Wainwright CL (1997) The role of nitric oxide in modulating ischaemia-induced arrhythmias in rats. J Cardiovasc Pharmacol 29: 554–562

Szabolcs M, Michler RE, Yang X, Aji W, Roy D, Athan E, Sciacca RR, Minanov OP, Cannon PJ (1996) Apoptosis of cardiac myocytes during cardiac allograft rejection. Relation to induction of nitric oxide synthase. Circulation 94:1665–1673

Szekeres M, Dezsi L, Monos E, Metsa-Ketela T (1997) Effect of a new nitric oxide donor on the biomechanical performance of the isolated ischaemic rat heart. Acta Physiol Scand 161:55–61

Thompson and Hess (1986) The oxygen free radical system: a fundamental mechanism in the production of myocardial necrosis. Prog Cardiovasc Dis 28:449–492

Tsao PS, Lefer AM (1990) Time course and mechanism of endothelial dysfunction in isolated ischaemic and hypoxic perfused rat hearts. Am J Physiol 259: H1660-H1666

Tsao PS, Aoki N, Lefer DJ, Johnson G, Lefer AM (1990) Time course of endothelial dysfunction and myocardial injury during myocardial ischaemia and reperfusion in the cat. Circulation 82:1402–1412

Van Benthuysen KM, McMurtry IF, Horowitz LD (1987) Reperfusion after acute coronary occlusion in dogs impairs endothelium-dependent relaxation to acetylcholine and augments contractile reactivity in vitro. J Clin Invest 79: 265–274

Vegh A, Szekeres L, Parratt JR (1992a) Preconditioning of the ischaemic myocardium; involvement of the L-arginine nitric oxide pathway. Br J Pharmacol 107:648–652

Vegh A, Gy Papp J, Szekeres L, Parratt JR (1992b) The local intracoronary administration of methylene blue prevents the pronounced anti-arrhythmic effect of ischaemic preconditioning. Br J Pharmacol 107:910–911

Vegh A, Gy Papp J, Szekeres L, Parratt JR (1993) Prevention by an inhibitor of the L-arginine-nitric oxide pathway of the anti-arrhythmic effects of bradykinin in anaesthetized dogs. Br J Pharmacol 110:18–19

Wang P, Zweier JL (1996) Measurement of nitric oxide and peroxynitrite generation in the post-ischaemic heart. J Biol Chem 271:29223–29230

Weselcouch EO, Baird AJ, Sleph P, Grover GJ (1995) Inhibition of nitric oxide synthesis does not affect ischaemic preconditioning in isolated perfused rat hearts. Am J Physiol 268:H242-H249

Weyrich AS, Ma X-L, Lefer AM (1992) The role of L-arginine in ameliorating reperfusion injury after myocardial ischaemia in the cat. Circulation 86:279–288

Wildhirt SM, Dudek RR, Suzuki H, Pinto V, Narayan KS, Bing RJ (1995) Immunohistochemistry in the identification of nitric oxide synthase isoenzymes in myocardial infarction. Cardiovasc Res 29:526–531

Williams MW, Taft CS, Ramnauth S, Zhao Z-Q, Vinten-Johansen J (1995) Endogenous nitric oxide (NO) protects against ischaemia–reperfusion injury in the rabbit. Cardiovasc Res 30:79–86

Wink DA, Hanbauer I, Krishna MC, DeGraff W, Gamson J, Mitchell JB (1993) Nitric oxide protects aginst cellular damage and cytotoxicity from reactive oxygen species. Proc Natl Acad Sci USA 90:9813–9817

Woolfson RG, Patel VC, Neild GH, Yellon DM (1995) Inhibition of nitric oxide synthesis reduces infarct size by an adenosine-dependent mechanism. Circulation 91:1545–1551

Xie Y-W, Shen W, Zhao G, Xu X, Wolin MS, Hintze TH (1996) Role of endothelium-derived nitric oxide in the modulation of canine myocardial mitochondrial respiration in vitro. Circ Res 79:382–387

Xie Y-W, Kaminski PM, Wolin MS (1998) Inhibition of rat cardiac muscle contraction and mitochondrial respiration by endogenous peroxynitrite formation during post-hypoxic reoxygenation. Circ Res 82:891–897

Yasmin W, Strynadka KD, Schulz R (1997) Generation of peroxynitrite contributes to ischaemia–reperfusion injury in isolated rat hearts. Cardiovasc Res 33:422–432

Yellon DM, Baxter GF, Garcia-Dorado D, Heusch G, Sumeray MS (1998) Ischaemic preconditioning: present position and future directions. Cardiovasc Res 37:21–33

Zhang X, Xie Y-W, Nasjletti A, Xu X, Wolin MS, Hintze TH (1997) ACE inhibitors promote nitric oxide accumulation to modulate myocardial oxygen consumption. Circulation 95:176–182

Zweier JL (1988) Measurement of superoxide-derived free radicals in the reperfused heart. Evidence for a free radical mechanism of reperfusion injury. J Biol Chem 263:1353–1357

CHAPTER 22
Nitric Oxide and Atherosclerosis

H. BULT, K.E. MATTHYS, and M.M. KOCKX

A. Introduction

Over the last decade, it has become clear that the functions of vascular endothelium are influenced by atherosclerosis. Conversely, atherogenesis is modulated by mediators released by the endothelial cells as well. Nitric oxide (NO) is one of those mediators, as are cytokines, prostacyclin, and an endothelium-derived hyperpolarizing factor (EDHF) whose identity is still debated. This review focuses on the alterations of NO signaling during atherosclerosis and discusses their consequences for the progression of the disease. First, the successive stages of human atherosclerosis, the corresponding animal models, and concepts about the pathogenesis of atherosclerosis are briefly described. The main text is focused on the complex interrelationships between NO and atherosclerosis.

B. Stages of Intimal Thickening and Atherosclerosis
I. The Physiological Intima: the Soil for Atherosclerosis

Human atherosclerotic lesions develop in the inner coat (tunica intima) of the large arteries (aorta, carotid, femoral, and coronary arteries). At birth, the intima of most arteries consists solely of endothelial cells but, soon after birth, focal and circumferential thickening occurs (STARY et al. 1992). This spontaneously developing intima consists of smooth-muscle cells, connective tissue and isolated macrophages and is considered an adaptation to mechanical wall stress (THUBRIKAR and ROBICSEK 1995). Although not pathologic at this stage, the thickened intima marks locations where atherosclerosis tends to develop later in life under the influence of atherogenic stimuli.

By creating a modest mechanical injury of the smooth muscle cells of the media, intimal thickening can be induced experimentally in species that do not develop intimal cushions. All models illustrate the three-wave paradigm for the involvement of smooth-muscle cells in the formation of intimal cushions (JACKSON 1994). The discrete injury of the media evokes smooth-muscle cell proliferation, followed by migration to the intima and an extended phase of proliferation in the intima. The most extensively investigated model involves balloon denudation of the intima of the rat carotid artery with an embolectomy catheter, which mimics physiological intimal thickening rather than restenosis after balloon angioplasty (JACKSON 1994). The removal of the endothelial cells, which hardly regenerate in this model, is not essential for the intimal hyperplasia. Placing a flexible collar around the rabbit carotid artery does not create direct endothelial injury but induces smooth-muscle cell proliferation in the media, followed by migration and prolonged proliferation in the intima (KOCKX et al. 1992; DE MEYER et al. 1997).

II. Successive Stages of Atherosclerosis

Early atherosclerosis is characterized by the deposition of intracellular and extracellular lipids and by the appearance of macrophages and T-lymphocytes in the intima. As macrophages and smooth-muscle cells underneath the endothelial cells accumulate lipid, they acquire a "foamy" appearance. Clusters of lipid-laden cells become macroscopically visible as fatty streaks (DAVIES and WOOLF 1993). These flat, fatty lesions may transform into raised fibrolipid plaques and finally into a fibroatheroma, which has a very characteristic microanatomy with a core of extracellular lipid covered at the luminal side by a thick fibrous cap. Surrounding the core are lipid-laden foam cells, while ischemia in the necrotic core initiates angiogenesis. This type of plaque may cause narrowing of the lumen once compensatory vascular remodeling (Chap. 26•, Sect. C.II) becomes exhausted. The ultimate stage, the complicated plaque, may arise either from fissure of the fibrous cap or from intra-plaque hemorrhage (Chap. 26•, Sect. C.III). If the thrombus is not occlusive, it becomes incorporated into the plaque and is organized by invading macrophages, endothelial cells and smooth-muscle cells, thereby further compromising the lumen of the vessel. The sequence of fissure, thrombus formation, organization and incorporation into the plaque may occur repeatedly.

Current knowledge of the initiation of the atherogenic process is largely based on rabbit, mouse, porcine or primate models of hypercholesterolemia. The hypercholesterolemia can be diet-induced or genetically determined, as in Watanabe Heritable Hyperlipidemic (WHHL) rabbits, certain transgenic mice or mice with a targeted gene disruption (LUSIS 1993), and provokes intravascular lipid infiltration leading to the formation of fatty streaks (DAVIES and WOOLF 1993). Protracted exposure to elevated plasma cholesterol levels eventually results in advanced fibrolipid plaques containing necrotic debris, as in advanced human disease.

III. Accelerated Atherosclerosis

The syndromes of accelerated atherosclerosis, namely heart-transplant atherosclerosis, coronary vein-graft disease and restenosis after percutaneous transluminal coronary angioplasty (PTCA), share etiologic mechanisms with spontaneous atherosclerosis by means of the "response to injury" hypothesis (Ross 1993) (Chap. 26, Sect. C.I). The vascular injury after PTCA stimulates intimal thickening (Chap. 22, Sect. B.I) but, in addition, the more severe damage of the media stimulates mural thrombus formation. The organization of the thrombus by smooth-muscle cells, macrophages, endothelial cells and neovascularization and the subsequent fibrosis accelerate neointima formation. Whether the intimal hyperplasia evolves to a premature occlusive process (restenosis) is, to a significant extent, determined by the capacity of the arteries to develop vascular remodeling (Chap. 22, Sect. C.II).

The inflation of an angioplasty balloon in normal arteries or in arteries with pre-existing (often cholesterol-induced) atherosclerotic lesions of rabbits, pigs or other experimental animals is used to mimic restenosis after PTCA (JACKSON 1994). Vessel-wall distension by the repeated inflation of a slightly oversized balloon creates more extensive injury of the media than the gentle passage of an embolectomy catheter used to induce intimal thickening and, thus, predisposes to thrombus formation. As in human restenosis, thrombus incorporation adds to the bulk of the neointima (JACKSON 1994). The deep balloon injury also activates fibroblasts in the adventitia; such fibroblasts differentiate to myofibroblasts. These may traverse the external elastic lamina and appear in the neointima at sites where the barrier function of the media has been destroyed (ZALEWSKI and SHI 1997). Thus, both adventitial myofibroblasts and smooth-muscle cells derived from the media may contribute to the intimal hyperplasia after PTCA. Further characteristics of balloon denudation are: the quick and often complete recovery of the endothelial cell layer through outgrowth from patches of cells that remained present after the angioplasty, and the major contribution of vascular remodeling to the final lumen size (BIRNBAUM et al. 1997; ZALEWSKI and SHI 1997).

C. Pathogenic Mechanisms

I. The Initiation of Atherosclerosis

The long-standing and continuously refined "response-to-injury" hypothesis (Ross 1993) considers the lesions as the result of an excessive inflammatory–fibroproliferative response to various forms of insults to the endothelium. Well-known risk factors for atherosclerosis include hypercholesterolemia, hypertension, diabetes and smoking. Impaired endothelial NO synthase (eNOS) activity and increased oxidative stress in the vascular wall (VOGEL 1997) are associated with these different conditions and, therefore, are implicated in the pathogenesis of the disease. Stimulation of the oxidative-stress-sensitive transcription factor nuclear factor κB (NF-κB), which acts as a promoter for genes, such as those coding for vascular cell adhesion molecule 1 (VCAM-1) and monocyte chemotactic protein-1 (MCP-1), results in inflammatory activation of the endothelium (COLLINS 1993) and the subsequent recruitment of mononuclear cells. These leukocytes are capable of producing numerous inflammatory mediators and growth factors for smooth-muscle cells, e.g., platelet-derived growth factor (PDGF), thereby contributing to lesion development. The dominant localization of plaques at flow dividers may be explained by the absence of the normal shear stress exerted by laminar blood flow, leading to downregulation of certain shear-stress-responsive endothelial genes involved in vessel homeostasis, such as eNOS and the antioxidant enzyme superoxide dismutase (SOD) (TOPPER et al. 1996).

In hypercholesterolemia-induced atherosclerosis, a causal role is attributed to oxidized low-density lipoprotein (oxLDL) (WITZTUM 1993; HAMILTON 1997). Oxidation of lipoproteins flooding the intima may result from the production of reactive oxygen intermediates or 15-lipoxygenase activity in the endothelial cells. OxLDL, in turn, is cytotoxic to endothelial cells due to the metal-catalyzed production of free radicals from lipid hydroperoxides contained in the modified lipoprotein particle (THOMAS et al. 1993). Furthermore, oxLDL is chemotactic for monocytes and T-lymphocytes. Newly formed epitopes in oxLDL elicit cell-mediated and humoral immune responses (LIBBY and HANSSON 1991). Minimally modified low-density lipoprotein (LDL) stimulates the endothelial cells and smooth-muscle cells to secrete MCP-1 and growth factors involved in the differentiation and proliferation of monocytes. In addition, oxLDL may, synergistically with cytokines, promote mononuclear leukocyte adhesion to the endothelium through the induction of VCAM-1 (KUME et al. 1992; KHAN et al. 1995). Monocyte-derived macrophages internalize oxLDL through scavenger receptors. As these receptors are not downregulated by the intracellular cholesterol level, massive cholesterol accumulation occurs, and the macrophages transform to foam cells. Thus, it appears that the vessel wall, through the oxidation of LDL, recruits macrophages to remove the invaded lipoprotein particles. This hypothesis also implies that a chronic inflammatory response will develop if the macrophages are unable to eliminate oxLDL sufficiently.

II. Remodeling of the Artery

The pioneering study by GLAGOV et al. (1987) demonstrated that arteries undergo a compensatory enlargement as atherosclerotic plaques develop (BIRNBAUM et al. 1997). It is now appreciated that vascular remodeling, i.e., compensatory enlargement, inadequate compensatory enlargement or shrinkage at sites of atherosclerotic lesions, is a major determinant of vessel lumen size both in native atherosclerosis and in restenosis after balloon angioplasty in animal models and human coronary arteries (BIRNBAUM et al. 1997; ZALEWSKI and SHI 1997). The intracellular and intercellular mechanisms of the compensatory enlargement, which proceeds until the plaque occupies more than 40% of the lumen (GLAGOV et al. 1987), are largely unclear. One hypothesis proposes that the endothelium senses changes in shear forces as the atheroma expands and that local release of mediators, among which NO could be a candidate, is involved in the rearrangement of cells in the media and adventitia, leading to a compensatory enlargement (BIRNBAUM et al. 1997).

III. Plaque Stability

The thromboembolic events following plaque fissure are a major cause of clinically manifest acute ischemic syndromes. Major mechanisms leading to coronary thrombosis include frank rupture of a plaque's fibrous cap, intra-plaque hemorrhage and superficial erosion of the endothelium. Plaque rupture occurs when the mechanical stresses in the fibrous cap exceed a critical level that the tissue can withstand (LEE and LIBBY 1997). Biological factors weakening the fibrous cap include infiltration with inflammatory macrophages and T cells and a reduction of the smooth-muscle cell number at critical locations. The macrophages can promote local expression or activation of matrix metalloproteinases (MMPs), which decrease the strength of the cap by degrading collagen and other matrix components. Furthermore, activated macrophages in atherosclerotic lesions kill smooth-muscle cells in their vicinity either via lytic damage leading to necrosis or by inducing apoptosis (KOCKX et al. 1996a, 1996b, 1998b). As smooth-muscle cells are central to the biosynthesis and maintenance of the fibrous cap, their number may become inadequate to repair the degradation. Generation of NO by activated macrophages could be one of the factors which tip the balance from a stable to an unstable plaque.

D. Dysfunction of eNOS Signaling in Atherosclerosis

Established atherosclerosis or the presence of risk factors like hypertension and hypercholesterolemia decrease the activity of the NO pathway. Endothelium-dependent dilation is lost progressively as atherogenesis continues. Conduit vessels with lesions as well as resistance vessels in which lesions do not develop are affected, the latter apparently to a lesser extent. In addition, impaired activity of endothelium-derived prostacyclin (BEETENS et al. 1986) and the elusive EDHF (URAKAMI-HARASAWA et al. 1997) may contribute to the loss of endothelial dilator function in atherosclerosis.

I. Impaired Relaxation in Isolated Arteries

It has been recognized for a long time that atherosclerotic blood vessels are very susceptible to vasospasm (SCHROEDER et al. 1977; GINSBURG et al. 1984; KASKI et al. 1986). Because coronary vasospasm can be provoked by several stimuli with different mechanisms of action, it has been proposed that dysfunction of the endothelium in atherosclerosis may contribute to that phenomenon by leaving constrictor responses unopposed (COCKS and ANGUS 1983; HEISTAD et al. 1984). Even before it was realized that NO accounts for the biological activity of endothelium-derived relaxing factor, it was indeed demonstrated that arteries from atherosclerotic animals show a loss of endothelium-dependent relaxation in organ-bath experiments (JAYAKODY et al. 1985; FREIMAN et al. 1986; HABIB et al. 1986; SREEHARAN et al. 1986; VERBEUREN et al. 1986). From then on, numerous in vitro studies confirmed the defect in the NO-signaling pathway in isolated atherosclerotic blood vessels in rabbits (BOSSALLER et al. 1987; CHAPPELL et al. 1987; JAYAKODY et al. 1987; KANAMURA et al. 1989; KOLODGIE et al. 1990; GALLE et al. 1991; TAGAWA et al. 1991), pigs (YAMAMOTO et al. 1987; COHEN et al. 1988; KOMORI et al. 1989; SHIMOKAWA and VANHOUTTE 1989), rats (YU et al. 1993) and primates (HARRISON et al. 1987), including humans (BERKENBOOM et al. 1987; BOSSALLER et al. 1987; FÖRSTERMANN et al. 1988; CHESTER et al. 1990). Both basal and stimulated NO release appeared to be affected (CHESTER et al. 1990; VERBEUREN et al. 1991; LEFER and MA 1993). Endothelial dysfunction in the rabbit aorta is strictly dependent on plaque size (VERBEUREN et al. 1986; KANAMURA et al. 1989; BULT et al. 1995). Several in vitro studies have demonstrated that incubation of vessel segments with lipoproteins and, in particular, with oxLDL inhibited endothelium-dependent relaxation in a way similar to that seen in hypercholesterolemia in vivo (FLAVAHAN 1992).

II. In vivo Studies of the eNOS Defect in Atherosclerotic Arteries

Urinary nitrate, an index metabolite for NO formation in vivo, is decreased in cholesterol-fed rabbits (BÖGER et al. 1995). Catheterization-based studies in patients with coronary-artery disease also demonstrated the impairment of endothelium-dependent coronary vasodilation to acetylcholine (HORIO et al. 1986; LUDMER et al. 1986; BOSSALLER et al. 1989; GORDON et al. 1989; NEWMAN et al. 1990), serotonin (VRINTS et al. 1992a) or increased flow (COX et al. 1989; DREXLER et al. 1989; NABEL et al. 1990), particularly at atherosclerosis-prone branch points (MCLENACHAN et al. 1990). The deterioration of endothelium-dependent vasodilation is an early event, as it can be observed in patients with typical angina or cardiac risk factors but angiographically smooth coronary arteries (WERNS et al. 1989; MCLENACHAN et al. 1990; VITA et al. 1990; YASUE et al. 1990; ZEIHER et al. 1991; VRINTS et al. 1992b; RASHEED and HODGSON 1993; KAWASHIMA et al. 1995). The current weight of evidence suggests that impaired endothelium-dependent vasodilation is the predominant mechanism underlying ischemic manifestations (VANHOUTTE and SHIMOKAWA 1989; VITA et al. 1990; MEREDITH et al. 1993). Unopposed vasoconstrictor responses in general and the loss of NO in the net reaction to some agonists, e.g., serotonin and norepinephrine in the pig and dog (COCKS and ANGUS 1983), may contribute to the occurrence of vasospasm in atherosclerotic vessels.

III. The Systemic Nature of the Defective eNOS Signaling

Several studies (ANDERSON et al. 1995) demonstrated that atherosclerosis in conduit vessels is accompanied by impaired endothelium-dependent vasodilation in the microcirculation, e.g., in the coronary (OSBORNE et al. 1989b; SELLKE et al. 1990; ZEIHER et al. 1991; KUO et al. 1992) and in peripheral resistance arteries (YAMAMOTO et al. 1988; CREAGER et al. 1990; LIAO et al. 1991;

CASINO et al. 1995). Also, the mere presence of cardiovascular risk factors was associated with dysfunctional microvascular endothelium (CELERMAJER et al. 1992; EGASHIRA et al. 1993). These findings have been confirmed in patients with hypertension (TADDEI et al. 1996; QUYYUMI et al. 1997), hypercholesterolemia (GOODE et al. 1997; O'DRISCOLL et al. 1997; QUYYUMI et al. 1997; STROES et al. 1997; TAMAI et al. 1997; TING et al. 1997; VERHAAR et al. 1998) or diabetes (NITENBERG et al. 1998; TIMIMI et al. 1998) and in smokers (HEITZER et al. 1996; ICHIKI et al. 1996), males (HASHIMOTO et al. 1995) and young, healthy subjects with a family history of premature coronary disease (CLARKSON et al. 1997). While the expected hypertensive effect might contribute to the progression of cardiovascular disease, this defect in microcirculation also implies that, in addition to a dysfunctional endothelial NO pathway, other factors are involved in the initiation and progression of atherosclerotic plaques, since lesions do not develop in these microvessels. The systemic nature of the endothelial dysfunction is now used for the non-invasive evaluation of endothelial function in readily accessible arteries (UEHATA et al. 1993). It should be kept in mind, however, that the contribution of EDHF to endothelium-dependent vasodilatation becomes more important in the peripheral circulation and that both the NO and the EDHF pathways are disturbed by hypercholesterolemia (URAKAMI-HARASAWA et al. 1997).

E. Explanations for the Defective eNOS-Signaling Pathway

The mechanisms underlying the dysfunctional endothelial NO-signaling pathway in atherosclerosis and hypercholesterolemia are multifactorial (FLAVAHAN 1992; HARRISON and OHARA 1995). Atherosclerotic arteries demonstrating disturbed endothelium-dependent relaxation are still capable of dilating to the NO donor nitroglycerin, which provides the smooth muscle with NO upon bioconversion (HARRISON and BATES 1993). As this demonstrates that the smooth muscle is still responsive to the dilatory action of NO, defective endothelial NO release or increased NO inactivation after release appear to be involved. This is supported by the decreased release of bioactive NO from isolated perfused atherosclerotic arteries, as assessed by a superfusion bioassay (SREEHARAN et al. 1986; SHIMOKAWA and VANHOUTTE 1989; MINOR et al. 1990; VERBEUREN et al. 1990; TAGAWA et al. 1991; BULT et al. 1995).

I. Endothelial Receptor Dysfunction

Endothelium-dependent dilation is lost in a progressive, hierarchical fashion (MEREDITH et al. 1993). Vasodilator responses to acetylcholine and serotonin are lost early, before impairment of the dilation due to other receptor agonists, to the receptor-independent Ca^{2+} ionophore A-23178 or to mechanical stimuli. The agonist specificity of the early dysfunction points to selective alterations in endothelial receptors or post-receptor effector pathways. In this respect, it has been demonstrated that the pertussis-toxin-sensitive G_i-protein-signaling pathway, which is employed by serotonin to elicit endothelium-dependent relaxation in the pig coronary artery, is impaired in the early stages of the atherosclerotic process (SHIMOKAWA et al. 1991). An immunohistochemical study of human coronary arteries using specific antibodies against the α subunits of G_{i1} and G_{i2} proteins demonstrated lower levels of G_i protein with aging, extent of atherosclerosis, hypertension and hyperlipidemia (TSUTSUI et al. 1994). Moreover, exposure of endothelial cells to oxLDL in vitro appeared to decrease the levels of G_i mRNA and protein (LIAO and CLARK 1995). Lysophosphatidylcholine, one of the many constituents of oxLDL, mimicked the effects of the whole particle in vascular-reactivity studies (KUGIYAMA et al. 1990; MANGIN et al. 1993; MUROHARA et al. 1994) and selectively inhibited G_i-protein-dependent signaling in porcine endothelial cells (FLAVAHAN 1993).

II. Expression of eNOS mRNA and Protein

The receptor selectivity in early atherosclerosis (Chap. 22, Sect. E.I) argues against a reduced expression of eNOS activity in endothelial cells, although this could occur at later stages. Monocyte-derived inflammatory cytokines, such as tumor necrosis factor α and interleukin (IL)-1α have been demonstrated to downregulate eNOS protein and NO release in cultured porcine and human aortic endothelial cells (MARCZIN et al. 1996). Paradoxically, in another study using human umbilical-vein cells, while steady state mRNA levels of eNOS fell after cytokine treatment, the overall eNOS activity appeared to be increased due to enhanced endogenous synthesis of the cofactor tetrahydrobiopterin (THB) (ROSENKRANZ et al. 1994). Exposure to cytokines also increased the transport of L-arginine into endothelial cells (DURANTE et al. 1996).

Exposure of endothelial cells in culture to low concentrations of oxLDL has been reported both to either increase (HIRATA et al. 1995) or decrease (LIAO et al. 1995) the expression of eNOS mRNA and protein. The discrepancy between the reports could be due to large variability among different preparations of oxLDL with respect to biological activities. Downregulation of eNOS mRNA by higher concentrations of oxLDL is a more consistent finding, whereas native LDL was without effect (HIRATA et al. 1995; LIAO et al. 1995). With respect to the in vivo situation, increased expression of eNOS mRNA and protein in endothelial cells overlying fatty streaks has been demonstrated in the aortas of hypercholesterolemic rabbits by in situ hybridization and immunohistochemistry (KANAZAWA et al. 1996). However, in a study of humans, a loss of eNOS expression in endothelial cells covering advanced lesions has been reported (WILCOX et al. 1997).

III. THB Deficiency

Deficiency of the cofactor THB in vitro caused uncoupling of the L-arginine–NO pathway and led to increased formation of oxygen radicals by NO synthase (NOS) itself (COSENTINO and KATUSIC 1995; WEVER et al. 1997). Although levels of THB have not been measured in endothelium of atherosclerotic blood vessels, physiological studies suggest that they may be deficient. In hypercholesterolemic patients, both the attenuated monomethyl L-arginine (L-NMMA)-induced vasoconstriction and the impaired serotonin-induced relaxation measured by forearm venous plethysmography were restored during infusion of THB or the folate precursor 5-methyltetrahydrofolate. Neither treatment had an effect in control subjects (STROES et al. 1997; VERHAAR et al. 1998). However, THB infusion may not be active by supplying NOS with its cofactor but simply by acting as an antioxidant (KOJIMA et al. 1995; VERHAAR et al. 1998).

IV. Arginine Availability

L-Arginine enters cells by facilitated diffusion via the y^+ transporter (BARBUL 1990). The y^+ transporter protein CAT1 co-localizes with eNOS in caveolae (McDONALD et al. 1997). Moreover, endothelial cells can resynthesize L-arginine from L-citrulline (HECKER et al. 1990). As exogenous L-arginine addition neither induces endothelium-dependent relaxation by itself in normal arteries nor enhances agonist-induced relaxations in isolated atherosclerotic vessel rings (VERBEUREN et al. 1993; TARRY and MAKHOUL 1994; BULT et al. 1995), the intracellular stores appear to be sufficient for maximal eNOS activity. Indeed, intracellular arginine concentrations in endothelial cells range from 0.1 mM to 0.8 mM, while the half-saturating concentration for eNOS is below $10\,\mu M$ (McDONALD et al. 1997). In view of the receptor selectivity of the endothelial dysfunction in early atherosclerosis (Chap. 22, Sect. D.I), arginine availability seems to be sufficient initially. However, in vivo studies showed that L-arginine may improve endothelium-dependent vasodilation, though the behavior of conduit arteries with overt atherosclerosis appears to be different from arterioles in the microcirculation, in which atherosclerosis does not develop.

1. Conduit Arteries with Atherosclerosis

Most authors agree that in vitro L-arginine addition fails to restore the endothelium-dependent relaxations in the atherosclerotic rabbit aorta (MÜGGE and HARRISON 1991; CAPAROTTA et al. 1993; BULT et al. 1995) and femoral artery (WHITE et al. 1994) or the output of NO (BULT et al. 1995). In patients with coronary-artery disease, peripheral-artery occlusive disease or diabetes, L-arginine also failed to improve endothelium-dependent dilation of the conduit arteries (DREXLER et al. 1991; BODE-BÖGER et al. 1996; NITENBERG et al. 1998).

In contrast, in vivo L-arginine administration to hypercholesterolemic rabbits improved the ex vivo endothelium-dependent relaxations of the isolated aorta; however, it should be noted that responses to nitroglycerin were affected to a very similar extent (COOKE et al. 1991). Also, prolonged in vivo L-arginine treatment ameliorated endothelium-dependent relaxations of isolated segments only marginally (SINGER et al. 1995). As the endothelial dysfunction in rabbit conduit arteries is strictly dependent on plaque size (Chap. 22, Sect. D.I), the marked anti-atherogenic effect of prolonged L-arginine supplementation (Chap. 22, Sect. G.II.1) most likely explains the improved endothelium-dependent relaxations.

2. Conduit Arteries Without Overt Atherosclerosis

Although the rabbit basilar artery develops neither atherosclerotic lesions nor a clear endothelial dysfunction after prolonged hypercholesterolemia (KANAMURA et al. 1989), an improvement of the endothelium-dependent relaxations has been reported after in vitro exposure to L-arginine (ROSSITCH et al. 1991). In addition, L-arginine attenuated the augmented vasoconstrictor responses to potassium chloride, serotonin and endothelin. However, cyclic guanosine 3′,5′-monophosphate (GMP)-mediated relaxations induced by endothelium-independent agonists were not studied and, hence, it is not entirely clear whether the actions of L-arginine can be attributed solely to enhanced endothelial NO production. Lefer and Ma measured constrictions evoked by the NOS inhibitor nitro-L-arginine methyl ester (L-NAME) as an index of basal NO release by the endothelial cells of lesion-free rabbit coronary arteries isolated after 3 weeks of cholesterol diet (LEFER and MA 1993). A reciprocal relationship existed between L-NAME-evoked contractions and plasma cholesterol, suggesting that basal NO release by the segments became compromised in the absence of overt atherosclerosis. However, as non-endothelial inducible NOS (iNOS) is induced in arteries of cholesterol-fed rabbits (Chap. 22, Sect. F.I), one cannot exclude the possibility that these vasoconstrictor responses to L-NAME resulted from inhibition of iNOS rather than eNOS.

3. Arterioles Without Overt Atherosclerosis

All studies except one (CASINO et al. 1994) reported that L-arginine infusion improved or even completely restored endothelium-dependent vasodilation in the coronary and peripheral microcirculation in hypercholesterolemic rabbits (GIRERD et al. 1990), pigs (KUO et al. 1992) and humans (DREXLER et al. 1991; CREAGER et al. 1992).

4. Possible Explanations for the Arginine Paradox

In view of plasma levels of L-arginine in the range of 150–250 μM and a K_m of 5–10 μM for NOS (GROSS and WOLIN 1995), it is surprising that L-arginine availability can ever limit NO biosynthesis. However, it is possible that the concentration of L-arginine in microdomains of the cell, e.g., in the caveolae, is not reflected in the total cellular concentration (McDonald et al. 1997). Another possibility is that the increase in membrane cholesterol associated with hypercholesterolemia impairs endothelial L-arginine transport, thus eventually depleting the intracellular stores. The latter could also result from increased L-arginine consumption due to iNOS expression (Chap. 22, Sect. F.I), as suggested by the raised output of inactive nitrogen oxides in the hypercholesterolemic rabbit aorta (Minor et al. 1990) or the enhanced activities of arginase I or II. Interestingly, arginase activities are inhibited by N^G-hydroxyarginine, the intermediate of the NOS pathway (Buga et al. 1996).

However, reversal (by L-arginine) of hypercholesterolemic endothelial dysfunction may not simply reflect the replenishment of the substrate for NO production (Jeremy et al. 1996). The positive effect of L-arginine could be based on its ability to scavenge superoxide anions (Wascher et al. 1997), thereby protecting NO from oxidative breakdown. Moreover, as the best results are obtained with L-arginine administration in vivo, other systemic effects of the amino acid, e.g., its secretagogue effects on the adrenals and pituitary gland, may prevail (Barbul 1990). This is illustrated by observations in healthy persons, where L-arginine infusion stimulated basal and acetylcholine-induced relaxation in the peripheral circulation (Imaizumi et al. 1992; Casino et al. 1994; Bode-Böger et al. 1994) and decreased the systemic blood pressure (Kanno et al. 1992; Bode-Böger et al. 1996). The concomitant increase in urinary nitrate and cyclic GMP could not simply be attributed to a direct stimulating effect of L-arginine on eNOS, as prostaglandin E_1-induced dilation also increased these parameters in urine (Bode-Böger et al. 1996). Indeed, the vascular effects seen after systemic L-arginine infusion in normal human volunteers is, to a substantial extent, mediated by the release of insulin (Giugliano et al. 1997). Therefore, further studies are needed to explain the beneficial effects of L-arginine.

V. Endogenous NOS Antagonists

The existence of endogenous NOS antagonists became evident with the detection of the inhibitors monomethyl arginine (L-NMMA) and asymmetric dimethylarginine (ADMA) in the plasma and urine of humans (Vallance et al. 1992). ADMA has been shown to accumulate in several human disease conditions associated with a loss of endothelium-mediated dilation, including peripheral arterial occlusive disease (Böger et al. 1997). The plasma of cholesterol-fed rabbits contains enhanced levels of ADMA, and higher intracellular concentrations were measured in regenerated endothelial cells after balloon angioplasty of the rabbit carotid artery. The latter was associated with a decreased intracellular arginine concentration (Azuma et al. 1995). The paradigm that an increased ratio of endogenous NOS inhibitors to NOS substrate can modulate NOS activity could offer another potential explanation for the beneficial effects of L-arginine on vasomotion (Chap. 22, Sect. E.IV). The human vasculature possesses dimethylarginase, which metabolizes L-NMMA and ADMA to L-citrulline (MACALLISTER et al. 1994). Inhibition of this enzyme might lead to higher intracellular levels of NOS inhibitors. However, intracellular and plasma levels of ADMA in the aforementioned vascular pathologies remain very low compared with L-arginine levels and the question of whether NOS activity may be affected by these endogenous inhibitors must be further investigated.

VI. Negative Feedback by NO Derived from iNOS

NOSs are vulnerable to feedback inhibition by NO (BUGA et al. 1993), possibly by the interaction of NO with the prosthetic heme group (GRISCAVAGE et al. 1994). Hence, high-output NO production by iNOS (Chap. 22, Sect. F.I) might downregulate eNOS activity. This is supported by the observation that chronic administration of large doses of a NO donor to rabbits depressed the ex vivo output of NO without reducing prostacyclin release (BULT et al. 1995).

VII. Superoxide Anion Inactivates NO

Under normal conditions, inactivation of NO by superoxide radicals is prevented by cytosolic CuZn SOD (MÜGGE et al. 1991a; LYNCH et al. 1997) and by extracellular SOD type C associated with heparan sulfate proteoglycans on the endothelial cell surface and in the interstitium (ABRAHAMSSON et al. 1992). However, hypercholesterolemia in the rabbit increased the production of reactive oxygen species by the intima (MINOR et al. 1990; OHARA et al. 1993) and the media beneath the atheromatous plaque (TAGAWA et al. 1991), resulting in increased degradation of NO (HARRISON and OHARA 1995). Increased vascular production of reactive oxygen species may result from enhanced xanthine oxidase activity in the endothelium (OHARA et al. 1993), from an increase in circulating plasma xanthine oxidase in response to hypercholesterolemia, which binds to endothelial cell glycosaminoglycans (WHITE et al. 1996) or from production by infiltrated monocytes (MÜGGE et al. 1994).

In addition to directly inactivating NO (CHIN et al. 1992), oxLDL has been shown to stimulate the respiratory burst in neutrophils (MAEBA et al. 1995), and lysophosphatidylcholine induces superoxide production in vascular smooth-muscle cells via protein kinase C activation (OHARA et al. 1994). Endothelial reduced nicotinamide adenine dinucleotide phosphate (NADPH) oxidase systems (ZULUETA et al. 1995) [activated by protein kinase C (SHARMA et al. 1991)] or smooth-muscle NADPH oxidase [activated by angiotensin II (RAJAGOPALAN et al. 1996)] may also be involved in the augmented production of oxygen radicals.

Furthermore, a striking feature of NOS is its ability to generate superoxide anion when either L-arginine or the cofactor THB is limiting (GROSS and WOLIN 1995). Under these circumstances, NADPH oxidation is uncoupled from synthesis of NO, and oxygen becomes the electron acceptor, resulting in superoxide formation. Whether the low arginine levels needed for superoxide biosynthesis occur in intact endothelial cells in vivo is unclear. However, protracted endothelial cell exposure to atherogenic native LDL concentrations increased the production of superoxide anion and peroxynitrite (Chap. 22, Sect. H.I), a process which was reversed by arginine supplementation. Superoxide was formed by three independent oxidative systems – cyclooxygenase, P_{450} isozyme and eNOS – of which the latter appeared to be the greatest source (PRITCHARD et al. 1995). iNOS did not seem to be involved; however, in conditions where iNOS (which has a much greater demand for substrate then eNOS) is induced, insufficient L-arginine might result in superoxide anion release. This would also explain the benefit of providing L-arginine, i.e., to promote re-coupling, thus reducing vascular superoxide production and prolonging the half-life of NO. These findings provide insight into the mechanisms by which hypercholesterolemia might both stimulate superoxide production and decrease functional NO levels.

The disturbed balance between vascular superoxide and endothelial NO production, resulting in the loss of functional NO, may be compensated for by iNOS activity in the vascular wall (Chap. 22, Sect. F.I) and/or by upregulation of endogenous SOD (HENRIKSSON et al. 1985; DEL BOCCIO et al. 1990; SHARMA et al. 1992). Addition of CuZn SOD (which does not penetrate cells) to the organ bath or preincubation with extracellular SOD type C (which binds extracellularly to vascular structures) also protected against the detrimental effects of superoxide radicals on endothelium-dependent relaxation (ABRAHAMSSON et al. 1992). Conversely, exhaustion of these protective mechanisms, which may be time-, species- or vessel-dependent, may tip over the balance towards a net decrease in functional NO.

In rabbits, but not in pigs, hypercholesterolemia alone did not impair the endothelial dilator function in large vessels; it only occurred in arteries with intimal plaques (VERBEUREN et al. 1986; KANAMURA et al. 1989; KOLODGIE et al. 1990; GALLE et al. 1991; TAGAWA et al. 1991; BULT et al. 1995), with the exception of the coronary arteries (OSBORNE et al. 1989a; LEFER and MA 1993). Apparently, the rabbit is capable of keeping the superoxide and NO production in balance, as long as lesions do not develop. Superoxide production in the media beneath the plaque (TAGAWA et al. 1991) or the presence of fatty streaks containing large amounts of macrophages and lipids may disturb the balance via high local superoxide production. Raising the antioxidant capacity in the vessel wall by the administration of CuZn SOD, polyethyleneglycolated (MÜGGE et al. 1991b) or liposome-entrapped (WHITE et al. 1994) to ensure cell entrance, partly restored the endothelium-dependent relaxation in the isolated aorta of the cholesterol-fed rabbit. In keeping with these findings, it has been shown that addition of antioxidant vitamins in the diets of cholesterol-fed rabbits preserved the endothelium-dependent dilation in the absence of an effect on lesion formation (KEANEY et al. 1994, 1995). Also, dietary correction of hypercholesterolemia in the rabbit normalized both the endothelial superoxide production and dramatically improved the vasodilator response to acetylcholine (OHARA et al. 1995).

Brachial-artery infusion of ascorbic acid to achieve millimolar concentrations improved acetylcholine-induced vasodilatation of the forearm vasculature both in subjects with hypercholesterolemia (TING et al. 1997) and in chronic cigarette smokers (HEITZER et al. 1996), but not in control subjects. Oral administration of 2 g ascorbic acid produced marked improvement in the forearm vascular response to hyperemia in patients with coronary artery-disease (LEVINE et al. 1996). However, oral administration of vitamin C producing only a modest increase in plasma concentrations did not improve endothelial function in hypercholesterolemic patients (GILLIGAN et al. 1994). Also, short-term treatment with the antioxidants probucol or vitamin E did not improve the forearm vascular responses to acetylcholine in patients with hypercholesterolemia (McDOWELL et al. 1994).

Eventually, atherosclerotic plaques, particularly when lipid-rich, may trap NO and may mechanically disturb the normal dilation of the medial smooth muscle. At this stage, relaxation in response to exogenous NO donors, e.g., nitroglycerin, and to endothelium-independent dilator substances, e.g., atrial natriuretic peptide, also becomes impaired (VERBEUREN et al. 1990).

F. Expression of iNOS

I. iNOS Expression in Atherosclerosis

Functional and biochemical evidence suggested that cholesterol feeding of rabbits induced iNOS expression in the aorta (VERBEUREN et al. 1993) and in the lungs (LANG et al. 1993). The former could account for the increased output of nitrogen oxides in the aortas of cholesterol-fed rabbits (MINOR et al. 1990). Histochemical studies confirmed the expression of iNOS mRNA and protein in cholesterol-induced atherosclerotic rabbit lesions, particularly in macrophages (ESAKI et al. 1997; LUOMA et al. 1998) and in human plaques, where it co-localizes with nitrotyrosine in macrophages and smooth-muscle cells (BUTTERY et al. 1996; WILCOX et al. 1997; LUOMA et al. 1998). In both species, the macrophages in the lesions also expressed high levels of extracellular SOD. Despite the abundant expression of the latter enzyme, epitopes characteristic of oxidized lipoproteins and nitrotyrosine residues implied that malondialdehyde, hydroxynonenal and peroxynitrite are important mediators of oxidative damage in iNOS-positive macrophage-rich lesions (LUOMA et al. 1998). A reduced glutathione content has been found in the aortas of hypercholesterolemic rabbits, and this renders the vascular tissue more susceptible to peroxynitrite toxicity (MA et al. 1997). Induction of iNOS was also observed in the endothelium and smooth muscle of intramyocardial vessels of patients with ischemic heart disease (HABIB et al. 1996) and in macrophages and smooth muscle cells in a rat model of cardiac-transplant atherosclerosis (RUSSELL et al. 1995).

Cytokines released by inflammatory cells may provide the stimulus for iNOS induction in smooth-muscle cells (FUKUMOTO et al. 1997). Also, oxLDL (MATTHYS et al. 1996) and LDL (POMERANTZ et al. 1993; MOHAN and DESAIAH 1994) have been shown to upregulate iNOS activity in macrophages and vascular smooth-muscle cells under certain conditions. However, mediators that inhibit iNOS induction, e.g., heat-shock proteins (WONG et al. 1995) or NO itself (COLASANTI et al. 1995), may determine the final output of NO. The observation that iNOS induction in vascular smooth muscle cells (as in macrophages) is accompanied by upregulation of L-arginine transport (WILEMAN et al. 1995) may contribute to the stimulating effect of L-arginine on vessel relaxation in some experimental settings.

II. Mechanical Injury and iNOS Expression

Vascular reactivity and histochemical studies pointed to iNOS induction after denudation of the rat carotid artery (JOLY et al. 1992; DOUGLAS et al. 1994; HANSSON et al. 1994; YAN and HANSSON 1998) and balloon angioplasty of rabbit arteries (BOSMANS et al. 1996; BULT et al. 1998) but not in collared rabbit arteries (DE MEYER et al. 1994). The immunohistochemical study, which suggested that the collar induced iNOS expression in the intima, suffers from a lack of specificity. In the balloon-injured rat carotid artery, iNOS mRNA became apparent within 24h post-injury, and in situ hybridization located iNOS mRNA in neointimal smooth-muscle cells, particularly at the luminal side of the vessel, conferring a non-thrombogenic surface (HANSSON et al. 1994).

G. NO: a Radical with Anti-Atherogenic Properties

Since the impairment in the eNOS pathway occurs early or may precede the development of visible atherosclerotic lesions, many authors have speculated on a causal role of this functional defect. This view is supported by a number of in vitro studies demonstrating the suppression by NO of several key processes involved in atherogenesis (Table 1).

I. In Vitro Studies

1. Interference with Oxidative Processes

Since superoxide anion contributes to oxidative stress, LDL modification and inflammatory gene transcription via the activation of NF-κB (COLLINS 1993), the decreased formation or inactivation of superoxide by NO may be considered protective. In this respect, it has been shown that the NO derived from iNOS inhibits xanthine oxidase in interferon-γ-stimulated macrophages (RINALDO et al. 1994) and in a cell-free system (CLANCY et al. 1992), possibly by reversible alteration of the flavin prosthetic site (FUKAHORI et al. 1994). NO

Table 1. Anti-atherogenic properties of nitric oxide in vitro

Property	References
Cytoprotection against oxidative stress	Wink et al. 1993, 1995; Motterlini et al. 1996
Inhibition of:	
Cell-mediated LDL oxidation	Jessup et al. 1992; Yates et al. 1992; Jessup and Dean 1993; Bolton et al. 1994; Malo-Ranta et al. 1994; Wang et al. 1994
Lipoxygenase activity	Maccarrone et al. 1996; Rubbo et al. 1995
oxLDL cytotoxicity	Struck et al. 1995
Xanthine oxidase	Clancy et al. 1992; Fukahori et al. 1994; Rinaldo et al. 1994
NADPH oxidase	Clancy et al. 1992
Endothelial permeability	Suttorp et al. 1996; Forster and Weinberg 1997
Endothelial activation	De Caterina et al. 1995
Monocyte chemotaxis	Bath et al. 1991
Monocyte adhesion	Bath et al. 1991; De Caterina et al. 1995; Tsao et al. 1995
Neutrophil adhesion	Siegfried et al. 1992; Lefer and Ma 1993; Ma et al. 1993; Niu et al. 1994; Mehta et al. 1995
MCP-1 expression	Zeiher et al. 1995
NF-κB and VCAM-1 activation	De Caterina et al. 1995; Zeiher et al. 1995; Spiecker et al. 1997
Smooth-muscle cell proliferation	Garg and Hassid 1989; Kariya et al. 1989; Nakaki et al. 1990; Assender et al. 1991; Scott-Burden and Vanhoutte 1993; Cornwell et al. 1994; Mooradian et al. 1995
Smooth-muscle cell migration	Dubey et al. 1995; Sarkar et al. 1996
T-cell proliferation	Fu and Blankenhorn 1992; Kawabe et al. 1992; Merryman et al. 1993
Platelet activation	Bassenge 1991
Macrophage apoptosis	Albina et al. 1993; Sarih et al. 1993; Messmer and Brüne 1996

LDL, low-density lipoprotein; *MCP-1*, monocyte chemotactic protein 1; *NADPH*, reduced nicotinamide adenine dinucleotide phosphate; *NF-κB*, nuclear phosphate κB; *VCAM-1*, vascular cell adhesion molecule 1.

also inhibits neutrophil superoxide anion production via direct action on the NADPH oxidase (Clancy et al. 1992). The NO donors known as NONOates abrogate the cytotoxic effects of superoxide on Chinese hamster lung fibroblasts (Wink et al. 1993). Moreover, NO also protected against cell damage by other reactive oxygen species (e.g., hydrogen peroxide and alkyl peroxides) by several mechanisms, such as inhibition of heme oxidation, Fenton-type oxidation of DNA and propagation of lipid peroxidation (Wink et al. 1995). Activation of the stress protein heme oxygenase by NO may contribute to its cytoprotective effect (Motterlini et al. 1996).

Furthermore, NO reduced the oxidative modification of LDL by macrophages (Jessup et al. 1992; Yates et al. 1992; Jessup and Dean 1993;

WANG et al. 1994), endothelial cells (MALO-RANTA et al. 1994) and lipoxygenase (RUBBO et al. 1995) by acting as terminator of radical chain-propagation reactions (RUBBO et al. 1995). As 15-lipoxygenase has been implicated in LDL oxidation (WITZTUM 1993), NO might also protect LDL by inhibiting lipoxygenase activity (MACCARRONE et al. 1996). Conversely, the decreased expression of iNOS activity in oxLDL-laden foam cells (JORENS et al. 1992; YANG et al. 1994) has been implicated in the accelerated oxidation of LDL by these macrophages (BOLTON et al. 1994). Furthermore, NO released by donor compounds inhibited the cellular toxicity of the lipid hydroperoxides contained in oxLDL, presumably by scavenging the propagatory free radicals generated during peroxidation of the endothelial cell membranes (STRUCK et al. 1995). NO donors also blocked (via a cyclic GMP-mediated mechanism) the hydrogen-peroxide-related increase in endothelial permeability (SUTTORP et al. 1996).

2. Maintenance of Endothelial Barrier Function

NO has been reported to oppose endothelial hyperpermeability (SUTTORP et al. 1996) and to determine the pattern of transport of albumin near branches of the normal rabbit aorta (FORSTER and WEINBERG 1997). These results imply that the influx of lipoproteins into the vascular wall, which is a hallmark of atherosclerosis, is modulated by endothelium-derived NO.

3. Interference with Leukocyte Recruitment

Inhibition of NOS in endothelial cells by L-NAME increased the intracellular oxidative stress, resulting in enhanced adhesion of neutrophils via CD18/intercellular adhesion molecule 1 (ICAM-1) interaction (NIU et al. 1994), upregulation of P selectin (MEHTA et al. 1995), VCAM-1 (DE CATERINA et al. 1995) and MCP-1 mRNA and protein (ZEIHER et al. 1995). Authentic NO gas inhibited monocyte adhesion and chemotaxis (BATH et al. 1991), and exposure to shear stress inhibited (via an NO-dependent mechanism) monocyte adhesion (TSAO et al. 1995). Furthermore, NO donors decreased the cytokine-induced expression of the endothelial adhesion molecules VCAM-1, ICAM-1 and E-selectin (DE CATERINA et al. 1995; SPIECKER et al. 1997) and decreased MCP-1 mRNA expression and secretion, presumably by suppressing a NF-κB-like transcriptional regulator (ZEIHER et al. 1995).

4. Antiproliferative Action of NO

As the proliferation of vascular smooth-muscle cells, macrophages and T-lymphocytes contributes to the progression of atherosclerosis, cell growth inhibition by NO could oppose lesion formation. NO has been shown to inhibit smooth-muscle cell proliferation (GARG and HASSID 1989; KARIYA et al. 1989; NAKAKI et al. 1990; ASSENDER et al. 1991; SCOTT-BURDEN and VANHOUTTE 1993; CORNWELL et al. 1994; MOORADIAN et al. 1995) and migration (DUBEY et al.

1995; SARKAR et al. 1996) in vitro. Both effects were cyclic GMP mediated. T-cell proliferation is also reduced by NO (FU and BLANKENHORN 1992; KAWABE et al. 1992; MERRYMAN et al. 1993). The mitotic effects of NO in endothelial cells are still controversial and require further studies. NO may induce endothelial cell growth and motility in vitro (MORBIDELLI et al. 1996; PAPAPETROPOULOS et al. 1997; ZICHE et al. 1997) or may terminate cell proliferation and promote endothelial cell differentiation (BABAEI et al. 1998).

5. Antiplatelet Effects of NO

Although the inhibitory effect of NO on platelet adhesion and aggregation (BASSENGE 1991) is often considered an anti-atherogenic effect of NO, platelets are minimally involved in early atherogenesis. However, the endothelial surface over advanced human plaques often shows focal loss of cells (DAVIES and WOOLF 1993), and platelet adhesion may promote the progression of those lesions, eventually leading to plaque fissuring and thrombosis. Also, following gross mechanical injury evoked by balloon angioplasty, platelet-derived products have been proposed to contribute to neointima formation (JACKSON 1994).

II. In Vivo Studies

Several in vivo studies support the concept that NO may suppress early atherosclerosis and intimal thickening evoked by mechanical injury (Table 2).

1. Inhibition of Experimental Atherosclerosis

Oral L-arginine supplementation caused a striking inhibition of fatty-streak formation in male (COOKE et al. 1992) but not in female (JEREMY et al. 1996) hypercholesterolemic rabbits. The enhanced endothelial adhesiveness for murine monocytes of isolated hypercholesterolemic aortas was significantly reduced if the rabbits had received supplemental dietary arginine, and it significantly increased in rabbits treated with L-NAME (TSAO et al. 1994). This was associated with, respectively, increased or decreased elaboration of vascular nitrogen oxides, as measured by chemiluminescence. The observation that the hypercholesterolemia-induced impairment in endothelium-dependent relaxation was only marginally improved by L-arginine treatment (SINGER et al. 1995) and was not sustained (JEREMY et al. 1996) suggests that the beneficial effect of L-arginine may primarily derive from the increased activity of iNOS. However, it should be noted again that arginine and other basic amino acids are potent hormonal secretagogues (BARBUL 1990; GIUGLIANO et al. 1997). Hence, part of the anti-atherosclerotic effect of systemic arginine administration might be related to the release of insulin, glucocorticoids or other immunosuppressive hormones, which are known to suppress intimal thickening and experimental atherosclerosis.

Table 2. Anti-atherogenic properties of nitric oxide (NO) in vivo

Modulation of leukocyte-endothelial interaction:	
Attenuated by L-arginine	TSAO et al. 1994
Attenuated by NO donors	GABOURY et al. 1993; Gauthier et al. 1994
Increased by NOS inhibitor	KUBES et al. 1991
Fatty streak formation in cholesterol-fed rabbits:	
Reduced by L-arginine	COOKE et al. 1992; SINGER et al. 1995
Reduced by NO donor	KOJDA and NOACK 1995
Increased by NOS inhibitors	CAYATTE et al. 1994; NARUSE et al. 1994
Intimal thickening evoked by balloon denudation:	
Reduced by L-arginine	MCNAMARA et al. 1993; TAGUCHI et al. 1993
Reduced by eNOS transfection	VON DER LEYEN et al. 1995; CHEN et al. 1998; JANSSENS et al. 1998
Reduced by exogenous NO	GUO et al. 1994; LEE et al. 1996
Increased by NOS inhibitors	FARHY et al. 1993; MCNAMARA et al. 1993; TAGUCHI et al. 1993; HANSSON et al. 1994; TARRY and MAKHOUL 1994; ELLENBY et al. 1996; VAN BELLE et al. 1996; YAN and HANSSON 1998
Intimal thickening evoked by perivascular collar or inflammation:	
Reduced by NO donors	DE MEYER et al. 1995; YIN and DUSTING 1997
Increased by eNOS deficiency	MOROI et al. 1998
Increased by NOS inhibitors	FUKUMOTO et al. 1997
Intimal hyperplasia developing in vein grafts:	
Reduced by L-arginine	DAVIES et al. 1996; OKAZAKI et al. 1997
Reduced by NO donors	CHAUX et al. 1998; FULTON et al. 1998
Inhibition of neo-intima and constrictive remodeling after balloon angioplasty:	
L-arginine	TARRY and MAKHOUL 1994; BOSMANS et al. 1999
eNOS gene transfer	VARENNE et al. 1998

eNOS, endothelial nitric oxide synthase; *NOS*, nitric oxide synthase.

The ambiguity of results obtained with L-arginine is avoided in studies with NO donor compounds or NOS inhibitors. L-NMMA and L-NAME increased leukocyte adhesion in vivo by a CD11/CD18-dependent mechanism (KUBES et al. 1991). Conversely, the NO donor SIN-1 prevented leukocyte adhesion. The observation that both SOD and SIN-1 inhibited leukocyte adhesion only under conditions associated with superoxide formation suggests that the anti-adhesive properties of NO may relate to its ability to inactivate superoxide anion (GABOURY et al. 1993). SPM-5185, another NO donor, attenuated leukocyte endothelial interaction in in vivo models of ischemia–reperfusion, in part through decreased endothelial P-selectin expression (SIEGFRIED et al. 1992; GAUTHIER et al. 1994).

Moreover, oral or parental treatment with L-NAME for 4–12 weeks (CAYATTE et al. 1994; NARUSE et al. 1994) enhanced fatty-streak formation significantly. Therefore, the data suggest that vascular NO, produced by eNOS or iNOS, inhibits de novo formation of intimal lesions. However, this conclusion is confounded by the observed interference with cholesterol resorption or metabolism, which leads to augmented plasma cholesterol levels in these rabbits, particularly after prolonged L-NAME treatment. Since hypercholesterolemia is the ultimate driving force for the lesions in this model, it is conceivable that this explains the accelerated atherosclerosis.

Pentaerythrityl tetranitrate, an organic nitrate, has been documented to inhibit cholesterol-induced fatty-streak formation in rabbits, but the effect was not seen with isosorbide mononitrate (KOJDA and NOACK 1995). This could be due to differences between the two organic nitrates with respect to the development of tolerance or the NO-releasing capacity. Conversely, treatment with molsidomine, prodrug of the spontaneous NO donor SIN-1, actually enhanced lesion formation in the hypercholesterolemic rabbit (BULT et al. 1995). This may relate to the generation of superoxide anion from SIN-1, which could abrogate the beneficial effects of the simultaneously released NO.

The inhibitory effect of NO on atherosclerosis may result from the above described in vitro actions, but NO-mediated decrease of endothelin production (LÜSCHER et al. 1993; BRUNNER et al. 1995) may also be involved. Endothelin is a potent mitogen (LÜSCHER 1994) and inducer of collagen synthesis (RIZVI et al. 1996) in vascular smooth-muscle cell cultures, and its production may be increased in atherosclerosis (LERMAN et al. 1991).

2. Inhibition of Intimal Thickening by NO

a) Neointima Formation after Balloon Denudation

Oral L-arginine supplementation suppressed intimal hyperplasia after balloon denudation of the rat carotid artery (McNAMARA et al. 1993; TAGUCHI et al. 1993). L-NAME reversed the effect of arginine (McNAMARA et al. 1993), indicating that the attenuation of the intimal hyperplasia was mediated by NO, presumably formed by iNOS (JOLY et al. 1992; DOUGLAS et al. 1994; HANSSON et al. 1994), since endothelial cells and eNOS activity do not recover in this model. Moreover, systemic (McNAMARA et al. 1993) or local perivascular (TAGUCHI et al. 1993) administration of L-NAME aggravated intimal thickening in response to balloon denudation. Increasing the flow in the injured carotid artery (by ligating the contralateral artery) significantly reduced intimal thickening, and this effect was, in part, mediated by endogenous NO (ELLENBY et al. 1996). The smooth-muscle cells in the media are more sensitive to the anti-proliferative effect of NO than those in the intima (YAN and HANSSON 1998). Likewise, the protective effect of angiotensin-converting-enzyme inhibitors (which also block kinin degradation) may be mediated, in part, by stimulation of the endogenous production of NO by bradykinin (FARHY et al. 1993; VAN BELLE et al. 1996).

Conversely, oral treatment with the cysteine-containing NO donor SPM-5185 (Guo et al. 1994) or chronic inhalation of NO (LEE et al. 1996) reduced intimal thickening in the denuded rat carotid artery. Nitroglycerin treatment after 3 weeks decreased early medial smooth-muscle cell proliferation without affecting intimal thickness (WOLF et al. 1995). This may be due to the development of tolerance associated with this class of nitrovasodilators (HARRISON and BATES 1993). A single local treatment of the denuded rabbit femoral artery with a protein adduct of NO inhibited platelet deposition and intimal proliferation (MARKS et al. 1995).

In vivo eNOS gene transfer to the denuded rat carotid artery provided further evidence for the inhibition of smooth-muscle cell accumulation by NO. Transfection of the eNOS gene in the media not only restored the calcium-dependent NO production and concomitant relaxations of the denuded artery, it also inhibited neointima formation at day 14 after balloon injury by 70% (VON DER LEYEN et al. 1995; JANSSENS et al. 1998). Similarly, when rat smooth-muscle cells transfected with the human eNOS gene were seeded onto the luminal surface of the balloon-denuded rat carotid artery, 37% inhibition of the neointimal hyperplasia was seen; orally administered nitro-L-arginine reversed the change (CHEN et al. 1998). Finally, percutaneous eNOS gene transfer after PTCA in the pig coronary artery also reduced the neointima formation (Chap. 22, Sect. G.II.4; VARENNE et al. 1998). These experiments provide direct evidence that NO is an endogenous inhibitor of vascular lesion formation in vivo. Furthermore, these experiments suggest the possibility of eNOS transfection [which has been successful in the human saphenous vein (CABLE et al. 1997)] or local delivery of long-lived NO adducts as potential therapeutic approaches to maintain patency after PTCA or in vein grafts.

b) Intimal Hyperplasia Due to Perivascular Manipulation

The inhibition of intimal thickening by NO is also seen when hyperplasia is induced by perivascular collars in rabbits or mice. Oral treatment of rabbits with the NO donor SPM-5185 (DE MEYER et al. 1995) or local treatment with spermine NONOate (YIN and DUSTING 1997) reduced the collar-induced intimal thickening, whereas only a tendency towards inhibition was observed upon oral treatment with molsidomine, prodrug of the NO donor SIN-1 (BULT et al. 1995). It is not clear whether the difference between the drugs was related to the dose, or characteristics of the NO donors, i.e., the presence of sulfydryl groups in SPM-5185 or the release of superoxide anion from SIN-1.

In mice with a targeted disruption of the eNOS gene, a perivascular collar around the femoral artery induced a much greater degree of intimal growth than in wild-type mice (MOROI et al. 1998). Finally, inhibitors of NOS increased the intimal hyperplasia in response to local inflammation evoked by IL-1 in the adventitia of pig coronary arteries (FUKUMOTO et al. 1997).

In these models, a clear relationship between inhibition of smooth-muscle cell mitosis and neointima formation is lacking. Smooth-muscle cell mitosis was influenced less than intimal thickening after eNOS gene transfer in denuded rat arteries (VON DER LEYEN et al. 1995) or after NO-donor treatment of rabbit collared arteries (DE MEYER et al. 1995). This suggests that NO exerts its major effect on smooth-muscle cell migration (DUBEY et al. 1995;

SARKAR et al. 1996), which is a crucial event in intimal thickening. Whether inhibition of migration is of importance to human atherosclerosis or restenosis remains to be determined, as atherosclerosis develops in an existing intima, and migration of smooth-muscle cells from media to intima is not considered a major determinant in atherogenesis (JACKSON 1994).

3. Inhibition of Intimal Hyperplasia in Vein Grafts

Oral L-arginine supplementation with L-arginine suppressed intimal hyperplasia in autologous vein grafts implanted in the arterial circulation of normal (DAVIES et al. 1996) and hypercholesterolemic (OKAZAKI et al. 1997) rabbits. Local perivascular application of NO donors also produced a reduction of intimal hyperplasia in vein grafts (FULTON et al. 1998). The mechanism of action was not entirely clear, and inhibition of proliferation, protein synthesis and production of matrix components could be involved (CHAUX et al. 1998).

4. Inhibition of Intimal Hyperplasia Induced by Balloon Angioplasty

Balloon angioplasty of lesion-free animal arteries creates more extensive injury of the media, and the subsequent thrombus formation accelerates intimal thickening. Although the endothelial cells regenerate quickly, the eNOS pathway remains dysfunctional (TARRY and MAKHOUL 1994), but iNOS is induced in non-endothelial vascular cells (BOSMANS et al. 1996), predominantly in macrophages in the adventitia and in organizing thrombi (BULT et al. 1998). Oral or local L-arginine supplementation suppressed the intimal hyperplasia after balloon angioplasty of rabbit arteries (TARRY and MAKHOUL 1994; BOSMANS et al. 1999), and eNOS gene transfer during the PTCA procedure exerted the same effect in porcine coronary arteries (VARENNE et al. 1998). In accordance with the finding in the collar model, treatment with the NO donor SIN-1 did not influence intimal thickening following porcine carotid angioplasty, in spite of a significant inhibition of medial smooth-muscle cell proliferation (GROVES et al. 1995).

5. Stimulation of Compensatory Remodeling

Recent reports suggest that NO released by eNOS mediates the compensatory enlargement in response to increased flow. In rabbits, long-term administration of L-NAME for 4 weeks caused a significant reduction of the adaptive increase of the vessel caliber, the medial cross-sectional area and the number of smooth-muscle cells. These effects were evoked by a chronic increase in blood flow due to an arteriovenous fistula between the common carotid artery and the jugular vein (TRONC et al. 1996). In response to external carotid artery ligation, mice with a targeted disruption of the eNOS gene did not remodel their ipsilateral common carotid arteries, whereas wild-type mice did show a compensatory enlargement (RUDIC et al. 1998). Rather, the eNOS mutant mice displayed a paradoxical increase in wall thickness, accompanied by a hyperplastic response of the arterial wall. These findings demonstrate a critical role for NO as an inhibitor of smooth-muscle proliferation in response to a remodeling stimulus (RUDIC et al. 1998). Furthermore, both studies imply that eNOS, whose activity (CORSON et al. 1996; FLEMING et al. 1998) and expression are raised by increased shear stress, provides the signal to involve adaptive enlargement of the vessel and its lumen.

Balloon angioplasty in normal rabbit carotid arteries evoked a complex process of remodeling, which had a greater impact on the lumen reduction than did the intimal thickening. Continuous subcutaneous infusion of L-arginine in the ventral neck area led to a significant inhibition of constrictive remodeling during the first 2 weeks (BOSMANS et al. 1999). In combination with the reduced intimal hyperplasia, this created a much greater vascular lumen in the arginine-treated animals when compared with the control group. Since the levels of circulating plasma L-arginine were not elevated by the treatment, these data may suggest that feeding the iNOS opposed wound-healing contraction by the myofibroblasts in the adventitia. Similarly, percutaneous eNOS gene transfer significantly reduced luminal narrowing in pig coronary arteries, most likely through a combined effect on neointima formation and on vessel remodeling after angioplasty (VARENNE et al. 1998).

In keeping with the beneficial effects of NO on lumen size after balloon angioplasty in the rabbit, treatment with linsidomine (SIN-1) followed by oral treatment with molsidomine during the PTCA procedure in patients with stable angina led to a modest improvement in long-term lumen patency (LABLANCHE et al. 1997). This beneficial effect was due to a better immediate result, without affecting clinical outcome or late restenosis.

H. NO: a Radical Promoter of Atherosclerosis

Under certain conditions, NO may activate processes that could promote atherosclerosis or increase the vulnerability of atherosclerotic plaques (Table 3).

I. Peroxynitrite Formation

Superoxide combines with NO to form the stronger oxidants peroxynitrite and its decomposition product, the hydroxyl radical (BECKMAN et al. 1990; RADI et al. 1991; GRAHAM et al. 1993; MILES et al. 1996). The rate of peroxynitrite formation appears to be critically dependent on the concentrations of NO, superoxide and iron (MILES et al. 1996). In the absence of iron, equimolar fluxes of NO and superoxide interact to yield potent oxidants, such as peroxynitrite, which oxidize organic compounds. Excess production of either radical remarkably inhibits these oxidative reactions. In the presence of redox-active iron complexes, NO may enhance or inhibit superoxide-dependent oxidation

Table 3. Pro-atherogenic properties of nitric oxide

Property	References
Oxidation of LDL	Darley-Usmar et al. 1992; Jessup et al. 1992; De Groot et al. 1993; Graham et al. 1993; Hogg et al. 1993; Chang et al. 1994; Wang et al. 1994; Leeuwenburgh et al. 1997
Cytotoxic effects	Beckman et al. 1990; Palmer et al. 1992; Fukuo et al. 1995; Gross and Wolin 1995; Forrester et al. 1996; Szabo et al. 1996; Shimizu et al. 1998
Smooth-muscle apoptosis	Fukuo et al. 1996
Stimulated matrix breakdown	Murrell et al. 1995; Frears et al. 1996; Trachtman et al. 1996

LDL, low-density lipoprotein.

and hydroxylation reactions, depending on their relative fluxes (Miles et al. 1996).

II. LDL Oxidation

NO may initiate lipid peroxidation in LDL in the absence (Chang et al. 1994; Wang et al. 1994) or presence (Darley-Usmar et al. 1992; Jessup et al. 1992; De Groot et al. 1993; Hogg et al. 1993; Leeuwenburgh et al. 1997) of superoxide anion. In the former case, the modified LDL demonstrated the biological properties of minimally oxidized LDL (Chang et al. 1994) with an increased lipid hydroperoxide content (Jessup et al. 1992) without further evolution to a high-uptake form recognized by the scavenger receptor (Darley-Usmar et al. 1992). More extensive LDL oxidation occurred if superoxide anion was present, e.g., during the decomposition of SIN-1 (Darley-Usmar et al. 1992).

III. Oxidative Cell Injury

The concept that peroxynitrite formation occurs in atherosclerosis is strongly supported by the immunohistochemical demonstration of extensive nitration of protein tyrosines in advanced atherosclerotic lesions of humans and rabbits (Beckmann et al. 1994; Buttery et al. 1996; Wilcox et al. 1997; Luoma et al. 1998). Excessive NO synthesis and peroxynitrite formation have been implicated in cytotoxic effects in endothelial cells (Beckman et al. 1990; Palmer et al. 1992; Shimizu et al. 1998), smooth-muscle cells (Fukuo et al. 1995) and macrophages (Szabo et al. 1996). Cell damage results from the inhibition of mitochondrial respiration, aconitase activity and DNA synthesis, as well as from iron loss (Gross and Wolin 1995). NO-induced p53 accumulation safeguarded against DNA damage through p53-mediated suppression of iNOS-gene expression, thus reducing the potential for NO-induced DNA damage

(FORRESTER et al. 1996). The release of basic fibroblast growth factor from damaged vascular smooth-muscle cells may counteract toxic effects on the endothelium by stimulating endothelial cell proliferation (FUKUO et al. 1995). In view of these findings, the protective effects of antioxidants in several models of atherosclerosis may, in part, derive from the prevention of NO breakdown by oxygen radicals.

IV. NO and Apoptosis

NO shares with tumor necrosis factor α the ability to initiate as well as block apoptosis, depending on multiple variables in the arterial wall (BRÜNE et al 1998). It is assumed that, in the normal arterial wall, NO is protective of endothelial and smooth-muscle cells. In atherosclerotic plaques, the situation may be fundamentally different, since the high-output isoform iNOS is expressed (BUTTERY et al. 1996; ESAKI et al. 1997; WILCOX et al. 1997; LUOMA et al. 1998) in an environment with oxidative stress. In this situation, NO or peroxynitrite (BECKMANN et al. 1994) could induce apoptotic smooth-muscle-cell death and destabilize the atherosclerotic plaque (KOCKX et al. 1998b; CROMHEEKE et al. 1999).

1. NO as an Inhibitor of Apoptosis in the Normal Arterial Wall

The low levels of NO formed by eNOS may protect against apoptosis via cyclic-GMP-dependent (POLTE et al. 1997) and -independent mechanisms. The latter include S-nitrosation of caspases, enzymes involved in the execution phase of apoptotic cell death. All caspases contain cysteine in their active center, and, recently, it has indeed been demonstrated that NO can nitrosate caspase-1 (IL-1β-converting enzyme) and caspase-3 (CPP-32), thereby affording protection against tumor-necrosis-factor-α-induced apoptotic cell death (DIMMELER et al. 1997, 1998).

NO can also activate cell protective mechanisms, e.g., upregulation of the expression of hemoxygenase 1 (KIM et al. 1995), heat-shock protein 70 (BELLMANN et al. 1996) or cyclooxygenase 2 (VON KNETHEN and BRÜNE 1997; BRÜNE et al. 1998). The upregulation of cyclooxygenase 2 may be important in our understanding of why iNOS-expressing macrophages survive the high doses of NO they produce. Induction of iNOS is associated with a strong increase in apoptotic cell death in the macrophage population (ALBINA et al. 1993). Re-activation of these macrophages in the presence of NOS inhibitors or prestimulation with a low dose of a NO donor protected the macrophages against the apoptotic effects of high (and thus apoptotic) NO concentrations (BRÜNE et al. 1998). The induction of cyclooxygenase 2 during the prestimulation period appeared to be a critical regulator for these protective effects (VON KNETHEN and BRÜNE 1997), and increased prostaglandin E_2 biosynthesis may represent a mechanism by which iNOS-expressing macrophages protect themselves against cytotoxic effects of NO (ZAMORA et al. 1998).

2. NO as an Inducer of Apoptosis

In initial studies, it was noted that NO targets naked DNA (WINK et al. 1991; NGUYEN et al. 1992) and causes both oxidative damage and deamination (DE ROJAS-WALKER et al. 1995). The damage elicits DNA-repair mechanisms and, in this concept, NO-induced apoptotic cell death is considered as a mechanism that is used by the cell if these DNA-repair mechanisms fail. Two repair mechanisms have been studied in this context: (1) the poly[adenosine diphosphate(ADP)–ribose] polymerase (PARP) pathway and (2) the p53–p21 system (BRÜNE et al. 1998).

a) PARP- and NO-Induced DNA Repair and Apoptosis

DNA damage will induce attachment of PARP to the strand breaks and extensive synthesis of short-lived polymers by the bound polymerase (ALTHAUS and RICHTER 1987; DE MURCIA and MENESSIER-DE MURCIA 1994). Although PARP has no direct role in DNA excision repair, the enzyme binds tightly to the DNA-strand breaks, and repair can be suppressed if PARP synthesis is prevented (WANG et al. 1997). Massive PARP activation following extensive DNA damage upon exposure to peroxynitrite leads to depletion of nicotinamide adenine dinucleotide ion (NAD+), the ADP–ribose donor (SZABO et al. 1996). In an effort to resynthesize NAD^+, adenosine triphosphate (ATP) becomes depleted, which ultimately leads to cell death due to energy deprivation. Moreover, inhibition of mitochondrial respiration (thereby affecting ATP synthesis) via destruction of Fe–S clusters has been noted (HENRY et al. 1993). It seems unlikely, however, that energy depletion due to PARP activation is a general pathway of NO-mediated apoptotic cell death, since apoptosis is an energy-requiring process, and PARP is cleaved in the early executive steps of the death pathway (BRÜNE et al. 1998).

b) p53/p21 and NO-Induced DNA Repair and Apoptosis

The tumor-suppressor gene p53 is regarded as a master guardian of the genome and a member of the DNA-damage response pathway (WHITE 1994). Therefore, NO-induced DNA damage will upregulate p53. Indeed, activation of iNOS in RAW 264.7 macrophages resulted in p53 accumulation, which preceded DNA fragmentation. Both apoptotic cell death and p53 accumulation were clearly downregulated by iNOS blockade (SARIH et al. 1993). p53 will arrest the cell cycle in G1 via p21 (WAF1/Cip 1), an inhibitor of cyclin-dependent kinases, to enable DNA repair. The induction of apoptosis via p53/p21 is less well understood but can be activated both by transactivation and by transactivation-independent pathways. Recent evidence suggests the involvement of the Bcl-2 family regulatory proteins that participate in the control of cell death (BRÜNE et al. 1998). It was demonstrated that the anti-apoptotic protein Bcl-2 was downregulated and the pro-apoptotic Bax was upregulated by NO (MESSMER et al. 1996). The promoter site of the Bax gene contains a p53-binding domain. Therefore, p53, as a sensor of DNA damage, could increase susceptibility to apoptosis through upregulation of Bax protein.

3. NO, Apoptosis and Plaque Stability

It has been demonstrated that vulnerable regions of advanced human plaques and the intima of saphenous vein grafts are characterized by a prominent apoptotic cell loss (ISNER et al. 1995; KOCKX et al. 1996a, 1998a). The apoptosis occurred predominantly in smooth-muscle cells situated near macrophage-derived foam cells. The frequency of apoptosis appeared to be low in the macrophages; instead, they showed signs of DNA repair or synthesis and expressed iNOS (BULT et al. 1998; CROMHEEKE et al. 1999). It is possible that the macrophages are resistant to the high levels of NO they produce (see above), whereas the smooth-muscle cells are not.

NO has been reported to cause programmed cell death in macrophages (ALBINA et al. 1993; SARIH et al. 1993) and smooth-muscle cells (FUKUO et al. 1996). The significance of apoptotic cell death of macrophages and smooth-muscle cells is very different (KOCKX 1998; KOCKX and HERMAN 1998). Apoptotic death of macrophages, which are the main source of MMP activity, could improve plaque stability, as it would decrease the breakdown of collagen fibers in the fibrous cap (LEE and LIBBY 1997). On the contrary, disappearance of smooth-muscle cells from the fibrous cap or other vulnerable regions could lead to plaque destabilization, since smooth-muscle cells are responsible for the synthesis of collagen fibers types I and III. Stabilization of atherosclerotic plaques could be achieved by inducing apoptosis in the macrophage population and by protecting the smooth-muscle cells. This concept could explain why clinical studies of cholesterol-lowering therapies have consistently shown a decrease in acute cardiovascular events in the absence of impressive reductions in coronary-artery stenosis severity by angiography (LEE and LIBBY 1997). This is illustrated by a study of atherosclerotic plaques in cholesterol-fed rabbits, in which cholesterol withdrawal did not reduce plaque size but induced important changes in the cell composition, DNA synthesis and apoptosis (KOCKX et al. 1998a). Macrophages disappeared from the plaques, whereas the smooth-muscle cells that remained lost their lipid accumulation and showed strong reductions of the expression of the pro-apoptotic Bax and the frequency of apoptosis. This indicated that the smooth-muscle cells became less susceptible to apoptosis after lipid lowering. Moreover, these cellular changes were accompanied by a drastic increase of cross-banded collagen fibers, pointing to an increased tensile strength of the plaque.

An obvious question is whether modulation of NO metabolism, e.g., with selective iNOS inhibitors, can be used to save the smooth-muscle cells. Interestingly, the majority of the iNOS-expressing cells in plaques are macrophages (LUOMA et al. 1998; CROMHEEKE et al. 1999). Therefore, the recent demonstration that iNOS-expressing macrophages become hypersensitive to the toxic effects of exogenously added NO donors (MOHR et al. 1998) could suggest another way to kill the macrophages while saving the smooth-muscle cells in advanced plaques. This could be important in our understanding of the potential benefits of NO donors in the treatment of atherosclerosis.

V. Matrix Breakdown

Enhanced matrix breakdown via the activation of MMPs by NO (Murrell et al. 1995; Trachtman et al. 1996) or the inactivation (by peroxynitrite) of the tissue inhibitor of metalloproteinase 1 (Frears et al. 1996) may contribute to plaque destabilization and may promote the development of a necrotic core in advanced plaques.

I. Summary

It has been known for a decade that the loss of endothelial NO production impairs endothelium-dependent dilation and promotes vasospasm in atherosclerotic arteries. More recent evidence indicates that dysfunction of the eNOS pathway may promote early atherosclerosis, in view of the protective effects of NO against leukocyte recruitment, oxidative processes and smooth-muscle cell migration and proliferation. It remains to be determined whether the improved eNOS signaling upon dietary supplementation with anti-oxidants or l-arginine or upon eNOS gene transfer after a PTCA procedure are accompanied by a suppression of atherosclerosis or restenosis in humans. However, it has been proposed that NO acts as a double-edged sword (Bult 1996), since there is ample evidence to consider NO as a molecular aggressor in chronic inflammatory processes like atherosclerosis. Finally, the finding that the adverse effects of NO are often associated with iNOS expression warrants studies of the effects of selective iNOS inhibitors in atherosclerosis.

Abbreviations

ADMA	asymmetric dimethylarginine
ADP	adenosine diphosphate
ATP	adenosine triphosphate
EDHF	endothelium-derived hyperpolarizing factor
eNOS	endothelial nitric oxide synthase (NOS-3)
GMP	guanosine 3′,5′-monophosphate
ICAM-1	intercellular adhesion molecule 1
IL	interleukin
iNOS	inducible nitric oxide synthase (NOS-2)
l-NAME	N^G-nitro-l-arginine methyl ester
l-NMMA	N^G-monomethyl-l-arginine
LDL	low-density lipoprotein
MCP-1	monocyte chemotactic protein 1

MMP	matrix metalloproteinase
NAD+	nicotinamide adenine dinucleotide ion
NADPH	reduced nicotinamide adenine dinucleotide phosphate
NF-κB	nuclear factor κB
NO	nitric oxide
NOS	nitric oxide synthase
OxLDL	oxidized low-density lipoprotein
PARP	poly(ADP–ribose) polymerase
PDGF	platelet-derived growth factor
PTCA	percutaneous transluminal coronary angioplasty
SOD	superoxide dismutase
THB	tetrahydrobiopterin
VCAM-1	vascular cell adhesion molecule 1

References

Abrahamsson T, Brandt U, Marklund SL, Sjöqvist P-O (1992) Vascular bound recombinant extracellular superoxide dismutase type C protects against the detrimental effects of superoxide radicals on endothelium-dependent arterial relaxation. Circ Res 70:264–271

Albina JE, Cui S, Mateo RB, Reichner JS (1993) Nitric oxide-mediated apoptosis in murine peritoneal macrophages. J Immunol 150:5080–5085

Althaus FR, Richter C (1987) ADP ribosylation of proteins-enzymology and biological significance. Mol Biol Biochem Biophys 37:1–125

Anderson TJ, Gerhard MD, Meredith IT, Charbonneau F, Delagrange D, Creager MA, Selwyn AP, Ganz P (1995) Systemic nature of endothelial dysfunction in atherosclerosis. Am J Cardiol 75:71B–74B

Assender JW, Southgate KM, Newby AC (1991) Does nitric oxide inhibit smooth muscle proliferation? J Cardiovasc Pharmacol 17 [Suppl. 3]:S104–S107

Azuma H, Sato J, Hamasaki H, Sugimoto A, Isotani E, Obayashi S (1995) Accumulation of endogenous inhibitors for nitric oxide synthesis and decreased content of L-arginine in regenerated endothelial cells. Br J Pharmacol 115:1001–1004

Babaei S, Teichert KK, Monge JC, Mohamed F, Bendeck MP, Stewart DJ (1998) Role of nitric oxide in the angiogenic response in vitro to basic fibroblast growth factor. Circ Res 82:1007–1015

Barbul A (1990) Physiology and pharmacology of arginine. In: Moncada S, Higgs EA (eds) Nitric oxide from L-arginine: a bioregulatory system. Elsevier, Amsterdam, pp 317–329

Bassenge E (1991) Antiplatelet effects of endothelium-derived relaxing factor and nitric oxide donors. Eur Heart J 12 [Suppl. E]:12–15

Bath PMW, Hassall DG, Gladwin A-M, Palmer RMJ, Martin JF (1991) Nitric oxide and prostacyclin: divergence of inhibitory effects on monocyte chemotaxis and adhesion to endothelium in vitro. Arterioscler Thromb 11:254–260

Beckman JS, Beckman TW, Chen J, Marshall PA, Freeman BA (1990) Apparent hydroxyl radical production by peroxynitrite: implications for endothelial injury from nitric oxide and superoxide. Proc Natl Acad Sci U S A 87:1620–1624

Beckmann JS, Ye YZ, Anderson PG, Chen J, Accavitti MA, Tarpey MM, White CR (1994) Extensive nitration of protein tyrosines in human atherosclerosis detected by immunohistochemistry. Biol Chem Hoppe Seyler 375:81–88

Beetens JR, Coene M-C, Verheyen A, Zonnekeyn LL, Herman AG (1986) Biphasic response of intimal prostacyclin production during the development of experimental atherosclerosis. Prostaglandins 32:319–334

Bellmann K, Jäättelä M, Wissing D, Burkart V, Kolb H (1996) Heat shock protein Hsp70 overexpression confers resistance against nitric oxide. FEBS Lett 391: 185–188

Berkenboom G, Depierreux M, Fontaine J (1987) The influence of atherosclerosis on the mechanical responses of human isolated coronary arteries to substance P, isoprenaline and noradrenaline. Br J Pharmacol 92:113–120

Birnbaum Y, Fishbein MC, Luo H, Nishioka T, Siegel RJ (1997) Regional remodeling of atherosclerotic arteries: a major determinant of clinical manifestations of disease. J Am Coll Cardiol 30:1149–1164

Bode-Böger SM, Böger RH, Creutzig A, Tsikas D, Gutzki FM, Alexander K, Frolich JC (1994) L-Arginine infusion decreases peripheral arterial resistance and inhibits platelet aggregation in healthy subjects. Clin Sci 87:303–310

Bode-Böger SM, Böger RH, Alfke H, Heinzel D, Tsikas D, Creutzig A, Alexander K, Frölich JC (1996) L-Arginine induces nitric oxide-dependent vasodilation in patients with critical limb ischemia. A randomized, controlled study. Circulation 93:85–90

Bolton EJ, Jessup W, Stanley KK, Dean RT (1994) Enhanced LDL oxidation by murine macrophage foam cells and their failure to secrete nitric oxide. Atherosclerosis 106:213–223

Bosmans JM, Bult H, Vrints CJ, Kockx MM, Herman AG (1996) Balloon angioplasty and induction of non-endothelial nitric oxide synthase in rabbit carotid arteries. Eur J Pharmacol 310:163–174

Bosmans JM, Vrints CJ, Kockx MM, Bult H, Cromheeke KMC, Herman AG (1999) Continuous perivascular L-arginine delivery increases total vessel area and reduces neointimal thickening after experimental balloon dilatation. Arterioscler Thromb Vasc Biol 19:767–776

Bossaller C, Habib GB, Yamamoto H, Williams C, Wells S, Henry PD (1987) Impaired muscarinic endothelium-dependent relaxation and cyclic guanosine 5′-monophosphate formation in atherosclerotic human coronary artery and rabbit aorta. J Clin Invest 79:170–174

Bossaller C, Hehlert-Friedrich C, Jost S, Rafflenbeul W, Lichtlen P (1989) Angiographic assessment of human coronary artery endothelial function by measurement of endothelium-dependent vasodilation. Eur Heart J 10 [Suppl. F]:44–48

Böger RH, Bode-Böger SM, Mügge A, Kienke S, Brandes R, Dwenger A, Frölich JC (1995) Supplementation of hypercholesterolaemic rabbits with L-arginine reduces the vascular release of superoxide anions and restores NO production. Atherosclerosis 117:273–284

Böger RH, Bode-Böger SM, Thiele W, Junker W, Alexander K, Frolich JC (1997) Biochemical evidence for impaired nitric oxide synthesis in patients with peripheral arterial occlusive disease. Circulation 95:2068–2074

Brüne B, Von Knethen A, Sandau KB (1998) Nitric oxide and its role in apoptosis. Eur J Pharmacol 351:261–272

Brunner F, Stessel H, Kukovetz WR (1995) Novel guanylyl cyclase inhibitor, ODQ reveals role of nitric oxide, but not of cyclic GMP in endothelin-1 secretion. FEBS Lett 376:262–266

Buga GM, Griscavage JM, Rogers NE, Ignarro LJ (1993) Negative feedback regulation of endothelial cell function by nitric oxide. Circ Res 73:808–812

Buga GM, Singh R, Pervin S, Rogers NE, Schmitz DA, Jenkinson CP, Cederbaum SD, Ignarro LJ (1996) Arginase activity in endothelial cells: inhibition by N^G-hydroxy-L-arginine during high-output NO production. Am J Physiol 40:H1988–H1998

Bult H (1996) Nitric oxide and atherosclerosis: possible implications for therapy. Mol Med Today 2:510–518

Bult H, De Meyer GRY, Herman AG (1995) Influence of chronic treatment with a nitric oxide donor on fatty streak development and reactivity of the rabbit aorta. Br J Pharmacol 114:1371–1382

Bult H, Cromheeke KM, Bosmans JM, Kockx MM, Vrints CJ, Herman AG (1998) Inducible NOS after experimental angioplasty and in human atherosclerosis. In:

Moncada S, Toda N, Maeda H, Higgs EA (eds) The biology of nitric oxide. Part 6. Portland, London, pp. 148

Buttery LDK, Springall DR, Chester AH, Evans TJ, Standfield N, Parums DV, Yacoub MH, Polak JM (1996) Inducible nitric oxide synthase is present within human atherosclerotic lesions and promotes the formation and activity of peroxynitrite. Lab Invest 75:77–85

Cable DG, O'Brien T, Schaff HV, Pompili VJ (1997) Recombinant endothelial nitric oxide synthase-transduced human saphenous veins: gene therapy to augment nitric oxide production in bypass conduits. Circulation 96:II–8

Caparotta L, Pandolfo L, Chinellato A, Ragazzi E, Froldi G, Aliev G, Fassina G (1993) L- arginine does not improve endothelium-dependent relaxation in in vitro Watanabe rabbit thoracic aorta. Amino Acids 5:403–411

Casino PR, Kilcoyne CM, Quyyumi AA, Hoeg JM, Panza JA (1994) Investigation of decreased availability of nitric oxide precursor as the mechanism responsible for impaired endothelium-dependent vasodilation in hypercholesterolemic patients. J Am Coll Cardiol 23:844–850

Casino PR, Kilcoyne CM, Cannon RO, Quyyumi AA, Panza JA (1995) Impaired endothelium-dependent vascular relaxation in patients with hypercholesterolemia extends beyond the muscarinic receptor. Am J Cardiol 75:40–44

Cayatte AJ, Palacino JJ, Horten K, Cohen RA (1994) Chronic inhibition of nitric oxide production accelerates neointima formation and impairs endothelial function in hypercholesterolemic rabbits. Arterioscler Thromb 14:753–759

Celermajer DS, Sorensen KE, Gooch VM, Spiegelhalter DJ, Miller OI, Sullivan ID, Lloyd JK, Deanfield JE (1992) Non-invasive detection of endothelial dysfunction in children and adults at risk of atherosclerosis. Lancet 340:1111–1115

Chang GJ, Woo P, Honda HM, Ignarro LJ, Young L, Berliner JA, Demer LL (1994) Oxidation of LDL to a biologically active form by derivatives of nitric oxide and nitrite in the absence of superoxide. Dependence on pH and oxygen. Arterioscler Thromb 14:1808–1814

Chappell SP, Lewis MJ, Henderson AH (1987) Effect of lipid feeding on endothelium dependent relaxation in rabbit aortic preparations. Cardiovasc Res 21:34–38

Chaux A, Ruan XM, Fishbein MC, Ouyang Y, Kaul S, Pass JA, Matloff JM (1998) Perivascular delivery of a nitric oxide donor inhibits neointimal hyperplasia in vein grafts implanted in the arterial circulation. J Thorac Cardiovasc Surg 115:604–612

Chen L, Daum G, Forough R, Clowes M, Walter U, Clowes AW (1998) Overexpression of human endothelial nitric oxide synthase in rat vascular smooth muscle cells and in balloon-injured carotid artery. Circ Res 82:862–870

Chester AH, O'Neil GS, Moncada S, Tadjkarimi S, Yacoub MH (1990) Low basal and stimulated release of nitric oxide in atherosclerotic epicardial coronary arteries. Lancet 336:897–900

Chin JH, Azhar S, Hoffman BB (1992) Inactivation of endothelial derived relaxing factor by oxidized lipoproteins. J Clin Invest 89:10–18

Clancy RM, Leszczynska-Piziak J, Abramson SB (1992) Nitric oxide, an endothelial cell relaxation factor, inhibits neutrophil superoxide anion production via a direct action on the NADPH oxidase. J Clin Invest 90:1116–1121

Clarkson P, Celermajer DS, Powe AJ, Donald AE, Henry RMA, Deanfield JE (1997) Endothelium-dependent dilatation is impaired in young healthy subjects with a family history of premature coronary disease. Circulation 96:3378–3383

Cocks TM, Angus JA (1983) Endothelium-dependent relaxation of coronary arteries by noradrenaline and serotonin. Nature 305:627–630

Cohen RA, Zitnay KM, Haudenschild CC, Cunningham LD (1988) Loss of selective endothelial cell vasoactive functions caused by hypercholesterolemia in pig coronary arteries. Circ Res 63:903–910

Colasanti M, Persichini T, Menegazzi M, Mariotto S, Giordano E, Caldarera CM, Sogos V, Lauro GM, Suzuki H (1995) Induction of nitric oxide synthase mRNA expression. Suppression by exogenous nitric oxide. J Biol Chem 270:26731–26733

Collins T (1993) Biology of disease. Endothelial nuclear factor-kB and the initiation of the atherosclerotic lesion. Lab Invest 68:499–508

Cooke JP, Andon NA, Girerd XJ, Hirsch AT, Creager MA (1991) Arginine restores cholinergic relaxation of hypercholesterolemic rabbit thoracic aorta. Circulation 83:1057–1062

Cooke JP, Singer AH, Tsao P, Zera P, Rowan RA, Billingham ME (1992) Antiatherogenic effects of L-arginine in the hypercholesterolemic rabbit. J Clin Invest 90:1168–1172

Cornwell TL, Arnold E, Boerth NJ, Lincoln TM (1994) Inhibition of smooth muscle cell growth by nitric oxide and activation of cAMP-dependent protein kinase by cGMP. Am J Physiol 36:C1405–C1413

Corson MA, James NL, Latta SE, Nerem RM, Berk BC, Harrison DG (1996) Phosphorylation of endothelial nitric oxide synthase in response to fluid shear stress. Circ Res 79:984–991

Cosentino F, Katusic ZS (1995) Tetrahydrobiopterin and dysfunction of endothelial nitric oxide synthase in coronary arteries. Circulation 91:139–144

Cox DA, Vita JA, Treasure CB, Fish RD, Alexander RW, Ganz P, Selwyn AP (1989) Atherosclerosis impairs flow-mediated dilation of coronary arteries in humans. Circulation 80:458–465

Creager MA, Cooke JP, Mendelsohn ME, Gallagher SJ, Coleman SM, Loscalzo J, Dzau VJ (1990) Impaired vasodilation of forearm resistance vessels in hypercholesterolemic humans. J Clin Invest 86:228–234

Creager MA, Gallagher SJ, Girerd XJ, Coleman SM, Dzau VJ, Cooke JP (1992) L-Arginine improves endothelium-dependent vasodilation in hypercholesterolemic humans. J Clin Invest 90:1248–1253

Cromheeke KM, Kockx MM, De Meyer GRY, Bosmans JM, Bult H, Vrints CJ, Herman AG (1999) Inducible nitric oxide synthase co-localizes with intracellular ceroid vesicles in human atherosclerotic plaques. submitted

Darley-Usmar VM, Hogg N, O'Leary VJ, Wilson MT, Moncada S (1992) The simultaneous generation of superoxide and nitric oxide can initiate lipid peroxidation in human low density lipoprotein. Free Rad Res Comms 17:9–20

Davies MG, Dalen H, Austerheim AMS, Gulbrandsen TF, Svendsen E, Hagen PO (1996) Suppression of intimal hyperplasia in experimental vein grafts by oral L-arginine supplementation and single ex vivo immersion in deferoxamine manganese. J Vasc Surg 23:410–420

Davies MJ, Woolf N (1993) Atherosclerosis: what is it and why does it occur? Br Heart J 69 [Suppl.]:S3–S11

De Caterina R, Libby P, Peng HB, Thannickal VJ, Rajavashisth TB, Gimbrone MA Jr, Shin WS, Liao JK (1995) Nitric oxide decreases cytokine-induced endothelial activation. Nitric oxide selectively reduces endothelial expression of adhesion molecules and pro-inflammatory cytokines. J Clin Invest 96:60–68

De Groot H, Hegi U, Sies H (1993) Loss of alpha-tocopherol upon exposure to nitric oxide or the sydnonimine SIN-1. FEBS Lett 315:139–142

De Meyer GRY, Bult H, Üstünes L, Kockx MM, Jordaens FH, Zonnekeyn LL, Herman AG (1994) Vasoconstrictor responses after neo-intima formation and endothelial denudation in the rabbit carotid artery. Br J Pharmacol 112:471–476

De Meyer GRY, Bult H, Üstünes L, Kockx MM, Feelisch M, Herman AG (1995) Effect of nitric oxide donors on neointima formation and vascular reactivity in the collared carotid artery of rabbits. J Cardiovasc Pharmacol 26:272–279

De Meyer GRY, Van Put DJM, Kockx MM, Van Schil P, Bosmans R, Bult H, Buyssens N, Vanmaele R, Herman AG (1997) Possible mechanisms of collar-induced intimal thickening. Arterioscler Thromb Vasc Biol 17:1924–1930

De Murcia G, Menessier-De Murcia J (1994) Poly(ADP-ribose) polymerase: a molecular nick-sensor. Trends Biochem Sci 19:172–176

De Rojas-Walker T, Tamir S, Ji H, Wishnok JS, Tannenbaum SR (1995) Nitric oxide induces oxidative damage in addition to deamination in macrophage DNA. Chem Res Toxicol 8:473–477

Del Boccio G, Lapenna D, Porreca E, Pennelli A, Savini F, Feliciani P, Ricci G, Cuccurullo F (1990) Aortic antioxidant defence mechanisms: time-related changes in cholesterol-fed rabbits. Atherosclerosis 81:127–135

Dimmeler S, Haendeler J, Nehls M, Zeiher AM (1997) Supression of apoptosis by nitric oxide via inhibition of interleukin-1β-converting enzyme (ICE)-like and cysteine protease protein (CPP)-32-like protease. J exp Med 185:601–607

Dimmeler S, Haendeler J, Sause A, Zeiher AM (1998) Nitric oxide inhibits APO-1/Fas-mediated cell death. Cell Growth Diff 9:415–422

Douglas SA, Vickery-Clark LM, Ohlstein EH (1994) Functional evidence that balloon angioplasty results in transient nitric oxide synthase induction. Eur J Pharmacol 255:81–89

Drexler H, Zeiher AM, Wollschläger H, Meinertz T, Just H, Bonzel T (1989) Flow-dependent coronary artery dilatation in humans. Circulation 80:466–474

Drexler H, Zeiher AM, Meinzer K, Just H (1991) Correction of endothelial dysfunction in coronary microcirculation of hypercholesterolaemic patients by L-arginine. Lancet 338:1546–1550

Dubey RK, Jackson EK, Lüscher TF (1995) Nitric oxide inhibits angiotensin II-induced migration of rat aortic smooth muscle cell. J Clin Invest 96:141–149

Durante W, Liao L, Iftikhar I, O'Brien WE, Schafer AI (1996) Differential regulation of L-arginine transport and nitric oxide production by vascular smooth muscle and endothelium. Circ Res 78:1075–1082

Egashira K, Inou T, Hirooka Y, Yamada A, Maruoka Y, Kai H, Sugimachi M, Suzuki S, Takeshita A (1993) Impaired coronary blood flow response to acetylcholine in patients with coronary risk factors and proximal atherosclerotic lesions. J Clin Invest 91:29–37

Ellenby MI, Ernst CB, Carretero OA, Scicli AG (1996) Role of nitric oxide in the effect of blood flow on neointima formation. J Vasc Surg 23:314–422

Esaki T, Hayashi T, Muto E, Yamada K, Kuzuya M, Iguchi A (1997) Expression of inducible nitric oxide synthase in T lymphocytes and macrophages of cholesterol-fed rabbits. Atherosclerosis 128:39–46

Farhy RD, Carretero OA, Ho KL, Scicli AG (1993) Role of kinins and nitric oxide in the effects of angiotensin converting enzyme inhibitors on neointima formation. Circ Res 72:1202–1210

Flavahan NA (1992) Atherosclerosis or lipoprotein-induced endothelial dysfunction. Potential mechanisms underlying reduction in EDRF/nitric oxide activity. Circulation 85:1927–1938

Flavahan NA (1993) Lysophosphatidylcholine modifies G protein-dependent signaling in porcine endothelial cells. Am J Physiol 264:H722–H727

Fleming I, Bauersachs J, Fisslthaler B, Busse R (1998) Ca^{2+}-independent activation of the endothelial nitric oxide synthase in response to tyrosine phosphatase inhibitors and fluid shear stress. Circ Res 82:686–695

Forrester K, Ambs S, Lupold SE, Kapust RB, Spillare EA, Weinberg WC, Felley Bosco E, Wang XW, Geller DA, Tzeng E, Billiar TR, Harris CC (1996) Nitric oxide-induced p53 accumulation and regulation of inducible nitric oxide synthase expression by wild-type p53. Proc Natl Acad Sci USA 93:2442–2447

Forster BA, Weinberg PD (1997) Changes with age in the influence of endogenous nitric oxide on transport properties of the rabbit aortic wall near branches. Arterioscler Thromb Vasc Biol 17:1361–1368

Förstermann U, Mügge A, Alheid U, Haverich A, Frölich JC (1988) Selective attenuation of endothelium-mediated vasodilation in atherosclerotic human coronary arteries. Circ Res 62:185–190

Frears ER, Zhang Z, Blake DR, O'Connell JP, Winyard PG (1996) Inactivation of tissue inhibitor of metalloproteinase-1 by peroxynitrite. FEBS Lett 381:21–24

Freiman PC, Mitchell GG, Heistad DD, Armstrong ML, Harrison DG (1986) Atherosclerosis impairs endothelium-dependent vascular relaxation to acetylcholine and thrombin in primates. Circ Res 58:783–789

Fu Y, Blankenhorn EP (1992) Nitric oxide-induced anti-mitogenic effects in high and low responder rat strains. J Immunol 148:2217–2222

Fukahori M, Ichimori K, Ishida H, Nakagawa H, Okino H (1994) Nitric oxide reversibly suppresses xanthine oxidase activity. Free Radic Res 21:203–212

Fukumoto Y, Shimokawa H, Kozai T, Kadokami T, Kuwata K, Yonemitsu Y, Kuga T, Egashira K, Sueishi K, Takeshita A (1997) Vasculoprotective role of inducible nitric oxide synthase at inflammatory coronary lesions induced by chronic treatment with interleukin-1beta in pigs in vivo. Circulation 96:3104–3111

Fukuo K, Inoue T, Morimoto S, Nakahashi T, Yasuda O, Kitano S, Sasada R, Ogihara T (1995) Nitric oxide mediates cytotoxicity and basic fibroblast growth factor release in cultured vascular smooth muscle cells. J Clin Invest 95:669–676

Fukuo K, Hata S, Suhara T, Nakahashi T, Shinto Y, Tsujimoto Y, Morimoto S, Ogihara T (1996) Nitric oxide induces up-regulation of Fas and apoptosis in vascular smooth muscle. Hypertension 27:823–826

Fulton GJ, Davies MG, Barber L, Gray JL, Svendsen E, Hagen PO (1998) Local effects of nitric oxide supplementation and suppression in the development of intimal hyperplasia in experimental vein grafts. Eur J Vasc Endovasc Surg 15:279–289

Gaboury J, Woodman RC, Granger DN, Reinhardt P, Kubes P (1993) Nitric oxide prevents leukocyte adherence: role of superoxide. Am J Physiol 265:H862–H867

Galle J, Busse R, Bassenge E (1991) Hypercholesterolemia and atherosclerosis change vascular reactivity in rabbits by different mechanisms. Arterioscler Thromb 11:1712–1718

Garg UC, Hassid A (1989) Nitric oxide-generating vasodilators and 8-bromo-cyclic guanosine monophosphate inhibit mitogenesis and proliferation of cultured rat vascular smooth muscle cells. J Clin Invest 83:1774–1777

Gauthier TW, Davenpeck KL, Lefer AM (1994) Nitric oxide attenuates leukocyte endothelial interaction via P selectin in splanchnic ischemia–reperfusion. Am J Physiol 30:G562–G568

Gilligan DM, Sack MN, Guetta V, Casino PR, Quyyumi AA, Rader DJ, Panza JA, Cannon ROI (1994) Effect of antioxidant vitamins on low density lipoprotein oxidation and impaired endothelium-dependent vasodilation in patients with hypercholesterolemia. J Am Coll Cardiol 24:1611–1617

Ginsburg R, Bristow MR, Davis K, Dibiase A, Billingham ME (1984) Quantitative pharmacologic responses of normal and atherosclerotic isolated human epicardial coronary arteries. Circulation 69:430–440

Girerd XJ, Hirsch AT, Cooke JP, Dzau VJ, Creager MA (1990) L-Arginine augments endothelium-dependent vasodilation in cholesterol-fed rabbits. Circ Res 67:1301–1308

Giugliano D, Marfella R, Verrazzo G, Acampora R, Coppola L, Cozzolino D, D'Onofrio F (1997) The vascular effects of L-arginine in humans. The role of endogenous insulin. J Clin Invest 99:433–438

Glagov S, Weisenberg E, Zarins CK, Stankunavicius R, Kolettis GJ (1987) Compensatory enlargement of human atherosclerotic coronary arteries. N Engl J Med 316:1371–1375

Goode GK, Garcia S, Heagerty AM (1997) Dietary supplementation with marine fish oil improves in vitro small artery endothelial function in hypercholesterolemic patients. A double-blind placebo-controlled study. Circulation 96:2802–2807

Gordon JB, Ganz P, Nabel EG, Fish RD, Zebede J, Mudge GH, Alexander RW, Selwyn AP (1989) Atherosclerosis influences the vasomotor response of epicardial coronary arteries to exercise. J Clin Invest 83:1946–1952

Graham A, Hogg N, Kalyanaraman B, O'Leary V, Darley-Usmar VM, Moncada S (1993) Peroxynitrite modification of low-density lipoprotein leads to recognition by the macrophage scavenger receptor. FEBS Lett 330:181–185

Griscavage JM, Fukuto JM, Komori Y, Ignarro LJ (1994) Nitric oxide inhibits neuronal nitric oxide synthase by interacting with the heme prosthetic group. J Biol Chem 269:21644–21649

Gross SS, Wolin MS (1995) Nitric oxide: pathophysiological mechanisms. Annu Rev Physiol 57:737–769

Groves PH, Banning AP, Penny WJ, Newby AC, Cheadle HA, Lewis MJ (1995) The effects of exogenous nitric oxide on smooth muscle cell proliferation following porcine carotid angioplasty. Cardiovasc Res 30:87–96

Guo J-P, Milhoan KA, Tuan RS, Lefer AM (1994) Beneficial effect of SPM-5185, a cysteine-containing nitric oxide donor, in rat carotid artery intimal injury. Circ Res 75:77–84

Habib FM, Springall DR, Davies GJ, Oakley CM, Yacoub MH, Polak JM (1996) Tumour necrosis factor and inducible nitric oxide synthase in dilated cardiomyopathy. The Lancet 347:1151–1155

Habib JB, Bossaller C, Wells S, Williams C, Morrisett JD, Henry PD (1986) Preservation of endothelium-dependent vascular relaxation in cholesterol-fed rabbit by treatment with the calcium blocker PN 200110. Circ Res 58:305–309

Hamilton CA (1997) Low-density lipoprotein and oxidised low-density lipoprotein: their role in the development of atherosclerosis. Pharmacol Ther 74:55–72

Hansson GK, Geng Y-J, Holm J, Hardhammar P, Wennmalm A, Jennische E (1994) Arterial smooth muscle cells express nitric oxide synthase in response to endothelial injury. J Exp Med 180:733–738

Harrison DG, Bates JN (1993) The nitrovasodilators. New ideas about old drugs. Circulation 87:1461–1467

Harrison DG, Ohara Y (1995) Physiologic consequences of increased vascular oxidant stresses in hypercholesterolemia and atherosclerosis: implications for impaired vasomotion. Am J Cardiol 75:75B-81B

Harrison DG, Freiman PC, Armstrong ML, Marcus ML, Heistad DD (1987) Alterations of vascular reactivity in atherosclerosis. Circ Res 61[suppl II]:II-74-II-80

Hashimoto M, Akishita M, Eto M, Ishikawa M, Kozaki K, Toba K, Sagara Y, Taketani Y, Orimo H, Ouchi Y (1995) Modulation of endothelium-dependent flow-mediated dilatation of the brachial artery by sex and menstrual cycle. Circulation 92:3431–3435

Hecker M, Sessa WC, Harris HJ, Änggard EE, Vane JR (1990) The metabolism of L-arginine and its significance for the biosynthesis of endothelium-derived relaxing factor: cultured endothelial cells recycle L-citrulline to L-arginine. Proc Natl Acad Sci USA 87:8612–8616

Heistad DD, Armstrong ML, Marcus ML, Piegors DJ, Mark AL (1984) Augmented responses to vasoconstrictor stimuli in hypercholesterolemic and atherosclerotic monkeys. Circ Res 54:711–718

Heitzer T, Just H, Münzel T (1996) Antioxidant vitamin C improves endothelial dysfunction in chronic smokers. Circulation 94:6–9

Henriksson P, Bergström K, Edhag O (1985) Experimental atherosclerosis and a possible generation of free radicals. Thromb Res 38:195–198

Henry Y, Lepoivre M, Drapier JC, Ducrocq C, Boucher JL, Guissani A (1993) EPR characterization of molecular targets for NO in mammalian cells and organelles. FASEB J 7:1124–1134

Hirata K-I, Miki N, Kuroda Y, Sakoda T, Kawashima S, Yokoyama M (1995) Low concentration of oxidized low-density lipoprotein and lysophosphatidylcholine up-regulate constitutive nitric oxide synthase mRNA expression in bovine aortic endothelial cells. Circ Res 76:958–962

Hogg N, Darley-Usmar VM, Wilson MT, Moncada S (1993) The oxidation of alpha-tocopherol in human low-density lipoprotein by the simultaneous generation of superoxide and nitric oxide. FEBS Lett 326:199–203

Horio Y, Yasue H, Rokutanda M, Nakamura N, Ogawa H, Takaoka K, Matsuyama K, Kimura T (1986) Effects of intracoronary injection of acetylcholine on coronary arterial diameter. Am J Cardiol 57:984–989

Ichiki K, Ikeda H, Haramaki N, Ueno T, Imaizumi T (1996) Long-term smoking impairs platelet-derived nitric oxide release. Circulation 94:3109–3114

Imaizumi T, Hirooka Y, Masaki H, Harada S, Momohara M, Tagawa T, Takeshita A (1992) Effects of L-arginine on forearm vessels and responses to acetylcholine. Hypertension 20:511–517

Isner JF, Kearney M, Bortman S, Passeri J (1995) Apoptosis in human atherosclerosis and restenosis. Circulation 91:2703–2711

Jackson CL (1994) Animal models of restenosis. Trends Cardiovasc Med 4:122–130

Janssens S, Flaherty D, Nong Z, Varenne O, Van Pelt N, Haustermans C, Zoldhelyi P, Gerard R, Collen D (1998) Human endothelial nitric oxide synthase gene transfer inhibits vascular smooth muscle cell proliferation and neointima formation after balloon injury in rats. Circulation 97:1274–1281

Jayakody RL, Senaratne MP, Thomson AB, Kappagoda CT (1985) Cholesterol feeding impairs endothelium-dependent relaxation of rabbit aorta. Can J Physiol Pharmacol 63:1206–1209

Jayakody L, Senaratne M, Thomson A, Kappagoda T (1987) Endothelium-dependent relaxation in experimental atherosclerosis in the rabbit. Circ Res 60:251–264

Jeremy RW, McCarron H, Sullivan D (1996) Effects of dietary L-arginine on atherosclerosis and endothelium-dependent vasodilatation in the hypercholesterolemic rabbit. Response according to treatment duration, anatomic site, and sex. Circulation 94:498–506

Jessup W, Dean RT (1993) Autoinhibition of murine macrophage-mediated oxidation of low-density lipoprotein by nitric oxide synthesis. Atherosclerosis 101:145–155

Jessup W, Mohr D, Gieseg SP, Dean RT, Stocker R (1992) The participation of nitric oxide in cell free- and its restriction of macrophage-mediated oxidation of low-density lipoprotein. Biochim Biophys Acta 1180:73–82

Joly GA, Schini VB, Vanhoutte PM (1992) Balloon injury and interleukin-1β induce nitric oxide synthase activity in rat carotid arteries. Circ Res 71:331–338

Jorens PG, Rosseneu M, Devreese A-M, Bult H, Marescau B, Herman AG (1992) Diminished capacity to release metabolites of nitric oxide synthase in macrophages loaded with oxidized low-density lipoproteins. Eur J Pharmacol 212:113–115

Kanamura K, Waga S, Tochio H, Nagatani K (1989) The effect of atherosclerosis on endothelium-dependent relaxation in the aorta and intracranial arteries of rabbits. J Neurosurg 70:793–798

Kanazawa K, Kawashima S, Mikami S, Miwa Y, Hirata K-I, Suematsu M, Hayashi Y, Itoh H, Yokoyama M (1996) Endothelial constitutive nitric oxide synthase protein and mRNA increased in rabbit atherosclerotic aorta despite impaired endothelium-dependent vascular relaxation. Am J Pathol 148:1949–1956

Kanno K, Hirata Y, Emori T, Ohta K, Eguchi S, Imai T, Marumo F (1992) L-Arginine infusion induces hypotension and diuresis/natriuresis with concomitant increased urinary excretion of nitrite/nitrate and cyclic GMP in humans. Clin Exp Pharmacol Physiol 19:619–625

Kariya K, Kawahara Y, Araki S, Fukuzaki H, Takai Y (1989) Antiproliferative action of cyclic GMP-elevating vasodilators in cultured rabbit aortic smooth muscle cells. Atherosclerosis 80:143–147

Kaski JC, Crea F, Meran D, Rodriguez L, Araujo L, Chierchia S, Davies G, Maseri A (1986) Local coronary supersensitivity to diverse vasoconstrictive stimuli in patients with variant angina. Circulation 74:1255–1265

Kawabe T, Isobe KI, Hasegawa Y, Nakashima I, Shimokata K (1992) Immunosuppressive activity induced by nitric oxide in culture supernatant of activated rat alveolar macrophages. Immunology 76:72–78

Kawashima T, Yashiro A, Nandate H, Himeno E, Oka Y, Kaku T, Nakashima Y, Kuroiwa A (1995) Increased susceptibility of angiographically smooth left anterior descending coronary artery to an impairment of vasoresponse to acetylcholine, and the relation between impaired vasoresponse and low-density lipoprotein cholesterol level. Am J Cardiol 75:1265–1267

Keaney JF Jr, Gaziano JM, Xu A, Frei B, Curran-Celentano J, Shwaery GT, Loscalzo J, Vita JA (1994) Low-dose α-tocopherol improves and high-dose α-tocopherol worsens endothelial vasodilator function in cholesterol-fed rabbits. J Clin Invest 93:844–851

Keaney JF Jr, Xu A, Cunningham D, Jackson T, Frei B, Vita JA (1995) Dietary probucol preserves endothelial function in cholesterol-fed rabbits by limiting vascular oxidative stress and superoxide generation. J Clin Invest 95:2520–2529

Khan BV, Parthasarathy SS, Alexander RW, Medford RM (1995) Modified low density lipoprotein and its constituents augment cytokine-activated vascular cell adhesion molecule-1 gene expression in human vascular endothelial cells. J Clin Invest 95:1262–1270

Kim YM, Bergonia H, Lancaster JR (1995) Nitrogen oxide-induced autoprotection in isolated rat hepatocytes. FEBS Lett 374:228–232

Kockx MM (1998) Apoptosis in the atherosclerotic plaque: quantitative and qualitative aspects. Arterioscler Thromb Vasc Biol 18:1519–1522

Kockx MM, Herman AG (1998) Apoptosis in atherogenesis: implications for plaque destabilization. Eur Heart J 19 [Suppl. G]:G23–G28

Kockx MM, De Meyer GRY, Jacob WA, Bult H, Herman AG (1992) Triphasic sequence of neointimal formation in the cuffed carotid artery of the rabbit. Arterioscler Thromb Vasc Biol 12:1447–1457

Kockx MM, De Meyer GRY, Bortier H, de Meyere N, Muhring J, Bakker A, Jacob W, Van Vaeck L, Herman A (1996a) Luminal foam cell accumulation is associated with smooth muscle cell death in the intimal thickening of human saphenous vein grafts. Circulation 94:1255–1262

Kockx MM, De Meyer GRY, Muhring J, Bult H, Bultinck J, Herman AG (1996b) Distribution of cell replication and apoptosis in atherosclerotic plaques of cholesterol-fed rabbits. Atherosclerosis 120:115–124

Kockx MM, De Meyer GRY, Buyssens N, Knaapen MWM, Bult H, Herman AG (1998a) Cell composition, replication, and apoptosis in atherosclerotic plaques after 6 months of cholesterol withdrawal. Circ Res 83:378–387

Kockx MM, De Meyer GRY, Buyssens N, Jacob W, Bult H, Herman AG (1998b) Apoptosis and related proteins in different stages of human atherosclerotic plaques. Circulation 97:2307–2315

Kojda G, Noack E (1995) Effects of pentaerythrityl tetranitrate and isosorbide-5-mononitrate in experimental atherosclerosis. Agents Actions Suppl 45:201–206

Kojima S, Ona S, Iizuka I, Arai T, Mori H, Kubota K (1995) Antioxidative activity of 5,6,7,8-tetrahydrobiopterin and its inhibitory effect on paraquat-induced cell toxicity in cultured rat hepatocytes. Free Radic Res 23:419–430

Kolodgie FD, Virmani R, Rice HE, Mergner WJ (1990) Vascular reactivity during the progression of atherosclerotic plaque. A study in Watanabe heritable hyperlipidemic rabbits. Circ Res 66:1112–1126

Komori K, Shimokawa H, Vanhoutte PM (1989) Hypercholesterolemia impairs endothelium-dependent relaxations to aggregating platelets in porcine iliac arteries. J Vasc Surg 10:318–325

Kubes P, Suzuki M, Granger DN (1991) Nitric oxide: an endogenous modulator of leukocyte adhesion. Proc Natl Acad Sci USA 88:4651–4655

Kugiyama K, Kerns SA, Morrisett JD, Roberts R, Henry PD (1990) Impairment of endothelium-dependent arterial relaxation by lysolecithin in modified low-density lipoproteins. Nature 344:160–162

Kume N, Cybulsky MI, Gimbrone MA Jr (1992) Lysophosphatidylcholine, a component of atherogenic lipoproteins, induces mononuclear leukocyte adhesion molecules in cultured human and rabbit arterial endothelial cells. J Clin Invest 90:1138–1144

Kuo L, Davis MJ, Cannon MS, Chilian WM (1992) Pathophysiological consequences of atherosclerosis extend into the coronary microcirculation. Restoration of endothelium- dependent responses by L-arginine. Circ Res 70:465–476

Lablanche JM, Grollier G, Lusson JR, Bassand JP, Drobinski G, Bertrand B, Battaglia S, Desveaux B, Juilliere Y, Juliard JM, Metzger JP, Coste P, Quiret JC, Dubois RJ, Crochet PD, Letac B, Boschat J, Virot P, Finet G, Le BH, Livarek B, Leclercq F, Beard T, Giraud T, Bertrand ME, et a (1997) Effect of the direct nitric oxide donors linsidomine and molsidomine on angiographic restenosis after coronary balloon angioplasty. The ACCORD Study. Angioplastic Coronaire Corvasal Diltiazem. Circulation 95:83–89

Lang D, Smith JA, Lewis MJ (1993) Induction of a calcium-independent NO synthase by hypercholesterolaemia in the rabbit. Br J Pharmacol 108:290–292

Lee JS, Adrie C, Jacob HJ, Roberts JD Jr, Zapol WM, Bloch KD (1996) Chronic inhalation of nitric oxide inhibits neointimal formation after balloon-induced arterial injury. Circ Res 78:337–342

Lee RT, Libby P (1997) The unstable atheroma. Arterioscler Thromb Vasc Biol 17:1859–1867

Leeuwenburgh C, Hardy MM, Hazen SL, Wagner P, Oh-Ishi S, Steinbrecher UP, Heinecke JW (1997) Reactive nitrogen intermediates promote low density lipoprotein oxidation in human atherosclerotic intima. J Biol Chem 272:1433–1436

Lefer AM, Ma X-L (1993) Decreased basal nitric oxide release in hypercholesterolemia increases neutrophil adherence to rabbit coronary artery endothelium. Arterioscler Thromb 13:771–776

Lerman A, Edwards BS, Hallett JW, Heublein DM, Sandberg SM, Burnett JC (1991) Circulating and tissue endothelin immunoreactivity in advanced atherosclerosis. New Engl J Med 325:997–1001

Levine GN, Frei B, Koulouris SN, Gerhard MD, Keaney JF, Vita JA (1996) Ascorbic acid reverses endothelial vasomotor dysfunction in patients with coronary artery disease. Circulation 93:1107–1113

Liao JK, Clark SL (1995) Regulation of G-protein alpha i2 subunit expression by oxidized low-density lipoprotein. J Clin Invest 95:1457–1463

Liao JK, Bettmann MA, Sandor T, Tucker JI, Coleman SM, Creager MA (1991) Differential impairment of vasodilator responsiveness of peripheral resistance and conduit vessels in humans with atherosclerosis. Circ Res 68:1027–1034

Liao JK, Shin WS, Lee WY, Clark SL (1995) Oxidized low-density lipoprotein decreases the expression of endothelial nitric oxide synthase. J Biol Chem 270:319–324

Libby P, Hansson GK (1991) Biology of disease. Involvement of the immune system in human atherogenesis: current knowledge and unanswered questions. Lab Invest 64:5–15

Ludmer PL, Selwyn AP, Shook TL, Wayne RR, Mudge GH, Alexander RW, Ganz P (1986) Paradoxical vasoconstriction induced by acetylcholine in atherosclerotic coronary arteries. New Engl J Med 315:1046–1051

Luoma JS, Strålin P, Marklund SL, Hiltunen TP, Särkioja T, Ylä-Herttuala S (1998) Expression of extracellular SOD and iNOS in macrophages and smooth muscle cells in human and rabbit atherosclerotic lesions. Colocalization with epitopes characteristic of oxidized LDL and peroxynitrite-modified proteins. Arterioscler Thromb Vasc Biol 18:157–167

Lüscher TF (1994) Endothelium in the control of vascular tone and growth: role of local mediators and mechanical forces. Blood Press Suppl 1:18–22

Lüscher TF, Boulanger CM, Yang Z, Noll G, Dohi Y (1993) Interactions between endothelium-derived relaxing and contracting factors in health and cardiovascular disease. Circulation 87 [Suppl. V]:V36–V44

Lusis AJ (1993) The mouse model for atherosclerosis. Trends Cardiovasc Med 3:135–143

Lynch SM, Frei B, Morrow JD, Roberts LJ, Xu A, Jackson T, Reyna R, Klevay LM, Vita JA, Keaney-JF J (1997) Vascular superoxide dismutase deficiency impairs endothelial vasodilator function through direct inactivation of nitric oxide and increased lipid peroxidation. Arterioscler Thromb Vasc Biol 17:2975–2981

Ma X-L, Weyrich AS, Lefer DJ, Lefer AM (1993) Diminished basal nitric oxide release after myocardial ischemia and reperfusion promotes neutrophil adherence to coronary endothelium. Circ Res 72:403–412

Ma X-L, Lopez BL, Liu GL, Christopher TA, Gao F, Guo Y, Feuerstein GZ, Ruffolo-RR J, Barone FC, Yue TL (1997) Hypercholesterolemia impairs a detoxification mechanism against peroxynitrite and renders the vascular tissue more susceptible to oxidative injury. Circ Res 80:894–901

MacAllister RJ, Fickling SA, Whitley GS, Vallance P (1994) Metabolism of methylarginines by human vasculature: implications for the regulation of nitric oxide synthesis. Br J Pharmacol 112:43–48

Maccarrone M, Corasaniti MT, Guerrieri P, Nistico G, Agro AF (1996) Nitric oxide-donor compounds inhibit lipoxygenase activity. Biochem Biophys Res Commun 219:128–133

Maeba R, Maruyama A, Tarutani O, Ueta N, Shimasaki H (1995) Oxidized low-density lipoprotein induces the production of superoxide by neutrophils. FEBS Lett 377:309–312

Malo-Ranta U, Ylä-Herttuala S, Metsä-Ketelä T, Jaakkola O, Moilanen E, Vuorinen P, Nikkari T (1994) Nitric oxide donor GEA 3162 inhibits endothelial cell-mediated oxidation of low density lipoprotein. FEBS Lett 337:179–183

Mangin EL Jr, Kugiyama K, Nguy JH, Kerns SA, Henry PD (1993) Effects of lysolipids and oxidatively modified low density lipoprotein on endothelium-dependent relaxation of rabbit aorta. Circ Res 72:161–166

Marczin N, Antonov A, Papapetropoulos A, Munn DH, Virmani R, Kolodgie FD, Gerrity R, Catravas JD (1996) Monocyte-induced down-regulation of nitric oxide synthase in cultured aortic endothelial cells. Arterioscler Thromb Vasc Biol 16:1095–1103

Marks DS, Vita JA, Folts JD, Keaney JF, Welch GN, Loscalzo J (1995) Inhibition of neointimal proliferation in rabbits after vascular injury by a single treatment with a protein adduct of nitric oxide. J Clin Invest 96:2630–2638

Matthys KE, Van Hove CE, Jorens PG, Rosseneu M, Marescau B, Herman AG, Bult H (1996) Dual effects of oxidized low-density lipoprotein on immune-stimulated nitric oxide and prostaglandin release in macrophages. Eur J Pharmacol 298:97–103

McDonald KK, Zharikov S, Block ER, Kilberg MS (1997) A caveolar complex between the cationic amino acid transporter 1 and endothelial nitric-oxide synthase may explain the "arginine paradox". J Biol Chem 272:31213–31216

McDowell IFW, Brennan GM, McEneny J, Young IS, Nicholls DP, McVeigh GE, Bruce I, Trimble ER, Johnston GD (1994) The effect of probucol and vitamin E treatment on the oxidation of low-density lipoprotein and forearm vascular responses in humans. Eur J Clin Invest 24:759–765

McLenachan JM, Vita J, Fish RD, Treasure CB, Cox DA, Ganz P, Selwyn AP (1990) Early evidence of endothelial vasodilator dysfunction at coronary branch points. Circulation 82:1169–1173

McNamara DB, Bedi B, Aurora H, Tena L, Ignarro LJ, Kadowitz PJ, Akers DL (1993) L-Arginine inhibits balloon catheter-induced intimal hyperplasia. Biochem Biophys Res Commun 193:291–296

Mehta A, Yang B, Khan S, Hendricks JB, Stephen C, Mehta JL (1995) Oxidized low-density lipoproteins facilitate leukocyte adhesion to aortic intima without affecting endothelium-dependent relaxation. Role of P selectin. Arterioscler Thromb Vasc Biol 15:2076–2083

Meredith IT, Yeung AC, Weidinger FF, Anderson TJ, Uehata A, Ryan TJ Jr, Selwyn AP, Ganz P (1993) Role of impaired endothelium-dependent vasodilation in ischemic manifestations of coronary artery disease. Circulation 87 [suppl V]:V-56-V-66

Merryman PF, Clancy RM, He XY, Abramson SB (1993) Modulation of human T-cell responses by nitric oxide and its derivative, S-nitrosoglutathione. Arthritis and Rheumatism 36:1414–1422

Messmer UK, Brüne B (1996) Nitric oxide (NO) in apoptotic versus necrotic RAW 264.7 macrophage cell death: the role of NO-donor exposure, NAD(+) content, and p53 accumulation. Arch Biochem Biophys 327:1–10

Messmer UK, Reed JC, Brüne B (1996) Bcl-2 protects macrophages from nitric oxide-induced apoptosis. J Biol Chem 351:261–272

Miles AM, Bohle DS, Glassbrenner PA, Hansert B, Wink DA, Grisham MB (1996) Modulation of superoxide-dependent oxidation and hydroxylation reactions by nitric oxide. J Biol Chem 271:40–47

Minor RL Jr, Myers PR, Guerra R Jr, Bates JN, Harrison DG (1990) Diet-induced atherosclerosis increases the release of nitrogen oxides from rabbit aorta. J Clin Invest 86:2109–2116

Mohan PF, Desaiah D (1994) Very low density and low density lipoproteins induce nitric oxide synthesis in macrophages. Biochem Biophys Res Commun 204:1047–1054

Mohr S, McCormick TS, Lapetina EG (1998) Macrophages resistant to endogenously generated nitric oxide-mediated apoptosis are hypersensitive to exogenously added nitric oxide donors: dichotomous apoptotic response independent of caspase 3 and reversal by the mitogen-activated protein kinase kinase (MEK) inhibitor PD 098059. Proc Natl Acad Sci USA 95:5045–5050

Mooradian DL, Hutsell TC, Keefer LK (1995) Nitric oxide (NO) donor molecules: effect of NO release rate on vascular smooth muscle cell proliferation in vitro. J Cardiovasc Pharmacol 25:674–678

Morbidelli L, Chang CH, Douglas JG, Granger HJ, Ledda F, Ziche M (1996) Nitric oxide mediates mitogenic effect of VEGF on coronary venular endothelium. Am J Physiol 39:H411–H415

Moroi M, Zhang L, Yasuda T, Virmani R, Gold HK, Fishman MC, Huang PL (1998) Interaction of genetic deficiency of endothelial nitric oxide, gender, and pregnancy in vascular response to injury in mice. J Clin Invest 101:1225–1232

Motterlini R, Foresti R, Intaglietta M, Winslow RM (1996) NO-mediated activation of heme oxygenase: endogenous cytoprotection against oxidative stress to endothelium. Am J Physiol 39:H107–H114

Mügge A, Harrison DG (1991) L-Arginine does not restore endothelial dysfunction in atherosclerotic rabbit aorta in vitro. Blood Vessels 28:354–357

Mügge A, Elwell JH, Peterson TE, Harrison DG (1991a) Release of intact endothelium-derived relaxing factor depends on endothelial superoxide dismutase activity. Am J Physiol 260:C219–C225

Mügge A, Elwell JH, Peterson TE, Hofmeyer TG, Heistad DD, Harrison DG (1991b) Chronic treatment with polyethylene-glycolated superoxide dismutase partially restores endothelium-dependent vascular relaxations in cholesterol-fed rabbits. Circ Res 69:1293–1300

Mügge A, Brandes RP, Böger RH, Dwenger A, Bode-Böger S, Kienke S, Frölich JC, Lichtlen PR (1994) Vascular release of superoxide radicals is enhanced in hypercholesterolemic rabbits. J Cardiovasc Pharmacol 24:994–998

Murohara T, Kugiyama K, Ohgushi M, Sugiyama S, Ohta Y, Yasue H (1994) LPC in oxidized LDL elicits vasocontraction and inhibits endothelium-dependent relaxation. Am J Physiol 267:H2441–H2449

Murrell GAC, Jang D, Williams RJ (1995) Nitric oxide activates metalloprotease enzymes in articular cartilage. Biochem Biophys Res Commun 206:15–21

Nabel EG, Selwyn AP, Ganz P (1990) Large coronary arteries in humans are responsive to changing blood flow: an endothelium-dependent mechanism that fails in patients with atherosclerosis. J Am Coll Cardiol 16:349–356

Nakaki T, Nakayama M, Kato R (1990) Inhibition by nitric oxide and nitric oxide-producing vasodilators of DNA synthesis in vascular smooth muscle cells. Eur J Pharmacol 189:347–353

Naruse K, Shimizu K, Muramatsu M, Toki Y, Miyazaki Y, Okumura K, Hashimoto H, Ito T (1994) Long-term inhibition of NO synthesis promotes atherosclerosis in the hypercholesterolemic rabbit thoracic aorta. Arterioscler Thromb 14:746–752

Newman CM, Maseri A, Hackett DR, el-Tamimi HM, Davies GJ (1990) Response of angiographically normal and atherosclerotic left anterior descending coronary arteries to acetylcholine. Am J Cardiol 66:1070–1076

Nguyen T, Brunson D, Crespi CL, Penman BW, Wishnok JS, Tannenbaum SR (1992) DNA damage and mutation in human cells exposed to nitric oxide in vitro. Proc Natl Acad Sci USA 89:3030–3034

Nitenberg A, Paycha F, Ledoux S, Sachs R, Attali JR, Valensi P (1998) Coronary artery responses to physiological stimuli are improved by deferoxamine but not by L-arginine in non-insulin-dependent diabetic patients with angiographically normal coronary arteries and no other risk factors. Circulation 97:736–743

Niu X-F, Smith CW, Kubes P (1994) Intracellular oxidative stress induced by nitric oxide synthase inhibition increases endothelial cell adhesion to neutrophils. Circ Res 74:1133–1140

O'Driscoll G, Green D, Taylor RR (1997) Simvastatin, an HMG-coenzyme A reductase inhibitor, improves endothelial function within 1 month. Circulation 95:1126–1131

Ohara Y, Peterson TE, Harrison DG (1993) Hypercholesterolemia increases endothelial superoxide anion production. J Clin Invest 91:2546–2551

Ohara Y, Peterson TE, Zheng B, Kuo JF, Harrison DG (1994) Lysophophatidylcholine increases vascular superoxide anion production via protein kinase C activation. Arterioscler Thromb 14:1007–1013

Ohara Y, Peterson TE, Sayegh HS, Subramanian RR, Wilcox JN, Harrison DG (1995) Dietary correction of hypercholesterolemia in the rabbit normalizes endothelial superoxide anion production. Circulation 92:898–903

Okazaki J, Komori K, Kawasaki K, Eguchi D, Ishida M, Sugimachi K (1997) L-arginine inhibits smooth muscle cell proliferation of vein-graft intimal thickness in hypercholesterolemic rabbits. Cardiovasc Res 36:429–436

Osborne JA, Lento PH, Siegfried MR, Stahl GL, Fusman B, Lefer AM (1989a) Cardiovascular effects of acute hypercholesterolemia in rabbits. Reversal with lovastatin treatment. J Clin Invest 83:465–473

Osborne JA, Siegman MJ, Sedar AW, Mooers SU, Lefer AM (1989b) Lack of endothelium-dependent relaxation in coronary resistance arteries of cholesterol-fed rabbits. Am J Physiol 256:C591–C597

Palmer RM, Bridge L, Foxwell NA, Moncada S (1992) The role of nitric oxide in endothelial cell damage and its inhibition by glucocorticoids. Br J Pharmacol 105:11–12

Papapetropoulos A, Garcia CG, Madri JA, Sessa WC (1997) Nitric oxide production contributes to the angiogenic properties of vascular endothelial growth factor in human endothelial cells. J Clin Invest 100:3131–3139

Polte T, Oberle S, Schröder H (1997) Nitric oxide protects endothelial cells from tumor necrosis factor-α-induced cytotoxicity: possible involvement of cyclic GMP. FEBS Lett 409:46–48

Pomerantz KB, Hajjar DP, Levi R, Gross SS (1993) Cholesterol enrichment of arterial smooth muscle cells up-regulates cytokine-induced nitric oxide synthesis. Biochem Biophys Res Commun 191:103–109

Pritchard KAJ, Groszek L, Smalley DM, Sessa WC, Wu M, Villalon P, Wolin MS, Stemerman MB (1995) Native low-density lipoprotein increases endothelial

cell nitric oxide synthase generation of superoxide anion. Circ Res 77:510–518
Quyyumi AA, Mulcahy D, Andrews NP, Husain S, Panza JA, Cannon RO (1997) Coronary vascular nitric oxide activity in hypertension and hypercholesterolemia. Comparison of acetylcholine and substance P. Circulation 95:104–110
Radi R, Beckman JS, Bush KM, Freeman BA (1991) Peroxynitrite oxidation of sulfhydryls. The cytotoxic potential of superoxide and nitric oxide. J Biol Chem 266:4244–4250
Rajagopalan S, Kurz S, Münzel T, Tarpey M, Freeman BA, Griendling KK, Harrison DG (1996) Angiotensin II-mediated hypertension in the rat increases vascular superoxide production via membrane NADH/NADPH oxidase activation. Contribution to alterations of vasomotor tone. J Clin Invest 1996:1916–1923
Rasheed Q, Hodgson JM (1993) Application of intracoronary ultrasonography in the study of coronary artery pathophysiology. J Clin Ultrasound 21:569–578
Rinaldo JE, Clark M, Parinello J, Shepherd VL (1994) Nitric oxide inactivates xanthine dehydrogenase and xanthine oxidase in interferon-gamma-stimulated macrophages. Am J Respir Cell Mol Biol 11:625–630
Rizvi MAD, Katwa L, Spadone DP, Myers PR (1996) The effects of endothelin-1 on collagen type I and type III synthesis in cultured porcine coronary artery vascular smooth muscle cells. J Mol Cellul Cardiol 28:243–252
Rosenkranz WP, Sessa WC, Milstien S, Kaufman S, Watson CA, Pober JS (1994) Regulation of nitric oxide synthesis by proinflammatory cytokines in human umbilical vein endothelial cells. Elevations in tetrahydrobiopterin levels enhance endothelial nitric oxide synthase specific activity. J Clin Invest 93:2236–2243
Ross R (1993) The pathogenesis of atherosclerosis: a perspective for the 1990s. Nature 362:801–809
Rossitch E Jr, Alexander EI, Black PM, Cooke JP (1991) L-Arginine normalizes endothelial function in cerebral vessels from hypercholesterolemic rabbits. J Clin Invest 87:1295–1299
Rubbo H, Parthasarathy S, Barnes S, Kirk M, Kalyanaraman B, Freeman BA (1995) Nitric oxide inhibition of lipoxygenase-dependent liposome and low-density lipoprotein oxidation: termination of radical chain propagation reactions and formation of nitrogen-containing oxidized lipid derivatives. Arch Biochem Biophys 324:15–25
Rudic RD, Shesely EG, Maeda N, Smithies O, Segal SS, Sessa WC (1998) Direct evidence for the importance of endothelium-derived nitric oxide in vascular remodeling. J Clin Invest 101:731–736
Russell ME, Wallace AF, Wyner LR, Newell JB, Karnovsky MJ (1995) Up-regulation and modulation of inducible nitric oxide synthase in rat cardiac allografts with chronic rejection and transplant arteriosclerosis. Circulation 92:457–464
Sarih M, Souvannavong V, Adam A (1993) Nitric oxide synthase induces macrophage death by apoptosis. Biochem Biophys Res Commun 191:503–508
Sarkar R, Meinberg EG, Stanley JC, Gordon D, Webb RC (1996) Nitric oxide reversibly inhibits the migration of cultured vascular smooth muscle cells. Circ Res 78:225–230
Schroeder JS, Bolen JL, Quint RA, Clark DA, Hayden WG, Higgins CB, Wexler L (1977) Provocation of coronary spasm with ergonovine maleate. New test with results in 57 patients undergoing coronary arteriography. Am J Cardiol 40:487–491
Scott-Burden T, Vanhoutte PM (1993) The endothelium as a regulator of vascular smooth muscle proliferation. Circulation 87 [Suppl. V]:V51–V55
Sellke FW, Armstrong ML, Harrison DG (1990) Endothelium-dependent vascular relaxation is abnormal in the coronary microcirculation of atherosclerotic primates. Circulation 81:1586–1593
Sharma P, Evans AT, Parker PJ, Evans FJ (1991) NADPH-oxidase activation by protein kinase C isotypes. Biochem Biophys Res Commun 177:1033–1040

Sharma RC, Crawford DW, Kramsch DM, Sevanian A, Jiao Q (1992) Immunolocalization of native antioxidant scavenger enzymes in early hypertensive and atherosclerotic arteries. Role of oxygen free radicals. Arterioscler Thromb 12: 403–415

Shimizu S, Nomoto M, Naito S, Yamamoto T, Momose K (1998) Simulation of nitric oxide synthase during oxidative endothelial cell injury. Biochemical Pharmacology 55:77–83

Shimokawa H, Vanhoutte PM (1989) Impaired endothelium-dependent relaxation to aggregating platelets and related vasoactive substances in porcine coronary arteries in hypercholesterolemia and atherosclerosis. Circ Res 64:900–914

Shimokawa H, Flavahan NA, Vanhoutte PM (1991) Loss of endothelial pertussis toxin-sensitive G protein function in atherosclerotic porcine coronary arteries. Circulation 83:652–660

Siegfried MR, Carey C, Ma X-L, Lefer AM (1992) Beneficial effects of SPM-5185, a cysteine-containing NO donor in myocardial ischemia–reperfusion. Am J Physiol 263:H771–H777

Singer AH, Tsao PS, Wang B-Y, Bloch DA, Cooke JP (1995) Discordant effects of dietary L-arginine on vascular structure and reactivity in hypercholesterolemic rabbits. J Cardiovasc Pharmacol 25:710–716

Spiecker M, Peng H-B, Liao JK (1997) Inhibition of endothelial vascular cell adhesion molecule-1 expression by nitric oxide involves the induction and nuclear translocation of IκBα. J Biol Chem 272:30969–30974

Sreeharan N, Jayakody RL, Senaratne MP, Thomson AB, Kappagoda CT (1986) Endothelium-dependent relaxation and experimental atherosclerosis in the rabbit aorta. Can J Physiol Pharmacol 64:1451–1453

Stary HC, Blankenhorn DH, Chandler AB, Glagov S, Insull W Jr, Richardson M, Rosenfeld ME, Schaffer SA, Schwartz CJ, Wagner WD, Wissler RW (1992) A definition of the intima of human arteries and of its atherosclerosis-prone regions. Circulation 85:391–405

Stroes E, Kastelein J, Cosentino F, Erkelens W, Wever R, Koomans H, Lüscher T, Rabelink T (1997) Tetrahydrobiopterin restores endothelial function in hypercholesterolemia. J Clin Invest 99:41–46

Struck AT, Hogg N, Thomas JP, Kalyanaraman B (1995) Nitric oxide donor compounds inhibit the toxicity of oxidized low-density lipoprotein to endothelial cells. FEBS Lett 361:291–294

Suttorp N, Hippenstiel S, Fuhrmann M, Krull M, Podzuweit T (1996) Role of nitric oxide and phosphodiesterase isoenzyme II for reduction of endothelial hyperpermeability. Am J Physiol 39:C778–C785

Szabo C, Zingarelli B, O'Connor M, Salzman AL (1996) DNA strand breakage, activation of poly(ADP–ribose) synthetase, and cellular energy depletion are involved in the cytotoxicity in macrophages and smooth muscle cells exposed to peroxynitrite. Proc Natl Acad Sci USA 93:1753–1758

Taddei S, Virdis A, Mattei P, Ghiadoni L, Sudano I, Salvetti A (1996) Defective L-arginine-nitric oxide pathway in offspring of essential hypertensive patients. Circulation 94:1298–1303

Tagawa H, Tomoike H, Nakamura M (1991) Putative mechanisms of the impairment of endothelium-dependent relaxation of the aorta with atheromatous plaque in heritable hyperlipidemic rabbits. Circ Res 68:330–337

Taguchi J, Abe J, Okazaki H, Takuwa Y, Kurokawa K (1993) L-Arginine inhibits neointimal formation following balloon injury. Life Sci 53:387–392

Tamai O, Matsuoka H, Itabe H, Wada Y, Kohno K, Imaizumi T (1997) Single LDL apheresis improves endothelium-dependent vasodilatation in hypercholesterolemic humans. Circulation 95:76–82

Tarry WC, Makhoul RG (1994) L-Arginine improves endothelium-dependent vasorelaxation and reduces intimal hyperplasia after balloon angioplasty. Arterioscler Thromb 14:938–943

Thomas JP, Geiger PG, Girotti AW (1993) Lethal damage to endothelial cells by oxidized low density lipoprotein: role of selenoperoxidases in cytoprotection against lipid hydroperoxide- and iron-mediated reactions. J Lipid Res 34:479–490

Thubrikar MJ, Robicsek F (1995) Pressure-induced arterial wall stress and atherosclerosis. Ann Thorac Surg 59:1594–1603

Timimi FK, Ting HH, Haley EA, Roddy M-A, Ganz P, Creager MA (1998) Vitamin C improves endothelium-dependent vasodilation in patients with insulin-dependent diabetes mellitus. J Am Coll Cardiol 31:552–557

Ting HH, Timimi FK, Haley EA, Roddy MA, Ganz P, Creager MA (1997) Vitamin C improves endothelium-dependent vasodilation in forearm resistance vessels of humans with hypercholesterolemia. Circulation 95:2617–2622

Topper JN, Cai J, Falb D, Gimbrone MA Jr (1996) Identification of vascular endothelial genes differentially responsive to fluid mechanical stimuli: cyclooxygenase-2, manganese superoxide dismutase, and endothelial cell nitric oxide synthase are selectively up-regulated by steady laminar shear stress. Proc Natl Acad Sci USA 93:10417–10422

Trachtman H, Futterweit S, Garg P, Reddy K, Singhai PC (1996) Nitric oxide stimulates the activity of a 72-kDa neutral matrix metalloproteinase in cultured rat mesangial cells. Biochem Biophys Res Commun 218:704–708

Tronc F, Wassef M, Esposito B, Henrion D, Glagov S, Tedgui A (1996) Role of NO in flow-induced remodeling of the rabbit common carotid artery. Arterioscler Thromb Vasc Biol 16:1256–1262

Tsao PS, McEvoy LM, Drexler H, Butcher EC, Cooke JP (1994) Enhanced endothelial adhesiveness in hypercholesterolemia is attenuated by L-arginine. Circulation 89:2176–2182

Tsao PS, Lewis NP, Alpert S, Cooke JP (1995) Exposure to shear stress alters endothelial adhesiveness. Role of nitric oxide. Circulation 92:3513–3519

Tsutsui M, Shimokawa H, Tanaka S, Kuwaoka I, Hase K, Nogami N, Nakanishi K, Okamatsu S (1994) Endothelial Gi protein in human coronary arteries. Eur Heart J 15:1261–1266

Uehata A, Gerhard MD, Meredith IT, Lieberman EL, Sylwyn AP, Creager M, Polak J, Ganz P, Yeung AC, Anderson TJ (1993) Close relationship of endothelial dysfunction in coronary and brachial artery. Circulation 88:I-618

Urakami-Harasawa L, Shimokawa H, Nakashima M, Egashira K, Takeshita A (1997) Importance of endothelium-derived hyperpolarizing factor in human arteries. J Clin Invest 100:2793–2799

Vallance P, Leone A, Calver A, Collier J, Moncada S (1992) Accumulation of an endogenous inhibitor of nitric oxide synthesis in chronic renal failure. Lancet 339:572–575

Van Belle E, Vallet B, Auffray JL, Bauters C, Hamon M, McFadden EI, Lablanche JM, Dupuis E, Bertrand ME (1996) NO synthesis is involved in structural and functional effects of ACE inhibitors in injured arteries. Am J Physiol 39:H298–H305

Vanhoutte PM, Shimokawa H (1989) Endothelium-derived relaxing factor and coronary vasospasm. Circulation 80:1–9

Varenne O, Pislaru S, Gillijns H, Van Pelt N, Gerard RD, Zoldhelyi P, Van de Werf F, Collen D, Janssens SP (1998) Local adenovirus-mediated transfer of human endothelial nitric oxide synthase reduces luminal narrowing after coronary angioplasty in pigs. Circulation 98:919–926

Verbeuren TJ, Jordaens FH, Van Hove CE, Van Hoydonck A-E, Herman AG (1990) Release and vascular activity of endothelium-derived relaxing factor in atherosclerotic rabbit aorta. Eur J Pharmacol 191:173–184

Verbeuren TJ, Jordaens FH, Zonnekeyn LL, Van Hove CE, Coene M-C, Herman AG (1986) Effect of hypercholesterolemia on vascular reactivity in the rabbit. I.

Endothelium-dependent and endothelium-independent contractions and relaxations in isolated arteries of control and hypercholesterolemic rabbits. Circ Res 58:552–564

Verbeuren TJ, Simonet S, Vallez M-O, Laubie A (1991) 5-HT$_2$-receptor blockade restores vasodilations to 5-HT in atherosclerotic rabbit hearts: role of the endothelium. J Cardiovasc Pharmacol 17 Suppl. 3]:S222–S228

Verbeuren TJ, Bonhomme E, Laubie M, Simonet S (1993) Evidence for induction of non-endothelial NO synthase in aortas of cholesterol-fed rabbits. J Cardiovasc Pharmacol 21:841–845

Verhaar MC, Wever RMF, Kastelein JJP, van Dam T, Koomans HA, Rabelink TJ (1998) 5-Methyltetrahydrofolate, the active form of folic acid, restores endothelial function in familial hypercholesterolemia. Circulation 97:237–241

Vita JA, Treasure CB, Nabel EG, McLenachan JM, Fish RD, Yeung AC, Vekshtein VI, Selwyn AP, Ganz P (1990) Coronary vasomotor response to acetylcholine relates to risk factors for coronary artery disease. Circulation 81:491–497

Vogel RA (1997) Coronary risk factors, endothelial function, and atherosclerosis: a review. Clin Cardiol 20:426–432

Von der Leyen HE, Gibbons GH, Morishita R, Lewis NP, Zhang L, Nakajima M, Kaneda Y, Cooke JP, Dzau VJ (1995) Gene therapy inhibiting neointimal vascular lesion: in vivo transfer of endothelial cell nitric oxide synthase gene. Proc Natl Acad Sci USA 92:1137–1141

Von Knethen A, Brüne B (1997) Cyclooxygenase-2: an essential regulator of NO-mediated apoptosis. FASEB J 11:887–895

Vrints CJM, Bult H, Bosmans J, Herman AG, Snoeck JP (1992a) Paradoxical vasoconstriction as result of acetylcholine and serotonin in diseased human coronary arteries. Eur Heart J 13:824–831

Vrints CJM, Bult H, Hitter E, Herman AG, Snoeck JP (1992b) Impaired endothelium-dependent cholinergic coronary vasodilation in patients with angina and normal coronary arteriograms. J Am Coll Cardiol 19:21–31

Wang JM, Chow SN, Lin JK (1994) Oxidation of LDL by nitric oxide and its modification by superoxides in macrophage and cell-free systems. FEBS Lett 342:171–175

Wang Z, Stingl L, Morrison C, Jantsch M, Los M, Schulze-Osthoff K, Wanger EF (1997) PARP is important for genomic stability but dispensable in apoptosis. Genes Dev 11:2347–2358

Wascher TC, Posch K, Wallner S, Hermetter A, Kostner GM, Graier WF (1997) Vascular effects of L arginine: anything beyond a substrate for the NO synthase? Biochem Biophys Res Commun 234:35–38

Werns SW, Walton JA, Hsia HH, Nabel EG, Sanz ML, Pitt B (1989) Evidence of endothelial dysfunction in angiographically normal coronary arteries of patients with coronary artery disease. Circulation 79:287–291

Wever RMF, Van Dam T, van Rijn HJ, De Groot F, Rabelink TJ (1997) Tetrahydrobiopterin regulates superoxide and nitric oxide generation by recombinant endothelial nitric oxide synthase. Biochem Biophys Res Commun 237:340–344

White CR, Brock TA, Chang LY, Crapo J, Briscoe P, Ku D, Bradley WA, Gianturco SH, Gore J, Freeman BA (1994) Superoxide and peroxynitrite in atherosclerosis. Proc Natl Acad Sci USA 91:1044–1048

White CR, Darley-Usmar V, Berrington WR, McAdams M, Gore JZ, Thompson JA, Parks DA, Tarpey MM, Freeman BA (1996) Circulating plasma xanthine oxidase contributes to vascular dysfunction in hypercholesterolemic rabbits. Proc Natl Acad Sci USA 93:8745–8749

White E (1994) Tumour biology. p53, guardian of Rb. Nature 371:21–22

Wilcox JN, Subramanian RR, Sundell CL, Tracey WR, Pollock JS, Harrison DG, Marsden PA (1997) Expression of multiple isoforms of nitric oxide synthase in normal and atherosclerotic vessels. Arterioscler Thromb Vasc Biol 17:2479–2488

Wileman SM, Mann GE, Baydoun AR (1995) Induction of L-arginine transport and nitric oxide synthase in vascular smooth muscle cells: synergistic actions of proinflammatory cytokines and bacterial lipopolysaccharide. Br J Pharmacol 116:3243–3250

Wink DA, Kasprzak KS, Maragos CM, Elespuru RK, Misra M, Dunams TM, Cebula TA, Koch WH, Andrews AW, Allen JS, Keefer LK (1991) DNA deanimating ability and genotoxicity of nitric oxide and its progenitors. Science 254:1001–1003

Wink DA, Hanbauer I, Krishna MC, Degraff W, Gamson J, Mitchell JB (1993) Nitric oxide protects against cellular damage and cytotoxicity from reactive oxygen species. Proc Natl Acad Sci USA 21:9813–9817

Wink DA, Cook JA, Pacelli R, Liebmann J, Krishna MC, Mitchell JB (1995) Nitric oxide (NO) protects against cellular damage by reactive oxygen species. Toxicol Lett 82:221–226

Witztum JL (1993) Role of oxidised low density lipoprotein in atherogenesis. Br Heart J 69, suppl: S12-S18

Wolf YG, Rasmussen LM, Sherman Y, Bundens WP, Hye RJ (1995) Nitroglycerin decreases medial smooth muscle cell proliferation after arterial balloon injury. J Vasc Surg 21:499–504

Wong HR, Finder JD, Wasserloos K, Pitt BR (1995) Expression of iNOS in cultured rat pulmonary artery smooth-muscle cells is inhibited by the heat shock response. Am J Physiol 13:L843–L848

Yamamoto H, Bossaller C, Cartwright J Jr, Henry PD (1988) Videomicroscopic demonstration of defective cholinergic arteriolar vasodilation in atherosclerotic rabbit. J Clin Invest 81:1752–1758

Yamamoto Y, Tomoike H, Egashira K, Nakamura M (1987) Attenuation of endothelium-related relaxation and enhanced responsiveness of vascular smooth muscle to histamine in spastic coronary arterial segments from miniature pigs. Circ Res 61:772–778

Yan Z, Hansson GK (1998) Overexpression of inducible nitric oxide synthase by neointimal smooth muscle cells. Circ Res 82:21–29

Yang X, Cai B, Sciacca RR, Cannon PJ (1994) Inhibition of inducible nitric oxide synthase in macrophages by oxidized low-density lipoproteins. Circ Res 74:318–328

Yasue H, Matsuyama K, Matsuyama K, Okumura K, Morikami Y, Ogawa H (1990) Responses of angiographically normal human coronary arteries to intracoronary injection of acetylcholine by age and segment. Possible role of early coronary atherosclerosis. Circulation 81:482–490

Yates MT, Lambert LE, Whitten JP, McDonald I, Mano M, Ku G, Mao SJT (1992) A protective role for nitric oxide in the oxidative modification of low density lipoproteins by mouse macrophages. FEBS Lett 309:135–138

Yin ZL, Dusting GJ (1997) A nitric oxide donor (spermine NONOate) prevents the formation of neointima in rabbit carotid artery. Clin Exp Pharmacol Physiol 24:436–438

Yu SM, Huang ZS, Wang CY, Teng CM (1993) Effects of hyperlipidemia on the vascular reactivity in the Wistar-Kyoto and spontaneously hypertensive rats. Eur J Pharmacol 248:289–295

Zalewski A, Shi Y (1997) Vascular myofibroblasts. Lessons from coronary repair and remodeling. Arterioscler Thromb Vasc Biol 17:417–422

Zamora R, Bult H, Herman AG (1998) The role of prostaglandin E_2 and nitric oxide in cell death in J774 murine macrophages. Eur J Pharmacol 349:307–315

Zeiher AM, Drexler H, Wollschläger H, Just H (1991) Modulation of coronary vasomotor tone in humans. Progressive endothelial dysfunction with different early stages of coronary atherosclerosis. Circulation 83:391–401

Zeiher AM, Fisslthaler B, Schray-Utz B, Busse R (1995) Nitric oxide modulates the expression of monocyte chemoattractant protein 1 in cultured human endothelial cells. Circ Res 76:980–986

Ziche M, Parenti A, Ledda F, Dell'Era P, Granger HJ, Maggi CA, Presta M (1997) Nitric oxide promotes proliferation and plasminogen activator production by coronary venular endothelium through endogenous bFGF. Circ Res 80:845–852

Zulueta JJ, Yu FS, Hertig IA, Thannickal VJ, Hassoun PM (1995) Release of hydrogen peroxide in response to hypoxia–reoxygenation: role of an NAD(P)H oxidase-like enzyme in endothelial cell plasma membrane. Am J Respir Cell Mol Biol 12:41–49

CHAPTER 23
Nitric Oxide in Brain Ischemia/Reperfusion Injury

M. Sasaki, T.M. Dawson, and V.L. Dawson

A. Introduction

The brain is a highly energetic organ, which requires a continuous supply of nutrients and oxygen from the circulatory system in order to function properly. Cessation of the blood supply to the brain for even a few minutes results in neuronal injury following initiation of a cascade of secondary mechanisms. Ischemia results in the reduction of resting membrane potential of glia and neurons in the brain. The leakage of potassium out of cells results in the depolarization of neurons, leading to a massive release of glutamate. Glutamate elicits its actions by acting on series of post-synaptic receptors, including the *N*-methyl-D-aspartate receptor (NMDA), non-NMDA receptors, and metabotropic glutamate receptors (Samdani et al. 1997). For nearly two decades, the actions of glutamate have been linked as important mediators of ischemic brain injury. The observation that activation of NMDA receptors generates nitric oxide (NO) in a calcium-dependent manner (Garthwaite et al. 1988; Bredt and Snyder 1989; Garthwaite et al. 1989) raised the possibility that NO might be an important mediator in regulating glutamate neurotoxicity. The observation that non-selective NO synthase (NOS) inhibitors could reduce glutamate neurotoxicity in vitro (Dawson et al. 1991b) and reduce infarct volume following transient focal ischemia in mice (Nowicki et al. 1991) suggested a role for NO as a neurotoxin. Immediately, there was controversy over the role of NO in neurotoxicity and ischemic damage (Dawson et al. 1994a; Dawson and Dawson 1996). Since then, a large body of scientific literature has evolved describing the role of NO in glutamate toxicity and ischemia-reperfusion injury. The early controversies were due to the important but opposing effects of NO generated from different NOS isoforms in the central nervous system (CNS), the use of non-selective NOS inhibitors, and the lack of understanding of the complex chemistry of NO in a biologic setting. Advances in pharmacology and chemistry, in addition to the generation of genetically engineered mice, have greatly expanded our understanding of NO biology in ischemic injury.

B. Neuronal NOS

Neuronal NOS (nNOS) is expressed in a small population of neurons scattered throughout the brain and spinal cord (Bredt et al. 1990; Dawson et al.

1991a). nNOS was the first isoform to be purified and cloned (BREDT et al. 1990). Under physiologic conditions, nNOS activation is believed to mediate synaptic plasticity and neuronal signaling and control, through the generation of NO, cerebral microcirculation during neural activity (DAWSON and SNYDER 1994; IADECOLA 1997). However, under pathologic conditions of ischemic injury, activation of nNOS contributes to neurotoxicity (DAWSON and DAWSON 1995; YUN et al. 1997).

A pathologic role for nNOS activation was first described in primary cortical neuronal cultures (DAWSON et al. 1991b). In these primary cortical cultures, nNOS is expressed at levels equivalent to in vivo expression – approximately 1–2% of the total neuronal population. In this culture model system, glutamate and NMDA neurotoxicity are mediated, in large part, by formation of NO. Co-exposure of primary cortical neurons to NMDA and arginine analogs, such as N^G-nitro-L-arginine (L-NNA) or N^G- nitro-L-arginine methyl ester (L-NAME), which are NOS inhibitors, attenuates NMDA neurotoxicity (DAWSON et al. 1991b, 1993). The neuroprotection afforded by NOS inhibitors can be competitively reversed by the nNOS substrate L-arginine. The stereoisomer, D-arginine, and the inactive analog, homo-arginine, are ineffective in reversing the neuroprotection afforded by l-NNA against NMDA neurotoxicity, confirming that l-NNA competes for the L-arginine binding site (DAWSON et al. 1991b). Concentration-response experiments were performed with a series of arginine-analog NOS inhibitors. The rank order of potency in inhibiting NMDA neurotoxicity paralleled their known rank order of potency in inhibiting nNOS (DAWSON et al. 1993).

Further evidence that activation of nNOS mediates a large component of NMDA neurotoxicity was provided by the use of inhibitors targeted towards regions other than the catalytic site. These inhibitors included flavoprotein inhibitors, calmodulin antagonists or agents that bind calmodulin, and calcineurin inhibitors. All of these agents decrease NOS catalytic activity and result in neuroprotection against NMDA neurotoxicity (DAWSON et al. 1993). nNOS neurons are resistant to excitotoxicity and make up only a fraction of the neurons present in primary cortical cultures, yet brief exposures to NMDA result in over 60–80% neuronal cell loss. Therefore, the NO generated in this small population of neurons must diffuse to surrounding cells to initiate a death cascade. Reduced hemoglobin scavenges NO but is too large to be cell permeant. Reduced hemoglobin completely prevents NMDA-induced neurotoxicity, confirming that NO generated in nNOS neurons diffuses to surrounding cells to initiate cell-death processes (DAWSON et al. 1991b). Additionally, NO generators, such as sodium nitroprusside or SIN-1, can induce neurotoxicity with kinetics similar to those of NMDA in primary cortical cultures (DAWSON et al. 1991b, 1993). Final confirmation of the importance of nNOS activation in NMDA-mediated neurotoxicity (DAWSON et al. 1996) and ischemic injury (HUANG et al. 1994; HARA et al. 1996) was provided by evaluating primary cortical cultures generated from genetically engineered nNOS-knockout mice. Cultures from nNOS-knockout mice are resistant to

NMDA and ischemic injury. However, the cell-death mechanisms recruited by generation of NO remain intact in the nNOS-knockout mice, since exposure to NO donors effectively elicits neurotoxicity in primary cultures from nNOS-knockout mice in a manner equivalent to that seen in wild-type mice (DAWSON et al. 1996). Therefore, the genetic manipulation has not impaired downstream death pathways but has merely uncoupled NMDA-receptor activation from NO production and thereby attenuated NMDA neurotoxicity.

While these initial observations have been replicated in a variety of cell-culture and slice-preparation systems, there are conflicting reports in the literature, in which NO-mediated neurotoxicity in culture models has not been observed (DAWSON and SNYDER 1994). In many cases, these differences can be attributed to insufficient expression of nNOS neurons or differences in experimental paradigms, which significantly change the chemical reaction pathways of NO. Recent studies indicate that NMDA-induced NO-mediated neurotoxicity is dependent on the number of nNOS neurons and the level of nNOS-protein expression (SAMDANI et al. 1997). The expression of nNOS in neurons is dependent on the culture conditions employed. Neurons grown on glial feeder layers contain relatively low levels of nNOS, which never mature into an adult expression pattern (SAMDANI et al. 1997). Neurons grown on poly-ornithine matrix mature and develop an adult pattern of nNOS expression. In neurons grown on a glial feeder layer, exposure to neurotrophins markedly increases the number of nNOS neurons, expression of nNOS protein, and nNOS catalytic activity (SAMDANI et al. 1997) and results in enhancement of NMDA neurotoxicity (KOH et al. 1995; SAMDANI et al. 1997) via NO-dependent mechanisms (SAMDANI et al. 1997). In contrast, nNOS expression in neurons grown on a poly-ornithine matrix and allowed to mature into an adult phenotype is unaffected by exposure to neurotrophins (KOH et al. 1995; SAMDANI et al. 1997). Neurotrophins do not influence the number of nNOS neurons in this culture system, and exposure to neurotrophins reduces NMDA neurotoxicity (SAMDANI et al. 1997). These results are reflective of observations made in numerous animal models in which neurotrophins protect against neurotoxicity. Confirmation that the differential effects observed between cultures grown in feeder layers versus cultures grown on a poly-ornithine matrix is due, in large part, to expression of nNOS in experiments in which primary cortical cultures were generated from mice lacking nNOS. Cultures from nNOS-knockout mice grown on glial feeder layers failed to produce a neurotrophin-mediated enhancement of neurotoxicity (SAMDANI et al. 1997). Thus, nNOS expression and NMDA/NO-mediated neurotoxicity are dependent on the culture paradigm.

The physiologic relevance of these in vitro observations was confirmed in animal models of focal ischemia. The use of non-selective nNOS inhibitors at concentrations that preferentially inhibit nNOS or the use of selective nNOS inhibitors result in markedly reduced infarct volume following middle cerebral artery occlusion (MCAO) in mice, rats, and cats (BUISSON et al. 1992; MONCADA et al. 1992; NISHIKAWA et al. 1993, 1994; CARREAU et al. 1994;

Ferriero et al. 1995; Quast et al. 1995; Zhang et al. 1995; Margaill et al. 1997; Matsui et al. 1997; Escott et al. 1998; Wei and Quast 1998). Investigators consistently observe neuroprotection with selective nNOS inhibitors, including 7-nitroindazol (Yoshida et al. 1994; Kelly et al. 1995; Escott et al. 1998; Nanri et al. 1998) and ARL17477 (Zhang et al. 1996), which effectively inhibit nNOS but not endothelial NOS (eNOS) activity (Babbedge et al. 1993; Moore et al. 1993a,b; Yoshida et al. 1994). The results following use of non-selective NOS inhibitors have varied widely. Some investigators have observed reduction in infarct volumes, other investigators observed no alteration in infarct volume (Dawson et al. 1994a), and some studies observe exacerbation of the injury (Yamamoto et al. 1992; Kuluz et al. 1993). In general, exacerbation of injury appears to occur at higher doses of non-selective NOS inhibitors, which effectively reduce eNOS activity, resulting in deleterious alterations of CBF and subsequent increased infarct volumes (Kuluz et al. 1993). When CBF is monitored and not compromised, the presence of the non-selective inhibitor L-NAME results in reduced neuronal injury (Nishikawa et al. 1993, 1994).

The importance of nNOS activation in the pathogenesis of infarct following MCAO is confirmed through experiments in nNOS null mice. Following permanent focal MCAO, nNOS-knockout mice have reduced infarct volumes when compared with age-matched wild-type control mice (Huang et al. 1994). Additionally, nNOS-knockout mice are resistant to focal ischemic injury in the reperfusion model of transient MCAO, and hippocampal damage is reduced in the model of global cerebral ischemia (Panahian et al. 1996). NO may mediate injury in these experimental stroke models through several pathways. It is thought that peroxynitrite, generated from NO and superoxide, mediates most of the cellular damage. However, NO itself can significantly contribute to glutamate overflow observed early in experimental stroke. NO makes these contributions through inhibition of glutamate transport (Pogun et al. 1994), stimulation of glutamate release (McNaught and Brown 1998), and alterations in spreading, depression-like waves of depolarization (Shimizu-Sasamata et al. 1998).

Although nNOS is constitutively expressed following pathologic insults, nNOS can be induced through new protein synthesis. Following MCAO in the rat, a rapid upregulation of nNOS mRNA, nNOS protein, and reduced nicotinamide adenine dinucleotide phosphate (NADPH)-diaphorase positive staining is observed in the ischemic lesion (Zhang et al. 1994a; Yoshida et al. 1995). It is not known whether this increase in nNOS expression correlates with increased NOS catalytic activity; however, it is possible that increased nNOS expression may contribute to the spread of neuronal damage following ischemic injury. The transcriptional elements responsible for the induction of nNOS are not known, as the promoter for nNOS has yet to be identified. It is possible that the upregulation of nNOS expression may be a stress mechanism used to elicit restorative function via activation of the Ras extracellular signal-regulated protein-kinase pathway (Yun et al. 1998). NO can activate the Ras pathway (Lander et al. 1997; Deora et al. 1998; Yun et al. 1998), leading to cyclic adenosine monophosphate responsive-element-binding protein phosphorylation. The Ras pathway has been implicated in long-term changes resulting in neuronal plasticity and activity-dependent survival (BARBACID 1987; SCHLESSINGER and ULLRICH 1992; LOWY and WILLUMSEN 1993; MACARA et al. 1996). It is possible that induction of nNOS under traumatic conditions may be an attempt by the brain to repair itself.

Immunohistochemical localization of nNOS correlates with the histochemical stain NADPH-diaphorase (DAWSON et al. 1991; HOPE et al. 1991). NADPH-diaphorase staining was first described by THOMAS and PEARSE in 1961 (THOMAS and PEARSE 1961, 1964). nNOS neurons are of extreme interest, because they are relatively spared from neuronal cell death after vascular stroke and excitotoxicity, as well as in Huntington's and Alzheimer's diseases (FERRANTE et al. 1985; KOH et al. 1986; KOH and CHOI 1988; UEMURA et al. 1990; HYMAN et al. 1992). Thirty years after the first description of NADPH-diaphorase staining, the discovery that nNOS is responsible for this histochemical stain, provided a curious paradox. nNOS neurons, which are spared from neuronal cell death, are the very neurons that are eliciting neuronal cell death in surrounding cells. This raises an important question regarding the intracellular strategies by which nNOS neurons are resistant to the neurotoxic environment which they create. Recent studies employing serial analysis of gene expression to identify putative candidates for cellular resistance between PC-12 cells sensitive to NO toxicity and PC-12 cells resistant to NO toxicity resulted in the identification of mitochondrial manganese superoxide dismutase (MnSOD) as a candidate gene product important in selective cell survival (GONZALEZ-ZULUETA et al. 1998). Immunohistochemical co-localization experiments confirmed that MnSOD is expressed at higher levels in nNOS neurons than in other neuronal cell types. Antisense knockdown of MnSOD renders nNOS neurons susceptible to NMDA neurotoxicity without influencing the overall susceptibility of non-nNOS cortical neurons to NMDA neurotoxicity. Knockout of MnSOD through genetic targeting results in exquisite sensitivity of nNOS neurons to NMDA neurotoxicity. This phenotype can be rescued by expression of MnSOD following adenovirus-mediated gene transfer. nNOS neurons from MnSOD-knockout mice in which MnSOD has been restored by adenovirus-mediated gene transfer are once again resistant to NMDA neurotoxicity. Furthermore, overexpression of MnSOD by adenovirus provides dramatic protection against NMDA and NO neurotoxicity to all cortical neurons (GONZALEZ-ZULUETA et al. 1998). These results suggest that superoxide anions generated by mitochondrial activity are important targets for NO, resulting in neurotoxicity. This is likely due to the reaction between superoxide anions and NO to form the toxic oxidant peroxynitrite (Fig. 1).

C. Endothelial NOS

eNOS is the major regulator of vascular hemodynamics. The role for endothelially derived NO in the regulation of blood flow was discovered by investigations into the mechanism of acetylcholine-induced relaxation of

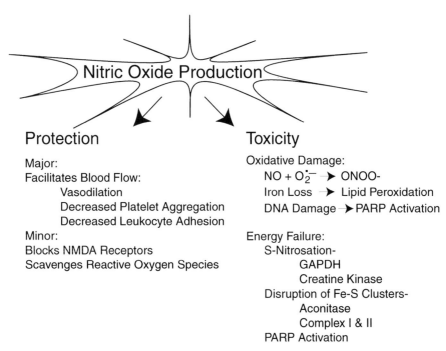

Fig. 1. Dual role of nitric oxide (NO) production in cerebral ischemic injury. Depending on the cellular compartment in which NO is produced, the time course and the local biochemical environment NO can either accelerate neurotoxicity or protect against neuronal damage following a reduction in cerebral blood flow. Protection is mediated in large part by production of NO by endothelial nitric oxide synthase (NOS) in the vascular compartment. Increased NO production results in vasodilation and maintenance of collateral cerebral blood flow and decreased platelet aggregation and leukocyte adhesion. NO can block the N-methyl-D-aspartate (NMDA) receptor, and some investigators have proposed that this blockade would provide protection. However, it is likely that, under conditions of ischemic injury, any regulatory effects of NO and other potential NMDA-receptor modulators would be overcome by the insult. In carefully controlled laboratory experiments, NO can limit oxidative damage and lipid peroxidation reactions. Since there are no markers to investigate these reactions in vivo, the physiologic relevance of these experiments is not yet known. Toxicity can be mediated by several possible pathways and is largely due to NO generated by neuronal NOS and inducible NOS. NO can promote oxidative damage by reacting with superoxide anion (O_2^-) to form peroxynitrite ($ONOO^-$). Peroxynitrite formation can affect iron metabolism, resulting in iron loss. Peroxynitrite can promote lipid peroxidation and DNA damage. Additionally, NO formation can impair energy metabolism at several sites. NO can inhibit the function of glyceraldehyde-3-phosphate dehydrogenase and creatine kinase and disrupt the iron–sulfur centers of aconitase and complexes I and II of the mitochondria, which can result in impaired mitochondrial respiration and increased superoxide-anion formation. Activation of poly(adenosine diphosphate–ribose) polymerase through DNA damage will result in consumption of nicotinamide adenine dinucleotide (NAD) and adensoine triphosphate during resynthesis of NAD. Both oxidative damage and energy failure likely contribute to the development of neurotoxicity following NO formation in ischemic injury

vascular smooth muscle and the vasodilatory effects of organic nitrates (ARNOLD et al. 1977; FURCHGOTT and ZAWADZKI 1980; IGNARRO et al. 1981, 1987; PALMER et al. 1987). eNOS is constitutively expressed and activated by increased intracellular calcium (MARLETTA 1993, 1994). eNOS is predominantly expressed in endothelial cells of the vascular system; however, in the brain, a small population of neurons also expresses eNOS (DINERMAN et al. 1994). Additionally, astrocytes express eNOS constitutively at low levels (TOGASHI et al. 1997; COLSANTI et al. 1998).

NO is the major regulator of vascular hemodynamics and is the primary regulator of blood-vessel relaxation. Inhibition of eNOS can result in large changes in CBF (NISHIKAWA et al. 1993, 1994; DALKARA et al. 1994; DALKARA and MOSKOWITZ 1997). Even partial inhibition of eNOS can have major effects on vascular hemodynamics. In studies evaluating non-specific NOS inhibitors in permanent and transient MCAO, the outcome of the studies was critically dependent on whether the dosing paradigm affected eNOS activity (DALKARA et al. 1994; NISHIKAWA et al. 1993, 1994; DALKARA and MOSKOWITZ 1997). The use of non-selective NOS inhibitors contributed to conflicting reports in the experimental stroke literature regarding the role of NO following MCAO. Administration of non-specific NOS inhibitors at concentrations which do not affect systemic blood flow are often sufficient to constrict pial arterioles, resulting in a reduction in CBF and subsequent increases in infarct volume following MCAO (IADECOLA et al. 1994). Supporting the importance of vascular NO in protecting against ischemic injury by maintaining blood flow, administration of NO donors or intra-arterial administration of L-arginine, both of which increase regional CBF, resulted in decreased infarct volumes distal to MCAO in rats (MORIKAWA et al. 1994; ZHANG et al. 1994).

These pharmacologic studies have been confirmed through genetically engineered mice lacking eNOS expression. In eNOS-knockout mice, infarct volumes are dramatically increased after MCAO in comparison with wild-type controls (HUANG et al. 1996). Deletion of eNOS renders these genetically engineered mice hypertensive (HUANG et al. 1995). However, hypertension does not account for the increased susceptibility of eNOS knockouts to MCAO, because infarct size is not altered following normalization of blood pressure with hydralazine (HUANG et al. 1996). eNOS-knockout mice present with a more pronounced reduction in regional CBF than wild-type mice in corresponding brain regions following MCAO. Dynamic CT scanning reveals that areas of hemodynamic penumbra are significantly smaller and the core relatively larger in eNOS knockouts than wild-type controls (Lo et al. 1996). Therefore, the susceptibility of eNOS knockouts to ischemic injury is likely due to a diminished capacity to adapt to reduced perfusion pressure at the margins of ischemic lesion. Additionally, it is likely that eNOS-knockout mice experience enhanced platelet and neutrophil adhesion, which would further render eNOS knockouts more susceptible to injury. Exposure to a non-specific NOS inhibitor, l-NNA, at concentrations that do not affect infarct volume in wild-type mice results in decreased infarct volume in eNOS-knockout mice following MCAO (HUANG et al. 1996). In eNOS-knockout mice, the only target for l-NNA is nNOS. The corollary experiment has also been performed. When exposed to the non-specific NOS inhibitor l-NNA, nNOS-knockout mice in which infarct volumes are reduced experience greatly increased infarct volumes (DALKARA and MOSKOWITZ 1997). In nNOS-knockout mice, the target for l-NNA is eNOS. In this animal, the protection afforded by eliminating nNOS production of NO is ameliorated by inhibiting eNOS-derived NO and altering cerebral vascular hemodynamics. These studies, combining genetically engineered mice with pharmacologic studies in experimental stroke models, elegantly highlight the dual actions of NO in focal ischemia. Overproduction of NO from activation of nNOS results in neurotoxicity. However, production of NO from eNOS is critically important in protecting the brain tissue by maintaining regional CBF (Fig. 1). eNOS protein and catalytic activity are both induced during the acute phase of ischemia. Upregulation of eNOS likely serves a protective role by enhancing the maintenance of CBF in the setting of ischemia.

Further evidence highlighting the importance of eNOS is provided from a surprising source. The cholesterol-lowering agents 3-hydroxy-3-methylglutaryl (HMG)-CoA reductase inhibitors protect against cerebral injury. This protection involves the selective upregulation of eNOS (ENDRES et al. 1998). Prophylactic treatment with HMG-CoA reductase inhibitors augments CBF, reduces cerebral infarct size, and improves neurological function in normocholesterolemic mice. The upregulation of eNOS by HMG-CoA reductase inhibitors is not associated with changes in serum cholesterol levels. The blood flow and neuroprotective effects of HMG-CoA reductase inhibitors are completely absent in eNOS-deficient mice, further implicating increased eNOS expression by HMG-CoA reductase inhibitors as the primary, if not the only, mechanism by which these agents protect against cerebral injury (ENDRES et al. 1998).

D. Immunologic NOS

Immunologic NOS (iNOS) is not normally expressed in healthy tissues. Under pathologic conditions, iNOS can be expressed in most tissues (NATHAN and XIE 1994; DAWSON and DAWSON 1996; THIEMERMANN 1997). In primary cortical cultures, induction of iNOS results in delayed neuronal cell death that develops over 3–7 days (CHAO et al. 1992; DAWSON et al. 1994c; ADAMSON et al. 1996). Induction of iNOS in primary cortical cultures can also exacerbate glutamate excitotoxicity (HEWETT et al. 1996). In human disease, iNOS expression has been implicated in CNS injury following ischemic insult, trauma, viral infection, bacterial infection, or other immunologic challenges. Expression of iNOS is observed predominantly in microglia, astrocytes, and the vasculature of the brain (VAN DAM et al. 1995; IADECOLA et al. 1996; OLESZAK et al. 1998).

Cerebral ischemia induces iNOS message and protein and enzymatic activity in the post-ischemic brain (IADECOLA et al. 1996; IADECOLA 1997). However, unlike nNOS and eNOS, in which catalytic activity is immediately induced and new protein is expressed within minutes to hours after MCAO occlusion, iNOS expression is delayed to 6–12h after MCAO. Following permanent MCAO, iNOS immunoreactivity is present in neutrophils infiltrating the ischemic brain while, in transient MCAO, iNOS is observed predominantly in vascular cells (ZHANG et al. 1995; IADECOLA et al. 1996). In global ischemia, iNOS expression is localized to reactive astrocytes (ENDOH et al. 1994). Although the expression of iNOS occurs many hours after the initial ischemic insult, the result of iNOS expression and NO production can be more severe and long lasting. Unlike nNOS and eNOS, in which rapid calcium influxes activate the enzyme, producing brief bursts of NO, iNOS causes a sustained and calcium-independent production of NO (NATHAN and XIE 1994; IADECOLA 1997). eNOS and nNOS produce NO in small and highly regulated bursts that are well suited for the molecular messenger function of NO. Under ischemic insults, higher levels of NO are generated; however, generation is limited. Once translated, iNOS monomers bind calmodulin (even at very low levels of intracellular calcium), dimerize, and are activated (MARLETTA 1993, 1994). In the presence of heme, tetrahydrobiopterin and L-arginine are formed and, in the presence of NADPH and oxygen, NO is formed. iNOS expression results in the production of large concentrations of NO continuously for long periods, which can result in neurotoxicity. Consistent with the notion that iNOS mediates a secondary phase of neurotoxicity following ischemic insult, treatment of mice with aminoguanidine 24h after infarct results in reduced infarct volume, and genetically engineered mice deficient in iNOS have smaller infarcts than wild-type mice following MCAO (IADECOLA et al. 1997).

E. The Role of NO in Focal Ischemic Brain Damage

Our current understanding of the impact of NO on ischemic brain injury (Fig. 2) suggests that the impact is dependent on the stage of evolution of tissue damage. At early time points following cerebral ischemia, i.e., under 2h, the vascular actions of NO are beneficial by promoting collateral circulation and microvascular flow (IADECOLA 1997). However, at these same time points, glutamate-induced calcium overload in ischemic neurons results in a persistent activation of nNOS and induction of nNOS, resulting in pathologic concentrations of NO production in neural tissue. Activation of the glutamate–NO pathway is likely to occur in the ischemic penumbra, a region at risk for infarction, in which glutamate excitotoxicity is most prominent. Administration of selective nNOS inhibitors ameliorates ischemic damage, in part, by impairing glutamate-induced NO synthesis without interfering with the beneficial vascular effects of endothelially derived NO (YOSHIDA et al. 1994; KELLY et al.

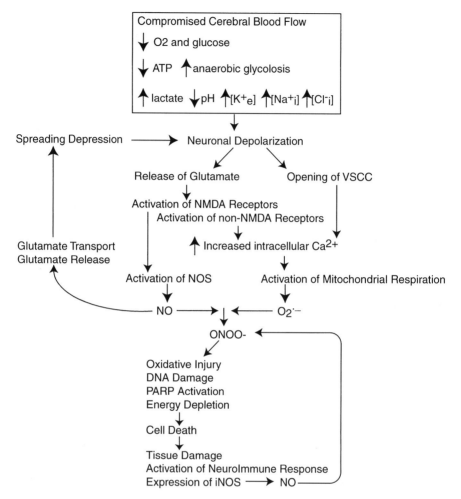

Fig. 2. A putative pathway towards neuronal injury following focal ischemia, with a focus on nitric oxide (NO). Blood-vessel occlusion compromises cerebral blood flow (CBF) and induces a multitude of cellular events. Ischemic events result in the reduction of the resting membrane potentials of glia and neurons in the brain and changes in pH, adenosine triphosphate (ATP) and lactate. Potassium leaks out of cells and depolarizes neurons, leading to a massive release of glutamate. Acting via N-methyl-D-aspartate (NMDA) receptors, glutamate triggers a release of NO, which combines with superoxide to form peroxynitrite. Peroxynitrite damages DNA, whose fragments activate poly[adenosine diphosphate (ADP)–ribose] polymerase (PARP). Massive activation of PARP leads to ADP-ribosylation and depletion of nicotinamide adenine dinucleotide (NAD). ATP is depleted in an effort to resynthesize NAD, leading to cell death by energy depletion. Energy depletion and oxidative damage can lead to cell death. The resultant tissue damage may activate a local inflammatory response, leading to expression of inducible NO synthase. Additionally, NO may participate in glutamate overflow, which occurs following ischemic insults by inhibiting glutamate transport, inducing glutamate release. NO is also implicated in spreading depression, which results in further neuronal depolarization, glutamate release, NMDA receptor activation and NO formation

1995; ZHANG et al. 1996; ESCOTT et al. 1998; NANRI et al. 1998). At later time points following cerebral ischemia, i.e., after 6h or more, iNOS is expressed, and large concentrations of NO that contribute to the further progression of tissue damage are produced (ZHANG et al. 1995). The iNOS inhibitor aminoguanidine, administered 6h following cerebral ischemia, can ameliorate some cerebral ischemic damage. Thus, in the late stages of cerebral ischemia, iNOS is expressed in the setting of post-ischemic inflammation. Therefore, the role of NO in ischemic injury is protective or destructive depending on the stage of evolution of the ischemic process and on the cellular compartment producing NO.

F. Targets of NO

The biochemical pathways by which excessive NO formation results in neurotoxicity are not known. However, neurotoxicity due to NOS activation is nearly always observed in a setting in which superoxide anion production is also increased. In biologic systems, the most permissive chemical reaction for NO is with superoxide anion to form peroxynitrite (BECKMAN 1991, 1994). It is likely that most of the cytotoxic outcomes attributed to increased NO formation are a result of peroxynitrite generation rather than a direct effect of NO. Under ischemic conditions, NO can also be generated non-enzymatically by the conversion of nitrates and nitrites (ZWEIER et al. 1995), which could contribute significantly to the amount of NO available to form peroxynitrite. Interestingly, a major target of peroxynitrite is mitochondrial MnSOD (MACMILLAN-CROW et al. 1996). Peroxynitrite inactivation of MnSOD would initiate a self-propagating cascade of neural injury through failure of mitochondrial scavenging of superoxide anion, leading to enhanced formation of peroxynitrite in a setting where NO production is elevated. NO and peroxynitrite both can mediate neurotoxicity by several mechanisms, including disruption of cellular metabolism through inhibition of aconitase, complex I and complex II of the mitochondrial electron transport chain, and glyceraldehyde-3-phosphate dehydrogenase (DAWSON and SNYDER 1994; GROSS and WOLIN 1995; DAWSON and DAWSON 1996). NO or peroxynitrite may mediate DNA damage from several possible mechanisms (TAMIR et al. 1996; CAULFIELD et al. 1998; WINK and MITCHELL 1998; WINK et al. 1998). Production of NO in an oxidative environment can result in formation of N_2O_3, which can deaminate DNA, leading to DNA-strand breaks. Peroxynitrite can oxidize DNA, resulting in DNA-strand breaks. Additionally, peroxynitrite can inactivate DNA ligase, further enhancing DNA damage (WINK and MITCHELL 1998; WINK et al. 1998).

DNA-strand breaks, particularly single-strand breaks, are potent activators of the nuclear enzyme poly[adenosine diphosphate (ADP)–ribose] polymerase (PARP) (LAUTIER et al. 1993; SZABO and DAWSON 1998). PARP activation appears to mediate a large component of NMDA–NO neurotoxicity (Zhang et al. 1994b; Eliasson et al. 1997). PARP is a nuclear enzyme which facilitates DNA repair and is important in maintaining genomic stability. It is important to note that PARP does not itself repair DNA and that DNA repair occurs in the absence of PARP (Satoh et al. 1994; Lindahl et al. 1995; Wang et al. 1995). Upon activation, PARP transfers the ADP–ribose moiety from nicotinamide adenine dinucleotide (NAD) to itself and other nuclear receptor proteins. PARP can transfer hundreds to thousands of ribose units within minutes of being activated (Lautier et al. 1993). For every 1 mol of ADP–ribose transferred, 1 mol of NAD is consumed and four free-energy equivalents of adenosine triphosphate (ATP) are required to regenerate NAD equivalents. Overactivation of PARP can, therefore, rapidly deplete cellular energy stores (Berger 1985). In the setting of ischemia, mitochondrial function may also be impaired, and the ability of the cells to replace NAD and ATP is compromised. NO can inhibit cis-aconitase, resulting in inhibition of glycolysis (Castro et al. 1994; Hausladen and Fridovich 1994). NO and peroxynitrite can inhibit mitochondrial respiration by competing with oxygen at cytochrome oxidase (Brudvig et al. 1980; Carr and Ferguson 1990; Brown 1997; Sharpe and Cooper 1998). Inhibition of complex I and complex II will lead to increased superoxide-anion production and increased peroxynitrite production. Additionally, NO can react with a thiol in creatinine kinase, thus inhibiting creatinine kinase (Gross et al. 1996) activity and leading to a decrement of ATP by inhibiting phosphoryl transfer between phosphoryl creatinine and ATP. Thus, excessive NO production and peroxynitrite formation at multiple points can inhibit ATP formation in a setting in which PARP activation is rapidly consuming NAD and ATP. Loss of NAD in a setting in which ATP generation is compromised can lead ultimately to energy failure and cell death.

Confirmation of this hypothesis is provided by both pharmacologic and genetic experiments. Neurotoxicity in primary cortical cultures exposed to excitotoxic concentrations of NMDA is attenuated by PARP inhibitors (Zhang et al. 1994b). The rank order of potency in inhibiting NMDA neurotoxicity parallels the rank order of potency of these agents in inhibiting PARP (Zhang ct al. 1994b). PARP inhibitors are also effective in preventing NO-mediated neurotoxicity and combined oxygen–glucose-mediated ischemic injury (Zhang et al. 1994; Eliasson et al. 1997). Primary cortical cultures generated from PARP-knockout mice are dramatically resistant to NMDA, NO, and ischemic injury (Eliasson et al. 1997). The physiologic relevance of these in vitro observations is confirmed in experimental stroke models (Eliasson et al. 1997; Endres et al. 1997, 1998). PARP-knockout mice have remarkably reduced infarct volumes (>80%) following transient MCAO as compared with the wild-type controls (Eliasson et al. 1997). This substantial reduction in infarct volume observed in PARP-knockout mice exceeds that observed with nNOS-knockout mice, superoxide-dismutase-overexpressing mice, other transgenic mice, or pharmacologic treatments. These data suggest that PARP activation may be a common point at which multiple insults converge to elicit neurotoxicity. These studies have identified PARP as a key target molecule for future therapeutic development for the treatment of stroke.

G. Summary

It is clear that NO plays major roles in modulating brain injury following ischemic events. The development of selective pharmacologic tools and the development of genetically engineered mice lacking each NOS isoform have greatly advanced our understanding of the diverse and opposing roles of NO in the CNS and the roles of NO in response to ischemic injury. Depending on the cellular compartment, the level of NO produced, and the setting in which it is produced, NO may be toxic or protective to the brain in ischemic conditions. Production of NO from either nNOS or iNOS leads to neurotoxicity, while NO production from eNOS protects brain tissue by maintaining regional CBF. These studies emphasize the necessity of developing truly selective inhibitors for nNOS and iNOS to adequately protect the brain from ischemic injury while not interfering with the function of eNOS in maintaining regional CBF. Furthermore, identification of NO targets that mediate neuronal cell death may yield important new therapeutic strategies for the treatment of stroke.

References

Adamson DC, Wildemann B, Sasaki M, McArthur J, Glass J, Dawson TM, Dawson VL (1996) Immunologic nitric oxide synthase is elevated in HIV infected individuals with dementia. Science 274:1917–1921

Arnold WP, Mittal CK, Katsuki S, Murad F (1977) Nitric oxide activates guanylate cyclase and increases guanosine 3′:5′-cyclic monophosphate levels in various tissue preparations. Proc Natl Acad Sci USA 74:3203–7

Babbedge RC, Bland-Ward PA, Hart SL, Moore PK (1993) Inhibition of rat cerebellar nitric oxide synthase by 7-nitro indazole and related substituted indazoles. Br J Pharmacol 110:225–8

Barbacid M (1987) Ras genes. Annu Rev Biochem 56:779–827

Beckman JS (1991) The double-edged role of nitric oxide in brain function and superoxide-mediated injury. J Dev Physiol 15:53–9

Beckman JS (1994) Peroxynitrite versus hydroxyl radical: the role of nitric oxide in superoxide-dependent cerebral injury. Ann N Y Acad Sci 738:69–75

Berger NA (1985) Poly(ADP-ribose) in the cellular response to DNA damage. Radiat Res 101:4–15

Bredt DS, Snyder SH (1989) Nitric oxide mediates glutamate-linked enhancement of cGMP levels in the cerebellum. Proc Natl Acad Sci USA 86:9030–3

Bredt DS, Hwang PM, Snyder SH (1990) Localization of nitric oxide synthase indicating a neural role for nitric oxide. Nature 347:768–70

Brown GC (1997) Nitric oxide inhibition of cytochrome oxidase and mitochondrial respiration: implications for inflammatory, neurodegenerative and ischaemic pathologies. Mol Cell Biochem 174:189–92

Brudvig GW, Stevens TH, Chan SI (1980) Reactions of nitric oxide with cytochrome c oxidase. Biochemistry 19:5275–85

Buisson A, Plotkine M, Boulu RG (1992) The neuroprotective effect of a nitric oxide inhibitor in a rat model of focal cerebral ischaemia. Br J Pharmacol 106:766–7

Carr GJ, Ferguson SJ (1990) Nitric oxide formed by nitrite reductase of *Paracoccus denitrificans* is sufficiently stable to inhibit cytochrome oxidase activity and is reduced by its reductase under aerobic conditions. Biochim Biophys Acta 1017:57–62

Carreau A, Duval D, Poignet H, Scatton B, Vige X, Nowicki JP (1994) Neuroprotective efficacy of N^ω-nitro-L-arginine after focal cerebral ischemia in the mouse and inhibition of cortical nitric oxide synthase. Eur J Pharmacol 256:241–9

Castro L, Rodriguez M, Radi R (1994) Aconitase is readily inactivated by peroxynitrite, but not by its precursor, nitric oxide. J Biol Chem 269:29409–15

Caulfield JL, Wishnok JS, Tannenbaum SR (1998) Nitric oxide-induced deamination of cytosine and guanine in deoxynucleosides and oligonucleotides. J Biol Chem 273:12689–95

Chao CC, Hu S, Molitor TW, Shaskan EG, Peterson PK (1992) Activated microglia mediate neuronal cell injury via a nitric oxide mechanism. J Immunol 149:2736–41

Colsanti M, Persichini T, Fabrizi C, Cavaliere E, Venturini Gm Ascenzi P, Lauro GM, Suzuki H (1998) Expression of a NOS-III-like protein in human astroglial cell culture. Biochem Biophys Res Commun 252:552–555

Dalkara T, Moskowitz MA (1997) Neurotoxic and neuroprotective roles of nitric oxide in cerebral ischaemia. Int Rev Neurobiol 40:319–36

Dalkara T, Yoshida T, Irikura K, Moskowitz MA (1994) Dual role of nitric oxide in focal cerebral ischemia. Neuropharmacology 33:1447–52

Dawson TM, Snyder SH (1994) Gases as biological messengers: nitric oxide and carbon monoxide in the brain. J Neurosci 14:5147–59

Dawson VL, Dawson TM (1995) Physiological and toxicological actions of nitric oxide in the central nervous system. Adv Pharmacol 34:323–42

Dawson VL, Dawson TM (1996) Nitric oxide neurotoxicity. J Chem Neuroanat 10:179–90

Dawson TM, Bredt DS, Fotuhi M, Hwang PM, Snyder SH (1991a) Nitric oxide synthase and neuronal NADPH diaphorase are identical in brain and peripheral tissues. Proc Natl Acad Sci USA 88:7797–801

Dawson VL, Dawson TM, London ED, Bredt DS, Snyder SH (1991b) Nitric oxide mediates glutamate neurotoxicity in primary cortical cultures. Proc Natl Acad Sci USA 88:6368–71

Dawson VL, Dawson TM, Bartley DA, Uhl GR, Snyder SH (1993) Mechanisms of nitric oxide-mediated neurotoxicity in primary brain cultures. J Neurosci 13:2651–61

Dawson DA, Graham DI, McCulloch J, Macrae IM (1994a) Anti-ischaemic efficacy of a nitric oxide synthase inhibitor and a N-methyl-D-aspartate receptor antagonist in models of transient and permanent focal cerebral ischaemia. Br J Pharmacol 113:247–53

Dawson TM, Zhang J, Dawson VL, Snyder SH (1994b) Nitric oxide: cellular regulation and neuronal injury. Prog Brain Res 103:365–9

Dawson VL, Brahmbhatt HP, Mong JA, Dawson TM (1994c) Expression of inducible nitric oxide synthase causes delayed neurotoxicity in primary mixed neuronal–glial cortical cultures. Neuropharmacology 33:1425–30

Dawson VL, Kizushi VM, Huang PL, Snyder SH, Dawson TM (1996) Resistance to neurotoxicity in cortical cultures from neuronal nitric oxide synthase deficient mice. J Neurosci 16:2479–2487

Deora AA, Win T, Vanhaesebroeck B, Lander HM (1998) A redox-triggered ras-effector interaction. Recruitment of phosphatidylinositol 3′-kinase to ras by redox stress. J Biol Chem 273:29923–8

Dinerman JL, Dawson TM, Schell MJ, Snowman A, Snyder SH (1994) Endothelial nitric oxide synthase localized to hippocampal pyramidal cells: implications for synaptic plasticity. Proc Natl Acad Sci USA 91:4214–8

Eliasson MJL, Sampei K, Mandir AS, Hurn PD, Traystman RJ, Jun Bao J, Pieper A, Wang Z-Q, Dawson TM, Snyder SH, Dawson VL (1997) Poly(ADP-Ribose) polymerase gene disruption renders mice resistant to cerebral ischemia. Nat Med 3:1–8

Endoh M, Maiese K, Wagner J (1994) Expression of the inducible form of nitric oxide synthase by reactive astrocytes after transient global ischemia. Brain Res 651:92–100

Endres M, Wang Z-Q, Namura S, Waeber C, Moskowitz MA (1997) Ischemic brain injury is mediated by the activation of Poly(ADP-ribose) polymerase. J Cerebral Blood Flow and Metabolism in press:

Endres M, Laufs U, Huang Z, Nakamura T, Huang P, Moskowitz MA, Liao JK (1998) Stroke protection by 3-hydroxy-3-methylglutaryl (HMG)-CoA reductase inhibitors mediated by endothelial nitric oxide synthase. Proc Natl Acad Sci USA 95:8880–5

Escott KJ, Beech JS, Haga KK, Williams SC, Meldrum BS, Bath PM (1998) Cerebroprotective effect of the nitric oxide synthase inhibitors, 1-(2- trifluoromethylphenyl) imidazole and 7-nitro indazole, after transient focal cerebral ischemia in the rat. J Cereb Blood Flow Metab 18:281–7

Ferrante RJ, Kowall NW, Beal MF, Richardson EP, Jr., Bird ED, Martin JB (1985) Selective sparing of a class of striatal neurons in Huntington's disease. Science 230:561–3

Ferriero DM, Sheldon RA, Black SM, Chuai J (1995) Selective destruction of nitric oxide synthase neurons with quisqualate reduces damage after hypoxia-ischemia in the neonatal rat. Pediatr Res 38:912–8

Furchgott RF, Zawadzki JV (1980) The obligatory role of endothelial cells in the relaxation of arterial smooth muscle by acetylcholine. Nature 288:373–6

Garthwaite J, Charles SL, Chess-Williams R (1988) Endothelium-derived relaxing factor release on activation of NMDA receptors suggests role as intercellular messenger in the brain. Nature 336:385–8

Garthwaite J, Garthwaite G, Palmer RM, Moncada S (1989) NMDA receptor activation induces nitric oxide synthesis from arginine in rat brain slices. Eur J Pharmacol 172:413–6

Gonzalez-Zulueta M, Ensz LM, Mukhina G, Lebovitz RM, Zwacka RM, Engelhardt JF, Oberley LW, Dawson VL, Dawson TM (1998) Manganese superoxide dismutase protects nNOS neurons from NMDA and nitric oxide mediated neurotoxicity. J Neurosci 18:2040–55

Gross SS, Wolin MS (1995) Nitric oxide: pathophysiological mechanisms. Annu Rev Physiol 57:737–69

Gross WL, Bak MI, Ingwall JS, Arstall MA, Smith TW, Balligand JL, Kelly RA (1996) Nitric oxide inhibits creatine kinase and regulates rat heart contractile reserve. Proc Natl Acad Sci USA 93:5604–9

Hara H, Huang PL, Panahian N, Fishman MC, Moskowitz MA (1996) Reduced brain edema and infarction volume in mice lacking the neuronal isoform of nitric oxide synthase after transient MCA occlusion. J Cereb Blood Flow Metab 16:605–11

Hausladen A, Fridovich I (1994) Superoxide and peroxynitrite inactivate aconitases, but nitric oxide does not. J Biol Chem 269:29405–8

Hewett SJ, Muir JK, Lobner D, Symons A, Choi DW (1996) Potentiation of oxygen glucose deprivation-induced neuronal death after induction of iNOS. Stroke 27:1586–91

Hope BT, Michael GJ, Knigge KM, Vincent SR (1991) Neuronal NADPH diaphorase is a nitric oxide synthase. Proc Natl Acad Sci USA 88:2811–4

Huang PL, Huang Z, Mashimo H, Bloch KD, Moskowitz MA, Bevan JA, Fishman MC (1995) Hypertension in mice lacking the gene for endothelial nitric oxide synthase. Nature 377:239–42

Huang Z, Huang PL, Panahian N, Dalkara T, Fishman MC, Moskowitz MA (1994) Effects of cerebral ischemia in mice deficient in neuronal nitric oxide synthase. Science 265:1883–5

Huang Z, Huang PL, Ma J, Meng W, Ayata C, Fishman MC, Moskowitz MA (1996) Enlarged infarcts in endothelial nitric oxide synthase knockout mice are attenuated by nitro-L-arginine. J Cereb Blood Flow Metab 16:981–7

Hyman BT, Marzloff K, Wenniger JJ, Dawson TM, Bredt DS, Snyder SH (1992) Relative sparing of nitric oxide synthase-containing neurons in the hippocampal formation in Alzheimer's disease. Ann Neurol 32:818–20

Iadecola C (1997) Bright and dark sides of nitric oxide in ischemic brain injury. Trends Neurosci 20:132–139

Iadecola C, Pelligrino DA, Moskowitz MA, Lassen NA (1994) Nitric oxide synthase inhibition and cerebrovascular regulation. J Cereb Blood Flow Metab 14:175–92

Iadecola C, Zhang F, Casey R, Clark HB, Ross ME (1996) Inducible nitric oxide synthase gene expression in vascular cells after transient focal cerebral ischemia. Stroke 27:1373–80

Iadecola C, Zhang F, Casey R, Nagayama M, Ross ME (1997) Delayed reduction of ischemic brain injury and neurological deficits in mice lacking the inducible nitric oxide synthase gene. J Neurosci 17:9157–64

Ignarro LJ, Lippton H, Edwards JC, Baricos WH, Hyman AL, Kadowitz PJ, Gruetter CA (1981) Mechanism of vascular smooth muscle relaxation by organic nitrates, nitrites, nitroprusside and nitric oxide: evidence for the involvement of S-nitrosothiols as active intermediates. J Pharmacol Exp Ther 218:739–49

Ignarro LJ, Buga GM, Wood KS, Byrns RE, Chaudhuri G (1987) Endothelium-derived relaxing factor produced and released from artery and vein is nitric oxide. Proc Natl Acad Sci USA 84:9265–9

Kelly PA, Ritchie IM, Arbuthnott GW (1995) Inhibition of neuronal nitric oxide synthase by 7-nitroindazole: effects upon local cerebral blood flow and glucose use in the rat. J Cereb Blood Flow Metab 15:766–73

Koh JY, Choi DW (1988) Cultured striatal neurons containing NADPH-diaphorase or acetylcholinesterase are selectively resistant to injury by NMDA receptor agonists. Brain Res 446:374–8

Koh JY, Peters S, Choi DW (1986) Neurons containing NADPH-diaphorase are selectively resistant to quinolinate toxicity. Science 234:73–6

Koh JY, Gwag BJ, Lobner D, Choi DW (1995) Potentiated necrosis of cultured cortical neurons by neurotrophins. Science 268:573–5

Kuluz JW, Prado RJ, Dietrich WD, Schleien CL, Watson BD (1993) The effect of nitric oxide synthase inhibition on infarct volume after reversible focal cerebral ischemia in conscious rats. Stroke 24:2023–9

Lander HM, Hajjar DP, Hempstead BL, Mirza UA, Chait BT, Campbell S, Quilliam LA (1997) A molecular redox switch on p21(ras). Structural basis for the nitric oxide-p21(ras) interaction. J Biol Chem 272:4323–6

Lautier D, Lagueux J, Thibodeau J, Menard L, Poirier GG (1993) Molecular and biochemical features of poly(ADP-ribose) metabolism. Mol Cell Biochem 122:171–93

Lindahl T, Satoh MS, Poirier GG, Klungland A (1995) Post-translational modification of poly(ADP-ribose) polymerase induced by DNA strand breaks. Trends Biochem Sci 20:405–11

Lo EH, Hara H, Rogowska J, Trocha M, Pierce AR, Huang PL, Fishman MC, Wolf GL, Moskowitz MA (1996) Temporal correlation mapping analysis of the hemodynamic penumbra in mutant mice deficient in endothelial nitric oxide synthase gene expression. Stroke 27:1381–5

Lowy DR, Willumsen BM (1993) Function and regulation of ras. Annu Rev Biochem 62:851–91

Macara IG, Lounsbury KM, Richards SA, McKiernan C, Bar-Sagi D (1996) The Ras superfamily of GTPases. Faseb J 10:625–30

MacMillan-Crow LA, Crow JP, Kerby JD, Beckman JS, Thompson JA (1996) Nitration and inactivation of manganese superoxide dismutase in chronic rejection of human renal allografts. Proc Natl Acad Sci USA 93:11853–8

Margaill I, Allix M, Boulu RG, Plotkine M (1997) Dose- and time-dependence of L-NAME neuroprotection in transient focal cerebral ischaemia in rats. Br J Pharmacol 120:160–3

Marletta MA (1993) Nitric oxide synthase structure and mechanism. J Biol Chem 268:12231–4

Marletta MA (1994) Nitric oxide synthase: aspects concerning structure and catalysis. Cell 78:927–30

Matsui T, Nagafuji T, Mori T, Asano T (1997) N^ω-nitro-L-arginine attenuates early ischemic neuronal damage of prolonged focal cerebral ischemia and recirculation in rats. Neurol Res 19:192–203

McNaught KS, Brown GC (1998) Nitric oxide causes glutamate release from brain synaptosomes. J Neurochem 70:1541–6

Moncada C, Lekieffre D, Arvin B, Meldrum B (1992) Effect of NO synthase inhibition on NMDA- and ischaemia-induced hippocampal lesions. Neuroreport 3:530–2

Moore PK, Babbedge RC, Wallace P, Gaffen ZA, Hart SL (1993a) 7-Nitro indazole, an inhibitor of nitric oxide synthase, exhibits anti-nociceptive activity in the mouse without increasing blood pressure. Br J Pharmacol 108:296–7

Moore PK, Wallace P, Gaffen Z, Hart SL, Babbedge RC (1993b) Characterization of the novel nitric oxide synthase inhibitor 7-nitro indazole and related indazoles: antinociceptive and cardiovascular effects. Br J Pharmacol 110:219–24

Morikawa E, Moskowitz MA, Huang Z, Yoshida T, Irikura K, Dalkara T (1994) L-arginine infusion promotes nitric oxide-dependent vasodilation, increases regional cerebral blood flow, reduces infarction volume in the rat. Stroke 25:429–35

Nanri K, Montecot C, Springhetti V, Seylaz J, Pinard E (1998) The selective inhibitor of neuronal nitric oxide synthase, 7-nitroindazole, reduces the delayed neuronal damage due to forebrain ischemia in rats. Stroke 29:1248–53

Nathan C, Xie QW (1994) Nitric oxide synthases: roles, tolls, controls. Cell 78:915–8

Nishikawa T, Kirsch JR, Koehler RC, Bredt DS, Snyder SH, Traystman RJ (1993) Effect of nitric oxide synthase inhibition on cerebral blood flow and injury volume during focal ischemia in cats. Stroke 24:1717–24

Nishikawa T, Kirsch JR, Koehler RC, Miyabe M, Traystman RJ (1994) Nitric oxide synthase inhibition reduces caudate injury following transient focal ischemia in cats. Stroke 25:877–85

Nowicki JP, Duval D, Poignet H, Scatton B (1991) Nitric oxide mediates neuronal death after focal cerebral ischemia in the mouse. Eur J Pharmacol 204:339–40

Oleszak EL, Zaczynska E, Bhattacharjee M, Butunoi C, Legido A, Katsetos CD (1998) Inducible nitric oxide synthase and nitrotyrosine are found in monocytes/macrophages and/or astrocytes in acute, but not in chronic, multiple sclerosis. Clin Diagn Lab Immunol 5:438–45

Palmer RM, Ferrige AG, Moncada S (1987) Nitric oxide release accounts for the biological activity of endothelium-derived relaxing factor. Nature 327:524–6

Panahian N, Yoshida T, Huang PL, Hedley-Whyte ET, Dalkara T, Fishman MC, Moskowitz MA (1996) Attenuated hippocampal damage after global cerebral ischemia in mice mutant in neuronal nitric oxide synthase. Neuroscience 72:343–54

Pogun S, Dawson V, Kuhar MJ (1994) Nitric oxide inhibits 3H-glutamate transport in synaptosomes. Synapse 18:21–6

Quast MJ, Wei J, Huang NC (1995) Nitric oxide synthase inhibitor N^G-nitro-L-arginine methyl ester decreases ischemic damage in reversible focal cerebral ischemia in hyperglycemic rats. Brain Res 677:204–12

Samdani AF, Dawson TM, Dawson VL (1997a) Nitric oxide synthase in models of focal ischemia. Stroke 28:1283–8

Samdani AF, Newcamp C, Resink A, Facchinetti F, Hoffman BE, Dawson VL, Dawson TM (1997b) Differential susceptibility to neurotoxicity mediated by neurotrophins and neuronal nitric oxide synthase. J Neurosci 17:4633–41

Satoh MS, Poirier GG, Lindahl T (1994) Dual function for poly(ADP-ribose) synthesis in response to DNA strand breakage. Biochemistry 33:7099–106

Schlessinger J, Ullrich A (1992) Growth factor signaling by receptor tyrosine kinases. Neuron 9:383–91

Sharpe MA, Cooper CE (1998) Interaction of peroxynitrite with mitochondrial cytochrome oxidase. Catalytic production of nitric oxide and irreversible inhibition of enzyme activity. J Biol Chem 273:30961–72

Shimizu-Sasamata M, Bosque-Hamilton P, Huang PL, Moskowitz MA, Lo EH (1998) Attenuated neurotransmitter release and spreading depression-like depolarizations after focal ischemia in mutant mice with disrupted type I nitric oxide synthase gene. J Neurosci 18:9564–71

Szabo C, Dawson VL (1998) Role of poly(ADP-ribose) synthetase in inflammation and ischaemia- reperfusion. Trends Pharmacol Sci 19:287–98

Tamir S, Burney S, Tannenbaum SR (1996) DNA damage by nitric oxide. Chem Res Toxicol 9:821–7

Thiemermann C (1997) Nitric oxide and septic shock. Gen Pharmacol 29:159–66

Thomas E, Pearse AGE (1961) The fine localization of dehydrogenases in the nervous system. Histochemistry 2:266–282

Thomas E, Pearse AGE (1964) The solitary active cells. Histochemical demonstration of damage-resistant nerve cells with a TPN-diaphorase reaction. Acta Neuropathology 3:238–249

Togashi H, Sasaki M, Frohman E, Taira E, Ratan RR, Dawson TM, Dawson VL (1997) Neuronal (type I) nitric oxide synthase regulates NFkB activity and immunologic (type II) nitric oxide synthase expression. Proc Natl Acad Sci USA 94:2676–2680

Uemura Y, Kowall NW, Beal MF (1990) Selective sparing of NADPH-diaphorase-somatostatin-neuropeptide Y neurons in ischemic gerbil striatum. Ann Neurol 27:620–5

Van Dam AM, Bauer J, Man AHWK, Marquette C, Tilders FJ, Berkenbosch F (1995) Appearance of inducible nitric oxide synthase in the rat central nervous system after rabies virus infection and during experimental allergic encephalomyelitis but not after peripheral administration of endotoxin. J Neurosci Res 40:251–60

Wang ZQ, Auer B, Stingl L, Berghammer H, Haidacher D, Schweiger M, Wagner EF (1995) Mice lacking ADPRT and poly(ADP-ribosyl)ation develop normally but are susceptible to skin disease. Genes Dev 9:509–20

Wei J, Quast MJ (1998) Effect of nitric oxide synthase inhibitor on a hyperglycemic rat model of reversible focal ischemia: detection of excitatory amino acids release and hydroxyl radical formation. Brain Res 791:146–56

Wink DA, Mitchell JB (1998) Chemical biology of nitric oxide: Insights into regulatory, cytotoxic, and cytoprotective mechanisms of nitric oxide. Free Radic Biol Med 25:434–56

Wink DA, Feelisch M, Fukuto J, Chistodoulou D, Jourd'heuil D, Grisham MB, Vodovotz Y, Cook JA, Krishna M, DeGraff WG, Kim S, Gamson J, Mitchell JB (1998a) The cytotoxicity of nitroxyl: possible implications for the pathophysiological role of NO. Arch Biochem Biophys 351:66–74

Wink DA, Vodovotz Y, Laval J, Laval F, Dewhirst MW, Mitchell JB (1998b) The multifaceted roles of nitric oxide in cancer. Carcinogenesis 19:711–21

Yamamoto S, Golanov EV, Berger SB, Reis DJ (1992) Inhibition of nitric oxide synthesis increases focal ischemic infarction in rat. J Cereb Blood Flow Metab 12:717–26

Yoshida T, Limmroth V, Irikura K, Moskowitz MA (1994) The NOS inhibitor, 7-nitroindazole, decreases focal infarct volume but not the response to topical acetylcholine in pial vessels. J Cereb Blood Flow Metab 14:924–9

Yoshida T, Waeber C, Huang Z, Moskowitz MA (1995) Induction of nitric oxide synthase activity in rodent brain following middle cerebral artery occlusion. Neurosci Lett 194:214–8

Yun HY, Dawson VL, Dawson TM (1997) Nitric oxide in health and disease of the nervous system. Mol Psychiatry 2:300–10

Yun HY, Gonzalez-Zulueta M, Dawson VL, Dawson TM (1998) Nitric oxide mediates N-methyl-D-aspartate receptor-induced activation of p21ras. Proc Natl Acad Sci USA 95:5773–8

Zhang F, White JG, Iadecola C (1994) Nitric oxide donors increase blood flow and reduce brain damage in focal ischemia: evidence that nitric oxide is beneficial in the early stages of cerebral ischemia. J Cereb Blood Flow Metab 14:217–26

Zhang F, Xu S, Iadecola C (1995) Time dependence of effect of nitric oxide synthase inhibition on cerebral ischemic damage. J Cereb Blood Flow Metab 15:595–601

Zhang ZG, Chopp M, Gautam S, Zaloga C, Zhang RL, Schmidt HH, Pollock JS, Forstermann U (1994a) Upregulation of neuronal nitric oxide synthase and mRNA, and selective sparing of nitric oxide synthase-containing neurons after focal cerebral ischemia in rat. Brain Res 654:85–95

Zhang J, Dawson VL, Dawson TM, Snyder SH (1994b) Nitric oxide activation of poly(ADP-ribose) synthetase in neurotoxicity. Science 263:687–9

Zhang ZG, Reif D, Macdonald J, Tang WX, Kamp DK, Gentile RJ, Shakespeare WC, Murray RJ, Chopp M (1996) ARL 17477, a potent and selective neuronal NOS inhibitor decreases infarct volume after transient middle cerebral artery occlusion in rats. J Cereb Blood Flow Metab 16:599–604

Zweier JL, Wang P, Samouilov A, Kuppusamy P (1995) Enzyme-independent formation of nitric oxide in biological tissues. Nat Med 1:804–9

CHAPTER 24
Therapeutic Potential of Nitric Oxide Synthase Gene Manipulation

H.E. von der Leyen and V.J. Dzau

A. General Principles of Gene Therapy

The introduction of recombinant DNA technology has led to a major paradigmatic shift in medicine (BERG 1981). Diseases are recognized now not only by their physiological/biochemical defects, but also by the molecular alterations at the level of gene expression (KATZ 1988). Novel therapeutic approaches have expanded from chemical synthesis to genetic engineering. The development of in vivo *gene-transfer* technology has created a powerful new tool to study disease mechanisms by providing the means to overexpress or to inhibit specific local factors which are believed to contribute to a pathological process. In addition, this technology provides the opportunity to develop novel therapeutic strategies, including gene replacement, gene correction, or gene augmentation, thereby paving the way for *gene therapy* (HAWKINS 1998; von der Leyen et al. 1998).

To improve the efficiency of gene transfer, a number of transduction systems have been developed recently. The encapsulation of the DNA of interest in artificial lipid membranes (liposomes) has resulted in successful gene transfer into certain cell types (FELGNER et al. 1987). Microinjection has been used effectively to introduce purified recombinant proteins, neutralizing antibodies, and functional oligonucleotides into cells, allowing a critical examination of the role of specific gene products in regulating and/or determining phenotypic features that can be assayed at the single cell level (CAPECCHI 1980; ADAMS et al. 1990). Furthermore, genes can be injected into cells in vivo (cardiac or skeletal muscle) for studies of regulated expression in the intact animal (KITSIS et al. 1991). Viral vectors have been shown to be efficient vehicles for in vivo gene transfer and have been used in cardiovascular research to transduce and express exogenous genes in vivo (NABEL et al. 1990; LEMARCHAND et al. 1993; MORISHITA et al. 1994). Currently, in vivo gene transfer techniques include (i) viral gene transfer using retrovirus or adenovirus, (ii) liposomal gene transfer with cationic liposomes (Lipofectamine), (iii) fusigenic liposomes using Sendai virus (hemagglutinating virus of Japan) complexed with neutral liposomes (fusigenic liposomes), (iv) direct (micro)injection, and (v) ex vivo genetic engineering, such as myoblast implantation. The characteristics of the different vector systems are described elsewhere (FELGNER and RHODES 1991; ANDERSON 1998; von der Leyen et al. 1998).

Table 1. Genetic engineering approaches to study nitric oxide

Gain of function: plasmid vector	Loss of function: antisense oligonucleotide
Overexpression of nitric oxide synthase (von der Leyen et al. 1995) – Intracellular and extracellular action – Autocrine and paracrine effect	Inhibition of transcription and/or translation of nitric oxide synthase gene (Cartwright et al. 1997) – Intracellular action

An important focus in cardiovascular medicine is the study of paracrine and autocrine regulation of cardiovascular function and potential aberrations in cardiovascular disease. Gene-transfer techniques have provided a useful tool for this area of research. Current approaches using experimental gene transfer to study genes involved with local physiological function include the application of oligonucleotides or recombinant DNA for gene replacement, inhibition, or augmentation (Table 1).

Nitric oxide (NO) has so many facets that it is uniting many fields of medicine, including immunology, neuroscience, and cardiovascular medicine. Being a reactive gas, NO functions both as a signaling molecule in endothelial and nerve cells and as a killer molecule by activated immune cells. Its ubiquitous distribution in the body and its multiple roles have influenced our understanding of how cells communicate and protect themselves (Gibaldi 1993; Änggard 1994; Schmidt and Walter 1994). The recent cloning and characterization of the different isoforms of the nitric oxide synthases (NOS) paved the way for a better understanding of the regulation of NO pathways and for the development of therapeutic gene transfer (Lamas et al. 1992; Lowenstein et al. 1992; Nathan and Xie 1994). Our chapter will focus mainly on reviewing the data of genetic engineering techniques using antisense oligonucleotides or DNA-expression vectors designed to modify the action of NOS in the vasculature.

B. Gain of Function
I. Overexpression of the NOS Gene

Vascular remodeling usually represents an adaptive process that occurs in response to long-term changes in hemodynamic and humoral conditions. However, this process can also contribute to the pathophysiology of vascular diseases (Gibbons and Dzau 1994). The active process of vascular remodeling involves changes in several cellular processes like cell growth, cell death, cell migration, and extracellular-matrix production or degradation. The potential therapeutic role of NO in vascular remodeling can be studied by the "gain of function" approach, i.e., the transfer of plasmid DNA (or genes) into cells or tissue in vivo to overexpress an isoform of NOS, with subsequent increased NO production as described below.

1. Overexpression of Endothelial Constitutive NOS

Injury to the endothelium plays an essential role in the pathogenesis of vascular disease (Ross 1993) and endothelium-derived relaxing factor/NO has an important regulatory function in maintaining vascular homeostasis (VANE et al. 1990). NO has been shown to inhibit vascular smooth-muscle cell proliferation in vitro and has been postulated to be an important local factor in modulating blood-vessel structures (GARG and HASSID 1989; DE CATERINA et al. 1995; MOORADIAN et al. 1995; TSAO et al. 1997). It has been hypothesized that the impairment of NO production and bioactivity may be the underlying mechanism for the various risk factors, such as hypercholesterolemia, hypertension, diabetes, and cigarette smoking, in the generation of vascular disease (HARRISON 1997). Experimental studies have shown that vascular injury induces the local expression of mitogens and chemotactic factors that stimulate vascular smooth muscle and leucocyte migration and proliferation. Thus, a deficiency in NO production or bioactivity and/or an excess of growth-promoting factors favors the development of vascular lesions. Vascular diseases, such as restenosis, atherosclerosis, bypass-graft failure, and transplant vasculopathy characterized by the above-mentioned imbalance of mediators, constitute unique opportunities for NOS gene transfer as the basis for therapy.

To determine the effect of overexpression of the constitutive isoform of NOS [endothelial-type NOS (eNOS); NOS III], we constructed a NOS expression vector in which the bovine eNOS complementary DNA (cDNA) was controlled by a *CAG* promoter [cytomegalovirus (CMV)-IE enhancer sequence connected to a modified (*AG*) chicken β-actin promoter], rabbit β-globin gene sequences including a polyadenylation signal, and a *SV40 ori* in a pUC13 vector (VON DER LEYEN et al. 1995). In vivo transfection of this construct using fusigenic liposomes (MANN et al. 1997) into balloon-injured rat carotid arteries not only restored NO production within the vessel wall but also significantly increased the vascular reactivity of the vessel (VON DER LEYEN et al. 1995). Furthermore, eNOS transgene expression resulted in a 70% inhibition of neointima formation after balloon injury (Fig. 1). Four complementary methods were used to verify successful in vivo eNOS gene transfer into the vessel wall: (i) transgene protein expression was documented by Western blot, (ii) localization of enzyme expression in the vessel wall was verified by in situ histochemical staining using the reduced nicotinamide adenine dinucleotide phosphate (NADPH) diaphorase assay, (iii) enzymatic activity of the transgene product was confirmed by measurement of increased NO generation from transfected vessel segments using the chemiluminescence methods, and (iv) biological effectiveness of the transgene product was assessed by changes in vascular reactivity induced by the increased local generation of NO, thereby

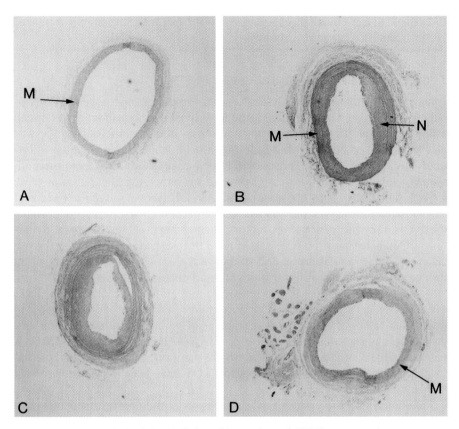

Fig. 1. Effect of endothelial cell nitric oxide synthase (eNOS) or control vector transfection on neointima formation in balloon-injured rat carotid arteries. Representative cross-sections (×25) from rat carotid arteries are shown: (A) uninjured, (B) injured untreated, (C) injured with control vector transfection, (D) injured with eNOS transfection. *M*, media; *N*, neointima. Reproduced with permission (VON DER LEYEN et al. 1995)

potentially counterbalancing vasospasm induced by vascular injury. In addition, at day 4 after eNOS transfection, a significant decrease in DNA synthesis (as measured by BrdU incorporation in eNOS-transfected vessels) was demonstrated (VON DER LEYEN et al. 1995). Given that the activation of mitogenic factors mediating the cellular processes essential for lesion formation occurs within the first couple of days after injury, the characterization of transgene expression was carried out during the first couple of days after balloon injury. It has been shown that interventions that inhibit cell proliferation and/or cell migration during this initial acute phase prevent neointima formation for at least 2 weeks after injury (SIMONS et al. 1992; MORISHITA et al. 1993; MANN et al. 1995). Thus, using in vivo gene transfer into the rat carotid injury model, the eNOS gene tranfer was a "proof of concept" that vascular-derived NO plays an important role in vivo in modulating vascular remodeling. To our knowledge, our study of NOS-III gene transfer was the first to document direct in vivo gene transfer of a cDNA encoding a functional enzyme in the vessel wall to achieve a therapeutic effect. In fact, eNOS gene transfer may be useful for gene therapy of neointimal hyperplasia and associated local vasospasm, thereby modulating vascular remodeling (Dzau and Horiuchi 1996).

By using adenovirus-mediated gene transfer, Janssens et al. (1998) extended the above-described genetic-engineering strategy and demonstrated successful transfer of the gene encoding human eNOS to balloon-injured rat carotid arteries. Thirty percent of medial smooth-muscle cells and foci of adventitial cells showed transgene expression that increased local NO production and restored cyclic guanosine monophosphate (cGMP) levels to those observed in uninjured vessels. After balloon injury, the neointimal surface area of NOS-transfected arteries was reduced by 70% compared with control injured arteries. Using a similiar adenoviral vector together with the Infiltrator device (Interventional Technologies, San Diego, Calif.) as delivery tool, the same group reported successful coronary gene transfer to injured porcine left anterior descending coronary arteries (Varenne et al. 1998). About 40% of medial smooth-muscle cells and 25% of adventitial cells were reportedly transduced with the transgene, suggesting highly efficient intramural adenovirus-mediated gene transfer brought about by the Infiltrator device. In this model of intracoronary gene transfer, overexpression of eNOS showed significant inhibition of neointimal lesion formation.

Recently, it was demonstrated that NO inhibits migration of mitogen-stimulated vascular smooth-muscle cells in vitro (Dubey et al. 1995). Stimulation of smooth-muscle cells of guinea pig coronary artery by platelet-derived growth factor (PDGF)-BB retards paxillin mobility in a time-dependent manner, as demonstrated by a mobility shift in sodium dodecyl sulfate polyacrylamide gel electrophoresis. This effect may be due to tyrosine phosphorylation of paxillin. Transfer of the eNOS (NOS III) gene by replication-deficient recombinant adenovirus in vascular smooth-muscle cells inhibited PDGF-BB-stimulated mobility shifts and tyrosine phosphorylation of paxillin (Fang et al. 1997). In addition, tyrosine phosphorylation of focal adhesion kinase (FAK) was inhibited. Considering the importance of FAK and paxillin in cell migration and proliferation, this gene transfer study suggested that the FAK–paxillin pathway is a target for NO action inhibiting vascular smooth-muscle cell migration and proliferation (Fang et al. 1997).

Several groups have also studied the overexpression of eNOS in cultured smooth-muscle cells or endothelial cells (Channon et al. 1996; Kullo et al. 1997c; Janssens et al. 1998) using adenovirus-mediated gene transfer. In extending their in vitro studies, Kullo et al. (1997b) showed a diminished contractile response and enhanced endothelium-dependent relaxation at day 4 after in vivo gene transfer of eNOS in rabbit carotid arteries. Endothelial transgene expression was confirmed by staining for β-galactosidase in *Lac Z*-

transduced arteries and eNOS-specific immunoreactivity in eNOS-transfected arteries. This gene-transfer strategy implies that vascular eNOS gene therapy may extend our therapeutic arsenal for the treatment of vasospasm and endothelial dysfunction (KULLO et al. 1997a, 1997b).

In an ex vivo study, a recombinant eNOS expression vector was successfully transferred to large canine cerebral arteries by adenovirus-mediated gene transfer (CHEN et al. 1997). Functional expression of eNOS resulted in increased basal production of cGMP, with a subsequent reduction in receptor-mediated contractile response and enhancement of endothelium-derived relaxation. The retrovirus-mediated overexpression of human eNOS in rat vascular smooth-muscle cells in culture with subsequent seeding of these transformed cells into denuded rat carotid arteries induced marked dilatation (threefold increase in vessel diameter) of the vessel at 2 weeks after seeding (CHEN et al. 1998). The perivascular application via cerebrospinal fluid, with subsequent expression of recombinant eNOS in cerebral arteries, may represent a potentially feasible therapeutic approach alleviating vasospastic conditions (CHEN et al. 1997; KULLO et al. 1997a). Incubation of rabbit carotid arteries (OOBOSHI et al. 1997) or porcine coronary arteries (CABLE et al. 1997a) in organ culture with a replication-deficient adenovirus carrying the cDNA for eNOS augmented vasorelaxation in response to stimuli that release NO. Gene transfer in organ culture resulted in eNOS transgene expression preferentially in adventitial cells, suggesting that adventitial gene transfer may be sufficient to alter vasomotor tone (OOBOSHI et al. 1997). In extending these gene-transfer studies in organ culture, CABLE et al. (1997b) transduced human saphenous veins with an adenovirus vector encoding bovine eNOS and demonstrated functional expression of recombinant NOS. This approach may provide a genetic engineering tool to reduce the risk of early thrombosis in saphenous vein grafts by providing increased vascular NO production.

Pulmonary gene transfer with delivery of recombinant eNOS adenovirus by a single aerosolization enabled diffuse transduction of bronchial and alveolar epithelial cells and vascular adventitial and endothelial cells, thereby mediating increased NO production (JANSSENS et al. 1996). During acute hypoxia, this local overexpression of eNOS in rat lungs significantly influenced pulmonary artery pressure and total pulmonary resistance without affecting systemic hemodynamics. Thus, aerosolized eNOS gene transfer can act as a selective pulmonary vasodilator, representing an attractive therapeutic approach to treat patients with pulmonary hypertension, extending the already established therapeutic application of NO inhalation (PEPKA-ZABA et al. 1991).

2. Overexpression of Inducible NOS

Gene transfer of human inducible NOS (iNOS) using a retroviral vector demonstrated significant protein expression in endothelial cells and smooth-muscle cells in vitro (TZENG et al. 1996a). Transfection of iNOS in vivo completely prevented myointimal thickness of porcine arteries induced by balloon-catheter injury (TZENG et al. 1996a). In smooth-muscle cells in vitro, NO synthesis after iNOS transfection was dependent on exogenous tetrahydrobiopterin (BH4). Unstimulated vascular smooth-muscle cells do not synthesize BH4 (TZENG et al. 1995). Recently, it was shown in NIH 3T3 cells constitutively expressing recombinant human iNOS, that iNOS subunits dimerize to form an active enzyme and that BH4 plays a critical role in promoting dimerization in intact cells (TZENG et al. 1995). A human expression plasmid encoding guanosine triphosphate (GTP) cyclohydrolase I (GTPCH), the rate-limiting enzyme for BH4 biosynthesis, was successfully cotransfected with iNOS into smooth-muscle cells to reconstitute iNOS activity (TZENG et al. 1996b). Thus, GPTCH gene transfer could be used to deliver a cofactor to targeted cells even if it is synthesized in neighboring cells, and may augment the production of NO after iNOS gene transfer.

One mechanism by which NO may prevent intimal hyperplasia may be protection of the endothelium. TZENG et al. (1997) constructed an adenoviral vector containing the iNOS (NOSII) and transfected cultured sheep arterial endothelial cells. The increased NO production after transfection did not affect the viability of the endothelial cells, and prolonged exposure to NO did not induce apoptosis of endothelial cells. Instead, NO inhibited lipopolysaccharide-induced apoptosis by reducing caspase-3-like protease activity (TZENG et al. 1997).

Recent experiments support a protective effect of NO in endothelial cells exposed to superoxide radicals (TSAO et al. 1997). Growing evidence suggests that atherosclerosis is an inflammatory disease and that monocyte adherence to the endothelium represents the earliest vessel-wall abnormality occurring within 1 week of the start of a high-cholesterol diet in experimental animals (Ross 1993). Superoxide radicals may play a pivotal role in early atherosclerotic-lesion formation by activating oxidative stress-inducible transcriptional factors (nuclear factor-κB), thereby upregulating the expression of inflammatory genes such as *MCP-1* and *VCAM-1* (MARUI et al. 1993). We have demonstrated that iNOS gene transfer reduced oxygen-radical production in vitro, thereby potentially influencing the adhesion of monocytes to vascular endothelium, as was demonstrated previously with NO donors (DE CATERINA et al. 1995; TSAO et al. 1997). It appears that iNOS gene transfer may provide a potentially powerful tool to modify vascular homeostasis beyond simple modulation of vasomotion (HARRISON 1997). Increased availability of NO within the vascular wall may counteract the detrimental effects of superoxide radicals and may reduce the degree of endothelial activation and dysfunction. Thus, iNOS gene transfer may be useful in preventing the development of atherosclerosis.

Alloimmune response in the rat aortic allograft model with subsequent endothelial dysfunction and injury results in the exposure of the underlying vascular smooth-muscle cells to mitogenic factors, with subsequent neointimal lesion formation (DAVIS et al. 1996). The development of allograft arteriosclerosis induced a sustained intramural expression of iNOS in rejecting aortic allografts (SHEARS et al. 1997). Interestingly, overexpression of an adenoviral vector encoding iNOS in the aortic grafts effectively inhibited intimal hyperplasia both in the presence and absence of cyclosporin A (SHEARS et al. 1997). Thus, overexpression of iNOS may allow iNOS expression to continue in the face of high cyclosporin and steroid doses, resulting in a suppression of allograft arteriosclerosis.

C. Loss of Function
I. Inhibition of NOS by Antisense Technology

Another strategy for the study of NOS function and its modulation for therapy is a "loss of function" approach by inhibiting the expression of specific NOS isoform gene(s) using antisense oligonucleotides. Oligonucleotides exert their molecular actions intracellularly either at the translational or transcriptional level (DAVIS 1994; FLANAGAN and WAGNER 1997; PHILLIPS and GYURKO 1997). Antisense oligonucleotides can bind specific mRNA and block ribosomal translocation, thereby inhibiting translation. The antisense DNA:mRNA duplex is also rapidly degraded by RNase H, thereby reducing the level of transcripts intracellularly (COLMAN 1990).

Oxidant stress is accompanied by the enhanced expression of iNOS, increased production of NO, and impaired cell viability in some cell systems (BECKMAN et al. 1990). With the description of a hypoxia-responsive element on the iNOS gene, a novel pathway for the activation of the iNOS gene was identified (MELILLO et al. 1995). In BSC-1 kidney tubular epithelial cells, selective inhibition of iNOS using phosphorothioate-modified antisense oligonucleotides dramatically improved BSC-1 cell viability after oxidant stress (PERESLENI et al. 1996). By inhibiting the hydrogen-peroxide-induced NO release, epithelial cells were rescued by reducing a possible detrimental effect of NO or NO-related superoxide radicals. Using this approach with transfer of antisense oligonucleotides in vivo into rats subjected to renal ischemia and concomitant hypoxia-induced oxidant stress, acute renal failure was attenuated, suggesting direct evidence for a cytotoxic effect of NO in this model of ischemic renal failure (NOIRI et al. 1996). The need to target the proximal tubular epithelium, the site of preferential injury in the acute-ischemia–reperfusion model, seems to have certain advantages for antisense oligonucleotides, since substantial accumulation of systemically injected phosphorothioate oligonucleotides occurs in the kidney and especially in the proximal tubular epithelium (RAPPAPORT et al. 1995). The nature of the interaction of NO and superoxide radicals and its subsequent pathophysiological consequences is still a matter of controversial discussion (STAMLER 1994; TAKANO et al. 1998). Because of differences in species, tissues, and pathophysiological models, NO may either scavenge superoxide radicals or may actually increase the cellular oxidant stress. These effects of NO are probably dose dependent, so massive NO formation may be toxic, but lower, more physiological formation of NO may be protective. In addition, the source of NO and the relationship between NO and tissue injury (TZENG and BILLIAR 1997) may be an important issue that needs to be addressed in future research (Sect. D).

Differences in the pathophysiological role of NO in different cell types may be explained by a tissue-specific transcriptional regulation of the iNOS promoter/enhancer as reported recently in smooth-muscle cells and macrophages (KOLYADA et al. 1996). CARTWRIGHT et al. (1997) generated a murine macrophage cell line expressing a 500-bp sequence of iNOS, in either the antisense or sense orientation, driven by the SV40 promoter/enhancer region. Adhesion of the antisense-treated cell line A10 to cytokine-stimulated murine endothelial cells was significantly higher than adhesion of the sense-treated cell lines. There was a negative correlation between the amount of NO produced and the level of adhesion, indicating an anti-adhesive role of NO.

Antisense oligonucleotides may also be used to study pulmonary vascular disease. In a recent study of cultured rat pulmonary artery smooth-muscle cells, the NO gene was induced in response to lipopolysaccharide and cytokines. Pre-incubation of the cells in the presence of an antisense oligonucleotide to the first 18 bases after the initiation codon of iNOS mRNA caused an significant decrease in cytokine-induced NO_2^- production in a concentration-dependent manner (THOMAE et al. 1993). Excessive NO production brought about by cytokine stimulation in septic shock contributes to hemodynamic instability and perhaps tissue leak. Thus, NOS inhibition in this pathophysiological context may confer benefit on clinical septic shock (PETROS et al. 1991).

D. Transgenic Animals with Disrupted NOS Gene

Transgenic mice with a disruption of the NOS gene represent a valuable tool to study the chronic functional role of NOS in various (patho)physiological conditions. GÖDECKE et al. (1998) constructed a knockout mouse with a defective eNOS gene (disruption of the NADPH-binding site) for a specific analysis of eNOS in the cardiovascular system. *Acute* inhibition of eNOS using the selective NOS inhibitor *N*-nitro-L-arginine methyl ester revealed that NO has an important function in setting the basal coronary artery tone and participates in reactive hyperemia and the response to acetylcholine. However, hemodynamic analysis in eNOS$^{-/-}$ knockout mice with *chronic* dysfunction of eNOS showed no changes in basal coronary flow and reactive hyperemia, suggesting the participation of compensatory mechanisms (GÖDECKE et al. 1998). These results are in contrast to the data of HUANG et al. (1995), who reported systemic hypertension in mice lacking the gene for eNOS. This apparent discrepancy could indicate that the signal transduction pathways, mediators, and/or compensatory responses may be different in large conduit vessels or small resistance vessels. Further studies will be necessary to elucidate the mechanism accounting for this interesting difference.

In order to develop a model of vascular remodeling in which the role of eNOS-derived NO can be investigated, RUDIC et al. (1998) generated eNOS knockout mice and induced vascular remodeling by external carotid-artery ligation. In the absence of eNOS (eNOS$^{-/-}$ knockout), not only was luminal remodeling impaired, vessel-wall thickness doubled due to proliferation of vascular smooth-muscle cells. Thus, these data support the recognition of NO as a major determinant of vessel-wall architecture responsible for vascular remodeling.

As discussed above, alloimmune injury to the donor vasculature is involved in the propagation of accelerated arteriosclerosis in chronic heart rejection after transplantation. Inducible NOS seems to play a protective role in transplant arteriosclerosis, since iNOS (NOS II) deficiency in iNOS$^{-/-}$ knockout mice caused a significant increase in severity and frequency of intimal thickening in response to alloimmune injury in a heterotopic cardiac-transplant model (KOGLIN et al. 1998).

Increased synthesis of the bioregulatory molecule NO is reported to be a part of the cellular and biochemical events activated during wound repair (SCHAFFER et al. 1996). In extending the series of pathophysiological studies in

Table 2. Potential clinical applications of nitric oxide (NO) synthase (NOS) gene transfer

Disease	NO effect	Genetic engineering approach
Vasospastic disorders		
Prinzmetal angina	Vasodilation	NOS overexpression
Vasculitis		
Inflammation	Destruction of microorganism	NOS overexpression
Vascular proliferation		
Restenosis	Antiproliferative effect	NOS overexpression
Vein-graft failure	Antiproliferative effect	NOS overexpression
Atherosclerosis	Antiproliferative effect	NOS overexpression
Transplant vasculopathy	Antiproliferative effect	NOS overexpression
Vascular access (hemodialysis)	Antiproliferative effect	NOS overexpression
Thrombosis	Antithrombogenic effect	NOS overexpression
Pulmonary hypertension	Pulmonary vasodilation	NOS overexpression
Oxidative injury	Scavenging of superoxide radicals	NOS overexpression
Reperfusion injury[a]	Scavenging or oxidative injury	NOS overexpression or NOS antisense
Septicemia	Vasodilation	Inhibition of NOS (antisense)

[a] Conflicting literature data (see text).

iNOS$^{-/-}$ knockout mice, YAMASAKI et al. (1998) demonstrated that the closure rate of excisional wounds was delayed by about 30% in iNOS-knockout mice. Transient iNOS expression achieved with adenoviral-mediated gene transfer using a simple, brief topical exposure to low-titer virus was adequate to reverse the delayed wound closure in the iNOS-deficient mice (YAMASAKI et al. 1998). The results in iNOS-deficient mice support the hypothesis that an increase in NO production resulting from upregulation of iNOS expression may be an important part of an endogenous physiological repair mechanism.

E. Potential Therapeutic Applications of NOS Gene Transfer

Potential applications for in vivo gene therapy using NOS overexpression or inhibition may cover a broad range of vascular or inflammatory disorders. However, there are still unresolved issues that remain to be addressed concerning the function and pathophysiology of NO. Clinical gene therapy for cardiovascular disease is now entering the stage of clinical pilot studies. Table 2 summarizes potential areas for NOS gene therapy. Given the recent rapid progress in technology development and transfer, we believe that the appropriate application of NOS gene therapy will bring new benefits to the treatment of specific diseases.

Acknowledgements. Heiko E. von der Leyen is supported by grants of the Deutsche Forschungsgemeinschaft (Le 567/3-1, 3-2, 6/1). Victor J. Dzau is the recipient of a National Institutes of Health (NIH) MERIT Award (HL 35610). Supported in part by NIH grants HL-35610 and HL-58516.

References

Adams BA, Tanabe T, Mikami A, Numa S, Beam KG (1990) Intramembrane charged movement restored in dysgenic skeletal muscle by injection of dihydropyridine receptor cDNAs. Nature 345:569–572

Anderson WF (1998) Human gene therapy. Nature 392:25–30

Änggard E (1994) Nitric oxide: mediator, murderer, and medicine. Lancet 343:1199–1206

Beckman JS, Beckman TW, Chen J, Marshall PA, Freeman BA (1990) Apparent hydroxyl radical production by peroxynitrite: implications for endothelial injury from nitric oxide and superoxide. Proc Natl Acad Sci U S A 87:1620–1624

Berg P (1981) Dissections and reconstructions of genes and chromosomes. Science 213:296–303

Cable DG, O'Brien T, Kullo IJ, Schwartz RS, Schaff HV, Pompili VJ (1997a) Expression and function of a recombinant endothelial nitric oxide synthase gene in porcine coronary arteries. Cardiovasc Res 35:553–559

Cable DG, O'Brien T, Schaff HV, Pompili VJ (1997b) Recombinant endothelial nitric oxide synthase-transduced human saphenous veins: gene therapy to augment nitric oxide production in bypass conduits. Circulation 96(Suppl):II-173–II-178

Capecchi M (1980) High efficiency transformation by direct microinjection of DNA into mammalian cells. Cell 22:479–488

Cartwright JE, Johnstone AP, Whitley GSJ (1997) Inhibition of nitric oxide synthase by antisense techniques: investigations of the roles of NO produced by murine macrophages. Br J Pharmacol 120:146–152

Channon KM, Blazing MA, Shetty GA, Potts KE, George SE (1996) Adenoviral gene transfer of nitric oxide synthase: high level expression in human vascular cells. Cardiovasc Res 32:962–972

Chen AFY, O'Brien T, Tsutsui M, Kinoshita H, Pompili VJ, Crotty TB, Spector DJ, Katusic ZS (1997) Expression and function of recombinant endothelial nitric oxide synthase gene in canine basilar artery. Circ Res 80:327–335

Chen L, Daum G, Forough R, Clowes M, Walter U, Clowes AW (1998) Overexpression of human endothelial nitric oxide synthase in rat vascular smooth muscle cells and in balloon-injured carotid artery. Circ Res 82:862–870

Colman A (1990) Antisense strategies in cell and developmental biology. J Cell Sci 97:399–409

Davis AR (1994) Current potential of antisense oligonucleotides as therapeutic drugs. Trends Cardiovasc Med 4:51–55

Davis SF, Yeung AC, Meredith IT, Charbonneau F, Ganz P, Selwyn AP, Anderson TJ (1996) Early endothelial dysfunction predicts the development of transplant coronary artery disease at 1 year post-transplant. Circulation 93:457–462

De Caterina R, Libby P, Peng HB, Thannickal J, Rajavashisth TB, Gimbrone MA, Shin WS, Liao JK (1995) Nitric oxide decreases cytokine-induced endothelial activation. J Clin Invest 96:60–68

Dubey R, Jackson E, Lüscher T (1995) Nitric oxide inhibits angiotensin II-induced migration of rat aortic smooth muscle cell. J Clin Invest 96:141–149

Dzau VJ, Horiuchi M (1996) In vivo gene transfer and gene modulation in hypertension research. Hypertension 28:1132–1137

Fang S, Sharma RV, Bhalla RC (1997) Endothelial nitric oxide synthase gene transfer inhibits platelet-derived growth factor-BB stimulated focal adhesion kinase and paxillin phosphorylation in vascular smooth muscle. Biochem Biophys Res Comm 236:706–711

Felgner PL, Rhodes G (1991) Gene therapeutics. Nature 349:351–352

Felgner PL, Gader TR, Holm M, Roman R, Chan HW, Wenz M, Northrop JP, Ringold GM, Danielsen M (1987) Lipofectin: a highly efficient, lipid mediated DNA-transfection procedure. Proc Natl Acad Sci U S A 84:7413–7417

Flanagan WM, Wagner RW (1997) Potent and selective gene inhibition using antisense oligodeoxynucleotides. Mol Cell Biochem 172:213–225

Garg UC, Hassid A (1989) Nitric-oxide generating vasodilators and 8-bromo-cyclic guanosine monophosphate inhibit mitogenesis and proliferation of cultured rat vascular smooth muscle cells. J Clin Invest 83:1774–1777

Gibaldi M (1993) What is nitric oxide and why are so many people studying it? J Clin Pharmacol 33:488–496

Gibbons GH, Dzau VJ (1994) The emerging concept of vascular remodeling. N Engl J Med 330:1431–1438

Gödecke A, Decking UKM, Ding Z, Hirchenhain J, Bidmon HJ, Gödecke S, Schrader J (1998) Coronary hemodynamics in endothelial NO synthase knockout mice. Circulation 82:

Harrison DG (1997) Cellular and molecular mechanisms of endothelial cell dysfunction. J Clin Invest 100:2153–2157

Hawkins JW (1998) A brief history of genetic therapy: gene therapy, antisense technology, and genomics. In: Wickstrom E (ed) Clinical trials of genetic therapy with antisense DNA and DNA vectors, edn. Marcel Dekker, New York, pp 1–38

Huang PL, Huang Z, Mashimo H, Bloch KD, Moskowitz MA, Bevan JA, Fishman MC (1995) Hypertension in mice lacking the gene for endothelial nitric oxide synthase. Nature 377:239–242

Janssens SP, Bloch KD, Nong Z, Gerard RD, Zoldhelyi P, Collen D (1996) Adenoviral-mediated transfer of the human endothelial nitric oxide synthase gene reduces acute hypoxic pulmonary vasoconstriction in rats. J Clin Invest 98:317–324

Janssens S, Flaherty D, Nong Z, Varenne O, van Pelt N, Haustermans C, Zoldhelyi P, Gerard R, Collen D (1998) Human endothelial nitric oxide synthase gene transfer inhibits vascular smooth muscle cell proliferation and neointima formation after balloon injury in rats. Circulation 97:1274–1281

Katz AM (1988) Molecular biology in cardiology, a paradigmatic shift. J Mol Cell Cardiol 20:355–366

Kitsis RN, Buttrick PM, McNally EM, Kaplan ML, Leinwand LA (1991) Hormonal modulation of a gene injected into rat heart in vivo. Proc Natl Acad Sci U S A 88:4138–4142

Koglin J, Glysing-Jensen T, Mudgett JS, Russell ME (1998) Exacerbated transplant arteriosclerosis in inducible nitric oxide-deficient mice. Circulation 97:2059–2065

Kolyada AY, Savikovsky N, Madias NE (1996) Transcriptional regulation of the human iNOS gene in vascular-smooth-muscle cells and macrophages: evidence for tissue specificity. Biochem Biophys Res Comm 220:600–605

Kullo IJ, Mozes G, Schwartz RS, Gloviczki P, Crotty TB, Barber DA, Katusic ZS, O'Brien T (1997a) Adventitial gene transfer of recombinant endothelial nitric oxide synthase to rabbit carotid arteries alters vascular reactivity. Circulation 96:2254–2261

Kullo IJ, Mozes G, Schwartz RS, Gloviczki P, Tsutsui M, Katusic ZS, O'Brien T (1997b) Enhanced endothelium-dependent relaxations after gene transfer of recombinant endothelial nitric oxide synthase to rabbit carotid arteries. Hypertension 30:314–320

Kullo IJ, Schwartz RS, Pompili VJ, Tsutsui M, Milstien S, Fitzpatrick LA, Katusic ZS, O'Brien T (1997c) Expression and function of recombinant endothelial NO synthase in coronary artery smooth muscle cells. Arterioscler Thromb Vasc Biol 17:2405–2412

Lamas S, Marsden PA, Li GK, Tempst P, Michel T (1992) Endothelial nitric oxide synthase: molecular cloning and characterization of a distinct constitutive enzyme isoform. Proc Natl Acad Sci U S A 89:6348–6352

Lemarchand P, Jones M, Yamada I, Crystal RG (1993) In vivo gene transfer and expression in normal uninjured blood vessels using replication-deficient recombinant adenovirus vectors. Circ Res 72:1132–1138

Lowenstein CJ, Glatt CS, Bredt DS, Snyder SH (1992) Cloned and expressed macrophage nitric oxide synthase contrasts with brain enzyme. Proc Natl Acad Sci U S A 89:7611–6715

Mann MJ, Gibbons GH, Kernoff RS, Diet FD, Tsao PS, Cooke JP, Kaneda Y, Dzau VJ (1995) Genetic engineering of vein grafts resistant to atherosclerosis. Proc Natl Acad Sci U S A 92:4502–4506

Mann MJ, Morishita R, Gibbons GH, von der Leyen HE, Dzau VJ (1997) DNA transfer into vascular smooth muscle using fusigenic Sendai virus (HVJ)-liposomes. Mol Cell Biochem 172:3–12

Marui N, Offerman M, Swerlick R, Kunsch C, Roxen CA, Ahmad M, Alexander RW, Medford RM (1993) Vascular cell-adhesion molecule-1 (VCAM-1) gene-transcription and expression are regulated through an antioxidant sensitive mechanism in human vascular endothelial cells. J Clin Invest 92:1866–1874

Melillo G, Musso T, Sica A, Taylor LS, Cox GW, Varesio L (1995) A hypoxia-responsive element mediates a novel pathway of the inducible nitric oxide. J Exp Med 182:1683–1693

Mooradian DL, Hutsell TC, Keefer LK (1995) Nitric oxide (NO) donor molecules: effect of NO release rates on vascular smooth muscle cell proliferation in vitro. J Cardiovasc Pharmacol 25:674–678

Morishita R, Gibbons GH, Ellison KE, Nakajima M, Zhang L, Kaneda Y, Ogihara T, Dzau VJ (1993) Single intraluminal delivery of antisense cdc2 kinase and proliferating-cell nuclear antigen oligonucleotides results in chronic inhibition of neointimal hyperplasia. Proc Natl Acad Sci U S A 90:8474–8478

Morishita R, Gibbons GH, Ellison KE, Nakajima M, von der Leyen H, Zhang L, Kaneda Y, Dzau VJ (1994) Intimal hyperplasia after vascular injury is inhibited by antisense cdk 2 kinase oligonucleotides. J Clin Invest 93:1458–1464

Nabel EG, Plautz G, Nabel GJ (1990) Site-specific gene expression in vivo by direct gene transfer into the arterial wall. Science 249:1285–1288

Nathan C, Xie QW (1994) Nitric oxide synthases: roles, tolls, and controls. Cell 78:915–918

Noiri E, Peresleni T, Miller F, Goligorsky MS (1996) In vivo targeting of inducible NO synthase with oligodeoxynucleotides protects rat kidney against ischemia. J Clin Invest 97:2377–2383

Ooboshi H, Chu Y, Rios CD, Faraci FM, Davidson BL, Heistad DD (1997) Altered vascular function after adenovirus-mediated overexpression of endothelial nitric oxide synthase. Am J Physiol 273(1 Pt 2):H265–H270

Pepka-Zaba J, Higgenbottam TW, Dinh-Xuan AT, Stone D, Wallwork J (1991) Inhaled nitric oxide as a cause of selective pulmonary vasodilatation in pulmonary hypertension. Lancet 338:1173–1174

Peresleni T, Noiri E, Bahou WF, Goligorsky MS (1996) Antisense oligodeoxynucleotides to inducible NO synthase rescue epithelial cells from oxidative stress. Am J Physiol 270:F971–F977

Petros A, Bennett D, Vallance P (1991) Effect of nitric oxide synthase inhibitors on hypotension in patients with septic shock. Lancet 338:

Phillips MI, Gyurko R (1997) Antisense oligonucleotides: new tools for physiology. News Physiol Sci 12:99–105

Rappaport J, Hanss B, Kopp JB, Copeland TD, Bruggeman LA, Coffman TM, Klotman PE (1995) Transport of phosphorothioate oligonucleotides in kidney: implications for molecular therapy. Kidney Int 47:1462–1469

Ross R (1993) The pathogenesis of atherosclerosis: a perspective for the 1990s. Nature 362:801–809

Rudic RD, Shesely EG, Maeda N, Smithies O, Segal SS, Sessa WC (1998) Direct evidence for the importance of endothelium-derived nitric oxide in vascular remodeling. J Clin Invest 101:731–736

Schaffer MR, Tantry U, Efron PA, Ahrendt GM, Thornton FJ, Barbul A (1996) Nitric oxide regulates wound healing. J Surg Res 63:237–240

Schmidt HHHW, Walter U (1994) NO at work. Cell 78:919–925

Shears LL, Kawaharada N, Tzeng E, Billiar TR, Watkins SC, Kovesdi I, Lizonova A, Pham SM (1997) Inducible nitric oxide synthase suppresses the development of allograft arteriosclerosis. J Clin Invest 100:2035–2042

Simons M, Edelman ER, DeKeyser JL, Langer R, Rosenberg RD (1992) Antisense c-myb oligonucleotides inhibit intimal arterial smooth muscle cell accumulation in vivo. Nature 359:67–70

Stamler JS (1994) Redox signaling: nitrosylation and related target interactions of nitric oxide. Cell 78:931–936

Takano H, Manchikalapudi S, Tang X, Qiu Y, Rizvi A, Jadoon AK, Zhang Q, Bolli R (1998) Nitric oxide synthase is the mediator of late preconditioning against myocardial infarction in conscious rabbits. Circulation 98:441–449

Thomae KR, Geller DA, Billiar TR, Davies P, Pitt BR, Simmons RL, Nakayama DK (1993) Antisense oligodeoxynucleotide to inducible nitric oxide synthase inhibits nitric oxide synthesis in rat pulmonary artery smooth muscle cells in culture. Surgery 114:272–277

Tsao PS, Wang B, Buitrago R, Shyy JY, Cooke JP (1997) Nitric oxide regulates monocyte chemotactic protein-1. Circulation 96:934–940

Tzeng E, Billiar TR (1997) Nitric oxide and the surgical patient. Arch Surg 132:977–982

Tzeng E, Billiar TR, Robbins PD, Loftus M, Stuehr DJ (1995) Expression of human inducible nitric oxide synthase in a tetrahydrobiopterin (H4B)-deficient cell line:

H4B promotes assembly of enzyme subunits into an active dimer. Proc Natl Acad Sci U S A 92:11771–11775

Tzeng E, Shears LL, Robins PD, Pitt BR, Geller DA, Watkins SC, Simmons RL, Billiar TR (1996a) Vascular gene transfer of the human inducible nitric oxide synthase: characterization of activity and effects on myointimal hyperplasia. Mol Med 2:211–225

Tzeng E, Yoneyama T, Hatakeyama K, Shears LL, Billiar TR (1996b) Vascular inducible nitric oxide synthase gene therapy: requirement for guanosine triphosphate cyclohydrolase I. Surgery 120:315–321

Tzeng E, Kim YM, Pitt BR, Lizonova A, Kovesdi I, Billiar TR (1997) Adenoviral transfer of the inducible nitric oxide synthase gene blocks endothelial cell apoptosis. Surgery 122:255–263

Vane JR, Änggård EE, Botting RM (1990) Regulatory functions of the endothelium. N Engl J Med 323:27–36

Varenne O, Pislaru S, Gillijns H, Van Pelt N, Gerard RD, Zoldhelyi P, Van de Werf F, Collen D, Janssens SP (1998) Local adenovirus-mediated transfer of human endothelial nitric oxide synthase reduces luminal narrowing after coronary angioplasty in pigs. Circulation 98:919–926

von der Leyen HE, Gibbons GH, Morishita R, Lewis NP, Zhang L, Nakajima M, Kaneda Y, Cooke JP, Dzau VJ (1995) Gene therapy inhibiting neointimal vascular lesion: In vivo gene transfer of endothelial-cell nitric oxide synthase gene. Proc Natl Acad Sci U S A 92:1137–1141

von der Leyen H, Mann MJ, Dzau VJ (1998) Gene therapy of cardiovascular disorders. In: Alexander RW, Schlant RC, Fuster V (eds) Hurst's the heart, 9th edn. McGraw-Hill, New York, pp 213–225

Yamasaki K, Edington HDJ, McClosky C, Tzeng E, Lizonova E, Kovesdi I, Steed DL, Billiar TE (1998) Reversal of impaired wound repair in iNOS-deficient mice by topical adenoviral-mediated iNOS gene transfer. J Clin Invest 101:967–971

Subject Index

A-453   43
acetylcholine   95, 278, 294, 344, 577
– coronary arteries, effect on   551
– EPSP   261
acetylsalicylic acid   247
aconitase   191
activator protein-1   77
active catalytic cycle   51
acute heart failure   365
acute inflammatory responses   498–499
acute lung injury   399, 404, 411
acute myocardial infarction   247, 371
– nitrovasodilators   375
acute respiratory distress syndrome (ARDS)   246, 399
– guidelines, use in adult ICU in UK   420(table)
– NO gas inhalation   400–412
– – acute effects   401–404
– – dependency   409–410
– – dose-response relationship   404–408
– – non-cardiac pulmonary edema   404
– – non-responders   409
– – right heart function   408–409
– – studies   410–412
adhesion   466
almitrine   402
α_1-adrenoceptors   278
α_1-antiprotease   419
α_1-synthrophin   75
α_2-adrenoceptors   278
α-fluoro-N-(3-aminomethylphenyl) acetamide   116(table), 120(fig.)
alveolar typeII cells   419
Alzheimer's disease   74, 623
amidine-containing inhibitors   119–124
2-amino-4-methyl-pyridine   121(fig.), 123
4-amino-H_4B 142–145
– effects on:
– – animals   144–145
– – cultured cells   144

– – purified enzymes   142–144
2-aminoazaheterocycles   124
aminoguanidine   115(table), 118–119, 421, 504
4-aminopteridines   145–149
– C6 chain   147–149
– function   145
– positions, 2, 5, 7   145–147
– pterine exosite   147–149
2-aminopyridines   123
AMPA receptors   259–260, 262
amyotrophic lateral sclerosis, neuronal NOS   111–112
Angeli's salts   19–20
angina, stable/instable   247, 365
– attack, prevention and treatment   374
– chronic, long-term management   374
– nitrovasodilators, effect   373–375
angiotensin-converting enzyme (ACE) inhibitors   333–334
angiotensin-II   319, 343, 377
antimicrobial defence mechanisms   508–510
antipterins   151–153, 154(fig.)
AP-1   461
apamin   298
Apo-1   532
apoptosis   159–160, 548(fig.), 575
– atherosclerosis, NO effect   595–598
– cardiac myocytes   506
apoptotic-signal transduction   162–163
arachidonic acid metabolites   501–502
arginase   444
arginine
– analogs   114–119
– atherosclerosis
– – availability   579
– – paradox, possible explanation   570–571
– conversion to citrulline   19
arginine vasopressin (AVP) see vasopressin

ARL-17477 123
artery, remodelling in atherosclerosis 574
ascorbic acid 584
aspirin 247–249
atherosclerosis 241–242, 571–598
– cholesterol (oxidized), levels 14
– eNOS:
– – defective pathway, explanation 577–584
– – dysfunction 575–577
– iNOS, expression 524–525
– – mechanical injury 585
– intimal thickening, stages 571–573
– pathogenic mechanisms 573–575
– radical promoter, NO 593–598
atrial-natriuretic peptide (ANP) 93, 190, 323, 584
atropine, resistance 278
autacoids 179
autocrine agent 317
autoxidation reaction 17–18
azo-bis-amidinopropane 14

Babylog 428
baloon angioplasty, atherosclerosis, studies 592
Bax proteins 164
Bcl-2 family members 160
– antagonism 164
bcl-2 gene 160
β-adrenoceptors 278
β-adrenergic
– agonists 211
– response
– – eNOS, regulation 220–221
– – iNOS, regulation 224
– stimulation, renin stimulation 343
β-cells (islets cells), insulin-producing 525–537
– NO-induced cell death, pathways 531–533
– – apoptotic pathway 531–532
– – mitochondrial damage 531
– NO-induced damage, cellular sources 528–530
– – β-cells 529–530
– – endothelial cells 529
– – macrophages 528
– – mitochondria 530
– – nuclear DNA 530–531
– NO-induced damage, protecting strategies 534–537

– – defence mechanism, improvement 535–536
– – NO formation, suppression 534–535
BHT-920 293
biological metals, reactions with NO 9–14
bis-indolylmaleimide I 81
blood flow
– control
– – NO and 20-HETE 197
– – NO and oxygen, interaction between 195–197
– lungs, oxygen tension 11
– tissues, heme binding site, NO/oxygen competition 11
blood pressure (mean), effect of E.coli endotoxin 387(fig.), 389(fig.)
blood-brain barrier 316
bovine aortic endothelial cells (BAEC) 79–80
bradykinin 95, 181, 219
– renovascular resistance 333
brain ischemia/reperfusion injury 619–631
brain natriuretic peptide (BNP) 93
bretylium 278

calcium
– eNOS 315
– – platelet-derived, activation 238
– membrane hyperpolarisation 298–299
– nitrergic nerves, activation 289–290
– nNOS 315
– NO, effect 187–189
– post-junctional transduction pathway 297
calcium binding protein 34
calmidazolium 81
calmodulin (CaM) 34–35, 315
– catalytic properties and response to NO 44–45
– eNOS, interaction with 181–182
– NOS activation, mechanism of action 46–47
– structural determinants, binding 47
cancer 244
carbinolamine dehydratase 138
carbon monoxide (CO) 96
– neurotransmitter 279
– soluble guanylyl cyclase, physiological activator 98
carboxy-PTIO 284–285, 295
cardiac function, regulation3 207–226

Subject Index

cardiac muscle cells (myocytes)
– biology, iNOS effect   224–225
– intracellular mechanism of action   214–219
– regulation, cGMP-dependent/independent mechanisms   216(table)
– targets for the NO effect   215(fig.)
cardiac output, septic shock   386
cardiomyopathy
– dilated   386
– – non-ischaemic   221
caspase(s)   1, 160, 463
– activation, p53 accumulation, apoptotic-signal transduction   162–163
caveolins   61, 82, 182, 186
– haemodynamic forces, effect   187
– interaction with eNOS   182
– skeletal and cardiac muscle   208
CD95   532
cell death   8, 13–14
– Angeli's salt cytotoxicity, NO and oxygen reaction   20
– apoptosis vs. necrosis   159–160
– NO and oxygen reaction   21
– programmed   159
central nervous system, physiological role   259–270
cerebral arteries, porcine   191
cerebral blood flow, local   264–265
cerebral ischaemic injury, NO dual role   624(fig.)
charybdotoxin   298
chelerythrine   81
chemiluminescence   429–430
chimeras   52–53
chlorzoxazone   126
cholecystokinin   316–317
circadian light-dark cycle, regulation   260
cis-4-methyl-5-pentylpyrollidin-2-imine   116(table)
CLA-510S   430
CLD-700AL   430
CLD-77AM   430
clonidine   293
clotrimazole   116(table), 126
CoASNO   286
cobalt   9
concanavalin (ConA)   503
conduit arteries, atherosclerosis   579–580
– without overt   580
copper   9, 14

coronary artery disease   241
– atherosclerotic   545
coronary blood flow/perfusion
– blood pressure reduction   371
– NO-dependent changes   550
– redistribution, nitrates induced   369–370
cortex, visual   261–262
corticosterone, NOS-I transcription, effect on   74
corticotrophin-releasing hormone (CRH)   316, 322(fig.)
– hypothalamo-pituitary-adrenal axis   318–320
critically ill patients   246
cyclic guanosine monophosphate (cGMP)   9
– dependent mechanisms   214–218
– – contraction-decreasing mechanisms   217–218
– – contraction-enhancing mechanisms   214–217
– synthesis   10
cyclic-AMP-responsive element-binding protein (CREB)   213
cyclooxygenase (COX), products   496–497
cyclooxygenase-2 (COX-2)
– macula densa cells   350
– renovascular hypertension, model   351
Cys residues   42–43
Cys-109, mutation to Ala-42   37–38(tables)
CYSNO   285
cytochrome p450, interaction with NO   10
cytokine(s)
– eNOS, effects   212–213
– inflammatory and NO   499–501
– iNOS (NOS II)
– – activation   449–451
– – expression   76
– responses, endogenous NO   463–465
cytotoxicity   8

DETCA   287–288
dexamethasone   444
diabetes mellitus (type I)
   immunopathogenesis   241–243, 525–537
– cellular immune reactions   527
– clinical characteristics   525–526

– nephropathy 329
– NO as pathogenic factor 528–533
– studies 526–527
dibutyryl cAMP 76
1,1-diethyl-2-hydroxy-2-nitroso-hydralasine 20
diethyl-dithiocarbonate 76, 287
dihydrorhodamine, oxidation 20–23
dimethylphenyl piperazine 289
dinitrosyl iron complexes 192–193
dipyridamol 369
dobutamine infusion 221
dopamine 323
drug(s)
– metabolism, liver cytochrome 450, inhibition 10
– nephrotoxicity 329
duroquinone 284

E-selectin 466, 497
EC-1.14.13.39 71
ECMO 400, 412, 414
electrochemical analyzers 430–431
enalapril 376
endothelial nitric oxide synthase (eNOS) 9, 11, 71–72, 180–187
– beating rate 210–211
– β-adrenergic agonists 211
– blood flow, control 194(fig.)
– brain ischaemia/reperfusion injury 623–626
– calcium dependent activation 181–185
– calcium independent activation 185–186
– cardiac function, regulation
– – basal, systolic/diastolic function 219–220
– – β-adrenergic response, regulation 220
– – endocardium 208
– – muscarinic cholinergic response 221–222
– – myocardium, oxygen consumption 192
– caveolin-1/3, interaction in cardiomyocytes 208
– cytokines, acute effect of 212–213
– expressional control 209–210
– fluid shear stress and NO production, link between 186–187
– hypoxia/hyperoxia, effects on 238
– kidney function 329–330
– mechanical forces 210

– muscarinic cholinergic agonists 211–212
– vascular haemodynamics 623–626
endothelial nitric oxide synthase-III (cNOS)
– activity, regulation 81–83
– cellular expression 79
– expression, regulation 79–81
endothelial NOS-associated protein-1 (ENAP) 183
endothelial receptor dysfunction, atherosclerosis 577–578
endothelin-1 179, 343
– blood pressure, role 190
endothelium-dependent relaxation 236
endothelium-derived hyperpolarising factor (EDHF) 179, 571
endothelium-derived relaxing factor (EDRF) 95, 236, 279–280, 283
– formation 95
enkephalin 317
ergotoxine 278
estradiol (pregnancy), nNOS expression induction 73–74
ethanol 295
ethylisothiourea 115(table)
7-etoxyresorufin 284
Evita 428
excitatory post-synaptic potential (EPSP) 261

fibrinogen receptor 249
fibroblast growth factor 76
fibronectin receptor 240
flavin adenine dinucleotide (FAD) 34–35, 137, 180
flavin mononucleotide (FMN) 34–35, 137, 180
– electron transfer in nNOS, role 46(fig.)
flavoproteins 181
fluid shear stress 184
– NO production, link between 186–187
focal ischaemic brain damage 627–629
forscolin 76, 347
Frank-Starling response 219, 368, 550

galanin 317
gap junctions 263–264
gastrointestinal disorders, nitrovasodilator effect in 376
GEA-3162 503–504
GEA-3175 501, 504

Subject Index

geldanamycin 183
gene therapy, principles 639–640
gene-transfer technology, in vivo 639–640
geniculate nucleus, retinal output 261–262
genistein 76
ginsenosides 290–291
GISSI-III study 247
glial cells 265
glibenclamide 298
glomerular filtration 329
– NO, role 335–336
glutamate 259
– receptors 264
glyceryl trinitrate 365, 366(fig.), 372(table)
– blood pressure 371
– intravenous administration 247
– preload 371
Go-6976 81
gonadotrophin-releasing hormone (GnRH) 317
G-protein(s) 286, 463
– coupled receptors 259
graft vs. host disease (GVHD) 505
green fluorescent protein 82
growth hormone 322
GSNO 285
guanethidine 278
guanylin 93
guanylyl cyclase 9, 11
– NO, activation by 9–10
– vascular tone, control 71

haemostasis, vascular 235
heart failure, vasodilators 375–376
heat shock protein(s) 61–62, 585
heat shock protein-70 167, 535
heat shock protein-90 61–62, 183–184
heme iron reduction 52
– control by H_4B and arginin 53, 54(fig.)
heme ligands 115–116(table), 124–127
heme oxygenase 167
– activation, pathophysiological conditions 10
heme proteins 7, 9
– NO, reactivity of 10(fig.)
heme-NO complex formation
– NOS catalysis, NO complex formation, impact 49–50
– partitioning into NO-bound form 49–50

hemoglobin 295
– NO binding 284
– purified, human 245–246
hepatocytes
– cytokine-stimulated 12
– heme oxygenase, activation 10
herbimycin A 76
20-HETE (hydroxyeicosatetraenoic acid) 197
high frequency oscillatory ventilation 413
high-density lipoproteins 242
histamine 95, 495
human umbilical vein endothelial cells (HUVEC) 79–80
Huntington's disease 623
hydralazine 376
hydraulic conductivity of filtration (Kf) 335
hydrogen peroxide (H_2O_2)
– cellular damage 12
– high-valence metal complexes 13(fig.)
hydroquinone 284
hydroxyarginine 19
hydroxycobalamin 284–285, 295
hydroxyphenylacetic acid (HPA), oxidation 20
5-hydroxytryptamine 259, 577
hypertension 241–242
– crisis 248
hypoglycaemia 319
hypotension, septic shock, predictor of outcome 386
hypothalamo-pituitary-adrenal axis 318–320
hypothalamo-pituitary-gonadal axis 320–322
hypothalamo-pituitary-thyroid axis 322
hypothalamus
– NO:
– – biosynthesis 315–317
– – physiology 317–323
hypoxia
– myocardium 546
– RNOS chemistry, cell death 20

I-Novevent 428
iberiotoxin 298
imidazole 124–126
2-iminobiotin 121(fig.), 124
2-iminopiperidine 124
immune system, NO function 443–468
– response 502–506
indazoles 126–127

indomethacin 195
– activity, regulation 78–79
– cellular expression 75
– expression, regulation 75–78
inducible nitric oxide syntase (iNOS) 9, 11, 71–75, 112–113, 213–214, 315
– brain ischaemia/reperfusion injury 626–627
– cardiac function, regulation 222–225
– – basal contractile function 223–224
– – β-adrenergic response, regulation 224
– – cardiomyocyte biology 208, 224–225
– immune system 444–467
– – cell types 444, 446
– – functions 453–467
– – induction and regulation 444–453
– induction in humans, septic shock 391
– inflammatory bowel disease 124
– kidney function 329–331
infection (chronic), NO production 10
inflammation, definition 494
inflammatory bowel disease, inducible NOS 112
inflammatory diseases, chronic 502–510
inflammatory glomerular disease 329
inhalation therapy, NO gas 399–431
– autoinhalation 415–417
– delivery 425–429
– – company-designed delivery 428
– – flow, continuous 426–428
– – inspiratory timed-injection technique 428
– – pre-mixing system 425–426
– exhaled 416(fig.)
– monitoring inhalation 429–431
– NO/nitrogen gas mixture 424–425
– use in adult ICU, UK guidelines 420(table)
inhibitory nerves 279
inositol 1,4,5-triphosphate (IP3) 187, 297–298
insulin-like growth factor I 76
integrins, fluid share stress 186
intercellular adhesion molecule-1 (ICAM-1) 466, 497, 550, 552
interferon-γ 76, 323, 444
– macrophages, non-inflammatory 447
– nNOS expression 74
– iNOS expression 76
– septic shock 385
interleukin(s), septic shock 385
interleukin-1 76, 315–316, 319–320, 507
interleukin-4 444

interleukin-6 499
interleukin-12 465–466
interleukin-13 444
intermittent mandatory ventilation, synchronised 425
iron
– containing proteins 191
– intracellular, target for NO 191
iron-nitrosyl (Fe-NO) complex 11
– formation, mechanisms 13, 49(table)
ischaemic heart, hypotension effect 371
ISIS-2 375
ISIS-4 247, 375
isoproterenol 347
isosorbide dinitrate, hydralazine combined with 376
isosorbide mononitrate 247, 366(fig.), 372(table)
isosorbide-2,5-dinitrate 365–366
isosorbide-5-nitrate 365–366
isothioureas 120–124

Janus kinases 463
junctional mechanisms 295–297
juxtaglomerular apparatus 329
kalirin 62
kidney function 329–352
kinin 495–496

L-arginine 238
– – active catalytic cycle 51–53
– – enzyme structural features 53–55
– – H$_4$B, role 56–57
– – heme, role 56–57
– – heme-NO complex formation 49–51
– oxidation 33
– platelet microenvironment 244–245
L-cysteine 285
L-NAME 119, 334–335, 340, 496–498, 504
L-NIL 115(table)
L-NIO 117–118
L-NMMA 119, 334, 496, 501, 507
L-NNA 115(table), 118
L-thiocitrullin 115(table), 118
lactation 317
lactoperoxidase 22
left ventricular pressure, end-diastolic (LVEDP) 219–220
leukocyte chemotaxis 466
leukotrienes 496
– production, NO effect on 14

linsidomine   593
lipid peroxidation   12
– inflammatory process   13–14
– reactive metal complexes   12
liposomes, gene transfer   639
lipoteichoic acid   387
lipoxygenase   14
lithium   74
losartan   352
low-density lipoproteins   242
LY-83583   286–288, 296
LY-83589   99

macrophage-deactivating factor   76
macrophage-stimulating protein (MSP)   449
macrophages
– atherosclerosis, mediated by   14
– NO production   9
– tumoricidal activity, NO role   12
macula densa-mediated renin secretion, NO effect   347–350
matrix metalloproteinase   575
matrix metalloproteinase-2   235
– dependent platelet aggregation   247
membrane hyperpolarization, following nitrergic nerve stimulation   298–299
membrane-bound guanylyl cyclase (mGCs)   93
– isoforms   93
met-enkephalin   295
metal complexes
– high-valence, NO chemistry   12, 13(fig.)
– reaction with NO   9–14
metal-oxygen complexes, interaction of NO   11–13
metalloproteins   21
methemoglobin   11–12, 284, 422–424
methemoglobinemia   417–418, 420(table)
– sodium nitroprusside, induced   378
methionine, oxidation   21
methotrexate   140–142
Michaelis-Menten kinetics   101
miconazole   126
mitochondrial respiration   191–192
mitogen-activated protein kinase (MAPK)   448–463
– cytotoxicity/apoptosis   161–162
– formation and signalling   160–161
– mediated toxicity, resistance against   164–167
– pro/anti-apoptotic molecule   169(fig.)

molsidomine   248, 365–367, 372(table)
– preload, effect on   368
monoamine(s)
– neural cells   259
– neurotransmitters   259
monocyte chemotactic protein-1   76
multi-organ failure (MOS), mechanism, NO overproduction   389, 390(fig.)
multiple inert-gas-elimination technique (MIGET)   400
multiple sclerosis, inducible NOS   112
muscarinic cholinergic
– agonists   211–212
– response, eNOS, regulation   221–222
mutants, NOS
– F-363   53
– F-470   36, 37(table), 40
– G-450   43
– R-365-A   40
– R-375   40
– W-188   53
– W-366   54
– W-455   37(table), 40
– W-457   36, 37(table), 40
myeloperoxidase   22
myoblast implantation   639
myocardial function, experimental studies   552–561
– arrhythmia, reperfusion induced   558–559
– blood/neutrophil perfused preparations   557
– myocardial infarctions   557–558
– post-ischaemic contractile function   552–557
myocardial ischaemia/reperfusion   545–563
– consequences   546
– iatrogenic   545
– interactions between NO and ROS   547
– ischaemic preconditioning   562–563
– NO and ONOO$^-$ influence   547–551
– pathophysiology, potential effect of NO   549(fig.)
– results, conflicting, reasons   561–562
– studies, experimental   551–561
myocardial oxygen supply/consumption   550–551
myoglobulin   21
myristoylation   81–82

N_2O_3 8
– formation, mechanism   17–18

– nitric oxide, indirect effects 15(fig.)
N-acetyl-L-cysteine 192–193
N-acetylcysteine 247
N-methyl-D-aspartate (NMDA) receptors 74–75, 259–260, 262
N-phenylisothioureas 122
N-terminal hairpin loop 40–42
natriuretic peptides 93
necrosis 159–160
necrotic death 159
nerve-derived hyperpolarising factor (NDHF) 283
nerve-nerve interactions 293–295
netergic-cholinergic interactions 293–294
neuroendocrine function 315–323
neuronal NOS (nNOS) 9, 11, 111–113
– brain ischaemia/reperfusion injury 619–623
– hypothalamus 316
– kidney function 329–330
– nitrergic nerves
– – localization 281–282
– – properties 281
– rat, domain arrangement 34(fig.)
neuronal nitric oxide synthase (nNOS) 73–75
– activity, regulation 74–75
– cellular expression 73
– expression, regulation 73–74
neuropeptide Y 284, 343
NF-κB 463, 499
nicotinamide adenine dinucleotide phosphate (NADPH) 33, 51
nifedipine 290
nimodipine 290
nitrate(s) 11–12
– contraindications 378
– organic 246–247, 386
– – coronary flow, redistribution 369
– – non-cardiovascular disorders 373(table)
– – side effects 365
– – vasodilation, preferential 368–369
– side effects 378
– tolerance 365, 377–378
nitrergic nerves 277, 279–284
– activation 289
– calcium, role in activation 289–290
– concept 279–281
– properties 281–284
– – anatomical distribution 282–283
– – localisation 281–282
– – physiological functions 282–283
– – transmission, unitary/dual/co 283–284
– transmission
– – NOS, blockade by inhibition 291–293
– – post-junctional 296–297
– – pre-junctional augmentation 290–291
nitrergic neurotransmitter, nature 284–288
nitrergic-adrenergic interactions 293
nitrergic-NANC interaction 294–295
nitric oxide (NO)
– antioxidant properties 12
– atherosclerosis
– – anti-atherogenic properties 585–593
– – pro-atherogenic properties 593–594
– β-cells, formation 529–530
– biosynthesis 493–494
– blood flow, control 193–197
– cardiac muscle cells 214–219
– cardiomyocyte contraction, targets for the effect 215(fig.)
– chemical biology 7–24
– – chemical reactivity 7, 8(fig.)
– – direct effect 7, 8(fig.), 9–14
– – indirect effect 7, 8(fig.), 14–15
– clearance 417
– cytostatic properties 13
– cytotoxicity 161–163
– deleterious agent (ischaemia/reperfusion), experimental studies 559–561
– discovery, endogenous mediator 7
– donors 246–249
– endogenously generated, hormone metabolism 10
– gas 246
– – inhalation therapy 399–431
– genetic engineering 640(table)
– immune system, role in tumoricidal processes 7
– immunoregulatory function 461–463
– inflammation 494–502
– inflammatory mediator 493–511
– inhibition, another type 60
– iron-containing proteins 191
– lusitropic effect 219
– mitochondrial respiration 191–192
– neurotoxicity, targets 629–631
– neurotransmission 71
– platelets 236–241
– – action, mechanism of 239–241
– – cell activation 237–238

Subject Index

– – co-factors, role 238
– – formation and action, pharmacological modulation 244–249
– – function, synergistic regulation 239
– – function in vitro/in vivo 238–239
– – rheology 238
– – substrate, role 238
– renal hemodynamics, shear-stress induced 333
– ROS, interactions with 547, 548(fig.)
– scavengers 245–246, 295
– signalling, invertebrates 260
– soluble guanylyl cyclase, physiological activator 94–95
– – mechanism 96–98
– uptake 417
– vascular smooth muscle, action of 187–193
– vascular tone, control 71, 179–180
nitric oxide (NO)/O2 chemistry 21–24
nitric oxide synthase (NOS) 9
– CAM activation
– – CAM binding, structural determinants 47
– – mechanism of action 46–47
– dimerization 57–59
– domain interactions 47–48
– domain organisations 33–35
– domain structure, modification 59–60
– isoenzymes 71
– – kinetic properties 72(table)
– kidney:
– – distribution 330–333
– – inhibitors 334–335
– – isoforms 329–330
– oxygenase domains and mutagenesis 35–43
– – cysteines and metal binding 42–43
– – H_4B-binding site 36–40
– – L-arg-binding site 36
– – N-terminal hairpin loop 40–42
– platelet, molecular biology 236–237
– reductase domains
– – catalytic properties and response to CAM 44–45
– – general features 43–44
– – mutagenesis 45–46
nitric oxide synthase (NOS) see also endothelial, neuronal and inducible nitric oxide synthases
nitric oxide synthase (NOS) inhibitors
– mechanism-based 114–129
– – heme ligands 124–127

– – rational design 127–129
– – substrate-based 114–124
– therapeutic concepts 111–112
nitric oxide synthase (NOS), therapeutic potentials 640–649
– gene manipulation, overexpression 640–646
– inhibition, antisense technology 646–647
– transgenic animals 647–649
nitric oxide/cyclic guanosine monophosphate (NO/cGMP) signalling cascade 94(fig.)
nitrogen dioxide 417–419
nitroglycerin 503, 584
– pulmonary vessels, effect on 399
7-nitroindazole 116(table), 127, 622
nitrosamines, production 16
nitrosation 16
– thiols 16(fig.)
– in vivo 8
nitrosative stress 8, 15–18
– effects 16(fig.)
nitrosothiol(s) 365
– containing proteins 192–193
nitrosyl 9
nitrovasodilators, therapeutic importance 365–378
– action, mechanism 367
– clinical use 373–377
– doses and action (onset and duration) 372(table)
– hemodynamic actions 368–372
– – blood pressure, effects 370–371
– – platelets 371–372
– – vasodilatation, preferential 368–369
– – vasodilatation, vessel-size-selective 369–370
– pharmacokinetics 372–373
nitroxyl 8, 19–20
– oxidative chemistry 21(fig.)
280-NOA 430
NOdomo 428–429
noformycin 121(fig.), 124
non-adrenergic neurotransmission 278–279
non-cholinergic neurotransmission 278–281
noradrenaline 190–191, 278, 347
NOS I see nNOS
NOS II see iNOS
NOS III see eNOS
NOS-knockout mice 112–114
nucleic acids, oxidation 19

ODQ 99, 241, 293, 296, 298–299, 344
C-agatoxin 290
C-conotoxin GVIA 290
C-conotoxin MVIIC 290
oxidative cell injury, atherosclerosis 524–525
oxidative stress 8, 15(fig.), 18–21, 545
oxo complexes 9, 10(fig.)
4-oxopteridines 149–154
– intact cells 153
oxygen, supply to tissue, NO as a sensor 11
oxyhemoglobin 280
– reaction with NO 11
oxyhydrazines 365
oxytocin 316–318, 377

p53 gene 160
– accumulation, apoptotic signal transduction 162–163
P-selectin 550, 552
pain, regulation 260
paracrine fashion 317
Parkinson's disease, neuronal NOS 112
PBITU 115(table)
PEEP 400
pentaerythriol tetranitrate 365, 366(fig.)
percutaneous baloon angioplasty 545
percutaneous transluminal coronary angioplasty (PTCA) 572–573
peripheral nervous system 277–299
peroxynitrite (ONOO$^-$) 8, 98, 191, 218–219, 241, 389, 419–422, 506, 547
– formation, atherosclerosis 593–594
– myocardium, effects on 550–551
– neurotoxocity, mediation 629–631
peroxynitrous acid (ONOOH) 22
persistent pulmonary hypertension of the newborn (PPHN) 399
– NO inhalation 412–414
phenylarsine oxide 81
1-phenylimidazole 116(table), 125
phosphatidyl inositol-3 kinase 449
phosphodiesterase (PDE) III 189–190
phosphofructokinase M 60
PHS-32 151–152
plaque 575
– stability, atherosclerosis 597–598
plasticity 266–270
platelet-activating factor (PAF) 385, 496
platelet-derived growth factor (PDGF) 76, 573

platelets
– adherence 235
– aggregation 235
– – inhibition, YC-1 93, 99–101
– antiplatelet effect, atherosclerosis 588
– control 235
– nitrovasodilators, effect 371–372
– rheology 235
– tumor methastasis, pathogenesis 244
poly (adenosine diphosphate-ribose) polymerase (PARP) 532–533
post-ischaemic endothelial dysfunction, experimental studies 551–552
post-junctional mechanisms 295–297
post-junctional transduction pathway
– calcium mobilisation, inhibition 297–298
– cyclic GMP, role 297
– membrane hyperpolarisation, role 298–299
post-synaptic density protein-93 (PSD-93) 60
post-synaptic density protein-95 (PSD-95) 60
prazosin 376
pre-eclampsia 241–243
pre-junctional mechanisms 289–292
pressure natriuresis, NO role 341
pressure support ventilation 425
probucol 584
prolactin, secretion 323
prostacyclin (PGI$_2$) 179, 372, 501
– haemostasis control 235
– pulmonary vessels, effect on 399
prostaglandin(s) 496
– E$_1$ 280
– E$_2$ 318, 501–502
– H$_2$ 179
– renin synthesis, NO effect 350–352
proteases 160
protective protein expression 166–167
protein inhibitor of NO synthase (PIN) 60–61
protein kinase(s)
– A 75
– AMP-activated 184–185
– C 75, 377, 448
– NO production, regulation 184
– G 75
protein phosphatases 1 and 2A 448
protein tyrosine kinases 463
protoporphyrin IX 96
pterins

- antagonists   137, 139(fig.)
- – approaches   141–142
- based inhibition of NOS   140–154
- – 4-amino-H₄B   142–145
- – 4-aminopteridines   145–149
- – H₄B levels, intracellular manipulating   140–141
- – 4-oxopteridines as inhibitors   149–154
- binding site   140
Pulmomix Mini   428
pulmonary hypertension, chronic   399
pulmonary vascular resistance   399
Puritan-Bennett-7200   426
pyrogallol   284
pyrrolidine dithiocarbamata (PDTC)   76

quinonoid dihydrobiopterin   139

ramiprilat   556
rauwolscine   293
reactive nitrogen oxide species (RNOS)   7, 17
- oxidative stress   8
- – mediation   19
reactive oxygen species (ROS)   13–14, 545, 548–549(fig.)
- interactions with NO myocardial function   547–548
redox-sensitive kinases   463
rejection, acute   505
relaxin   245
renal autoregulation   329
- NO, role of   336–340
- – myogenic response   338
- – tubuloglomerular feedback   338–340
renal blood flow   329
- regulation, NO role   333–335
- – medullary   341
- – release, endogenous mediators   333–334
renal failure   329
renal nerves, NO effect   346–347
renin
- NO effect on:
- – release/pressure control   344–347
- – secretion   329, 333–334
- – synthesis   343, 350–352
renin-angiotensin system, nitrate tolerance   377–378
reperfusion, myocardial ischaemia   545–563

- arrhythmia   546
- – studies   558–559
rheology, platelets   235
- endothelium-derived NO   238
Ro-31-8220   81
ryanodine receptor   218

S-methylisothiourea   421
S-nitroso-cysteine   191
S-nitroso-glutathione   248–249
- platelet function, inhibition   241
S-nitroso-N-acetylcysteine (SNAC)   286
S-nitroso-N-acetylpenicillamine (SNAP)   286, 501
S-nitrosohemoglobin   191
S-nitrosothiols   16, 422
sarcoplasmic/endoplasmic reticulum Ca-ATPase (SERCA)   188
sepsis
- clinical, NO in   391–394
- *see also* septic shock
septic shock   385–395
- clinical features   385–387
- – cardiovascular changes   386
- – tissue oxygenation   386
- – tissue/organ damage   386–387
- experimental models   387–390
- NO production   10
- outcome studies   394–395
septicaemia   243–244
serotonin   259, 495, 577–578
Servo-300   429
Siemens Servo 900C   426, 428
signal transducer and activator of transcription (STAT)   77, 447
signal transduction   8
single-nephron glomerular filtration rate (SNGFR)   335
sleep-waking cycle, regulation   260
smoking, platelet NOS activity   245
smooth muscle, relaxation   93
sodium nitroprusside   184, 247–248, 260, 344, 366(fig.), 503, 620
- effect on:
- – blood pressure   370–1
- – platelets   371–372
- – preferential vasodilatation   368
- – pulmonary vessels   399
sodium trioxodinitrate (Na₂N₂O₃)   19–20
soluble adenylyl cyclase   95(fig.)
- catalytic domains   104(fig.)
soluble guanylyl cyclase (sGC)   259

- blockade 295–296
- endothelin-1, production 190
- heme group 97(fig.)
- platelet fractions 239
- regulation 93–101
- – carbon monoxide, physiological activator 98
- – modulators 99–101
- – redox regulation 98–99
- structure 101–105
- – catalytic domain 103–105
- – isoforms and tissue distribution 101–102
- – regulatory heme-binding domain 102–103
- – subunits, primary structure/homology 102

somatostatin 317, 322–323
Sp-1 463
steroid hormones, nNOS expression 73
streptozotocin, effect on β-cells 526
substance P 219–220, 294, 316
- intracoronary infusion 221
superoxide dismutase (SOD) 21–22, 98, 179, 195, 287, 419–422, 545, 574
- NO inactivation, atherosclerosis 582–584
sydnonimine SIN-1 20, 248, 344, 365, 503, 620
synaptic transmission 261–263
syntrophins 60

T-helper cells, deviation 466–467
tacrine 74
TATA box 76–77
testosterone synthesis, NO role 10
tetahydrobiopterin (H_4B) 35, 111, 137, 142(table), 180, 444, 578
- administration 141
- binding site 36–40, 42
- deficiency, atherosclerosis 579
- dependence of the NOS reaction 137–140
- heme reduction, control 53
- intracellular levels, manipulation 140–141
- nitric oxide synthase, functioning 57(table)
- structural formulae 139(fig.), 148(fig.)
tetradecanoyl phorbol-13 acetate 76
tetramethrin 290
tetrodotoxin 289
thiocitrulline 122
thiols, nitrosation 16(fig.)

thrombosis 235
- coronary 246
- NOS inhibitors 245
thromboxane A_2 235, 412
thrombus formation 239
Sn-protoporphyrin IX 167
transforming growth factor 2 76, 78, 444
triacyl glycerol 245
trilinolein 247
TRIM 292
Triton X-100-insoluble (cytoskeletal) proteins 81
troponin-1 217
tubular function 342–343
tubulo glomerular feedback 337–340
tumour necrosis factor α 76, 209, 449
- apoptosis, atherosclerosis 595
- septic shock 385
tyrphostin 76

UK-14304 293
uraemia 244
uterus, nitrovasodilators, effects on 376–377

vascular cell adhesion molecule-1 (VCAM-1) 466, 497, 574
vascular disorders, pathogenesis 241–244
vascular endothelial growth factor (VEGF) 183–184
vascular response
- chemical mediators 495–496
- injury 496–498
vascular smooth muscle, NO action 187–193
- effect of NO on:
- – calcium 187–189
- – cyclic nucleotide phosphodiesterase III 189–190
- – vascular tone, control 190–193
vascular tone, regulation 179–180
vasoactive intestinal polypeptide 317
vasodilator-stimulated phosphoprotein (VASP) 240
vassopressin 316–318
ventilators, adult/pediatric 425–426
- pressure control 427(fig.)
- volume control 427(fig.)
verapamil 290
vitamin C 584
vitamin E 574

Subject Index

voltage operated channel (VOC) 297
– calcium role 290
von Willebrand factor 240

1400-W 115(table), 120(fig.), 122–123
Windkessel function 371

xanthine oxidase (XO) 14, 22–23, 195, 547

YC-1 99–101, 249
yohimbine 293

zinc 9